Lecture Notes in Computer Science　9490

Commenced Publication in 1973
Founding and Former Series Editors:
Gerhard Goos, Juris Hartmanis, and Jan van Leeuwen

More information about this series at http://www.springer.com/series/7407

Sabri Arik · Tingwen Huang
Weng Kin Lai · Qingshan Liu (Eds.)

Neural
Information Processing

22nd International Conference, ICONIP 2015
Istanbul, Turkey, November 9–12, 2015
Proceedings, Part II

 Springer

Editors
Sabri Arik
University of Istanbul
Istanbul
Turkey

Tingwen Huang
University at Qatar
Doha
Qatar

Weng Kin Lai
Tunku Abdul Rahman University College
Kuala Lumpur
Malaysia

Qingshan Liu
University of Science Technology
Wuhan
China

ISSN 0302-9743   ISSN 1611-3349   (electronic)
Lecture Notes in Computer Science
ISBN 978-3-319-26534-6   ISBN 978-3-319-26535-3   (eBook)
DOI 10.1007/978-3-319-26535-3

Library of Congress Control Number: 2015954339

LNCS Sublibrary: SL1 – Theoretical Computer Science and General Issues

Springer International Publishing AG Switzerland is part of Springer Science+Business Media
(www.springer.com)

Preface

This volume is part of the four-volume proceedings of the 22nd International Conference on Neural Information Processing (ICONIP 2015), which was held in Istanbul, Turkey, during November 9–12, 2015. The ICONIP is an annual conference of the Asia Pacific Neural Network Assembly (APNNA; which was reformed in 2015 as the Asia Pacific Neural Network Society, APNNS). This series of ICONIP conferences has been held annually since 1994 in Seoul and has become one of the leading international conferences in the areas of artificial intelligence and neural networks.

ICONIP 2015 received a total of 432 submissions by scholars coming from 42 countries/regions across six continents. Based on a rigorous peer-review process where each submission was evaluated by an average of two qualified reviewers, a total of 301 high-quality papers were selected for publication in the reputable series of *Lecture Notes in Computer Science* (LNCS). The selected papers cover major topics of theoretical research, empirical study, and applications of neural information processing research. ICONIP 2015 also featured the Cybersecurity Data Mining Competition and Workshop (CDMC 2015), which was jointly held with ICONIP 2015. Nine papers from CDMC 2015 were selected for the conference proceedings.

In addition to the contributed papers, the ICONIP 2015 technical program also featured four invited speakers, Nik Kasabov (Auckland University of Technology, New Zealand), Jun Wang (The Chinese University of Hong Kong), Tom Heskes (Radboud University, Nijmegen, The Netherlands), and Michel Verleysen (Université catholique de Louvain, Belgium).

We would like to sincerely thank to the members of the Advisory Committee and Program Committee, the APNNS Governing Board for their guidance, and the members of the Organizing Committee for all their great efforts and time in organizing such an event. We would also like to take this opportunity to express our deepest gratitude to all the reviewers for their professional review that guaranteed high-quality papers.

We would like to thank Springer for publishing the proceedings in the prestigious series of *Lecture Notes in Computer Science*. Finally, we would like to thank all the speakers, authors, and participants for their contribution and support in making ICONIP 2015 a successful event.

November 2015

Sabri Arik
Tingwen Huang
Weng Kin Lai
Qingshan Liu

Organization

General Chair

Sabri Arik   Istanbul University, Turkey

Honorary Chair

Shun-ichi Amari   Brain Science Institute, RIKEN, Japan

Program Chairs

Tingwen Huang	Texas A&M University at Qatar, Qatar
Weng Kin Lai	School of Technology, Tunku Abdul Rahman College (TARC), Malaysia
Qingshan Liu	Huazhong University of Science Technology, China

Advisory Committee

P. Balasubramaniam	Deemed University, India
Jinde Cao	Southeast University, China
Jonathan Chan	King Mongkut's University of Technology, Thailand
Sung-Bae Cho	Yonsei University, Korea
Tom Gedeon	Australian National University, Australia
Akira Hirose	University of Tokyo, Japan
Tingwen Huang	Texas A&M University at Qatar, Qatar
Nik Kasabov	Auckland University of Technology, New Zealand
Rhee Man Kil	Korea Advanced Institute of Science and Technology (KAIST), Korea
Irwin King	Chinese University of Hong Kong, SAR China
James Kwok	Hong Kong University of Science and Technology, SAR China
Weng Kin Lai	School of Technology, Tunku Abdul Rahman College (TARC), Malaysia
James Lam	The University of Hong Kong, Hong Kong, SAR China
Kittichai Lavangnananda	King Mongkut's University of Technology, Thailand
Minho Lee	Kyungpook National University, Korea
Andrew Chi-Sing Leung	City University of Hong Kong, SAR China
Chee Peng Lim	University Sains Malaysia, Malaysia
Derong Liu	The Institute of Automation of the Chinese Academy of Sciences (CASIA), China

Chu Kiong Loo	University of Malaya, Malaysia
Bao-Liang Lu	Shanghai Jiao Tong University, China
Aamir Saeed Malik	Petronas University of Technology, Malaysia
Seichi Ozawa	Kobe University, Japan
Hyeyoung Park	Kyungpook National University, Korea
Ju. H. Park	Yeungnam University, Republic of Korea
Ko Sakai	University of Tsukuba, Japan
John Sum	National Chung Hsing University, Taiwan
DeLiang Wang	Ohio State University, USA
Jun Wang	Chinese University of Hong Kong, SAR China
Lipo Wang	Nanyang Technological University, Singapore
Zidong Wang	Brunel University, UK
Kevin Wong	Murdoch University, Australia

Program Committee Members

Syed Ali, India
R. Balasubramaniam, India
Tao Ban, Japan
Asim Bhatti, Australia
Jinde Cao, China
Jonathan Chan, Thailand
Tom Godeon, Australia
Denise Gorse, UK
Akira Hirose, Japan
Lu Hongtao, China
Mir Md Jahangir Kabir, Australia
Yonggui Kao, China
Hamid Reza Karimi, Norway
Nik Kasabov, New Zealand
Weng Kin Lai, Malaysia
S. Lakshmanan, India
Minho Lee, Korea
Chi Sing Leung, Hong Kong, SAR China
Cd Li, China

Ke Liao, China
Derong Liu, USA
Yurong Liu, China
Chu Kiong Loo, Malaysia
Seiichi Ozawa, Japan
Serdar Ozoguz, Turkey
Hyeyoung Park, South Korea
Ju Park, North Korea
Ko Sakai, Japan
Sibel Senan, Turkey
Qianqun Song, China
John Sum, Taiwan
Ying Tan, China
Jun Wang, Hong Kong, SAR China
Zidong Wang, UK
Kevin Wong, Australia
Mustak Yalcin, Turkey
Enes Yilmaz, Turkey

Special Sessions Chairs

Zeynep Orman	Istanbul University, Turkey
Neyir Ozcan	Uludag University, Turkey
Ruya Samli	Istanbul University, Turkey

Publication Chair

Selcuk Sevgen   Istanbul University, Turkey

Organizing Committee

Emel Arslan	Istanbul University, Turkey
Muhammed Ali Aydin	Istanbul University, Turkey
Eylem Yucel Demirel	Istanbul University, Turkey
Tolga Ensari	Istanbul University, Turkey
Ozlem Faydasicok	Istanbul University, Turkey
Safak Durukan Odabasi	Istanbul University, Turkey
Sibel Senan	Istanbul University, Turkey
Ozgur Can Turna	Istanbul University, Turkey

Contents – Part II

Motor Imagery Task Classification Using a Signal-Dependent Orthogonal Transform Based Feature Extraction

Mostefa Mesbah[1,2], Aida Khorshidtalab[3], Hamza Baali[4], and Ahmed Al-Ani[5(✉)]

[1] Department of Electrical and Computer Engineering College of Engineering,
Sultan Qaboos University, P O Box: 33 Muscat 123, Sultanate of Oman
mmesbah@ieee.org
[2] School of Computer Science and Software Engineering,
The University of Western Australia, 35 Stirling Highway, Perth, WA 6009, Australia
[3] Intelligent Mechatronics System Research Unit, Department of Mechatronics Engineering,
International Islamic University Malaysia (IIUM), Kuala Lumpur, Malaysia
ida.khorshidtalab@gmail.com
[4] Malaysia Industry Transformation, Technology Park Malaysia,
57000 Kuala Lumpur, Malaysia
baaliha@yahoo.fr
[5] Faculty of Eng and IT, University of Technology Sydney,
Ultimo, NSW 2007, Australia
Ahmed.Al-Ani@uts.edu.au

Abstract. In this paper, we present the results of classifying electroencephalographic (EEG) signals into four motor imagery tasks using a new method for feature extraction. This method is based on a signal-dependent orthogonal transform, referred to as LP-SVD, defined as the left singular vectors of the LPC filter impulse response matrix. Using a logistic tree based model classifier, the extracted features are mapped into one of four motor imagery movements, namely left hand, right hand, foot, and tongue. The proposed technique-based classification performance was benchmarked against those based on two widely used linear transform for feature extraction methods, namely discrete cosine transform (DCT) and adaptive autoregressive (AAR). By achieving an accuracy of 67.35 %, the LP-SVD based method outperformed the other two by large margins (+25 % compared to DCT and +6 % compared to AAR-based methods).

Keywords: Brain-computer interface · SVD · Feature extraction · Linear prediction · Orthogonal transform

1 Introduction

The aim of Brain-computer interface (BCI) is to set a direct communication link between the brain and external electronic devices whereby brain signals are translated into useful commands. Such communication link would assist people suffering from severe muscular (motor) disabilities with an alternative means of communication and control

© Springer International Publishing Switzerland 2015
S. Arik et al. (Eds.): ICONIP 2015, Part II, LNCS 9490, pp. 1–9, 2015.
DOI: 10.1007/978-3-319-26535-3_1

that bypass the normal output pathways [1–3]. In this paper, we focus on an important sub-component of BCI systems, namely feature extraction. This sub-component's aim is to identify a set of features that are effective in discriminating between different classes of interest.

Transform based approaches form an important class of feature extraction techniques. Their aim is to find a more compact lower-dimensional representation in which most of the signal's information is packed in a few number of uncorrelated coefficients. By eliminating irrelevant features (transform coefficients), these methods allow extracting effective features that preserve the generalization capability while lessening the computational complexity associated with the classification stage [4]. These transform-based approaches can be subdivided into linear and nonlinear, supervised and unsupervised, and signal dependent and signal independent methods. The most widely used linear techniques are PCA and LDA. The first one is unsupervised and aims at maximizing the variance of the projected data, using the eigenvectors of the sample covariance matrix, onto a low-dimensional subspace called principal subspace. In contrast, the latter is supervised and attempts to find a linear mapping that maximizes linear class separability of the data in a low-dimensional space [5].

Recently, the authors introduced a signal-dependent linear orthogonal transform, referred to as LP-SVD transform [6]. The transform has the advantage of forming the transformation matrix using only the AR model parameters, instead of the data samples as in the case of PCA. This transform is used in this paper to map EEG data into a new domain where only a few spectral coefficients contain most of the signal's energy. A subset of these transform coefficients, in conjunction with the LP coefficients and the error variance, were used as features in the classification of EEG into four class motor imagery tasks. The feature extraction method was validated using BCI IIIa competition dataset and its classification capability was assessed against two state-of-the-art methods based on DCT and AAR transforms.

The rest of the paper is organized as follows. Section 2 (*a*) describes the EEG data, its acquisition, and its pre-processing. Section 2 introduces the LP-SVD transform and how it is used in feature extraction. Section 3 compares the classification performance of the proposed LP-SVD based technique against two methods based on two of the most widely used linear transform for feature extraction. Section 4 concludes the paper.

2   Methodology

2.1   Data Acquisition and Pre-processing

The dataset IIIa from the BCI competition III (2005) [7] was used to evaluate the effectiveness of the proposed feature extraction method. It is a widely used benchmark dataset of multiclass motor imagery tasks recorded from three subjects; referred to as K3b, K6b and L1b. The multichannel EEG signals were recorded using a 64-channel Neuroscan EEG amplifier (Compumedics, Charlotte, North Carolina, USA). Only 60 EEG channels were actually recorded from the scalp of each subject using the 10–20 system and referential montage. The left and right mastoids served as reference and ground respectively. The recorded signal was sampled at 250 Hz and filtered using a bandpass filter with 1 and 50 Hz cut-off frequencies. A notch filter was then applied to suppress the interference originated from power lines. During the experiments, each subject was instructed to perform imagery movements associated with visual cues. Each trial started with an empty black screen at $t = 0$ s. At time point $t = 2$ s, a short beep tone was presented and a cross '+' appeared on the screen to raise the subject's attention. At $t = 3$ s, an arrow pointed to one of the four main directions (left, right, upwards or downwards) was presented. Each of the four directions, indicated by this arrow, instructed the subject to imagine one of the following four movements: left hand, right hand, tongue or foot, respectively. The imagination process was performed until the cross disappeared at $t = 7$ s. Each of the four cues was randomly displayed ten times in each run. No feedback was provided to the subject. The recorded dataset from subject K3b consists of 9 runs, while the ones from K6b and L1b consist of 6 runs each, which resulted in 360 trials for subject K3 and 240 trials for each of the other two subjects.

2.2 The LP-SVD Transform

The LP-SVD transform is constructed using a two-step process, namely the estimation of LPC filter coefficients and the computation of the left singular vectors of LPC filter impulse response matrix using singular value decomposition (SVD).

Linear prediction (LP) consists of computing the current signal observation, $y(n)$, using a linear combination of its P past samples, namely, $y(n - i)$ for $i = 1, \ldots, P$. This can be expressed mathematically by [8]

$$y(n) = - \sum_{i=1}^{P} a_i y(n - i) + e(n),\tag{1}$$

where, a_i are the linear prediction coefficients (LPCs), P is the prediction order and $e(n)$ is the prediction error. Equation (1) can be written in a more compact form using the following matrix notations:

$$y = He,\tag{2}$$

where $y = [y(1), \ldots, y(N)]^T$ and $e = [e(1), \ldots, e(N)]^T$ are respectively the $N \times 1$ columns vectors of the data samples and the prediction residual, while H is the $N \times N$ impulse response matrix of the synthesis filter (also called LPC filter) whose entries are completely determined by the linear prediction coefficients a_i. The matrix H is lower triangular and Toeplitz. Applying the SVD to H gives:

$$y = UDV^T e\tag{3}$$

U and V are the $N \times N$ orthogonal matrices containing the left and right eigenvectors of H and D is the $N \times N$ diagonal matrix of singular values [9].

We define the transformation that maps the measurement vector (y) to a feature vector (θ) by [6]:

$$\theta = U^T y\tag{4}$$

It is important to note that the transform operation $(U^T y)$ by itself does not achieve any dimensionality reduction. It only decorrelates and packs a large fraction of the signal energy into a relatively few transform coefficients as shown in Fig. 1.

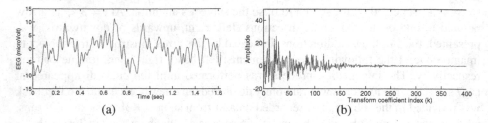

(a)   (b)

Fig. 1. Signal transformation using LP-SVD: (a) original EEG signal trace from subject L1b, (b) transform coefficients with AR(1) as a signal model.

2.3 LP-SVD-Based Feature Extraction

Our approach involves extracting features from each EEG segment. These features include the estimated LP coefficients (a_i), the prediction error variance (Vr), and a subset of the most significant transform coefficients θ. These features are described below.

According to the above LP analysis, the EEG vector is described in terms of all-poles filter coefficients and the prediction error. There are two classical approaches used to estimate the LP parameters, namely the autocorrelation and the covariance methods. In this study, we used the autocorrelation method as it guarantees the stability of the filter and allows the efficient Levinson-Durbin recursion to be used to estimate the model parameters [8]. Once the coefficients are estimated, the prediction error sequence can be computed using (1). The estimate of the prediction error $e(n)$ variance is given by:

$$Vr = \frac{1}{N-1} \sum_{n=1}^{N} (e(n) - \bar{e})^2 , \qquad (5)$$

where $\bar{e}$ is the arithmetic mean of the prediction error vector e and N is its length.

The data vector y is presented in the new coordinates $\{u_i\}$ by the transform coefficients or scores θ_i. The transform coefficients corresponding to the K largest singular values are selected as features:

$$\hat{\theta} = \hat{U}y, , \text{ The columns of } \hat{U} \text{ are } \{u_1, u_2, \dots, u_K\} . \qquad (6)$$

a. *DCT–based feature extraction procedure.*

The DCT is a signal independent, real-valued, orthogonal transform that is asymptotically equivalent to the optimal principal component analysis (PCA) for highly correlated first-order stationary autoregressive signals [10]. The orthonormal basis vectors w_k of an N points discrete cosine transform (DCT-II) are giving by:

$$w_k = \begin{cases} \frac{1}{\sqrt{N}} (1, 1, \dots, 1)^T & \text{for } k = 1 \\ \frac{2}{\sqrt{N}} \left(\cos \frac{k\pi}{2N}, \cos \frac{3k\pi}{2N}, \dots, \cos \frac{(2N-1)k\pi}{2N} \right)^T & \text{for } k = 2, \dots, N \end{cases} \tag{7)}$$

The $N \times N$ orthogonal DCT matrix is then defined as $W = (w_1, \dots, w_N)$. It follows immediately that the relation between a data vector y and its DCT transform Y is given by:

$$Y = W^T y \tag{8}$$

The resulting DCT coefficients represented by the vector Y are concentrated in the low-frequency subspace as shown in Fig. 2. Dimensionality reduction using DCT is realized by using only these low frequency coefficients as features and discarding the remaining high frequency coefficients. This is illustrated by the following linear mapping.

$$\hat{Y} = \hat{W}y, \tag{9}$$

The columns of $\hat{W}$ are $\{w_1, w_2, \dots, w_K\}$

Figure 2 shows an exemplary DCT coefficients vector of the EEG data of Fig. 1. The energy of the transformed data is packed into the first few low frequency coefficients while all high frequency coefficients are relatively small.

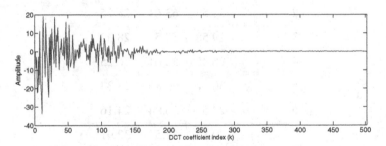

Fig. 2. DCT-II transform of the EEG signal shown in Fig. 1(a)

3 Experimental Results and Discussion

This section is divided into two parts. The first part is devoted to the AR model order selection. The second part evaluates the performance of the LP-SVD-based feature extraction method against two well-known related feature extraction methods. The classifier used to measure the performance is a logistic model tree implemented as part of the Weka software package with its default parameters [11]. This classifier, that uses SimpleLogistic, has a merit over other classifiers due to its use of LogitBoost. To evaluate the classification results, we used 10 fold cross-validation where the data is randomly split into 10 folds of equal size.

3.1   AR Model Selection

To investigate the appropriate AR model order and the number of transform coefficients to be retained as features, we performed a series of simulations. In this part, only the parameters characterizing the LP-SVD transform are used as features, namely, a subset of transform coefficients ($\hat{\theta}$), the LP coefficients (a_i) and the prediction error variance (*Vr*). The features were extracted from the electrode sites over the primary motor area C3, CZ, and C4. These are widely considered to be the most informative channels associated with motor imagery tasks [12].

We varied the AR model order from one to seven using the EEG segments from t = 3.5 s to t = 5.5 s (501 samples) from each trial. The best model order was selected based on the resulting classification accuracy. This criterion is more suitable, in the present context, than the commonly used one in signal representation (modeling), namely the tradeoff between the model order and the prediction error variance. Table 1 shows the classification results as function of the order of the AR model.

Table 1. AR model order selection

AR model order	Subject		
	L1b	K3b	K6b
	Accuracy (%)		
1	**42.08**	**66.11**	**47.5**
2	32.5	63.61	28.75
3	29.58	57.5	28.33
4	26.25	58.61	36.25
5	30	60.83	28.33
6	22.5	56.94	24.16
7	20.83	57.5	32.91

For all subjects, the highest classification accuracy, on average, was obtained with first order AR model and using a subset of four transform coefficients with results ranging from 42.08 % for subject l1b to 66.11 % for subject K3b. Therefore, this model order and number of transform coefficients were used in subsequent analysis.

3.2   Feature Extraction Evaluation

This part compares the performance of the feature extraction method to those using similar approaches, which are based on signal modeling and orthogonal transform. These techniques are based on adaptive autoregressive (AAR) model [12] and discrete cosine transform (DCT). In particular, Schlögl et al. [12] applied a third order adaptive autoregressive (AAR) model for EEG signal analysis. The extracted AAR coefficients, which provide dynamic information about the signal spectrum, served as features. The authors used three different classifiers namely, neural network based on k-nearest neighbour (kNN), support vector machines (SVM), and linear discriminant analysis (LDA) to classify the EEG signal into one of the four classes described earlier. The results showed that the SVM-based classifier achieved the best accuracies followed by LDA and then kNN. The authors also reported that the best results were obtained when using the features extracted from all 60 monopolar channels. In this evaluation, we used these same channels to provide a fair comparison between the methods.

To find the adequate number of DCT coefficients that achieve the highest classification performance for the different subjects, we varied the number of retained DCT coefficients from 5 to 50 with a step size of 5. Table 2 summarizes the obtained classification results as a function of the number of retained DCT coefficients. The number of coefficients required, for subjects K6b, L1b and K3b, to achieve the highest classification accuracies were 15, 40, and 20, respectively.

Table 2. Performance (classification accuracy) of DCT-based feature extraction using 60 Monopolar Channels

Number of coefficients retained	Subject		
	K6b	L1b	K3b
5	37.91	38.75	33.61
10	43.75	41.25	36.94
15	**45.833**	40.41	37.77
20	40.00	42.08	**38.05**
25	42.50	42.50	33.88
30	41.25	41.66	35.83
35	40.41	42.91	36.38
40	40.00	**43.75**	33.611
45	39.16	42.50	35.00
50	38.75	39.16	34.44

The performances of the three feature extraction approaches mentioned above are summarized in Table 3. It can be seen that when only the transform coefficients were used as features, the proposed approach outperformed the DCT-based one by up to 23 % in terms of accuracy (for subject L1b) with 10 times fewer number of features. Meanwhile, when the LP coefficient and the residual error variance were added to the LP-SVD transform coefficients, our technique performed better than the two methods for subjects L1b and K6b and achieved comparable results to the AAR-based method for subject K3b. On average, the improvement, in terms of accuracy was about +25 % compared to DCT and +6 % compared to AAR-based methods. It is pertinent to point out that, unlike DCT which results only in the transform coefficients as features, our method results in other features, LPC coefficients and residual signal variance, that led to a better characterization of the signal. In addition, the DCT is signal independent while our proposed transform is signal dependent. These two facts explain the difference in performance between the two methods.

Table 3. Comparative analysis of different features extraction approaches

Subject	DCT coefficients (best results)	AAR(3) Best SVM results [12]	LP-SVD coefficients (4 coefficients)	LP-SVD coefficients (4) + AR(1) + Error variance
	ACC %			
L1b	43.75	53.90	52.50	58.75
K3b	38.05	77.20	55.00	76.66
K6b	45.83	52.40	67.08	66.66
Avg	42.54	61.16	58.19	**67.35**

4   Conclusion

In the present study, we presented a feature extraction approach based on the combination of autoregressive modeling and orthogonal transformation. Results of classification experiments, using a benchmark dataset from the BCI competition III, and comparison against closely related approaches, namely DCT and AAR, demonstrates that the proposed feature set is compact and offers a significant improvement in performance as judged by the classification accuracy. The number of transform coefficients was kept constant during all the experiments. It would be interesting to address the issue of parameter tuning in future studies. Future work will also include adding more features to improve the performance beyond the one obtained in this study.

References

1. Lotte, F., Congedo, M., Lécuyer, A., Lamarche, F., Arnaldi, B.: A review of classification algorithms for EEG-based brain–computer interfaces. J. Neural Eng. **4**, R1-R13 (2007)
2. Zander, T.O., Kothe, C., Jatzev, S., Gaertner, M.: Enhancing human-computer interaction with input from active and passive brain-computer interfaces. In: Tan, D.S., Nijholt, A. (eds.) brain-computer interfaces. Human-Computer Interaction Series, pp. 181–199. Springer, London (2010)
3. Saa, J.F.D., Cetin, M.: Discriminative methods for classification of asynchronous imaginary motor tasks from EEG data. IEEE Trans. Neural Syst. Rehabil. Eng. **21**(5), 716–724 (2013)
4. Ozertem, U., Erdogmus, D., Jenssen, R.: Spectral feature projections that maximize Shannon mutual information with class labels. Pattern Recogn. **39**(7), 1241–1252 (2006)
5. Bishop, C.M.: Pattern Recognition and Machine Learning. Springer, New York (2006)
6. Baali, H., Akmeliawati, R., Salami, M.J.E., Khorshidtalab, A., Lim, E.-G.: ECG parametric modeling based on signal dependent orthogonal transform. IEEE Signal Process. Lett. **21**(10), 1293–1297 (2014)

7. Blankertz, B., Müller, K.-R., Krusienski, D.J., Schalk, G., Wolpaw, J.R., Schlögl, A., Gert Pfurtscheller, JdR, Millan, M.S., Birbaumer, N.: The BCI competition III: validating alternative approaches to actual BCI problems. IEEE Trans. Neural Syst. Rehabil. Eng. **14**(2), 153–159 (2006)
8. Vaidyanathan, P.P.: The theory of linear prediction. Synth. Lect. Sign. Proces. **2**(1), 1–184 (2007)
9. Strang, G.: Computational Science and Engineering, vol. 1. Wellesley-Cambridge Press, Wellesley (2007)
10. Ahmed, N., Milne, P.J., Harris, S.G.: Electrocardiographic data compression via orthogonal transforms. IEEE Trans. Biomed. Eng. **6**(BME-22), 484–487 (1975)
11. Hall, M., Frank, E., Holmes, G., Pfahringer, B., Reutemann, P., Witten, I.H.: The WEKA data mining software: an update. ACM SIGKDD Explor. Newsl. **11**(1), 10–18 (2009)
12. Schlögl, A., Lee, F., Bischof, H., Pfurtscheller, G.: Characterization of four-class motor imagery EEG data for the BCI-competition 2005. J. Neural Eng. **2**(4), L14 (2005)

Robust Ensemble Classifier Combination Based on Noise Removal with One-Class SVM

Ferhat Özgür Çatak[✉]

TÜBİTAK BİLGEM, Cyber Security Institute, Kocaeli/Gebze, Turkey
ozgur.catak@tubitak.gov.tr

Abstract. In machine learning area, as the number of labeled input samples becomes very large, it is very difficult to build a classification model because of input data set is not fit in a memory in training phase of the algorithm, therefore, it is necessary to utilize data partitioning to handle overall data set. Bagging and boosting based data partitioning methods have been broadly used in data mining and pattern recognition area. Both of these methods have shown a great possibility for improving classification model performance. This study is concerned with the analysis of data set partitioning with noise removal and its impact on the performance of multiple classifier models. In this study, we propose noise filtering preprocessing at each data set partition to increment classifier model performance. We applied Gini impurity approach to find the best split percentage of noise filter ratio. The filtered sub data set is then used to train individual ensemble models.

Keywords: One-class SVM · Data partitioning · Noise filtering · Gini impurity · Large scale data classification

1 Introduction

It's clear that we collect and store larger amounts of data in databases. The need for efficiently and effectively analyzing and utilizing the information contained in the data has been increasing. Just as big data technologies evolved, the quantity and variety of data has also increased, and becoming more focused on storing every type of data. The main purpose of the storing of the data is intended to obtain information from data using a variety of machine learning methods. One of the primary machine learning techniques is classification, which labels the new samples based on a training set whose class labels are provided [1,2]. Classification methods are applied in various areas such as bioinformatics, pattern recognition, text mining, social network analysis, etc.

In Big Data age, traditional classifier algorithms have new challenges to scaling up in order to address the large-scale data set training. Most of existing classification algorithms assume that the data can fit in a memory in training phase of learning. These algorithms cannot be comfortably implemented to data sets that larger than computer memory capacity. Data partitioning strategy is

© Springer International Publishing Switzerland 2015
S. Arik et al. (Eds.): ICONIP 2015, Part II, LNCS 9490, pp. 10–17, 2015.
DOI: 10.1007/978-3-319-26535-3_2

one of the methods that can be applied to the training of high-dimensional data sets that are used for the building of classification model in order to overcome the input data complexity. In order to prevent the building of weak classification model that emerged from the data chunks, the input set needs to be strengthened through various methods.

In this study, the noise filtering approach is applied to each individual sub data set to clean noisy input data, then, AdaBoost ensemble method is used to strength the classification model at each data partition. We applied one-class Support Vector Machine (SVM) method to filter noisy instances from each individual data partition and then AdaBoost ensemble based classification method is used to each individual data partition to increase the model accuracy.

The overall contributions of the study are listed as follows:

1. Using data partitioning method, the complexity of input matrix, which is quite high for the single memory, is reduced in this manner.
2. Each individual sub-set of input matrix is reinforced with noise filtering method using one-class SVM and Gini impurity.
3. Each sub-set of input matrix is used in the training phase of the different ensemble classifier, so that each instances are considered when building a global classification model.

Gini impurity is used to calculate the uncertainty about source of input data set. This measure is applied to estimate the degree of information diversity provided by cleaned partition of sub data set.

The remainder of this paper is organized as follows: Sect. 2 briefly explains the methods that are used in this work. Section 3 describes the proposed data cleaning and partitioning method. Section 4 gives the experimental results. In Sect. 5, we give conclusion and future works.

2 Preliminaries

The approach presented in this paper uses one-class SVM algorithm to remove noisy instances, AdaBoost to build ensemble classifier models, and data partitioning to train over all data set instances. All elements are introduced here briefly.

2.1 One-Class SVM

SVM [3] method is used to find classification models using the maximum margin separating hyper plane. Schölkopf et al. [4] proposed a training methodology that handles only one class classification called as "one-class" classification.

One-class SVM algorithm is a method used to detect the outliers in the data. Basically the method finds soft boundaries of the data set, and then, model determines whether new instance belongs to this data set or not. Suppose, we are given a data set, $\mathbf{x}_1, \ldots \mathbf{x}_m \in X$ drawn from an unknown underlying probability distribution P. We are interested in estimating a set S such that the probability

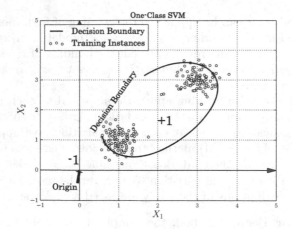

Fig. 1. One-class SVM. The origin, $(0,0)$, is the single instance with label -1.

that a test point from P lies inside in S with an a priori specified probability value. As shown in Fig. 1, origin is labeled as -1, and the all training instances are labeled as $+1$.

Let $S = \{(\mathbf{x}_i, y_i) | \mathbf{x}_i \in \mathbb{R}^n, y \in \{1, \ldots, K\}\}_{i=1}^m$ be input instances in $\mathbb{R}^n$, $\phi : X \to H$ be a kernel function that maps the input instances to another space. Then standard SVM method tries to find a hyper plane that solves the separation problem with an optimization problem. The objective function of the SVM classifier is formulated as follows.

$$min_{\mathbf{w}, \xi, \rho} \left(\frac{1}{2} ||\mathbf{w}||^2 + \frac{1}{mC} \sum_i \xi_i - \rho \right)$$

$$subject\, to \tag{1}$$

$$(w.\phi(\mathbf{x}_i)) \geq \rho - \xi_i$$

$$\xi_i \geq 0, \forall i = 1, \ldots, m$$

where $\mathbf{w}$ is orthogonal to the separating hyper plane, C is smoothness parameter, $\mathbf{x}_i$ is the i-th input instances, m is the total number of input instances, ξ_i are the slack variables, ρ is the distance between origin and separating hyper plane.

By using Lagrange techniques, $\mathbf{w}$ and ρ are obtained, then the decision function becomes:

$$f(\mathbf{x}) = sign\left((\mathbf{w}.\phi(\mathbf{x})) - \rho\right). \tag{2}$$

2.2 AdaBoost

The AdaBoost [5] is a supervised learning algorithm designed to solve classification problems [6]. The algorithm takes as input a training set $(\mathbf{x}_1, y_1), ..., (\mathbf{x}_n, y_n)$ where the input sample $\mathbf{x}_i \in R^p$, and the output value, y_i, in a finite space

$y \in 1, ...K$. AdaBoost algorithm assumes a set of training data sampled independently and identically distributed (i.i.d.) from some unknown distribution $\mathcal{X}$.

Given a space of feature vectors X and two possible class labels, $y \in \{-1, +1\}$, AdaBoost goal is to learn a strong classifier $H(\mathbf{x})$ as a weighted ensemble of weak classifiers $h_t(\mathbf{x})$ predicting the label of any instance $\mathbf{x} \in X$ [7].

$$H(\mathbf{x}) = sign(f(\mathbf{x})) = sign\left(\sum_{t=1}^{T} \alpha_t h_t(\mathbf{x})\right). \tag{3}$$

2.3 Data Partitioning Strategies

The use of multiple classifiers, learning methods are applied to base classifiers with different methods. Data partitioning is used a variety of reasons. First reason is the diversity that means uncorrelated base classifiers [8,9]. Another reason is the reducing the input complexity of large-scale data sets [10]. Last one is to build classifier models for the specific part of the input instances [11].

Data partitioning is basically divided into two different groups; filter based data partitioning and wrapper based data partitioning [12]. In wrapper based data partitioning, sub-data sets are created using base classifier outputs [13]. In filter based data partitioning, sub-data sets are created before individual classifiers are trained [14].

3 Proposed Approach

In this section we provide the details of the proposed noise filter based sub data set training method. The basic idea of noise removing based on one-class SVM technique is introduced in Sect. 3.1. The analysis of proposed method is described in Sect. 3.2.

3.1 Basic Idea

Our main task is to partition the input data set into sub-data sets, (X_m, Y_m), and, create local classifier ensembles for each sub data chunk. Noise removing process is applied to each individual sub-data set as pre-processing. Weighted voting method is used to combine the each ensemble classifier, and then, a single classifier model is created. Overall of the proposed method is shown in Fig. 2.

3.2 Analysis of the Proposed Algorithm

Kragh et al. showed that ensemble methods of neural networks gets better accuracy performance over unseen examples [15]. The main motivation of this work is the idea that small size classifier ensembles can obtain more accurate classifier model that are comparable to individual classifiers.

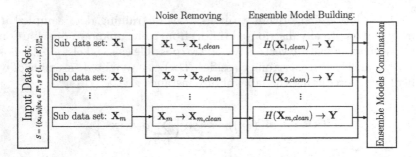

Fig. 2. Overall of proposed approach.

In the proposed model, at every sub-data set, there is a set of classifier functions (ensemble classifier), $H^{(m)}$, that acts as a single classification model. The single model at every sub-data set, m, is defined as follows:

$$H^{(m)}(\mathbf{x}) = \arg \max_k \sum_{t=1}^{M} \alpha_t h_t(\mathbf{x}) \qquad (4)$$

The selected ensemble classifier models from last phase of our algorithm are combined into one single classification model, $\hat{H}(\mathbf{x})$, using accuracy based majority voting method.

$$\hat{H}(\mathbf{x}) = \arg \max_k \sum_{i=1}^{m} \beta H^{(m)}(\mathbf{x}) \qquad (5)$$

where β is the accuracy of ensemble classifier.

4   Experiments

In this section, we perform experiments on real-world data sets from the public available data set repositories. Public data sets are used to evaluate the proposed learning method. Classification models of each data set are compared for accuracy results without removing noisy samples from them.

4.1   Experimental Setup

In this section, our approach is applied to five different data sets to verify its model effectivity and efficiency. The data sets are summarized in Table 1, including cod-rna, ijcnn1, letter, shuttle and SensIT Vehicle. We choose 50 as the data split size, m, and 3 different classification methods including Extra Trees [16], k-nn and SVM.

Table 1. Description of the testing data sets used in the experiments.

Data set	#Train	#Test	#Classes	#Attributes
cod-rna	59,535	157,413	2	8
ijcnn1	49,990	91,701	2	22
letter	15,000	5,000	26	16
shuttle	43,500	14,500	7	9
SensIT vehicle	78,823	19,705	3	100

4.2 Effect of Noise Removing on Input Matrix

In this section, we show the impact of noise removal pre-processing on the sample data sets. In order to show the noise removing affects, we used the "Gini Impurity" to measure the quality of procedure. Gini approaches deal appropriately with data diversity of a data. The Gini measures the class distribution of variable $\mathbf{y} = \{y_1, \cdots, y_m\}$. The Gini impurity can be written as:

$$g = 1 - \sum_k p_j^2 \tag{6}$$

where p_j is the probability of class k, in data set $\mathcal{D}$.

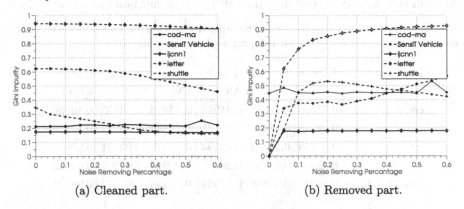

(a) Cleaned part.   (b) Removed part.

Fig. 3. The impact of one-class SVM on the performance on selected data sets in terms of Gini impurity.

The cleaning results are shown in Fig. 3a and b. As expected, the Gini impurity value decreases with the cleaning of the noisy instances from the data, and increases on separated noisy data. Our aim is to minimizing the Gini impurity on clean data set, $\mathbf{X}_{clean}$, and maximizing the value on noisy data set $\mathbf{X}_{noisy}$. As a result, this division ratio which minimizes the ratio between the two values was regarded as the optimum value.

$$Split\,Percantage = \arg\max_p \frac{Gini(\mathbf{X}_{clean}, p)}{Gini(\mathbf{X}_{noisy}, p)} \tag{7}$$

Table 2 shows the best Gini impurity performances of each data set used in our experiments.

Table 2. The best noise removal percentages of each data sets.

Data sets	Percentage	$Gini(\mathbf{X}_{clean})$	$Gini(\mathbf{X}_{noisy})$	$\frac{Gini(\mathbf{X}_{clean})}{Gini(\mathbf{X}_{noisy})}$
cod-rna	0,55	0,331344656	0,56831775	0,583027112
SensIT vehicle	0,60	0,461155751	0,568847982	0,810683638
ijcnn1	0,60	0,167652825	0,179397223	0,934534117
letter	0,60	0,9039108	0,926430562	0,975691905
shuttle	0,30	0,214142833	0,506938229	0,422423918

4.3 Simulation Results

The process of the experiments are as follows: Firstly, we trained our data sets without using noise removal. Then we perform classification on test data sets, and calculate the accuracy of classifiers. We repeated the experiments 50 times, and average accuracy is calculated. Table 3 shows the average accuracy of each example data sets with and without noise removing using one-class SVM method.

As can be seen on Table 3, the noise removing based partitioned proposed algorithm significantly outperforms the splitted classifier building in most cases.

Table 3. Classification performance on example datasets using One-Class SVM noise removing and without removing for the proposed learning algorithm.

Data set	Extra trees		K-nn		SVM	
	All	Clean	All	Clean	All	Clean
cod-rna	0,75929	0,78652	0,91955	0,93513	0,88553	0,89806
ijcnn1	0,69758	0,72175	0,75533	0,77833	0,82989	0,84326
letter	0,91255	0,90853	0,90594	0,90318	0,92121	0,9199
shuttle	0,47801	0,44792	0,20262	0,26974	0,62301	0,63939
SensIT vehicle	0,9602	0,90558	0,87904	0,88266	0,99232	0,99174

5 Conclusions

In this paper, we have introduced a novel data partitioning based classifier building method, which improves the sub data sets with removing the noisy instances using one-class SVM and find best noise removing ratio with Gini impurity value. We carried out a series of computer experiments to find a global ensemble classifier and the performance of our proposed method. The training process of a partitioned data set is simple, fast and final classifier model handle overall training instances. Our experimental results show that the memory requirement of training phase reduced remarkably, and the accuracy increased by using the noise removal process. The proposed method is a practical multiple ensemble classifier training model to classify large-scale data sets.

In the future work, our plan is to study different noise removing methods to clean sub data set. We plan adaptive noise removing ratio to make our method as autonomous as possible.

References

1. Anderson, J.R., Michalski, R.S., Carbonell, J.G., Mitchell, T.M.: Machine Learning: An Artificial Intelligence Approach, vol. 2. Morgan Kaufmann, San Mateo (1986)
2. Ramakrishnan, R., Gehrke, J.: Database Management Systems. Osborne/McGraw-Hill, Berkeley (2000)
3. Vapnik, V.: The Nature of Statistical Learning Theory. Springer Science and Business Media, New York (2000)
4. Schölkopf, B., Platt, J.C., Shawe-Taylor, J., Smola, A.J., Williamson, R.C.: Estimating the support of a high-dimensional distribution. Neural Comput. $13(7)$, 1443–1471 (2001)
5. Freund, Y., Schapire, R.E.: A desicion-theoretic generalization of on-line learning and an application to boosting. In: Vitányi, P. (ed.) Computational Learning Theory. Lecture Notes in Computer Science, vol. 904, pp. 23–37. Springer, Heidelberg (1995)
6. Freund, Y., Schapire, R., Abe, N.: A short introduction to boosting. J.-Jpn. Soc. Artif. Intell. $14(771$–$780)$, 1612 (1999)
7. Landesa-Vzquez, I., Alba-Castro, J.L.: Double-base asymmetric AdaBoost. Neurocomputing 118, 101–114 (2013)
8. Kuncheva, L.I.: Using diversity measures for generating error-correcting output codes in classifier ensembles. Pattern Recogn. Lett. $26(1)$, 83–90 (2005)
9. Dara, R.A., Makrehchi, M., Kamel, M.S.: Filter-based data partitioning for training multiple classifier systems. IEEE Trans. Knowl. Data Eng. $22(4)$, 508–522 (2010)
10. Chawla, N.V., Moore, T.E., Hall, L.O., Bowyer, K.W., Kegelmeyer, W.P., Springer, C.: Distributed learning with bagging-like performance. Pattern Recogn. Lett. $24(1)$, 455–471 (2003)
11. Woods, K., Bowyer, K., Kegelmeyer Jr., W.P.: Combination of multiple classifiers using local accuracy estimates. In: 1996 IEEE Computer Society Conference on Computer Vision and Pattern Recognition, Proceedings CVPR 1996, pp. 391–396. IEEE (1996)
12. Blum, A.L., Langley, P.: Selection of relevant features and examples in machine learning. Artif. Intell. $97(1)$, 245–271 (1997)
13. Freund, Y., Schapire, R.E., et al.: Experiments with a new boosting algorithm. ICML 96, 148–156 (1996)
14. Breiman, L.: Bagging predictors. Mach. Learn. $24(2)$, 123–140 (1996)
15. Krogh, A., Vedelsby, J.: Neural network ensembles, cross validation, and active learning. In: Advances in Neural Information Processing Systems, pp. 231–238. MIT Press (1995)
16. Geurts, P., Ernst, D., Wehenkel, L.: Extremely randomized trees. Mach. Learn. $63(1)$, 3–42 (2006)

Soil Property Prediction: An Extreme Learning Machine Approach

Dina Masri, Wei Lee Woon, and Zeyar Aung[✉]

Department of Electrical Engineering and Computer Science,
Masdar Institute of Science and Technology, Abu Dhabi, UAE
{dmasri,wwoon,zaung}@masdar.ac.ae

Abstract. In this paper, we propose a method for predicting functional properties of soil samples from a number of measurable spatial and spectral features of those samples. Our method is based on Savitzky-Golay filter for preprocessing and a relatively recent evolution of single hidden-layer feed-forward network (SLFN) learning technique called extreme learning machine (ELM) for prediction. We tested our method with Africa Soil Property Prediction dataset, and observed that the results were promising.

Keywords: Soil property · Prediction · Neural network · Extreme learning machine (ELM) · Kernel-based ELM

1 Introduction

Computational prediction of soil properties is an important task in modern agricultural and environmental studies. It allows us to perform low-cost analysis on the measurable soil features in order to forecast the soil's functional properties like primary productivity, nutrient and water retention, and resistance to soil erosion. These properties are important for planning sustainable agricultural intensification and natural resources management. The best available low cost analysis can be done using diffuse reflectance infrared spectroscopy measurements and geo-referencing of soil samples. Using spectroscopy, the amount of light absorbed by a soil sample is measured at different wavelengths to provide an infrared spectrum of the soil sample.

In this work, we propose a method to first preprocess the data using Savitzky-Golay filter, and then build an effective predictive model using a newly emerging single hidden-layer feed-forward neural network learning (SLFN) technique called "extreme learning machine (ELM)". We tested our method with "Africa Soil Property Prediction Challenge" dataset [9], and observed that the results were promising with low prediction error rates and low standard deviations of errors.

© Springer International Publishing Switzerland 2015
S. Arik et al. (Eds.): ICONIP 2015, Part II, LNCS 9490, pp. 18–27, 2015.
DOI: 10.1007/978-3-319-26535-3_3

2 Problem Definition

Our objective is to predict of 5 target soil "functional properties" from the 16 spatial and the 3,566 spectral features which are relatively easy to measure. (Note: we follow the same objective as in the Africa Soil Property Prediction Challenge [9], which was sponsored by Africa Soil Information Service [1].)

The explanatory input variables are the spectral and spatial features of the soil sample. The spectral features consists of a range of 3,578 Near-infrared (NIR)

Table 1. Variables of soil samples.

Variable Name	Description
16 Spatial Variables	
BSAN	Near-infrared average long-term Black Sky Albedo measurement
BSAS	Shortwave average long-term Black Sky Albedo measurement
BSAV	Visible average long-term Black Sky Albedo measurement
CTI	Compound topographic index
ELEV	Elevation
EVI	Average long-term enhanced vegetation index
LSTD	Average long-term land surface day time temperature
LSTN	Average long-term land surface day night time temperature
REF1	Average long-term Reflectance measurement for blue
REF2	Average long-term Reflectance measurement for red
REF3	Average long-term Reflectance measurement for near-infrared
REF7	Average long-term Reflectance measurement for mid-infrared
RELI	Topographic relief
TMAP	Mean annual precipitation (rainfall)
TMFI	Modified Fournier index of rainfall
Depth	Depth of soil sample ("Topsoil" for 0–20 cm depth and "Subsoil" for 20–50 cm depth)
3,578 Spectral Variables	
m599.76–m7497.96	Mid-infrared absorbance measurements at different wavelengths (599.76–7497.96 cm^{-1}) [Note: 12 of them are removed later.]
5 Target Variables (Soil Functional Properties)	
CA	Mehlich-3 extractable Calcium
P	Mehlich-3 extractable Phosphorus
pH	pH value
SOC	Soil organic carbon
Sand	Sand content

absorbance measurements at different wavelengths ranging from (599.76–7497.96 cm^{-1}). (Note: among them, 12 CO_2 spectral features ranging between 2352.76 and 2379.76 cm^{-1} are removed as suggested by the experts since this area of the spectrum picks up atmospheric CO_2 absorption features that are not due to the soil sample [9]. After this removal, 3,566 spectral features remain.) The spatial features are from environmental and remote sensing data sources as well as the depth from which the soil sample is taken.

The five target variables are (1) Mehlich-3 extractable Calcium (Ca), (2) Mehlich-3 extractable Phosphorus (P), (3) pH value (pH), (4) Soil organic carbon (SOC), and (5) Sand content (Sand). A summary of the variables are presented in Table 1.

The dataset we used was the one provided in the Africa Soil Property Prediction Challenge [9]. It contains 1,158 training instances (soil samples) and 728 test instances. We did not use the test instances in our study because the values of their target variable were not publicly available. It should be also noted that geographical clustering was observed in the raw dataset [10], which was due to the spatially stratified multi-level sampling design that was used during the assembling phase of the data. For that reason, the data were randomized (shuffled) before the training and cross validation processes.

3   Proposed Method

3.1   Data Preprocessing

As mentioned in Sect. 2, the training dataset is of very high dimensionality (3,582 input features) with respect to the number of training instances (1,158). This calls for preprocessing of the data before entering it to the predictive model training process. Since the spectral features comprise 3,566 NIR variables, our main focus is to preprocess them and reduce their number.

According to Rinnan et al. [17], preprocessing of NIR spectral data is an integral part of the any predictive modeling involving this type of data. It is essential to remove any physical phenomena in the NIR spectra in order to retrieve the most important information within. In general, the spectra is highly influenced by non-linearities caused by light scatter of the sample under study. There are two main methods used for NIR preprocessing [17], namely, scatter correction and spectral derivatives. The first method involves statistical analysis such as multiplicative scatter correction. The second method, which is the simplest, involves smoothing the spectra then performing a derivative (usually of first or second order) to decrease the signal to noise ratio.

One of the most famous smoothing filters used for spectral derivative preprocessing is called Savitzky-Golay smoothing filter [18]. This filter basically fits the NIR spectra to a polynomial, then average out the resulting fitted model. It uses a windowed data with odd number of points to fit the polynomial. Up until this step, only the measurement noises are smoothed out, as shown in Fig. 1, where a sample spectral raw data is plotted with its Savitzky-Golay smoothed version plotted over it. To change the NIR data from absolute values to relative

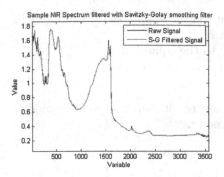

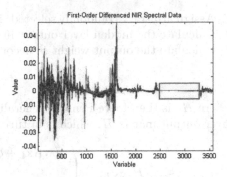

Fig. 1. Savitzky-Golay smoothed NIR sample spectra.

Fig. 2. First order differenced NIR training data (Color figure online).

rate of change values, a first order derivative (differencing) is performed after the smoothing step. The resulting differenced spectral data is presented in Fig. 2. It can be noticed in the figure that some parts of the spectra, like the area in the red box, are of no use and most probably carry no information. Removing such areas of the spectra serves as a means of discarding irrelevant features, thus achieving dimensionality reduction.

3.2   Prediction Using Extreme Learning Machine (ELM)

Neural network has proven to provide effective performances in diverse application areas over the years [15,16,19]. Single hidden layer feed-forward network (SLFN) is a powerful category of neural network. A SLFN with any continuous bounded and nonlinear activation function can form decision regions with arbitrary shapes in multi-dimensional cases. That means an SLFN can approximate any continuous function and implement any classification application [5]. To overcome the slowness of learning in neural networks and long iterative process for parameter tuning, based on the findings in [5], Huang et al. [7] proposed a new evolutionary learning technique called extreme learning machine (ELM). It can train any SLFN with exceptional speed (thousands of times faster than the traditional technique) and extreme generalization abilities.

These improvements in ELM are based on the observation that the mapping parameters between the input and the hidden layer are not correlated to the output, so they are set randomly at the beginning of the training process. Moreover, this learning technique tends to result in the smallest training errors as well as the smallest norm of weights. That means not only it is targeted at the minimization of error, but also on the minimization of the norm of the tuning parameters. This specific property is in accordance with Bartlett's theory [2] that the generalization ability of any feed-forward neural network is much better with smaller weights or tuning parameters.

Given a training set $X = [(\mathbf{x}_i, \mathbf{y}_i) \mid \mathbf{x}_i \in \mathbb{R}^n, \mathbf{y}_i \in \mathbb{R}^m, i = 1, \ldots, N]$, any non-zero activation function $g(\mathbf{x})$, and the number of hidden nodes (neurons) L, the ELM algorithm can be summarized in the following three steps [7].

1. Assign arbitrary (randomly selected) input weight $\mathbf{w}_i$ and bias $b_i, i = 1, \ldots, L$.
2. Calculate the hidden layer output matrix H.
3. Calculate the output weight β, according to the equation:

$$\beta = H^\dagger Y \tag{1}$$

where $H^\dagger$ is the Moore-Penrose generalized inverse of matrix of SLFN's hidden layer output matrix H, which is in turn defined as:

$$H_{(N \times L)} = \begin{pmatrix} g(\mathbf{w}_1.\mathbf{x}_1 + b_1) & \cdots & g(\mathbf{w}_L.\mathbf{x}_1 + b_L) \\ \vdots & \ddots & \vdots \\ g(\mathbf{w}_1.\mathbf{x}_N + b_1) & \cdots & g(\mathbf{w}_L.\mathbf{x}_N + b_L) \end{pmatrix}$$

and

$$\beta_{(L \times m)} = \begin{pmatrix} \beta_1^T \\ \vdots \\ \beta_L^T \end{pmatrix}, \quad Y_{(L \times m)} = \begin{pmatrix} \mathbf{y}_1^T \\ \vdots \\ \mathbf{y}_L^T \end{pmatrix}$$

H is constant throughout the training since the input mapping is done randomly and is fixed. So, training an SLFN using ELM can simply be done by finding a least-squares solution $\widehat{\beta}$ of the linear system $H\beta = Y$, and this solution is equivalent to $\widehat{\beta} = H^\dagger Y$. This learning technique is not iterative and the solution is unique. Thus, the risk of local minima does not exist.

To improve the generalization performance robustness, a regularization parameter (C) can be added to Eq. 1 [6]. Now, the new solution is:

$$\widehat{\beta} = (\frac{I}{C} + H^\dagger H)^{-1} H^\dagger H \tag{2}$$

ELM can also be extended to kernel learning [6] because it can use any type of feature mapping (between the hidden layer and the output) including kernels. Here, the output function of ELM becomes:

$$f(\mathbf{x}) = h(\mathbf{x})H^\dagger(\frac{I}{C} + H^\dagger H)^{-1}Y = \begin{pmatrix} K(\mathbf{x}, \mathbf{x}_1) \\ \vdots \\ K(\mathbf{x}, \mathbf{x}_N) \end{pmatrix}^T (\frac{I}{C} + \Omega_{ELM})^{-1}Y \tag{3}$$

where $\Omega_{ELM} = HH^\dagger : \Omega_{ELM}[i, j] = h(\mathbf{x}_i) \cdot h(\mathbf{x}_j) = K(\mathbf{x}_i, \mathbf{x}_j)$, with Ω_{ELM} being the kernel matrix. $K(\mathbf{x}_i, \mathbf{x}_j)$ can be any kernel function like Gaussian, polynomial, linear, etc. Here, the feature mapping function $h(\mathbf{x})$ is not needed to be known by the users, instead its kernel is given. Moreover, unlike the basic ELM where the feature mapping from the hidden layer to the output $h(\mathbf{x})$ is done using any activation function, here the number of hidden nodes (L) is not needed to be known either.

4   Results and Discussions

In order to implement the ELM-based prediction model to solve the soil property prediction task at hand, we used the codes [8] provided by the same research group who developed ELM. Two major ELM algorithms were tested: (1) basic ELM and (2) kernel-based ELM.

For the basic ELM, the Sigmoid activation function was used since it is the most prominent. (It was concluded in [6] that the training time spent by ELM with Sigmoid additive nodes increases very slowly when the number of training data increases.) We varied two main parameters, namely, (1) the number of nodes (neurons) in the hidden layer, L and (2) the ELM regularization parameter, C.

For the kernel-based ELM, the Gaussian kernel was chosen. We tuned two main parameters, namely, (1) the Gaussian kernel parameter, γ and (2) the ELM regularization parameter, C.

For each unique parameter setting, ten-fold cross validation (CV) was performed on the training set of 1,158 instances from the Africa Soil Property Prediction Challenge [9] dataset. For the basic Sigmoid-based ELM, the parameter values tested are: $L = \{10, 50, 100, 200, 500, 700, 1000\}$ and $C = \{0.001, 0.01, 0.1, 1, 10, 50, 100, 1000, 5000, 10000\}$, thus 70 different CV trials were performed. For the Gaussian kernel-based ELM, the following values were used: $C = \{0.001, 0.01, 0.1, 0.2, 0.5, 1, 5, 10, 20, 50, 100, 1000, 10000\}$ and $\gamma = \{0.001, 0.01, 0.1, 0.2, 0.4, 0.8, 1, 5, 10, 50, 100, 1000, 10000\}$, requiring 169 different CV trials.

The results were evaluated using the mean column-wise root mean squared error (MCRMSE) metric, where the errors of each ten-fold CV trials were averaged over $T = 115$ or 116 test instances across $m = 5$ target variables. y_{ij} and $\hat{y}_{ij}$ stand for the actual and the predicted values respectively.

$$\text{MCRMSE} = \frac{1}{m} \sum_{j=1}^{m} \sqrt{\frac{1}{T} \sum_{i=1}^{T} (y_{ij} - \hat{y}_{ij})} \qquad (4)$$

Moreover, we used the mean cross validation standard deviation (MCVSTD) metric, where the average standard deviation of each ten-fold CV test was measured by averaging over all the target variables for each parameter change in the prediction models.

The results of the ten-fold CV tests over the different parameter values are depicted in Figs. 3 and 4. A Matlab 3-D plot was created for each of the basic Sigmoid-based and the Gaussian kernel-based models showing the effect of the regularization parameter as well as the model specific parameter (i.e., L and γ, respectively). Tables 2 and 3 show samples of the best CMRMSE values obtained for the both models given specific parameter combinations, as well as the values of MCVSTD.

In Fig. 3, it can be observed that there is no strong dependency of the performance on the combination of (C, L) in the basic Sigmoid-based ELM. As long as L is relatively high, the model performs well. (This conclusion was also obtained in [6].) On the contrary, for the Gaussian kernel-based ELM, Fig. 4 shows that the model performance is much more sensitive on the combination of (C, γ).

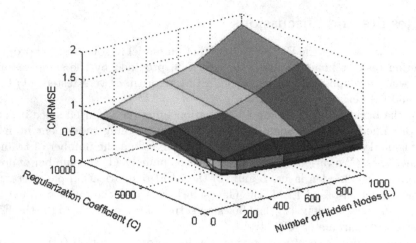

Fig. 3. Basic Sigmoid-based ELM's results for different values of C and L.

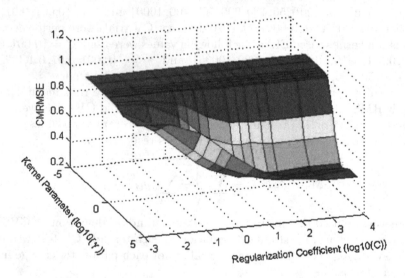

Fig. 4. Gaussian kernel-based ELM's results for different values of C and γ.

Regarding the prediction accuracy, under the provided tuning of parameters, the Gaussian kernel-based model performed better with the CMRMSE value as low as 0.3937 with 0.1529 standard deviation of the ten-fold CV. The average value of CMRMSE in the table is 0.4199. This beats the leaderboard top result in Africa Soil Property Prediction Challenge [11], which was 0.4689. However, we acknowledge that this is not an accurate comparison because our result is just a ten-fold CV error on the training data, but not on the test data used in the competition (because their actual target variable values are not publicly available). Nonetheless, our results do show the potential effectiveness for our proposed predictive model.

Table 2. Sample results obtained by different combinations of (L, C) parameters for basic Sigmoid-based ELM. The minimum (best) values are highlighted.

L	C	CMRMSE	MCVSTD
1000	0.01	0.5077	0.1786
1000	0.10	**0.4724**	**0.1342**
1000	1.00	0.4985	0.1419
1000	10.00	0.5792	0.2490
700	0.01	0.5331	0.1675
700	0.10	0.4899	0.1586
700	1.00	0.4987	0.1494
700	10.00	0.5435	0.1402
700	50.00	0.5967	0.1640
500	0.01	0.5533	0.2202
500	0.10	0.5014	0.1511
500	1.00	0.5149	0.1385
500	10.00	0.5438	0.1615
500	50.00	0.5893	0.2212
500	100.00	0.6162	0.2677
200	0.10	0.5720	0.1855
200	1.00	0.5660	0.1952
200	10.00	0.5703	0.1529
200	100.00	0.6035	0.1541

Table 3. Sample results obtained by different combinations of (γ, C) parameters for Gaussian kernel-based ELM. The minimum (best) values are highlighted.

γ	C	CMRMSE	MCVSTD
10000	1000	0.4430	0.1637
1000	1000	0.4265	0.1294
1000	100	0.4185	0.1356
1000	50	0.4276	0.1696
1000	20	0.4432	0.1433
100	10000	0.4033	0.1528
100	1000	0.4017	**0.1285**
100	100	**0.3937**	0.1529
100	50	0.4073	0.1542
100	20	0.4199	0.1476
100	10	0.4214	0.1534
100	5	0.4143	0.1842
50	10000	0.4219	0.1376
50	1000	0.4234	0.1421
50	100	0.4129	0.1296
50	50	0.4116	0.1444
50	20	0.4299	0.2140
50	10	0.4291	0.2060
50	5	0.4293	0.1969

5   Related Work

Different machine learning algorithms have been tried and tested to solve the problem of soil property prediction [9]. These comprise multi-layer neural networks, Bayesian additive regression trees (BART), support vector machine (SVM), ridge regression, lasso regression, elastic net lasso, gradient boosting regressor, and many others [10]. The BART model was provided as an example with a benchmark CMRMSE of 0.56551. However, to our best knowledge, no one has tried ELM to solve this problem before.

ELM as a learning technique has a number of advantages over other state-of-the-art classification and regression algorithms. The main reasons are its generalization ability, unique solutions with minimized training error, and the ability to theoretically model any function or decision boarder no matter how complex it is. For example, as mentioned in [6], compared to two widely-used variants of support vector machine for regression, namely, least-square support vector machine (LS-SVM) and proximal support vector machine (P-SVM), ELM is subject to fewer and milder optimization constraints. Moreover, SVM sometimes may provide sub-optimal solutions, unlike ELM where it has a unique solution. ELM exhibits both better scalability and generalization performance.

Different integration attempts between SVM and the concept of ELM has been proposed in the literature [4,13], where the concept of randomized feature spaces for SVM algorithms was introduced in order to enhance SVM's generalization performance. The basic idea is to use ELM to compute a kernel of the first layer of a SLFN, which in turn is used train the SVM.

ELM was also compared to another newly emerged neural network-based learning algorithm called deep learning or deep networks [3]. Deep learning outperforms all traditional classification/regression methods like multi-layer neural networks, SLFNs, and SVMs for big data analysis. However, the training process of deep learning is very slow. A new structure of ELM was proposed in [3] to resemble the deep learning process, but with a much faster learning speed. This structure consists of a multi-layer ELM (ML-ELM) which performs layer-by-layer unsupervised learning like deep learning does. This new learning algorithm is much faster than the existing deep learning techniques while maintaining a comparable predictive performance.

ELM was extended into an online sequential learning algorithm in [14]. The proposed algorithm can learn data one-by-one or by chunks with variable sizes. In [12], ELM was used in an application to recognize human actions with incremental learning using a very minimal number of video frames at a high speed.

ELM is still a newly emerging technique that needs a lot of exploration and enhancements, and its concepts are applicable to various learning techniques. In short, it is like a gold mine to dig into.

6   Conclusion

In this paper, we have proposed a predictive modeling technique for predicting the functional properties of soil samples based on their spatial and spectral features. First, the spectral part of the data was pre-processed using a smoothing step with Savitzky-Golay filter, followed by a first order differencing step to get rid of the physical effects and emphasize on the spectral properties. Then, a relatively new and very promising SLFN algorithm called extreme learning machine (EML) was used for predictive modeling of the soil properties. We have tried two variations of EML, namely, basic Sigmoid-based EML and Gaussian kernel-based EML. When tested on the Africa Soil Property Prediction Challenge dataset, both methods offer good prediction results with low prediction error rates and low standard deviations of errors. Therefore, we believe that our proposed method can be practically useful in many agricultural and environmental applications, which have to deal with the soil's functional properties.

References

1. AfSIS: Africa soil information service (2014). http://africasoils.net/
2. Bartlett, P.L.: The sample complexity of pattern classification with neural networks: the size of the weights is more important than the size of the network. IEEE Trans. Inf. Theory **44**(2), 525–536 (1998)
3. Cambria, E., Huang, G.B., Kasun, L.L.C., et al.: Extreme learning machines [trends & controversies]. IEEE Intell. Syst. **28**(6), 30–59 (2013)
4. Frénay, B., Verleysen, M.: Using SVMs with randomised feature spaces: an extreme learning approach. In: Proceedings of the 2010 European Symposium on Artificial Neural Networks, Computational Intelligence and Machine Learning, pp. 315–320 (2010)
5. Huang, G.B., Chen, Y.Q., Babri, H.A.: Classification ability of single hidden layer feedforward neural networks. IEEE Trans. Neural Netw. **11**(3), 799–801 (2000)
6. Huang, G.B., Zhou, H., Ding, X., Zhang, R.: Extreme learning machine for regression and multiclass classification. IEEE Trans. Syst. Man Cybern. Part B Cybern. **42**(2), 513–529 (2012)
7. Huang, G.B., Zhu, Q.Y., Siew, C.K.: Extreme learning machine: a new learning scheme of feedforward neural networks. In: Proceedings of the 2004 IEEE International Joint Conference on Neural Networks, vol. 2, pp. 985–990. IEEE (2004)
8. Huang, G.B., et al.: ELM: H2O R Interface. Nanyang Technological University, Singapore (2004). http://www.ntu.edu.sg/home/egbhuang/elm_codes.html
9. Kaggle: Africa soil property prediction challenge (2014). https://www.kaggle.com/c/afsis-soil-properties/
10. Kaggle: Africa soil property prediction challenge - forums (2014). https://www.kaggle.com/c/afsis-soil-properties/forums
11. Kaggle: Africa soil property prediction challenge - leaderboard (2014). https://www.kaggle.com/c/afsis-soil-properties/leaderboard
12. Liang, N.Y., Huang, G.B., Saratchandran, P., Sundararajan, N.: A fast and accurate online sequential learning algorithm for feedforward networks. IEEE Trans. Neural Netw. **17**(6), 1411–1423 (2006)
13. Liu, Q., He, Q., Shi, Z.: Extreme support vector machine classifier. In: Washio, T., Suzuki, E., Ting, K.M., Inokuchi, A. (eds.) PAKDD 2008. LNCS (LNAI), vol. 5012, pp. 222–233. Springer, Heidelberg (2008)
14. Minhas, R., Mohammed, A.A., Wu, Q.M.J.: Incremental learning in human action recognition based on snippets. IEEE Trans. Circuits Syst. Video Technol. **22**(11), 1529–1541 (2012)
15. Neupane, B., Perera, K.S., Aung, Z., Woon, W.L.: Artificial neural network-based electricity price forecasting for smart grid deployment. In: Proceedings of the 2012 IEEE International Conference on Computer Systems and Industrial Informatics, pp. 1–6 (2012)
16. Oentaryo, R., Lim, E.P., Finegold, M., et al.: Detecting click fraud in online advertising: a data mining approach. J. Mach. Learn. Res. **15**(1), 99–140 (2014)
17. Rinnan, Å., van den Berg, F., Engelsen, S.B.: Review of the most common preprocessing techniques for near-infrared spectra. TrAC Trends Anal. Chem. **28**(10), 1201–1222 (2009)
18. Savitzky, A., Golay, M.J.E.: Smoothing and differentiation of data by simplified least squares procedures. Anal. Chem. **36**(8), 1627–1639 (1964)
19. Wong, Y.K., Woon, W.L.: An iterative approach to enhanced traffic signal optimization. Expert Syst. Appl. **34**(4), 2885–2890 (2008)

Recent Advances in Improving the Memory Efficiency of the TRIBE MCL Algorithm

László Szilágyi[1,2,3]([✉]), Lajos Loránd Nagy[2], and Sándor Miklós Szilágyi[1,4]

[1] Department of Control Engineering and Information Technology, Budapest
University of Technology and Economics, Budapest, Hungary
[2] Faculty of Technical and Human Sciences,
Sapientia University of Transylvania, Tîrgu-Mureş, Romania
lalo@ms.sapientia.ro
[3] Canterbury University of Christchurch, Christchurch, New Zealand
[4] Department of Informatics, Petru Maior University of Tîrgu-Mureş,
Tîrgu-Mureş, Romania

Abstract. A fast and highly memory-efficient implementation of the TRIBE-MCL clustering algorithm is proposed to perform the classification of huge protein sequence data sets using an ordinary PC. Improvements compared to previous versions are achieved through adequately chosen data structures that facilitate the efficient handling of symmetric sparse matrices. The proposed algorithm was tested on huge synthetic protein sequence data sets. The validation process revealed that the proposed method extended the data size processable on a regular PC from previously reported 250 thousand to one million items. The algorithm needs 10–20 % less time for processing the same data sizes than previous efficient Markov clustering algorithms, without losing anything from the partition quality. The proposed solution is open for further improvement via parallel data processing.

Keywords: Protein sequence clustering · Markov clustering · Markov processes · Efficient computing · Sparse matrix

1 Introduction

Markov clustering performs a hierarchical grouping of input data based on a graph structure and its associated connectivity matrix. It has a series of successful applications in the field of protein sequence and interaction network analysis [6], video processing [8], image processing [4], language modeling [9], community detection [12], human action categorization [18], and FPGA circuit design [3].

When the input data consists of protein sequences, each sequence will be associated to a node of the graph, and edge weights will be the pairwise similarity values computed with existing alignment methods like: Needleman-Wunsch [11],

Research supported by the Hungarian National Research Funds (OTKA), Project no. PD103921. S. M. Szilágyi is a Bolyai Fellow of the Hungarian Academy of Sciences.

Smith-Waterman [13], BLAST [1], and PRIDE [7]. In case of large-scale data sets, the BLAST similarity measures are preferred due to its sparse nature, which allows for quick and memory-efficient processing.

TRIBE-MCL [6] is a clustering method based on Markov chain theory [5], which assigns a graph structure to the protein set such a way that each protein has a corresponding node. Edge weights are stored in the so-called similarity matrix $\mathbf{S}$, which acts as a stochastic matrix. At any moment, edge weight s_{ij} reflects the posterior probability that protein i and protein j have a common evolutionary ancestor. TRIBE-MCL is an iterative algorithm, performing in each loop two main operations on the similarity matrix: inflation and expansion. Inflation raises each element of the similarity matrix to power $r > 1$, which is a previously established fixed inflation rate, favoring higher similarity values in the detriment of lower ones. Expansion, performed by raising matrix $\mathbf{S}$ to the second power, is aimed to favor longer walks along the graph. Further operations like column or row normalization, and matrix symmetrization are included to serve the stability and robustness of the algorithm, and to enforce the probabilistic constraint. Similarity values that fall below a previously defined threshold value ε are rounded to zero. Clusters are obtained as connected subgraphs in the graph.

Handling matrices of several hundreds of thousand rows and columns is prohibitively costly in both runtime and storage space. Recent fast TRIBE-MCL implementations (e.g. [16,17]) significantly reduced runtime, but the memory limitations still exist. The main goal of this paper is to introduce a novel memory efficient TRIBE-MCL approach that uses only sparse matrices to store similarity values and a one-dimensional array to store intermediate values of a single row during expansion. This change can significantly upgrade the size of processable data sets, and may also improve processing speed. The proposed method will be validated using large synthetic protein data sets derived from the SCOP95 database [2,10,14] using our method presented in [15].

The remainder of this paper is structured as follows. Section 2 presents the details of the proposed memory efficient TRIBE-MCL algorithm. Section 3 evaluates the behavior of the proposed method and discusses the achieved results and outlines the role of each parameter, while Sect. 4 concludes this study.

2 Methods

Making the TRIBE-MCL processing of graphs of up to $n = 10^6$ nodes accessible for ordinary PCs requires adequate data structures to store the nonzero values on the similarity matrix (SM), and also an adequate organization of the algorithmic steps. Compared to our previous solution, it is necessary to redefine the operations performed during each cycle. The main proposals in this order are listed in the following.

At any moment of the processing, the stored SM is symmetric. The input SM is symmetric. Inflation, consisting in raising each element of the matrix to a given power, does not influence matrix symmetry. Expansion, consisting in raising the SM to the second power, also gives symmetric output if the input

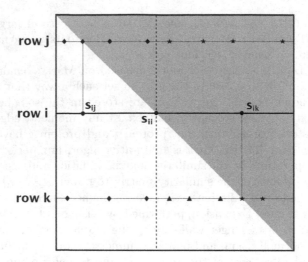

Fig. 1. Expansion using the employed data structures. Computing row i of the expanded matrix requires reading all rows with index $j < i$ and $k > i$ for which nonzeros s_{ij} or s_{ik} exist. Values indicated by stars ($\star$) are directly accessible from the stored SM. Those drawn as triangles ($\blacktriangle$) need information on the lower half of the matrix. Nonzeros indicated by diamonds ($\blacklozenge$) are irrelevant at this step, as only that part of row i is to be computed, which falls in the upper diagonal half of the SM.

was symmetric. The only operation that pushes similarity values s_{ij} out of symmetry is the normalization of rows. However, each normalization is followed by a symmetrization, and by unifying these two steps, the intermediary output of normalization does not need to be stored.

Having the stored SM always symmetric, it is not necessary to store the whole matrix, only one half of it including the diagonal, as all further elements are deductible from these ones. Our choice was to store the upper diagonal half of sparse SM, thus in each row the first stored element is the one situated on the diagonal. All nonzero elements are stored in a row-wise order, together with their column coordinate. The number of nonzeros in each row, and a pointer to the first element of each row are stored separately.

Inflation is performed by parsing the array of stored nonzero elements and raising them to the power indicated by the inflation rate.

The unified normalization and symmetrization is performed in two steps. In a first step, the array of nonzeros is parsed and the sum of each row (Σ_i, $i = 1 \ldots n$) is computed. Non-diagonal elements s_{ij} $(i \neq j)$ in the stored upper half matrix are added to the sum of both rows i and j. In the second step, for each stored element s_{ij}, the normalized values become $\overline{s}_{ij} = s_{ij}/\Sigma_i$ and $\overline{s}_{ji} = s_{ij}/\Sigma_j$, respectively. The new s_{ij} value after symmetrization will be $\sqrt{\overline{s}_{ij}\overline{s}_{ji}}$, and will only be stored if it exceeds the similarity threshold ε.

Expansion is the only operation of TRIBE-MCL, which requires two instances of the similarity matrix: one for the input and one for the output of expansion.

Data: Initial similarity matrix **S**, inflation rate $r > 1$, similarity threshold ε
Result: Final similarity matrix **S**
repeat
     Inflate(S)
     Normalize-and-Symmetrize with elimination(S)
     Normalize-and-Symmetrize(S)
     Build auxiliary data structure for the lower diagonal half matrix
     Expand(S)
until *convergence*;
Identify clusters

       Algorithm 1. The steps of the proposed algorithm

Similarly to our solution given in [16], the expanded matrix is computed row by row, and during the expansion of a row, the computed output needs non-sparse storage. This requires a float valued array of n elements, denoted by E initialized with zero values for each row.

During expansion, the new row with index i, denoted by $\overline{\mathbf{S}}_i$ is obtained as follows:

$$E = \left(\sum_{j \in \text{row}_i} s_{ij} s_{j1} \quad \sum_{j \in \text{row}_i} s_{ij} s_{j2} \quad \cdots \quad \sum_{j \in \text{row}_i} s_{ij} s_{jn} \right), \qquad (1)$$

which in sparse matrix notations becomes

$$\overline{\mathbf{S}}_i = \sum_{j \in \text{row}_i} s_{ij} \mathbf{S}_j. \qquad (2)$$

When the row is computed, its nonzero values falling in the upper half of the SM are transferred to the output sparse matrix.

The application for the expansion formula given in Eq. (2) is not trivial, due to the structural properties of the stored SM. Explanation in this order is given in Fig. 1. During the computation of row i of the squared matrix, the area of interest is only the part of row i situated behind the diagonal element s_{ii}, as the others will not be stored anyway. Consequently all nonzeros outside row i, on the left side of the vertical dotted line are irrelevant.

The algorithm parses all nonzeros in row i. Those situated in columns with index $j \leq i$ will require reading row j, whose elements of interest are all accessible in the upper half sparse matrix structure. On the other hand, nonzeros situated in columns with index $k > i$, imply reading row k whose elements of interest are partly outside the stored zone. This is why, before proceeding to the expansion operation, another sparse matrix structure is created to store information on the elements in the lower diagonal half of the matrix. The latter structure does not store the actual value of similarity, but stores information on the existence and position of the corresponding element in the upper diagonal half matrix structure. Creating the lower diagonal half matrix structure in each iteration requires two parsings over the upper half matrix structure. During the first,

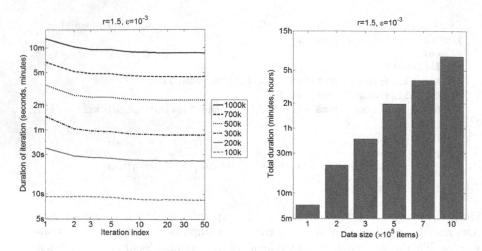

Fig. 2. Benchmark figures for a median density data sets of variable number of items, at constant inflation rate $r = 1.5$ at a constant value of similarity threshold $\varepsilon = 10^{-3}$: (left) duration of individual iterations; (right) overall runtime on fifty iterations.

we can count how many nonzeros are present in each row, and can allocate memory correspondingly and set row heads and element counts for all rows. During the second parsing, the actual information is extracted from the upper half nonzero elements (column information, and offset position with respect to row head) and added to the lower diagonal half matrix. This structure makes the lower part values, drawn in Fig. 1 as triangles (▲), directly accessible, facilitating efficient data processing.

The proposed TRIBE-MCL algorithm is summarized in Algorithm 1.

3   Results and Discussion

The proposed method underwent a series of benchmark tests using synthetic test data sets of sizes ranging from 10 thousand to one million items. Data sets were generated using the method indicated in [15]. For each data size, 21 instances were created and the one with median density was chosen for the test.

Figure 2(left) exhibits the duration of each of the first fifty iterations in case of various matrix sizes, at inflation rate fixed at $r = 1.5$ and similarity threshold $\varepsilon = 10^{-3}$. As long as most nodes of the graph are connected together, namely in the first 5–6 loops, the computational load is somewhat higher and considerably falls thereafter, being virtually constant and low from the 10th loop. This difference between the duration of initial and late iterations gets more relevant as the data size grows. Figure 2(right) indicates the total runtime necessary to perform 50 loops in case of various data sizes. A comparison with our previous algorithms version [16] reveals that the main improvement is achieved in the size of data an ordinary PC can deal with. The new algorithm can easily handle a one-million-node graph, while the previous one was limited at 250 thousands. The execution time for the same amount of nodes in the graph is also reduced by 10–20 %.

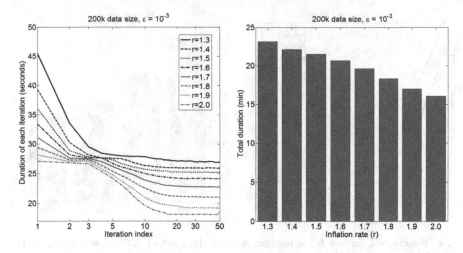

Fig. 3. Benchmark figures for a median density data set of 200k items, showing the influence of inflation rate r at a constant value of similarity threshold ε: (left) duration of individual iterations; (right) overall runtime on fifty iterations.

Figure 3 exhibits the effect of the inflation rate on the computational load of the algorithm. The input data here consisted of 200 thousand items having a similarity matrix of median density. Figure 3(left) shows the duration of individual iterations, while Fig. 3(right) indicates the total runtime of clustering performed in 50 loops. As the inflation rate grows, the similarity matrix becomes sparser and thus the total runtime and also the length of late iterations are shorter.

Figure 4 shows the influence of the similarity threshold ε on the computational load of the algorithm, using median density data sets of 200 thousand items. Figure 4(left) shows the duration of single iterations, while Fig. 4(right) exhibits the total runtime of clustering performed in 50 loops. A lower value of the similarity threshold keeps small similarity values longer in the matrix, and consequently the processing needs more time. The final outcome of clusters, and consequently the accuracy of clusters is hardly influenced by ε. Consequently ε should be kept high enough to support efficient data processing. However, if ε is too high, it may determine the algorithm the eliminate all non-zeros at once from a row of the matrix, leading to serious damage in the obtained partition.

All efficiency tests were carried out on a PC with quad core i7-4770 processor running at 3.4 GHz frequency and 16 GB RAM memory, using a single core of the microprocessor. The upper limit of processable data size is constrained by the memory of the employed computer. The main determining factor is the maximum number of nonzero values in the similarity matrix, usually reached after the first expansion operation. With the current version of the algorithm, an ordinary PC with 4 GB RAM can easily process a graph of 350,000 nodes, while one million nodes require an upper class PC with 16 GB RAM.

In case of huge input data sets, the approximative memory requirement is 11 bytes multiplied by the maximum number of nonzeros in the similarity matrix at

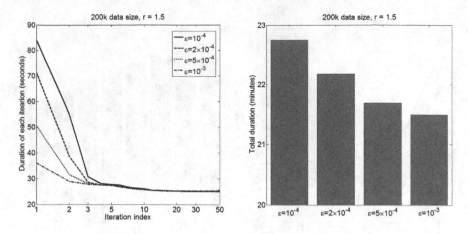

Fig. 4. Benchmark figures for a median density data set of 200k items, showing the influence of r and ε: (left) duration of individual iterations; (right) overall runtime on fifty iterations.

any time during the data processing. At a reasonably high inflation rate (e.g. $r \geq 1.3$), the maximum is likely to be reached right after the first expansion. Further improvement of the processing speed is achievable via parallel processing either with CPU or GPU. Further extension of the processable data size is achievable by temporary storage of data on solid state drives.

4   Conclusions

In this paper we have proposed an ultimate efficient approach to the graph-based TRIBE-MCL clustering method, a useful tool in protein sequence classification. The proposed approach proved extremely quick, and its memory needs are strongly reduced since previous versions. This novel implementation represents a major step of TRIBE-MCL towards handling huge data sets in reasonable time. Further enhancement of the algorithm's efficiency may be achieved via parallel implementation in CPUs or GPUs. Our future efforts will be focused on developing TRIBE-MCL algorithm versions to efficiently handle huge networks described by sparse matrices of unloadable size.

References

1. Altschul, S.F., Madden, T.L., Schaffen, A.A., Zhang, J., Zhang, Z., Miller, W., Lipman, D.J.: Gapped BLAST and PSI-BLAST: a new generation of protein database search program. Nucl. Acids Res. **25**, 3389–3402 (1997)
2. Andreeva, A., Howorth, D., Chadonia, J.M., Brenner, S.E., Hubbard, T.J.P., Chothia, C., Murzin, A.G.: Data growth and its impact on the SCOP database: new developments. Nucl. Acids Res. **36**, D419–D425 (2008)

3. Dai, H., Zhou, Q., He, O., Bian, J.: Markov clustering based placement algorithm for island-style FPGAs. In: IEEE International Conference on Green Circuits and Systems, pp. 123–128. IEEE Press, New York (2010)
4. Dhara, M., Shukla, K.K.: Characteristics of restricted neighbourhood search algorithm and Markov clustering on modified power-law distribution. In: 1st International Conference on Recent Advances in Information Technology, pp. 520–525. IEEE Press, New York (2012)
5. Eddy, S.R.: Profile hidden Markov models. Bioinformatics **14**, 755–763 (1998)
6. Enright, A.J., van Dongen, S., Ouzounis, C.A.: An efficient algorithm for large-scale detection of protein families. Nucl. Acids Res. **30**, 1575–1584 (2002)
7. Gáspári, Z., Vlahovicek, K., Pongor, S.: Efficient recognition of folds in protein 3D structures by the improved PRIDE algorithm. Bioinformatics **21**, 3322–3323 (2005)
8. Hospedales, T., Gong, S.G., Xiang, T.: A Markov clustering topic model for mining behaviour in video. In: 12th IEEE International Conference on Computer Vision, pp. 1156–1172. IEEE Press, New York (2009)
9. Keensub, L., Ellis, D.P.W., Loui, A.C.: Detecting local semantic concepts in environmental sounds using Markov model based clustering. In: IEEE International Conference on Acoustics Speech and Signal Processing, pp. 2278–2281. IEEE Press, New York (2010)
10. Lo Conte, L., Ailey, B., Hubbard, T.J., Brenner, S.E., Murzin, A.G., Chothia, C.: SCOP: a structural classification of protein database. Nucl. Acids Res. **28**, 257–259 (2000)
11. Needleman, S.B., Wunsch, C.D.: A general method applicable to the search for similarities in the amino acid sequence of two proteins. J. Mol. Biol. **48**, 443–453 (1970)
12. Pons, P., Latapy, M.: Computing communities in large networks using random walks. In: Yolum, I., Güngör, T., Gürgen, F., Özturan, C. (eds.) ISCIS 2005. LNCS, vol. 3733, pp. 284–293. Springer, Heidelberg (2005)
13. Smith, T.F., Waterman, M.S.: Identification of common molecular subsequences. J. Mol. Biol. **147**, 195–197 (1981)
14. Structural Classification of Proteins database. http://scop.mrc-lmb.cam.ac.uk/scop
15. Szilágyi, L., Kovács, L., Szilágyi, S.M.: Synthetic test data generation for hierarchical graph clustering methods. In: Loo, C.K., Yap, K.S., Wong, K.W., Teoh, A., Huang, K. (eds.) ICONIP 2014, Part II. LNCS, vol. 8835, pp. 303–310. Springer, Heidelberg (2014)
16. Szilágyi, L., Szilágyi, S.M., Hirsbrunner, B.: A fast and memory-efficient hierarchical graph clustering algorithm. In: Loo, C.K., Yap, K.S., Wong, K.W., Teoh, A., Huang, K. (eds.) ICONIP 2014, Part I. LNCS, vol. 8834, pp. 247–254. Springer, Heidelberg (2014)
17. Szilágyi, S.M., Szilágyi, L.: A fast hierarchical clustering algorithm for large-scale protein sequence data sets. Comput. Biol. Med. **48**, 94–101 (2014)
18. Zhu, X., Li, H.: Unsupervised human action categorization using latent Dirichlet Markov clustering. In: 4th International Conference on Intelligent Networking and Collaborative Systems, pp. 347–352. IEEE Press, New York (2012)

Weighted ANN Input Layer for Adaptive Features Selection for Robust Fault Classification

Muhammad Amar[1], Iqbal Gondal[1,2(✉)], and Campbell Wilson[1]

[1] FIT, Monash University, Melbourne, Australia
{muhammad.amar,iqbal.gondal,campbell.wilson}@monash.edu
[2] ICSL, Federation University Australia, Ballarat, Australia
iqbal.gondal@federation.edu.au

Abstract. Model based feature selection for identification of diverse faults in rotary machines can significantly cost time and money and it is nearly impossible to model all faults under different operating environments. In this paper, feed-forward ANN input-layer-weights have been used for the adaptive selection of the least number of features, without fault model information, reducing the computations significantly but assuring the required accuracy by mitigating the noise. In the proposed approach, under the assumption that presented features should be translation invariant, ANN uses entire set of spectral features from raw input vibration signal for training. Dominant features are then selected using input-layer-weights relative to a threshold value vector. Different instances of ANN are then trained and tested to calculate F1_score with the reduced dominant features at different SNRs for each threshold value. Trained ANN with best average classification accuracy among all ANN instances gives us required number of dominant features.

Keywords: Machine health monitoring (MHM) · Adaptive feature selection · Features reduction · Artificial neural networks (ANNs) · Fault diagnosis

1 Introduction

Frequency signatures of different defective components in the machines can be captured in the form of vibrations signals [1, 2]. Recorded vibrations contain raw signal and noise components. For different applications and surroundings, noise sources can conceal these raw vibrations. Mathematical models are used to estimate characteristic frequencies for faults under consideration [1, 3, 4], and during condition monitoring process, vibration spectrum is observed around these characteristic frequencies to detect faults. Authors in [1] used basic motion model of the bearing to determine characteristic frequencies for features extraction for ANN classifier. Zhang et al. in [4] have used weighted window around the frequency of interest for features calculation. Thus in literature most of the existing techniques use mathematical models to obtain robust features. In industrial environments, with increased number of diverse machine applications it is slow, tedious, and financially inefficient to model each machine and all of its component for feature extraction. Thus, for a large variety of machines with diverse applications, mathematical models

© Springer International Publishing Switzerland 2015
S. Arik et al. (Eds.): ICONIP 2015, Part II, LNCS 9490, pp. 36–43, 2015.
DOI: 10.1007/978-3-319-26535-3_5

and machine's surroundings independent fault diagnostic solution is needed. To avoid modelling the system, entire spectrum of the captured vibrations can be used as a feature set. But in worst SNR scenarios, using the entire spectrum of feature set will result in noise accumulation and could mislead the classifiers [5]. In vibration spectral features, each frequency contains two constituent components: noise and fault information. Lesser the number of features, lesser would be the noise accumulated. Therefore minimum frequency features are preferred to reduce the noise sources and run time computations. But the use of very few frequency features can result in faster computations but lead to inefficient learning of the classifiers [1, 5] and poor fault detection. Thus for the robust and faster fault classification, least number of frequency features with significant fault information are advisable.

Features can be learned using different transform and classifiers. For instance, Fourier Transform (FT) of vibration signals can be used to train ANN classifiers to detect bearing faults [1, 6]. FFT is among the most widely used transforms for feature selection and extraction purposes [1, 7]. FFT has advantage of representing translation variant time domain signals into translation invariant frequencies. These translation invariant frequency features are preferred for classifier like ANN. But in some cases, FFT can be inefficient to calculate the exact amplitude of short duration transients as compared to FFT window length [5, 6]. To incorporate transients of different duration and amplitude, WT can be useful as it decomposes signals into different frequency sub-bands called nodes [7, 8]. Signals at each node of WT are translation variant and thus are unable to readily provide required feature set.

In this paper an adaptive weight based features selection (AWLF) method using ANN has been proposed. To train ANN, spectral contents of normal and fault vibrations have been used. Relative weight of input layer of trained ANN provides corresponding strength of frequencies for dominant features selection.

The rest of the paper is organized as follows: Sect. 2 explains proposed framework for adaptive features Selection, Sect. 3 describes time segmentation and spectral features of the vibration signal, Sect. 4 presents artificial neural network training and weight profile, Sect. 5 explains dominant features selection using ANN, Sect. 6 presents bearing fault case study to validate AWLF and Sect. 7, finally, concludes the paper.

2 Adaptive Features Selection Framework

The Proposed AWLF framework for adaptive feature selection using ANN input layer weights is shown in Fig. 1 and this proposed AWLF framework can be explained stepwise as:

1)   ANN training using all usual classes.
2)   Determine Feature Weight Profile (FWP) and Mean of Feature Profile (MFP)
3)   for i←0:step:stop
4)   thr ← i*MFP
5)   RFWP ← FWP > thr
6)   Train ANN with RFWP raw vibrations
7)   Avg F1_Score over all classes under different SNRs
8)   End

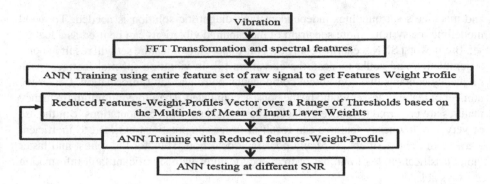

Fig. 1. Framework of the proposed AWLF scheme.

At the first stage, feedforward ANN is trained with full frequency features (step 1) to get Feature Weight Profile (FWP) and mean of feature weight profile (MFP) from ANN input layer weights (step 2). In an iterative process, starting at (step 3), a thr is selected (step 4) iteratively to get a Reduced Feature Weight Profile (RFWP) by selecting only the features having their weights greater than respective *thr* (step 5) in the FWP. ANN is trained using raw vibrations with this RFWP (step 6) and is tested for different SNRs (steps 7) to get average F1_Score for each thr level over all classes. This procedure is repeated (steps 3–7) for each *thr* level in the threshold vector. At the end, based upon the average F1_score, least number of features meeting the required accuracy are selected. For implementation purposes, the AWLF is explained in details from Sects. 3, 4 and 5.

3　Time Segmentation and Spectral Features

First step in the AWLF is to time segment a vibration signal X of length l into m sub-signals using a window of length of w samples as given by (1):

$$m = \frac{l}{w} \tag{1}$$

An arbitrary segment x_j can be given by (2):

$$x_j = X\left(((j-1)\,w + 1):jw\right) \tag{2}$$

Using FFT, x_j is represented in feature vector form F with R number of features (3):

$$F = \{f1, f2, f3, \dots, fR\} \tag{3}$$

The features obtained will be used by an ANN for dominant feature selection.

4   Artificial Neural Network Training and Weight Profile

Biologically inspired ANNs are capable of classifying different feature patterns using stored information in weighted connections between different neurons [9]. Feedforward ANNs, used in this paper, consist of input layer, one output layer and fixed one hidden layer, with the linear activation function in the output layer and sigmoid in the hidden layer. Minimum number of neurons giving the best classification accuracy are preferred in step (1) for hidden layer. Once the ANN architecture is decided then training phase starts. Before training, the dataset is divided into training, validation and test sets. In AWLF, during supervised learning, the weights are tuned using feedforward and back propagation algorithm such that ANN minimizes the mean square error (MSE) between current output and target output.

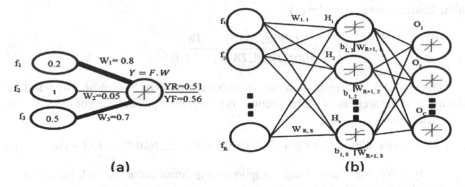

Fig. 2. (a) Single neuron and (b) Feedforward ANN architecture

In a trained ANN, features with higher weight are indicative of containing more information as compared to others. With a variable threshold values, least number of these dominant features, meeting given classification accuracy, can be extracted using FWP of ANN. Thus, ANN can be used to select relatively information-rich features while ignoring least affective features and thus resulting in reduced noise sources for better classification accuracy under worst SNRs. Figure 2.a shows a single neuron with input feature vector F, having three features, output Y and respective connection-weights-vector W for each feature using linear activation function. From Fig. 2.a, for the trained neuron, connection 1 and connection 3 have high weight profile as compared to connection 2. For any threshold, features with weight < threshold can be ignored resulting in reduced features. With threshold = 0.1, input feature f2 will be discarded and the retained dominant features will be f1 and f3, resulting in 1/3 reduction of the features. Results show that output YR, with reduced features, is not much different from output YF, with full features. The output difference can be minimized by retraining the ANN with new reduced features. Thus, trained ANN FWP can be used for dominant feature selection and effectively resulting in noise filtration for better classification accuracy.

In Fig. 2.b a general feedforward ANN is shown. This ANN contains input layer, output layer with linear activation function, and one hidden layer with sigmoid activation function. It has feature vector F, output layer class vector O and H hidden layer neurons vector with their respective connections and weights (W). In figure, R is the number of input features, S is the number of hidden layer neurons and b is a bias of a neuron. Any hidden layer neuron, including bias, will have R + 1 connections. As bias can be ignored for feature selection, thus, individual weight profile (IWP) for any hidden layer neuron S is given by (4):

$$IWP^S = \{W_{1,S}, W_{2,S}, W_{3,S}, \dots, W_{R-1,S}, W_{R,S}\} \tag{4}$$

Here, $W_{1,S}$ corresponds to connection weight between hidden layer neuron s and 1st input feature, f_1. To better understand and visualize the relative weight strength of features of different neurons, normalized all-weight profile (NAWP) for S number of hidden layer neurons using the IWP of connections between input layer and each neuron in the hidden layer is given by (5):

$$NAWP = \frac{\{IWP^1, IWP^2, \dots, IWP^S\}}{Max(\{IWP^1, IWP^2, \dots, IWP^S\})} \tag{5}$$

Individual features weight (IFW) for any input feature r can be measured by selecting maximum of the weights of all connections between rth input and each of the S hidden layer neurons (6):

$$IFW(r) = max(NAWP(r), NAWP(r+R), NAWP(r+2R), \dots, NAWP(r+R(S-1))) \tag{6}$$

Finally, FWP, the relative weight importance information of the each input feature based on the trained ANN is given by (7):

$$FWP = \{IFW(1), IFW(2), IFW(3), \dots, IFW(R)\} \tag{7}$$

FWP obtained from (7) can be used to find mean of feature profile (MFP) by (8):

$$MFP = Mean(FWP) \tag{8}$$

In FWP, the important features will have relatively higher weightage as compared to less important ones. MFP will be, later, used for the selection of the dominant features using multiple of MFP as threshold.

5   Features Selection Using ANN

To find the weight based least number of appropriate features, we follow the steps mentioned in Fig. 1. After getting MFP, we follow the AWLF algorithm from step 4 to step 7. Based on MFP a threshold vector is used to obtain RFWP which is then used for different ANN instance training. Trained ANN instances are tested with different fault classes under different SNR conditions and the best performing ANN, giving us required averaged classification accuracy with least number of features, can be selected. In conclusion, if we provide AWLF framework with baseline data then it will automatically find out the least number of features to achieve required classification accuracy at any given SNR. In the next section a bearing fault classifier is discussed to validate the proposed framework.

6   Bearing Fault Classifier

In this paper, a case study of bearing fault classification, with one normal and three fault classes, has been presented to validate the real time applicability of AWLF.

6.1   Experimental Setup

In the experimental setup given in [10], one normal and three faulty bearings with inner race, ball and outer race fault, were used one by one to support the shaft of the motor. Faults created by electro-discharge machining with diameter of 0.007 in has been used for the three of the faults classes. At 1750 rpm, with 2HP load the vibration were recorded at 12000 samples per second. Additive white Gaussian noise (AWGN) has been added to the recorded vibrations at different SNRs to see the robustness of the AWLF [5, 7].

6.2   Feature Set

ANN with architecture of $(IN \times 3 \times 4)$ has been used for AWLF. Here, IN, 3 and 4 corresponds to the number of input features, number of hidden layer neurons and output classes respectively. Least number of neurons, 3, meeting the required classification has been used in hidden layer [5]. The recorded vibrations for each class were time segmented with window size of 1024 [5] and then transformed into frequency domain using FFT to get a set of features for the four of the fault classes using (1) – (3). These four feature sets were combined and randomly shuffled to get a training data set. After splitting this data set into 70 % training, 15 % testing and 15 % validation sets; ANN was trained using supervised learning. After training, NAWP was obtained using (5). NAWP contains WP (weight profile) for each of the hidden layer neuron of the ANN concatenated into one profile. FWP and MFP obtained from (7) and (8) using maximum

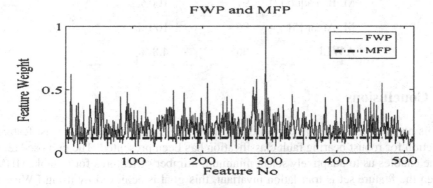

Fig. 3. FWP and MFP.

weight criterion in (6) for a trained ANN is shown in Fig. 3. It is clear from the figure that some of the features have very high IFW as compared to others. If a relatively high threshold is selected then most of the features will be discarded because of their lower IFW and vice versa.

With the use of multiple of MFP threshold, the infinitesimal features can be filtered out and the dominant features can be retained for further training purposes. Figure 4 shows percentage features reduction, for different thresholds with averaged classification accuracy over four classes. It's clear from the figure that initial features reduction results in decreased number of noise sources and, thus, improves the classification accuracy till a critical point. But further reduction of features, beyond this point, results in deteriorated accuracy because the reduction of information sources. On −10 dB SNR curve in Fig. 4, best accuracy of 96 % was achieved using only 4.8 % of features. Table 1 shows that AWLF gives 96 % accuracy equaling that of in [5] at worst case SNR of −10 dB with 4.8 % features only. This feature reduction from 100 % [5] to 4.8 % with comparable accuracy of 96 % significantly reduces run time computations and thus validates the applicability and novelty of AWLF in adaptive feature selection.

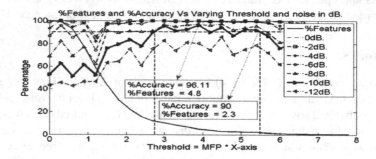

Fig. 4. %features reduction and classification accuracy

Table 1. Classification accuracy comparison of AFSR at SNR = −10 dB

Scheme	Accuracy %	Features Used
M. F. Yaqub [7]	91	100 %
M. Amar [5]	96	100 %
AWLF	**96**	**4.8 %**

7   Conclusions

In this paper artificial neural network input layer weight based adaptive least features selection for robust bearing fault classification has been presented. The proposed technique enables us to adaptively select minimum number of features for reliable MHM. Given the feature set is translation invariant, this goal is achieved by using FWP and

MFP of the ANN trained with full feature set of raw vibrations. The process is repeated with different, MFP multiple, thresholds to reduce the feature set size meeting required classification accuracy even under worst SNRs. Results show that best case classification accuracy of 96 % can be achieved using only 4.8 % features. Thus, these results validate that the input layer weights of trained feed forward ANN can be successfully used to adaptively select the least number of dominant features for successful fault identification by eliminating the need of mathematical model and by reducing the run time computations while keeping the classification accuracy intact.

References

1. Li, B., Chow, M.Y., Tipsuwan, Y., Hung, J.C.: Neural-network-based motor rolling bearing fault diagnosis. IEEE Trans. Ind. Electron. **47**(5), 1060–1069 (2000)
2. Bellini, A., Immovilli, F., Rubini, R., Tassoni, C.: Diagnosis of bearing faults of induction machines by vibration or current signals: a critical comparison. In: Industry Applications Society Annual Meeting, pp. 1–8 (2008)
3. Yiakopoulos, C.T., Gryllias, K.C., Antoniadis, I.A.: Rolling element bearing fault detection in industrial environments based on a K-means clustering approach. Expert Syst. Appl. **38**(3), 2888–2911 (2011)
4. Bin, Z., Sconyers, C., Byington, C., Patrick, R., Orchard, M., Vachtsevanos, G.: A probabilistic fault detection approach: application to bearing fault detection. IEEE Trans. Ind. Electron. **58**(5), 2011–2018 (2011)
5. Amar, M., Gondal, I., Willson, C.: Vibration spectrum imaging: a bearing fault classification approach. IEEE Trans. Ind. Electron. **62**(1), 494–502 (2015)
6. Amar, M., Gondal, I., Willson, C.: Multi-size-window spectral augmentation: neural network bearing fault classifier. In: The 8th IEEE Conference on Industrial Electronics and Applications, pp. 261–266 (2013)
7. Yaqub, M.F., Gondal, I., Kamruzzaman, J.: Inchoate fault detection framework: adaptive selection of wavelet nodes and cumulant orders. IEEE Trans. Instrum. Meas. **61**(3), 685–695 (2012)
8. Bouzida, A., Touhami, O., Ibtiouen, R., Belouchrani, A., Fadel, M., Rezzoug, A.: Fault diagnosis in industrial induction machines through discrete wavelet transform. IEEE Trans. Ind. Electron. **58**(9), 4385–4395 (2011)
9. Su, H., Chong, K.T.: Induction machine condition monitoring using neural network modeling. IEEE Trans. Ind. Electron. **54**(1), 241–249 (2007)
10. Bearing Data Center, http://www.eecs.case.edu/laboratory/bearing/welcome_overview.htm

Neural Network with Evolutionary Algorithm for Packet Matching

Zelin Wang[1,2], Zhijian Wu[1(✉)], Xinyu Zhou[1], Ruimin Wang[3], and Peng Shao[1]

[1] State Key Laboratory of Software Engineering, School of Computer,
Wuhan University, Wuhan 430072, China
zhijianwu@whu.edu.cn
[2] School of Computer Science and Tecnnology, Nantong University,
JangSu 226019, China
[3] International School of Software, Wuhan University, Wuhan 430072, China

Abstract. The performances of network forwarding device are determined by the efficiency of packet matching algorithm. It is difficult for
network device based on traditional algorithm acting as software core to
achieve linear forwarding. This paper proposes a new packet matching
algorithm, which achieves packet matching by combination of evolutionary algorithms and neural networks. Firstly, the evolutionary algorithm
is used to evolve weight value and activation function of neural networks.
Secondly, influence factor is applied to prune the neurons of the hidden
layer. And finally, back-propagation algorithm is utilized to fine-tune the
neural network. So a compact and efficient neural network structure to
solve packet matching is created by the creative procedure. Data experiments show that this new algorithm effectively improves the performance
of packet matching compared with the classical algorithms. And it can
completely solve the problem of large-scale rule packet matching.

Keywords: Packet matching · Neural network · Evolutionary algorithm

1 Introduction

With the popularity of optic fiber, the interconnection equipments such as
routers, switches and firewalls have become the bottleneck of network and their
performances are influenced by the packet matching speed especially. related
works in [1–5] have put forward many important methods by many scholars, but which could not solve the problem of packet linear speed forwarding.
Recently, researchers have proposed to use evolutionary algorithms to preprocess
the packet matching in [6,7]. Although these algorithms solve approximatively
packet linear forwarding of a multidimensional and large-scale rule library, the
assumption of a linear relationship between data features and mapping domain
of these features provokes a lot of restrictions in algorithm application and deteriorates the distribution performance of mapping spatial. So these algorithms
are sometimes not feasible and lack of stability in practice.

© Springer International Publishing Switzerland 2015
S. Arik et al. (Eds.): ICONIP 2015, Part II, LNCS 9490, pp. 44–51, 2015.
DOI: 10.1007/978-3-319-26535-3_6

Evolutionary algorithm is widely, successfully and effectively used in machine learning, process control, and other fields, which can be concluded from [8–10]. In recent years, it has show a trend of integration of kinds of intelligent algorithm. Design of an ANN is to actually design its structure. If the practical problem which need solve is more complex, it is more difficult or even infeasible to design ANN by artificial means. So we should design neural network with efficient automatic method instead of inefficient manual method. Intelligent algorithm in [11] provides a good thinking. In addition, the application of evolutionary algorithm can also merge the neural network structure and weights to learn together.

The paper proposes an algorithm combining evolutionary algorithm with neural network to solve packet matching, which is abbreviated by GNNPM. In which, Firstly, the evolutionary algorithm is not only used to evolve weight value but activation function of neural networks. Secondly, influence factor is applied to prune the neurons of the hidden layer. And finally, back-propagation algorithm is utilized to fine-tune the neural network. So a compact and efficient neural network structure to solve packet matching is created by the creative procedure.

2   The Basic Idea of GNNPM

In this paper, the source IP address, destination IP address, source port, destination port and the upper transport layer protocol type as the characteristics of the data were extracted. In Firewall and routing table, IPv6 address is represented in the form of $\times\times\times\times$: $\times\times\times\times$: $\times\times\times\times$: $\times\times\times\times$: $\times\times\times\times$: $\times\times\times\times$: $\times\times\times\times$: $\times\times\times\times$.

Source IP address is identified from a high bit to low bit, and each section is identified as the $X_{1i}, i \in \{1, 2, 3......8\}$. Similarly, the each section of destination IP address is identified as $X_{2i}, i \in \{1, 2, 3...8\}. X_{ji}, j \in \{1, 2\}, i \in \{1, 2, 3...8\}$, is transformed from the hexadecimal form into the decimal form.

$$X_i = Extract(IP) = \sum_{j=1}^{8} X_{ij}, i \in \{1, 2\} \tag{1}$$

Source port and destination port are represented in the form of 16-bit binary, and they are identified in the form of corresponding decimal Y_3 and Y_4. $X_i = Y_i \bmod 256, i \in (3, 4)$. The upper transport layer protocol is represented in the form of eight bit, and are identified in the form of corresponding decimal X_5. X represents the vector $(X_1, X_2, X_3, X_4, X_5,)$. In this paper, X domain space of the mapping function and $F(X)$ mapping space are known, and algorithm uses three-layer ANN to construct $F(X)$. Evolutionary algorithm is used to search for the optimal combination of link weights and activation functions, and influence factor is used to prune hidden layer neurons, which influence factor value is less than a certain performance threshold, and finally, BP is used to fine-tune the ANN structure. Finally, the traditional sequential search strategy is used to search a certain rule in the appropriate classification obtained by the neural network.

2.1   Set ANN Weights and Activation Functions

Data Normalization: Because difference between each element of the vector X is very large, each element of the vector X needs data normalization processing. This article uses the mean and variance of the estimated value of each element to normalize. According to the following formula, normalized processing is done.

$$\bar{X}_k = \frac{1}{N}\sum_{i=1}^{N} X_{ik}, k \in \{1,2,3,4,5\}, \sigma_k^2 = \frac{1}{N-1}\sum_{i=1}^{N}(X_k - \bar{X}_k)^2, \tilde{X}_{ik} = \frac{X_{ik} - \bar{X}_k}{\sigma_k}$$

(2)

where $\tilde{X}_{ik}$ is a normalized results of the k-th data characteristics of the i-th data.

Weight value is at the range of $(-\hat{w}, \hat{w})$. The $\hat{w}$ value depends on specific condition. When the weight between the input layer and hidden layer is initialized, the $\hat{w}$ is equail to $1/\sqrt{5}$. The input weights should be selected randomly in the $(-1/\sqrt{5}, 1/\sqrt{5})$. When the weight between the output layer and hidden layer is initialized, the $\hat{w}$ is equail to $1/\sqrt{Cn}$. The Cn is the number of connection of the output layer and hidden layer. The output weights should be selected randomly in the $(-1/\sqrt{Cn}, 1/\sqrt{Cn})$.

Activation Function: For three layers of ANN, the algorithm proposed applies a linear activation function in hidden layer, but it uses evolutionary algorithms to select activation function in the set (Binary sigmoid, Linear, Tanh, threshold, logb, gaussian) in output layer. So, the algorithm gives not only full consideration to the linear relationship, and takes into account the nonlinear relationship between the input and output. Encoding scheme: In this paper, evolutionary algorithm applies the real coding, which is relatively natural to express the problem, and reduces the trouble of mutual conversion between binary encoding and real number. When the parameter space is continuous, using the binary encoding to express the real numbers, must carry out discretion, which would result in errors. Encoding is as $(f_v\, w_1\, w_2 \ldots w_N)$. Where f_v is activation function of output layer, and $w_1\, w_2 \ldots w_N$ are weight values. N is computed by $N = (N_i + N_o)\, N_h$ Where N_i, N_o and N_h are the number of neurons of the input layer, and output layer and hidden layer, respectively.

The Evolution Operator: Selection Operator: In the course of operation, both of before operation and after operation, must be controlled. It is necessary to prevent the destruction of the fine individuals, but also to ensure the diversity of the population. In this paper, evolutionary algorithm apply the $\mu+\lambda+1$ mating pool strategy of tournament selection. The μ is mating pool capacity, the λ is the number of new individual, and 1 is the optimal solution retained by each cycle. These m individuals are randomly selected from the parent population, and then uses tournament selection to select an optimal individual into the mating pool from these m individuals. The procedure is recycled to operate, until the number of individuals of the mating pool reach μ.

Crossover Operator: The evolutionary algorithm Chooses a pair of individual from the mating pool and implements cross-operating. The algorithm assumes as follows: $X(x_1, x_2...x_n)$ and $Y(y_1, y_2...y_n)$ two individuals are selected from the mating pool; $X_c(x_{c1}, x_{c2}...x_{cn})$ and $Y_c(y_{c1}, y_{c2}...y_{cn})$ are generated by crossover operator. The algorithm generates n random probability value corresponding to each item of X and Y, and sets crossover probability. When the probability value corresponding to location c_i is less than crossover probability, crossover operation is implemented by the $x_{ci} = x_i + (x_i - y_i)$ and $y_{ci} = x_i - (x_i - y_i)$.

Mutation Operator: The two individuals are implemented by crossover operation, at the same time they are all implemented by Mutation operator. A mutation probability labeled by Pmutuation is set up, and generates a random number correspoinding to individual $X(x_1, x_2...x_n)$. If the random number is less than Pmutuation, the mutation operation is implemented, according to the $x_m = x_i + \gamma\sigma_i$ Where γ is the coefficient of dynamic change, and $\gamma = (\frac{1}{t})\gamma$, and t is the number of evolution cycles. The σ_i is the standard deviation of the i-th element of all individuals in the mating pool.The Table 1 shows that γ of dynamic change can improve the performances of packet matching.

Table 1. Different setting of γ leads to different performance

Rule number	1 k		1.8 k		2.6 k	
	Initialize (s)	Searching (s)	Initialize (s)	Searching (s)	Initialize (s)	Searching (s)
$\gamma = 0.5$	355	0.00000083	726	0.00000091	2130	0.00000108
$\gamma = 0.1$	972	0.00000102	2169	0.00000121	6827	0.00000184
$\gamma = 0.9$	1010	0.00000098	1983	0.00000124	6906	0.00000179
Dynamicγ	128	0.00000069	140	0.00000071	251	0.00000075

Fitness Function: Fitness selected must have the ability to measure the distribution performance of the mapping point, and excellent fitness must make certain that point aggregation is strong in class, and the distance between the class is relatively far.

$$fitness1 = \sqrt{\sum_{i=1}^{N_O}(NUM_i - \bar{E})^2} \qquad (3)$$

Where the N_O is the number of classification, and the NUM_i is the number of individuals in class i, and the $\bar{E}$ is the average of point number of all classification.

$$\bar{E} = \frac{1}{M}\sum_{i=1}^{M}N_i \qquad (4)$$

The NUM_i is the number of individuals in class i. fitness1 value is smaller, then the number of points of each class more and more tends to average value.

$$fitness2 = -\sum_{i=1}^{M}\sum_{j=i}^{M}\frac{(\mu_i - \mu_j)^2}{\sigma_i^2 + \sigma_j^2} \qquad (5)$$

The μ_i and μ_j are the individual deviation of class i and j, respectively, and the σ_i and σ_j are variance of the class i and class j, respectively. The value of fitness2 is the smaller, then the gathered condition of the points within the class is better. So, the points between classes is better dispersed between classes. So the fitness $= \alpha$ fitness1 $+ \beta$ fitness2, Where the α and β are the adjustment coefficient, and their values are restricted in (0, 1). the value of fitness is smaller, then the structure of the neural network will be better.

2.2   Pruning Redundant Neurons in the Hidden Layer

When we use evolutionary algorithms to set up ANN weights, and activation function, the number of hidden layer neurons are set by

$$N_h = \lfloor \sqrt{N_i + N_o} \rfloor + 1 \tag{6}$$

Where the N_i and N_o are the neuron number of the input layer and output layer, respectively. This section is to prune neuron of hidden layer. According to the I-Factor $= \sigma_i^2$, Where the σ_i is the variance for all input samples neurons in the output, the algorithm proposed calculates the influence factor (I-Factor) of each neuron of the hidden layer, it ranks the neurons according to the value of neuron I-Factor, and it searches for the neuron with smallest I-Factor.

The smallest I-Factor is compared with the threshold TIV. If it is less than the threshold, the neuron with the smallest I-Factor will be pruned, and the algorithm calculate the fitness value of new ANN. If fitness values is only minor changes, the change is accepted. Otherwise, according to $w_j = w_{ij} \mu_i$, Where μ_i is output expectations of the hidden layer pruned neuron, the corresponding deviation item is appended to output neuron j, which is connected with i pruned neuron.

The algorithm calculate the fitness value of ANN again. If fitness value is only minor deterioration (the deterioration value is les than TDV), the deletion is accepted, and cycles to consider the hidden layer neurons with smaller I-Factor. Otherwise this deletion neuron is restored, and stop pruning.

2.3   Fine-Tune ANN Structure

The ideal output of the sample is that samples were averagely distributed to each class, namely fitness1 value is equal to zero. While the reality is not the case. We may apply the difference of actual output value (AV) and desired output value (DV) to adjust appropriate weights, and search for the optimal combination of weights, when $\|AV - DV\| \leq Tv$. Back propagation algorithm is used to calculate the effective error for each hidden layer unit, and deduce the learning rule of an input layer to hidden layer weights.

$$\Delta \omega_{kj} = \eta(t_k - z_k)f'(net_k)y_j \tag{7}$$

Where the $\Delta \omega_{kj}$ is the weight adjustment value of output layer k-th unit and the hidden layer j-th unit, and η is the learning rate. The t_k is the desired output value, and the z_k is actual output value. The $f'(net_k)$ is the derivative of activation function between the output layer and the hidden layer, and $y_j = f(net_j)$.

$$\Delta\omega_{ji} = \eta[\sum_{k=1}^{M} \omega_{kj}(t_k - z_k)f'(net_k)]\zeta'(net_j)x_i \tag{8}$$

Where the $\Delta\omega_{ji}$ is weight adjustment value of input layer i-th unit and hidden layer j-th unit, and η is the learning rate. The $f'(net_k)$ is the derivative of activation function between the output layer and the hidden layer, and $\zeta'(net_j)$ is the derivative of activation function between the input layer and hidden layer.

3   Algorithm Design

The algorithm is designed as follows:

INPUT: $\mu, \lambda, \gamma, \alpha, \beta$, TIV, TDV, TV, TRN

OUTPUT: ANN structure

Step 1:initialization

Step 2: evolve ANN weights and activation functions

   Step 2.1:select μ individuals by tournament selection

   Step 2.2:rank μ individuals, ν is the best individual

   Step 2.3:Choose two individuals from the mating pool

   Step 2.4:implement crossover and mutation

      if the new individuals number reaches λ, goto the next step

      otherwise, continue to implement Step 2.3

   Step 2.5:select the superior into new mating pool

      the best replace ν, if it is superior to ν

      if ν is not changed consecutively for three times,

      or evolution cycle Number reaches threshold, goto Step 3

      otherwise, goto Step 2.3

Step 3: Using the I-Factor to prune hidden layer neurons

Step 4: using BP to fine-tuning the ANN structure

Step 5: organize all individuals of each class in order.

---

4   Experiments and Results

In this section, the algorithm proposed is compared with the algorithm Grid-of-the Trie [1–4] and HiCuts [1,4], and the algorithm (NLMPM) [6]. Because the Grid-of-the Trie algorithm is the most typical algorithm based on tries, and the same HiCuts is the most typical based on the space partitioning. The NLMPM is the best algorithm based on intelligence, at present.

Table 2. Comparison of performance of three kinds of algorithm

	Initialize (s)	Searching (s)	Memory (MB)	Initialize (s)	Searching (s)	Memory (MB)
	600 rule number			1000 rule number		
GNNPM	123	0.00000068	0.54	128	0.00000069	0.62
NLMPM	69	0.00000094	0.75	107	0.00000097	0.99
Grid-of-Trie	54	0.0000012	1.23	88	0.0000022	1.78
HiCuts	58	0.0000022	2.74	112	0.0000028	3.41
	1400 rule number			1800 rule number		
GNNPM	134	0.00000069	0.71	140	0.00000071	0.78
NLMPM	163	0.00000099	1.47	245	0.00000104	1.58
Grid-of-Trie	138	0.0000034	2.35	182	0.0000045	3.76
HiCuts	149	0.0000038	4.32	172	0.0000046	4.87
	2200 rule number			2600 rule number		
GNNPM	245	0.00000103	1.27	251	0.00000105	1.29
NLMPM	312	0.00000112	1.61	394	0.00000117	2.02
Grid-of-Trie	234	0.0000068	4.12	283	0.0000092	6.53
HiCuts	212	0.0000056	7.87	267	0.0000087	16.45

Table 2 shows the average time consumption of initialization, and the average time consumption of packet matching of the four kinds of algorithms, and the average of memory consumption, for the four kinds of algorithms.

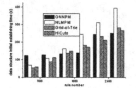
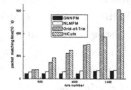
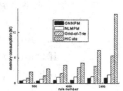

(a) the average time-consuming for initialization

(b) the average time-consuming of each packet matching

(c) the average memory consumption

The experimental results show the algorithm proposed on the average packet matching performance is better than the Grid-of-the Trie and the HiCuts. In addition, the NLMPM algorithm is to determine the parameters of the function, while the proposed algorithm assumes the functions and parameters are unknown, so, this algorithm can make the data distribution better. So, the performance is slightly better than the NLMPM algorithm.

The experimental results display that the algorithm proposed, on the terms of average memory consumption, is more superior than other algorithms, because the algorithm needs just to keep the weight parameters and the class of activation functions, while the Grid-of-the Trie and the HiCuts need to store the tree structure, and the NLMPM algorithm can not guarantee a good distribution of the results, so, will cause the spatial redundancy.

5   Conclusions

The algorithm proposed solves constructively the problem of the packet linear forwarding. The speed of packet matching, compared to traditional methods, is greatly increased. Comparied with the traditional methods, the proposed algorithm has a weak relationship between rule number and matching speed. Comparied with the NLMPM algorithm, the stability of the algorithm proposed is better, and the performance is also slightly improved, especially, the time consumption of initial process and the memory consumption of the packet matching are both better. When the number of rules increases, the algorithm proposed can use ANN cascade technical to solve problem.

Acknowledgements. This work was supported by the National Natural Science Foundation of China (No.F0207, No.61070008, No.61364025), Application research project of Nantong science and Technology Bureau (No.BK2014057), the Science and Technology Foundation of Jiangxi Province (No.20151BAB217007).

References

1. Hazem, H., Adel, E.-A., Ehab, A.-S.: Adaptive statistical optimization techniques for firewall packet filtering. School of Computer Science, DePaul University, Chicago, USA (2006)
2. Kencl, L., Schwarzer, C.: Traffic-adaptive packet filtering of denial of service attacks. In: Proceedings of the 2006 International Symposium on World of Wireless, Mobile and Multimedia Networks, pp. 485–489 (2006)
3. Acharya, S., Abliz, M., Mills, B., Znati, T.F.: OPTWALL: a traffic-aware hierarchical firewall optimization. In: Proceedings of the Network and Distributed Systems Symposium (2007)
4. Tsai, C.-H., Chu, H.-M., Wang, P.-C.: Packet classification using multi-iteration RFC. In: 2013 IEEE Annual Computer Software and Application Conference Workshop (2013)
5. Sreelaja, N.K., Vijayalakshmi Pai, G.A.: Ant colony optimization based approach for efficient packet filtering in firewall. Appl. Soft Comput. **10**, 1222–1236 (2010)
6. Wang, Z.L., Wu, Z.J., Yin, L.: Hign-dimension large-scale patcket matching algorithm in IPV6. Acta Eletronica Sinica **11**, 2181–2186 (2013)
7. Wang, Z.L., Wu, Z.J.: Packet matching algorithm based on improving differential evolution. Wuhan Univ. J. Natural Sci. **17**(5), 447–453 (2012)
8. Padhy, N.P.: Artificial Intelligence and Intelligent Systems. Oxford University Press, Oxford (2005)
9. Salameh Walid, A.: Detection of intrusion using neural networks: a customized study. Stud. Inf. Control **13**(2), 137–145 (2004)
10. Yao, X., Yong, X.: Recent advances in evolutionary computation. J. Comput. Sci. Technol. **21**(1), 1–18 (2006)
11. Carvalho, A.R., Ramos, F.M., Chaves, A.A.: Metaheuristics for the feedforward artificial neural network (ANN) architecture optimization problem. Neural Comput. Appl. **20**, 1273–1284 (2011)

Trading Optimally Diversified Portfolios in Emerging Markets with Neuro-Particle Swarm Optimisation

Pascal Khoury[1,2] and Denise Gorse[1(✉)]

[1] Department of Computer Science, University College London,
Gower Street, London, WC1E 6BT, UK
{P.Khoury,D.Gorse}@cs.ucl.ac.uk
[2] Charlemagne Capital, 39 St. James's Street, London, SW1A 1JD, UK

Abstract. In previous work the authors have developed trading models using both particle swarm optimisation and neural networks for specific emerging markets industry sectors. Here, a more flexible model is developed that is effective across a wide range of sectors. It is discovered there is a strong dependence of the quality of returns on the minimum number of trades allowed within a given time period (a risk-minimisation measure used to maintain portfolio diversity) and that in the case of emerging markets the optimal value for this parameter may be different to the standard investment industry recommendation. Learning is then extended to include this parameter, with out-of-sample testing demonstrating very promising results.

Keywords: Particle swarm optimisation · Ensemble classifiers · Emerging markets · Trading model

1 Introduction

Emerging markets (EM) such as those of Asia and Latin America are increasingly attractive to investors. However as well as offering the opportunity for high returns, these markets can also expose investors to high risks. It is common in trading and investment strategies to limit the minimum number of open positions during a given time period in order to preserve the diversity of the portfolio, as diversity is widely believed to be a key factor in the management of risk [1]. However recommendations may need to be appropriately modified for less developed markets; it would be counterproductive to pursue diversity at the expense of adding largely unprofitable or risky elements to an EM portfolio.

In previous work by the authors [2], which used a mixture of neural networks (NN) and particle swarm optimisation (PSO) to construct a trading model for consumer industry equities, the diversity of the portfolio was determined by a factor we will here call *mintrades*, the minimum number of recommended open positions per time period. In that work *mintrades* was set to an industry-conventional value of 30; it will be shown here the NN/PSO trading system can be extended to learn an appropriate value for *mintrades*, and that for EM it may be preferable to hold a somewhat less diverse portfolio than usually recommended.

© Springer International Publishing Switzerland 2015
S. Arik et al. (Eds.): ICONIP 2015, Part II, LNCS 9490, pp. 52–60, 2015.
DOI: 10.1007/978-3-319-26535-3_7

2 Background

2.1 Trading in Emerging Markets

Attempting to forecast the future is an enterprise inevitably fraught with risk. Even when the direction of movement of a stock price has been predicted correctly there is no guarantee of profit, as transaction costs may cancel out a modest gain. These observations are particularly true for emerging markets, where prediction of equity price movements is more difficult due to a range of factors: uncertain economic growth; policies associated with central banks, especially in the wake of the global financial crisis of 2008; market risks such as those associated with fluctuations in the strength of the US dollar and in the oil price, and in any lessening of investors' appetites for EM equities. In addition transaction costs in emerging markets tend to be higher than in more developed markets, making it especially important to avoid initiating a trade whose potential return would be too small to outweigh these costs.

2.2 Multifactor Stock Selection Model

Alongside the well-known strategy of 'buy-and-hold' (buy all assets at the start of the test period, make no changes to the portfolio during this period, and sell all holdings at the end), we will benchmark our results against an industry-standard trading model referred to here as the 'multifactor model'. This model is based on a consideration of both company fundamentals and past share price performance [1, 3]. It looks at a linear combination of relevant factors, treating these dynamically in a way that allows it to adjust to economic cycles without explicit human intervention, and is considered overall to be a good choice for use in emerging markets, though it tends to perform better in periods when markets are driven by company fundamentals than when they are driven by news flows and sentiment, such as the year-end holiday season.

3 Data Set

Previous NN/PSO work by the authors has focused on a particular area of EM economic activity, consumer companies in [2] or steel manufacturers in [4, 5]. Here, the data used represent a broad range of industry sectors, as shown in Table 1. Trading with this diverse portfolio is a more challenging problem, as some of the economic factors among the NN inputs may not be equally relevant across all sectors, or may affect companies operating in these sectors in different ways.

Data were gathered from February 2007 to February 2015, for 49 consecutive (non-overlapping) two-monthly time periods. Periods 1–39 (4836 examples) were used for training and validation, and periods 40–49 (1240 examples) for testing, with summary statistics given in Tables 2 and 3. It is notable that while there are two to three times as many negative returns than positive ones, the average positive return is nearly three times as large as the average negative return, this being caused by the presence of a small number of companies which exhibit particularly strong growth.

Table 1. Breakdown by sector of the 124 emerging markets companies considered (regional breakdown: Africa 4.8 %; Asia 77.4 %; Europe 5.6 %; LatAm 12.1 %)

Economic sector	%	Economic sector	%
Commercial services	1.6	Health technology	3.2
Communications	8.1	Industrial services	0.8
Consumer durables	4.0	Non-energy minerals	12.9
Consumer non-durables	12.9	Process industries	8.1
Consumer services	3.2	Producer manufacturing	7.3
Distribution services	2.4	Retail trade	6.5
Electronic technology	8.1	Technology services	0.8
Energy minerals	9.7	Transportation	4.0
Finance	0.8	Utilities	5.6

Table 2. Summary statistics for training/validation set (periods 1–39)

	Training/Validation: 4836 Examples		
	Class 1 (negative)	Class 2 (negligible)	Class 3 (positive)
% of total	43.11	33.13	23.76
Min return	−64.98	0.00	10.02
Max return	0.00	9.99	105.40
Av return	−8.64	4.45	20.02

Table 3. Summary statistics for testing set (periods 40–49)

	Testing: 1240 examples		
	Class 1 (negative)	Class 2 (negligible)	Class 3 (positive)
% of total	44.44	42.42	13.15
Min return	−40.32	0.00	10.02
Max return	0.00	9.94	70.44
Av return	−6.55	4.16	16.67

For each pattern in the training, validation, or test sets the task will be to predict its return class—negative (sell), negligible (do nothing), or positive (buy)—on the basis of a range of fundamental variables known to impact company performance, a practice that has a sound theoretical basis [6]. Thirty such inputs were used, examples being the per share value of a company on paper relative to the equity share price, a company's ability to pay dividends, or to generate earnings and recurring cash inflows so to maintain its business, the amount of capital it invests, or the sales it generates. In addition we used a 31st input representing past (one month) share price volatility, quantised into the signal set {−1 (low), 0 (medium), +1 (high)}; this input was found helpful in our previous work [2] focused on trading EM consumer company equities.

4   Methods

4.1   Particle Swarm Optimisation

Particle swarm optimisation (PSO) [7] is a biologically inspired, population based search algorithm with many applications, including to finance [8]. At time t each particle $i = 1 \ldots N$ in the swarm represents a solution $x_{i,t}$ to a problem, in the current case in the form of a trading strategy as embodied by the weights of a neural network. Particles search the parameter (weight) space by a mechanism that combines a tendency to return to a personal best position $p_{i,t}$ (*cognitive* contribution weighted by φ_1) with one to follow the swarm's best performing member (*social* contribution weighted by φ_2). The equations used to update the velocity and position are

$$v_{i,t+1} = \chi \left(v_{i,t} + \varphi_1 \beta_1 \left(p_{i,t} - x_{i,t} \right) + \varphi_2 \beta_2 \left(g_t - x_{i,t} \right) \right), \tag{1a}$$

$$x_{i,t+1} = x_{i,t} + v_{i,t}, \tag{1b}$$

where β_1, β_2 are random numbers drawn uniformly from [0,1], and as in [2] we have adopted the constriction coefficient method [9] to encourage swarm convergence, which with $\varphi_1 = \varphi_2 = 2.05$ gives a constriction coefficient $\chi \approx 0.7289$.

4.2   Fitness Measure and the Role of *Mintrades*

The fitness measure used here for training, validation, and testing will be the profit function P_t used in [2] for consumer company trading:

$$P_t = \left[\begin{array}{l} 0 \text{ if } \sum_i |c_{i,t}| < mintrades \\ \frac{100}{\sum_i |c_{i,t}|} \left(c_{i,t} \sum_i R_t - 2 |c_{i,t}| \delta \right) \text{ otherwise} \end{array} \right. \tag{2}$$

where i indexes the company (1...124), and R_t is the aggregate stock return at time t across all the 124 companies considered; $c_{i,t} \in \{−1, 0, +1\}$ is a class designator related by $c_{i,t} = \text{Class}_{i,t} - 2$ to the three return classes (Class 1, low returns; Class 2, negligible returns; Class 3, high returns) described above, the sum over i of the absolute values | $c_{i,t}$| thus being the number of trading actions recommended at time t; and δ is the percentage transaction cost, set here to 0.5.

This measure includes a crucial parameter *mintrades*: if for a given time period less than this number of trades are recommended, it is considered the portfolio is insufficiently diverse and no trading is carried out. In [2] *mintrades* was set to the value 30, the same value as is conventionally used with the multifactor stock selection model described in Sect. 2.2. Initial experiments described in Sect. 5.1 use this same value; however it will later be shown in Sect. 5.2 that profitability depends strongly on this parameter, and in Sect. 5.3 that it is additionally possible to use the NN/PSO system to discover a more appropriate value for *mintrades*, alongside its learning of the trading decision network weights, from training and validation data.

4.3   Training Methodology

The results to be described used a swarm size of 100 particles and 25 PSO training iterations. While the latter number is relatively few, overfitting of the data was as in [2] a significant problem, such that optimal results on the validation data sets were usually obtained within 5–10 iterations, so longer runs only increased computational load without any corresponding gain. Individual particles were here single layer nets with three softmax output neurons, with class targets orthogonally encoded such that Class 1 (negative returns) = (1,0,0); Class 2 (negligible returns) = (0,1,0); Class 2 (positive returns) = (0,0,1). A number of experiments (not shown) with multilayer nets were carried out, but though these better fitted the training data, overfitting quickly occurred, such that parameter sets extracted for application to the test data performed no better than those derived from single layer nets.

As in [2] validation was used both to monitor for overtraining and as a means to construct a committee which to make classification decisions for the test set (the final 10 time periods from 40 to 49, equivalent to 1240 patterns). Validation sets were here sequences of 8 consecutive time periods (equivalent to 992 patterns) and to build in maximum robustness all possible consecutive 8-period selections from the 39 training/ validation periods were potentially used. For each set the global best weights that did best on the validation set were used to create a committee, with admission being dependent on obtaining a profit of 10 % on the validation set. The weight sets thus selected (up to a maximum of 31) were then used to classify the test set by taking an average of their outputs for each of the three output neurons of the network.

5   Results

Each NN/PSO experiment to be described below consisted of 25 runs, with average performance given along with standard deviations in the relevant figures or tables.

5.1   Trading with *Mintrades* = 30

In this first experiment a *mintrades* value of 30 was used, as normally utilised in the multifactor model of Sect. 2.2, and as used in [2] by an NN/PSO system developed specifically for use with EM consumer companies. As Table 4 shows, not only does the NN/PSO system achieve nearly double the test profit of the multifactor model, it also shows a similar increase in Sortino ratio, a measure of risk-adjusted profit, used here with a 2 % base rate target. The NN/PSO system also substantially outperforms buy-and-hold, a strategy difficult to beat in an upward-trending market as here.

Table 4. Comparative test period performance of buy-and-hold, multifactor model, and NN/PSO trading system with *mintrades* = 30

	% Profit	Sortino ratio
Buy-and-hold	9.45	0.264
Multifactor model	6.47	0.191
NN/PSO	12.39 ± 2.85	0.451 ± 0.184

5.2 Dependence of Test Profit on *Mintrades* Value

It was decided to explore the role of the mintrades parameter in determining the reprofitability of both the NN/PSO system and the multifactor model (it has no application to buy-and-hold as by definition all assets are traded). Figure 1 shows that while the conventional setting of *mintrades* = 30 is not unreasonable for the multifactor model, it appears a value ~ 10 would be preferable for the NN/PSO trading system.

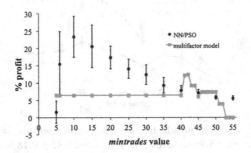

Fig. 1. Test data % profit as a function of *mintrades* for NN/PSO and the multifactor model

However it would be inappropriate to choose a preferred value of *mintrades* from the above results, as they refer to test data performance. If *mintrades* is to be assigned a value other than its conventional setting it must one be acquired from consideration of the training/validation data only, to which topic we now turn.

5.3 Learning *Mintrades* from Training/Validation Data

A adaptive value for *mintrades* is implemented by adding a further weight z_i to each particle i's parameter list, unconstrained and real-valued, as is the case for those parameters representing the influence of economic and other factors on the neural network's output classifications, but from which an integer value for the minimum number of trades per time period can be derived via (3) below:

$$\text{mintrades}_i = \text{nint}(5\,(1-f_i) + 25f_i), \quad \text{where } f_i = (1 + \exp(-z_i))^{-1} \tag{3}$$

(in which the function nint(x) denotes the nearest integer to x). *Mintrades* was bounded between 5 and 25 as from the experiments summarised in Fig. 3 this seemed to be a useful working range; initial z_i values were selected so that initial *mintrades* values were random in this interval. The NN/PSO training methodology was otherwise as described in Sect. 4.3, apart from there now being a need to define a *mintrades* value for the committee decision from those values learned by its component members. It was decided this should be chosen to be the minimum of those *mintrades* values found by the committee members, else otherwise a strong vote for a small subset could be effectively excluded.

Figure 2 compares the cumulative % profit over the test period achieved by the three methods of Table 4, together with that from the adaptive mintrades variant. It is clear this last, which had a run- and time period-averaged discovered *mintrades* value of 10.60 ± 1.11—considerably less than the value of 30 conventionally recommended, and similar to the value that might have been deduced from Fig. 1—was the most effective of the four methods in terms of both profitability and Sortino ratio.

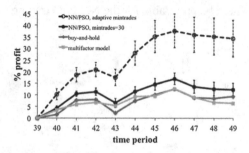

Fig. 2. Cumulative % test profit from adaptive *mintrades* and *mintrades* = 30 NN/PSO models, averaged over 25 runs (average Sortino ratios 2.41 ± 1.71 and 0.45 ± 0.18 respectively), compared to buy-and-hold (Sortino ratio 0.26) and multifactor model (Sortino ratio 0.19)

It was of interest to discover whether when using an adaptive *mintrades* the probability of selection for trading was relatively uniform across the 124 EM equities, or whether certain stocks were more often favoured. As Figs. 3a and b show, the latter was the case, with five EM companies, 87 (Nine Dragons Paper Company; Hong Kong; Process Industries), 30 (China Petroleum; China; Energy Minerals), 10 (Arcelik; Turkey; Consumer Durables), 120 (Tencent Holdings; China; Technology Services), and 51 (Glow Energy; Thailand; Utilities) especially often picked.

These five companies represent a broad range of industry sectors, and are notable for their growth potential; Arcelik for example is not only a market leader in consumer durables in its home market of Turkey but also exports to faster growing emerging markets such as South Africa, while Tencent is the leading online gaming operator in China and runs the most popular messaging and social platform in that country. The sector diversity displayed by the trading model means it is likely decisions are based more on fundamental factors rather than price momentum; this is good news as strategies driven by the latter are more vulnerable to market fluctuations based on news and sentiment. The broad spread of industry sectors displayed by the NN/PSO portfolio is also

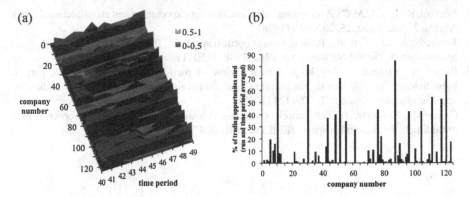

Fig. 3. (a) 3D plot where height denotes the run-averaged proportion of times a given company makes a trade in a given test period; (b) Overall % of trading opportunities used during testing for each of the 124 companies, averaged over 25 runs and test periods 40–49; most favored stocks are (in decreasing order of use) numbers 87, 30, 10, 120, 51

welcome in that this sector spread would offset the lessened diversity that comes with the selection of fewer companies to be traded.

6   Discussion

It has been shown that in addition to learning which of a number of input fundamental factors can best inform a trading decision for a particular equity, it is possible, using particle swarm optimisation, to train a neural network to at the same time optimise the composition of the portfolio. For the data set considered the companies most favoured by the NN/PSO system represented a wide range of industry sectors, containing some companies already known to be valuable assets in an emerging markets portfolio along with others less frequently traded. It is planned to investigate the NN/PSO recommendations in more detail in order to determine which features of the favoured companies predominantly drive their selection, and also to discover whether the methods described here could also be used for more developed markets.

References

1. Fabozzi, F.J., Markowitz, H.M.: Multifactor Equity Risk Models. Wiley, New Jersey (2011)
2. Khoury, P., Gorse, G.: Investing in emerging markets using neural networks and particle swarm optimisation. In: Proceedings of IJCNN 2015 (July 2015, to appear)
3. Grinold, R.C., Kahn, R.N.: Active Portfolio Management. McGraw-Hill, New York (1999)
4. Khoury, P., Gorse, D.: Identification of factors characterising volatility and firm-specific risk using ensemble classifiers. In: Huang, T., Zeng, Z., Li, C., Leung, C.S. (eds.) ICONIP 2012, Part IV. LNCS, vol. 7666, pp. 450–457. Springer, Heidelberg (2012)
5. Khoury, P., Gorse, D.: Investigation of the predictability of steel manufacturer stock price movements using particle swarm optimisation. In: Lee, M., Hirose, A., Hou, Z.-G., Kil, R.M. (eds.) ICONIP 2013, Part II. LNCS, vol. 8227, pp. 673–680. Springer, Heidelberg (2013)

6. Frankel, R., Lee, C.M.C.: Accounting valuation, market expectation and cross-sectional stock returns. J. Acc. Econ. **25**, 283–319 (1998)
7. Kennedy, J., Eberhart, R.: Particle swarm optimization. In: IEEE International Conference Symposium on Neural Networks, pp. 1942–1948. IEEE Press, New York (1995)
8. Banks, A., Vincent, J., Anyakoha, C.: An review of particle swarm optimization. part II: hybridisation, combinatorial, multicriteria and constrained optimization, and indicative applications. Nat. Comput. **7**, 109–124 (2008)
9. Clerc, M., Kennedy, J.: The particle swarm—explosion, stability, and convergence in a multidimensional complex space. IEEE Trans. Evol. Comp. **6**, 58–73 (2002)

Generalized Kernel Normalized Mixed-Norm Algorithm: Analysis and Simulations

Shujian Yu[1]([✉]), Xinge You[2], Xiubao Jiang[2], Weihua Ou[3], Ziqi Zhu[4], Yixiao Zhao[1], C.L. Philip Chen[5], and Yuanyan Tang[2,5]

[1] Department of Electrical and Computer Engineering,
University of Florida, Gainesville, FL, USA
yusjlcy9011@ufl.edu
[2] School of Electronic Information and Communications,
Huazhong University of Science and Technology, Wuhan, Hubei, China
[3] School of Mathematics and Computer Science, Guizhou Normal University,
Guiyang, Guizhou, China
[4] School of Computer Science and Technology,
Wuhan University of Science and Technology, Wuhan, Hubei, China
[5] The Faculty of Science and Technology, University of Macau, Macau, China

Abstract. This paper is a continuation and extension of our previous research where kernel normalized mixed-norm (KNMN) algorithm, a combination of the *kernel trick* with the mixed-norm strategy, was proposed to demonstrate superior performance for system identification under non-Gaussian environment. Meanwhile, we also introduced a naive adaptive mixing parameter (AMP) updating mechanism to make KNMN more robust under nonstationary scenarios. The main contributions of this paper are threefold: firstly, the ℓ_p-norm is substituted for the ℓ_4-norm in the cost function, which can be viewed as a generalized version to the form of mixed-norms; secondly, instead of using the original AMP proposed in our previous work, a novel time-varying AMP is employed to provide better tracking behavior to the nonstationarity; and thirdly, the mean square convergence analysis is conducted, where the second moment behavior of weight error vector is elaborately studied. Simulations are conducted on two benchmark system identification problems, and different kinds of additive noises are added respectively to verify the effectiveness of improvements.

Keywords: KNMN algorithm · Generalized KNMN algorithm · Adaptive mixing parameter · System identification

1 Introduction

In the era of data deluge, adaptive online learning (AOL) is well motivated as the computation complexity of processing the entire data set as a batch is prohibitive and the learning algorithms must be performed without revisiting the past entries [15]. In this direction, kernel-based AOL becomes one of the

© Springer International Publishing Switzerland 2015
S. Arik et al. (Eds.): ICONIP 2015, Part II, LNCS 9490, pp. 61–70, 2015.
DOI: 10.1007/978-3-319-26535-3_8

most emerging and promising research areas [3]. The main advantage of kernel methods is their ability to represent nonlinear functions linearly and universally in Reproduce Kernel Hilbert Space (RKHS) by taking advantage of a Mercer kernel, which in general results in a simple convex optimization problem with a unique global optimum [3,10].

Successful examples of kernel-based AOL algorithms include kernel least mean square (KLMS) [9], kernel affine projection algorithm (KAPA) [11], and kernel normalized least mean square (KNLMS) [13], etc. Almost all the current kernel-based AOL algorithms assume additive Gaussian distributed noises for the sake of mathematical simplicity [7,8]. However, it is well known that Gaussian model alone cannot accurately describe the noise statistics [5]. As a result, algorithms developed under the Gaussian noise hypothesis always suffer from performance degradation [8].

To address this issue, kernel robust mixed norm (KRMN) [7] algorithm and kernel normalized mixed norm (KNMN) [16] algorithm were proposed separately. Instead of using the standard squared error based adaptation criterion, the strategy of combining squared error and other error power in the cost function is employed in both KRMN and KNMN, and the corresponding weight update function in RKHS can be represented as (1) and (2), respectively:

$$\mathbf{\Omega}(k) = \mathbf{\Omega}(k) + \mu \left[2\lambda e(k) + (1 - \lambda)\text{sign}(e(k)) \right] \varphi(\mathbf{x}(k)) \tag{1}$$

$$\mathbf{\Omega}(k) = \mathbf{\Omega}(k) + \mu \left[2\lambda e(k) + 4(1 - \lambda)\text{sign}(e^3(k)) \right] \varphi(\mathbf{x}(k)) \tag{2}$$

where $\mathbf{\Omega}(k)$ denotes the estimated weight vector in RKHS for the kth iteration, μ is the step size, $e(k)$ is the error term, and $\lambda \in [0, 1]$ is the mixing parameter of convex optimization functions. In addition, $\varphi(\mathbf{x}(k))$ stands for mapping the original data vector $\mathbf{x}(k)$ into RKHS.

In this paper, our previously proposed KNMN is extended. The performance of KNMN is substantially improved thanks to the following three factors. First, the ℓ_p-norm is substituted for the ℓ_4-norm in the cost function, which can be viewed as a generalized version to the form of mixed-norms. Second, instead of using two alternative mixing parameters to loosely track the nonstationary operating environment, a novel time-varying adaptive mixing parameter (AMP) is employed to provide better tracking behavior in the same scenario. Third, the mean square convergence analysis is conducted, where the second moment behavior of weight vector is elaborately studied.

This paper is organized as follows. KNMN is described in Sect. 2. The generalized KNMN (GKNMN) is introduced in Sect. 3, where the mechanism for time-varying AMP is also presented. Section 4 analyzes the mean square convergence and Sect. 5 shows the simulation results for system identification. This paper is concluded in Sect. 6.

2   Kernel Normalized Mixed-Norm (KNMN) Algorithm

2.1   KNMN

KNMN is proposed to cope with the problem of AOL under nonstationary operation environment. It has superior performance for nonlinear system identification, especially when the additive noises follow non-Gaussian statistics [16]. The key idea of KNMN is to combine the linear mixed-norm algorithm with the so-called *kernel trick*, while introducing a normalized step size and an adaptive mixing parameter to deal with the nonstationarity. Equation (3) provides the learning rule of KNMN in RKHS:

$$
\begin{cases}
\mathbf{\Omega}(0) = 0 \\
e(k) = d(k) - \mathbf{\Omega}^T(k-1)\varphi(\mathbf{x}(k)) \\
\mathbf{\Omega}(k) = \mathbf{\Omega}(k-1) + \mu(k)\left[2\lambda e(k) + 4(1-\lambda)e^3(k)\right]\varphi(\mathbf{x}(k))
\end{cases}
\tag{3}
$$

where $\varphi(\cdot)$ and $\mathbf{\Omega}$ have the same meaning as in (1) and (2), and $\mu(k)$ is the normalized step size. In [16], $\mu(k)$ is formulated as:

$$
\mu(k) = \frac{e(k)}{\left[2\lambda e(k) + 4(1-\lambda)e^3(k)\right]\kappa\langle\mathbf{x}(k),\mathbf{x}(k)\rangle}
\tag{4}
$$

Therefore, the detailed KNMN is summarized as Algorithm 1.

Algorithm 1. The Kernel Normalized Mixed-Norm (KNMN) Algorithm.

Input: $\{x(k), d(k)\}, k = 1, 2, \cdots$
Initialization: Choose initial step size $\mu(1)$, mixing parameter λ, kernel function κ
  $\alpha_1 = \mu(1)[2\lambda d(1) + 4(1-\lambda)d^3(1)]$;
Computation:
1: **while** $\{x(k), d(k)\}(k > 1)$ available, **do**
2:   evaluate the output $y(k) = \sum_{j=1}^{k}\alpha_j\kappa\langle\mathbf{x}(j),\mathbf{x}(k)\rangle$;
3:   compute the error term $e(k) = d(k) - y(k)$;
4:   compute the normalized step size $\mu(k) = \frac{e(k)}{[2\lambda e(k)+4(1-\lambda)e^3(k)]\kappa\langle\mathbf{x}(k),\mathbf{x}(k)\rangle}$;
5:   update the coefficient $\alpha_k = \mu(k)[2\lambda e(k) + 4(1-\lambda)e^3(k)]$
6: **end while.**

2.2   KNMN with Adaptive Mixing Parameter (AMP) $\lambda(k)$

There are two approaches to set AMP $\lambda(k)$ for mixed-norm algorithms: a constant value or an adaptive value which is self-adaptive to the nonstationary operating environment [8,16]. Theoretically, the optimal AMP is given by [2,12]

$$
\lambda(k) = \text{Prob}\{d > |d(k)| \cup d < -|d(k)|\}
\tag{5}
$$

where d models the desired signal $d(k)$ in the absence of additive noises.

In our previous work, the well-known outliner trimmed sliding window [6] strategy is employed to provide a simple but effective simplification to (5). More specifically, a threshold τ is set as:

$$
\tau = \frac{1.5\sum_{i=1}^{n}|d(i)|}{n}
\tag{6}
$$

where n is the window length (set to 10 in [16]). If the incoming desired output d is larger than the threshold, a small $\lambda(k)$ is selected, whereas if d is smaller than the threshold, then $\lambda(k)$ close to 1 is selected.

The underlying concept for this simple strategy is that the probability of the desired signal contains outliers can be approximated as $[1 - \lambda(k)]$, i.e., the LMS is progressively replaced by LMF in RKHS as the probability of an outlier increases [2].

3   Generalized Kernel Normalized Mixed-Norm (GKNMN) Algorithm

3.1   GKNMN

Adaptive filtering algorithms designed via employing a convex mixing of ℓ_2-norm and ℓ_4-norm have a disadvantage when the absolute value of the error is greater than one [17]. This will make (3) unstable, except when a relatively smaller step size or a larger mixing parameter are chosen. On the other hand, ℓ_p-norm ($p \leq 2$) based minimization algorithms for signal recovery, parameter estimation or other related signal processing areas have been extensively studied in the literature.

The proposed GKNMN in RKHS is thus defined as minimizing the following cost function:

$$J(k) = \lambda(k)E[e(k)^2] + (1 - \lambda(k))E[|e(k)|^p] \tag{7}$$

with $p \in [1, 4]$, and the corresponding weight update function in RKHS can be represented as:

$$\mathbf{\Omega}(k) = \mathbf{\Omega}(k - 1) + \mu(k)[2\lambda(k)e(k) + p(1 - \lambda(k))|e(k)|^{(p-1)}\text{sign}(e(k))]\varphi(\mathbf{x}(k)) \tag{8}$$

The following sections focus on the derivation of the normalized step size $\mu(k)$ as well as the time-varying adaptive mixing parameter $\lambda(k)$.

Derivation of $\mu(k)$.
Multiplying $\varphi(\mathbf{x}(k))$ on both sides of (8), we have:

$$\mathbf{\Omega}(k)^T\varphi(\mathbf{x}(k)) = \mathbf{\Omega}(k - 1)^T\varphi(\mathbf{x}(k)) + \mu(k)[2\lambda(k)e(k)$$
$$+ p(1 - \lambda(k))|e(k)|^{(p-1)}\text{sign}(e(k))]\kappa\langle\mathbf{x}(k), \mathbf{x}(k)\rangle \tag{9}$$

Similar to [16], if we define posterior error as

$$e_{post}(k) = d(k) - \mathbf{\Omega}(k)^T\varphi(\mathbf{x}(k)) \tag{10}$$

After subtracting $d(k)$ from both sides of (9), we have:

$$e_{post}(k) = e(k)\left\{1 - \mu(k)\kappa\langle\mathbf{x}(k), \mathbf{x}(k)\rangle[2\lambda(k) + p(1 - \lambda(k))|e(k)|^{(p-2)}]\right\} \tag{11}$$

To minimize the square of the posterior error $e_{post}(k)$, the term inside above large bracket in (11) is set to zero, hence yielding the normalization formula for $\mu(k)$, which is given by:

$$\mu(k) = \frac{e(k)}{[2\lambda(k)e(k) + p(1 - \lambda(k))|e(k)|^{(p-1)}\text{sign}(e(k))]\kappa\langle\mathbf{x}(k), \mathbf{x}(k)\rangle}. \quad (12)$$

Mechanism for Time-Varying AMP $\lambda(k)$. Equation (6) provides a simple scheme for updating $\lambda(k)$ between two alternative values. To further improve the performance of algorithm tracking behavior in nonstationary operating environment, motivated with the modified variable step-size (MVSS) strategy proposed in [1], we employ a similar mechanism to update the time-varying AMP $\lambda(k)$ in GKNMN. The updating function for $\lambda(k)$ is thus defined as:

$$\lambda(k + 1) = \delta\lambda(k) + \gamma\rho(k)^2 \quad (13)$$

and

$$\rho(k) = \beta\rho(k - 1) + (1 - \beta)e(k)e(k - 1) \quad (14)$$

where $\delta \in [0, 1], \beta \in [0, 1]$ are the exponential weighting parameters which govern the averaging time constant, and $\gamma > 0$. For more descriptions on δ, β and γ, the interested reader is referred to [1].

According to [1], the introduction of $\rho(k)$ serves two objectives. Firstly, the error autocorrelation between $e(k)$ and $e(k-1)$ is generally a good measurement of the proximity to the optimum. Secondly, it has better performance to sense the adaption process, especially in the presence of ambient noises. Note that, in our following experiments, $\rho(1)$ is set to 0, and $\lambda(k)$ is set to λ_{min} or λ_{max} when it falls below or above the lower and upper bounds, respectively.

Therefore, the detailed GKNMN is summarized as Algorithm 2.

3.2   Relation to Previous Works

It is worth noting that GKNMN provides a high-level generalization to previous works in this area, which features generalization mechanisms to normalized step size $\mu(k)$, time-varying AMP $\lambda(k)$, as well as the form of mixed norms which is characterized by parameter p.

If $p = 2$, the cost function defined with (7) degrades to the classical KLMS [9] algorithm (without normalization strategy) whatever the value of $\lambda(k)$ in the range $[0, 1]$, for which the unimodality of the cost function is preserved.

If $p = 1$, the algorithm degrades to KRMN [7] algorithm (without normalization strategy). Moreover, KRMN algorithm further reduces to the kernel least absolute deviation (KLAD) [7] algorithm for $\lambda(k) \triangleq 0$.

If $p < 2$, ℓ_p gives less weight to larger errors and thus tends to reduce the influence of impulse interface or noises follow long-tailed distributions [17]. On the contrary, if $2 < p \leq 4$, the algorithm has better tracking capability when the environment is corrupted with short-tailed noises [16]. More specifically, the algorithm degrades to KNMN [16] and KNMN-AMP [16] for $p = 4$.

Algorithm 2. The Generalized Kernel Normalized Mixed-Norm (GKNMN) Algorithm

---

Input: $\{x(k), d(k)\}, k = 1, 2, \cdots$
Initialization: Choose initial step size $\mu(1)$, initial mixing parameter $\lambda(1)$, mixed-norm parameter p, exponential weighting parameters δ and β, and kernel function κ
$\qquad \alpha_1 = \mu(1)[2\lambda(1)d(1) + p(1 - \lambda(1))|d(1)|^{p-1}\text{sign}(d(1))]$;
Computation:
1: **while** $\{x(k), d(k)\}(k > 1)$ available, **do**
2:    evaluate the output $y(k) = \sum_{j=1}^{k} \alpha_j \kappa \langle \mathbf{x}(j), \mathbf{x}(k) \rangle$;
3:    compute the error term $e(k) = d(k) - y(k)$;
4:    compute the normalized step size
$\qquad \mu(k) = \dfrac{e(k)}{[2\lambda(k)e(k) + p(1 - \lambda(k))|e(k)|^{(p-1)}\text{sign}(e(k))]\kappa \langle \mathbf{x}(k), \mathbf{x}(k) \rangle}$;
5:    update the coefficient $\alpha_k = \mu(k)[2\lambda(k)e(k) + p(1 - \lambda(k))|e(k)|^{(p-1)}\text{sign}(e(k))]$
6:    update the time-varying adaptive mixing parameter for next iteration
$\qquad \rho(k) = \beta\rho(k - 1) + (1 - \beta)e(k)e(k - 1)$
$\qquad \lambda(k + 1) = \delta\lambda(k) + \gamma\rho(k)^2$
7: **end while**

---

4   Mean Square Convergence Analysis

This section provides a brief mean square convergence analysis to GKNMN based on the fundamental energy conservation relation (ECR) which is initially proposed in [14]. According to recently works [4,7,16], if we define the weight error vector as $\widetilde{\mathbf{\Omega}}(k) = \mathbf{\Omega}^*(k) - \mathbf{\Omega}(k)(\mathbf{\Omega}^*(k)$ is the optimal weight vector in RKHS), *a priori* error as $e_a(k) = \widetilde{\mathbf{\Omega}}(k - 1)\varphi(\mathbf{x}(k))$, and a *posteriori* error[1] as $e_p(k) = \widetilde{\mathbf{\Omega}}(k)\varphi(\mathbf{x}(k))$, then the standard ECR in RKHS can be represented as:

$$\|\widetilde{\mathbf{\Omega}}(k)\|_{\mathbb{F}}^2 + e_a^2(k) = \|\widetilde{\mathbf{\Omega}}(k - 1)\|_{\mathbb{F}}^2 + e_p^2(k) \tag{15}$$

here, $\|\widetilde{\mathbf{\Omega}}(k)\|_{\mathbb{F}}^2 = \widetilde{\mathbf{\Omega}}(k)^T \widetilde{\mathbf{\Omega}}(k)$ and $\mathbb{F}$ refers to the high dimensional feature space. Subtracting $\mathbf{\Omega}^*(k)$ from both side of (8), and multiplying $\varphi(\mathbf{x}(k))$, we have:

$$e_p(k) = e_a(k) + \mu(k)\kappa\langle \mathbf{x}(k), \mathbf{x}(k) \rangle[2\lambda(k)e(k)$$
$$+ p(1 - \lambda(k))|e(k)|^{(p-1)}\text{sign}(e(k))] \tag{16}$$

Substituting (16) into (15) and after straightforward derivation:

$$E\left[\left\|\widetilde{\mathbf{\Omega}}(k)\right\|_{\mathbb{F}}^2\right] = E\left[\left\|\widetilde{\mathbf{\Omega}}(k - 1)\right\|_{\mathbb{F}}^2\right] - 2\mu(k)E\left[e_a(k)\left(2\lambda(k)e(k)\right.\right.$$
$$\left.\left. + p(1 - \lambda(k))|e(k)|^{(p-1)}\text{sign}(e(k))\right)\right]$$
$$+ \mu^2(k)E\left[\left(2\lambda(k)e(k)\right.\right.$$
$$\left.\left. + p(1 - \lambda(k))|e(k)|^{(p-1)}\text{sign}(e(k))\right)^2\right] \tag{17}$$

---

[1] The *posterior* error $e_p(k)$ defined here is different from $e_{post}(k)$ in (10).

To make GKNMN converge after finite iterations, we have:

$$E\left[\left\|\widetilde{\mathbf{\Omega}}(k)\right\|_{\mathbb{F}}^2\right] \le E\left[\left\|\widetilde{\mathbf{\Omega}}(k-1)\right\|_{\mathbb{F}}^2\right]$$

$$\Longleftrightarrow -2\mu(k)E\left[e_a(k)\Big(2\lambda(k)e(k)+p(1-\lambda(k))|e(k)|^{(p-1)}\text{sign}(e(k))\Big)\right]$$

$$+\mu^2(k)E\left[\Big(2\lambda(k)e(k)+p(1-\lambda(k))|e(k)|^{(p-1)}\text{sign}(e(k))\Big)^2\right]$$

$$\le 0$$

$$\Longleftrightarrow \mu(k) \le \frac{2E\left[e_a(k)\Big(2\lambda(k)e(k)+p(1-\lambda(k))|e(k)|^{(p-1)}\text{sign}(e(k))\Big)\right]}{E\left[\Big(2\lambda(k)e(k)+p(1-\lambda(k))|e(k)|^{(p-1)}\text{sign}(e(k))\Big)^2\right]} \quad (18)$$

Therefore, if the normalized step $\mu(k)$ satisfies (18), the weight error vector in $\mathbb{F}$ will be monotonically decreased (and hence convergent). Interested readers can refer to [7] for simplification of (18).

5   Simulations and Results Analysis

The performance of our proposed GKNMN is evaluated in the nonlinear system identification setting. Four algorithms are re-implemented for comparison purpose: KAPA [11], KNLMS [13], KRMN [7], and KNMN [16]. The reasons behind the choice of selected comparison methods are described in [16]. Two benchmark systems are considered herein, where our goal is to model them using the input sequence x_k and the desired sequence d_k (corrupted with ambient noises n_k), which is given by $d_k = \psi(x_k)$. In the following two independent examples, both x_k and d_k consist of 3000 samples and we perform 300 Monte-Carlo simulations.

The first example features a nonlinear system which is described by a difference equation (System I):

$$\begin{cases} x_k = (0.8 - 0.5\exp(-x_{k-1}^2))x_{k-1} - (0.3 + 0.9\exp(x_{k-1}^2))x_{k-2} + 0.1\sin(x_{k-1}\pi) \\ x_0 = x_1 = 0.1 \\ y_k = x_k \\ d_k = y_k + n_k \end{cases} \quad (19)$$

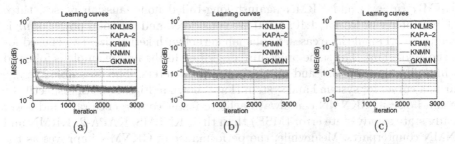

(a)   (b)   (c)

Fig. 1. Learning curves for system I, where the power ratios of Gaussian noise and Bernoulli noise are set to be (a) 1 : 1, (b) 1 : 10, and (c) 1 : 100, respectively. The corresponding steady-state MSEs ($\times 10^{-3}$) over the last 100 samples for GKNMN are 17.36, 7.85, and 7.61.

these data were used to estimated the nonlinear model $d_k = \psi(x_{k-1}, x_{k-2})$, where n_k is a linear combination of Gaussian noises and Bernoulli noises with different power ratios. To verify the superiority of GKNMN ($p = 4$) and KNMN against short-tailed non-Gaussian noises, three different power ratios 1:1, 1:10, 1:100 are considered and the total power of n_k is fixed to be 0.1 in all three cases. Gaussian kernel with size 0.38 is selected for all the aforementioned algorithms. To provide a fair comparison, the step size is tuned so that all the algorithms exhibit similar initial convergence rate. Other parameters are determined via grid search with increment $10^{-\ell}$ within the corresponding searching range $[10^{-\ell}, 10^{-\ell+1}]$ ($\ell = 1, 2, 3$)).

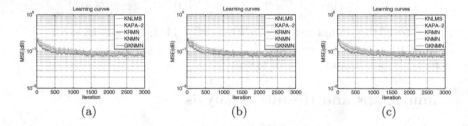

(a)   (b)   (c)

Fig. 2. Learning curves for system II, where the power ratios of Gaussian noise and Laplacian noise are set to be (a) 1:1, (b) 1:10, and (c) 1:100, respectively. The corresponding steady-state MSEs ($\times 10^{-2}$) over the last 100 samples for GKNMN are 7.65, 7.64, and 7.60.

The second example considers another discrete-time nonlinear dynamical system (system II):

$$\begin{cases} v_k = 1.1 \exp(-|v|_{k-1}) + x_k \\ y_k = v_k^2 \\ d_k = y_k + n_k \end{cases} \tag{20}$$

where v_0 is 0.5, x_k is $i.i.d.$ Gaussian distributed input sequence with zero-mean and standard deviation 0.25, n_k is a linear combination of Gaussian noises and Laplacian noises with different power ratios. Again, to verify the superiority of GKNMN ($p = 1$) and KRMN against long-tailed non-Gaussian noises, three different power ratios 1:1, 1:10, 1:100 are considered and the total power of n_k is fixed to be 0.1 in all three cases. Laplacian kernel with kernel size 0.35 is used for all the algorithms. The same procedure described in the first example is followed to initialize the step size and all the other parameters. The ensemble-average learning curves for system I and system II are shown in Figs. 1 and 2, respectively. As can be seen, GKNMN can always achieve faster convergence rate and lower steady state steady state error (MSE) than their KNLMS, KAPA-2, KRMN and KNMN counterparts. Meanwhile, the performance of GKNMN improves as the percentage of the non-Gaussian noise increases with the total noise power fixed.

To further assess the performance of proposed GKNMN for different values of p under different noisy environments. Three different scenarios are further

analyzed in detail, i.e., $p = 1, p = 2, p = 4$. Figure 3 illustrates the corresponding learning behavior of GKNMN to System I, under Uniform noise (representative short-tailed noise), Gaussian noise and Laplacian noise (representative
long-tailed noise) environment (the noise power is fixed to be 0.1), respectively.
As we can see, the best result is obtained with different p under different noise
environment, and $p = 4$ demonstrates superior performance against short-tailed
noises, while $p = 1$ favors long-tailed noises.

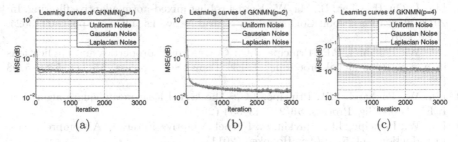

(a)   (b)   (c)

Fig. 3. Learning curves of GKNMN for system I under different noise environment
when (a) $p = 1$, (b) $p = 2$, and (c) $p = 4$.

6 Conclusion

In this paper, based on our previously proposed kernel normalized mixed-norm
(KNMN) algorithm for nonlinear system identification under non-Gaussian non-
stationary operating environment, we present a generalized version of KNMN
(GKNMN) to further improve its performance. Compared with standard KNMN,
we use the ℓ_p-norm to substitute for ℓ_4-norm, to provide a generalization to the
form of mixed norms. Meanwhile, we also introduce a novel updating mechanism to the time-varying adaptive mixing parameter (AMP), which also gives a
high-level generalization to different AMPs used in previous works. The mean
square convergence analysis is conducted. Simulations are also conducted, the
final results validate the theory part developed in this paper.

Acknowledgments. This work is supported partially by the National Natural Science Foundation of China (no.61402122) and the 2014 Ph.D. Recruitment Program of
Guizhou Normal University.

References

1. Aboulnasr, T., Mayyas, K.: A robust variable step-size lms-type algorithm: analysis
   and simulations. IEEE Trans. Sig. Process. **45**(3), 631–639 (1997)
2. Chambers, J., Avlonitis, A.: A robust mixed-norm adaptive filter algorithm. Sig.
   Process. Lett. IEEE **4**(2), 46–48 (1997)
3. Chen, B., Yuan, Z., Zheng, N., Príncipe, J.C.: Kernel minimum error entropy
   algorithm. Neurocomputing **121**, 160–169 (2013)

4. Chen, B., Zhao, S., Zhu, P., Príncipe, J.C.: Mean square convergence analysis for kernel least mean square algorithm. Sig. Process. **92**(11), 2624–2632 (2012)
5. Chuah, T.C., Sharif, B.S., Hinton, O.R.: Robust adaptive spread-spectrum receiver with neural net preprocessing in non-gaussian noise. IEEE Trans. Neural Netw. **12**(3), 546–558 (2001)
6. Lee, C.H., Lin, C.R., Chen, M.S.: Sliding-window filtering: an efficient algorithm for incremental mining. In: Proceedings of the Tenth International Conference on Information and Knowledge Management, pp. 263–270. ACM (2001)
7. Liu, J., Qu, H., Chen, B., Ma, W.: Kernel robust mixed-norm adaptive filtering. In: 2014 International Joint Conference on Neural Networks (IJCNN), pp. 3021–3024. IEEE (2014)
8. Liu, W., Pokharel, P.P., Príncipe, J.C.: Correntropy: properties and applications in non-gaussian signal processing. IEEE Trans. Sig. Process. **55**(11), 5286–5298 (2007)
9. Liu, W., Pokharel, P.P., Principe, J.C.: The kernel least-mean-square algorithm. IEEE Trans. Sig. Process. **56**(2), 543–554 (2008)
10. Liu, W., Principe, J.C., Haykin, S.: Kernel Adaptive Filtering: A Comprehensive Introduction, vol. 57. Wiley, Hoboken (2011)
11. Liu, W., Príncipe, J.: Kernel affine projection algorithms. EURASIP J. Adv. Sig. Process. **2008**(1), 784292 (2008)
12. Mandic, D.P., Papoulis, E.V., Boukis, C.G.: A normalized mixed-norm adaptive filtering algorithm robust under impulsive noise interference. In: 2003 IEEE International Conference on Acoustics, Speech, and Signal Processing, 2003. Proceedings, (ICASSP 2003), vol. 6, pp. VI-333. IEEE (2003)
13. Richard, C., Bermudez, J.C.M., Honeine, P.: Online prediction of time series data with kernels. IEEE Trans. Sig. Process. **57**(3), 1058–1067 (2009)
14. Sayed, A.H.: Fundamentals of Adaptive Filtering. Wiley, Hoboken (2003)
15. Slavakis, K., Giannakis, G., Mateos, G.: Modeling and optimization for big data analytics: (statistical) learning tools for our era of data deluge. Sig. Process. Mag. IEEE **31**(5), 18–31 (2014)
16. Yu, S., You, X., Zhao, K., Ou, W., Tang, Y.: Kernel normalized mixed-norm algorithm for system identification. In: 2015 International Joint Conference on Neural Networks (IJCNN). IEEE (2015, in press)
17. Zidouri, A.: Convergence analysis of a mixed controlled ℓ_2-ℓ_p adaptive algorithm. EURASIP J. Adv. Sig. Process. **2010**, 103 (2010)

A Parallel Sensitive Area Selection-Based Particle Swarm Optimization Algorithm for Fast Solving CNOP

Shijin Yuan, Feng Ji[✉], Jinghao Yan, and Bin Mu

School of Software Engineering, Tongji University, Shanghai China
{yuanshijin,0423jfeng,13alexyan,binmu}@tongji.edu.cn

Abstract. Recently, more and more researchers apply intelligent algorithms to solve conditional nonlinear optimal perturbation (CNOP) which is proposed to study the predictability of numerical weather and climate prediction. The difficulty of solving CNOP using intelligent algorithm is the high dimensionality of complex numerical models. Therefore, previous researches either are just tested in ideal models or have low time efficiency in complex numerical models which limited the application of CNOP. This paper proposes a sensitive area selection-based particle swarm optimization algorithm (SASPSO) for fast solving CNOP. Meanwhile, we adopt the self-adaptive dynamic control swarm size strategy to SASPSO method and parallel SASPSO with MPI. To demonstrate the validity, we take Zebiak-Cane (ZC) numerical model as a case. Experimental results show that the proposed method can obtain a better CNOP more efficiently than SAEP [1] and PCAGA [2] which are two latest researches on intelligent algorithms for solving CNOP.

Keywords: Sensitive area selection · PSO · CNOP · ZC model · SaDC · MPI

1 Introduction

Conditional nonlinear optimal perturbation (CNOP) is proposed to study the predictability of numerical weather and climate prediction. It has been applied successfully to ENSO predictability [3], spring predictability barrier (SPB) of El Niño events [4], and targeted observations, etc. The most popular method for solving CNOP is the adjoint-based method [5] which usually selects the spectral projecting gradient (SPG2) as the optimization method. It requires two inputs: the objective function and associated gradient. In particular, the objective function is calculated by the nonlinear models while the gradient by the adjoint models [6]. However, many modern numerical models have no corresponding adjoint models, and it is a huge engineering to implement one which it limited the application of CNOP. Therefore, more and more researchers focus on intelligence algorithm for solving CNOP (IA-CNOP) because it is free of gradient information and can deal with the non-differential object function [7, 8]. Zheng et al. [7] and Ye et al. [8] tried the intelligent algorithm to solving CNOP. However, these works are just validated on ideal models with low dimension. For the high dimension numerical model, it is inefficient to apply intelligent algorithm to calculate CNOP directly due to the large amount of calculation. Therefore, Mu et al. [2] and Wen et al. [1] proposed PCAGA

© Springer International Publishing Switzerland 2015
S. Arik et al. (Eds.): ICONIP 2015, Part II, LNCS 9490, pp. 71–78, 2015.
DOI: 10.1007/978-3-319-26535-3_9

(principal component analysis based-GA method) and SAEP (simulated annealing based ensemble projecting) method, respectively to solve CNOP. They reduce high dimensions space to low feature space firstly, and then solve CNOP in the low feature space with intelligent algorithm. These two works are significant to apply IA-CNOP to solve CNOP for high dimension numerical model. However, how to get enough data samples, which can represent the global characteristics, is a challenge. In addition, the efficiency needs to be improved.

Therefore, this paper proposes a sensitive area selection-based PSO algorithm (SASPSO) for fast solving CNOP. The process of SASPSO is as follows: (1) firstly selects a sensitive area based on the contribution of each dimension; (2) then searches solution in original space and strengthens search in the sensitive area by improved PSO with self-adaptive dynamic control strategy. In addition, we implement the parallel SASPSO with MPI. To demonstrate the validity, we apply it to solve CNOP in ZC model. Experiment results show that parallel SASPSO can obtain the similar results to ADJ-CNOP, better results than SAEP and PCAGA, and consumes less time.

The structure of this paper is arranged as follows: Sect. 2 introduce the concept of CNOP and give a description for SASPSO method as well as self-adaptive control dynamic control strategy. Experiments are presented in Sect. 3. Finally, this paper ends with the conclusion and future work in Sect. 4.

2   Sensitive Area Selection-Based Particle Swarm Optimization Algorithm (SASPSO) for Fast Solving CNOP

2.1   CNOP

CNOP is the perturbation δX_0^* which makes the target function $J_{TE}(\delta X_0)$ achieve the maximum with a condition of $\|\delta X_0\| \leq \beta$, i.e.

$$J(\delta X_0^*) = \max_{\|\delta X_0\| \leq \beta} \|J_{TE}(\delta X_0)\|^2 \tag{1}$$

$$J_{TE}(\delta X_0) = M_{t_0 \to t}(X_0 + \delta X_0) - M_{t_0 \to t}(X_0) \tag{2}$$

Here, $M_{t_0 \to t}$ is the discrete nonlinear propagator of a nonlinear model from the initial time t_0 to the prediction time t, J_{TE} is the nonlinear evolution of the initial perturbation. β is the constraint radius of the initial perturbation, $\|\cdot\|$ is used to measure the norm of the evolution of the initial perturbation. To apply existing optimization algorithms conveniently, we transfer the maximum problem into the minimum problem.

$$f(\delta X_0) = \min_{\|\delta X_0\| \leq \beta}(-\|J_{TE}(\delta X_0)\|^2) \tag{3}$$

Function f usually is considered as the target function which called adaptive function in intelligence optimization algorithm. Therefore, we can use intelligent algorithm which is free of the adjoint model to solve the target function to obtain CNOP.

2.2 SASPSO Method

In the SASPSO method, there are three important parts: dimension sensitivity map generation, sensitive area selection and improved PSO method.

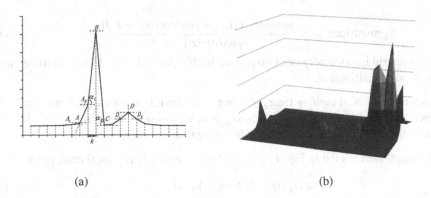

(a)   (b)

Fig. 1. Example of sensitivity map

Dimension Sensitivity Map Generation. The aim of dimension sensitivity map generation is to get the map which can measure the degree of sensitivity of each dimension. Firstly, we add a perturbation to a dimension, then we call the function M which is the discrete nonlinear propagator provided by the numerical model's corresponding nonlinear model. It can obtain the evolution of the perturbation which can represent the sensitivity of the corresponding dimension. Therefore, all the evolution of perturbation in each dimension consist the dimension sensitivity map.

Sensitive Area Selection Based on Immersion Simulation. After obtaining the sensitivity map, we need to recognize the sensitive area. Due to the influence of geographical information, the evolution of the perturbation is not always very large in each dimension. Therefore, we put forward that the sensitive area may not only be the region meets the technical definition of the extremum area but also contain the area surrounding the extremum region and some local extremum different from the perimeter zone represent a region of interest. Therefore, we proposed an automatic recognition algorithm inspired by the ideas of immersion simulation method [9] to recognize peak area which satisfies the demands of specific recognition. In this case, we regard the sensitivity map as a real 3D map whose value corresponding to altitude. Figure 1(a) and (b) are the example of the 2-d and 3-d image of the sensitive map, respectively. We take the 2-d image for example to show how it works. We assume that each point has the following parameters: α_L, α_R, growthRate$_\alpha$, $\Delta 1$, $\Delta 2$. α_L and α_R are showed in Fig. 1(a). A_L, A_R are the points on the left and right slide of A, $\Delta 1$, $\Delta 2$ are the physical constraint conditions. growthRate$_\alpha$ is utilized to evaluate the growth rate of the sensitivity index. We give the tangent transformation of α_L and α_R, and the mathematical formula of growthRate$_\alpha$ for point i as follow:

$$\tan\alpha_L(i) = \frac{senMatrix(i,j) - senMatrix(i-k,j)}{k} (i = 1, \dots n_x - k, j = 1, \dots n_y - k) \quad (4)$$

$$\tan\alpha_R(i) = \frac{senMatrix(i,j) - senMatrix(i+k,j)}{k} (i = 1, \dots n_x - k, j = 1, \dots n_y - k) \quad (5)$$

$$growthRate_{\alpha(i)} = \frac{senMatrix(i,j) - senMatrix(i-k,j)}{senMatrix(i-k,j)} \quad (6)$$

k is the grid interval between two points, In this paper, k = 1. There are three basic types of point distribution:

- Such as point A, if $\tan\alpha_L * \tan\alpha_R < 0$, we consider it is neither a peak nor a trough.
- Such as point B and D, $\tan\alpha_L > 0, \tan\alpha_R > 0$, we consider this point as a peak.
- Such as point C, $\tan\alpha_L < 0, \tan\alpha_R < 0$, we treat this point as a trough.

If a point satisfies the in Eqs. (7), (8) and (9), we take it as a local peak point.

$$\tan\alpha_L(i) > 0, \ \tan\alpha_R(i) > 0 \quad (7)$$

$$\tan\alpha_L(i) > \Delta 1 \quad (8)$$

$$growthRate_{\alpha(i)} > \Delta 2 \quad (9)$$

After we got the local peak point, we flooding from the local peak points, unlike the immersion simulation algorithm, the proposed method does not need to generate a watershed. In the process of submerged, we take constraints to determine whether each region needs to immerse in the next step, the constraints are set according to the practical situation and your recognize requirements, so that we can get the sensitive area which meets the requirements of the characteristics.

Improved PSO with Self-adaptive Dynamic Control. Particle swarm optimization (PSO) algorithm is an evolutionary computation technique developed by Kennedy and Eberhart in 1995 [10]. Original PSO algorithm has a fixed number of population size, which is not wise due to the requirement of population size may vary across periods. Oversized population will decrease the computational efficiency, while, if the population size is too little, search ability will be weakened. Therefore, this paper adopts self-adaptive dynamic control (SaDC) population size strategy to PSO algorithm to make the individuals work more efficiently [11]. The details of the PSO with SaDC is showed as pseudo-code in Algorithm 1. Here, PS^0 represents the initial population size, PS_{max} and PS_{min} are the maximum and minimum of the population size, S_{dec}, S_{inc} and k are the control parameters of SaDC strategy. In formula (10) and (11), the superscript represents the iterative step, w is the inertia coefficient, c_1 is the self-awareness to track p_{id} the historically optimal position, c_2 is the social-awareness of the particle swarm to track g_{id} the globally optimal particle, x_{id} is current solution, r_1 and r_2 are random numbers satisfying uniform distribution in the range of [0,1], and r is the restraint factor to control the speed.

Algorithm 1. PSO with SaDC Strategy

Initialization:
1: Set the parameters w, c_1, c_2, r, PS_{max}, PS_{min};
   $PS^0 \leftarrow (PS_{max} + PS_{min})/2$, $S_{dec} \leftarrow 0$, $S_{inc} \leftarrow 0$
PSO:
2: Random initialize a particle in the sensitive area;
3: $project(particle)$;
   ▷ project the particle back onto the boundary
4: while the termination is not satisfied do
5:    $adaptFunction(particle)$ ▷ Calculate the adaption values
6: update globally and locally optimal positions;
   SaDC strategy (step7~step12):
7:    if the globally optimal position is update, then
8:        $S_{dec} \leftarrow S_{dec} + 1$, $S_{inc} \leftarrow 0$
9:    else
10:       $S_{inc} \leftarrow S_{inc} + 1$, $S_{dec} \leftarrow 0$
11:   if $S_{dec} > 2k$ then delete the poor performance particles
12:   if $S_{inc} > k$ then add more particles
13:   $advanceStep(particle)$;
   ▷ move by standard PSO movement formula (10)(11)[10]

$$V_{id}^g = wV_{id}^{g-1} + c_1 r_1\left(p_{id}^{g-1} - x_{id}^{g-1}\right) + c_2 r_2\left(g_{id}^{g-1} - x_{id}^{g-1}\right) \qquad (10)$$

$$x_{id}^g = x_{id}^{g-1} + rV_{id}^g \qquad (11)$$

14:   $project(particle)$;
15: end while
Output: CNOP (The optimal solution)

---

3 Experiments

Our experiments run on our server with 8 GB memory and two Intel Xeon CPU E5645, each with 6 cores. In order to verify the validity, we apply it to solve CNOP in ZC model [12]. In this case, the initial perturbation δX_0 consists of the sea surface temperature anomalies (SSTA) and thermohaline anomalies (THA), the related variables form 1080-dimensional vectors, and the time span of evolution is 9 months. First, we analysis effectiveness of SaDC strategy. Then, we compare the proposed method with the adjoint-based method (for short, ADJ-CNOP), SAEP [1] and PCAGA [2]. In addition, we parallel SASPSO algorithm. Experiment results show that parallel SASPSO can obtain the similar results to ADJ-CNOP, better results than SAEP and PCAGA, and consumes less time.

3.1 Effectiveness Analysis of SaDC Strategy

We now compare the magnitude of CNOP fitness value and time consuming given by SASPSO with and without SaDC strategy. The two experiments have the same initial parameter, and the number of the parallel cores is 8. From Fig. 2(a) we can know that the fitness value of CNOP get by SASPSO is a little better than SASPSO without SaDC and the time consuming of SASPSO method is much faster than SASPSO without SaDC. We can know that the SaDC strategy makes particles works more efficiently.

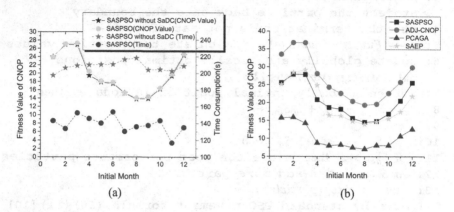

Fig. 2. (a) Fitness value of CNOP and time consumption in SASPSO with and without SaDC. (b) Magnitude of CNOP fitness value of SASPSO, PCAGA, SAEP and ADJ-CNOP method.

3.2 Comparison Analysis

Generally, there are three aspects to judge whether a method is good or not for solving CNOP: the fitness value of CNOP, the spatial structure of CNOP and the time efficiency, respectively. Typically, for the fitness value of CNOP, our inspection standard is the higher the better and the values of 12 months in a year should show up the same tendency with ADJ-CNOP method. However, the spatial structure of CNOP is the most important aspect, because it is the direct evidence of El Nino events (or La Nina events) when the SSTA components of tropical Pacific shows east "positive" and the west "negative" (or east "negative" west "positive"), and the THA components that show a sinking (or up) across the whole equatorial. For the time efficiency, obviously, the lower time consumption is better when achieve the similar results.

About Optimal Fitness Values. Figure 2(b) shows the proposed method can achieve a larger CNOP value than PCAGA. Therefore, the SASPSO method is better than PCAGA. The SASPSO performs better in most months while the SAEP method works better in other months. Therefore, the SASPSO method is similar with SAEP method in the effectiveness. In addition, The SASPSO follows the same trend with the ADJ-CNOP on the whole. Therefore, the CNOP given by the SASPSO method is a good approximation of ADJ-CNOP.

About Spatial Structure. Figure 3 depicts the optimal precursors and corresponding 9 month SSTA evolutions and initial optimization month is October. The results demonstrate that the SASPSO can be used to determine the optimal precursors of ENSO, and these optimal precursors can evolve into an El Niño event. It can also be seen from the figures given by SASPSO method retain the main large-scale spatial structure of those from the ADJ-CNOP. Compared with other methods, the result obtained by SASPSO method achieves closer spatial features to ADJ-CNOP method.

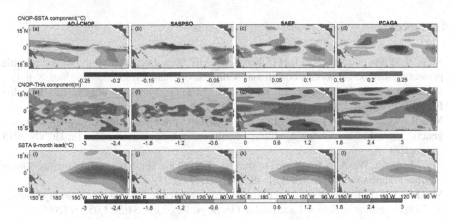

Fig. 3. Four column from left to right are the CNOP pattern of October obtain by ADJ-CNOP, SASPSO, SAEP and PCAGA in order. Three rows of subfigures are SSTA, THA and nonlinear evolution of SSTA after 9 months, the time span of 9 months.

About Efficiency. Since the SASPSO algorithm is a global search swarm intelligence algorithm, it is an effective way to reduce the time-consuming through parallel. Therefore, we adopt the MPI technique to speed up the proposed method. Our method can achieve 5.4 times speedup under 8 cores.

Here are the time consumption comparison of the four methods. For ADJ-CNOP method, it will take 361 s with 10 initial guess fields. However, when we use 8 cores to parallel the SASPSO, and the initial population size is 300, the time consumption is only 142 s average. In the same experimental environment, the SAEP and PCAGA are much slower than the parallel SASPSO.

4   Conclusion and Future Work

In this paper, we propose a parallel sensitive area selection-based PSO algorithm (SASPSO) for fast solving CNOP. Experiment results show that parallel SASPSO can obtain the similar results to ADJ-CNOP, better results than SAEP and PCAGA, and consumes less time. Compared with ADJ-CNOP, SASPSO method is free of the adjoint model and do not need the gradient information, so that it can handle the non-differentiable objective function. Compared with SAEP method and PCAGA method, SASPSO method searches solution in original space and reduces the difficulty of selecting data samples according to different physical problem because it does not need so many representative data samples to reduce dimensions. In the future, we will verify the effectiveness of SASPSO method against more complex numerical models with high dimensions, such as MM5 model and WRF model. In addition, we will try sensitive area selection with other intelligent algorithm to solve CNOP.

Acknowledgements. This work is supported by the National Natural Science Foundation of China (Grant No. 41405097).

References

1. Wen, S., Yuan, S., Mu, B., Li, H., Chen, L.: SAEP: simulated annealing based ensemble projecting method for solving conditional nonlinear optimal perturbation. In: Sun, X-h, Qu, W., Stojmenovic, I., Zhou, W., Li, Z., Guo, H., Min, G., Yang, T., Wu, Y., Liu, L. (eds.) ICA3PP 2014, Part I. LNCS, vol. 8630, pp. 655–668. Springer, Heidelberg (2014)
2. Mu, B., Zhang, L.L., Yuan, S.J., Li, H.Y.: PCAGA: principal component analysis based genetic algorithm for solving conditional nonlinear optimal perturbation. In: IJCNN 2015 (Accepted)
3. Mu, M., Duan, W.: A new approach to studying ENSO predictability: conditional nonlinear optimal perturbation. Chin. Sci. Bull. **48**, 1045–1047 (2003)
4. Mu, M., Xu, H., Duan, W.: A kind of initial errors related to "spring predictability barrier" for El Niño events in Zebiak-Cane model. Geophys. Res. Lett. **34**(3), L03709 (2007)
5. Xu, H., Duan, W.S., Wang, J.C.: The tangent linear model and adjoint of a coupled ocean-atmosphere model and its application to the predictability of ENSO. In: IEEE International Conference on Geoscience and Remote Sensing Symposium, pp. 640–643 (2006)
6. Duan, W.S., Feng, X., Mu, M.: Investigating a nonlinear characteristic of El Niño events by conditional nonlinear optimal perturbation. Atmos. Res. **94**(1), 10–18 (2009)
7. Zheng, Q., et al.: On the application of a genetic algorithm to the predictability problems involving "on-off" switches. Adv. Atmos. Sci. **29**(2), 422–434 (2012)
8. Ye, F.H., Zhang, L., Gong, X.W., Zheng, Q.: Applications of an improved particle swarm optimization to conditional nonlinear optimal perturbation. J. Jiangnan Univ. (Nat. Sci. Ed.) **10**(4), 1–5 (2011)
9. Vincent, L., Soille, P.: Watersheds in digital spaces: an efficient algorithm based on immersion simulations. IEEE Trans. Pattern Anal. Mach. Intell. **13**(6), 583–598 (1991)
10. Hsieh, S.-T., et al.: Solving large scale global optimization using improved particle swarm optimizer. In: IEEE Congress on Evolutionary Computation, CEC 2008 (IEEE World Congress on Computational Intelligence). IEEE (2008)
11. Wang, R.F., et al.: Nature computation with self-adaptive dynamic control strategy of population size. J. Softw. **23**(7), 1760–1772 (2012)
12. Zebiak, S.E., Cane, M.A.: A model El Nino-Southern osillation. Mon. Weather Rev. **115**, 2262–2278 (1987)

Realization of Fault Tolerance for Spiking Neural Networks with Particle Swarm Optimization

Ruibin Feng, Chi-Sing Leung$^{(\boxtimes)}$, and Peter Tsang

Department of Electronic Engineering, City University of Hong Kong,
Kowloon Tong, Hong Kong
rfeng4-c@my.cityu.edu.hk, eeleungc@cityu.edu.hk

Abstract. The spiking neural network (SNN) model has been an important topic in the past two decades. Many training algorithms, such as SpikeProp, were designed and applied to various applications. However, the fault tolerant ability in SNNs was not fully understood. Based on our study, the SNN model with the classical training objective function cannot even handle the single fault situation, in which one of the hidden neurons is damage. To improve the fault tolerant ability, we design an objective function and utilize the particle swarm optimization approach to minimize it. Simulation results show that our approach is much better than the classical objective function.

Keywords: Spiking neural networks · Fault tolerance · Particle swarm optimization

1 Introduction

Spiking neural networks [1] (SNNs), the 3th generation neural networks, have been widely investigated in the past two decades. Instead of using threshold logic units or sigmoidal neurons, in SNNs the information propagates through time in spiking neurons. A spiking neuron receives spikes from other neurons. It generates a spike (or saying fire), when its incoming spikes reach the threshold at certain points in time. An important feature is that instead of the magnitude of spikes, the time when a neuron fires is mainly affected by the number of the incoming spikes received. Theoretical analysis demonstrates that SNNs can simulate arbitrary feedforward sigmoidal neural networks [2].

Many algorithms, including gradients based methods, were proposed in [3,4] for SNNs. Besides, temporal encoding schemes were designed to encode an analog signal into spike times in [5,6]. Based on these, SNNs were successfully applied in many real applications, including lipreading [4] and face detection [7].

While numerous works have been conducted in the mathematical modeling of SNNs, equal effort has been spent on the integration of SNNs with the biological systems. The technology has cast light on the possibility that ultimately, the human neural systems can be partly replaced with semiconductor based SNNs. One important and interesting property in the human brain is the inherent fault

© Springer International Publishing Switzerland 2015
S. Arik et al. (Eds.): ICONIP 2015, Part II, LNCS 9490, pp. 79–86, 2015.
DOI: 10.1007/978-3-319-26535-3_10

tolerant ability [8]. However, according to [9–11], without proper training, the performance of the second generation neural networks could degrade drastically when network fault appears. Hence, it is essential to investigate the fault tolerant ability of SNNs. As showed in Sect. 5 later, SNNs trained by the classical objective function cannot handle even one fault hidden neuron. Then methods for SNNs to resist structure damages are needed to design. One of the straightforward approaches is to replicate the neurons and synaptic connections, which is borrowed from classical artificial neural network [12]. But it requires more resources and makes SNN structure considerably more complicated than its original form. In this paper, instead of replicating the network elements, we embed the fault tolerant requirement to the training process. Because of the existence of local minima, gradient descent type of learning algorithm may not be the proper choice in training fault tolerant SNNs, as the search paths are prone to being trapped in local solutions. Instead, we introduce an evolution algorithm, particle swarm optimization [13] (PSO) algorithm, to train fault tolerant SNNs.

Organization of the paper is given as follows. After the introduction, a brief description of the SNN is given in Sect. 2. Section 3 demonstrates the fault model and objective function and PSO algorithm for training SNN is shown in Sect. 4. Simulation results in Sect. 5 are used to evaluate our proposed method. Finally, we have a conclusion summarizing the essential findings.

2 Spiking Neural Network

As introduced in [14], we consider a SNN with a feedforward structure. A neuron fires when the internal neuron state, or "membrane potential" reaches a threshold. The internal state changes based on the incoming spikes following spike response model (SRM), which is described in [15]. For the purpose of simplification, we assume each neuron fires at most one time in the given time window.

Given a neuron j, it will receive the spikes with fire times $t_i, i \in \Gamma_j$, where Γ_j is the set of presynaptic neurons. Every presynaptic neurons $i \in \Gamma_j$ connects neurons i using a fixed number of m synaptic terminals. Every synaptic terminal has a different delay time $d^k, k = 1, \ldots, m$ as well as a tunable weight w_{ij}^k. One incoming spike from a synaptic terminal k connected with the presynaptic neuron i at time t is given by

$$u_i^k(t) = \varepsilon(t - t_i - d^k), \tag{1}$$

where t_i is the fire time of neuron i, d^k is the time decay of synaptic terminal k, and $\varepsilon(t)$ is the spike-response function described as

$$\varepsilon(t) = \frac{t}{\tau} \exp^{1 - \frac{t}{\tau}} \mathcal{H}(t), \tag{2}$$

where $\mathcal{H}(t)$ is the well-know heavyside function which generates 0 for $t \leq 0$ and 1 otherwise. τ is the membrane potential decay time which is assumed to be constant in the whole structure. Armed with (1) and (2), the dynamic of internal state for neuron j can be defined as the weighted sum of the pre-synaptic spikes:

$$x_j(t) = \sum_{i \in \Gamma_j} \sum_{k=1}^{m} w_{ij}^k u_i^k(t). \tag{3}$$

After $x_j(t)$ reaches the threshold θ, neuron j will fire a spike to the postsynaptic neurons. Then the state of the neuron $x_j(t)$ is reset.

The SNNs in this paper have a 3-layer structure with one input layer, one hidden layer, and one output layer. The number of input, hidden and output neurons are denoted as $\mathbb{I}$, $\mathbb{H}$ and $\mathbb{O}$, respectively. The SNN receives an input pattern $t(n)_j^I, j = 1, \ldots, \mathbb{I}$ and tries to learn the corresponding desire output pattern $t(n)_j^D, j = 1, \ldots, \mathbb{O}$ where $n = 1, \ldots, N$ and N is the number of training pattern. The goal is to minimize the following classical objective function:

$$E = \frac{1}{2N} \sum_{n=1}^{N} \sum_{j=1}^{\mathbb{O}} (t(n)_j^O - t(n)_j^D)^2, \tag{4}$$

and $t(n)_j^O$ is network output of output neuron j for training pattern n. Note that the input values and the output values are required to encode the spike time.

3   Fault Model and Objective Function

As mentioned in [10,16], in hardware implementations, component failure and open circuit are unavoidable. One common model to formulate the failure is stuck-at-zero fault [16]. In stuck-at-zero fault, the internal state variable of fault neuron j is tied at zero $x_j(t) = 0$ and the neuron cannot fire for all t. We assume that only one fault hidden neuron exists.

In order to increase the tolerant ability, we propose a new objective function:

$$E_f = E + \lambda \sum_{h=1}^{\mathbb{H}} E_h, \tag{5}$$

where E is given in (4) and E_h is the mean square error when hidden neuron h is faulty. E_h is given by

$$E_h = \frac{1}{2N} \sum_{n=1}^{N} \sum_{j=1}^{\mathbb{O}} (t(n)_{j,h}^O - t(n)_j^D)^2. \tag{6}$$

$t(n)_{j,h}^O$ is the fire time of output neuron j for training pattern n when $x_h(t) = 0$ for all t. λ is used to balance the faultless and faulty performance of SNN.

4   Particle Swarm Optimization

As mention before, gradient based algorithm may not suit to minimize the objective function (5) because of the existence of local minima. Hence, we proposed the use of PSO to train a faulty SNN. PSO is one kind of evolution algorithms [13]. The search process is done by a swarm of particles, with initial position vectors and velocity vectors with the same dimensions. Each element of a vector represents a tunable weight in our problem. The value of the objective function 5 corresponding to each particle reflects its relative fitness in the swarm. Each particle updates their position according to the best candidate in the whole swarm, as well as the personal best position of the particle itself.

For a particle i, the update from position $\boldsymbol{p}_i(n)$ to $\boldsymbol{p}_i(n+1)$ is given by

$$\boldsymbol{p}_i(n+1) = \boldsymbol{p}_i(n) + \boldsymbol{v}_i(n+1), \tag{7}$$

where $\boldsymbol{v}_i(n+1)$ is the velocity of particle i and is given by

$$\boldsymbol{v}_i(n+1) = w(n) \times \boldsymbol{p}_i(n) + c_1(n) \times U_1 \times (pbest_i - \boldsymbol{p}_i(n)) + c_2(n) \times U_2 \times (Gbest - \boldsymbol{p}_i(n)). \tag{8}$$

In Eq. (8), U_1 and U_2 are random real numbers in the range of $[0, 1]$. Parameter $w(n)$ is called the inertia weight. The variables $pbest_i$ and $Gbest$ are the best position (measured by (5)) from the particle i personal history and the whole swarm experience, respectively.

Various methods are proposed to update $w(n)$, $c_1(n)$ and $c_2(n)$. In this paper, we use linear declining function [17] to get the parameters for each iteration as following:

$$w(n) = w_{start} - \frac{w_{start} - w_{end}}{MAXITER} \times n, \tag{9}$$

and

$$c_i(n) = c_{i,start} - \frac{c_{i,start} - c_{i,end}}{MAXITER} \times n, \text{for } i = 1, 2, \tag{10}$$

where $MAXITER$ is the maximum number of iteration before PSO process is terminated.

As recommended by [17], w_{start} and w_{end} are set to 0.9 and 0.4. Besides, $[c_{1,start}, c_{1,end}]$ and $[c_{2,start}, c_{2,end}]$ are [2.5,0.5] and [0.5,2.5].

5   Simulation Results

In this section, we evaluate our proposed method based on two classical tasks: XOR-problem and Iris dataset.

XOR-Problem: The purpose of this simple example to investigate how the fault situation affects the output spike time. In the XOR-problem, there two input variables and one output. We encode input-output pairs as spike time patterns according to Table 1 [3]. The first spike time is the bias. Hence, the network structure is $3 \times 5 \times 1$. The synaptic terminal number is 16 and the delay time $d^k = k, k = 1, \ldots, 16$. It means the number of tunable weights is $3 \times 5 \times 16 + 5 \times 1 \times 16 = 320$. Besides, the membrane potential decay time τ in (2) is 7. The gradient based algorithm "SpikeProp" [3] is used for comparison.

Table 1. The input and the target spike patterns for the XOR problem.

Input pattern	Input spike time $[t(n)_1^I, t(n)_2^I, t(n)_3^I]$	Desired output $t(n)^D$
00	[0,0,0]	16
01	[0,0,6]	10
10	[0,6,0]	10
11	[0,6,6]	16

Its objective function is the classical objective function given by (4). For the PSO, we try two different λ values: $\lambda = 0$ and $\lambda = 1$.

When we set $\lambda = 0$, the training objective is identical to (4). For PSO, the swarm population is 100 and the maximum iteration is 200. The comparison between proposed method with SpikeProp are shown in Tables 2, 3 and 4.

From Table 2, it can be seen that the performance of the SNN trained by SpikeProp can produce the correct outputs when there are no faulty neurons. However, for example, as shown in Table 2, when hidden node 2 is faulty, the output spike time is around 20 regardless of the input pattern. It means the SNN trained SpikeProp with the classical objective function does not have any fault tolerant ability.

From Table 3, it can be seen that the SNN trained by PSO with $\lambda = 0$ can produce the correct outputs too, when there are no faulty neurons. When fault happens at hidden neuron 1 or hidden neuron 5, the output spike times are far away from the desire output. That means, when the classical objective function is used, the SNN trained by PSO cannot improve the fault tolerant ability.

From Table 4, the SNN trained by PSO with $\lambda = 1$ can produce the correct outputs when there are no faulty neurons. Besides, when fault happens, the output spike times are still very close to the desire spike times. That means, with the modified objective function, we can improve the fault tolerant ability.

Table 2. XOR: output spike times from SpikeProp.

Input pattern	Desire output spike time	Faultless output spike time	Faulty output spike time stuck-at-zero at hidden neuron				
			1	2	3	4	5
00	16	15.99	18.52	20.55	50	20.19	8.53
01	10	10.02	17.98	20.46	21.81	19.91	9.4
10	10	10.04	17.84	20.28	22.17	20.23	9.35
11	16	16.04	20.87	22.81	50	22.72	11.7

Iris Dataset: The Iris dataset contains 150 samples, where each example contains **4 input variables**. There are three classes in this dataset. Each class has

Table 3. XOR: output spike times from PSO with $\lambda = 0$.

Input pattern	Desire output spike time	Faultless output spike time	Faulty output spike time stuck-at-zero at hidden neuron				
			1	2	3	4	5
00	16	16	15.99	14.69	16	15.99	10.20
01	10	10	10.21	8.86	10	13.70	10
10	10	10	15.81	9.09	10	13.5	7.85
11	16	16	16	16.03	16	16	10.23

Table 4. XOR: output spike times from PSO with $\lambda = 1$.

Input pattern	Desire output spike time	Faultless output spike time	Faulty output spike time stuck-at-zero at hidden neuron				
			1	2	3	4	5
00	16	15.86	16.45	15.85	16.01	16.62	15.56
01	10	9.93	10.23	10.46	9.93	9.93	10.54
10	10	9.85	9.85	10.62	11.05	10.26	9.85
11	16	16	16.24	16.12	16.64	15.73	16.33

50 samples. We randomly select 25 examples from each class as the training set. The rest samples are used as the test set.

Followed the practice in [3], each input variable is encoded as **12 Gaussian receptive fields**, uniformly quantized to 11 levels and use integer from 0 to 10 to represent. The number of input spiking neurons is equal to $4*12+1 = 49$ (one is the bias). Since the number of classes is equal to 3, the SNN has three output neurons. If the sample belongs to the jth class, then the desire spike time for the jth output neuron is set to 10 and the desire spike times for other output neurons are set to 16. The synaptic terminal number is 16. The number of hidden neurons is 10. With this arrangement, the network structure is $49 \times 10 \times 3$. The number of tunable weights is $49*10*16+10*3*16 = 8320$. Our preliminary simulation found that the standard SpikProp is easily stuck at the local minimum. Even for the fault-free case, the classification rate is very low (around 60–70 % only). Therefore, we use PSO to train the SNN. We try two different λ values: $\lambda = 0$ and $\lambda = 1$. When we set $\lambda = 0$, the training objective is identical to that of the classical SpikeProp. Also, we repeat the experiment 10 times with different initial conditions. The results are shown in Tables 5 and 6.

From Table 5, with the classical objective function, when one hidden neuron is faulty, the average test set classification rate over the ten trials is dropped by 0.090. In addition, when the classical objective function is used, in many trials some hidden neurons are very important. If they are faulty, then the classification rate is greatly reduced. For instance, in Trail No. 2, if either hidden neuron 1 or hidden neuron 9 is faulty, the classification rate is greatly dropped. In other trials,

Table 5. PSO with the classical objective function: Iris test set classification rate.

Trial no.	No fault	Node fault case										mean
		stuck-at-zero at hidden neuron										
		1	2	3	4	5	6	7	8	9	10	
1	0.920	0.920	0.747	0.920	0.640	0.840	0.787	0.907	0.92	0.667	0.813	0.816
2	0.893	0.653	0.893	0.893	0.840	0.907	0.840	0.893	0.893	0.613	0.653	0.808
3	0.867	0.880	0.813	0.853	0.867	0.653	0.720	0.867	0.867	0.613	0.867	0.800
4	0.853	0.413	0.693	0.827	0.867	0.720	0.680	0.827	0.707	0.787	0.747	0.727
5	0.853	0.880	0.827	0.653	0.853	0.573	0.800	0.853	0.760	0.840	0.853	0.789
6	0.947	0.933	0.640	0.813	0.960	0.947	0.947	0.880	0.947	0.960	0.947	0.897
7	0.867	0.733	0.573	0.453	0.853	0.573	0.693	0.867	0.760	0.893	0.880	0.728
8	0.840	0.747	0.840	0.787	0.627	0.840	0.520	0.800	0.840	0.853	0.760	0.761
9	0.867	0.573	0.560	0.853	0.693	0.853	0.813	0.880	0.867	0.827	0.867	0.779
10	0.880	0.760	0.760	0.827	0.467	0.893	0.880	0.867	0.853	0.707	0.853	0.787
Average	**0.879**	–	–	–	–	–	–	–	–	–	–	**0.789**

Table 6. PSO with the modified objective function: Iris test set classification rate.

Trial no.	No fault	Node fault case										mean
		stuck-at-zero at hidden neuron										
		1	2	3	4	5	6	7	8	9	10	
1	0.907	0.893	0.84	0.8	0.907	0.907	0.827	0.907	0.907	0.827	0.853	0.867
2	0.88	0.88	0.787	0.88	0.84	0.84	0.853	0.893	0.827	0.893	0.667	0.836
3	0.827	0.8	0.827	0.787	0.813	0.853	0.733	0.72	0.813	0.787	0.827	0.796
4	0.893	0.853	0.667	0.893	0.893	0.893	0.813	0.867	0.827	0.747	0.893	0.835
5	0.827	0.827	0.76	0.8	0.787	0.76	0.667	0.827	0.827	0.813	0.813	0.788
6	0.933	0.92	0.907	0.72	0.933	0.933	0.933	0.933	0.92	0.933	0.867	0.9
7	0.907	0.907	0.84	0.92	0.88	0.88	0.907	0.92	0.827	0.907	0.867	0.885
8	0.853	0.88	0.84	0.827	0.76	0.867	0.84	0.813	0.853	0.707	0.84	0.823
9	0.813	0.787	0.667	0.787	0.707	0.773	0.813	0.667	0.84	0.733	0.787	0.756
10	0.947	0.893	0.893	0.947	0.933	0.907	0.92	0.853	0.893	0.907	0.92	0.907
Average	**0.879**	–	–	–	–	–	–	–	–	–	–	**0.839**

we also observe the similar behavior. That means, when the classical objective function is used, some hidden neurons dominates the approximation ability. If they are out of order, the performance of the trained SNN becomes very poor.

On the other hand, from Table 6, with the modified objective function, when one hidden neuron is faulty, the average test set classification rate over the ten trials is dropped by 0.040 only. In addition, after we investigate the results of each trial, we found out that in each trial, hidden neurons are likely to be equally important.

6   Conclusion

In this paper, we study the fault tolerant ability of SNNs. We found that with the classical training objective, the fault tolerant ability of the trained SNN is poor and some hidden neurons dominate the approximation ability. To improve the fault tolerant ability, we design a modified objective function and utilize the PSO approach to minimize the objective function. Simulation results show that our approach is able to improve the average fault tolerant ability. Besides, with the modified objective function, we found out that the hidden neurons are more likely to be equally important.

Acknowledgement. The work was supported by the General Research Fund from Hong Kong (Project No.: CityU 116511).

References

1. Maass, W.: Networks of spiking neurons: the third generation of neural network models. Neural Networks **10**(9), 1659–1671 (1997)
2. Maass, W.: Fast sigmoidal networks via spiking neurons. Neural Comput. **9**(2), 279–304 (1997)
3. Bohte, S.M., Kok, J.N., Poutre, H.L.: Error-backpropagation in temporally encoded networks of spiking neurons. Neurocomputing **48**(1), 17–37 (2002)
4. Booij, O.: Temporal pattern classification using spiking neural networks. Unpublished master's thesis, University of Amsterdam (2004)
5. Lewicki, M.S.: Efficient coding of natural sounds. Nat. Neurosci. **5**(4), 356–363 (2002)
6. Maass, W., Bishop, C.M.: Pulsed Neural Networks. MIT Press, Cambridge (2001)
7. Matsugu, M., Mori, K., Ishii, M., Mitarai, Y.: Convolutional spiking neural network model for robust face detection. In: Proceedings of ICONIP 2002, vol. 2, pp. 660–664 (2002)
8. Zhang, X., Song, S., Wu, C.: Robust Bayesian classification with incomplete data. Cogn. Comput. **5**(2), 170–187 (2013)
9. Zhou, Z.-H., Chen, S.F.: Evolving fault-tolerant neural networks. Neural Comput. Appl. **11**(3–4), 156–160 (2003)
10. Leung, C.S., Sum, J.: A fault-tolerant regularizer for RBF networks. IEEE Trans. Neural Networks **19**(3), 493–507 (2008)
11. Leung, C.S., Sum, J.: RBF networks under the concurrent fault situation. IEEE Trans. Neural Netw. Learn. Syst. **23**(7), 1148–1155 (2012)
12. Oya, T., Asai, T., Amemiya, Y., Schmid, A., Leblebici, Y.: Single-electron circuit for inhibitory spiking neural network with fault-tolerant architecture. In: Proceedings of IEEE International Symposium on Circuits and Systems, pp. 2535–2538 (2005)
13. Venter, G., Sobieszczanski-Sobieski, J.: Particle swarm optimization. AIAA J. **41**(8), 1583–1589 (2003)
14. Natschlager, T., Ruf, B.: Spatial and temporal pattern analysis via spiking neurons. Network: Comput. Neural Syst. **9**(3), 319–332 (1998)
15. Gerstner, W.: Time structure of the activity in neural network models. Phys. Rev. E **51**(1), 738 (1995)
16. Ho, K., Leung, C.S., Sum, J.: Convergence and objective functions of some fault/noise-injection-based online learning algorithms for RBF networks. IEEE Trans. Neural Networks **21**(6), 938–947 (2010)
17. Ratnaweera, A., Halgamuge, S., Watson, H.C.: Self-organizing hierarchical particle swarm optimizer with time-varying acceleration coefficients. IEEE Trans. Evol. Comput. **8**(3), 240–255 (2004)

Parallel Cooperative Co-evolution Based Particle Swarm Optimization Algorithm for Solving Conditional Nonlinear Optimal Perturbation

Shijin Yuan, Li Zhao[✉], and Bin Mu

School of Software Engineering, Tongji University, Shanghai, China
{yuanshijin,0821frank_zhao,binmu}@tongji.edu.cn

Abstract. Conditional nonlinear optimal perturbation (CNOP) is proposed to study the predictability of numerical weather and climate prediction. Recent researches show that evolutionary algorithms (EAs) could solve CNOP efficiently, such as SAEP and PCAGA. Both of them use dimension reduction methods with EAs to solve CNOP. But these methods always need large scale data samples and their data information are usually incomplete, which sometimes may cause the result unsatisfactory. Another way is to use cooperative co-evolution (CC) method, it adopts multi populations to change the mode of traditional searching optimum solutions. The CC method is applied in the original solution space which could avoid the defects that dimension reduction method has. In this paper, we propose cooperative co-evolution based particle swarm optimization algorithm (CCPSO) for solving CNOP. In our method, we make improvements on PSO with tabu search algorithm. Then we parallelize our method with MPI (PCCPSO). To demonstrate the validity, we compare our method with adjoint-based method, SAEP and PCAGA in ZC model. Experimental results of CNOP magnitudes and patterns show PCCPSO has the satisfactory results that are approximate to the adjoint-based method and better than SAEP and PCAGA. The time consumption of PCCPSO is about 5 min. It is approximate to the adjoint-based method with 15 initial guess fields and faster than SAEP and PCAGA. Our method can reach the speedup of 7.6 times with 12 CPU cores.

Keywords: Cooperative co-evolution · CNOP · PSO · ZC model

1 Introduction

CNOP has played an important role in predictability and sensitivity studies of meteorology and oceanography [1]. The common method for solving CNOP is the adjoint-based method (for short, XD method) which is always referred to as the benchmark [2]. Its corresponding gradient is obtained by the adjoint model. Unfortunately, many numerical models have no corresponding adjoint models and new implementations cost tremendous engineering work. Recently, many researchers applied EAs to solve CNOP. However, numerical models usually have tens of thousands dimensions in solution spaces, which costs long computation time. This is referred to as the curse of dimensionality [3].

© Springer International Publishing Switzerland 2015
S. Arik et al. (Eds.): ICONIP 2015, Part II, LNCS 9490, pp. 87–95, 2015.
DOI: 10.1007/978-3-319-26535-3_11

For better efficiency, some researchers adopt dimension reduction methods with EAs to solve CNOP. Wen proposed simulated annealing algorithm through PCA to solve CNOP (SAEP) [4]. The genetic algorithm was applied by Zhang for solving CNOP based on PCA (PCAGA) [5]. Both of them are valid methods for solving CNOP. Sometimes, however, when the data sets lose some data information, the results become less precise. Moreover, numerical models like MM5 and WRF [6] have much higher dimensions, which means dimension reduction methods may not obtain the accurate result. Another way to improve the efficiency is cooperative co-evolution (CC), which was proposed by Potter [7]. The most common models of CC are 'island model' and 'domain model' [8]. In these models, multi populations are adopted and individual in the population represents a feasible solution. Each population searches for the optimal solution and the best individuals are transferred among the populations, which is considered to be the shared information to help the co-evolution. Besides, the co-evolution of multi populations makes itself easy to be parallelized.

In this paper, we propose a cooperative co-evolution based PSO algorithm (CCPSO) for solving CNOP, then we use MPI to parallelize it (PCCPSO). Due to the fact that PSO has the shortcoming of premature convergence and its local optimization ability is weak, we combined it with tabu search (TS), which has the advantage of good local search ability. We adopt multi populations that have coevolution relationship to obtain CNOP. Each population will deliver its best fitness value and optimal particle to next population. To demonstrate the validity, the Zebiak-Cane model is utilized as a case to compare our method with XD method, SAEP and PCAGA.

The rest of the paper is organized as follows: Sect. 2 introduces the background. In Sect. 3, we introduce the method CCPSO and then we show the parallelization of it (PCCPSO). Experiments and analysis are showed in Sect. 4. This paper ends with conclusions and future works in Sect. 5.

2 Background

2.1 CNOP

CNOP is the perturbation $u_{0\delta}^{*}$ that makes the target function $J(u_0)$ achieve the maximum at the prediction time, τ under the constraint of $\|u_0\|_{\delta} \leq \delta$, i.e.

$$J(u_{0\delta}^{*}) = \max_{\|u_0\|_{\delta} \leq \delta} J(u_0),$$
$$J(u_0) = \left\| M_{\tau}(U_0 + u_0) - M_{\tau}(U_0) \right\|_{\delta} \tag{1}$$

where M_{τ} is the propagator of a nonlinear model from the initial time to the prediction time τ, $J(u_0)$ is the nonlinear evolution of the initial perturbation, u_0 is the initial perturbation, and δ is the magnitude of uncertainty. $\|\bullet\|$ means the L^2 norm. For the convenience of computing, Eq. (1) can be converted into a minimum problem:

$$J(u_{0\delta}^{*}) = \min_{\|u_0\|_{\delta} \leq \delta} -J(u_0). \tag{2}$$

2.2   Zebiak-Cane (ZC) Model

The Zebiak-Cane model [9] is a mesoscale air-sea coupled model for the tropical pacific. It has been widely applied to the research of the predictability and dynamics of El Niño/ Southern Oscillation (ENSO). In ZC model, u_0 in Eq. (1) is a perturbation vector consists of the sea surface temperature anomalies (SSTA) and thermocline depth anomalies (THA). The dimensions of ZC model grid is 30*34 (for SSTA and THA). As there is no value in the marginal area, we reduce the size of data set to 20*27*2.

2.3   Particle Swarm Optimization (PSO) and Tabu Search (TS)

Particle swarm optimization (PSO) [10] is one of the most effective optimization algorithms. The PSO is described as:

$$v_i^{t+1} = \omega * v_i^t + c_1 * r_1 * (pbest_i^t - x_i^t) + c_2 * r_2 * (pbest_i^t - x_i^t)$$
$$x_i^{t+1} = x_i^t + v_i^{t+1}$$
(3)

Where $\omega \geq 0$ is an inertia weight coefficient, $c_1 \geq 0$ and $c_2 \geq 0$ are acceleration coefficients, r_1 and r_2 are random numbers in the range [0, 1]. x_i^t denotes the location of the i-th particle on the t-th iteration, and v_i^t denotes a velocity vector of the i-th particle on the t-th iteration. $pbest_i^t$ is personal best location that gives the best value of the evaluation function of the i-th particle on the t-th iteration, $gbest_i^t$ is global best location that means gives the best fitness value on the t-th iteration.

$$C_j = rand(-1, 1), \; S_{i,j} = rand(-1, 1), \; i = 1, 2, \ldots, m; j = 1, 2, \ldots, n$$
(4)

$$S_{i,j} = \begin{cases} S_{i,j}, & ||C - S_i|| \leq r \\ C_j - \dfrac{r}{||C - S_i||} * (C_j - S_{i,j}), & ||C - S_i|| > r \end{cases}, \; i = 1, 2, \ldots, m; j = 1, 2, \ldots, n$$
(5)

Tabu search [11, 12] employs local search methods. Local searches take a potential solution to a problem and check its immediate neighbors to find an improved solution. Firstly, it will initialize the current point C, and generate m neighbors $(S_1, S_2, \ldots, S_m)$ by Eq. (4). $S_{i,j}$ is the value of j-th dimension of i-th neighbor. Then use Eq. (5) to keep every neighbor in the neighborhood hyper sphere. The TS then compute the adaptive values of C and copy $f(C)$ to the best solution. If a potential solution has been previously visited, it is marked as "tabu" (forbidden) so that the algorithm does not consider that possibility repeatedly.

3   PCCPSO: Parallel Cooperative Co-evolution Based Particle Swarm Optimization Algorithm

3.1   Improved Particle Swarm Optimization Algorithm with Tabu Search

The main reason PSO has the shortcoming of premature convergence and weak optimization ability is lacking of population diversity or trapping in local optimum [13]. To avoid it, we make improvements with tabu search. During the search period of PSO, optimal particle is perturbed with a linear increasing probability that makes the particles maintain vitality which could prevent the population from lacking diversity. Gaussian disturbance was put into in the personal best positions in Eq. (6), which could prevent particles falling into local minima and improve the convergence speed and accuracy.

$$v_i^{t+1} = \omega * v_i^t + c_1 * r_1 * (pbest_i^t + r_2 * gauss_i^t - x_i^t) + c_2 * r_3 * (pbest_i^t - x_i^t)$$
$$gauss_i^t = r_4 * gaussian(\mu, \sigma^2) \tag{6}$$

r_1, r_2, r_3, r_4 are random numbers in range [0, 1]. $gaussian_i^t$ is the Gaussian disturbance of the i-th particle on the t-th iteration. μ is the mean value and σ^2 is the variance. Mutation probability p has lots of randomness. When iterations increases, we take p to add diversity. We set p rises from pp_{begin} to pp_{end}, which is linear increasing. The definition of p is as Eq. (7)

$$p = pp_{begin} + (pp_{end} - pp_{begin}) * \frac{Curcnt}{Loopcnt} \tag{7}$$

When the global search of PSO converges to a certain degree, the algorithm will call tabu search to carry on the fine local search. When the rate of fitness change in latest T steps is less than threshold ε, the population could be considered to converge to a certain extent. Then the TS begins. The starting condition is defined as Eq. (8). Where $Curcnt$ is the current number of iterations, $f(Curcnt)$ and $f(Curcnt-T)$ are current fitness value and fitness value T steps before.

$$\frac{|f(Curcnt) - f(Curcnt - T)|}{|f(Curcnt - T)|} \le \varepsilon. \tag{8}$$

3.2   CCPSO

To avoid the disadvantages that the dimension reduction methods have, in this paper, we apply CC strategy with multi populations to search for CNOP. Multi populations are generated in the solution space to search for the optimal solution. Each population contains two stages, one is PSO and the other is TS. The interrelation of the two stages are based on Sect. 3.1. We call our method CCPSO. Its process is shown in Algorithm 1. Firstly, we make initialization and set parameters. Then update the velocity and position of PSO. Mutation possibility p and threshold ε are computed to decide whether to invoke TS. Then TS will run the local search and get the $gbest$. The global best fitness value and optimal particle will be delivered to the next population. Finally, the optimal solution *gbest* (CNOP) will be output when the termination condition is satisfied. The process of our method is shown in Algorithm 1.

3.3 PCCPSO

In this paper, we are to parallelize CCPSO (PCCPSO) without changing the co-evolutionary feature and a master-slave model is designed. During the computation of CNOP, the time consumption increases with the rise of times of calling adaptive function of ZC model linearly. In this paper, we adopt MPI to speed up our method.

In our parallel model, there are two kind of processes, the master process and the slave process. The master process is to control the main CCPSO algorithm and distribute computing tasks to slave processes and collect results after they finish. For slave processes, they are to execute the work sent to them and send results back. In PSO and TS, which are very convenient to divide subtasks, we send particles to *n* slave processes that make the computation of adaptive values compute in parallel. After each slave process finishes their computation, they will find out the best particle and adapt value, and then send them back to the main process.

Algorithm 1. CCPSO

1:	**Input**: SSTA
2:	**While** (termination is not satisfied)
3:	Initialization: set parameters *max_iter, p, ε, pbest, gbest, n*
4:	Update particle velocity and position according to Eq. (3)
5:	if *rand*() < *p*, add disturbance according to Eq. (6)
6:	if fitness < ε, go to step 3
7:	Initialize neighbors according to Eq. (4)
8:	Use Eq. (5) to keep every neighbor in the neighborhood hyper sphere
9:	TS compute the adaptive value.
10:	Deliver the *gbest* and optimal particle, go to step 4.
11:	**End while**
12:	**Output**: *gbest*

4 Experiments and Analysis

Our experiments run on a HP Z800 workstation with 8 GB memory and 2 Intel Xeon E5645 2.4 GHz CPU, each with 6 cores. The operating system is Fedora 13. We perform the comparison among PCCPSO and the XD method, SEAP and PCAGA.

4.1 Parameter Selection

The NINO3.4 index is one of several ENSO indicators based on sea surface temperatures. NINO3.4 is the average sea surface temperature anomaly in the region bounded by 5°N to 5°S, from 170°W to 120°W. This region has large variability on El Niño time scales, and is close to the region where changes in local sea surface temperature are important for shifting the large region of rainfall typically located in the far western Pacific. In ZC model, the center area along the equator is where the NINO3.4 locates. The variation of ENSO mainly converge on the area around equator [14] and the value of THA almost keeps unchanged throughout the year. Therefore, we set the grid of SSTA as 10*27 and choose the SSTA as the input of ZC model.

In [15], we take their recommended parameters of PSO. We take ω as 0.729, $c_1 = c_2 = 1.49$. And we set pp_{begin} and pp_{end} are 0.1 and 0.9 respectively. Based on our pre-experiments, we set other parameters as shown in Table 1.

Table 1. Parameter settings in experiments

Name	Meaning	Value
N	Number of dimensions in solution space	270
ε	Threshold of PSO and TS	0.25
max_iter1	Number of iterations in PSO	100
max_iter2	Number of iterations in TS	150
m	Number of neighbors in TS	30
n	Number of particles in PSO	300
R	Radius of neighborhood	0.1
M	Swarm numbers	3

4.2 Comparison Analysis

In this paper, we compare the magnitudes and patterns of CNOP with the XD method, SAEP and PCAGA. We set each month as the initial month and its optimization time span as 9 months. The magnitudes of CNOP in each initial month is shown in Fig. 1. It is obvious that all lines have the same trend. The magnitudes of PCCPSO are less than SAEP in June, July and August, but higher in other months. It has greater magnitudes than PCAGA throughout the year. Difference between the XD method and PCCPSO is shown in dash line called diff. The biggest difference is in June while the smallest one in April. Without loss of generality, we choose the two months as the examples to show the pattern comparison.

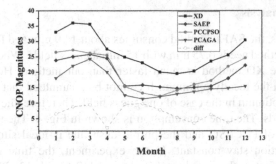

Fig. 1. The CNOP magnitudes of four methods. The black line is XD method, the red line is PCCPSO, the blue line is SAEP, and the green line is PCAGA. The dash line is the difference value between XD and PCCPSO (Color figure online).

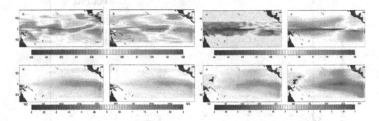

Fig. 2. CNOP pattern of June. The first column on the left is the pattern of SAEP, the second is PCAGA, the third is PCCPSO and the fourth is XD method

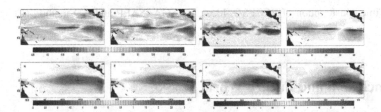

Fig. 3. CNOP pattern of April. The first column on the left is the pattern of SAEP, the second is PCAGA, the third is PCCPSO and the fourth is XD method

In Fig. 2, two rows of sub-figures are the patterns of the SSTA and the SSTA evolution. In the first row, it is obvious that the left part is negative while the right part is positive, which is the character of precursor of El Nino. In the second row, the patterns of SSTA evolution in these methods look similar. All of them show that the water in east pacific have the phenomenon of abnormal high temperature, which is the main feature of El Nino. The result of April is shown in Fig. 3. The CNOP pattern generated by PCCPSO looks similar with XD method in both cases that have the biggest and smallest magnitude difference. Therefore, we could draw the conclusion that PCCPSO can be treated as an approximate solution for CNOP.

4.3   Efficiency Analysis

With 80 dimensions, the SAEP method consumes about 14.9 min and PCAGA is about 21 min. The XD method needs 6.75 min with 15 initial guess fields. When just one guess field is utilized, the XD method is much faster than our method. However, the XD method may obtain the locally optimum. It cannot be guaranteed that the XD method obtain the global optimum in the case of one guess field. Therefore, the XD method use 15 initial guess fields. The time consumption is shown in Fig. 4. The red line is actual time consumption of PCCPSO and the purple dash line is ideal situation. Methods without parallelization stay constant. In the experiment, the time consumption of PCCPSO is about 5 min, the speedup of it can reach 7.6 with 12 CPU cores, which improves the efficiency of solving CNOP greatly.

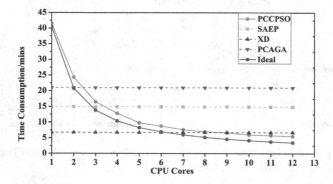

Fig. 4. Time consumption of PCCPSO, Ideal, PCAGA, SAEP and XD method (Color figure online).

5   Conclusion and Future Work

This paper proposes a parallel cooperative co-evolution particle swarm optimization algorithm in ZC model to solve CNOP. Multi populations of PSO combined with TS are applied to search for the optimum solution. On one hand, the proposed method has similar magnitudes, patterns and speed as the XD method, and it is free of adjoint model. On the other hand, our method has better magnitudes and patterns than SAEP and PCAGA. Moreover our method is free of dimensionality reduction methods. Therefore, our method could be treated as an approximate solution to CNOP. That is, we can apply the proposed method to CNOP of other numerical models like MM5 or WRF that have high dimensions without adjoint models, which improves the capability of the applications of the CNOP. It is a main task of our future works.

References

1. Duan, W.S., Mu, M.: Conditional nonlinear optimal perturbation as the optimal precursors for El Nino Southern oscillation events. Geogr. Res. **109**, 1–12 (2004)
2. Birgin, E.G., Martinez, J.E.M., Raydan, M.: No monotone spectral projected gradient methods on convex sets. Soc. Ind. Appl. Math. J. Optim. **10**, 1196–1211 (2000)
3. Goldberg, D., Korb, B., Deb, K.: Messy genetic algorithms: motivation, analysis, and first results. Complex Syst. **3**(5), 493–530 (1989)
4. Wen, S., Yuan, S., Mu, B., Li, H., Chen, L.: SAEP: simulated annealing based ensemble projecting method for solving conditional nonlinear optimal perturbation. In: Sun, X.-h., Qu, W., Stojmenovic, I., Zhou, W., Li, Z., Guo, H., Min, G., Yang, T., Wu, Y., Liu, L. (eds.) ICA3PP 2014, Part I. LNCS, vol. 8630, pp. 655–668. Springer, Heidelberg (2014)
5. Mu, B., Zhang, L.L.: PCAGA: principal component analysis based genetic algorithm for solving conditional nonlinear optimal perturbation. In: International Joint Conference on Neural Networks, Ireland (2015, Accepted)
6. Xu, H.: The tangent linear model and adjoint of a coupled ocean-atmosphere model and its application to the predictability of ENSO. In: Geoscience and Remote Sensing Symposium, pp. 640–643 (2006)
7. Potter, M.A.: A cooperative co-evolutionary approach to function optimization. In: International Conference of Parallel Problem Solving from Nature, vol. 2, pp. 249–257 (1994)
8. Tao, X.M., Xu, J.: Multi-species cooperative particle swarm optimization algorithm. Control Decis. **24**(9), 1406–1411 (2009)
9. Zebiak, S.E.: A model El Nino-Southern osillation. Mon. Weather Rev. **115**, 2262–2278 (1987)
10. Kennedy, J.: Particle swarm optimization. In: IEEE International Conference of Neural Networks, pp. 1942–1948 (1995)
11. Fred, G.: Tabu search - part 1. ORSA J. Comput. **1**(2), 190–206 (1989)
12. Fred, G.: Tabu search - part 2. ORSA J. Comput. **2**(1), 4–32 (1990)
13. Li, Y.G., Deng, Y.Q.: Improved particle swarm optimization algorithm with tabu search. Comput. Eng. **38**(18), 155–157 (2012)
14. Duan, W.S., Feng, X.: Investigating a nonlinear characteristic of El Niño events by conditional nonlinear optimal perturbation. Atmos. Res. **94**, 8–10 (2009)
15. Clerc, K.: The particle swarm-explosion, stability, and convergence in a multi- dimensional complex space. IEEE Trans. Evol. Comput. **6**, 58–73 (2002)

Discovery of Interesting Association Rules Using Genetic Algorithm with Adaptive Mutation

Mir Md. Jahangir Kabir[✉], Shuxiang Xu, Byeong Ho Kang,
and Zongyuan Zhao

School of Engineering and ICT, University of Tasmania, Hobart, Australia
{mmjkabir, Shuxiang.Xu, Byeong.Kang,
Zongyuan.Zhao}@utas.edu.au

Abstract. Association rule mining is the process of discovering useful and interesting rules from large datasets. Traditional association rule mining algorithms depend on a user specified minimum support and confidence values. These constraints introduce two major challenges in real world applications: exponential search space and a dataset dependent minimum support value. Data analyzers must specify suitable dataset dependent minimum support value for mining tasks although they might have no knowledge regarding the dataset and these algorithms generate a huge number of unnecessary rules. To overcome these kinds of problems, recently several researchers framed association rule mining problem as a multi objective problem. In this paper, we propose ARMGAAM, a new evolutionary algorithm, which generates a reduced set of association rules and optimizes several measures that are present in different degrees based on the datasets are used. To accomplish this, our method extends the existing ARMGA model for performing an evolutionary learning, while introducing a reinitialization process along with an adaptive mutation method. Moreover, this approach maximizes conditional probability, lift, net confidence and performance in order to obtain a set of rules which are interesting, useful and easy to comprehend. The effectiveness of the proposed method is validated over a few real world datasets.

Keywords: Data mining · ARMGA · Positive association rules · Genetic algorithm · Conditional probability

1 Introduction

Data mining techniques or knowledge discovery in datasets (KDD) define the extraction of novel, valid, interesting, useful, and understandable knowledge or patterns from data. Association rule mining plays a vital role in advancing the research, applications and development of data mining techniques. An association rule is an implication between two item sets A and B, A $\rightarrow$ B, which is used to define the dependencies between the item sets in a dataset. The problem of mining association rules has been considered by many researchers and a large number of algorithms have been developed for extracting association rules from different type of datasets [1, 2].

© Springer International Publishing Switzerland 2015
S. Arik et al. (Eds.): ICONIP 2015, Part II, LNCS 9490, pp. 96–105, 2015.
DOI: 10.1007/978-3-319-26535-3_12

Most of the existing classical algorithms for mining association rules are based on a support-confidence framework. This framework consists of two sub processes: finding all frequent item sets and generating rules from those frequent item sets based on a user defined support value and a confidence value, respectively. Several authors have noted that these algorithms raised the following major challenges: (1) Users need to specify an appropriate threshold value for mining rules although they have no information regarding the dataset, and (2) Association rule mining is an NP-Hard problem because searching all frequent item sets satisfying a minimum support value reveals an exponential search space of size 2^n, where n is the number of item sets [3, 4]. Finally, it generates a huge number of unnecessary rules from frequent item sets, resulting in weak mining performance [5, 6].

Recently, a large number of research papers have used evolutionary algorithms for mining association rules. These studies have found that evolutionary algorithms (EAs) particularly genetic algorithms based approaches are efficient tools especially when the search space is too large to use deterministic search methods [6, 7]. Because of inherent parallel structure, GA based methods are effective for automatic processing of large amount of data and discovering meaningful and significant information [8]. In real world applications, datasets not only use quantitative or numeric values but also contain categorical values. For this reason, several studies have been proposed for mining Boolean association rules (BARs) from datasets with categorical values [3, 9, 10].

Recent GA based approach named ARMGA [3, 11], uses conditional probability as a fitness function to extract high quality BARs without generating unnecessary rules. This algorithm uses only one evaluation criterion to measure the quality of the rules. Recently, some researchers have framed the association rule mining problem as a multiobjective problem rather than mono objective problem and jointly optimize different measures, in order to extract a set of rules which are easy to understand and interesting [6, 12, 13]. These approaches remove some of the drawbacks of mono-objective algorithms and mine high quality rules from the datasets with quantitative or numerical values [6, 14, 15].

Motivated by the features of multi-objective approaches, in this paper we propose a new genetic algorithm based approach, named ARMGAAM, which jointly optimize multiple objectives to mine a reduced set of BARs without considering user defined support and confidence values. The generated rules are easy to understand, interesting and having a good trade off among the number of rules, support, confidence and other objectives of the datasets. To accomplish this, our approach extends the recent ARMGA in order to perform an evolutionary learning and condition selection and maximizes three other objectives: lift, net confidence and conditional probability. Moreover, this approach introduces re-initialization process and an adaptive mutation method in order to increase the diversity in the population.

In order to evaluate the performance of the proposed method, this experimental study is carried out on four real world datasets, with the number of items and records ranging from 21 to 73 and 277 to 5456, respectively. The following studies are performed as follows: First, we compare the performance of our method with another GA based approach for mining BARs named ARMGA [3, 11]. Second, We compare the obtained results with two other classical algorithms Apriori [16, 17] and Eclat [1, 18]. Third, the scalability of the proposed method is studied and finally we have analyzed some of the rules that are obtained by our method.

2 Preliminaries

Initially association rules were used in market-basket analysis but these days its application has been extended to different real world fields including e-commerce, telecommunication, intrusion detection, bioinformatics, web mining, etc. [2].

Let $I = \{i_1, i_2, \ldots, i_r, \ldots, i_{n-1}, i_n\}$ be an item set which contains n-numbers of items and the transaction dataset be $T = \{t_1, t_2, \ldots, t_{k-1}, t_k\}$ which contains k-numbers of transactions. Each transaction of T, i.e. t_i where $i \in k$, is a subset of an item set I is such that $t_i \subseteq I$. If A and B are two item sets, then an association rule between these item sets is defined as $A \rightarrow B$, where A is antecedent and B is its consequent and $A \subseteq I$, $B \subseteq I$, $A \cap B = \varphi$. Support and confidence are the two quality measurement factors for evaluating the validity of an association rule $A \rightarrow B$, which are defined as follows: The support and confidence values of a rule $A \rightarrow B$ is defined by the term, $supp(A \cup B) = |(A \cup B)|/|T|$ and $conf(A \rightarrow B) = supp(A \cup B)/supp(A)$, respectively, the total number of records in a dataset is defined by the term $|T|$. That is, support means the occurring frequency of an item set in a dataset and strength of a rule is measured by confidence. A rule $A \rightarrow B$ is valid if $supp(A \cup B) \geq$ min_supp and $conf(A \rightarrow B) \geq$ min_supp, where min_supp and min_conf are an user defined support and confidence value [1, 16]. However, several researchers have noted that support-confidence framework has led to the generation of a huge number of misleading rules. A rule $A \rightarrow B$ is misleading, if supp(B) > confidence $(A \rightarrow B)$ i.e. there is a negative correlation between the item sets of antecedent and consequent. High support based item sets are the source of misleading rules, since they exist in most of the records and therefore any items may seem to be a good predictor because of the presence of the high support based item sets. On the other hand, confidence measure does not take into account the consequent part of a rule. For this reason it does not identify negative dependence or statistical independence between item sets [6].

In recent years, several authors have proposed different measures according to the potential interest of the users [6, 19]. We briefly explain some of those that are used in the current literature for mining BARs.

The conditional probability measure of a rule analyzes the dependence between A and B and it is defined as:

$$CP(A|B) = \{supp(A \cup B) - supp(A)supp(B)\}/\{supp(A)(1 - supp(B))\} \quad (1)$$

Its obtain values in $[-\infty, \infty]$, where misleading rules are represented by $0 > value > -\infty$, $0 < value < \infty$ represents positive association rules, and value $= 0/ -\infty/\infty$ represents trivial rules. The ratio between the confidence and the expected confidence of a rule is measured by lift and it is defined as,

$$lift(A \rightarrow B) = supp(A \cup B)/\{ supp(A)supp(B)\} \quad (2)$$

The netconf measure is used to evaluate a rule based on the support value of that rule and its consequent and antecedent support. Its domain range is $[-1,1]$, where positive values, negative values and zero represent positive dependence, negative dependence and independence, respectively. Netconf of a rule $A \rightarrow B$ is defined as follows:

$$netconf(A \rightarrow B) = [supp(A \cup B) - \{supp(A)supp(B)\}]/[supp(A)(1 - supp(A))]. \quad (3)$$

3 Methods

This section will portrays our proposed model named ARMGAAM for mining a reduced set of BARs with a good tradeoff between the number of generated rules and good coverage of the dataset, considering three objectives: lift, net confidence and conditional probability, including support and confidence values. In order to perform an evolutionary learning, this approach extends the ARMGA algorithm and introduces two new components: the reinitialization process and an adaptive mutation method to its evolutionary model. In the following, we will briefly describe all the characteristics (see Sects. 3.1 and 3.2) and represent a flowchart of the algorithm (see Sect. 3.3).

3.1 Reinitialization and Adaptive Mutation Technique Within ARMGA

We extend the existing ARMGA algorithm, which uses conditional probability as a fitness function and a genetic algorithm to optimize the association rule mining problem. In this study we use an adaptive mutation method [20]. The mutation operator is used to keep the diversity from one generation of a population to the next one. Mutation changes one or more genes of a chromosome with respect to a mutation probability, mp. Existing ARMGA follows fixed mutation probability for mutation operation. Normally, the value of mutation probability (mp) is set to low. The search will become a random search if it set too high. Adaptive mutation based approach provides better performance than fixed mutation [20]. In the proposed approach, the rate of mutation is reduced with the increasing of the generation number. Mutation rate is adapted with respect to the fitness value of the offspring. Initially, a large mutation rate is applied for more exploration on the search space (to explore more on the search space). Based on the fitness value, the offspring are categorized into different ranks. The rate of mutation is assigned for an offspring based on its rank. Top rank offspring is mutated at a lower rate than lower rank one.

The reinitialization process is used to move away from local optima and increase the diversity in the population. This process is only applied if the number of new chromosomes of a population is less than α % of a current population.

3.2 Objectives and Evolutionary Operators

Three objectives: lift, net confidence and conditional probability are maximized in order to get interesting knowledge from a dataset. We are interested in generating only strong rules which show a strong dependence among the item sets and avoid the problem of the support-confidence framework (see Sect. 1). Notice that, positive association rules represent positive dependence, thus we are interested in those rules that have CP > 0. It is important to note that existing ARMGA generates those BARs which have higher CP value but low values in other objectives. In this approach we are focusing on those BARs which maintain a good trade-off between the number of generated rules and in all other measures of a dataset. A chromosome is defined as a gene vector which represents the attributes and an indicator. Given an association rule of k length means that, a rule contains k items. Figure 1 shows the configuration of the chromosome which contains k-genes, i.e. k number of items. The first place of the chromosome contains a number which acts as an indicator for separation of the antecedent from the consequent. So, the k-rule represents a chromosome of length $k + 1$. For example a rule is $A \rightarrow B$, where antecedent A contains $item_1$ to $item_n$ and the consequent B contains $item_{n+1}$ to $item_k$, where $0 < n < k$.

Given a rule length k, the size of the population, we use a random function to initialize a population. By using selection operator, an individual chromosome is chosen from a given population. This operator acts as a filter to choose an individual chromosome based on the given selection probability (sp). The value of probability, sp set high to explore more on the search space. Crossover is one of the significant features of genetic algorithms. This operator is applied on two chromosomes of a given population called parent chromosomes to reproduce two new offspring chromosomes by exchanging parts of the parent chromosomes. The ARMGAAM model uses a two-point crossover mechanism where the two crossover points are generated randomly. That is, any segment of parent chromosomes are chosen.

3.3 Flowchart of the Algorithm

According to the above description, the proposed approach for mining BARs can be summarized through the following flowchart:

PROCEDURE 1. ARMGAAM
Input D: Dataset D, sizeof_population, selection probability sp, crossover probability cp, rule length k
Output: Positive association rules with potential interest
(0) Categorical attributes of a dataset are mapped into Boolean attributes
(1) **begin** i ← 0
(2) pop[i] ← initialization (sizeof_population)
(3) **while** not terminate_func(pop[i]) **do**
(4) **begin** temp_pop_after_selection[0] ← 0
(5) temp_pop_after_crossover [0] ← 0
(6) **for** $\forall chrom$ ∈ pop[i] **do**
(7) **if** select(chrom, sp) **then**
(8) temp_pop_after_selection[0] ← chrom
(9) temp_pop_after_crossover[0] ← crossover (temp_pop_after_selection, cp)
(10) **for** $\forall chrom$ ∈ temp_pop_after_crossover[0] **do**
(11) pop[i] ← rank_based_adaptive_mutation (chrom)
(12) i ← i+1
(13) If the number of new chromosomes in a current population is less than α% of the previous population, then reinitialize the population.
(14) end
(15) **return** pop [i]
(16) end

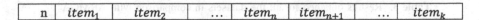

n	$item_1$	$item_2$	...	$item_n$	$item_{n+1}$	...	$item_k$

Fig. 1. A chromosome of an association rule of length k

This proposed approach is stopped if the maximum number of evaluation is reached.

4   Experimental Results

For evaluating the performance of the proposed approach, we have conducted a set of experiments on four real world datasets. These datasets were taken from the University of California at Irvine (UCI) machine learning repository (http://archive.ics.uci.edu/ml/datasets.html).

The specifications of these datasets are summarized in Table 1, the number of attributes is represented by Attributes and the number of examples in a dataset is represented by Examples. In these experiments, we compare the performance of the proposed approach with three other algorithms, named ARMGA [3, 11], Apriori [16, 17], and Eclat [1, 18]. Table 2 shows the parameters of the analyzed algorithms. In this experiment, instead of searching specific values we used standard common parameters which work well for facilitating the comparisons. The parameters of the remaining algorithms are selected based on the recommendations of corresponding authors of each approach. For all the experiments conducted in this study, the results shown in the table for the GA based approaches always refer to those non-dominated rules which consider positive dependence among the item sets with potential interest. For developing the different experiments, we have considered the average results of three runs for each dataset.

Table 1. Datasets that are considered for the experimental analysis

	Chess (Kr vs Kp)	Breast cancer	Car Eval.	Plant Cell Sig.
Attributes	73	51	21	43
Examples	3196	277	1728	5456

Table 2. Parameters considered for different algorithms

ARMGA	popsize = 100, P_{sp} = 0.95, P_{cp} = 0.85, P_{mp} = 0.01, α = 0.01, maxloop = 5 ~ 25, k = 3
Apriori	min_supp = 0.01, min_conf = 0.1
Eclat	min_supp = 0.01, min_conf = 0.1
ARMGAAM	popsize = 100, P_{sp} = 0.95, P_{cp} = 0.85, P_{mp} = variable, α = 5 %, maxloop = 5 ~ 25, k = 3

The performance of our approach against an evolutionary algorithm for mining BARs, the ARMGA [3, 11] algorithm, is shown in Table 3, where #R represents the number of generated BARs, Av_{supp}, Av_{conf}, Av_{lift}, $Av_{netconf}$, Av_{CP} are average support,

Table 3. Results obtained for all the datasets in comparison with ARMGA

Algorithms	#R	Av_{sup}	Av_{conf}	Av_{lift}	$Av_{netconf}$	Av_{CP}
Chess (King-Rook vs King-Pawn)						
ARMGA	47	0.09	0.85	2.71	0.34	0.72
ARMGAAM	32	**0.10**	**0.86**	**3.39**	**0.41**	**0.78**
Car Evaluation						
ARMGA	38	**0.02**	0.08	1	**0.0001**	**0.0001**
ARMGAAM	26	**0.02**	**0.1**	1	**0.0001**	**0.0001**
Plant Cell Signaling						
ARMGA	18	**0.63**	**0.86**	1.81	0.54	0.63
ARMGAAM	15	0.6	0.84	**1.97**	**0.56**	**0.65**
Breast Cancer						
ARMGA	9	**0.03**	0.17	3.19	0.1	0.1
ARMGAAM	5	0.02	**0.29**	**8.36**	**0.27**	**0.26**

confidence, lift, netconf, conditional probability, respectively. As with ARMGAAM, it generates a reduced set of BARs than ARMGA for all the datasets.

From an analysis of the results shown in Table 3, it can be concluded that for all datasets the rules obtained by our approach show improvements in almost all the interestingness measures over those obtained rules generated by ARMGA. The comparison of our approach with other two classical rule mining algorithms, Apriori [16, 17] and Eclat [1, 18], is shown in Table 4. In most datasets, Apriori and Eclat generate a large set of BARs, have high support and confidence values, but a low value for each of the remaining measures due to the fact those classical rule mining algorithms

Table 4. Results obtained for all the datasets in comparison with the classical algorithms

Algorithms	#R	Av_{sup}	Av_{conf}	Av_{lift}	$Av_{netconf}$	Av_{CP}
Chess(King-Rook vs King-Pawn)						
Apriori	17592	**0.68**	**0.88**	1.01	0.04	0.05
Eclat	17592	**0.68**	**0.88**	1.01	0.04	0.05
ARMGAAM	32	0.10	0.86	**3.39**	**0.41**	**0.78**
Car Evaluation						
Apriori	5082	**0.02**	**0.18**	1	−4.17E-09	−5.05E-09
Eclat	5082	**0.02**	**0.18**	1	−4.17E-09	−5.05E-09
ARMGAAM	26	**0.02**	0.1	1	**0.0001**	**0.0001**
Plant Cell Signaling						
Apriori	12147	**0.9**	**0.96**	1.02	∞	∞
Eclat	12147	**0.9**	**0.96**	1.02	∞	∞
ARMGAAM	15	0.6	0.84	**1.97**	0.56	0.65
Breast Cancer						
Apriori	94	**0.29**	**0.85**	1.14	0.16	**0.4**
Eclat	94	**0.29**	**0.85**	1.14	0.16	**0.4**
ARMGAAM	5	0.02	0.29	**8.36**	**0.27**	0.26

generate a huge number of misleading rules (see Sect. 2). For plant cell signaling dataset, the value ∞ shown in the table represents the maximum value in some measures. This value generates because of the presence of a huge number of trivial rules (see Sect. 2). By contrast, our approach allows us to obtain a reduced set of BARs which have similar or low values for support and confidence measures but high or similar values for the rest of the measures.

Table 5. Expended runtime (in seconds) of all the algorithms when the number of attributes is increased within a dataset Chess (King-Rook vs King-Pawn)

Number of attributes					
Algorithms	15	25	35	55	73
ARMGA	5.1	6.3	7.1	6.97	6.8
Apriori	0.6	1.01	2.06	4.6	56.74
Eclat	0.6	1.05	2.76	4.9	59.16
ARMGAAM	5.87	4.36	4.92	6.81	5.78

Table 6. Expended runtime (in seconds) of all the algorithms when the number of examples is increased within a dataset Chess (King-Rook vs King-Pawn)

Number of examples					
Algorithms	20 %	40 %	60 %	80 %	100 %
ARMGA	2.61	3.1	3.8	4.1	6.8
Apriori	18.61	28.68	44.79	48.12	56.74
Eclat	22.16	30.56	45.78	44.19	59.16
ARMGAAM	2.85	2.81	4.1	4.43	5.78

Several experiments have been carried out for analyzing the scalability of the proposed approach for Chess (King-Rook vs King-Pawn) dataset. The experiments were performed on an Intel(R) core i5-3210 M CPU @2.50 GHz, 4 GB RAM running on Windows 7 Enterprise. The average runtime expended by the algorithms, when the number of attributes and examples are increased is shown in Tables 5 and 6,

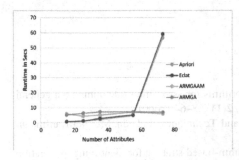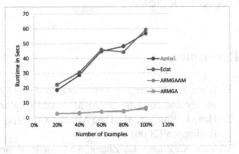

Fig. 2. Required runtime for different algorithms for different number of attributes and examples in a Chess (King-Rook vs King-Pawn) dataset

respectively. From Fig. 2 we can see that all the evolutionary algorithms scale quite linearly whereas the classical algorithms, Apriori and Eclat, increase exponentially especially when the number of attributes is increased. Some useful and interesting rules which are generated by our proposed approach are shown in Table 7. The number of generated rules is shown in Table 4, two of these are interpreted in Table 7.

Table 7. Some of the obtained Rules of a car evaluation dataset

Dataset	Rules
Car evaluation	R1: The buying price of a car is high only if it can carry 2 persons and the size of the luggage boot is big
	R2: The car of 5 or more doors has a medium safety only if it carries more persons

5 Summary

In this paper we proposed a new genetic algorithm based approach named ARM-GAAM to mine a reduced set of Boolean association rules. The generated BARs are interesting, easy to understand and maximizing three objectives: lift, net confidence and conditional probability. To accomplish this, this algorithm extends the ARMGA algorithm for performing evolutionary learning and selection of a condition of each rule. This proposed approach introduces the reinitialization process and the adaptive mutation method to its evolutionary model in order to perform evolutionary learning and to improve the diversity of the obtained set of rules. Moreover, the obtained rules are strong, showing a strong relationship among the item sets and solving the problem of the support-confidence framework.

From the experimental results obtained over four real world datasets, it can be concluded that our approach allows us to mine a reduced set of BARs with good trade off among the number of generated rules, support, confidence, lift, net confidence and conditional probability of all the datasets. Finally, the proposed approach have a good computational cost and scalability when the problem size increases.

Acknowledgements. This research work was funded by School of Engineering and ICT, University of Tasmania, Australia, and website: http://www.utas.edu.au/cricos, under CRICOS Provider Code 00586B.

References

1. Hipp, J., Güntzer, U., Nakhaeizadeh, G.: Algorithms for association rule mining—a general survey and comparison. ACM sigkdd Explor. **2**(1), 58–64 (2000)
2. Han, J., Kamber, M.: Data Mining: Concepts and Techniques, 2nd edn. Morgan Kaufmann, Burlington (2006)
3. Yan, X., Zhang, C., Zhang, S.: Genetic algorithm-based strategy for identifying association rules without specifying actual minimum support. Expert Syst. Appl. **36**(2), 3066–3076 (2009)

4. del Jesus, M.J., Gámez, J.A., González, P., Puerta, J.M.: On the discovery of association rules by means of evolutionary algorithms. Wiley Interdiscip. Rev. Data Min. Knowl. Discov. **1**(5), 397–415 (2011)
5. Berzal, F., Blanco, I., Sánchez, D.: Measuring the accuracy and interest of association rules: a new framework. Intell. Data Anal. **6**(3), 221–235 (2002)
6. Martin, D., Rosete, A., Alcala-Fdez, J., Herrera, F.: A new multiobjective evolutionary algorithm for mining a reduced set of interesting positive and negative quantitative association rules. IEEE Trans. Evol. Comput. **18**(1), 54–69 (2014)
7. Mukhopadhyay, A., Maulik, U., Bandyopadhyay, S., Coello, C.A.C.: A survey of multiobjective evolutionary algorithms for data mining: Part i. IEEE Trans. Evol. Comput. **18**(1), 4–19 (2014)
8. Maulik, U., Bandyopadhyay, S., Mukhopadhyay, A.: Multiobjective Genetic Algorithms for Clustering: Applications in Data Mining and Bioinformatics. Springer, Berlin (2011)
9. Shenoy, P., Srinivasa, K., Venugopal, K., Patnaik, L.: Evolutionary approach for mining association rules on dynamic databases. In: Proceeding of the 7th Pacific-Asia Conference on Advances in Knowledge Discovery and Data Mining, PAKDD, pp. 325–336 (2003)
10. Shenoy, P., Srinivasa, K., Venugopal, K., Patnaik, L.: Dynamic association rule mining using genetic algorithms. Intell. Data Anal. **9**(5), 439–453 (2005)
11. Yan, X., Zhang, C., Zhang, S.: ARMGA: identifying interesting association rules with genetic algorithms. Appl. Artif. Intell. Int. J. **19**(7), 677–689 (2005)
12. Alatas, B., Akin, E.: MODENAR: multi-objective differential evolution algorithm for mining numeric association rules. Appl. Soft Comput. **8**(1), 646–656 (2008)
13. Ghosh, A., Nath, B.: Multi-objective rule mining using genetic algorithms. Inf. Sci. (Ny) **163**(1–3), 123–133 (2004)
14. Salleb-aouissi, A., Vrain, C., Nortet, C., Kong, X., Cassard, D.: QuantMiner for mining quantitative association rules. Mach. Learn. Res. **14**(1), 3153–3157 (2013)
15. Webb, G.I.: Discovering associations with numeric variables. In: Proceedings of the Seventh ACM SIGKDD International Conference on Knowledge Discovery and Data Mining, pp. 383–388 (2001)
16. Agrawal, R., Srikant, R.: Fast algorithms for mining association rules. In: 20th International Conference on Very Large Data Bases, pp. 487–499 (1994)
17. Borgelt, C.: Efficient implementations of Apriori and Eclat. In: IEEE ICDM Workshop on Frequent Item Set Mining Implementations, pp. 280–296 (2003)
18. Zaki, M.J.: Scalable algorithms for association mining. IEEE Trans. Knowl. Data Eng. **12**(3), 372–390 (2000)
19. Geng, L., Hamilton, H.J.: Interestingness measures for data mining. ACM Comput. Surv. **38**(3), 1–32 (2006)
20. Kannimuthu, S., Premalatha, K.: Discovery of high utility itemsets using genetic algorithm with ranked mutation. Appl. Artif. Intell. **28**(4), 337–359 (2014)

Semi-supervised Non-negative Local Coordinate Factorization

Cherong Zhou[1], Xiang Zhang[1], Naiyang Guan[1,2 (✉)], Xuhui Huang[3], and Zhigang Luo[1,2 (✉)]

[1] Science and Technology on Parallel and Distributed Processing Laboratory, Changsha, China
rongchezhou@163.com, zhangxiang_43@aliyun.com
[2] Institute of Software, College of Computer, Changsha, China
{ny_guan,zgluo}@nudt.edu.cn
[3] Department of Computer Science and Technology, College of Computer, National University of Defense Technology, Changsha, Hunan 410073, People's Republic of China
xhhuang@nudt.edu.cn

Abstract. Non-negative matrix factorization (NMF) is a popular matrix decomposition technique that has attracted extensive attentions from data mining community. However, NMF suffers from the following deficiencies: (1) it is non-trivial to guarantee the representation of the data points to be sparse, and (2) NMF often achieves unsatisfactory clustering results because it completely neglects the labels of the dataset. Thus, this paper proposes a semi-supervised non-negative local coordinate factorization (SNLCF) to overcome the above deficiencies. Particularly, SNLCF induces the sparse coefficients by imposing the local coordinate constraint and propagates the labels of the labeled data to the unlabeled ones by indicating the coefficients of the labeled examples to be the class indicator. Benefit from the labeled data, SNLCF can boost NMF in clustering the unlabeled data. Experimental results on UCI datasets and two popular face image datasets suggest that SNLCF outperforms the representative methods in terms of both average accuracy and average normalized mutual information.

Keywords: Non-negative matrix factorization · Local coordinate coding · Semi-supervised learning

1 Introduction

Unsupervised [1] and supervised subspace learning [3,8] have been well-studied and successfully applied in computer vision and signal processing. Clustering is an unsupervised subspace learning which seeks the grouping relationships among examples, and have been widely applied in data mining due to its effectiveness. Many representative methods such as K-means [2] and probabilistic latent semantic indexing (PLSI, [4]) have achieved quite good results. K-means iteratively assigns the object to the cluster whose center is close to the object,

© Springer International Publishing Switzerland 2015
S. Arik et al. (Eds.): ICONIP 2015, Part II, LNCS 9490, pp. 106–113, 2015.
DOI: 10.1007/978-3-319-26535-3_13

while PLSI utilizes the statistical latent class models. Different from above methods, NMF imposes the non-negative constraints to learn the parts-based representation, and induces more robust features than non-sparse, global ones. Thus, NMF has drawn extensive attentions from machine learning community [5,11,16].

Non-negative matrix factorization (NMF, [6]) is a powerful matrix decomposition technique that decomposes the non-negative matrix into two non-negative factors, i.e., the cluster centers and the coefficients. Since NMF can learn effective cluster centers, it has been widely applied in clustering tasks. More importantly, recent works [4,7] show that NMF theoretically equals to the state-of-the-art clustering methods including both K-means and PLSI. However, traditional NMF methods does not always guarantee any factor of the decomposition results to be sparse [1]. This leads to the performance degeneration in clustering [9]. To overcome this deficiency, Li *et al.* proposed local NMF (LNMF, [10]) imposes sparse constraint which leads to sparser representation than that of traditional NMF. Hoyer *et al.* [12] incorporated the sparseness constraints over both factors to present the sparse NMF method (SNMF). Besides, Yuan *et al.* [13] developed the projective NMF (PNMF) to induce parts-based representation by implicitly enforcing the orthogonal constraint over the basis. Cai *et al.* [14] proposed the non-negative local coordinate factorization (NLCF) to guarantee its sparseness by adding the local coordinate constraint. Since NLCF requires that the learned basis vector be close to the original data points, each data point can be approximated by a linear combination of as few basis vectors as possible. However, NLCF completely neglects the labels of the datasets. Although several supervised NMF methods has been proposed to address this issue, these supervised methods are more inconvenient for practical applications because the labels are expensive in practice.

The most effective strategy to solve the above problem can resort to the semi-supervised learning [15] methods which can make full use of both few labeled and large volume of unlabeled examples. For instance, Cho *et al.* [17] incorporated support vector machine (SVM) into NMF to propose the NMF-α method. Besides, Liu *et al.* [18] proposed the constrained NMF method (CNMF) that assumes that the similar input patterns owning the similar lower-dimensional coefficients. However, each class of one labeled sample will degenerate CNMF into conventional NMF. Chen *et al.* [19] denoted a semi-supervised NMF (SS-NMF) to provide pairwise constraints on a small percentage of the data points. These constrains specify whether any two data points should belong to the same cluster or should strictly belong to different clusters. For clustering tasks, the learned coefficients should be sparse enough to approximate the class indicator. Unfortunately, these methods cannot completely solve the aforementioned problems [17].

To overcome the above deficiencies, we propose a semi-supervised NLCF (SNLCF) which employs both unlabeled and labeled examples in the dataset. Particularly, SNLCF propagates the labels of the labeled data points to the unlabeled data points by imposing the coefficients of the labeled data points to be as close to their class indicators as possible. Inherited from NLCF, SNLCF can boost the representation ability of the learned basis by NMF meanwhile guarantee sparsity of the learned coefficients. To optimize SNLCF, we developed the multiplicative update rule (MUR) and further proved its convergence. Experiments on UCI datasets and two popular face image datasets including UMIST [20] and FERET [21] suggest that SNLCF provides a better representation and achieves a higher accuracies than the representative methods in quantities.

2   Semi-supervised Non-negative Local Coordinate Factorization

To utilize the discriminative information provided by labeled data points, we propose a semi-supervised non-negative local coordinate factorization (SNLCF) which incorporates the unlabeled and labeled data points into NLCF [14]. Given the data matrix $X = [X^l, X^u]$, $X \in R^{m \times n}$ where n denotes the number of all samples, X^l and X^u signify the labeled and unlabeled examples, respectively.

For the unlabeled data points, SNLCF incorporates the local coordinate constraint [22,23] over the basis and the coefficients into NMF as follows:

$$\min_{U \geq 0, V^u \geq 0} \|X^u - UV^u\|^2 + \beta \sum_{i=1}^{n_u} \sum_{t=1}^{k} |v_{ti}^u| \|u_t - x_i^u\|^2, \qquad (1)$$

where β signifies a positive regularization parameter, k is the cluster number, and n_u means the total number of unlabeled data points. U represents the basis matrix. The entry v_{ti}^u denotes the t-th row and i-th column of matrix V^u, and the vector x_i^u denotes the i-th unlabeled data point.

For the labeled data points, different from (1), we impose the coefficients to be as close to the class indicator as possible. This induces effective group relationships among the original examples. Thus we can obtain the following objective:

$$\min_{U \geq 0, V^l \geq 0} \|X^l - UV^l\|_F^2 + \alpha \|V^l - Q\|_F^2 \qquad (2)$$

where α signifies a positive regularization parameter, x_i^l denotes the i-th sample in matrix X^l. We set $Q_{ti} = 1$ if x_i^l belongs to cluster s_t; otherwise, $Q_{ti} = 0$.

By incorporating (1) and (2), we obtain the objective of SNLCF as follows:

$$\min_{U \geq 0, V \geq 0} \|X - UV\|_F^2 + \alpha \|V^l - Q\|_F^2 + \beta \sum_{i=1}^{n_u} \sum_{t=1}^{k} v_{ti}^u \|u_t - x_i^u\|_2^2 \qquad (3)$$

Note that (2) can impose the basis bias to the labeled data while (1) makes the learned basis preferable to the unlabeled data. This results in a balance between the representative capacities on both labeled and unlabeled data points in (3). Thus, we do not impose the local coordinate constraint over the labeled data. In summary, SNLCF can effectively represent each data point with only few nearby basis vectors and thus lead to sparse representation. Moreover, benefit from the discriminative information of the labeled data points, SNLCF can boost both NLCF and NMF in clustering.

2.1   Optimization Algorithm

It is impractical to yield the global solution to (3) in polynomial time because the objective function is non-convex with respect to both factor matrices. Fortunately, we can obtain the local solution of (3) by alternating optimization.

Here we develop the multiplicative update rule (MUR) to (3) and establish the following theorem:

Theorem 2.1. The objective function (3) is non-increasing under the following multiplicative update rules:

$$U_{jt} \leftarrow U_{jt} \frac{\left(XV^T + \beta X^u (V^u)^T\right)_{jt}}{\left(UVV^T + \beta UH\right)_{jt}}, \tag{4}$$

$$V_{ti}^l \leftarrow V_{ti}^l \frac{\left(U^T X^l + \alpha Q\right)_{ti}}{\left(U^T UV^l + \alpha V^l\right)_{ti}}, \tag{5}$$

$$V_{ti}^u \leftarrow V_{ti}^u \frac{\left(2\left(\beta + 1\right) U^T X^u\right)_{ti}}{\left(2U^T UV^u + \beta\left(C + D\right)\right)_{ti}}, \tag{6}$$

where H is a diagonal matrix whose entries are row sums of V^l. The column vector $c = diag\left((X^u)^T X^u\right) \in R^{n_u}$ and $C = (c, c, \dots, c)^T \in R^{k \times n_u}$. The column vector $d = diag\left(U^T U\right) \in R^k$ and $D = (d, d, \dots, d) \in R^{k \times n_u}$.

The proof is left in **Appendix A**. The time complexity of the MUR-based algorithm mainly lies in (4), (5) and (6), respectively. Its total time overhead takes $O(mnk + mk^2 + nk^2 + mn + mk + nk)$.

3   Experiments

This section conducts the clustering experiments on UCI datasets and two face image datasets including UMIST [20] and FERET [21] to verify the effectiveness of SNLCF by comparing SNLCF with K-means [2], NMF [6,24], CNMF [18] and NLCF [14]. For each dataset, the pixel values are normalized into the range from 0 to 1. For semi-supervised methods, we randomly select two instances as the labeled samples. We independently conducts each experiment 100 trails to remove the influence of randomness and then evaluate the compared methods in terms of both accuracy [21] and normalized mutual information (NMI, [21]).

3.1   UCI Datasets

This section verifies the effectiveness of SNLCF on four biological datasets including breast cancer, diabetes, heart and liver disorders from UCI machine learning repository. All these datasets contain two categories. We first remove those samples with missing values and some redundant attributions from four datasets, and thus respectively obtain 683 instances with 10 attributions, 768 instances with 8

Table 1. The accuracy and NMI of the compared methods on breast cancer, diabetes, heart and liver disorders datasets, respectively. The bold black numbers highlight the best results on each dataset.

Datasets / Algorithms	Accuracy(NMI)% breast-cancer	diabetes	heart	liver-disorders
K-means	95.01(70.12)	**65.83(3.32)**	59.87(4.840)	54.44(0.029)
NMF	95.79(73.08)	63.51(1.44)	53.33(0.019)	51.61(0.099)
CNMF	95.62(72.63)	63.64(1.48)	53.36(0.028)	51.86(0.120)
NLCF	93.02(62.92)	52.40(0.10)	53.92(0.450)	54.65(0.730)
SNLCF	**96.42(76.41)**	55.42(0.26)	**67.24(10.100)**	**58.53(2.140)**

attributions, 270 instances with 13 attributions and 345 instances with 6 attributions in our experiments. We set parameters $\alpha = 0.2, 2, 2.5, 5$ and $\beta = 1, 2.5, 1.5, 2$ for SNLCF on four datasets, respectively. Simultaneously, we set $\mu = 0.01$ in NLCF on both datasets.

Table 1 suggests that K-means obtains higher values of both accuracy and NMI than that of SNLCF on the diabetes dataset. However, SNLCF outperforms K-means, NMF, CNMF, and NLCF on breast-cancer, heart and liver-disorders dataset in terms of both accuracy and NMI.

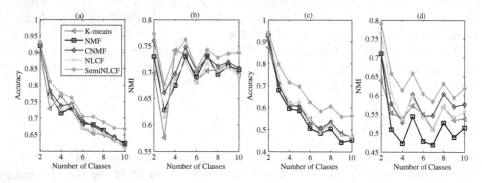

Fig. 1. The accuracy and NMI versus the number of class of SNLCF, NLCF, CNMF, NMF and K-means on (a) (b) UMIST and (c) (d) FERET datasets.

3.2 Face Image Datasets

Experiment Settings. The UMIST database [20] contains images from 20 individuals, totally 575 different images. All images are grayscale and cropped to 40×40 pixels. Each subject has fifteen right profile images, and the subset consists of 300 images in this experiment. The FERET database [21] contains images from 40 individuals. Each subject has 10 different images from various facial expressions such as smiling. All images are grayscale and cropped to 40×40 pixels. We set parameters $\alpha = 0.001$ and $\beta = 1.8$ for SNLCF, and $\mu = 0.01$ in NLCF on both datasets. Figure 1 reports that SNLCF consistently outperforms

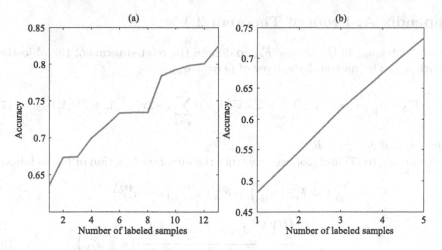

Fig. 2. Accuracy versus the number of labeled samples on (a) UMIST and (b) FERET dataset.

K-means, NMF, CNMF and NLCF in terms of accuracy and NMI on both datasets. It confirms that using labels of the dataset can improve the effectiveness of NLCF and NMF in clustering.

The number of the labeled data points is an important factor in semi-supervised clustering. To study the effect of the number of labeled data points, we repeated this above experiments by varying the number of labeled data points. Figure 2 shows the result of SNLCF when varing the numbers of labeled data points on UMIST and FERET dataset, respectively. It also shows that the clustering accuracy improves as the number of labeled data points. It confirms the enhancement of the discriminative information of the labeled data points to clustering in our SNLCF model. This model is both simple and effective, and the optimization is efficient, it can be tenderised in the future referring to [26].

4   Conclusion

This paper proposes a novel semi-supervised non-negative local coordinate factorization (SNLCF). SNLCF can take full advantages of the labels of few labeled samples to boost NMF in clustering unlabeled samples. Moreover, SNLCF incorporates the local coordinate constraint into NMF to guarantee the coefficients to be sparse. Besides, we developed the multiplicative update rule (MUR) to optimize SNLCF and proved its convergence. Experiment results suggest the effectiveness of SNLCF on UCI datasets and two face image datasets.

Acknowledgements. This work was partially supported by National High Technology Research and Development Program ("863" Program) of China (under grant No. 2015AA020108) and National Natural Science Foundation of China (under grant No. 61502515).

Appendix A: Proof of Theorem 2.1

For any entry u_{ab} in U, we use F_{ab} to denote the related terms of the objective O with u_{ab}. The partial derivatives of O over u_{ab} is:

$$F'_{ab} = \frac{\partial O}{\partial U_{ab}} = ((2UVV^T - 2XV^T) + \beta \sum_{i=1}^{n_u} (-2x_i^u 1^T \Lambda_i + 2U\Lambda_i))_{ab}, \quad (7)$$

where $\Lambda_i = diag(v_i^u) \in R^{k \times k}$.

According to (7) and [25], we construct the auxiliary function of F_{ab} as follows

$$G\left(u, u_{ab}^{(t)}\right) = F_{ab}\left(u_{ab}^{(t)}\right) + F'_{ab}\left(u_{ab}^{(t)}\right)\left(u - u_{ab}^{(t)}\right)$$

$$+ \frac{\left(UVV^T\right)_{ab} + \beta \sum_{i=1}^{n_u} (U\Lambda_i)_{ab}}{u_{ab}^{(t)}} \left(u - u_{ab}^{(t)}\right)^2. \quad (8)$$

By setting the derivatives of (8) to zero, we can obtain the update rule (4). Similarly, we can obtain the update rules (5) and (6) for V_u and V_l, respectively. According to the property of the auxiliary function [25], we can obtain

$$O\left(U^{(t+1)}, V^{(t+1)}\right) \leq O\left(U^{(t)}, V^{(t)}\right) \leq O\left(U^{(t-1)}, V^{(t-1)}\right). \quad (9)$$

Based on (9), this MUR can guarantee the objective to be non-increased. Meanwhile, the equalition (9) holds when their sub-gradients of objective (3) are zeros. Thus, this MUR can converge to local minima of SNLCF. This completes the proof.   □

References

1. Zhang, D., Zhu, P., Hu, Q.: A linear subspace learning approach via sparse coding. In: 2011 IEEE International Conference on Computer Vision (ICCV), pp. 755–761. IEEE (2011)
2. Punj, G., Stewart, D.W.: Cluster analysis in marketing research: review and suggestions for application. J. Market. Res. **20**(2), 134–148 (1983)
3. Tao, D., Tang, X., Li, X., Wu, X.: Asymmetric bagging and random subspace for support vector machines based relevance feedback in image retrieval. IEEE Trans. Pattern Anal. Mach. Intell. **28**(7), 1088–1099 (2006)
4. Hofmann, T.: Probabilistic latent semantic indexing. In: Proceedings of the 22nd Annual International ACM SIGIR Conference on Research and Development in Information Retrieval, pp. 50–57. ACM (1999)
5. Guan, N., Tao, D., Luo, Z., Yuan, B.: Non-negative patch alignment framework. IEEE Trans. Neural Netw. **22**(8), 1218–1230 (2011)
6. Lee, D.D., Seung, H.S.: Learning the parts of objects by non-negative matrix factorization. Nature **401**(6755), 788–791 (1999)
7. Ding, C., Li, T., Peng, W.: On the equivalence between non-negative matrix factorization and probabilistic latent semantic indexing. Comput. Stat. Data Anal. **52**(8), 3913–3927 (2008)

8. Tao, D., Li, X., Wu, X., Maybank, S.J.: Geometric mean for subspace selection. IEEE Trans. Pattern Anal. Mach. Intell. **31**(2), 260–274 (2009)
9. Hoyer, P.O.: Non-negative sparse coding. In: Proceedings of the 2002 12th IEEE Workshop on Neural Networks for Signal Processing, pp. 557–565. IEEE (2002)
10. Li, S.Z., Hou, X., Zhang, H., Cheng, Q.: Learning spatially localized, parts-based representation. In: Proceedings of the 2001 IEEE Computer Society Conference on Computer Vision and Pattern Recognition, CVPR 2001, vol. 1, p. I-207. IEEE (2001)
11. Guan, N., Zhang, X., Luo, Z., Tao, D., Yang, X.: Discriminant projective non-negative matrix factorization. PLoS ONE **8**(12), e83291 (2013)
12. Hoyer, P.O.: Non-negative matrix factorization with sparseness constraints. J. Mach. Learn. Res. **5**, 1457–1469 (2004)
13. Yuan, Z., Oja, E.: Projective nonnegative matrix factorization for image compression and feature extraction. In: Kalviainen, H., Parkkinen, J., Kaarna, A. (eds.) SCIA 2005. LNCS, vol. 3540, pp. 333–342. Springer, Heidelberg (2005)
14. Chen, Y., Zhang, J., Cai, D., Liu, W., He, X.: Nonnegative local coordinate factorization for image representation. IEEE Trans. Image Process. **22**(3), 969–979 (2013)
15. Wang, C., Yan, S., Zhang, L., Zhang, H.: Non-negative semi-supervised learning. In: International Conference on Artificial Intelligence and Statistics, pp. 575–582 (2009)
16. Guan, N., Tao, D., Luo, Z., Yuan, B.: NeNMF: an optimal gradient method for nonnegative matrix factorization. IEEE Trans. Signal Process. **60**(6), 2882–2898 (2012)
17. Cho, Y., Saul, L.K.: Nonnegative matrix factorization for semi-supervised dimensionality reduction (2011). arXiv preprint arXiv:1112.3714
18. Liu, H., Wu, Z.: Non-negative matrix factorization with constraints. In: Twenty-Fourth AAAI Conference on Artificial Intelligence (2010)
19. Chen, Y., Rege, M., Dong, M., Hua, J.: Non-negative matrix factorization for semi-supervised data clustering. Knowl. Inf. Syst. **17**(3), 355–379 (2008)
20. Wechsler, H.: Face Recognition: From Theory to Applications, vol. 163. Springer, Heidelberg (1998)
21. Phillips, P.J., Moon, H., Rizvi, S.A., Rauss, P.J.: The FERET evaluation methodology for face-recognition algorithms. IEEE Trans. Pattern Anal. Mach. Intell. **22**(10), 1090–1104 (2000)
22. Yu, K., Zhang, T., Gong, Y.: Nonlinear learning using local coordinate coding. In: Advances in Neural Information Processing Systems, pp. 2223–2231 (2009)
23. Wang, J., Yang, J., Yu, K., Lv, F., Huang, T., Gong, Y.: Locality-constrained linear coding for image classification. In: 2010 IEEE Conference on Computer Vision and Pattern Recognition (CVPR), pp. 3360–3367. IEEE (2010)
24. Lee, D.D., Seung, H.S.: Algorithms for non-negative matrix factorization. In: Advances in Neural Information Processing Systems, pp. 556–562 (2001)
25. Dempster, A.P., Laird, N.M., Rubin, D.B.: Maximum likelihood from incomplete data via the EM algorithm. J. Royal Stat. Soc. Ser. B (Methodol.) **39**, 1–38 (1977)
26. Tao, D., Li, X., Wu, X., Maybank, S.J.: General tensor discriminant analysis and gabor features for gait recognition. IEEE Trans. Pattern Anal. Mach. Intell. **29**(10), 1700–1715 (2007)

A Strength Pareto Evolutionary Algorithm for Live Migration of Multiple Interdependent Virtual Machines in Data Centers

Tusher Kumer Sarker and Maolin Tang[✉]

School of Electrical Engineering and Computer Science,
Queensland University of Technology, 2 George Street,
Brisbane, QLD 4001, Australia
{t.sarker,m.tang}@qut.edu.au

Abstract. Although live VM migration has been intensively studied, the problem of live migration of multiple interdependent VMs has hardly been investigated. The most important problem in the live migration of multiple interdependent VMs is how to schedule VM migrations as the schedule will directly affect the total migration time and the total downtime of those VMs. Aiming at minimizing both the total migration time and the total downtime simultaneously, this paper presents a Strength Pareto Evolutionary Algorithm 2 (SPEA2) for the multi-VM migration scheduling problem. The SPEA2 has been evaluated by experiments, and the experimental results show that the SPEA2 can generate a set of VM migration schedules with a shorter total migration time and a shorter total downtime than an existing genetic algorithm, namely Random Key Genetic Algorithm (RKGA). This paper also studies the scalability of the SPEA2.

Keywords: Live VM migration · Scheduling · Migration time · Downtime · Strength Pareto Evolutionary Algorithm

1 Introduction

The emergence of virtualization technology and live Virtual Machine (VM) migration technology has brought enormous benefits to data centers. Through virtualization a number of VMs can be created on the top of a Physical Machine (PM). Since the VMs are logically independent from the PMs on which those VMs are created, a VM can be migrated from one PM to another while the VM is running, which is called *live VM migration*. Live VM migration technology enables many new online management and maintenance activities, such as server consolidation [1], load balancing [2], and proactive fault tolerance [3].

Although the problem of live migration of a single VM has been studied intensively, the problem of live migration of multiple interdependent VMs has hardly been investigated. The performance of live migration of multiple VMs is measured by the total migration time and the total downtime of the VMs,

© Springer International Publishing Switzerland 2015
S. Arik et al. (Eds.): ICONIP 2015, Part II, LNCS 9490, pp. 114–121, 2015.
DOI: 10.1007/978-3-319-26535-3_14

both of which depend on the scheduling of the multi-VM migrations. The total migration time is the time difference between the time when the last VM migration is completed and the time when the first migration is initiated, and the total downtime is the sum of the time during which a VM remains in a down state internally.

A pioneering research on the live multi-VM migration scheduling problem was done by Ghorbani et al., who proposed a heuristic algorithm to solve the *VM sequence planning problem* [4]. In their work, they found a migration order for a given set of interdependent VMs to their designated target PMs in a bandwidth constrained network so that maximum number of VMs are migrated without violating the link bandwidth capacity. However, they did not provide detailed experimental results, such as total migration time and total downtime in their paper. Nus et al. [5] developed and investigated several VM migration scheduling algorithms for minimizing the total migration time for migrating multiple identical VMs. They did not consider the network topology and assumed that each migration took a predefined amount of migration time. However, VM migration time varies with the bandwidth and deployed VMs in a data center are heterogeneous. The problem of scheduling multiple heterogeneous interdependent VMs were initially tackled in [6,7]. In [6], a heuristic algorithm was proposed to schedule multi-VM migrations. The strategy used by the heuristic algorithm was to give the highest priority to the VM that would result in the least increase in the network congestion incurred by the data flow between the VM and its dependent VMs after migrating the VM. A Random Key Genetic Algorithm (RKGA) was proposed in [7] to provide a migration schedule with a minimum combined minimum total migration time and the minimum total downtime. Since the multi-VM migration scheduling problem is a multi-objective optimization problem, by its nature, this paper attempts to solve it using an evolutionary Pareto-based multi-objective optimization algorithm, namely *Strength Pareto Evolutionary Algorithm 2*, or SPEA2 [8].

The remainder of this paper is organized as follows. Section 2 formulates the problem. Section 3 details how to design a SPEA2 for the multi-VM migration scheduling problem. The evaluation of the SPEA2 is conducted in Sect. 4. Finally, the research work is concluded in Sect. 5.

2   Problem Statement

The multi-VM migration scheduling problem refers to finding a sequence of VM migrations for a given set of interdependent VMs in a data center such that both the total migration time and the total downtime are minimized.

A data center, $G_D = (N, L)$, consists of a set of nodes, N, including PMs, P, and switches, W, and a set of communication links, L, which interconnect these PMs and switches. The weight on a link $l_{i,j} \in L$ represents the bandwidth capacity of the link. A PM is characterized by its resources. Two types of resources are considered in this research – CPU and memory. The interdependency among the VMs is represented by a weighted graph, $G_V = (V, E)$. An undirected edge

$e_{i,j} \in E$ between two VMs, $vm_i \in V$ and $vm_j \in V$, represents the inter-VM dependency and the weight on the edge, $w_{i,j}$, is the amount of traffic flow rate between them. The migration of vm_i, from $pm_j \in P$ to $pm_k \in P$, represented by a 3-tuple, $\langle vm_i, pm_j, pm_k \rangle$, requires that the available bandwidth between pm_j and pm_k, $b_{j,k}$, must be enough to carry the inter-VM traffic flow between pm_j and pm_k. Moreover, for the migration tuple, the CPU and memory requirements of vm_i, denoted by C_{vm_i} and M_{vm_i}, respectively, must be met by the respective available resources at pm_k, C_{pm_k} and M_{pm_k}. Therefore, the migration of vm_i to pm_k is subject to the following resource constraints:

$$
\begin{aligned}
C_{vm_i} &\leq C_{pm_k} \\
M_{vm_i} &\leq M_{pm_k} \\
w_{i,j} &\leq b_{j,k}
\end{aligned}
\tag{1}
$$

As this research problem deals with the migration of multiple VMs and multiple VM migrations can be performed in parallel, the time for migrating a set of parallel VMs, h_i, denoted as $T_m^{h_i}$, is the time when the first VM in h_i starts migrating to the time when the last VM in h_i finishes migrating. The total migration time, T_m, is the time required to complete all the VM migrations, which is ultimately the summation of the migration times of q sets of parallel VMs, where q is the total number parallel VMs.

$$
T_m = \sum_{i=1}^{q} T_m^{h_i}.
\tag{2}
$$

The total downtime, T_d, is the summation of downtimes experienced by migrating each VM vm_i, $t_d^{vm_i}$. Thus, the total downtime for migrating n VMs is given by (3). The migration time and downtime of a VM is calculated according to the procedure given in [7].

$$
T_d = \sum_{i=1}^{n} t_d^{vm_i}.
\tag{3}
$$

Given a set of VM migrations and the inter-VM dependency between the VMs, G_V, in a data center, G_D, the multi-VM migration scheduling problem is to find a sequence of parallel VM migration sets, $\langle h_1, h_2, \ldots, h_n \rangle$, such that both T_m and T_d are minimized.

3 The SPEA2

The SPEA2 generates a set of nondominated solutions. A solution is defined as a schedule of VM migrations. The quality of a solution, X_i, is determined by an objective vector, $\langle g_1^{X_i}, \ldots, g_l^{X_i}, \ldots, g_r^{X_i} \rangle$, where $g_l^{X_i}$ is an objective value. In minimization problem solution, X_i, dominates solution, X_j, i.e. $X_i \succ X_j$, if $\forall l, g_l^{X_i} \leq g_l^{X_j}$ and $\exists l, g_l^{X_i} < g_l^{X_j}$. In this research, the objective vector is $\langle T_m, T_d \rangle$.

Algorithm 1. The SPEA2 for scheduling VMs migration

1: randomly generate an initial population
2: initialize the archive to an empty set
3: **while** termination criteria are false **do**
4:    **for** each individual in the population **do**
5:       find the schedule of migrations as described in Sect. 3.1
6:       calculate the total migration time and total downtime
7:    **end for**
8:    calculate the fitness values of all individuals in the population and archive
9:    do environment selection
10:   do mating selection to fill mating pool
11:   apply crossover and mutation operators
12: **end while**
13: output the set of nondominated solutions

The SPEA2 works in four steps: (1) fitness calculation; (2) environment selection; (3) mating selection; and (4) genetic operations. These four steps iterate until a certain termination condition is met. In fitness calculation, the solutions in the current archive and current population are evaluated with respect to their objective vectors, $\langle T_m, T_d \rangle$. The environment selection updates a fixed size archive with the solutions from the current archive and current population. In this regard, the archive is filled up by the nondominated solutions first. If the number of nondominated solutions is less than a pre-defined archive size, then the rest of the archive is filled up by the dominated solutions with better fitnesses. On the contrary, if there are more nondominated solutions than the archive size, then archive truncation is required. In the archive truncation procedure, the nondominated solutions in the dense area are considered. A solution which has minimum Euclidean distance to another solution is chosen for removal, and if more than one solution exist with the same minimum distance then the second nearest neighbour is chosen and so forth. The removal process iterates until the size of the archive is not truncated to the pre-defined size. The mating pool selection fills the mating pool by the better solutions in the current archive and current population. Genetic operators are applied on the solutions in the mating pool to get a new generation. Algorithm 1 is the pseudocode of SPEA2.

3.1   Chromosome Representation

A chromosome represents a schedule of migrations and the random-key presentation [7] is used to obtain such a schedule. A random number in the range [0, 1] is generated against each gene, where each gene corresponds to a VM. The VMs are prioritized for migration from the lowest to the highest value of random numbers, and VMs with the same random number get consecutive migration priorities. The VMs are scanned from the highest to the lowest migration priorities to check the migration feasibilities to their target PMs. The VMs that satisfy the resource constraints as specified in (1) during a scan is scheduled to migrate in parallel. The scan procedure iteratively schedules all the migrations.

3.2 Genetic Operators

The crossover and mutation operators used in [7] are adopted in the SPEA2. In the crossover, an offspring is produced from two randomly selected parents in the mating pool. A random number in the range [0, 1] is generated against each gene and if the value of this random number is below a predefined value then the gene from the second parent is chosen; otherwise the gene of first parent is selected. In the mutation, a gene of an arbitrary chromosome is altered with 20 % probability to be changed by the new value in the range [0, 1].

3.3 Fitness Calculation

The fitness value of a solution, X_i, is determined by the strengths of its domina-tors. The strength of X_i, $S(X_i)$, refers to the number of solutions it dominates in both the archive, A_t, and the population, P_t, i.e.

$$S(X_i) = |\{X_j \mid X_j \in A_t \cup P_t, X_i \succ X_j\}|. \tag{4}$$

The symbol $\succ$ denotes the dominance relation. The raw fitness value of X_i, $R(X_i)$, is the added strengths of the solutions that dominate X_i.

$$R(X_i) = \sum_{X_j \in A_t \cup P_t, X_j \succ X_i} S(X_j). \tag{5}$$

Equation (5) indicates that the raw fitness values of the solutions that do not dominate each other are zeros. Hence, density information is taken into account to calculate actual fitness value. In density estimation technique, the Euclidean distance of the k-th nearest neighbour of X_i is calculated and denoted as $\sigma_{X_i}^k$. In our experiments we set $k = 1$. Then density corresponding to X_i is normalized as $D(X_i) = \frac{1}{\sigma_{X_i}^k + 2}$ and added to $R(X_i)$ to obtain the actual fitness value of X_i as follows:

$$F(X_i) = R(X_i) + D(X_i), X_i \in A_t \cup P_t. \tag{6}$$

Equation (6) indicates that a better solution gets lower fitness value.

4 Evaluation

We show the effectiveness of the SPEA2 for different size test problems. A test problem size is defined by the number of VMs designated for migration. We con-ducted the experiments for test problems of migrating between 20 VMs and 200 VMs with an increment of 20 VMs. As it is quite difficult to perform such large scale experiments in real data centers, we performed the simulation experiments. We evaluate the SPEA2 by comparing through the RKGA [7]. In the following subsections, we describe the simulation setup and experimental results.

4.1   Simulation Setup

In our experiments, we simulated a data center comprising 686 PMs, connected through the CLOS topology, and 1000 VMs. The capacity of each link was 1 Gbps. The virtual network consisted of several VM clusters with the maximum cluster size of 5 VMs and the traffic flow rate between two dependent VMs was randomly picked up from the set $\{1, 2, 3, 4, 5\}$ Mbps. For each test problem the VMs were arbitrarily chosen from a pool of 200 VMs. Each PM was configured with 50 CPU and 64 GB memory. The attributes of a VM were arbitrarily chosen from the Amazon EC2 instance types [9]. Memory modification rate of a VM was in the range of $[1, 100]$ Mbps. In all experiments the simulation setup, i.e. the number of PMs and VMs, the CPU and memory capacities of each PM and VM, the link capacities and the initial placement of VMs remained unchanged.

The parameters for the SPEA2, i.e. the archive size, population size, mating pool size and maximum number of generations were respectively 40, 100, 100 and 150; and 90 % and 10 % of offspring were reproduced by the crossover and mutation operations respectively. This set of parameters was chosen through trials. The SPEA2 terminated when either 150 generations were explored or when there was no improvement in 50 consecutive generations. Both the SPEA2 and the RKGA were implemented in Java, and all the experiments were conducted on a 2.80 GHz Intel Core i7-2640M CPU and 8.00 GB RAM desktop computer.

4.2   Experimental Results

We compared the solutions generated by the SPEA2 and the RKGA for each of the test problems through the relative dominance of their solutions. As the SPEA2 provides a set of nondominated solutions unlike the RKGA, which provides only one solution, we evaluated the relative dominance between them. We define the relative dominance as the number of solutions of the SPEA2 dominate the solution of the RKGA and vice versa. Therefore, this criterion has been chosen for comparison to show the superiority of the SPEA2 over the RKGA and vice versa. Due to stochastic nature of the SPEA2 and the RKGA, 10 experiments were conducted for each test problem and the average of this 10 runs was taken for comparison.

Comparison of Relative Dominance of the SPEA2 and the RKGA: Table 1 shows the relative dominance between the SPEA2 and the RKGA. For each test problem the results presented in Table 1 is the average of 10 runs. The experimental results show that some of the nondominated solutions generated by the SPEA2 dominate the solution generated by the RKGA for all test problems starting from 40 VMs. For smaller size test problems (20 VMs) as small number of solutions are generated by the SPEA2, it gets less chance that solutions in the SPEA2 dominate that in the RKGA. However, for this test problem some solutions in the SPEA2 overlap the solution in the RKGA, i.e. some solutions in the SPEA2 are identical to that of the RKGA. The experimental results also illustrate that the solution in the RKGA does not dominate any solution

Table 1. Relative dominance between SPEA2 and RKGA

#VM	No. of solutions (SPEA2)	No. of solutions (RKGA)	No. of solutions of SPEA2 dominating solutions of RKGA	No. of solutions of RKGA dominating solutions of SPEA2
20	6	1	0	0
40	8.8	1	0.2	0
60	10.5	1	0.4	0
80	8.9	1	0.7	0
100	9.9	1	0.2	0
120	8.5	1	2.2	0
140	6.8	1	1.9	0
160	8	1	3.2	0
180	10.2	1	2.9	0
200	9.4	1	4.6	0

in the SPEA2 for any test problem. This concludes that no solution in the SPEA2 is worse than that of the RKGA and for larger test problems the SPEA2 outperforms the RKGA in terms of their relative dominance.

Comparison of Solution Qualities of the SPEA2 and the RKGA: The quality of a solution is measured by the total migration time and total downtime. The comparison graphs for total migration time and total downtime are respectively shown in Figs. 1 and 2. Albeit the solutions generated by the SPEA2 dominate the solution in the RKGA, the average total migration time calculated for each test problem in the RKGA is less than that of the SPEA2. The reason for that, for each test problem a significant number of solutions produced for the SPEA2 by 10 experiments, for example, for test problem of size 200 VMs 94 solutions were generated from 10 runs, and for the RKGA only one solution is generated in each run. Therefore, the average total migration time calculated for the SPEA2 becomes more than that of the RKGA. Moreover, the solution

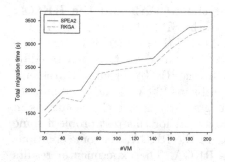

Fig. 1. Total migration time graphs

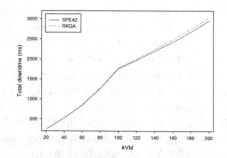

Fig. 2. Total downtime graphs

set of the SPEA2 contains a wide range of nondominated solutions, and some of these nondominated solutions are extremely diverse, i.e. total migration time is very large compared to total downtime. This gives the chance of minimizing the average total downtime with an increase of total migration time; and Fig. 2 depicts this scenario where the SPEA2 shows slight improvement over the RKGA in terms of total downtime. Both the graphs of total migration time and total downtime, show the near linear trend with the increased number of migrating VMs indicating good scalability of the SPEA2. The fluctuations are observed due to heterogeneity of VMs, as different configuration of VMs result in different migration time and downtime.

5   Conclusion

In this paper, we have developed a SPEA2 for the multi-VM migration scheduling problem. We have also evaluated the SPEA2 through by experiments. The experimental results have revealed that the SPEA2 produces better solutions than the RKGA, an existing genetic algorithm for the multi-VM migration scheduling problem. In addition, the experimental results have demonstrated the good scalability of the SPEA2.

References

1. Wood, T., Shenoy, P., Venkataramani, A., Yousif, M.: Sandpiper: black-box and gray-box resource management for virtual machines. Comput. Netw. **53**, 2923–2938 (2009)
2. Bobroff, N., Kochut, A., Beaty, K.: Dynamic placement of virtual machines for managing SLA violations. In: Proceedings of the 10th IFIP/IEEE International Symposium on Integrated Network Management, Munich, pp. 119–128 (2007)
3. Engelmann, C., Vallee, G.R., Naughton, T., Scott, S.L.: Proactive fault tolerance using preemptive migration. In: Proceeding of the 17th Euromicro International Conference on Parallel, Distributed and Network-based Processing, Weimar, pp. 252–257 (2009)
4. Ghorbani, S., Caesar, M.: Walk the line: consistent network updates with bandwidth guarantees. In: Proceedings of the 1st Workshop on Hot Topics in Software Defined Networks, Helsinki, pp. 67–72 (2012)
5. Nus, A., Raz, D.: Migration plans with minimum overall migration time. In: Proceedings of the 2014 IEEE/IFIP Network Operations and Management Symposium, Krakow, pp. 1–9 (2014)
6. Sarker, T.K., Tang, M.: Performance-driven live migration of multiple virtual machines in datacenters. In: Proceedings of the 2013 IEEE International Conference on Granular Computing, Beijing, pp. 253–258 (2013)
7. Sarker, T.K., Tang, M.: A random key genetic algorithm for live migration of multiple virtual machines in data centers. In: Proceedings of the 21st International Conference on Neural Information Processing, Kuching, pp. 212–220 (2014)
8. Zitzler, E., Laumanns, M., Thiele, L.: SPEA2: improving the strength pareto evolutionary algorithm for multiobjective optimization. In: Proceeding of the EUROGEN2001 Conference, Athens, pp. 95–102 (2001)
9. http://aws.amazon.com/de/ec2/instance-types/

A Survey of Applying Machine Learning Techniques for Credit Rating: Existing Models and Open Issues

Xiang Wang[1], Min Xu[1(✉)], and Özgür Tolga Pusatli[2]

[1] Global Big Data Technologies Centre,
University of Technology Sydney, Sydney, Australia
Xiang.Wang-1@student.uts.edu.au, Min.Xu@uts.edu.au
[2] Department of Mathematics and Computer Science,
Cankaya University, Ankara, Turkey
pusatli@cankaya.edu.tr

Abstract. In recent years, machine learning techniques have been widely applied for credit rating. To make a rational comparison of performance of different learning-based credit rating models, we focused on those models that are constructed and validated on the two mostly used Australian and German credit approval data sets. Based on a systematic review of literatures, we further compare and discuss about the performance of existing models. In addition, we identified and illustrated the limitations of existing works and discuss about some open issues that could benefit future research in this area.

Keywords: Credit rating · Single classifier models · Hybrid learning models · Literature survey

1 Introduction

Credit risk management is of a key importance for banks and financial institutions. To effectively manage risk exposure, banks and financial institutions should be able to correctly identify the creditworthiness of a client as it represents his/her capability to meet the obligation of paying back interests and principal as have been agreed.

Scoring card technique is widely used by financial practitioners for assessing customer creditworthiness. Under this scheme, credit experts carry out detailed analysis to discriminate 'good' and 'bad' borrowers based on available information of customers [1]. The assessment of a customer's creditworthiness depends on credit experts' judgments, which can be very time consuming and financially costly. In order to obtain rating results in an automatic and efficient manner, researchers have constructed various credit rating models. Some studies focus on statistical technique based models, including linear regression [2–4], probit regression (PR) and logit regression (LR) [5–8]. Recently, with the big breakthrough of machine learning technique development, a great number of researchers have attempted to construct learning-based models for credit rating, for example, Multivariate Discriminant Analysis (MDA) [9–11], Decision Trees (DT) [12, 13], Neural Networks (NN) [14–17],

© Springer International Publishing Switzerland 2015
S. Arik et al. (Eds.): ICONIP 2015, Part II, LNCS 9490, pp. 122–132, 2015.
DOI: 10.1007/978-3-319-26535-3_15

K Nearest Neighbour (KNN) [18, 19], Support Vector Machine (SVM) [20–24]. Furthermore, these techniques can be combined with each other to generate hybrid models [5, 25–28].

In this paper, we review existing learning-based models that are constructed and validated based on two most commonly used datasets. We compare credit rating accuracy generated by different models on the above two datasets in order to provide meaningful evaluations on the performance of different learning-based models. Besides the above-mentioned contribution, this paper also discusses about the open issues that could benefit researchers and financial practitioners in the area.

The rest of the paper is organized as follow. Section 2 introduces two most commonly used credit data sets. Section 3 reviews existing learning-based models used for credit rating. Section 4 compares and discusses performance of different models. Section 5 illustrates the limitation of existing research and discusses open issues for further research.

2 Introduction of Two Most Commonly Used Data Sets

Machine learning techniques have been tested on several real world credit data sets, for instance, Korean personal credit data set [29], farmers petty credit loan data set [30] and British credit card approval data set [20]. However, those data sets are used exclusively in specific literatures only. In order to make rational comparison of model performance, we recognized two data sets available from UCI database [31], which have been widely used by researchers in learning-based credit analysis.

The Australian credit data set contains 690 instances (45 % positive and 55 % negative) of credit card application approvals. For each instance, there are 14 attributes including 6 numerical and 8 categorical attributes. While, the German credit data set contains 1000 credit card approval instances, with 700 positive approvals (70 % of the total instances) and 300 negative instances. It consists of 20 attributes (7 numerical and 13 categorical attributes).

The two data sets have some properties in common, which bring challenges for developing an automatic credit rating system. Firstly, both datasets are mixture of numerical and categorical attributes. This requests the automatic system to be able to deal with both data types. Moreover, some numerical attributes have wide interval of values. As a result, data standardization is very likely to cause information lose. Lastly, the two data sets are weakly non-linear; therefore, a suitable form of machine learning model should be chosen [32]. In order to deal with those challenges, many research efforts have been attracted to construct and validate learning-based models based on these two datasets. The details of these researches are introduced in the next section.

3 Existing Learning-Based Models for Credit Rating

Predicting creditworthiness with machine learning techniques can be treated as a classification task. This is due to the factor that the credit grade system use categorical class labels to represent creditworthiness of objects. Machine learning-based techniques could not only remarkably improve productivity compare to expert's manual scoring, their advantage also lies on the improvement on predictive accuracy [33]. Most of the existing learning-based models are built based on popular machine learning techniques, such as DT, NN, SVM, KNN and so on. In addition, there are hybrid models which are constructed by combining two or more clusters and/or classifiers [15, 21, 34–39]. To make meaningful comparison of model efficiency, we focus on those that are constructed and validated on the two selected data sets.

3.1 Single Classifier Models

Decision tree, neural network and support vector machine are the most investigated learning-based classifiers for credit rating. In this following, we will discuss about the existing single classifier models used for credit rating.

3.1.1 Decision Tree

Decision tree is one of the most successful supervised classification techniques. Many research efforts have been put into developing tree construction algorithms. The most popular algorithms, such as ID3 (Iterative Dichotomiser 3) [40], C4.5 (successor of ID3) [13], CART (Classification And Regression Tree) [41], and CHAID (CHi-squared Automatic Interaction Detector) [42]. Quinlan [13] construct decision tree credit rating models based on Australian and German data sets. According to their report, the C4.5 decision tree achieves better predictive accuracy than most of statistical methods. Compare to other learning-based techniques, learning results from DT model are easier to be explained.

3.1.2 Neural Networks

A significant advantage of NN model for credit rating is that it deals well with complicated problem, where a large number of attributes are simplified in a model.

There are several factors that affect performance of NN classifiers. Firstly, network architectures decide effectiveness of NN model for credit rating. West [14] investigate five neural network architectures(multilayer perceptron (MLP), mixture of experts (MOE), radial basis function (RBF), learning vector quantization (LVQ) and fuzzy adaptive resonance (FAR) networks) and conclude that only MOE and RBF NN models are suitable for credit scoring applications; Secondly, number of neurons and hidden layers impact NN model performance. In [15], Tsai et al. construct NN models with different number of neurons and illustrate that model performance is related to the number of neurons; Thirdly, NN model performance can be affected by training and validation ratios. Khashman [17] tests NN classifiers with different training to validation data ratios and argues that the more balanced the training and validation ratio the better the accuracy achieved.

To further enhance a NN model, Khashman [43] constructs a NN model that takes human emotions into consideration. He argues the proposed emotional NN outperforms conventional models in terms of both predictive accuracy and training speed.

3.1.3 Support Vector Machine

Support Vector Machine (SVM) has caught eyeballs of researchers for its outstanding performance in solving practical classification problems [46]. Huang and Day [33] compare performance of SVM model with LR, MDA, KNN, NN and DT classifiers and conclude that SVM yield very good performance on the credit data sets.

By employing different kernel functions, SVM technique can be applied to classify non-linearly separable data sets. Chen and Li [34] test SVM models with three different kernel function forms (e.g. Linear, Polynomial and Radial Basis Functions (RBF)) and conclude that model with linear kernel function has highest accuracy. Besides kernel function forms, SVM parameters, e.g. SVM penalty parameter C and RBF parameter γ, have significant impact on accuracy of SVM classifier. Grid search approach is an effective way to find the optimal parameters for SVM model [47]. Moreover, Huang et al. [22] propose a model which combines a genetic algorithm (GA) to perform model parameters optimization and feature selection. Their experiment shows that GA-SVM model generates more robust accuracy than NN, Generic Programming (GP) and C4.5. Furthermore, in order to incorporate domain knowledge into model construction, Lin [25] propose a monotonicity-constrained model, which imposes constrains on the monotonicity of selected attributes to align model design with expert knowledge.

3.1.4 Other Models

Besides DT, NN and SVM, different learning based models are also applied and proved to be effective for credit rating. We summaries some existing researches in this subsection.

Leung et al. [48] construct a genetic programming (GP) model for credit scoring. They compare GP model with NN, DT, rough sets (RST), and LR and conclude that GP yield better performance than others. Bellotti and Crook [20] applies the KNN classifier on German and Australian data sets and concludes that KNN classifier generates plausible result and requires simpler computation. Lan et al. [49] achieved 86.2 % predictive accuracy on Australian data set with a simple artificial immune system model (SAIM), which handles missing attributes well. In [50], an artificial immune network (AINE) based classifier is proposed. Experimental results indicate that the proposed AINE classifier is a competitive classifier for credit scoring. Even though many learning-based models have been proved to be effective for credit rating, Lan et al. [49] test classification based on associations (CBA) models for credit rating. They conclude that CBA model is not a good choice for credit analysis.

3.2 Hybrid Models

Hybrid learning model combines different machine learning classifiers and/or clusters to construct a more plausible classifier. By doing so, it either improves the quality of input data, or advances performance of classification model.

In this section, we will illustrate existing work of applying hybrid models to improve credit rating accuracy through two major ways. One is performing feature selection before classification. The other one is ensemble learning, which combine results of multiple base learning classifiers to obtain better predictive performance.

3.2.1 Hybrid Feature Selection Models

Feature selection method generates feature subset for use in model construction. The process helps to save unnecessary computation and improve model interpretability.

Chen and Li [34] proposed two stage models, which use LDA, DT, RST and F-score as feature selection methods to generate subsets before training SVM classifiers. The proposed hybrid feature selection models achieve higher accuracy than single classifiers. Moreover, Huang et al. [21] use GA as feature selection method and compare results with GP, C4.5 and F-score feature selection approaches before training SVM classifiers. The GA-SVM model enables the search for optimal feature subsets and SVM parameters to be done simultaneously; therefore, improving model accuracy and efficiency.

Various feature selection methods are also tested on NN and KNN classifiers. In [35], a group of neural network classifiers are trained with feature subsets selected by ten feature selection methods (e.g. best first search, info-gain, wrapper-subset, gain-ratio, random-search, chi-square, consistency-subset, principal-component, relief-attribute and SVM-attribute). Li [19] train KNN classifiers with feature subsets selected by LDA, DT, RST and F-score feature selection methods. The results from above mentioned experiments suggest that feature selection can improve performance of hybrid models; however, the effectiveness of feature selection methods varies.

3.2.2 Ensemble Models

Ensemble methods combine two or more classifiers and make them to complement each other. Ensemble method could effectively reduce the influence of bias and errors of single classifiers in order to achieve a better classification performance. In credit rating literatures, there are three ensemble strategies that are most commonly used for credit rating: Bagging, Boosting and Stacking.

In [36], Wang et al. compare performances of the three popular ensemble methods experimentally with LR, NN, DT and SVM techniques employed as base classifiers. West et al. [39] test cross-validation, bagging and boosting ensemble methods with neural network base classifiers. In their experiments, bagging and boosting methods promote diversity through re-sampling techniques, in which subsets are randomly drawn for training classifiers; whereas, stacking method aggregates predictions of four diversified learning techniques.

Diversity is crucial for the success of an ensemble model [44]. An alternative way to boost diversity is to manipulate subsets in the attribute space (rather than instance space). Wang et al. [38] propose two ensemble trees: RS-bagging DT and bagging-RS DT, which promote diversity in feature space. The proposed models improve performance of DT by reducing the influence of the noise data and the redundant attributes to the accuracy of DT.

On the whole, experiments on the two data sets suggests ensemble models in general outperform standing alone classifiers.

4 Performance Comparison on Different Models

In previous sections, we have introduced different learning-based models that are constructed and validated on the two data sets. In this section, we compare the performance of existing models in three groups: (1) models with a single classifier; (2) models with a feature selection and (3) ensemble models.

4.1 Performance Comparison on Models with a Single Classifier

Table 1 summarizes the performance of 11 models with a single classifier based on different classification algorithms. All single classifier models are generated and validated by our selected credit rating data sets. There is no evidence that any classifier is proved to be able to dominate others in providing optimum credit rating prediction. It is clear that the predictive accuracy varies in response to different classifiers and data sets. For instance, KNN classifier has the best performance on Australian dataset with 89.1 % instances are correctly classified, but it gets the lowest accuracy on German data set.

Table 1. Performance of different machine learning techniques in terms of credit rating accuracy for the two data sets

Machine-learning algorithms	Credit rating accuracy		Reference
	Australian	German	
KNN	**89.1**	72.2	[19]
GP	88.27	77.44	[45]
NN	86.23	**79.5**	[46]
AINE-based	85.36	77.1	[47]
SVM	85.34	75.4	[21]
MC-SVM	–	75.8	[24]
SAIS	85.2	75.4	[48]
CBA	85.65	73.3	[49]
LR	84.06	76	[46]
C4.5	82.5	72.4	[13]
LDA	77.97	75.5	[46]

4.2 Performance Comparison on Hybrid Feature Selection Models

Feature selection is an effective way to improve model efficiency. Table 2 shows that classifiers trained with selected feature subset perform at least as well as those trained with full dataset. Training model with less attributes helps to improve classification accuracy and computation cost by reducing noises and homogeneity of the data.

Table 2. Performance of hybrid models after feature selection (best scenarios)

Selection methods and learning algorithms	Australia		Germany		Reference
	Attributes selected	Credit rating accuracy	Attributes selected	Credit rating accuracy	
SVM + Grid search + F-score	7.6	84.2	20.4	77.5	[21]
GA + SVM	7.3	86.9	13.3	77.92	
GP + SVM	8.2	87	13.2	78.1	
C4.5 + SVM	12.2	85.9	20.3	73.6	
LDA + SVM	7	86.52	12	76.1	[34]
DT + SVM	7	86.29	12	73.7	
RST + SVM	7	85.22	12	75.6	
F-score + SVM	7	85.1	12	76.7	
BFS + NN	6	88.08	3	72.9	[35]
Info-Gain + NN	14	80.42	20	73.8	
Wrapper-Subset + NN	7	66.38	1	68.8	
Gain-Ratio + NN	14	85.3	20	73.8	
Random-Search + NN	6	79.42	6	71.5	
Chi-square + NN	14	79.42	20	71.6	
Consistency-subset + NN	12	84.78	14	71.8	
Principal-component + NN NNNNNANN	12	84.78	20	71.6	
Relief-Attribute + NN	14	79.42	20	72.6	
SVM-Attribute + NN	14	79.42	20	73.6	
LDA + KNN	7	88.27	12	74.5	[19]
DT + KNN	7	88.41	12	71.9	
RST + KNN	7	88.54	12	72.3	
F-score + KNN	7	90.4	12	74.3	

Another motivating finding is that the optimum number of features selected is related to the base classification technique rather than the feature selection methods.

4.3   Performance Comparison on Ensemble Models

Ensemble overcomes deficiency of weak classifiers by introducing diversity, which helps to reduce model bias and variance. Experimental results presented in Table 3 show that the ensemble method in general perform better than single classifier models, while the bagging method is the most efficient ensemble strategy.

Ensemble methods yield better performance than single classifiers for three reasons: first of all, the training data may not provide satisfactory information for selecting an accurate hypothesis; moreover, the learning processes of a classifier might not be complete enough; lastly, the target function may not be contained in the hypothesis space. In these circumstances, an ensemble classifier can provide a good approximation [50].

Table 3. Performance of ensemble methods based on ANN, LRA, DT and SVM techniques

Learning algorithm + ensemble methods	Credit rating accuracy		Reference
	Australian	German	
ANN	86.77 %	74.69 %	[39]
ANN + Cross Validation	86.90 %	75.81 %	
ANN + Bagging	87.20 %	74.87 %	
ANN + Boosting	85.23 %	74.53 %	
LRA	86.56 %	76.14 %	[36]
LRA + Bagging	86.64 %	76.08 %	
LRA + Boosting	86.56 %	76.14 %	
DT	84.39 %	72.10 %	
DT + Bagging	86.30 %	74.92 %	
DT + Boosting	85.22 %	72.77 %	
SVM	85.67 %	76.28 %	
SVM + Bagging	85.71 %	75.93 %	
SVM + Boosting	84.19 %	76.30 %	
Stacking (LRA, DT, ANN, SVM)	86.57 %	76.97 %	
DT + Random space	86.93 %	76.12 %	[38]
Random Forest	86.89 %	77.05 %	
Rotation Forest	86.55 %	77.00 %	
DT + Random space-Bagging	88.17 %	78.36 %	
DT + Bagging-RS	88.01 %	78.52 %	

5   Limitations of Existing Researches and Open Issues

More and more researchers and financial practitioners are keen to witness a revolutionary breakthrough in applying learning based models for real world business practice. Through comparing and analyzing existing models generated and tested on two major datasets, we identify two major limitations of previous works, which could indicate some potential directions for future investigations:

- Limitations of the datasets: The two data sets have several defects: firstly, the sample sizes and number of attributes are far less than datasets used in business practice; secondly, the two data sets contains only static data, which are not dynamic enough to represent of the fast changing financial business. Even though many plausible results have been obtained by researchers, there still exist some uncertainties whether machine learning techniques are reliable enough to be employed in business practice.
- Limitations of incorporation of domain knowledge: Some research have attempted to incorporate expert knowledge into model construction and the outcome tends to be superior to other models. Incorporate domain knowledge into machine learning model is an effective way to improve model efficiency and interpretability. Therefore, this yet largely understudied research direction has a great potential in its application.

References

1. Thomas, L.C.: A survey of credit and behavioural scoring: forecasting financial risk of lending to consumers. Int. J. Forecast. **16**, 149–172 (2000)
2. Horrigan, J.O.: The determination of long-term credit standing with financial ratios. J. Acc. Res. **4**, 44–62 (1966)
3. Pogue, T.F., Soldofsky, R.M.: What's in a bond rating. J. Financ. Quant. Anal. **4**, 201–228 (1969)
4. West, R.R.: An alternative approach to predicting corporate bond ratings. J. Acc. Res. **8**, 118–125 (1970)
5. Hwang, R.C., Cheng, K., Lee, C.F.: On multiple-class prediction of issuer credit ratings. Appl. Stochast. Models Bus. Ind. **25**, 535–550 (2009)
6. Grunert, J., Norden, L., Weber, M.: The role of non-financial factors in internal credit ratings. J. Bank. Finance **29**, 509–531 (2005)
7. Martin, D.: Early warning of bank failure: a logit regression approach. J. Bank. Finance **1**, 249–276 (1977)
8. Kaplan, R.S., Urwitz, G.: Statistical models of bond ratings: a methodological inquiry. J. Bus. **52**, 231–261 (1979)
9. Altman, E.I.: Financial ratios, discriminant analysis and the prediction of corporate bankruptcy. J. Finance **23**, 589–609 (1968)
10. Michel, A.J.: Municipal bond ratings: a discriminant analysis approach. J. Financ. Quant. Anal. **12**, 587–598 (1977)
11. Shi, Y., Peng, Y., Xu, W., Tang, X.: Data mining via multiple criteria linear programming: applications in credit card portfolio management. Int. J. Inf. Technol. Decis. Mak. **1**, 131–151 (2002)
12. Hongxia, W., Xueqin, L., Yanhui, L.: Enterprise credit rating model based on fuzzy clustering and decision tree. In: Information Science and Engineering (ISISE) International Symposium on, pp. 105–108. IEEE (2010)
13. Quinlan, J.R.: C4.5: Programs for Machine Learning. Morgan Kaufmann Publishers Inc., Burlington (1993)
14. West, D.: Neural network credit scoring models. Comput. Oper. Res. **27**, 1131–1152 (2000)
15. Tsai, C.-F., Wu, J.-W.: Using neural network ensembles for bankruptcy prediction and credit scoring. Expert Syst. Appl. **34**, 2639–2649 (2008)
16. Atiya, A.F.: Bankruptcy prediction for credit risk using neural networks: a survey and new results. IEEE Trans. Neural Netw. **12**, 929–935 (2001)
17. Khashman, A.: Neural networks for credit risk evaluation: investigation of different neural models and learning schemes. Expert Syst. Appl. **37**, 6233–6239 (2010)
18. Henley, W., Hand, D.J.: A k-nearest-neighbour classifier for assessing consumer credit risk. The Statistician **45**, 77–95 (1996)
19. Li, F.-C.: The hybrid credit scoring strategies based on knn classifier. In: Sixth International Conference on Fuzzy Systems and Knowledge Discovery, FSKD 2009, pp. 330–334. IEEE (2009)
20. Bellotti, T., Crook, J.: Support vector machines for credit scoring and discovery of significant features. Expert Syst. Appl. **36**, 3302–3308 (2009)
21. Huang, C.-L., Chen, M.-C., Wang, C.-J.: Credit scoring with a data mining approach based on support vector machines. Expert Syst. Appl. **33**, 847–856 (2007)
22. Huang, Z., Chen, H., Hsu, C.-J., Chen, W.-H., Wu, S.: Credit rating analysis with support vector machines and neural networks: a market comparative study. Decis. Support Syst. **37**, 543–558 (2004)

23. Kim, K.-J., Ahn, H.: A corporate credit rating model using multi-class support vector machines with an ordinal pairwise partitioning approach. Comput. Oper. Res. **39**, 1800–1811 (2012)
24. Chen, C.-C., Li, S.-T.: Credit rating with a monotonicity-constrained support vector machine model. Expert Syst. Appl. **41**, 7235–7247 (2014)
25. Lin, S.L.: A new two-stage hybrid approach of credit risk in banking industry. Expert Syst. Appl. **36**, 8333–8341 (2009)
26. Zhou, L., Lai, K.K., Yu, L.: Least squares support vector machines ensemble models for credit scoring. Expert Syst. Appl. **37**, 127–133 (2010)
27. Chen, Y.-S., Cheng, C.-H.: Hybrid models based on rough set classifiers for setting credit rating decision rules in the global banking industry. Knowl. -Based Syst. **39**, 224–239 (2013)
28. Yeh, C.-C., Lin, F., Hsu, C.-Y.: A hybrid KMV model, random forests and rough set theory approach for credit rating. Knowl. -Based Syst. **33**, 166–172 (2012)
29. Kweon, S.-C., Sawng, Y.-W., Kim, S.-H.: An integrated approach using data mining and genetic algorithm in customer credit risk prediction of installment purchase financing. In: International Symposium on Collaborative Technologies and Systems, CTS 2006, pp. 125–131. IEEE (2006)
30. Hai, L., Shi, B., Peng, G.: A credit risk evaluation index system establishment of petty loans for farmers based on correlation analysis and significant discriminant. J. Softw. **8**, 2344–2351 (2013)
31. Lichman, M.: {UCI} Machine Learning Repository (2013)
32. Baesens, B., Van Gestel, T., Viaene, S., Stepanova, M., Suykens, J., Vanthienen, J.: Benchmarking state-of-the-art classification algorithms for credit scoring. J. Oper. Res. Soc. **54**, 627–635 (2003)
33. Huang, S.-C., Day, M.-Y.: A comparative study of data mining techniques for credit scoring in banking. In: 2013 IEEE 14th International Conference on Information Reuse and Integration (IRI), pp. 684–691. IEEE (2013)
34. Chen, F.-L., Li, F.-C.: Combination of feature selection approaches with SVM in credit scoring. Expert Syst. Appl. **37**, 4902–4909 (2010)
35. Raghavendra, B., Simha, J.B.: evaluation of feature selection methods for predictive modeling using neural networks in credits scoring. Int. J. Adv. Netw. Appl. **2**, 714–718 (2010)
36. Wang, G., Hao, J., Ma, J., Jiang, H.: A comparative assessment of ensemble learning for credit scoring. Expert Syst. Appl. **38**, 223–230 (2011)
37. Maclin, R., Opitz, D.: Popular ensemble methods: an empirical study. J. Artif. Intell. Res. **11**, 169–198 (1999)
38. Wang, G., Ma, J., Huang, L., Xu, K.: Two credit scoring models based on dual strategy ensemble trees. Knowl. -Based Syst. **26**, 61–68 (2012)
39. West, D., Dellana, S., Qian, J.: Neural network ensemble strategies for financial decision applications. Comput. Oper. Res. **32**, 2543–2559 (2005)
40. Quinlan, J.R.: Induction of decision trees. Mach. Learn. **1**, 81–106 (1986)
41. Chipman, H.A., George, E.I., McCulloch, R.E.: Bayesian CART model search. J. Am. Stat. Assoc. **93**, 935–948 (1998)
42. Kass, G.V.: An exploratory technique for investigating large quantities of categorical data. App. Stat. **29**, 119–127 (1980)
43. Khashman, A.: Credit risk evaluation using neural networks: emotional versus conventional models. Appl. Soft Comput. **11**, 5477–5484 (2011)
44. Kuncheva, L.I., Whitaker, C.J.: Measures of diversity in classifier ensembles and their relationship with the ensemble accuracy. Mach. Learn. **51**, 181–207 (2003)

45. Ong, C.-S., Huang, J.-J., Tzeng, G.-H.: Building credit scoring models using genetic programming. Expert Syst. Appl. **29**, 41–47 (2005)
46. Chuang, C.L., Huang, S.T.: A hybrid neural network approach for credit scoring. Expert Syst. **28**, 185–196 (2011)
47. Chang, S.-Y., Yeh, T.-Y.: An artificial immune classifier for credit scoring analysis. Appl. Soft Comput. **12**, 611–618 (2012)
48. Leung, K., Cheong, F., Cheong, C.: Generating compact classifier systems using a simple artificial immune system. IEEE Trans. Syst. Man Cybern. B Cyber. **37**, 1344–1356 (2007)
49. Lan, Y., Janssens, D., Chen, G., Wets, G.: Improving associative classification by incorporating novel interestingness measures. Expert Syst. Appl. **31**, 184–192 (2006)
50. Dietterich, T.G.: Machine-learning research. AI Mag. **18**, 97 (1997)

A New Evolutionary Algorithm for Extracting a Reduced Set of Interesting Association Rules

Mir Md. Jahangir Kabir[✉], Shuxiang Xu, Byeong Ho Kang,
and Zongyuan Zhao

School of Engineering and ICT, University of Tasmania, Hobart, Australia
{mmjkabir,Shuxiang.Xu,Byeong.Kang,
Zongyuan.Zhao}@utas.edu.au

Abstract. Data mining techniques involve extracting useful, novel and interesting patterns from large data sets. Traditional association rule mining algorithms generate a huge number of unnecessary rules because of using support and confidence values as a constraint for measuring the quality of generated rules. Recently, several studies defined the process of extracting association rules as a multi-objective problem allowing researchers to optimize different measures that can present in different degrees depending on the data sets used. Applying evolutionary algorithms to noisy data of a large data set, is especially useful for automatic data processing and discovering meaningful and significant association rules. From the beginning of the last decade, multi-objective evolutionary algorithms are gradually becoming more and more useful in data mining research areas. In this paper, we propose a new multi-objective evolutionary algorithm, MBAREA, for mining useful Boolean association rules with low computational cost. To accomplish this our proposed method extends a recent multi-objective evolutionary algorithm based on a decomposition technique to perform evolutionary learning of a fitness value of each rule, while introducing a best population and a class based mutation method to store all the best rules obtained at some point of intermediate generation of a population and improving the diversity of the obtained rules. Moreover, this approach maximizes two objectives such as performance and interestingness for getting rules which are useful, easy to understand and interesting. This proposed algorithm is applied to different real world data sets to demonstrate the effectiveness of the proposed approach and the result is compared with existing evolutionary algorithm based approaches.

Keywords: Data mining · Association rules · Multi-objective evolutionary algorithms · Conditional probability · Interestingness

1 Introduction

Data mining techniques are the most important tools for discovering valid, novel, useful and interesting patterns from large data sets. Nowadays, a huge amount of data is collected and stored inexpensively and is easy to access due to the digital revolution. This limitless growth of data, makes the knowledge extraction process more difficult and in most cases the problems become complex [1]. Therefore, designing an efficient deterministic method is not suitable. Because of inherent parallel structures,

© Springer International Publishing Switzerland 2015
S. Arik et al. (Eds.): ICONIP 2015, Part II, LNCS 9490, pp. 133–142, 2015.
DOI: 10.1007/978-3-319-26535-3_16

evolutionary algorithms have been found to be effective for automatic processing of large data sets through optimal operator settings and to extract meaningful and important information [2]. Mining association rules is one of the most common data mining tasks for extracting interesting and hidden knowledge from big data sets [3]. Association rules are used to define and represent relationships between item sets in a data set. If X and Y are item sets where $X \cap Y = \emptyset$, then an association rule between the item sets is represented by $X \rightarrow Y$.

A large number of research studies for mining association rules are based on a support-confidence framework [4, 5]. This framework consists of two sub processes and most of the existing association rule mining algorithms follow these factors to measure the interestingness of a rule. These factors are as follows: (1) finding all frequent item sets from a large database which satisfies a user defined support value and (2) generating rules from those frequent item sets which satisfies a user defined confidence value. These particulars raise two major issues. The first one is that, users need to specify a suitable threshold value although they have no knowledge regarding the real world data sets. Another one is an exponential search space of size 2^n, where n is the number of item sets [6].

Many evolutionary algorithms [7], particularly genetic algorithms, are proposed in the literature for extracting a reduced set of Boolean association rules (BAR) [6, 8–10]. Genetic algorithm based methods are considered to be one of the most effective search techniques for solving complex problems and have proved to be a successful approach, especially when the size of the search spaces are large enough for using deterministic methods. These algorithms removes some of the limitations of the above mentioned challenges and generate high quality rules as well as few unnecessary rules due to using only one evaluation criteria [11].

In this paper, we propose MBAREA, a new genetic algorithm based approach which is using multiple evaluation criteria in order to mine a reduced set of high quality rules without generating unnecessary rules.

In order to assess the performance of the proposed method, we are presenting an experimental analysis using six real world data sets, with the number of attributes ranging from 23 to 118 and the number of records ranging from 267 to 12,960. We have performed the following studies: First, we have compared the performance of our results with two other evolutionary approaches, ARMGA [6, 10] and ARMMGA [8]. Second, we have explained the scalability of the proposed approach and finally, we have analyzed some of the rules obtained by our proposed method.

2 Preliminaries

In real world applications the data sets consist of nominal attributes. For mining Boolean association rules, nominal attributes are mapped into Boolean attributes on which association rule mining algorithms are applied. For instance, in a solar flare data set, the attributes named evolution and activity are categorized into {decay, no growth, growth} and {reduced, unchanged} respectively. These two attributes are mapped into a set of items $I = \{I_1 = decay, I_2 = no\ growth, I_3 = growth, I_4 = reduced, I_5 = unchanged\}$. The aim of this study is to find interesting association rules from this mapped data set to find Boolean association rules (BARs). For example, a Boolean association rule can be defined as, buys (X,"Data Mining Book") ∧ buys (X,"EA Book") → buys (X,"EARM Book"). A common practice is to use association rules for solving a wide range of different real world problems, such as health, medical science, etc [12]. The classical algorithms only focus on generating association rules if these satisfy user defined support and confidence values. A commonly used method is to generate frequent item sets from a data set based on a user defined support value where supp(A∪B) = | (A∪B)|/|D|, here A and B are two item sets of a data set D and A ⊆ I, B ⊆ I and A ∩ B = φ [3, 4]. After generating frequent item sets, the next step is to generate rules from those frequent patterns. A rule is valid if it satisfies a user defined confidence value. A confidence value of a rule A → B is defined as supp(A∪B)/supp(A).

Most of these algorithms were only focused on those rules which frequently appeared in a data set that is high support based rules [11, 13]. However several authors have noticed some drawbacks of this structure which leads to the generation of a huge number of misleading and trivial rules [8, 11]. A rule is misleading if supp(Y) > confidence (X → Y) i.e. the item set in the antecedent is negatively correlated with the item sets in the consequent, since the buying of one of these items actually decreases the probability of purchasing the other [14].

Many researchers used different measures for evaluating the quality of a rule and those approaches significantly reduced the generation of misleading rules [6, 8, 15]. However, still those approaches suffered from generating misleading as well as trivial rules. A rule X → Y is trivial, if supp(X) = 0 or supp(Y) = 0, then supp(X∪Y) = 0 for any item sets of X or Y, respectively.

In recent years, several studies have proposed other measures according to the interest of users [11]. We briefly describe a few of those which will be used in the current study.

The conditional probability [16] of a rule analyzes the dependence between X and Y and it is defined as:

$$CP(X|Y) = \{supp(X \cup Y) - supp(X)supp(Y)\}/\{supp(X)(1 - supp(Y))\} \quad (1)$$

Its domain range is $[-\infty, \infty]$, where $0 >$ value $> -\infty$ represents misleading rules, $0 <$ value $< \infty$ represents positive association rules, and value $= 0/-\infty/\infty$ represents trivial rules. The ratio between the confidence and the expected confidence of the rule is measured by lift [11] and it is defined as,

$$lift(X \to Y) = supp(X \cup Y)/\{supp(X)supp(Y)\} \quad (2)$$

For finding interesting rules, new rules are generated based on each item present in the consequent part of a rule. Since a number of items are present in the consequent part of a rule and it is not predefined, this approach may not be suitable for an association rule mining task. Recall the definition of interesting [17], a new expression for measuring the interestingness of a rule A → B is defined as follows:

$$I = [supp(A \cup B)/supp(A)] \times [supp(A \cup B)/supp(B)] \times [supp(A \cup B)/|D|] \quad (3)$$

Here I is the interestingness constraint of a rule A → B and the total number of records in a database is defined by the term |D|. Its domain range is [0, ∞], where 0, ∞ and 0 < value < ∞ represents independence, trivial rules and positive dependence, respectively.

3 Methods

This section describes our method for obtaining a reduced set of interesting association rules with a good trade-off between the coverage and the number of generated rules, considering three objectives conditional probability, lift and interestingness. This proposal extends the existing ARMGA and ARMMGA algorithms for performing an evolutionary learning and introduces two new components: class based variable adaptation operator and best population. The characteristic and flowchart of the algorithm are presented through the following sections.

3.1 Class Based Mutation and Best Population

In order to store all the non-dominated rules which are generated in the intermediate generation of a population, provoking the diversity of the population, and increasing the coverage of data sets we have to introduce class based mutation approach along with best population method. The mutation operator is used to keep the diversity from one generation of a population to the next one. Mutation changes one or more genes of a chromosome with respect to a mutation probability, mp. Existing GA based approaches such as ARMGA and ARMMGA, followed fixed mutation probability and randomly mutated the chromosomes. Although these methods used low mutation probability, it mutated few high quality chromosomes due to random function. For this reason, some top quality chromosomes get less chance for future generation of a population. To prevent this problem and to give more chance to the best chromosomes for future generation of a population, we classified the whole population into δ, based on a fitness value of each chromosome. Top class chromosomes have a higher fitness value but assign with a low mutation ratio whereas low class chromosomes are mutated with high mutation probability. Through this approach high class chromosomes take part for future generation of a population. Best population (BP) keeps all the non-dominated rules which are generated in intermediate generation of a population. Moreover, BP will be updated with the generation of a new population following the non-dominance criteria. This process helps us to increase the coverage of a data set and for performing enhanced exploration of the search space.

3.2 Objectives and Genetic Operators

Three objectives are maximized for this problem: conditional probability (CP), lift and interestingness. We are only interested in mining very strong rules which have positive dependence between items and avoid the problem of support-confidence framework based methods. Notice that positive association rules allow us to represent positive dependence, thus we are interested in those rules which have CP > 0. Thus, a rule

$X \rightarrow Y$ must satisfy the following conditions: (i) CP > 0; (ii) supp (X∪Y) > 0; (iii)-supp(X) $\neq$ 0 and supp (Y) $\neq$ 0. In this study, CP will act as a fitness function of a valid rule for filtering out from misleading and trivial rules. A rule with a CP value near to one means high degree of positive dependence between item sets and may be more important to the users. Interestingness is a measure of a rule through which we can say how interesting a rule to the users. Here we have used the well-known interestingness measure (see Sect. 2). Since its range is not bounded, allows us to better value denotes the difference between the rules and reduces the number of generation.

A chromosome is a gene vector which represents the attributes and an indicator for separation between item sets. Given an association rule of k length means that, a rule contains k items which is shown by Fig. 1. For example a rule is $A \rightarrow B$, where antecedent A contains $item_1$ to $item_n$ and the consequent B contains $item_{n+1}$ to $item_k$, where $0 < n < k$. The first place of a chromosome is an indicator for separation from antecedent to consequent.

By using selection operator, an individual chromosome is chosen from a given population. This operator acts as a filter to choose an individual chromosome based on the fitness function and selection probability (sp). Crossover operator is applied on two chromosomes of a given population called parent chromosomes to reproduce two new offspring chromosomes by exchanging parts of the parent chromosomes. An example of how this operator works is shown in Fig. 2. Mutation operator is explained in 3.1.

3.3 MBAREA Algorithm

According to the above description, the MBAREA algorithm is summarized through the following structure.

PROCEDURE: MBAREA	
Input:	Database D, size of population popSize, selection probability sp, crossover probability cp, class_size δ, mutation probability mp[δ], rule length k
Output:	Best Population (Association rules with potential interest)
(0)	categorized attributes of a database D, are mapped into Boolean attributes
(1)	i ← 0, bestPopulation[i] ← 0, mp[δ] ← mutation_function(δ)
(2)	population[i] ← initialization (popSize)
(3)	**while** not termination_condition_reached (population[i])
(4)	pop_after_selection[i] ← 0
(5)	pop_after_crossover[i] ← 0
(6)	**for** ∀chromosome ∈ population[i] do
(7)	pop_after_selection[i] ← selection(population[i], sp)
(8)	pop_after_selection[i]←sort_population(pop_after_selection[i])
(9)	pop_after_crossover[i] ← crossover (pop_after_selection[i], cp)
(10)	pop_after_crossover[i]←sort_population(pop_after_crossover[i])
(11)	population[i] ← class_based_mutation (pop_after_crossover[i], mp[δ])
(12)	best_population ← non-dominated_rules(population[i])
(13)	remove redundant rules from best_population
(14)	i ← i+1
(15)	**end**
(16)	**The best_population is returned.**

n	$item_1$	$item_2$		...	$item_n$	$item_{n+1}$		...	$item_k$

Fig. 1. A chromosome of an association rule of k length

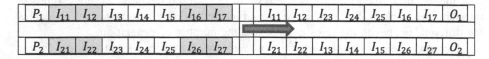

Fig. 2. Two-point crossover example

This process is continued until it satisfies any of the following conditions occur: (i) if the maximum number of evaluation is reached, or (ii) if the average value of fitness function of current population is less than the value α of previous population.

4 Experimental Analysis

Several experiments have been carried out on different data sets for analyzing the performance of the proposed algorithm. For testing the proposed algorithm and comparing the result with ARMGA and ARMMGA approaches, we have considered six real world data sets, which are available in the UCI machine learning repository (http://archive.ics.uci.edu/ml/datasets.html).

Table 1 summarizes the specifications of those data sets, where Attributes (B) represents the number of Boolean attributes and Records is the number of records.

The parameters, which are used for running the algorithms are shown in Table 2. For ARMMGA and ARMGA, the parameters are selected according to the recommendations of each proposal. As we described in Sect. 3.1, a class based mutation method is applied for our approach and the probability of mutation ratio is decreased with respect to the class of chromosomes in a population.

Table 1. Data sets considered for the experimental analysis

	Mushroom	Balance scale	Nursery	Monk's problems	Solar flare	SPECT heart
Attributes (B)	118	23	32	19	50	46
Records	8124	625	12960	431	1066	267

Table 2. Parameters considered for running the algorithms

Algorithms	Parameters
ARMMGA	Popsize = 100, P_{sel} = 0.95, P_{cro} = 0.85, P_{mut} = 0.01, db = 0.01, k = 3
ARMGA	Popsize = 100, P_{sel} = 0.95, P_{cro} = 0.85, P_{mut} = 0.01, α = 0.01, k = 3
MBAREA	Popsize = 100, P_{sel} = 0.95, P_{cro} = 0.85, P_{mut} = [100 − {(100/δ)*n}] %, δ = 5, n = 1 ∼ δ, k = 3, α = 0.01

Because of using a weak constraint function [8], ARMMGA generates positive association rules including misleading and trivial rules, which are shown in Fig. 3. In this experiment we considered single evaluation result for different popsizes. The performance of our approach against other algorithms is shown in Table 3, where #Rules is the number of generated rules and Av_{supp}, AV_{conf}, AV_{lift}, Av_{int} and CP are average support, confidence, lift, interest and conditional probability, respectively. In order to develop the different experimental analysis, we have considered the average results of three runs for each data set.

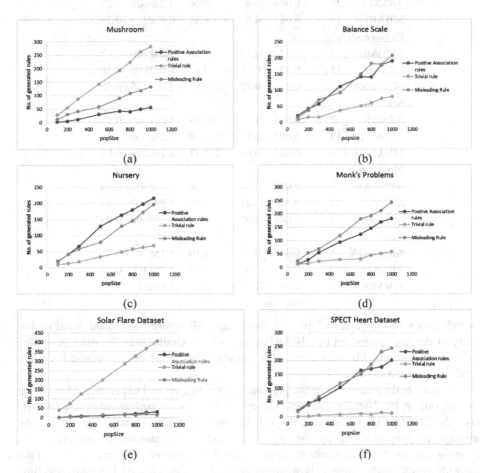

(a)   (b)

(c)   (d)

(e)   (f)

Fig. 3. Different types of rules for different data sets are generated by ARMMGA because of using weak constraint

The rules obtained by our approach presents better or similar values for different measures than the rules obtained by other algorithms. As with ARMMGA, it generates a smaller number of rules but some of those are misleading or trivial rules. For some data sets, ARMGA obtain good average support but low values for the rest of the measures.

Table 3. Results obtained by evolutionary algorithms for different data sets

Algorithms	#Rules	Av$_{supp}$	Av$_{conf}$	Av$_{lift}$	Av$_{int}$	CP
SPECT heart						
ARMMGA	9	**0.29**	**0.87**	1.25	**0.0005**	0.2
ARMGA	37	0.25	0.6	1.5	0.0004	0.3
MBAREA	11	0.1	0.83	**1.88**	**0.0005**	**0.68**
Monk's problems						
ARMMGA	7	0.04	0.64	1.88	8.84E-06	0.19
ARMGA	23	**0.06**	0.4	3.06	4.27E-05	0.24
MBAREA	15	**0.06**	**0.8**	**6.46**	**0.0001**	**0.76**
Balance scale						
ARMMGA	5	0.02	0.72	1.56	2.24E-06	0.35
ARMGA	15	**0.03**	0.51	**2.82**	**6.02E-06**	0.38
MBAREA	8	**0.03**	**0.77**	1.678	2.60E-06	**0.58**
Solar flare						
ARMMGA	7	**0.005**	1	236.88	5.86E-07	0.5
ARMGA	12	0.003	0.59	113.19	2.16E-06	0.6
MBAREA	10	**0.005**	0.947	**237.31**	**4.10E-06**	**0.94**
Mushroom						
ARMMGA	10	**0.028**	0.38	12.09	3.75E-07	0.40
ARMGA	29	0.018	0.47	6.26	2.29E-07	0.37
MBAREA	18	0.002	**0.97**	**12.74**	**1.47E-06**	**0.94**
Nursery						
ARMMGA	4	**0.028**	0.52	1.19	1.15E-07	0.2
ARMGA	14	0.02	0.5	3.54	**3.28E07**	0.37
MBAREA	6	0.01	**1**	**8**	1.47E-11	**1**

For all data sets, the values of average confidence, lift, interest and conditional probability of the rules generated by MBAREA are more or similar than the rules generated by other algorithms. Moreover, the rules generated by the proposed approach are not misleading or trivial.

To analyze the scalability of the proposed algorithm, several experiments have been carried out on the Nursery data set. The experiments were performed on an Intel(R) core i5-3210M CPU @2.50 GHz, 4 GB RAM running on Windows 7 Enterprise. The average runtime required by the algorithms, when the number of attributes and examples are increased is shown in Tables 4 and 5, respectively. We can see that all the algorithms scale quite linearly, however in most of the cases MBAREA takes less time than other algorithms. Some useful and interesting BARs which are generated by MBAREA are shown in Table 6. These are the rules with positive dependence among the item sets and that have maximum value of other objectives such as lift, CP and so on. For example, the rule 28 → 19, 12 of the Nursery data set in Table 6 could be interpreted as follows: the decision will be to "recommend" the application only if the financial condition of a parent is "convenient" and the number of children is "one".

Table 4. Runtime (in secs) needed for different attributes of the nursery data set

Algorithms	Number of attributes				
	8	12	20	25	32
ARMMGA	10.68	8.34	11.4	10.87	13.13
ARMGA	9.15	10.88	12.25	13.37	13.85
MBAREA	9.1	9.43	10.09	14.35	10.88

Table 5. Runtime (in secs) needed for increasing number of examples of the nursery data set

Algorithms	Number of examples				
	20 %	40 %	60 %	80 %	100 %
ARMMGA	8	9.45	10	10.4	13.13
ARMGA	3.77	9.17	10.46	13.24	13.85
MBAREA	3.09	8.32	8.58	12.84	10.88

Table 6. Rules obtained by our proposal for different data sets

Data	Rules	Conf	Lift	CP
SPECT Heart	37, 26 → 16	1	1.74	1
Mushroom	98, 86 → 34	1	1.19	1
Nursery	28 → 19, 12	1	8	1

5 Conclusion

We have proposed MBAREA, a new EA for mining a reduced set of positive BARs. The generated rules are interesting, easy to understand and maximize two objectives performance and interestingness. To accomplish this, this approach extends the existing ARMGA and ARMMGA for evolutionary learning and selection of a condition of each rule. This algorithm introduces class based mutation method to the evolutionary model and a best population technique to improve the diversity of the generated rules and to store all the non-dominated rules which are generated in the intermediate generation of a population. Analyzing the results obtained over six real world data sets, it can be concluded that the generated rules maintain the good trade-off among the number of rules, confidence, conditional probability, interest and lift values in all the data sets. Moreover, the generated rules are very strong which indicates a strong relationship between the item sets and solves the drawback of support dependent methods. Finally, the experimental results show that our proposed algorithm has a good computational cost and scales well when the problem size is increased.

References

1. Van Renesse, R., Birman, K.P., Vogels, W.: Astrolabe: a robust and scalable technology for distributed system monitoring, management, and data mining. ACM Trans. Comput. Syst. **21**(2), 164–206 (2003)

2. Maulik, U., Bandyopadhyay, S., Mukhopadhyay, A.: Multiobjective Genetic Algorithms for Clustering: Applications in Data Mining and Bioinformatics. Springer, Berlin (2011)
3. Han, J., Kamber, M.: Data Mining: Concepts and Techniques, 2nd edn. Morgan Kaufmann, Burlington (2006)
4. Aggarwal, C.C., Yu, P.S.: A new framework for itemset generation. In: PODS Conference, pp. 18–24 (1998)
5. Han, J., Pei, J., Yin, Y.: Mining frequent patterns without candidate generation. ACM SIGMOD 29(2), 1–12 (2000)
6. Yan, X., Zhang, C., Zhang, S.: Genetic algorithm-based strategy for identifying association rules without specifying actual minimum support. Expert Syst. Appl. 36(2), 3066–3076 (2009)
7. Eiben, A.E.: Introduction to Evolutionary Computing. Springer, Berlin (2003)
8. Qodmanan, H.R., Nasiri, M., Minaei-Bidgoli, B.: Multi objective association rule mining with genetic algorithm without specifying minimum support and minimum confidence. Expert Syst. Appl. 38(1), 288–298 (2011)
9. Kannimuthu, S., Premalatha, K.: Discovery of high utility itemsets using genetic algorithm with ranked mutation. Appl. Artif. Intell. 28(4), 337–359 (2014)
10. Yan, X., Zhang, C., Zhang, S.: ARMGA: identifying interesting association rules with genetic algorithms. Appl. Artif. Intell. Int. J. 19(7), 677–689 (2005)
11. Martin, D., Rosete, A., Alcala-Fdez, J., Herrera, F.: A new multiobjective evolutionary algorithm for mining a reduced set of interesting positive and negative quantitative association rules. IEEE Trans. Evol. Comput. 18(1), 54–69 (2014)
12. Ampan, A.C.: A programming interface for medical diagnosis prediction. Artif. Intell. LI(1), 21–30 (2006)
13. del Jesus, M.J., Gámez, J.A., González, P., Puerta, J.M.: On the discovery of association rules by means of evolutionary algorithms. Wiley Interdisc. Rev. Data Min. Knowl. Discov. 1(5), 397–415 (2011)
14. Zhou, L., Yau, S.: Efficient association rule mining among both frequent and infrequent items. Comput. Math Appl. 54(6), 737–749 (2007)
15. Mukhopadhyay, A., Maulik, U., Bandyopadhyay, S., Coello, C.A.C.: A survey of multiobjective evolutionary algorithms for data mining: part i. IEEE Trans. Evol. Comput. 18(1), 4–19 (2014)
16. Piatetsky-Shapiro, G.: Discovery, analysis, and presentation of strong rules. In: Knowledge Discovery in Databases, pp. 229–248. AAAI/MIT Press, Menlo Park (1991)
17. Wakabi-Waiswa, P.P., Baryamureeba, V.: Extraction of interesting association rules using genetic algorithms. Int. J. Comput. ICT Res. 2(1), 101–110 (2008)

Non-negative Spectral Learning for Linear Sequential Systems

Hadrien Glaude[1,2]([⊠]), Cyrille Enderli[1], and Olivier Pietquin[2,3]

[1] Thales Airborne Systems, Elancourt, France
[2] University Lille, CRIStAL, UMR 9189, SequeL Team, Villeneuve d'Ascq, France
hadrien.glaude@gmail.com
[3] Institut Universitaire de France (IUF), Paris, France

Abstract. Method of moments (MoM) has recently become an appealing alternative to standard iterative approaches like Expectation Maximization (EM) to learn latent variable models. In addition, MoM-based algorithms come with global convergence guarantees in the form of finite sample bounds. However, given enough computation time, by using restarts and heuristics to avoid local optima, iterative approaches often achieve better performance. We believe that this performance gap is in part due to the fact that MoM-based algorithms can output negative probabilities. By constraining the search space, we propose a non-negative spectral algorithm (NNSpectral) avoiding computing negative probabilities by design. NNSpectral is compared to other MoM-based algorithms and EM on synthetic problems of the PAutomaC challenge. Not only, NNSpectral outperforms other MoM-based algorithms, but also, achieves very competitive results in comparison to EM.

1 Introduction

Traditionally, complex probability distributions over structured data are learnt through generative models with latent variables (*e.g* Hidden Markov Models (HMM)). Maximizing the likelihood is a widely used approach to fit a model to the gathered observations. However, for many complex models, the likelihood is not convex. Thus, algorithms such as Expectation Maximization (EM) and gradient descent, which are iterative procedures, converge to local minima. In addition to being prone to get stuck into local optima, these algorithms are computationally expensive, to the point where obtaining good solutions for large models becomes intractable. A recent alternative line of work consists in designing learning algorithms for latent variable models exploiting the so-called Method of Moments (MoM). The MoM leverages the fact that low order moments of distributions contain most of the distribution information and are typically easy to estimate. The MoM have several pros over iterative methods. It can provide extremely fast learning algorithms as estimated moments can be computed in a time linear in the number of samples. MoM-based algorithms are often consistent with theoretical guarantees in form of finite-sample bounds. In addition, these algorithms are able to learn a large variety of models [1,3,11,15].

© Springer International Publishing Switzerland 2015
S. Arik et al. (Eds.): ICONIP 2015, Part II, LNCS 9490, pp. 143–151, 2015.
DOI: 10.1007/978-3-319-26535-3_17

In a recent work, [15] showed that numerous models are encompassed into the common framework of linear Sequential Systems (SSs) or equivalently Multiplicity Automata (MA). Linear SSs represent real functions over a set of words $f_{\mathcal{M}} : \Sigma^{\star} \to \mathbb{R}$, where Σ is an alphabet. In particular, they can be used to represent probability distributions over sequences of symbols. Although, for the sake of clarity, we focus on learning stochastic rational languages (defined in Sect. 2), our work naturally extends to other equivalent or encompassed models: Predictive State Representations (PSRs), Observable Operator Models (OOMs), Partially Observable Markov Decision Processes (POMDPs) and HMMs.

Beyond all the appealing traits of MoM-based algorithms, a well-known concern is the negative-probabilities problem. Actually, most of MoM-based algorithms do not constrain the learnt function to be a distribution. For some applications requiring to learn actual probabilities, this is a critical issue. For example, a negative and unnormalized measure does not make sense when computing expectations. Sometimes, an approximation of a probability distribution is enough. For instance, computing the Maximum a Posteriori (MAP) only requires the maximum to be correctly identified. But even for these applications, we will show that constraining the learned function to output non-negative values helps the learning process and improves the model accuracy. A second concern is the observed performance gap with iterative algorithms. Although they usually need restarts and other heuristics to avoid local minima, given enough time to explore the space of parameters, they yield to very competitive models in practice. Recently, an empirical comparison [5] have shown that MoM-based algorithms still perform poorly in comparison to EM.

In this paper, we propose a new MoM-based algorithm, called non-negative spectral learning (NNSpectral) that constraints the output function to be non-negative. Inspired by theoretical results on MA defined on non-negative commutative semi-rings, NNSpectral uses Non-negative Matrix Factorization (NMF) and Non-Negative Least Squares (NNLS). An empirical evaluation on the same twelve synthetic problems of the PAutomaC challenge used by [5] allows us fairly comparing NNSpectral to three other MoM-based algorithms and EM. Not only, NNSpectral outperforms previous MoM-based algorithms, but also, it is the first time to our knowledge that a MoM-based algorithm achieves competitive results in comparison to EM.

2   Multiplicity Automata

2.1   Definition

Let Σ be a set of symbols, also called an alphabet. We denote by $\Sigma^{\star}$, the set of all finite words made of symbols of Σ, including the empty word ε. Words of length k form the set Σ^{k}. Let u and $v \in \Sigma^{\star}$, uv is the concatenation of the two words and $u\Sigma^{\star}$ is the set of finite words starting by u. In the sequel, K is a commutative semi-ring, in particular $\mathbb{R}$ or $\mathbb{R}^{+}$. We are interested in mapping of $\Sigma^{\star}$ into K called formal power series. Let f be a formal power series, for a set of words S, we define $f(S) = \sum_{u \in S} f(u) \in K$. Some formal power series can be represented by compact models, called MA.

Definition 1 (Multiplicity Automaton). *Let K be a semi-ring, a K-multiplicity automaton (K-MA) with n states is a structure $\langle \Sigma, Q, \{A_o\}_{o \in \Sigma}, \alpha_0, \alpha_\infty \rangle$, where Σ is an alphabet and Q is a finite set of states. Matrices $A_o \in K^{n \times n}$ contain the transition weights. The vectors $\alpha_\infty \in K^n$ and $\alpha_0 \in K^n$ contain respectively the terminal and initial weights. A K-MA $\mathcal{M}$ defines a rational language $r_\mathcal{M} : \Sigma^* \to \mathbb{R}$,*

$$r_\mathcal{M}(u) = r_\mathcal{M}(o_1 \ldots o_k) = \alpha_0^\top A_u \alpha_0 = \alpha_0^\top A_{o_1} \ldots A_{o_k} \alpha_\infty.$$

A function f is said *realized* by a K-MA $\mathcal{M}$, if $r_\mathcal{M} = f$.

Definition 2 (Rational Language). *A formal power series r is rational over K iff it is realized by a K-MA.*

Two K-MA that define the same rational language are *equivalent*. A K-MA is *minimal* if there is not an equivalent K-MA with strictly fewer states. An important operation on K-MA is the *conjugation* by an invertible matrix. Let $R \in K^{n \times n}$ be invertible and $\mathcal{M} = \langle \Sigma, Q, \{A_o\}_{o \in \Sigma}, \alpha_0, \alpha_\infty \rangle$ be a K-MA of dimension n, then

$$\mathcal{M}' = \left\langle \Sigma, Q, \{R^{-1} A_o R\}_{o \in \Sigma}, R^\top \alpha_0, R^{-1} \alpha_\infty \right\rangle,$$

defines a conjugated K-MA. In addition, $\mathcal{M}$ and $\mathcal{M}'$ are equivalent. In this paper, we are interested in learning languages representing distributions over finite-length words that can be compactly represented by a K-MA. When $K = \mathbb{R}$, this forms the class of stochastic rational language.

Definition 3 (Stochastic Rational Language). *A stochastic rational language p is a rational language with values in $\mathbb{R}^+$ such that $\sum_{u \in \Sigma^*} p(u) = 1$.*

Stochastic rational languages are associated to the following subclass of $\mathbb{R}$-MA.

Definition 4 (Stochastic Multiplicity Automaton). *A stochastic multiplicity automaton (SMA) $\mathcal{M}$ is a $\mathbb{R}$-MA realizing a rational stochastic language.*

2.2 Hankel Matrix Representation

Let $f : \Sigma^* \to K$ be a formal power series, we define $H_f \in K^{\Sigma^* \times \Sigma^*}$ the bi-infinite Hankel matrix whose rows and columns are indexed by Σ^* such that $H_f[u, v] = f(uv)$,

$$H_f = \begin{array}{c} \\ \varepsilon \\ a \\ b \\ aa \\ \\ \end{array} \begin{array}{c} \varepsilon \quad\quad a \quad\quad\; b \quad\quad\; aa \quad\quad \cdots \\ \begin{pmatrix} f(\varepsilon) & f(a) & f(b) & f(aa) & \cdots \\ f(a) & f(aa) & f(ab) & f(aaa) & \cdots \\ f(b) & f(ba) & f(bb) & f(baa) & \cdots \\ f(aa) & f(aaa) & f(aab) & f(aaaa) & \cdots \\ \vdots & \vdots & \vdots & \vdots & \ddots \end{pmatrix} \end{array}$$

When f is a stochastic language, H_f contains occurring probabilities that can be estimated from samples by empirical counts. Details about matrices defined over semi-rings can be found in [12]. We note H the Hankel matrix when its formal power series can be inferred from the context. Let for all $o \in \Sigma$, $H_o \in K^{\Sigma^* \times \Sigma^*}$, $\mathbf{h}_S \in K^{\Sigma^*}$ and $\mathbf{h}_P \in K^{\Sigma^*}$ be such that $H_o(u,v) = f(uov)$, $\mathbf{h}_S(u) = \mathbf{h}_P(u) = f(u)$. These vectors and matrices can be extracted from H. The Hankel representation of formal series lies in the heart of all MoM-based learning algorithms, because of the following fundamental theorem.

Theorem 1 (See [7]). *Let r be a rational language over K and $\mathcal{M}$ a K-MA with n states that realizes it, then rank$(H_r) \leq n$. Conversely, if the Hankel matrix H_r of a formal power series r has a finite rank n, then r is a rational language over K and can be realized by a minimal K-MA with exactly n states.*

Note that, the original proof assumes that K is a field but remains true when K is a commutative semi-ring, as ranks, determinants and inverses are well defined in semi-modules. In addition, the proof gives also the construction of H from a K-MA and *vice-versa*. For a K-MA with n states, observe that $H[u,v] = (\boldsymbol{\alpha}_0^\top A_u)(A_v \boldsymbol{\alpha}_\infty)$. Let $P \in K^{\Sigma^* \times n}$ and $S \in K^{n \times \Sigma^*}$ be matrices defined as follows,

$$P = ((\boldsymbol{\alpha}_0^\top A_u)^\top)_{u \in \Sigma^*}^\top, \qquad\qquad S = (A_v \boldsymbol{\alpha}_\infty)_{v \in \Sigma^*},$$

then $H = PS$. Moreover, we have that,

$$H_o = PA_oS, \quad \mathbf{h}_S^\top = \boldsymbol{\alpha}_0^\top S, \quad \mathbf{h}_P = P\boldsymbol{\alpha}_\infty. \tag{1}$$

So the K-MA parameters can be recovered by solving Eq. 1. Hopefully, we do not need to consider the bi-infinite Hankel matrix to recover the underlying K-MA. Given a basis $\mathcal{B} = (\mathcal{P}, \mathcal{S})$ of prefixes and suffixes, we denote by $H_\mathcal{B}$ the sub-block of H. A basis $\mathcal{B}$ is *complete* if $H_\mathcal{B}$ has the same rank than H. A basis is *suffix-closed* if $\forall u \in \Sigma^*, \forall o \in \Sigma, ou \in \mathcal{S} \Rightarrow u \in \mathcal{S}$. In [4], the author shows that if $\mathcal{B} = (\mathcal{P}, \mathcal{S})$ is a suffix-closed complete basis, by defining P over $\mathcal{P}$, S over $\mathcal{S}$ and H over $\mathcal{B}$, we can recover a MA using Eq. 1.

2.3 Spectral Learning

This section reviews the Spectral Learning algorithm to learn $\mathbb{R}$-MA from samples generated by a stochastic rational language. In the literature, severals methods are used to build a suffix-closed basis from data. For example, one can use all prefixes and suffixes that appear in the training set. In addition, we require that sets of prefixes and suffixes contain the empty word ε. Once a basis is chosen, the Spectral algorithm first estimates the probabilities in $H_\mathcal{B}$ by empirical counts. Then, it recovers a factorized form of $H_\mathcal{B} = UDV^\top$ through a truncated Singular Value Decomposition (SVD). Finally, setting $P = UD$ and $S = V^\top$, the algorithm solves Eq. 1. More precisely, let $\mathbf{1}_\varepsilon^S$ and $\mathbf{1}_\varepsilon^P$ be vectors filled with 0s with a single 1 at the index of the empty word in the basis, we have that $\mathbf{h}_S = (\mathbf{1}_\varepsilon^P)^\top H_\mathcal{B}$, $\mathbf{h}_P = H_\mathcal{B} \mathbf{1}_\varepsilon^S$. Let, for all $o \in \Sigma^*$, $T_o \in \mathbb{R}^{|S| \times |S|}$ be matrices such that, $T_o[u,v] = \delta_{u=ov}$ then we have that $H_o = H_{\mathcal{B}}T_o$. Using these identities in Eq. 1, one obtains Algorithm 1. In the experiments, following the advices of [8], we normalized the feature-variance of the coefficients of the Hankel matrix by independently scaling each row and column by a factor $c_u = \sqrt{|\mathcal{S}|/(\#u + 5)}$, where $\#u$ is the number of occurrences of u. In addition, depending on the problem, it can be better to work with other series derived from p. For example, the substring-based series $p^{substring}(u) = \sum_{w,v \in \Sigma^*} p(wuv)$ is related to p^{string}. According to [4], if p^{string} is realized by a SMA, then $p^{substring}$ is too. In addition, he provides an explicit conversion between string-based SMA and substring-based SMA preserving the number of states. For all algorithms compared in Sect. 4, we used the series leading to the best results for each problem.

Algorithm 1. Spectral algorithm for $\mathbb{R}$-MA.

---

1. Choose a set of prefixes $\mathcal{P} \subset \Sigma^*$ and suffixes $\mathcal{S} \subset \Sigma^*$ both containing ε
2. Using $\mathcal{S}$, build the matrices T_o for all $o \in \Sigma$ such that $T_o[u,v] = \delta_{u=ov}$
3. Estimate $H_{\mathcal{B}}$ by empirical counts.
4. $U, D, V = \mathcal{SVD}_n(H_{\mathcal{B}})$ using the truncated SVD, where n is a parameter of the algorithm
5. For all $o \in \Sigma$ do $A_o = V^\top T_o V$
6. $\alpha_0^\top = (1_\varepsilon^P)^\top H_{\mathcal{B}} V$
7. $\alpha_\infty = V^\top 1_\varepsilon^S$

---

3   Non-negative Spectral Learning

As mentioned in the introduction, Algorithm 1 is designed to learn $\mathbb{R}$-MA and is very unlikely to return a SMA. This is a major drawback when a probability distribution is required. In this case, one has to rely on heuristics like thresholding the values to be contained in $[0, 1]$, which introduces errors in predictions. A natural enhancement of the Spectral algorithm would be to constraint the return model to be a SMA. Unfortunately, this is not likely to be feasible due to the underlying complexity between $\mathbb{R}$-MA and SMA. Indeed, although $\mathbb{R}$-MA are strictly more general than SMA, checking whether a $\mathbb{R}$-MA is stochastic is undecidable [9]. In terms of algorithms, constraining the return model to be a SMA would require adding an infinite number of constraints. As a matter of fact, a $\mathbb{R}$-MA realizing a language r requires the non-negativity ($\forall u \in \Sigma^*, r(x) \geq 0$) and the convergence to 1 ($\lim_{L \to +\infty} \sum_{l=0}^{L} \sum_{u \in \Sigma^l} r(u) = 1$) of the series to be a SMA. The undecidability comes only from the non-negativeness. Note that only the existence of the limit is really required as a convergent series can always be normalized. Thus, we propose to restrict learning to $\mathbb{R}^+$-MA, which by definition produces non-negative series. Although, SMA are not included in $\mathbb{R}^+$-MA, there are $\mathbb{R}^+$-MA realizing probability distributions and called Probabilistic Finite Automata (PFA). Relations between all these classes of MA are summed up in Fig. 1.

Definition 5 (Probabilistic (Deterministic) Finite Automaton). *A probabilistic finite automaton (PFA)* $\mathcal{M} = \langle \Sigma, Q, \{A_o\}_{o \in \Sigma}, \alpha_0, \alpha_\infty \rangle$ *is a SMA with non-negative weights verifying* $\mathbf{1}^\top \alpha_0 = 1$, $\alpha_\infty + \sum_{o \in \Sigma} A_o \mathbf{1} = 1$. *The weights of PFA are in* $[0, 1]$ *and can be viewed as probabilities over transitions, initial states and terminal states. A Probabilistic Deterministic Infinite Automata (PDFA) if a PFA with deterministic transitions.*

Thus, in contrast to the Spectral algorithm which returns a $\mathbb{R}$-MA, NNSpectral returns a $\mathbb{R}^+$-MA avoiding the non-negativity probability problem. The NNSpectral algorithm, given in Algorithm 2, also uses the decomposition in Theroem 1, but applied to MA defined on the semi-ring $\mathbb{R}^+$ instead of the field $\mathbb{R}$. Thus, there exists a low-rank factorization with non-negative factors of the Hankel matrix representing a $\mathbb{R}^+$-MA. Finding such a decomposition is a well-known problem, called NMF. It aims at finding a low rank decomposition of a given non-negative data matrix $H_\mathcal{B} \in \mathbb{R}^{n \times m}$ such that $\|H_\mathcal{B} - PS\|_F$ where $P \in \mathbb{R}^{n \times r}$ and $S \in \mathbb{R}^{r \times m}$ are component-wise non-negative and $\|\cdot\|_F$ is the Frobenius norm. Unfortunately, NMF is NP-Hard in general [16] and the decomposition is not unique which makes the problem ill-posed. Hence, in practice, heuristic algorithms are used which have only the guarantee to converge to a stationary point. Most of them run in $\mathcal{O}(nmr)$. We refer the reader to [10] for a comprehensive survey of these algorithms. In the experiments, the alternating non-negative least squares algorithm has shown to be a good trade-off between convergence speed and quality of the approximation. This algorithm iteratively optimizes P and S by solving NNLS problems. NNLS is a non-negative constrained version of the least squares method. Being equivalent to a quadratic programming problem under linear constraints, it is convex and algorithms converge to the optimum. In the experiments, NNLS problems were solved using the projected gradient algorithm of [14]. So, given a non-negative factorization of $H_\mathcal{B}$, solving Eq. 1 is done by NNLS to ensure the non-negativity of the weights. Although NNSpectral cannot learn SMA (or the equivalent OOMs, PSRs), a SMA can be arbitrarily well approximated by a PFA (or the equivalent HMMs, POMDPs) with a growing number of states [13].

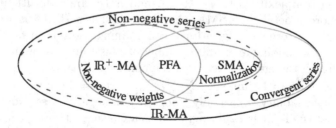

Fig. 1. Solid lines are polynomially decidable. Dashed line are undecidable. See [2,9].

Algorithm 2. NNSpectral algorithm for $\mathbb{R}^+$-MA

1. Choose a set of prefixes $\mathcal{P} \subset \Sigma^*$ and suffixes $\mathcal{S} \subset \Sigma^*$ both containing ε
2. Using $\mathcal{S}$, build the matrices T_o for all $o \in \Sigma$ such that $T_o[u,v] = \delta_{u=ov}$
3. Estimate $H_\mathcal{B}$ by empirical counts.
4. $P, S \leftarrow \text{argmin}_{PS} \|H_\mathcal{B} - PS\|_F$ s.t. $P \in \mathbb{R}^{|\mathcal{P}| \times n}, S \in \mathbb{R}^{n \times |\mathcal{S}|} \geq 0$ $\triangleright$ by NMF
5. For all $o \in \Sigma$ do $A_o \leftarrow \text{argmin}_{A \geq 0} \|AS - ST_o\|_F$ $\triangleright$ by NNLS
6. $\alpha_0 \leftarrow \text{argmin}_{\alpha \geq 0} \|\alpha^\top S - (\mathbf{1}_\varepsilon^P)^\top H_\mathcal{B}\|_F$ $\triangleright$ by NNLS
7. $\alpha_\infty \leftarrow S \mathbf{1}_\varepsilon^S$

4   Numerical Experiments

The Probabilistic Automata learning Competition (PAutomaC) is dealing with the problem of learning probabilistic distributions from strings drawn from finite-state automata. From the 48 problems available, we have selected the same twelve problems than in [5], to provide a fair comparison with other algorithms. The generating model can be of three kinds: PFA, HMMs or PDFA. Four models have been selected from each class. A detailed description of each problem can be found in [17]. Table 1 compares the best results of NNSpectral between learning from strings or substrings to EM and the best results among the following MoM-based algorithms: CO [6] using strings, Tensor [1] using strings and Spectral using strings and substrings. A description and comparison of these algorithms can be found in [5].

The quality of a model can be measured by the quality of the probability distribution it realizes. The objective is to learn a MA realizing a series p close

Table 1. Comparison with other algorithm for Perplexity (left table) and WER (right table). "MoM" stands for the best of MoM-based algorithms. Model sizes are listed in parentheses.

ID	Perplexity				WER			
	NNSpectral	EM	MoM	True model	NNSpectral	EM	MoM	True model
HMM 1	**30.54**(64)	500.10	44.77	29.90(63)	72.7(30)	75.7	**71.3**	68.8(63)
14	116.98(11)	**116.84**	128.53	116.79(15)	**68.8**(7)	68.8	70.0	68.4(15)
33	32.21(14)	**32.14**	49.22	31.87(13)	**74.3**(6)	74.3	76.7	74.1(13)
45	**24.08**(2)	107.75	31.87	24.04(14)	78.24(2)	**78.1**	80.1	78.1(14)
PDFA 6	76.99(28)	**67.32**	95.12	66.98(19)	**47.1**(21)	47.4	50.2	46.9(19)
7	**51.26**(12)	51.27	62.74	51.22(12)	48.41(42)	**48.1**	50.6	48.3(12)
27	**43.81**(46)	94.40	102.85	42.43(19)	**73.9**(20)	83.0	75.5	73.0(19)
42	**16.12**(20)	168.52	23.91	16.00(6)	**56.6**(8)	58.1	61.4	56.6(6)
PFA 29	25.24(35)	**25.09**	34.57	24.03(36)	47.6(36)	49.2	**47.3**	47.2(36)
39	**10.00**(6)	10.43	11.24	10.00(6)	**59.4**(19)	63.3	62.0	59.3(6)
43	**32.85**(7)	461.23	36.61	32.64(67)	**76.8**(25)	77.4	78.0	77.1(67)
46	12.28(44)	**12.02**	25.28	11.98(19)	**78.0**(20)	77.5	79.4	77.3(19)

to the distribution $p_\star$, which generated the training set $\mathcal{T}$. The quality of p is measured by the perplexity that corresponds to the average number of bits needed to represent a word using the optimal code given by $p_\star$.

$$\text{Perplexity}(\mathcal{M}) = 2^{-\sum_{u \in \mathcal{T}} p_\star(u) \log(p_\mathcal{M}(u))}$$

The quality of model can also be evaluated by the Word Error Rate (WER) metric. It measures the fraction of incorrectly one-step-ahead predicted symbols.

In the simulations, we perform a grid search to find the optimal rank for each of the performance metric. For each problem, the best algorithm is indicated by a bold number and the best score between NNSpectral and other MoM-based algorithms is underlined. The score of the true model is reported for comparison. For the perplexity, NNSpectral outperforms other MoM-based algorithm on the 12 problems and does better that EM for 7 problems. For the other 5 problems, NNSpectral achieves performances very close to EM and to the true model. For the WER metric, NNSpectral beats other MoM-based algorithms on 10 problems and scores at top on 7 problems. Also, for all problems NNSpectral produces perplexity close to the optimal ones, whereas, on 5 problems EM fails to produce an acceptable solution.

5 Conclusion

In this paper, we proposed a new algorithm inspired by the theoretical developments of MA defined on semi-rings. NNSpectral works by constraining the search space to ensure non-negativity of the rational language. Like other MoM-based algorithms, NNSpectral is able to handle large-scale problems much faster than EM. Here, the consistency is lost due to the use of heuristics to solve the NMF problem. Experimentally, this does not seem to be a major problem because NNSpectral outperforms other MoM-based algorithms and is competitive with EM that sometimes produces aberrant solutions. Thus, NNSpectral provides a good alternative to the EM algorithm with a much lower computational cost. In further works, we would like to investigate how using NNSpectral to initialize an EM algorithm and how adding constraints in NNSpectral to ensure the convergence of the series. In addition, relations with NMF could be further exploit to produce tensor [1] or kernel-based algorithms.

References

1. Anandkumar, A., Ge, R., Hsu, D., Kakade, S.M., Telgarsky, M.: Tensor decompositions for learning latent variable models (2012). arXiv preprint arXiv:1210.7559
2. Bailly, R., Denis, F.: Absolute convergence of rational series is semi-decidable. Inf. Comput. **209**(3), 280–295 (2011)
3. Bailly, R., Habrard, A., Denis, F.: A spectral approach for probabilistic grammatical inference on trees. In: Hutter, M., Stephan, F., Vovk, V., Zeugmann, T. (eds.) Algorithmic Learning Theory. LNCS, vol. 6331, pp. 74–88. Springer, Heidelberg (2010)

4. Balle, B.: Learning finite-state machines: algorithmic and statistical aspects. Ph.D. thesis (2013)
5. Balle, B., Hamilton, W., Pineau, J.: Methods of moments for learning stochastic languages: unified presentation and empirical comparison. In: Proceedings of ICML-14, pp. 1386–1394 (2014)
6. Balle, B., Quattoni, A., Carreras, X.: Local loss optimization in operator models: a new insight into spectral learning. In: Proceedings of ICML-12 (2012)
7. Carlyle, J.W., Paz, A.: Realizations by stochastic finite automata. J. Comput. Syst. Sci. 5(1), 26–40 (1971)
8. Cohen, S.B., Stratos, K., Collins, M., Foster, D.P., Ungar, L.H.: Experiments with spectral learning of latent-variable PCFGs. In: Proceedings of HLT-NAACL-13, pp. 148–157 (2013)
9. Denis, F., Esposito, Y.: On rational stochastic languages. Fundamenta Informaticae 86(1), 41–77 (2008)
10. Gillis, N.: The Why and How of nonnegative matrix factorization. ArXiv e-prints, January 2014
11. Glaude, H., Pietquin, O., Enderli, C.: Subspace identification for predictive state representation by nuclear norm minimization. In: Proceedings of ADPRL-14 (2014)
12. Guterman, A.E.: Rank and determinant functions for matrices over semirings. In: Young, N., Choi, Y. (eds.) Surveys in Contemporary Mathematics, pp. 1–33. Cambridge University Press, Cambridge (2007)
13. Gybels, M., Denis, F., Habrard, A.: Some improvements of the spectral learning approach for probabilistic grammatical inference. In: Proceedings of ICGI-12, vol. 34, pp. 64–78 (2014)
14. Lin, C.J.: Projected gradient methods for nonnegative matrix factorization. Neural Comput. 19(10), 2756–2779 (2007)
15. Thon, M., Jaeger, H.: Links between multiplicity automata, observable operator models and predictive state representations – a unified learning framework learning framework. J. Mach. Learn. Res. (2015, to appear)
16. Vavasis, S.A.: On the complexity of nonnegative matrix factorization. SIAM J. Optim. 20(3), 1364–1377 (2009)
17. Verwer, S., Eyraud, R., de la Higuera, C.: Results of the pautomac probabilistic automaton learning competition. J. Mach. Learn. Res. - Proc. Track 21, 243–248 (2012)

Optimism in Active Learning
with Gaussian Processes

Timothé Collet[1,2]([✉]) and Olivier Pietquin[3]

[1] MaLIS Research group, CentraleSupélec, Metz, France
timothe.collet@centralesupelec.fr
[2] GeorgiaTech-CNRS UMI 2958, Metz, France
[3] IUF (Institut Universitaire de France), University Lille - CRIStAL Lab
(UMR 9189), Lille, France
olivier.pietquin@univ-lille1.fr

Abstract. In the context of Active Learning for classification, the classification error depends on the joint distribution of samples and their labels which is initially unknown. Online estimation of this distribution, for the purpose of minimizing the error, involves a trade-off between exploration and exploitation. This is a common problem in machine learning for which multi-armed bandit theory, building upon the paradigm of Optimism in the Face of Uncertainty, has been proven very efficient. We introduce two novel algorithms that use Optimism in the Face of Uncertainty along with Gaussian Processes for the Active Learning problem. Evaluations lead on real-world datasets show that these new algorithms compare positively to state-of-the-art methods.

Keywords: Active learning · Classification · Optimism in the face of uncertainty · Gaussian processes

1 Introduction

Classification is a supervised learning framework consisting in learning a mapping from instances to labels, existing in finite number. It uses a training set, being composed of instances already associated to labels. In order to get a high prediction accuracy, a large amount of instances should be available in the training set. In the recent years, it has become easier to collect, store and process large data sets. However, the association of each instance of the training set to labels requires manual annotations by an expert which is time-consuming.

In Active Learning, the goal is to minimize the number of requests to the expert needed to achieve a targeted classification performance. Or, equivalently, to maximize its performance under a fixed budget of requests. Hence, such an algorithm dynamically builds the training set by sequentially deciding which instance to present to the expert. The instance is chosen considering all the previously received labels, such that its inclusion in the training set leads to the best classifier performance. In the sequel, we work under the pool-based

© Springer International Publishing Switzerland 2015
S. Arik et al. (Eds.): ICONIP 2015, Part II, LNCS 9490, pp. 152–160, 2015.
DOI: 10.1007/978-3-319-26535-3_18

sampling scheme in which a set of unlabeled instances is available. The Active Learning algorithm successively selects instances from the pool to be labeled.

Among the many ways existing to formulate this problem [8], this work focuses on the Error Minimization class of methods. In this framework, the selection criterion is derived from the true risk, or misclassification rate. It is thus directly linked to the (mostly used) measure of performance for classifiers (without using a proxy as often). We distinguish two strategies in this framework: the function to be minimized is either the maximum or the averaged true risk across the instance space. In the former, the strategy is to estimate the true risk related to each unlabeled instance and to sample the one for which it is maximum. In the latter, the strategy is to simulate the sampling of each instance and to effectively sample the one resulting in the maximum decrease of the true risk. This second strategy has the advantage of representativeness [3], which means that an instance surrounded by unlabeled data is preferred to an isolated one.

The true risk cannot be known exactly, since it requires the knowledge of the true label distribution over the whole instance space which has to be learned. But it can be estimated, along with a measure of uncertainty, using the labels received so far. Among the methods that exist in Error Minimization, the difference lies in how to manage the uncertainty on the true risk. In [9], the true risk is directly replaced by its estimation. In [4,5], the binary loss is expected over the posterior distribution of class proportion. In [3], a min-max approach is used to ensure a minimum decrease of the risk.

In this paper, we propose a new solution to manage the uncertainty in the Error Minimization problem based on the Optimism in the Face of Uncertainty paradigm. Indeed, querying an instance may be used in order to directly increase the classifier performance, but also to improve future decisions. This relates to the exploration/exploitation dilemma for which the Optimism in the Face of Uncertainty is a well-renown approach, particularly for finite budget problems. Optimism in the Face of Uncertainty has already been successfully applied to Active Learning. In [2], the authors focused on the reduction of the version space. In [1], the instance space is partitioned and an allocation strategy is defined.

As for many of such methods, ours relies on the establishment of a probability bound on the value of the ideal criterion. Then it selects the instance for which the bound is maximum. We apply this solution to Gaussian Processes. Indeed, they provide a simple Bayesian learning framework which outputs a probability distribution on the conditional density of the class labels. It is thus well fitted for Active Learning, and particularly for Optimistic methods.

We briefly review Gaussian Processes in Sect. 2. Then, we present two novel algorithms: OLRM (Optimistic Local Risk Minimization) and OGRM (Optimistic Global Risk Minimization) for both strategies in Sects. 3 and 4. Experiments and results are provided in Sects. 5 and 6.

2   Gaussian Processes

A stochastic process is a generalization of a random vector to functions, each input is associated with a probability distribution on the output space. In the case of Gaussian Processes (GP), this distribution is Normal and can therefore be characterized only by its mean and variance. It thus provides a simple Bayesian framework for the supervised learning problem.

Let us denote the instance space by X and the label set by Y. In binary classification, the label set is composed of two elements, here $Y = \{-1, 1\}$. In the case of noisy classification, the expert decisions about labels can be modeled by a Bernoulli distribution:

$$P(y \in Y = 1 | x \in X) = \frac{\mu(x) + 1}{2},$$

where $\mu(x)$ is the mean label for instance x. We use a GP to estimate the values of $\mu(x)$ from S.

Let us denote $\mathcal{D}_p$ the set containing all the instances from the pool, it is split into $\mathcal{D}_l$ and $\mathcal{D}_u$, respectively the sets of labeled and unlabeled examples of size n_l and n_u. Suppose that we are given a training set $S = \{(x_i, y_i)\}_{i \in [\![1, n_l]\!]}$ containing all the instances in $\mathcal{D}_l$ and the associated labels given by the expert.

A GP makes use of a kernel function $k(.,.)$ to specify the covariance between outputs. Let $K_{l,l} = [k(x_i, x_j)]_{n_l \times n_l}$ be the covariance matrix of labeled instances. Let $K_{x,l} = [k(x, x_i)]_{x_i \in \mathcal{D}_l}$ be the covariance of an instance $x \in X$ and the labeled instances.

If the underlying expert distribution is reduced to Gaussian $N(\mu(x), \sigma^2)$, conditioning the joint Gaussian distribution of $\mu(x)$ on the observed labels gives the following joint posterior distribution:

$$\mu(x) | S \sim \mathcal{N}(m_S(x), \sigma_S^2(x)).$$

with $m_S(x) = K_{x,l}(K_{l,l} + \sigma^2 I)^{-1} \mathbf{y}_l$ and $\sigma_S^2(x) = k(x, x) - K_{x,l}(K_{l,l} + \sigma^2 I)^{-1} K_{x,l}{}^\top$.

The GP classifier predicts labels according to the sign of the posterior distribution's mean, according to Eq. (1). Note that this mean is also the solution of the Regularized Least Square Regression (RLSR). The GP adds a confidence measure to this. It is particularly useful in Active Learning, as it provides hints about the region of the instance space where the model should be refined.

$$l(x) = sign(m_S(x)). \tag{1}$$

3   Local Risk Minimization

In this section, we propose OLRM, an algorithm based on Optimism in the Face of Uncertainty that minimizes the maximum local risk of the classifier. We first define the loss function, then derive a criterion that tends to minimize it.

Suppose that we are given a budget n. At each time step $t \in [\![1, n]\!]$, the Active Learning algorithm selects an unlabeled instance $x_t \in \mathcal{D}_u$, submits it to the expert, receives a label y_t and adds the pair to the training set. The sets of labeled and unlabeled instances automatically adapt to the reality of instances, and should be indexed by t. But we prefer to avoid overloading notations.

At any time step t, the current training set $\mathcal{S}_t$ is used to train the GP. This one is then used to predict the label $l_t(x)$ for any instance $x \in X$ from Eq. (1). The expected prediction error, or local risk, is the probability that the expert would give a different label:

$$r_t(x) = \begin{cases} \frac{1+\mu(x)}{2} & \text{if } l_t(x) = -1 \\ \frac{1-\mu(x)}{2} & \text{if } l_t(x) = +1. \end{cases} \tag{2}$$

We want to minimize the maximum risk one would encounter by presenting an unknown instance $x \in X$. We therefore define the following loss:

$$L_t = \arg \max_{x \in X} r_t(x). \tag{3}$$

In order to minimize this kind of loss, the solution is to select the instance for which the local risk is maximum.

$$x_{t+1} = \arg \max_{x \in \mathcal{D}_u} r_t(x).$$

Indeed, sampling an instance will necessarily lower the risk at its location, sampling where the maximum risk is attained guarantees to decrease this maximum.

However, this loss cannot be used as such. Indeed, this supposes knowing the true value of $\mu(x)$ at least for every unlabeled instances, which is unrealistic. If it was known, prediction would be trivial. We propose to use Optimism in the Face of Uncertainty in order to approach this solution estimating only a distribution over $\mu(x)$ by a GP. In this paradigm, a confidence bound on the ideal criterion is established relatively to a probability δ, then its upper bound is used as a heuristic. We show how to use the distribution of $\mu(x)$ to derive the distribution of the local risk.

Suppose that the label $l_t(x) = -1$ is predicted. Then,

$$\mathbb{P}(r_t(x) \geq \epsilon_t(x)) = \mathbb{P}(\mu(x) \geq 2\epsilon_t(x) - 1).$$

The GP trained with labeled instances outputs the following distribution:

$$\mu(x)|\mathcal{S}_t \sim \mathcal{N}(m_{\mathcal{S}_t}(x), \sigma^2_{\mathcal{S}_t}(x)).$$

Thus, with Φ the cumulative distribution function of the Normal distribution,

$$\mathbb{P}(r_t(x) \geq \epsilon_t(x)) = \delta \iff 2\epsilon_t(x) - 1 = \Phi^{-1}(\delta, m_{\mathcal{S}_t}(x), \sigma^2_{\mathcal{S}_t}(x)).$$

This applies symmetrically for $l_t(x) = 1$. Let us remind that $l_t(x) = sign$ $(m_{\mathcal{S}_t}(x))$ from Eq. (1), then:

$$\epsilon_t(x) = \frac{1 + \Phi^{-1}(\delta, |m_{\mathcal{S}_t}(x)|, \sigma^2_{\mathcal{S}_t}(x))}{2}. \tag{4}$$

Thus, ϵ_t bounds r_t with probability $1 - \delta$. Following an optimistic, approach OLRM selects at each time step the unlabeled instance for which ϵ_t is maximum:

$$x_{t+1} = \arg\max_{\mathcal{D}_u} \epsilon_t(x_p). \qquad (5)$$

Minimizing the local risk is a solution that has shown good results in state-of-the-art. However, it does not consider the density of instances around the considered instances. Indeed, the criterion is computed using only their own prediction from the GP. In the next section, we change the loss to consider this.

4   Global Risk Minimization

In this section, we propose OGRM, an algorithm based on Optimism in the Face of Uncertainty that minimizes the global risk of the classifier. The main improvement of OGRM compared to OLRM is that it considers the representativeness of an instance. We first define the loss function being used and then show a criterion that tends to minimize it.

We now consider the expected error one would encounter by presenting an instance $x \in X$ drawn from the instance distribution which is the global risk of the classifier:

$$R_t(\mathcal{S}_t) = \int_X r_t(x) dP(x).$$

It is impractical because of the integral and because we do not known exactly the instance distribution. However, we can estimate the global risk from the available pool $\mathcal{D}_p$ which follows this distribution. We therefore define the empirical loss:

$$\tilde{R}_t(\mathcal{S}_t) = \sum_{x \in \mathcal{D}_p} r_t(x).$$

Defining a strategy that optimally minimizes this loss is NP-hard. However, it is common to make a myopic approximation. Thus the ideal strategy is to simulate the sampling of every unlabeled instances in $\mathcal{D}_u$. Then, to effectively sample the one which results in the highest decrease of the risk:

$$x_{t+1} = \arg\max_{x \in \mathcal{D}_u} \Delta_{x_s} \tilde{R}_t(\mathcal{S}_t), \qquad (6)$$

where

$$\Delta_{x_s} \tilde{R}_t(\mathcal{S}_t) = \tilde{R}_t(\mathcal{S}_t) - \tilde{R}_t(\mathcal{S}_t \cup \{x_s, y_s\}),$$

where y_s is the true label of x_s.

This supposes to know the true values of $\mu(x)$ and y_s, which is unrealistic. We propose to use again the Optimism in the Face of Uncertainty approach in order to sample the best instance considering only a Bayesian distribution of $\mu(x)$, given by the GP. We show how to compute a confidence bound on the criterion from Eq. (6) relatively to a probability δ.

Let us consider the simulation of sampling $x_s \in \mathcal{D}_u$ and that the label given by the expert is y_s. Let $l_t^+(x) = sign(m_{\mathcal{S}_t \cup \{x_s, y_s\}}(x))$. Then,

$$\forall x \in \mathcal{D}_p, \quad r_t(x) = \mathbb{1}_{l_t(x) \neq sign(\mu(x))} - \mathbb{1}_{l_t^+(x) \neq sign(\mu(x))}$$

$$= \begin{cases} \mathbb{1}_{l_t(x) \neq -1} - \mathbb{1}_{l_t^+(x) \neq -1} \text{ with prob. } \mathbb{P}(\mu(x) < 0|\mathcal{S}_t \cup \{x_s, y_s\}) \\ \mathbb{1}_{l_t(x) \neq 1} - \mathbb{1}_{l_t^+(x) \neq 1} \text{ with prob. } \mathbb{P}(\mu(x) \geq 0|\mathcal{S}_t \cup \{x_s, y_s\}). \end{cases}$$

We can then combine the cases for every instance in the pool to deduct the probability: $\mathbb{P}(\Delta_{x_s} R_t(\mathcal{S}_t)|\mathcal{S}_t \cup \{x_s, y_s\})$. We can also combine the cases for y_s, given that $\mathbb{P}(y_s = +1) = \mathbb{P}(\mu(x_s) \geq 0|\mathcal{S}_t)$, to deduct: $\mathbb{P}(\Delta_{x_s} R_t(\mathcal{S}_t)|\mathcal{S}_t)$. We then compute the probability bound e_t relatively to a fixed probability δ as:

$$\mathbb{P}(\Delta_{x_s} R_t(\mathcal{S}_t) \leq e_t(x_s)|\mathcal{S}_t) = 1 - \delta. \tag{7}$$

Following an optimistic approach, OGRM selects at each time step the unlabeled instance for which e_t is maximum:

$$x_{t+1} = \arg\max_{x_s \in \mathcal{D}_u} e_t(x_s). \tag{8}$$

5   Experiments

We evaluated both algorithms on several datasets from the UCI repository [7]. Each experiment consisted in a series of 1,000 runs. For each of these runs, the dataset was randomly divided into two equal parts: one was used as the pool of instances in which the algorithms were allowed to pick, the other one was used as a test set, hidden from the algorithms. Each algorithm is initialized with a single labeled instance randomly drawn from the pool. At each time step, the accuracy of the classifier is recorded as the proportion of well-classified instances on the test set. The global performance displayed on figures results from averaging the accuracy of every runs.

We compare OLRM and OGRM with the following methods: (1) Random Sampling: each instance to be labeled is randomly drawn from the pool, (2) Uncertainty Sampling (US) [6] samples instance closest to the boundary, (3) QUerying Informative and Representative Examples (QUIRE) [3] follows a min-max strategy for minimization of the global risk, (4) Expected Value-of-information (VOI) [5] minimizes the expected global risk, and (5) GPAL (Uncertainty) [4] minimizes of the expected local risk.

All these algorithms use GP or RLSR classifier with a Gaussian kernel. Indeed, different classifiers have different intrinsic performances. Thus, it is not meaningful to compare Active Learning algorithms which are based on different classifiers. The optimal variance parameter of the kernel has been found using grid-search.

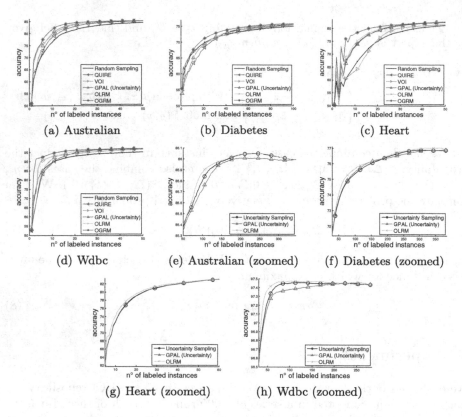

Fig. 1. Evaluation of algorithms

6   Results

Let us first bring out elements of discussion. Formally, if the δ parameter in OLRM is set to 0.5, it is strictly equivalent to US. Thus, OLRM appears as a generalization of US. And thus, OGRM is a generalized extension of US including representativeness. Notice that OGRM considers the estimation of the decrease of the risk and VOI the decrease of the estimation of the risk, which is different.

Figure 1[a–d] display the classification accuracy for each evaluated methods as a function of the budget of labeled instances. Notice that even though the marker only appears ten times on each curve, the accuracy is plotted for every integer number of labeled instances. As some curves are hard to differentiate, Fig. 1[e–h] show a zoomed version of, respectively, Fig. 1[a–d], focusing on those methods.

The δ parameter in OLRM and OGRM controls the exploration/exploitation trade-off. In the case of OGRM, this parameter was tuned using a gridsearch. It appeared that any value lower than 0.5 leads to almost exactly the same performance, with a slight tendency for values closer to 0.5.

In the case of OLRM, no value of the parameter had an outstanding performance. The optimal parameter always depended on the number of labeled instances considered. However, optimistic algorithms are often studied through a finite budget framework. This supposes that the budget is known from the beginning and the parameter may be chosen accordingly. The displayed curves for this algorithm shows the maximum performance OLRM would have for each budget if the parameter was chosen w.r.t. to it. A deeper theoretical analysis could provide the optimal parameter given the budget.

We see on Fig. 1[a–d] that the three algorithms based on minimization of the local risk have close performance. By looking at the zoomed versions on Fig. 1[e–h], we see that they are not exactly equivalent. While GPAL is designed as an improvement of US considering the variance provided by the GP, it is most of the time outperformed by it. We also see that, for a majority of budgets, there exists a parameter of ORLM that achieves better performance than US.

We observe on Fig. 1[a–d] that OGRM outperforms all the other methods. Using the global risk loss function inherently makes the performance increase sooner. Indeed, it considers the density of instances, and has an *a priori* about where to sample with no observations. It is therefore surprising that QUIRE behaves worse than local risk methods. On Fig. 1[a–b], VOI starts with the same performance as OGRM but does not follow. This may be due to the aforementioned reason.

7   Conclusion

In this paper, we introduced a new way to deal with uncertainty in the Error Minimization problem based on Optimism in the Face of Uncertainty. We present two algorithms that work with two different losses. One can be seen as a generalization of Uncertainty Sampling, while the second extends the former to consider the instances distribution. Evaluations on real-world datasets showed that both algorithms compared positively to state-of-the-art methods. Future work will focus on theoretical analysis and on the adaptation to the multi-class case. We will also consider improving the modeling of noise distribution in the GP to better stick to the expert's one.

References

1. Collet, T., Pietquin, O.: Active learning for classification: an optimistic approach. In: 2014 IEEE Symposium on Adaptive Dynamic Programming and Reinforcement Learning (ADPRL), pp. 1–8. IEEE (2014)
2. Ganti, R., Gray, A.G.: Building bridges: viewing active learning from the multi-armed bandit lens (2013). arXiv preprint arXiv:1309.6830
3. Huang, S.J., Jin, R., Zhou, Z.H.: Active learning by querying informative and representative examples. In: Advances in Neural Information Processing Systems, pp. 892–900 (2010)
4. Kapoor, A., Grauman, K., Urtasun, R., Darrell, T.: Active learning with gaussian processes for object categorization. In: IEEE 11th International Conference on Computer Vision, ICCV 2007, pp. 1–8. IEEE (2007)
5. Kapoor, A., Horvitz, E., Basu, S.: Selective supervision: guiding supervised learning with decision-theoretic active learning. IJCAI **7**, 877–882 (2007)

6. Lewis, D.D., Gale, W.A.: A sequential algorithm for training text classifiers. In: Croft, B.W., van Rijsbergen, C.J. (eds.) Proceedings of the 17th Annual International ACM SIGIR Conference on Research and Development in Information Retrieval, pp. 3–12. Springer, New York (1994)
7. Lichman, M.: UCI machine learning repository (2013). http://archive.ics.uci.edu/ml
8. Settles, B.: Active learning literature survey. Univ. Wis. Madison **52**(11), 55–66 (2010)
9. Zhu, X., Lafferty, J., Ghahramani, Z.: Combining active learning and semi-supervised learning using gaussian fields and harmonic functions. In: ICML 2003 Workshop on the Continuum from Labeled to Unlabeled Data in Machine Learning and Data Mining, pp. 58–65 (2003)

A Penalty-Based Genetic Algorithm for the Migration Cost-Aware Virtual Machine Placement Problem in Cloud Data Centers

Tusher Kumer Sarker and Maolin Tang$^{(\boxtimes)}$

School of Electrical Engineering and Computer Science,
Queensland University of Technology, 2 George Street,
Brisbane, QLD 4001, Australia
{t.sarker,m.tang}@qut.edu.au

Abstract. In the past few years, the virtual machine (VM) placement problem has been studied intensively and many algorithms for the VM placement problem have been proposed. However, those proposed VM placement algorithms have not been widely used in today's cloud data centers as they do not consider the migration cost from current VM placement to the new optimal VM placement. As a result, the gain from optimizing VM placement may be less than the loss of the migration cost from current VM placement to the new VM placement. To address this issue, this paper presents a penalty-based genetic algorithm (GA) for the VM placement problem that considers the migration cost in addition to the energy-consumption of the new VM placement and the total inter-VM traffic flow in the new VM placement. The GA has been implemented and evaluated by experiments, and the experimental results show that the GA outperforms two well known algorithms for the VM placement problem.

Keywords: VM placement · VM migration · Penalty-based genetic algorithm

1 Background and Related Work

Virtual Machine (VM) placement is a key management activity in today's data centers. VMs, which are created on the top of a Physical Machine (PM), can be resized when the resource requirements by the applications deployed on the VMs are changed or can be migrated to another PM when the current PM cannot meet the resource requirements. Moreover, the resource utilization of a PM by its VMs changes over time due to creation of new VMs in the data center and the removal of existing VMs. Therefore, the current VM placement, which was optimal, may not be optimal any more, and a new optimal VM placement needs to be found in order to minimize the total energy consumption in the data center and to minimize the inter-VM traffic.

A study by Le et al. reveals that the energy costs of a data center at its average utilization is more than \$15M per year [1]. Thus, an extensive research has

© Springer International Publishing Switzerland 2015
S. Arik et al. (Eds.): ICONIP 2015, Part II, LNCS 9490, pp. 161–169, 2015.
DOI: 10.1007/978-3-319-26535-3_19

been conducted to optimize the data center resource usage aiming at minimizing the data center energy consumption [1–6].

In addition, the VM placement requires to consider the inter-VM traffic flow, the data flow between two dependent VMs, through the data center. A data center hosts heterogeneous applications ranging from enterprise applications to scientific applications. The VMs that execute the components of such a composite application exhibit a significant amount of traffic flow among them. Therefore, an arbitrary placement of VMs may result in network congestion due to the inter-VM traffic flow through the data center network. A recent work by Meng et al. [7] explored network traffic-aware VM placement for various network architectures while the energy consumption was not considered. Shrivastava et al. [8] proposed a heuristic algorithm, AppAware, which considered the application dependency during the placement of VMs to minimize the inter-VM traffic flow through the data center.

However, only few efforts have been given that combine both the energy-aware and traffic-aware VM placement. More importantly, the migration cost, which we define in this paper as total number of VM migrations required to transform from the current VM placement to a new VM placement, has not been considered in the existing work on the VM placement. Each VM live migration degrades the performance of the application residing in that VM [5,9]. As a result, the gain from optimizing VM placement may be less than the loss of the migration cost from current VM placement to the new VM placement, and therefore, it is essential to minimize the number of VM migrations to minimize the negative effect on the performance of the applications in the VMs.

To address the above issues, this paper presents a penalty-based genetic algorithm (GA) for the VM placement problem (VMPP) that considers the migration cost in addition to the energy-consumption by PMs and the total inter-VM traffic flow of the new VM placement. The GA has been implemented and evaluated by experiments, and the experimental results show that the GA outperforms two popular algorithms for the VM placement problem: the First Fit Decreasing (FFD) algorithm, which focuses on minimizing the total energy consumption, and the AppAware algorithm, which concentrates on minimizing the inter-VM traffic.

The organization of this paper is as follows: Sect. 2 formulates the VM placement problem; Sect. 3 details our penalty-based GA; Sect. 4 presents the evaluation of our GA; and finally Sect. 5 concludes this research and discusses future work.

2   Problem Formulation

Let V and P be the set of VMs and the set of PMs, respectively, in a data center, and V_{pm_s} be the set of VMs deployed on PM, $pm_s \in P$. Thus, the set, $\{V_{pm_1}, V_{pm_2}, \cdots, V_{pm_{|P|}}\}$, is a VM placement of the data center. To transform from one VM placement to another, we need to use the live VM migration technology to migrate VMs from their current PM to their target PM, if their current and target PMs are different; and N number of migrations is required for that, $N \geq 1$.

This research work considers the inter-VM traffic flow among the dependent VMs, therefore, only data dependency among the VMs has been considered. The VM data dependency is represented by a $|V| \times |V|$ matrix, D. An element of D, d_{ij}, is a positive value which represents the amount of data flow between two VMs, vm_i and vm_j, i.e. $d_{ij} \in \mathbb{R}^+, 1 \leq i, j \leq |V|$. An inter-VM traffic flow takes place through the data center network when two dependent VMs are deployed in two different PMs. Let h_{st} is the network topology distance between two PMs, pm_s and pm_t. A topology distance of a pair of PMs is the number of network tiers required to connect these PMs. Then the inter-VM traffic flow between pm_s and pm_t for the VMs, $vm_i \in V_{pm_s}$ and $vm_j \in V_{pm_t}$, is $2 \times d_{ij} \times h_{st}$. Thus, the total inter-VM traffic flow for each pair of PMs, (pm_s, pm_t), is calculated by (1).

$$T_{st} = 2 \times \sum_{\forall vm_i \in V_{pm_s}, \forall vm_j \in V_{pm_t}} d_{ij} \times h_{st}. \qquad (1)$$

It is assumed that there are k types of resources in a PM, pm_s. Let $C_{pm_s}^k$ and $L_{pm_s}^k$ are respectively the capacity and consumed resource of k-type in pm_s. The utilization of each type of resource in pm_s is calculated as follows:

$$\mu_{pm_s}^k = \frac{L_{pm_s}^k}{C_{pm_s}^k}. \qquad (2)$$

The energy consumption by a PM, pm_l, is calculated using (3) in the light of the model proposed in [6] as follows:

$$E_{pm_l} = (e_{pm_l}^{max} - e_{pm_l}^{idle}) \times \mu_{pm_l}^{cpu} + e_{pm_l}^{idle} \times \lceil \mu_{pm_l}^{cpu} \rceil. \qquad (3)$$

where $e_{pm_l}^{max}$ and $e_{pm_l}^{idle}$ are the energy consumed by pm_l in full utilization and idle state respectively, $\mu_{pm_l}^{cpu}$ is the CPU utilization of pm_l, and $\lceil . \rceil$ is the ceiling function. Equation (3) ensures that a PM without any VM is switched-off.

The placement of vm_i into pm_t is subject to satisfaction of the resource constraints and bandwidth constraint of pm_t. To place vm_i into pm_t each type of resource requirement of vm_i must be satisfied by the available resource of pm_t. The migration of vm_i to pm_t, however, changes the inter-VM traffic flow between pm_t and other PMs, pm_q, on which dependent VMs, vm_j, of vm_i are residing. The placement of vm_i into pm_t, is therefore, requires to satisfy that the available bandwidth between pm_q and pm_t, B_{qt}, must be adequate to carry inter-VM traffic between vm_i and vm_j, i.e. $d_{ij} \leq B_{qt}$.

The VMPP problem is thus stated as: given a set of VMs, V, with their resource requirements and inter-VM dependency, D; a data center network, consists of a set of PMs, P, and a set of switches, W; the configurations of PMs; and the current placement of VMs, it is to find a target PM for each and every VM of V such that the costs– total number of migrations required to obtain the target placement (N), total energy consumption calculated using (5) and total inter-VM traffic flow in the data center as given by (6) are minimized, i.e. minimize:

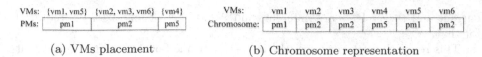

(a) VMs placement   (b) Chromosome representation

Fig. 1. Chromosome encoding scheme

$$N \tag{4}$$

$$E = \sum_{l=1}^{|P|} E(pm_l). \tag{5}$$

$$T = \sum_{\forall(pm_s,pm_t)} T_{st}. \tag{6}$$

subject to:

$$\forall k \forall t, \mu_{pm_t}^k \leq 1. \tag{7}$$

$$\forall vm_i \in V_{pm_s}, \forall vm_j \in V_{pm_t}, d_{ij} \leq B_{st}. \tag{8}$$

3   The GA for the VMPP

As the VMPP is computationally large and complex, we have developed a penalty-based GA for the VMPP. The following subsections discuss the design of our GA.

3.1   Chromosome Representation

In our GA, a VM placement is represented by a chromosome of $|V|$ genes where each gene represents the target PM of a VM. For example, Fig. 1a shows the target placement of VMs where vm_1 and vm_5 are planned to be placed on pm_1. The corresponding chromosome for this plan is represented in Fig. 1b.

3.2   Genetic Operators

The new generation is obtained through an elitism scheme and two genetic operations: crossover and mutation. With the elitism scheme, a proportion of the current population with better fitness values are copied to the next generation.

Through the crossover operation two offspring are produced by applying two-point crossover in two parent chromosomes. The first parent is selected arbitrary from the chromosomes that are copied to next generation through the elitism and the second parent is randomly chosen from the current population. Then the portions between two arbitrary points of these parent chromosomes are swapped to produce new offspring. In the mutation operation, 10 % of genes of an arbitrarily selected chromosome from the set of chromosomes found through the elitism scheme are altered arbitrarily to form new chromosomes.

3.3 Fitness Calculation

This research aims to minimize three objectives – total number of migrations, total energy consumption and total inter-VM traffic flow in the data center simultaneously. Equation (9) gives the fitness function of our GA.

$$F(X) = \begin{cases} 0.5 + \sum\limits_{i=1}^{3} \frac{Q_i^{max} - Q_i}{2 \times Q_i^{max}} \times w_i & \text{if solution is feasible} \\ \frac{|V| - |V_O|}{2 \times |V| + 1} & \text{otherwise} \end{cases} \tag{9}$$

where Q_i and Q_i^{max} are respectively a cost incurred due to a placement plan and the maximum possible cost of that type for a feasible solution, $0 \leq Q_i \leq Q_i^{max}$; w_i is the assigned weight to each cost and $\sum w_i = 1$; and V_O is the set of VMs for which no target PMs are found. A solution is feasible when for all VMs in V, corresponding target PMs are found, i.e. $V_O = \emptyset$.

As one of the objectives of this research is to minimize the number of migrations, it is constrained that a VM can be migrated for once. Therefore, the Q_i and Q_i^{max} for total number of migrations cost are respectively $N, 0 \leq N \leq |V|$, and $|V|$. The Q_i^{max} value for the energy cost is calculated for the placement plan that causes maximum energy consumption by the PMs. This is obtained by placing the VM with most CPU requirement to the least energy efficient PM and the process iterates by placing the following VMs with most CPU requirements to the next least energy efficient PMs. The energy efficiency of a PM, pm_l, is determined by per CPU unit energy consumption, i.e. $Energy\ efficieny\ (pm_l) = \frac{e_{pm_l}^{max}}{C_{pm_l}^{cpu}}$, and a PM with the most per CPU unit energy consumption is the least energy efficient. Q_i^{max} is thus obtained by aggregating the energy consumption by each active PM for the placement plan of the most CPU requirement VM to the least energy efficient PM. The Q_i value for total energy cost is found from (5). The maximum amount of inter-VM traffic flow through the data center is obtained when any pair of dependent VMs are placed in two distinct PMs. The total inter-VM traffic flow for a feasible placement is computed using (6). Equation (9) ensures that any feasible solution always gets the fitness value between 0.5 and 1.0, i.e. $0.5 \leq F(X) \leq 1.0$; and the solution with minimum costs, gets the highest fitness value.

On the contrary, a penalty is assigned to any infeasible solution. An infeasible solution is a chromosome in which some VMs cannot be placed according to their gene values, i.e. $V_O \neq \emptyset$. The penalty assigned to such an infeasible solution ensures the fitness value is less than 0.5, i.e. $0 \leq F(X) < 0.5$.

3.4 Our GA

In our GA, a heuristic algorithm has been developed to generate the initial population. The idea behind this heuristic algorithm is trying to keep the VMs to their current source PMs. The VM that cannot be retained in its source PM due to violation of resource constraints is arbitrarily mapped to another target PM. The initial population generated in this way contains the solutions with less number of migrations requirements, which is one of our research objectives. Afterward, the following steps are repeated until a termination condition is met: (1) calculate the fitness value of each chromosome in the population using (9); (2) use the elitism scheme to copy some fitter chromosomes from the current population to the new population; and (3) use the genetic operators to reproduce new chromosomes and add them into the new population.

4 Evaluation

4.1 Experimental Design

We simulated a data center with 128 PMs connected using CLOS topology. The capacity of each link was 1 Gbps. The CPU and RAM capacities of a PM were chosen in the ranges [20, 40] and [30, 50] GB respectively. The maximum energy consumption by a PM in full CPU utilization was chosen arbitrarily from the range [100, 200] kWh; and in idle state 40 % of maximum energy was consumed.

Five test problems with different number of VMs were randomly generated. The numbers of VMs in the five randomly generated test problems were 20, 40, 60, 80 and 100. A VM was a randomly chosen Amazon EC2 instance type [10]. The inter-VM traffic flow rate between two dependent VMs was a random value in the range [1, 10] Mbps. A VM cluster size was chosen arbitrarily with maximum 8 VMs. The initial VM placement was obtained through the bin packing approach.

The parameters of our GA were set as follows. The population size and the number of generations of our GA were 300 and 400, respectively. The probabilities for crossover rate and mutation rate were 88 % and 7 %, respectively. The elitism rate was 5 %. The GA terminated when 400 generations have evolved.

All the algorithms– the GA, FFD and AppAware, were implemented in Java on a desktop computer of Intel Core i7-3770 CPU of 3.40 GHz and 8.00 GB RAM. For each of the randomly generated test problems, we used our GA, the FFD and the AppAware to solve it. Due to the stochastic nature of our GA, we repeatedly used it to tackle each of the test problems for 30 times and used the average of this 30 runs to compare with the solutions generated by the FFD and the AppAware.

Four experiments were conducted. The first experiment was to test the capacity of our GA in minimizing the total energy consumption objective, which was done by setting $w_1 = 1.0$, $w_2 = w_3 = 0.0$, where w_1, w_2 and w_3 are the weight for the total energy consumption objective, the inter-VM traffic flow objective, and the migration cost objective, respectively, in the fitness function.

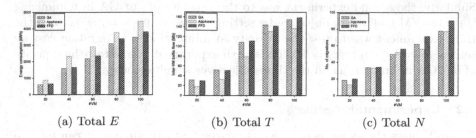

Fig. 2. Weights for E, T and N are respectively 1.0, 0.0 and 0.0

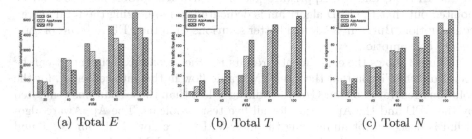

Fig. 3. Weights for E, T and N are respectively 0.0, 1.0 and 0.0

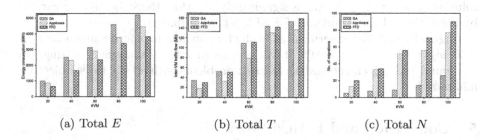

Fig. 4. Weights for E, T and N are respectively 0.0, 0.0 and 1.0

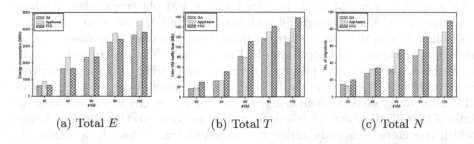

Fig. 5. Weights for E, T and N are respectively 0.85, 0.1 and 0.05

Similarly, the second experiment was to the capacity of our GA in minimizing the inter-VM traffic flow objective by setting $w_2 = 1.0$, $w_1 = w_3 = 0.0$, and the third experiment was to test the capacity in minimizing the migration cost by setting $w_3 = 1.0$, $w_1 = w_2 = 0.0$. The fourth experiment was to test the capacity of our GA in minimizing all the three objectives simultaneously.

4.2   Experimental Results

Figure 2 shows the experimental results for the first experiment. It can be seen from the Fig. 2a that the total energy consumption by the PMs, E, for the solutions generated by our GA is significantly less than those solutions generated by the FFD [5] and the AppAware [8] for all five test problems. It should be pointed out that the FFD algorithm is dedicated to minimizing the total energy consumption. But, our GA found better solutions than the FFD in terms of total energy consumption.

Figure 3 shows the experimental results for the second experiment. It can be seen from the Fig. 3a that the inter-VM traffic flow in the data center, T, for the solutions generated by our GA is significantly less than those solutions generated by the FFD and the AppAware for all five test problems. The AppAware algorithm is designed for minimizing the inter-VM traffic flow. But, our GA found better solutions than the AppAware in terms of inter-VM traffic flow.

Figure 4 shows the experimental results for the third experiment. It can be seen from the Fig. 4c that the VM migration cost in the data center, N, for the solutions generated by our GA is significantly less than those solutions generated by the FFD and the AppAware for all five test problems.

Figure 5 shows the experimental results for the fourth experiment where $w_1 = 0.85$, $w_2 = 0.10$, and $w_3 = 0.05$. It can be seen that using the weight setting, our GA found VM placements for the five test problems with all the three objectives minimized.

5   Conclusion and Future Work

In this paper, we have developed a penalty-based GA for the VM placement problem that considers the migration cost in addition to the energy consumption and total inter-VM traffic flow for the new VM placement plan. We have evaluated the GA by comparing it with two popular VM placement algorithms, FFD and AppAware, against three criteria – total number of migrations, total energy consumption and total inter-VM traffic flow. The experimental results have shown that when the priority is given to minimizing the total energy consumption our GA can beat the FFD algorithm, which focuses on minimizing the total energy consumption, and when the priority is given to minimizing the inter-VM traffic flow our GA is able to find better solutions than the AppAware, which concentrates on minimizing the inter-VM traffic flow. In addition, the experimental results have shown that by selecting a set of appropriate weights for the three objectives, our GA can find solutions that are better than those solutions generated by the FFD in terms of total energy consumption and also better than those solutions created by the AppAware in terms of inter-VM flow with a smaller number of VM migrations at the same time.

The research results presented in this paper are our preliminary research work on the VM placement problem considering the VM migration cost in which the VM migration cost is measured by the number of VM migrations from the current VM placement to a new VM placement. In the future, we adopt more sophisticated measurements for the migration cost, such as total VM migration time and total migration downtime in the migration-aware VM placement problem. We will also further improve the performance of our GA.

References

1. Le, K., Bianchini, Zhang, R.J., Jaluria, Y., Meng, J., Nguyen, T.D.: Reducing electricity cost through virtual machine placement in high performance computing clouds. In: Proceedings of the 2011 International Conference for High Performance Computing, Networking, Storage and Analysis, Seattle, pp. 22–33 (2011)
2. Wu, Y., Tang, M., Tian, Y.-C., Li, W.: A simulated annealing algorithm for energy-efficient virtual machine placment. In: Proceedings of the 2012 IEEE International Conference on Systems, Man, and Cybernetics, Seoul, pp. 1245–1250 (2012)
3. Wu, G., Tang, M., Tian, Y.-C., Li, W.: Energy-efficient virtual machine placement in data centers by genetic algorithm. In: Huang, T., Zeng, Z., Li, C., Leung, C.S. (eds.) ICONIP 2012, Part III. LNCS, vol. 7665, pp. 315–323. Springer, Heidelberg (2012)
4. Tang, M., Pan, S.: A hybrid genetic algorithm for the energy-efficient virtual machine placement problem in data centers. Neural Process. Lett. **41**, 211–221 (2015). Springer
5. Verma, A., Ahuja, P., Neogi, A.: pMapper: power and migration cost aware application placement in virtualized systems. In: Issarny, V., Schantz, R. (eds.) Middleware 2008. LNCS, vol. 5346, pp. 243–264. Springer, Heidelberg (2008)
6. Beloglazov, A., Anton, A., Abawajy, J., Buyya, R.: Energy-aware resource allocation heuristics for efficient management of data centers for cloud computing. Future Gener. Comput. Syst. **28**, 755–768 (2012)
7. Meng, X., Pappas, V., Zhang, L.: Improving the scalability of data center networks with traffic-aware virtual machine placement. In: Proceedings of the 2010 IEEE INFOCOM, San Diego, pp. 1–9 (2010)
8. Shrivastava, V., Zerfos, P., Lee, K., Jamjoom, H., Liu, Y., Banerjee, S.: Application-aware virtual machine migration in data centers. In: Proceedings of the 2011 IEEE INFOCOM, Shanghai, pp. 66–70 (2011)
9. Voorsluys, W., Broberg, J., Venugopal, S., Buyya, R.: Cost of virtual machine live migration in clouds: a performance evaluation. In: Jaatun, M.G., Zhao, G., Rong, C. (eds.) Cloud Computing. LNCS, vol. 5931, pp. 254–265. Springer, Heidelberg (2009)
10. http://aws.amazon.com/de/ec2/instance-types/

Frequency Decomposition
Based Gene Clustering

Md Abdur Rahman[1,2], Madhu Chetty[1]([✉]), Dieter Bulach[3],
and Pramod P. Wangikar[4]

[1] School of Science and IT, Federation University, Ballarat, Australia
{ma.rahman,madhu.chetty}@federation.edu.au
[2] Faculty of Engineering,
American International University Bangladesh, Dhaka, Bangladesh
arahman@aiub.edu
[3] Life Sciences Computation Centre,
Victorian Life Sciences Computation Initiative, Carlton, Australia
dieter.bulach@unimelb.edu.au
[4] Department of Chemical Engineering,
Indian Institute of Technology, Mumbai, India
wangikar@iitb.ac.in

Abstract. Gene expressions have been commonly applied to understand the inherent underlying mechanism of known biological processes. Although the microarray gene expressions usually appear aperiodic, with proper signal processing techniques, its periodic components can be easily obtained. Thus, if expressions of interconnected (regulatory and regulated) genes are decomposed, at least one common frequency component will appear in these genes. Exploiting this novel concept, we propose a frequency decomposition approach for gene clustering to better understand the gene interconnection topology. This method, based on Hilbert Huang Transform (HHT) enables us to segregate every periodic component of the gene expressions. Next, a multilevel clustering is performed based on these frequency components. Unlike existing clustering algorithms, the proposed method assimilates a meaningful knowledge of the gene interactions topology. The information related to underlying gene interactions is vital and can prove useful in many existing evolutionary optimisation algorithms for genetic network reconstruction. We validate the entire approach by its application to a 15-gene synthetic network.

Keywords: Frequency decomposition · Gene expression · HHT · Clustering

1 Introduction

At a molecular level, genes are interconnected to perform specific tasks according to the necessity of the organism. These interconnected genes may be broadly categorised as *regulated* or *regulatory* genes. Regulated genes are defined as genes

© Springer International Publishing Switzerland 2015
S. Arik et al. (Eds.): ICONIP 2015, Part II, LNCS 9490, pp. 170–181, 2015.
DOI: 10.1007/978-3-319-26535-3_20

which perform activities under the influence of a specific regulatory gene or a subset of genes. Regulated genes exist in all living organisms (e.g. eukaryotes, prokaryotes and archaea) to accomplish the functioning of molecular machinery of life. Recent advancements in microarray technologies have provided opportunities for seeking deeper insight into the gene interactions via the observation of gene expressions. Typically, the time series gene expression data are obtained by measurements carried out in a non-uniform manner at irregular time intervals. The time series data suffers from temporal sparsity—insufficient time samples for any meaningful observations—dimensionality curse—many genes and fewer samples—noisiness in data, or biological variation—samples obtained from different sources and environmental conditions using variety of techniques for sample preparation and measurement [1].

Current gene expression based clustering methods [2–5] use distinguishable features such as Pearson Correlation Coefficient (PCC), Euclidean distance and Entropy. Few works used the differential prioritization to determine the optimal combination of feature selection and classifier aggregation [6]. All these matrices are mathematical in nature and do not necessarily reflect causality of any biological functionality [7]. Thus GRN models based on these clustering algorithms fail to reflect the actual causality of biological connections among the genes. From signal processing knowledge, we know that if a linear system gets multiple signals with different frequencies, the output will be a combination of all the input signals. Hence, we propose that clustering genes with identical or closer frequency components in the time course of their gene expression profile provides a novel, biologically relevant grouping of expressed genes. This is because it is quite possible that, input signals of regulatory genes leave their footprint on the expressions of the regulated genes. In other words, the frequency (or frequencies) of the input signals may manifest on the output of the regulated genes. Thus, a network topology developed based on this concept will imbibe more biologically relevant interconnections.

Due to the non-linearity and non-stationarity of the gene expressions, extraction of periodic components from the time series is a big challenge. Traditional signal processing techniques such as fast Fourier transform (FFT) [8,9], spectrogram [10] and wavelet transforms [11] require interpolations that may compromise the integrity of the expressions [12]. Consequently, these methods are not exactly suitable for reliable decomposition of the gene expressions. Hence, we propose to apply Hilbert Huang Transform (HHT) [13,14] to decompose gene expressions and obtain the frequency components. Being data driven, the proposed method does not require any pre-processing of data and makes it easier to calculate the energy and frequency of each periodic component of the gene expressions. We can safely assume that the dominating frequency components (component with highest energy) are strongly connected. The genes grouped using these frequencies will result in clusters of genes of similar frequencies. Thus, we obtain a prior knowledge of intra-cluster gene interaction with strong links. Repeating the process based on second highest frequency components for clustering will provide next level of gene connections, i.e. inter-cluster connections. With increasing number of components, we can acquire more knowledge about gene interactions, albeit that which gets progressively weaker. However, incorporation of this prior knowledge in any of the known evolutionary optimisation algorithms for genetic network can help build a strong biologically relevant network. Eventually, design of GRN [15,16] can be aided significantly by applying the clustering of genes.

The rest of the paper is structured as follows. Section 2 idea of gene interconnections based on multiple frequency components obtained by HHT algorithm. Section 3 explains the underlying principles of the proposed frequency based method for the clustering of genes. Results and analysis are presented in Sect. 4. Finally, Sect. 5 concludes the paper.

2   Background

Many biological processes such as the cell cycle, cardiac excitation-contraction and the circadian clock are inherently periodic in nature and most of these responses are controlled by a gene or a set of genes which possess periodic expression profiles. A gene expression can be viewed as a non-linear oscillatory signal with the expression being represented as a combination of one or more periodic functions.

2.1   Gene Interactions

For any genetic network, we propose that, *"if gene expressions of interconnected (regulatory and regulated) genes are decomposed, they will exhibit the presence of at least one common frequency component"*. Stated simply, this means that the "activation" or "inhibition" signals of the regulatory genes are superimposed on the gene expressions of the regulated genes. Hence, a regulated gene receives these control signals as inputs from multiple regulatory genes and its observed gene expression is essentially a combination of several oscillatory inputs. This is illustrated, in Fig. 1, where we can see a regulated gene $G3$ receiving multiple regulatory inputs, one from gene $G1$ and another from gene $G2$.

For sake of analysis, let us consider that these two genes $G1$ and $G2$ are producing control signals which are approximated as sinusoid with frequencies $f1$ and $f2$ and a phase difference ϕ. For the expression of $G3$, two scenarios are possible, (i) If $G3$ generates a frequency $f3$, the expression of $G3$ will be a combination of three frequencies $f1$, $f2$ and $f3$. In this case, gene $G3$ is both a regulated gene (by upstream genes $G1$ and $G2$) and a regulatory gene (for the down stream genes) and (ii) if G3 does not generate a frequency, then the expression of $G3$ consists of only two input frequencies, $f1$ and $f2$. In either case, an observed microarray expression of any regulated gene ($G3$) can be represented by (1).

$$x(t) = \sum_{i=1}^{k} A_i . \sin(2\pi f_i t \pm \phi_i) \qquad (1)$$

Here, A_i is the amplitude indicating the signal strength, f_i is the frequency and ϕ_i is phase of the i-th component of the gene expression $x(t)$. The variable $i = 1, 2, \ldots n$; where n is total number of frequencies present.

2.2 Gene Expression Decomposition

For the reasons given earlier, Hilbert Huang Transform (HHT) approach is most suitable for the decomposition of microarray gene expressions. This method is performed in two steps, (i) Decomposing the non-linear and non-stationary signals into a series of intrinsic mode functions (IMF) and (ii) Extracting instantaneous frequencies for each IMF. The first step is accomplished by application of empirical mode decomposition (EMD) method. Since EMD is completely data driven, it depends entirely on the data and preserves the non-uniformity and non-stationarity characteristics of the target signal. Furthermore, being applicable to non-uniform sampling, EMD method is especially suitable for the microarray data (which typically has fewer samples). For a given gene expression $x(t)$, the EMD is performed as shown in the flowchart of Fig. 2.

At the end of the EMD process, we are left with the residue $r(t)$ and n number of IMFs $c_i(t), i = 1, 2, \ldots n$. The original signal $x(t)$ can be reconstructed by (2).

$$x(t) = \sum_{i=1}^{n} c_i(t) + r(t) \qquad (2)$$

The IMFs, $c_i(t)$ are extracted from the higher frequency component toward the lower frequency components. An IMF signal can be written as AM-FM signal as $c_i(t) = a_i(t)q_i(t)$. Where, $q_i(t) = e^{(-j\theta_i(t))}$, amplitude modulated part $a_i(t)$ is the envelop and $\theta i(t)$ is the phase of FM signal. The residue $r(t)$ shows the trend of the input signal. After obtaining the IMF components, the instantaneous frequency components are calculated by Hilbert Transform (HT) [14]. Again, the original gene expression $x(t)$ can be obtained from the real part of the instantaneous frequencies calculated by HT according to (3).

$$x(t) = \text{Re} \sum_{i=1}^{n} a_i(t)e^{j \int \omega_i(t)dt} \qquad (3)$$

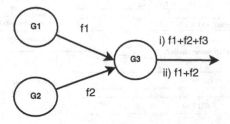

Fig. 1. Frequency based interpretation of gene regulation

where, a_i and ω_i are the instantaneous amplitude and frequency of each IMFs respectively. The instantaneous frequency is calculated by $\omega_i(t) = d\theta_i(t)/dt$.

3   Proposed Method

Although the microarray expression of any regulated gene will usually appear aperiodic with the aid of (2), we obtain the periodic components of this aperiodic signal.

3.1   Gene Representative and Significant Frequencies

The flowchart of the proposed method is shown in Fig. 3. At first, a specific fold change with appropriate threshold (typically 2 fold change) is applied discard the genes with low amplitude of oscillations and thus not significantly contributing to the system dynamics [17]. Subsequently, by applying the HHT method, we find the major frequency components in each gene expression. If there are n IMF components, different frequencies of the gene expression constitute to the energy of the vector $[E_1, E_2, \ldots, E_n]$. Then the frequency component with highest energy is regarded as the representative frequency (F_R) of the gene. If the energy of other IMFs are within a certain range of F_R, these are regarded as significant frequency $[F_{S1}, F_{S2}, \ldots, F_{Sn}]$ components. The value of this range depends on specific application, the organism under consideration and the measurement techniques. In this paper, to include reasonably strong signals, we set

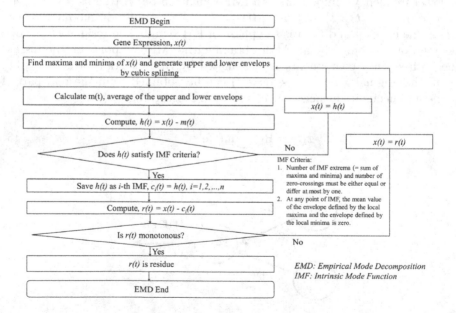

Fig. 2. EMD algorithm flowchart

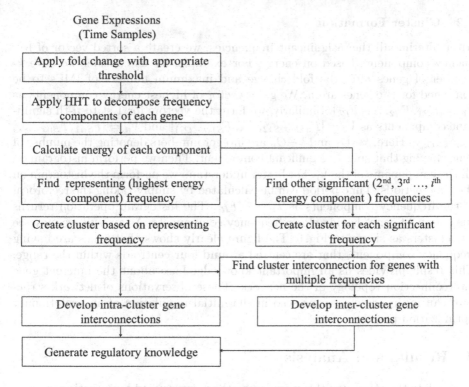

Fig. 3. Flowchart of Proposed method

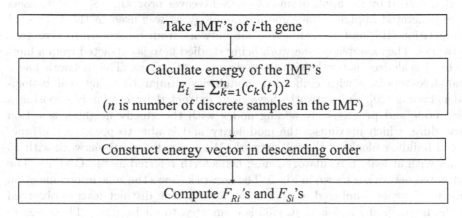

Fig. 4. Calculation process of representative and significant frequencies

the range value to 40 %. A higher threshold value will decrease the number of components of the gene and vice versa. This process is shown schematically in the flowchart of Fig. 4.

3.2   Cluster Formation

After filtering all the insignificant frequencies, we create a sorted vector of frequency components based on energy. For example, let us consider we have k-number of genes after the fold change and maximum number of IMFs to be combined for the genes are n. We get a vector of highest energy components as $V_1 = [F_{R1}, F_{R2}, \ldots, F_{Rk}]$. Similarly, we form the vectors for 1-st to n-th significant components as $V_2 = [F_{S11}, F_{S12}, \ldots, F_{S1(k-O_1)}]$ and $V_n = [F_{Sn1}, F_{Sn2}, \ldots, F_{Sn(k-O_n)}]$. Here, $k - O_1$ and $k - O_n$ are integer numbers denoting the number of genes having that specific significant component. Then we perform histogram on all the components of the V_1. For better understand we perform the histogram on the time periods. Time period can be calculated by simply taking the reciprocal of the frequency components as $T_i = 1/F_{R_i}$. The histogram operation returns the counts of genes having closer frequency components. As a result we may see few clusters as shown in Fig. 6. The figure clearly shows that genes are having frequency components that are centred around four centroids within the range. This result provides us an important knowledge-base about the inherent gene interconnection topology of the network. These observations of network structure can incorporate a strong seed in the initial population of an evolutionary optimization process.

4   Results and Analysis

To validate the efficacy of the proposed method, we studied a synthetic gene network obtained by the application of GeneNetWeaver program [18]. The datasets were generated applying the same methodology as was used in the DREAM4 competition [19], and extracting the sub-network from the known in-vivo gene networks. The example sub-network being studied here is extracted from a hierarchical scale-free network with 64 nodes and 207 edges. This network has a scale-free topology with embedded modularity similar to many real biological networks [20]. Sub-networks have been extracted by randomly choosing a seed node and progressively adding nodes with the greedy neighbor selection procedure, which maximises the modularity and is able to preserve the functional building blocks of the full network. We generated a sub-network with 15 genes with at least 5 regulatory genes; henceforth referred as "15-GNET". The extracted network is shown in Fig. 5. The network shows the gene interconnection among 15 genes numbered as $G32, G33, \ldots G47$. One distinct feature observed in the network of Fig. 5 is that, $G32$ is connected to all 14 genes. Thus, we are expected to observe the presence of a common frequency among all 15 genes in the network.

Now we apply the proposed method to the expressions of 15 genes and form the clusters. The components are sorted according to the signal strengths where "Comp1" is the highest energy and "Comp4" is the lowest energy components of the respective gene. For example, we can see that "Comp1" of 8 genes are members of cluster CL-1. Similarly, based on "Comp1", clusters CL-2, CL-4 and CL-5 have 1, 2 and 4 member genes respectively. Then we perform the same clustering operation (with the obtained centroids and range from "Comp1") on

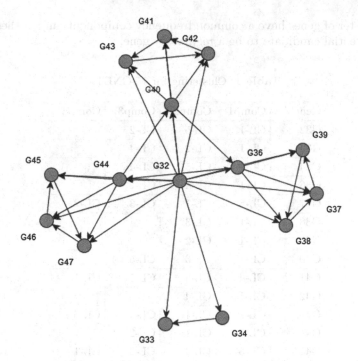

Fig. 5. 15 gene network

"Comp2" to "Comp4" and find the corresponding cluster numbers. Finally we obtain Table 1 that shows component-wise clustering of 15 genes with maximum of 4 significant components. Each entry in the table represents the cluster number of the corresponding component.

Table 1 shows the few interesting behaviour of the genes under investigation. Especially, we can see the frequency component from CL-1 is present in all the genes. This means at least one gene from CL-1 to be regulating every other gene in the network. Now, we can also observe that $G32$ is a member of cluster CL-1. While the method cannot reveal that gene $G32$ is the regulator, it still helps understand, the underlying structure of gene interconnections to be of one of the eight genes of CL-1 to be regulating other genes. This clearly indicates a highly probable network topology of these 15 genes, which can be considered as seed network in any known evolutionary optimization approach [21,22]. Results on other networks produced similar results but are not produced here due to space constraints. Due to space limitations these results are not included in the paper. Based on these observations, following rules can be proposed for generating seed networks incorporating this information as prior network knowledge.

1. Genes having common frequency component in the gene expression are inter-connected.
2. Genes possessing the highest energy in the gene expressions have higher probability to be regulatory genes.

3. If a cluster of genes have a common frequency component, any of these genes is a potential candidate to be a regulatory gene.

Table 1. Clustering for 15-GNET

Gene	Comp1	Comp2	Comp3	Comp4
G32	CL-1	CL-5	CL-2	
G33	CL-1	CL-2	CL-4	
G34	CL-1	CL-5	CL-4	
G36	CL-1	CL-4	CL-2	
G37	CL-5	CL-1	CL-3	CL-2
G38	CL-1	CL-3		
G39	CL-1	CL-2		
G40	CL-1	CL-3	CL-5	
G41	CL-4	CL-2	CL-5	CL-1
G42	CL-1	CL-4		
G43	CL-2	CL-1	CL-4	CL-5
G44	CL-5	CL-1	CL-2	
G45	CL-5	CL-4	CL-2	CL-1
G46	CL-5	CL-1	CL-2	
G47	CL-4	CL-1		

Cluster centroids				
CL-1	CL-2	CL-3	CL-4	CL-5
172.4763	334.9893	497.5024	741.2719	903.7849

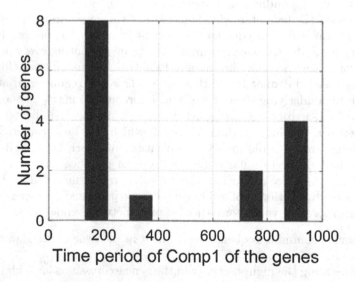

Fig. 6. Histogram based clustering of 15-GNET

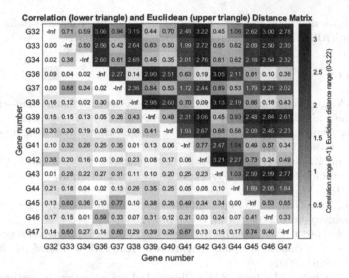

Fig. 7. Pair-wise Correlation and Euclidean distance among the genes

To further study how the approach helps identify probable gene inter-connections, we compare the proposed method with two popular approaches currently used in gene clustering, i.e. Pearson Correlation Coefficient (PCC) and Euclidean distance (E_d) [4,23]. The pair-wise PCC and E_d for the expressions 15 genes as shown in Fig. 7. The lower triangle of the figure shows the correlation coefficients while the upper triangle shows the E_d. Now, according to original network of Fig. 5, we expect to see high PCC and low E_d among the $G32$ and all other genes. However, both these popular approaches fail to detect any of these interconnections. In fact, we can only observe 10 pairs with high correlation and lower E_d. Hence, Seed gene networks based on consideration of PCC and E_d, would not be realistic. On the other hand, our method based on frequency decomposition correctly predicted the presence of an interconnection among all the genes.

5 Conclusion

We proposed a novel method of gene clustering based on frequency decomposition of gene expression to understand the underlying genetic interactions which are biologically relevant, i.e. follow the requirements of underlying regulatory and regulated interconnections. Clustering is done based on the frequency components observed within the genes. Compared with the known popular methods based on correlations and Euclidean distance, we observed that the gene interactions closely resembles the actual interactions in the network. Our future work is to utilise the prior knowledge about gene interactions in any existing evolutionary optimization algorithm in gene regulatory network reconstruction.

Acknowledgment. This research is supported by the Australia-India strategic research fund (AISRF) having the reference number BF040037.

References

1. Smith, A.A.: Classification and alignment of gene-expression time-series data. Ph.D. thesis, UNIVERSITY OF WISCONSIN-MADISON (2009)
2. Aach, J., Church, G.M.: Aligning gene expression time series with time warping algorithms. Bioinformatics **17**(6), 495–508 (2001)
3. Bar-Joseph, Z., Gerber, G.K., Gifford, D.K., Jaakkola, T.S., Simon, I.: Continuous representations of time-series gene expression data. J. Comput. Biol. **10**(3–4), 341–356 (2003)
4. Ernst, J., Nau, G.J., Bar-Joseph, Z.: Clustering short time series gene expression data. Bioinformatics **21**(suppl 1), i159–i168 (2005)
5. Déjean, S., Martin, P.G., Baccini, A., Besse, P.: Clustering time-series gene expression data using smoothing spline derivatives. EURASIP J. Bioinform. Syst. Biol. **2007**(1), 70561 (2007)
6. Ooi, C.H., Chetty, M., Teng, S.W.: Differential prioritization in feature selection and classifier aggregation for multiclass microarray datasets. Data Min. Knowl. Disc. **14**(3), 329–366 (2007)
7. Feng, J., Barbano, P.E., Mishra, B.: Time-frequency feature detection for time-course microarray data. In: Proceedings of the 2004 ACM symposium on Applied computing, pp. 128–132. ACM (2004)
8. Du, L., Wu, S., Liew, A.W.-C., Smith, D.K.: Spectral analysis of microarray gene expression time series data of plasmodium falciparum. Int. J. Bioinform. Res. Appl. **4**(3), 337–349 (2008)
9. Rosa, B.A., Jiao, Y., Oh, S., Montgomery, B.L., Qin, W., Chen, J.: Frequency-based time-series gene expression recomposition using priism. BMC Syst. Biol. **6**(1), 69 (2012)
10. Sussillo, D., Kundaje, A., Anastassiou, D.: Spectrogram analysis of genomes. EURASIP J. Appl. Sig. Process. **2004**, 29–42 (2004)
11. Dodin, G., Vandergheynst, P., Levoir, P., Cordier, C., Marcourt, L.: Fourier and wavelet transform analysis, a tool for visualizing regular patterns in dna sequences. J. Theor. Biol. **206**(3), 323–326 (2000)
12. Bar-Joseph, Z.: Analyzing time series gene expression data. Bioinformatics **20**(16), 2493–2503 (2004)
13. Huang, N.E., Shen, Z., Long, S.R., Wu, M.C., Shih, H.H., Zheng, Q., Yen, N.-C., Tung, C.C., Liu, H.H.: The empirical mode decomposition and the hilbert spectrum for nonlinear and non-stationary time series analysis. In: Proceedings of the Royal Society of London. Series A: Mathematical, Physical and Engineering Sciences, vol. 454, no. 1971, pp. 903–995 (1998)
14. Huang, N.E., Shen, S.S.: Hilbert-Huang Transform and Its Applications, vol. 5. World Scientific, Singapore (2005)
15. Chowdhury, A.R., Chetty, M., Vinh, N.X.: Incorporating time-delays in s-system model for reverse engineering genetic networks. BMC Bioinform. **14**(1), 196 (2013)
16. Ram, R., Chetty, M.: A markov-blanket-based model for gene regulatory network inference. IEEE/ACM Trans. Comput. Biol. Bioinform. **8**(2), 353–367 (2011)
17. Mariani, T.J., Budhraja, V., Mecham, B.H., Gu, C.C., Watson, M.A., Sadovsky, Y.: A variable fold change threshold determines significance for expression microarrays. FASEB J. **17**(2), 321–323 (2003)

18. Marbach, D., Schaffter, T., Mattiussi, C., Floreano, D.: Generating realistic in silico gene networksfor performance assessment of reverse engineering methods. J. Comput. Biol. **16**(2), 229–239 (2009)
19. Stolovitzky, G., Monroe, D., Califano, A.: Dialogue on reverse-engineering assessment and methods. Ann. N. Y. Acad. Sci. **1115**(1), 1–22 (2007)
20. Schaffter, T., Marbach, D., Roulet, G.: Genenetweaver user manual (2012)
21. Chowdhury, A.R., Chetty, M.: An improved method to infer gene regulatory network using s-system. In: 2011 IEEE Congress on Evolutionary Computation (CEC), pp. 1012–1019, IEEE (2011)
22. Xuan, N., Chetty, M., Coppel, R., Wangikar, P.P.: Gene regulatory network modeling via global optimization of high-order dynamic bayesian network. BMC Bioinform. **13**(1), 131 (2012)
23. Jiang, D., Tang, C., Zhang, A.: Cluster analysis for gene expression data: a survey. IEEE Trans. Knowl. Data Eng. **16**(11), 1370–1386 (2004)

Using Genetic Algorithm in Profile-Based Assignment of Applications to Virtual Machines for Greener Data Centers

Meera Vasudevan[1], Yu-Chu Tian[1]([✉]), Maolin Tang[1], Erhan Kozan[2], and Jing Gao[3]

[1] School of Electrical Engineering and Computer Science, Queensland University of Technology, GPO Box 2434, Brisbane, QLD 4001, Australia
y.tian@qut.edu.au
[2] School of Mathematical Sciences, Queensland University of Technology, GPO Box 2434, Brisbane, QLD 4001, Australia
[3] College of Computer and Information Engineering, Inner Mongolia Agricultural University, 306 Zhaowuda Road, Hohhot 010018, Inner Mongolia, China
gaojing@imau.edu.cn

Abstract. The increase in data center dependent services has made energy optimization of data centers one of the most exigent challenges in today's Information Age. The necessity of green and energy-efficient measures is very high for reducing carbon footprint and exorbitant energy costs. However, inefficient application management of data centers results in high energy consumption and low resource utilization efficiency. Unfortunately, in most cases, deploying an energy-efficient application management solution inevitably degrades the resource utilization efficiency of the data centers. To address this problem, a Penalty-based Genetic Algorithm (GA) is presented in this paper to solve a defined profile-based application assignment problem whilst maintaining a trade-off between the power consumption performance and resource utilization performance. Case studies show that the penalty-based GA is highly scalable and provides 16 % to 32 % better solutions than a greedy algorithm.

Keywords: Data center · Energy efficiency · Application assignment · Resource scheduling · Genetic algorithm · Profiling

1 Introduction

Data centers are facing an escalation of services related to high-powered technologies such as artificial intelligence, IPv6, Remote Direct Memory Access (RDMA), virtualizations and cloud solutions. This in turn predictably increases the energy consumption and operation costs to power and maintain these systems at an alarming pace. A report from the Natural Resources Defense Council (NRDC) indicates that data centers consumed 91 billion kWh of electrical energy in 2013. This statistics is projected to increase by 53 % [1] by year 2020.

The necessity for green and energy-efficient measures has become very real and emerging for reducing carbon footprint and the exorbitant energy costs.

© Springer International Publishing Switzerland 2015
S. Arik et al. (Eds.): ICONIP 2015, Part II, LNCS 9490, pp. 182–189, 2015.
DOI: 10.1007/978-3-319-26535-3_21

Energy and cost distribution studies, e.g., Le *et al.* [2], have demonstrated that deploying green initiatives at data centers reduces the carbon footprint by 35 % at only a 3 % cost increase. However, energy-aware measures with simultaneous maximum performance efficiency and minimum energy consumption [3] are difficult to achieve. In most cases, deploying an energy-efficient solution inevitably degrades the resource utilization efficiency of the data centers.

To tackle this challenging issue, this paper presents a penalty-based genetic algorithm to solve the profile-based application assignment problem. The concepts of profiles and profile-based assignment of applications to Virtual Machines (VMs) have been recently established in our previous work [4]. A greedy algorithm has been proposed in our previous work [4] to solve the profile-based assignment problem. The work of this paper significantly improves our previous work by developing a penalty-based genetic algorithm (GA) for deriving a better solution for reducing energy consumption and increasing resource utilization.

The paper is organized as follows. Section 2 reviews related work and motivates the research. Section 3 describes and formulates the profile-based application assignment problem. The penalty-based genetic algorithm is presented in Sect. 4. Case studies are conducted in Sect. 5 to demonstrate the algorithm. Finally, Sect. 6 concludes the paper.

2   Related Work

Evolutionary algorithms such as genetic algorithms (GA) have been applied for job scheduling in data centers and cloud computing. A GA based task and VM scheduler is presented in a paper [5] for cloud systems with a bi-objective of makespan and average CPU utilization. The paper indicates that a good scheduling algorithm should satisfy both application and resource centric objectives. It proposes a penalty-based GA that satisfies energy, CPU and memory utilization objectives. Also, the problem size considered is significantly large.

An energy-efficient resource allocation method is presented in [6], which uses an open source GA framework called jMetal. The allocation objectives also include optimizing task completion times whilst satisfying computational and networking task requirements. The method ensures scalability and performance efficiency for a large number of tasks. Our approach in the present paper utilizes profiles built for both applications and VMs, allowing the penalty-based GA for very large problem sizes without compromising performance efficiency.

VM placement problems, which are NP-complete, have been solved successfully using GA. In [7], the authors minimize the energy consumption of servers and the communication network within the data centers using GA based VM placement. The work is extended in [8] to significantly improve the energy and performance efficiency with an enhanced hybrid genetic algorithm. Our work in the present paper proposes an energy-efficient penalty-based GA for allocating applications to VMs using a profiling method. The scope of this paper is on the application placement management of data centers.

3   Problem Formulation

The original problem of assigning applications to virtual machines is transformed into a constrained combinatorial optimization as below:

A binary decision variable $x_{ij}, i \in I, j \in J$ represents the assignment of an application a_i, $i \in I$, onto a VM V_j, $j \in J$:

$$x_{ij} = \begin{cases} 1 & \text{if } a_i \text{ is allocated to } V_j; \; i \in I, j \in J, \\ 0 & \text{otherwise.} \end{cases} \tag{1}$$

The CPU utilization of a VM V_j is denoted by $\mu[V_j]$ and is derived from a ratio of the CPU busy time to the time interval ($\lambda = 15 \, \text{min}$). The total number of instructions to execute application a_i is given by IC_i.

$$\mu[V_j] = \frac{1}{\lambda} \cdot \sum_{i=1}^{N} \left[\frac{x_{ij} \cdot IC_i}{\mu_j^{CPU}} \right] \tag{2}$$

The total power consumption associated with a data center:

$$P = l[P_{idle} + (E_{usage} - 1)P_{peak} + (P_{peak} - P_{idle})U_{avg}] \tag{3}$$

where P_{peak} and P_{idle} represents the power consumed at the maximum and idle server utilization respectively. The Power Usage Efficiency (PUE) is represented by E_{usage}. U_{avg} is the average CPU utilization of all VMs across the data center for the time interval under consideration. $l \in L$ represents the number of active servers in the data center. The Energy Cost C_{ij} of executing application $a_i, i \in I$, on VM $V_j, j \in J$, is calculated as the product of the ratio of peak and idle power of the host physical machine and the execution time of application a_i on VM V_j:

$$C_{ij} = \frac{P_{peak}}{P_{idle}} \cdot \frac{IC_i}{\mu_j^{CPU}} \tag{4}$$

The constrained combinatorial optimization model for the assignment of a set of applications to VMs is given as:

$$F(obj) = \min \sum_{j=1}^{M} \sum_{i=1}^{N} C_{ij} \cdot x_{ij} \tag{5}$$

$$\text{s.t.} \qquad IC_j/\lambda \le \mu_j^{CPU}, \; \forall j \in J; \tag{6}$$

$$\sum_{i=1}^{N} x_{ij} \eta_i^{mem} \le \eta_j^{mem}, \; \forall j \in J; \tag{7}$$

$$\sum_{j=1}^{M} x_{ij} = 1, \; \forall i \in I; \tag{8}$$

$$x_{ij} = 0 \text{ or } 1, \; \forall i \in I, j \in J. \tag{9}$$

The constraints in Eqs. (6) and (7) ensure that the allocated resources are within the total capacity of the VM. Constraint (8) restricts an application from running on more than one VM. The binary constraint of the allocation decision variable x_{ij} is given by (9).

4   Penalty-Based Genetic Algorithm

Our assignment problem is a combinatorial optimization problem, which is NP-hard. Thus, a steady-state genetic algorithm can be used to solve the problem. This section presents a penalty-based genetic algorithm for the profile-based application assignment problem. The objectives of the penalty-based GA include: minimizing energy consumption; and maximizing resource utilization. Every iteration of the algorithm creates a population consisting of a set of chromosomes representing a possible assignment solution. The initial population consists of chromosomes generated by random allocation of applications to VMs. The following is a description of the genetic operators in the genetic algorithm. Figure 1 represents the working of the genetic operators. The chromosomes are represented by value encoding and parent chromosomes are derived from the roulette wheel selection. Uniform crossover and mutation by selecting and exchanging two genes is applied to the parent solutions to produce the offspring solutions. The termination condition is that cycle is repeated for each generation until a maximum number of generations is reached or an individual is found which adequately solves the problem.

Fitness. The fitness function determines the quality of the solution when compared to an optimal solution. The fitness function effectively penalises an allocation solution that violates the CPU and memory constraints discussed in Eqs. (6) and (7). The lower the energy cost and penalty in terms of resource utilization

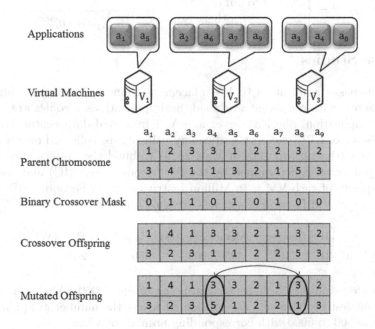

Fig. 1. Value encoding, uniform crossover using binary mask and mutation by selection and exchange of two genes

efficiency, the higher the fitness function. Feasible solutions have a positive fitness value, whereas infeasible solutions incur a negative fitness. The fitness function is derived as:

$$F(X) = w_1 \cdot \bar{F}_{obj} - \frac{w_2}{M} \cdot \sum_{j=1}^{M} \left[\phi_j^{cpu} + \phi_j^{mem} \right] \tag{10}$$

The weights $[w_1, w_2]$ associated with the fitness function is currently set to $[2, 1]$. The multiplicative inverse of the objective function discussed in Eq. (5) is represented by $\bar{F}_{obj}$. In order to normalise and scale the objective function $\bar{F}_{obj}$ to a range of $[1, 10]$, we use:

$$\bar{F}_{obj} = \left[\frac{F_{worst} - F_{obj}}{F_{worst} - F^{\star}} \right] \cdot \left[\frac{F^{\star}}{F_{obj}} \right] \cdot r + 1 \tag{11}$$

Where, the range $r = 9$. The best (minimized) and worst objective function is represented by $F^{\star}$ and F_{worst}, respectively. The penalty for CPU and memory constraint violations are derived as follows:

$$\phi_j^{cpu} = \begin{cases} 0, & \text{if } U_{avg} = 1 \\ \lambda \cdot \mu_j^{cpu}/IC_j, & \text{if } 0 < U_{avg} < 1 \\ 2 & \text{if } U_{avg} = 0 \end{cases} \tag{12}$$

$$\phi_j^{mem} = \begin{cases} 2(1 - 1/\alpha), & \text{if } \alpha \geqslant 1, \\ 2, & \text{otherwise}, \end{cases} \quad \alpha = \frac{\eta_j^{mem}}{\sum_{i=1}^{N} x_{ij} \cdot \eta_i^{mem}}. \tag{13}$$

5   Case Studies

The profile-based application to VM placement framework targets a big class of data centers with consistent workloads and applications. Profiles are created for every application, physical server and VM from real data center workload logs consisting of CPU, memory and energy utilizations, collected over a period of seven days (the 12th to 19th of May, 2014) to build the profiles. The length of each application is determined by the Instruction Count (IC) and the computing capacity of each VM is in Million Instructions Per Second (MIPS). The application and VM parameter settings are shown below:

IC (instr)	IPS (inst/sec)	Memory (bytes)	P_{peak} (W)	P_{idle} (W)	E_{usage}
$[5, 10] \times 10^9$	$[1, 2] \times 10^9$	$[1000, 5000]$	350	200	2

In our case studies, a data center consisting of upto 2000 VMs is considered. Six different test problem sets are considered where the number of applications ranges from 500 to 5000 with corresponding number of VMs:

The implementation of our profile dependent penalty-based genetic algorithm is carried out with a pre-set population size of 200 individuals in each generation.

Problem	1	2	3	4	5	6
VMs	100	400	800	1200	1600	2000
Applications	500	1000	2000	3000	4000	5000

The termination condition is reached when there is no change in the average and maximum fitness values of strings for 10 generations. The number of maximum generations is set to be 200. The probabilities for crossover and mutation are configured to be 0.75 and 0.02, respectively.

The high scalability of the GA is established by solving the allocation problem for upto 2000 VMs and 5000 applications. Figure 2 displays the algorithm solution time with respect to the increasing problem size. As the test problem size $[M * N]$ increases, the solution time of the GA increases linearly.

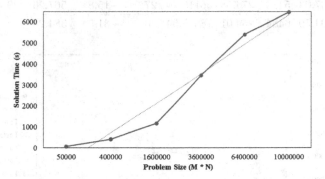

Fig. 2. Scalability of the penalty-based GA.

In order to evaluate the quality of solutions produced by our GA, we solve the test problems using a simple greedy algorithm. The genetic algorithm is stochastic in nature. The quality of allocation solutions in terms of energy consumption are assessed by using GA to solve 30 configurations of each of the test problems as shown in Fig. 3. The resulting mean of energy consumption and solution times is given in Table 1. According to the results, the GA produces 16 % to 32 % better solutions in terms of energy consumption than the greedy algorithm. Although the solution times are higher compared to the greedy approach for the increasing number of applications, the GA maintains an efficient trade-off with energy consumption.

A paired t-test is conducted for the two independent means provided by the GA and greedy algorithm for each of the six test problems. The null hypothesis is that there is no difference between the GA and greedy energy consumption means. The confidence interval is set at 95 % and a two-tailed hypothesis is assumed. The t-stat values are recorded in Table 1 and the p-values are all significantly less than 0.05. The results show that the difference between the means are significant and thus, the null hypothesis is rejected.

As shown in Fig. 4, the average sum of CPU and memory utilization efficiency of the penalty-based GA is 3 % to 22 % more efficient when compared to the greedy approach. Also the variance of the CPU utilization of the GA (0.56) is lower than that of the Greedy algorithm (0.90). This indicates the GA is more consistent in resource allocation.

Table 1. Energy and solution time performance (Energy unit: W; time unit: sec).

Genetic Algorithm				Greedy		T-Test			
Energy	SD	Time	SD	Energy	Time	t-stat	std. err	DF	crit 2-tail
12878.47	1227.90	69	7.53	15017.28	2	−9.54	224.18	29	2.045
21379.27	2404.85	412	38.98	28234.01	6	−15.61	439.06	29	2.045
27113.47	1086.97	1189	50.39	33482.86	9	−32.091	198.45	29	2.045
32001.33	1264.83	3459	341.61	38416.58	18	−27.778	230.92	29	2.045
47149.83	3107.03	5412	276.58	56115.70	27	−15.81	567.26	29	2.045
65904.90	2104.70	6484	250.01	78025.54	40	−31.54	384.26	29	2.045

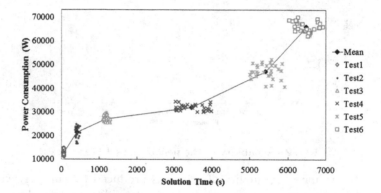

Fig. 3. GA energy consumption and solution time for 30 configurations of each test problem set.

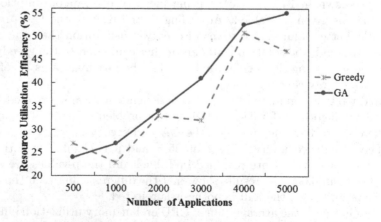

Fig. 4. Resource utilization efficiency.

6   Conclusion

A penalty-based genetic algorithm has been presented for profile-based assignment of applications to VMs. Improving our previous work significantly, it optimizes the energy consumption of data centers while maintaining utilization performance efficiency in terms of CPU and memory. The case studies have demonstrated that the algorithm is highly scalable and provides significantly better solutions than a greedy application placement approach.

Acknowledgments. Author J. Gao would like to acknowledge the support from the National Natural Science Foundation of China under the Grant Number 61462070, and the Inner Mongolia Government under the Science and Technology Plan Grant Number 20130364.

References

1. Whitney, J., Delforge, P.: Scaling Up Energy Efficiency Across the Data Center Industry: Evaluating Key Drivers and Barriers. Issue Paper, Natural Resources Defense Council (NRDC) (2014)
2. Le, K., Bianchini, R., Martonosi, M., Nguyen, T.: Cost- and energy-aware load distribution across data centers. In: Proceedings of HotPower, pp. 1–5 (2009)
3. Greenberg, A., Hamilton, J., Maltz, D.A., Patel, P.: The cost of a cloud: research problems in data center networks. ACM SIGCOMM Comput. Commun. Rev. **39**, 68–73 (2008)
4. Vasudevan, M., Tian, Y.-C., Tang, M., Kozan, E.: Profiling: an application assignment approach for green data centers. In: 40th IEEE Annual Conference of the Industrial Electronics Society, pp. 5400–5406. IEEE Press, Dallas (2014)
5. Sindhu, S., Mukherjee, S.: A genetic algorithm based scheduler for cloud environment. In: 4th International Conference on Computer and Communication Technology (ICCCT), pp. 23–27 (2013)
6. Portaluri, G., Giordano, S., Kliazovich, D., Dorronsoro, B.: A power efficient genetic algorithm for resource allocation in cloud computing data centers. In: 3rd International Conference on Cloud Networking (CloudNet), pp. 58–63. IEEE Press (2014)
7. Wu, G., Tang, M., Tian, Y.-C., Li, W.: Energy-efficient virtual machine placement in data centers by genetic algorithm. In: Huang, T., Zeng, Z., Li, C., Leung, C.S. (eds.) ICONIP 2012, Part III. LNCS, vol. 7665, pp. 315–323. Springer, Heidelberg (2012)
8. Tang, M., Pan, S.: A hybrid genetic algorithm for the energy-efficient virtual machine placement problem in data centers. Neural Process. Lett. **41**, 211–221 (2015). Springer, US

EDL: An Extended Delay Learning Based Remote Supervised Method for Spiking Neurons

Aboozar Taherkhani[1]([✉]), Ammar Belatreche[1], Yuhua Li[2], and Liam P. Maguire[1]

[1] University of Ulster, Londonderry, UK
Taherkhani-a@emaile.ulster.ac.uk,
{a.belatreche,lp.maguire}@ulster.sc.uk
[2] University of Salford, Manchester, UK
y.li@salford.ac.uk

Abstract. This paper presents an Extended Delay Learning based Remote Supervised Method, called EDL, which extends the existing DL-ReSuMe learning method previously proposed by the authors for mapping spatio-temporal input spiking patterns into desired spike trains. EDL merges the weight adjustment property of STDP and anti-STDP with a delay shift method similar to DL-ReSuMe but also introduces the following distinct features to improve learning performance. Firstly, EDL adjusts synaptic delays more than once to find more precise value for each delay. Secondly, EDL can increase or decrease the current value of delays during a learning epoch by initialising the delays at a value higher than zero at the start of learning. Thirdly, EDL adjusts the delays related to a group of inputs instead of a single input. The ability of multiple changes of each delay in addition to the adjustment of a group of delays helps the EDL method to find more appropriate values of delays to produce a desired spike train. Finally, EDL is not restricted to adjusting only one type of inputs (inhibitory or excitatory inputs) at each learning time. Instead, it trains the delays of both inhibitory and excitatory inputs cooperatively to enhance the learning performance.

Keywords: Delay shift learning · Spiking neuron · Spatiotemporal pattern · Supervised learning · Synaptic delay

1 Introduction

It is broadly agreed that spikes (a.k.a pulses or action potentials), which represents short and sudden increases in voltage of a neuron, are used to transfer information between neurons. The principle of encoding information through spikes is a controversial matter. Previously, it was supposed that the brain encode information through spike rates [1]. However, neurobiological research findings have shown high speed processing in the brain that cannot be performed by rate coding scheme alone [2]. It has shown that human's visual processing can perform a recognition task in less than 100 ms by using neurons in multiple layers (from the retina to the temporal lobe). It takes about 10 ms processing time for each neuron. The time-window is too small for rate coding to occur [2, 3]. High speed processing tasks can be performed using precise timing of spikes [3, 4]. The ability of capturing the

© Springer International Publishing Switzerland 2015
S. Arik et al. (Eds.): ICONIP 2015, Part II, LNCS 9490, pp. 190–197, 2015.
DOI: 10.1007/978-3-319-26535-3_22

temporal information of input signals enables spiking neurons to be more powerful than their non-spiking predecessors. However, the intrinsic complexity of spiking neurons needs the development of robust and biologically more plausible learning rules [3, 5].

SpikeProp [6] is one of the first supervised learning methods for spiking neurons. It is inspired by the classical back propagation algorithm. The multi-spike learning algorithm [7] is another example of gradient based methods which are based on the estimation of the gradient of an error function. The gradient based methods suffer from the problems of silent neuron and local minima. In [8] an evolutionary strategy based supervised method is proposed. Unlike the gradient-based methods, this is a derivative-free optimization method and has achieved better performance than SpikeProp. However, it is time consuming due to the iterative nature of evolutionary algorithms.

ReSuMe is a biologically plausible supervised learning algorithm for spiking neurons that can train a neuron to produce multiple spikes. It uses Spike-timing-dependent plasticity (STDP) and anti-STDP to adjust synaptic weights [9]. The STDP is driven by using a remote teacher spikes train to enhance appropriate synaptic weights to force the neuron to fire at desired times. Using remote supervised teacher spikes enables ReSuMe to overcome the silent neuron problem existing in the gradient based learning methods such as SpikeProbe. The ability of on-line processing and locality are two remarkable properties of ReSuMe. It works based on weight adjustment.

Delay Learning Remote Supervised Method (DL-ReSuMe) [4] integrates both the weight adjustment method and the delay shift approach to improve the learning performance. DL-ReSuMe uses STDP and anti-STDP of synapses to adjust synaptic weights similar to ReSuMe [9]. DL-ReSuMe learns the precise timing of a desired spike train in a shorter time with a higher accuracy compared to the well-known ReSuMe algorithm. ReSuMe has problem with silent window, a short time window in the input spatiotemporal pattern without any input spikes, around a desired spike. However, DL-ReSuMe uses delay adjustment to shift nearby input spikes to the appropriate time in the silent window at a desired spike. An extended version of DL-ReSuMe called Multi-DL-ReSuMe has been proposed in [10] to train multiple neurons for classification of spatiotemporal input patterns.

We have shown in our previous work [4] that the performance of DL-ReSuMe drops for high values of total simulation time T_t and comparatively high frequencies of input spike train or desired spike train. Because, the number of input spikes and desired output spikes are increased when Tt is increased, and consequently increases the negative effect of a delay learning related to a desired spike on the other desired spikes. In DL-ReSuMe each delay is only adjusted once by considering one of the desired spikes and input spikes shortly before the desired spike. The adjusted delay could have negative effects on the other learned desired spikes. In DL-ReSuMe, there are no mechanisms to compensate the negative effect of a delay adjustment by considering various desired spikes and refine a delay by multiple adjustments. When the number of spikes is increased by increasing T_t, the negative effect is also increased and the performance of DL-ReSuMe drops as shown in [4].

In this paper, we extend the DL-ReSuMe by adding some new characteristics. The extended version of DL-ReSuMe, called EDL, can achieve higher performance than DL-ReSuMe and ReSuMe for different values of Tt.

2   Synaptic Weight and Delay Learning in EDL

The structure of a neuron in EDL is shown in Fig. 1. Like DL-ReSuMe, EDL trains both the synaptic delays and weights. It is a heuristic method which trains a spiking neuron to map a spatiotemporal input pattern into a desired output spike train. The delay training is performed at the time of a desired spike and at the time of an undesired actual output spike.

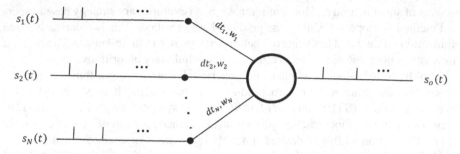

Fig. 1. The neuron structure used in DL-ReSuMe. There are N input synapses each of which is associated with a weight and a delay.

EDL adjusts the delays to generate a set of desired spikes by increasing the neuron PSP at the instances of the desired spikes. EDL increases the delays of excitatory inputs that have input spikes shortly before a desired time to increase the PSP at the desired time. The increment in the delays brings the positive PSP produced by the excitatory input close to the time of the desired spike which consequently increases the level of the total PSP of the neuron at the time of the desired spike and may cause the neuron to hit the firing threshold and fire at the desired time. Additionally, EDL reduces the delays of inhibitory inputs that have input spikes shortly before the desired times. The reduction of the delays increases the distances of the negative PSP produced by the inhibitory inputs from the desired times. Thus the total produced PSP is increased at the desired time and help the neuron to produce the desired spike.

EDL also tries to remove undesired actual output spikes by reducing the neuron PSP at the instances of the undesired actual output spikes. This is done by reducing the delays of excitatory inputs that have spikes shortly before the undesired output spikes. The delay reduction reduces the PSP produced by the excitatory inputs at time of an undesired output spike time which makes the neuron less likely to reach the firing threshold, hence cancelling the undesired output spike. Furthermore, EDL increases appropriately the delays of inhibitory inputs that have spikes shortly before an undesired output spike to reduce the neuron PSP at the time of the undesired output spike. The delay adjustment continues over a number of learning epochs and a delay could be adjusted many times to train a neuron to fire at the desired spiking times.

EDL, similar to DL-ReSuMe, uses STDP to increase the neuron PSP at the time of desired spikes, i.e. it increases the weights of (both excitatory and inhibitory) input synapses that have spikes shortly before a desired time. EDL also uses anti-STDP at the time of an undesired output spikes to reduce the weights of input synapses that have spikes shortly before an undesired spike.

A local variable called spike trace is used to calculate the value of a delay adjustment. The trace shows the effect of an input spike in a synapse in a time interval. The effect of an input spike train is represented by a local variable called trace. The trace of the ith input spike train, $s_i(t)$, is called $x_i(t)$. $x_i(t)$ is described in (1). $x_i(t)$ jumps to a saturated value 'A' whenever a presynaptic spike arrives and then decreases exponentially according to the saturated model [4, 11]:

$$x_i(t) = \begin{cases} Ae^{-(t-t_i^f-del_i)/\tau_x} & \text{for } t_i^f < t < t_i^{f+1} \\ A & \text{for } t = \dots, t_i^{f-1}, t_i^f, t_i^{f+1}, \dots \end{cases} \quad (1)$$

where τ_x is the time constant of the exponential decay, amplitude A is a constant value, and del_i is the delay corresponding to the ith input. The level of $x_i(t)$ at the moment t can be used to determine the time of the last previous input spike. $x_i(t)$ can also be used to find the input synapse that has the nearest spike before the current time, t. Suppose that the mth synapse ($m = 1, 2, \dots$ or N) has the nearest spike before the current time t and consequently, out of all (input) synapses, has the maximum trace at time t, $x_m(t)$, and it should be less than A, because the spike should occur before the current time. The delay, dt_m, which is necessary to be applied to the synapse m, to shift the effect of its spike to the time t can be calculated using the inverse of the trace function given in (1). The result is shown in (2).

$$dt_m = t - t_m^f - del_m = -\tau_x Ln\left(\frac{x_m(t)}{A}\right) \quad (2)$$

where t_m^f is firing time of the fth spike of the input m before current time t, del_m is the delay related to the mth input. The delays are adjusted at the times of desired spikes and actual output spikes in proportion to the time interval of the closest input spike before a desired spike or an actual output spike.

The delay of ith input is adjusted at the instances of desired spikes, t^{fd}, and at the instances of the neuron actual output spikes, t^{fa}. Suppose that the ith input synapse is an excitatory synapse. The delay adjustment of the delay related to the ith excitatory input, $dte_i(t)$, is calculated by (3).

$$dte_i(t) = \begin{cases} +dte_m(t)x_i(t)/xe_m(t), & t = t^{fd} \\ -dte_m(t)x_i(t)/xe_m(t), & t = t^{fa} \\ 0, & Otherwise \end{cases} \quad (3)$$

where t^{fd} is the firinig time of the fd^{th} desired spike in the desired spike train i.e. $fd = 1, 2, \dots$ or N_d, and N_d is the number of spikes in the desired spike train. t^{fa} is the firing time of the fa^{th} spike at the neuron actual output spike train (fa = 1, 2, \dots or N_a, and N_a is the number of spikes in the actual output spike train). $xe_m(t)$ (when $t = t^{fd}$) is the trace of the excitatory input that has the closest input spike before the desired spike at t^{fd}. The time interval of the closest input spike and the desired spike at t^{fd}, $dte_m(t = t^{fd})$, can be calculated by substituting $xe_m(t^{fd})$ in (2). The positive delay at the desired time ($t = t^{fd}$) in (2) brings the excitatory PSP produced by the ith input close to the desired spike and consequently increases the total PSP at the desired time and it can produce an actual output spike at the desired time. Similarly, the negative adjustment of the delay of excitatory inputs in (2) moves the produced excitatory PSP away from the output spike at $t = t^{fa}$ and tries to cancel the actual output spike. When the learning algorithm trains the neuron to produce an actual output spike at the time of a desired spike i.e. $t = t^{fd} = t^{fa}$, the enhancement and the reduction of the delay is equal according to (3), and the net delay adjustment is zero. Therefore, the delay adjustment is stopped when the neuron learns to fire the desired spike train. Equation (3) also shows that the maximum delay adjustment is related to the input that has the closest spike before the time t and it is equal to $dte_m(t)$ (when $x_i(t) = xe_m(t)$) and the delays of other inputs are adjusted in proportion to their distance from the time t (depending to $x_i(t)$).

If the ith input synapse is an inhibitory synapse, the delay of the ith input is reduced at the times of the desired spikes such that the distance of the inhibitory PSPs produced by the input spikes before the desired time is increased. The delays are increased at the times of actual output spikes to bring the produced inhibitory PSPs by the ith input close to the actual output times, according to (4).

$$dti_i(t) = \begin{cases} -dti_m(t)\, x_i(t)\, /xi_m(t), & t = t^{fd} \\ +dti_m(t)\, x_i(t)\, /xi_m(t), & t = t^{fa} \\ \qquad 0, & Otherwise \end{cases} \qquad (4)$$

where $dti_i(t)$ is the delay adjustment of the ith inhibitory input at the time t, $xi_m(t)$ is the trace of the mth inhibitory input synapse that have the closest input spike before time t between all inhibitory inputs that have spikes before the time t. $dti_m(t)$ is the time distance between the closest inhibitory input spike and the time t that is calculated by putting $xi_m(t)$ in (2). Equation (4) shows that the delay adjustment of the ith input is reduced by decreasing $x_i(t)$ or by increasing the time distance of its input spike before time t. The amount of the delay increment and decrement at the time t is equal and they cancel each other when an actual output is produced at the time of a desired spike, i.e. $t = t^{fd} = t^{fa}$. In this situation the delay adjustment is stabilized. The delay adjustments are limited by a maximum amount $dtMax$.

3   Experimental Setup

The learning task is to train a neuron to fire a desired spike train in response to a spatio-temporal input pattern. The proposed learning algorithm trains the neuron in a number of learning epochs. The input and desired spikes are generated by a Poisson process. Once the input and the output pair are generated they are frozen during a learning run. In each learning run that is composed of a number of learning epochs the learning parameters (synaptic weights and delays) are adjusted to train the neuron to produce the desired spike train. Learning parameters are initialised at the start of a learning run and are then continuously updated during a number of learning epochs. The average performance over a number of runs is reported. Each run is performed using a different pair of a spatiotemporal input pattern and a single desired output spike train. The different spike trains for the different runs are drawn from random Poisson processes. At the start of each learning run the weights and delays are reset. The weights are initialized randomly, and the delays are initialized to $dtMax/2$.

During a learning epoch the delay adjustments of synapse i which are calculated by (3) or (4) are accumulated and the accumulated delay adjustment is added to the ith input delay, del_i. The new delay should be within $[0, dtMax]$.

The weights are updated by the delayed version of ReSuMe rule which is governed by (5) where the delays, del_i, are applied to the ReSuMe weight update.

$$\frac{dw_i(t)}{dt} = \left[s_d(t) - s_o(t)\right] \left[a + \int_0^{+\infty} Tw(s)\, s_i\left(t - del_i - s\right) ds\right] \tag{5}$$

where del_i is the delay related to input i. In the delayed version of ReSuMe, the term $s_i\left(t - del_i - s\right)$ is used instead of $s_i(t - s)$ used in ReSuMe. At the start of the training procedure, 20 % of the weights are considered inhibitory and 80 % of the weights are considered excitatory as in [12].

4 Results and Discussion

The performance of EDL, DL-ReSuMe and ReSuMe are compared in Fig. 2. A spatio-temporal input pattern with 400 spike trains is used. Each spike train is generated by a Poisson process. The mean frequency of the input pattern is 5 HZ. A desired spike train is also produced by a Poisson process with 100 Hz mean frequency. The total time duration, T_t, of input and output spike trains are changed from 100 to 1000 ms. The simulation is done for 20 independent runs and the median value of the 20 run is shown in Fig. 2.

The results show that the performance of EDL is higher than the performance of ReSuMe for all the values of T_t (up to 1000 ms). In addition, the performances of EDL and DL-ReSuMe are identical when the simulation time is shorter than 500 ms but the performance of EDL then becomes higher for T_t longer than 500 ms. Like DL-ReSuMe, the performance of EDL also drops after 500 ms but less sharply than DL-ReSuMe. The weights are adjusted based on STDP and anti-STDP. In the STDP and anti-STDP the time of an input spike is important. For example, during STDP, if an input spike occurs shortly before and close to a desired spike time, the corresponding weight is potentiated by a high value. When a delay is applied to the input after the weight adjustment, the PSP produced by the input spike is delayed. A high enough delay can shift the produced PSP beyond the desired time and because of the high weight of the input (related to the previous weight learning) it may cause an undesired output spike after the desired time, and this can be considered as the negative effect of the delay adjustment on the weight learning. A high value of T_t can result in each delay adjustment affecting a higher number of input spikes and consequently increases the negative effect of delay adjustment which can contribute to a decrease in the learning performance as shown in Fig. 2.

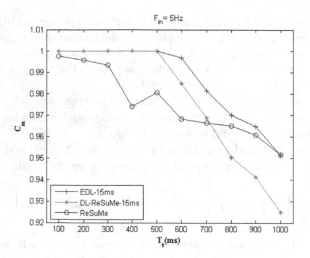

Fig. 2. Comparison of the performances of EDL, DL-ReSuMe and ReSuMe for different values of T_t. The input spatiotemporal pattern and the desired spike train are produced by Poisson processes with frequencies of 5 Hz and 100 Hz, respectively. **dtMax** is set to 15 ms for both EDL and DL-ReSuMe. The synaptic delays are initialized to dtMax/2 = 7.5 ms and 0 ms for EDL and DL-ReSuMe, respectively.

5   Conclusion and Future Work

In this paper, DL-ReSuMe is extended to improve the learning performance resulting in the proposed EDL (Extended Delay Learning) method. EDL trains a neuron to fire at desired times in response to an input spatiotemporal spiking pattern using the synergy between weights and delays training. This interplay between weights and delays helps the neuron fire at desired times while at the same time cancel its spikes fired at undesired times. Unlike DL-ReSuMe, delays are changed for both inhibitory and excitatory inputs and a synaptic delay can be changed several times in EDL. Multiple adjustments of a delay at various times give the EDL method the ability to assess the effect of a delay on various desired spikes and actual output spikes and then to reconcile the various effects by finding an appropriate delay adjustment.

The learning performance of EDL was compared against DL-ReSuMe, developed by the authors, and the well-established ReSuMe method and the obtained results have shown an overall increased accuracy for different durations of Poisson input spatiotemporal spiking pattern and desired spike train.

Future work will explore the application of EDL to a network of spiking neurons and its viability in learning classes of spatiotemporal spiking patterns.

References

1. Masquelier, T., Deco, G.: Learning and coding in neural networks. Principles of Neural Coding Anonymous, pp. 513–526. CRC Press, Boca Raton (2013)

2. Thorpe, S., Delorme, A., Van Rullen, R.: Spike-based strategies for rapid processing. Neural Netw. **14**, 715–725 (2001)
3. Vreeken, J.: Spiking neural networks, an introduction, Technical Report UU-CS, pp. 1–5 (2003)
4. Taherkhani, A., Belatreche, A., Li, Y., Maguire, L.P.: DL-ReSuMe: a delay learning based remote supervised method for spiking neurons. IEEE Trans. Neural Netw. Learn. Syst. **PP**, 1 (2015)
5. Taherkhani, A., Belatreche, A., Li, Y., Maguire, L.: A new biologically plausible supervised learning method for spiking neurons. In: Proceedings of ESANN, Bruges (Belgium) (2014)
6. Bohte, S.M., Kok, J.N., La Poutre, H.: Error-backpropagation in temporally encoded networks of spiking neurons. Neurocomputing **48**, 17–37 (2002)
7. Xu, Y., Zeng, X., Han, L., Yang, J.: A supervised multi-spike learning algorithm based on gradient descent for spiking neural networks. Neural Netw. **43**, 99–113 (2013)
8. Belatreche, A., Maguire, L., McGinnity, M.: Advances in design and application of spiking neural networks. Soft Comput. Fusion of Found. Methodol. Appl. **11**, 239–248 (2007)
9. Ponulak, F., Kasiński, A.: Supervised learning in spiking neural networks with ReSuMe: sequence learning, classification, and spike shifting. Neural Comput. **22**, 467–510 (2010)
10. Taherkhani, A., Belatreche, A., Yuhua, L., Liam, P.M.: Multi-DL-ReSuMe: Multiple neurons delay learning remote supervised method. IN: The 2015 International Joint Conference on Neural Networks (IJCNN), Killarney, Ireland (2015)
11. Morrison, A., Diesmann, M., Gerstner, W.: Phenomenological models of synaptic plasticity based on spike timing. Biol. Cybern. **98**, 459–478 (2008)
12. Izhikevich, E.M.: Polychronization: computation with spikes. Neural Comput. **18**, 245–282 (2006)

Enhanced Genetic Algorithm Applied
for Global Optimization

Fadzil Ahmad[1,2(✉)], Nor Ashidi Mat Isa[2], Zakaria Hussain[1],
Saiful Zaimy Yahaya[1], Rozan Boudville[1], Mohamad Faizal Abdul Rahman[1],
Aini Hafiza Mohd Saod[1], and Zuraidi Saad[1]

[1] Faculty of Electrical Engineering, Universiti Teknologi MARA (UiTM), Pulau Pinang,
13500 Permatang Pauh, Penang, Malaysia
[2] Imaging and Intelligent Systems Research Team (ISRT),
School of Electrical and Electronic Engineering, Universiti Sains Malaysia (USM),
14300 Nibong Tebal, Penang, Malaysia
fadzilahmad@ppinang.uitm.edu.my

Abstract. Conventional genetic algorithm (GA) has several drawbacks such as premature convergence and incapable of fine tuning around potential region. Thus, new enhanced GA that focuses on new search, crossover and elitism strategy is proposed in this study. It involves solution enhancement phase by performing search among high quality chromosomes via new crossover operator. A modified elitism operation is devised to ensure that the performance of enhanced GA not getting worse than the standard GA in case of solution enhance phase fails to find better chromosomes. In modified elitism, best chromosomes resulted from the enhancement phase and normal population will have to compete among each other to survive in next generation. The enhanced GA has been applied for solving global optimization of benchmark test functions and compared with standard GA. Based on the occurrences of the algorithms produce the best result across different test functions and elitism size; it is proven that the proposed method outperforms standard GA.

Keywords: Genetic algorithm · Global optimization · Crossover · Elitism

1 Introduction

Global optimization is a process of obtaining the best solution of a given problem that would have either minimized or maximized the problem objective and to satisfy all the given constrains. This concept is widely used by mankind of modern civilization for solving various problems in many research areas such as in finance [1], engineering design [2] and biomedical [3].

Genetic Algorithm (GA) is a suitable candidate in solving global optimization problem. As a family of global search technique, it performs a global multi-directional search and capable of quickly locating near optimal solution even in complicated search space. It has been successfully applied for global optimization of neural network [4],

© Springer International Publishing Switzerland 2015
S. Arik et al. (Eds.): ICONIP 2015, Part II, LNCS 9490, pp. 198–205, 2015.
DOI: 10.1007/978-3-319-26535-3_23

aerodynamic shapes and the composite material of a car [5], microarray technology [6] and structural and operational design of skyscrapers, mega cities, highways, factories and sophisticated machines [7].

Unfortunately, GA is unable to fine tune the solution around the potential region due to the incapability of local search. Besides, the pre-mature convergence is another often-observed problem in GA [8]. It is a phenomenon in which the quality of the generated offspring is always being inferior to their parent. Therefore in certain optimization problem where the search space is too complex, conventional GA may not able to produce promising result. In order to attain better optimization performances, modification to conventional GA is needed. Furthermore, it is broadly acknowledged that the optimization performances of a GA can be further improved either by integrating new search strategy into the GA or modifying the standard genetic operator such as selection, reproduction, crossover, mutation and elitism scheme.

In a standard GA procedure, high quality chromosomes will be destroyed if they are not selected as parent and not involved in elitism process. In this work, new enhanced GA that focuses search around high potential search region related with high quality chromosomes is proposed. It is also based on the theory that the most highly fit parents (both male and female) are most likely to produce healthiest offspring. The new search procedure involves genes exchange among high quality chromosomes via newly proposed crossover operator. It differs from conventional crossover in a way that the destruction of near optimal genetic information in the chromosome is avoided during the crossover operation.

It is not guaranteed that the GA enhancement phase via new crossover operator will always produce a better quality chromosome. This is due to the probabilistic nature of the GA approach. If low quality chromosomes are allowed to survive, the search capability of the GA will deteriorate. Therefore a modified elitism strategy is introduced whereby the chromosomes before and after the enhancement phase will have to compete among each other to survive in next generation. By implementing the modified elitism, in the case of the enhancement phase fail to produce better chromosomes the performance of enhanced GA will not get worse than the standard GA.

2 Methodology

The general idea of the new GA enhancement phase and modified elitism is visualized in Fig. 1. It can be observed that the chromosomes are ranked based on their fitness value and put in a sorted list with the highest rank chromosome be on top of the list. Then, the enhanced GA exploits a group of top ranked chromosomes of size k through selection and modified crossover. It is can be clearly observed that the modified elitism involves two groups before (normal population) and after enhancement (enhanced population) phase. The implementation of the algorithm is summarized as follows:

(1) **Initial population**: The initial population of enhanced GA is randomly generated. Each chromosome in the population consists of 2 components or gene segments; each of them varies within a pre-specified range. Based on the quantified interval, each gene segment is coded using binary value.

(2) **Fitness calculation**: The gene segments for each chromosome in binary are decoded to obtain their equivalent decimal values and substituted into the test function equation that we intend to obtain its global minimum. Chromosome that returns the smallest value represents the best solution.

(3) **Selection**: Based on the fitness value, the selection method is executed using Roulette rank-based selection [9]. The fittest chromosome occupies the largest area in the Roulette wheel.

(4) **Modified crossover**: The crossover location is controlled so that it only occurs at location that will not split the gene segment. This is to avoid the destruction of near optimal information contain in gene segment. Top quality chromosomes (size k) are obtained from GA population. First group of first gene segments (size l) is created via rank based Roulette selection using the ranking of their respective chromosome. Second group of second gene segments is created in the same way of first gene segment group is created. The two groups are combined to form a new group of enhanced chromosomes.

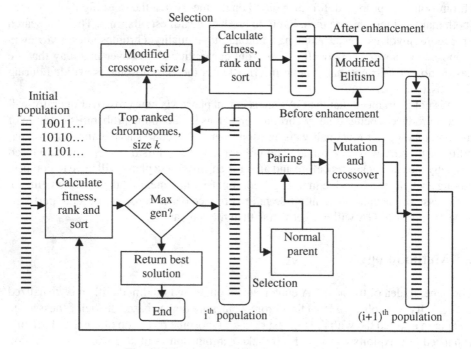

Fig. 1. The architecture of the enhanced GA.

(5) **Modified elitism**: The best chromosomes before and after enhancement phase will have to compete each other to survive in next generation. E chromosomes are selected from both normal and improved population. The two groups are combined and reordered based on fitness value so that the best chromosome being on the top of the list. The top E chromosomes are then selected and moved to next generation.

(6) **Standard crossover and mutation**: The parents are selected (Roulette ranked selection) from normal population and paired for the crossover operation. It is followed by mutation operation. These are based on standard crossover and mutation operation as described in [10].

(7) **New generation**: The offspring obtain from Step 6 is then combined with the elitist chromosomes (the chromosomes that survive the modified elitism) to form a new generation. All Step 2 until Step 7 are repeated until maximum number of generation is reached (Fig. 2).

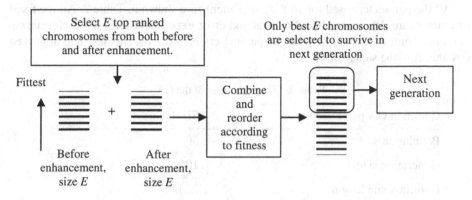

Fig. 2. The operation of modified elitism.

2.1 The Test Functions

The proposed algorithm is employed for optimizing highly complicated and multimodal benchmarking test functions – Ackley, Rastrigin and Schaffer2 and Shubert. The solution domain of these test functions consist of many local optima solutions. Only the algorithms that have reached a certain level of robustness are capable to avoid the local optimal and obtain the global optimal solution.

The formulation and search surface of all the test functions can be found in [11]. Their characteristics are tabulated in Table 1.

Table 1. The characteristics of the test functions the evaluated range.

Function name	Global minima values	Global minima locations	Evaluated range
Ackley	0	0,0	[−32.768; 32.768]
Schaffer N.2	0	0,0	[−100; 100]
Rastrigin	0	0,0	[−5.12; 5.12]
Eggholder	−959.6407	512,404.2319	[−512; 512]

2.2 Experimental Setup

In order to evaluate the performance of the enhanced GA and to compare with the standard GA, two set of experiments for finding global minimum is conducted by varying the elitism size, E. First set uses standard GA and second set uses enhanced GA. Standard GA does not have the enhancement phase and modified elitism. As a member of stochastic computational techniques, GA does not guarantee to produce consistent result every time it is executed. Therefore, the performance of both the standard and improved GA is evaluated based on the average and best result produced from 100 repetitions of the algorithm.

All the parameters used for all the experiments are shown in Table 2. All the fixed parameters are determined based on trial and error experiment for good performance. For a fair comparison, all the common parameters between the standard and improved GA are set to the same values.

Table 2. The parameter of the GA.

Common GA parameters	Size/value
Population size	50
Generation size	100
Chromosome length	32 bits
Elitism size, E	2, 4 and 6
Selection scheme	Rank based roulette selection
Crossover probability	0.8
Mutation probability	0.05
Improved GA parameters	
Top ranked chromosome size, k	20
Enhanced crossover chromosomes size, l	20

3 Results and Discussion

Table 3 tabulates the results of the best global optima solution found from the 100 repetitions of the algorithms for all the test functions and elitism size, E. The algorithms aim to find the minimum value in the solution space of the test functions and the smallest values is the best. At $E = 2$, it is obvious that, for all the test functions the enhanced GA outperforms the standard GA. At $E = 4$, enhanced GA produces better results on Ackley and Rastrigin, however on Schaffer2 standard GA outperforms enhanced GA. At $E = 6$, both enhanced and standard GA score better results on 2 out of 4 test functions. From Table 3, number of occurrences the enhanced and standard produces better results is 8 and 3 respectively. From the results it is proven that the enhanced GA outperforms standard GA in obtaining best global minimum location.

Table 3. The best performance over 100 repetitions of the algorithm. The best results are highlighted in bold font.

Elitism Size, E	2	4	6
Ackley			
Standard GA	2.24E-05	1.05E-05	1.75E-06
Enhanced GA	**7.54E-06**	**5.96E-06**	**1.60E-06**
Rastrigin			
Standard GA	2.55E-09	2.04E-11	**2.25E-12**
Enhanced GA	**1.38E-09**	**4.84E-12**	1.41E-11
Schaffer2			
Standard GA	3.15E-11	**9.98E-13**	**2.60E-14**
Enhanced GA	**1.41E-11**	2.30E-12	3.55E-14
Eggholder			
Standard GA	−959.6385	−959.6404	−959.6406
Enhanced GA	**−959.6404**	−959.6404	**−959.6407**

Table 4. The average performance over 100 repetitions of the algorithm

Elitism Size, E	2	4	6
Ackley			
Standard GA	2.1E-03	4.3E-04	2.3E-04
Enhanced GA	**2.0E-03**	**3.4E-04**	**1.7E-04**
Rastrigin			
Standard GA	4.5E-05	2.1E-06	1.0E-02
Enhanced GA	**8.6E-06**	**1.9E-06**	1.0E-02
Schaffer2			
Standard GA	1.4E-03	4.0E-04	**2.3E-04**
Enhanced GA	**6.3E-04**	**2.5E-04**	5.2E-04
Eggholder			
Standard GA	−933.01	−931.00	−931.44
Enhanced GA	**−941.36**	**−939.67**	**−938.26**

Table 4 compares the average performance of the algorithms for all the test functions and elitism size. It is demonstrated that, at $E = 2$ and 4 the enhanced GA produces better results for all the test function. At $E = 6$, the results are mixed. The occurrences enhanced and standard GA produces better results is 10 and 1 respectively. To determine the best algorithm considering both the best and the best averaged global optima found, the occurrences for Tables 3 and 4 are total up. Enhanced GA produces 18 occurrences, meanwhile standard GA only scores 4 occurrences.

At $E = 2$, for the best and best averaged performance, the enhanced GA produces better result for all the test functions. This result proves that at low elitism size the search capability of enhanced GA is better than standard GA.

4   Conclusion

This paper introduces enhanced GA that involves solution enhancement phase by performing search among high quality chromosomes via new crossover operator. A modified elitism operation is devised to ensure that the performance of enhanced GA not getting worse that the standard GA in case of solution enhance phase fails to find better results. In modified elitism, best chromosomes resulted from the enhancement phase and normal population will have to compete among each other to survive in next generation. The enhanced GA is applied for solving global optimization problem of highly completed and multimodal test function. The results prove that the capability of enhanced GA in finding global minimum location is better than standard GA.

References

1. Pan, W.-T.: A new fruit fly optimization algorithm: taking the financial distress model as an example. Knowl.-Based Syst. **26**, 69–74 (2012)
2. Grossmann, I.E.: Global Optimization in Engineering Design, vol. 9. Springer-Science +Business Media, Dordrecht (2013)
3. Liu, X., Yang, Q., Qing, H., Long, N.: A new global optimization strategy for medical image elastic registration. In: 2012 IEEE-EMBS International Conference on Biomedical and Health Informatics (BHI), pp. 337–340. IEEE (2012)
4. Ahmad, F., Isa, N.A.M., Hussain, Z., Osman, M.K., Sulaiman, S.N.: A GA-based feature selection and parameter optimization of an ANN in diagnosing breast cancer. In: Pattern Analysis and Applications, pp. 1–10 (2014)
5. Takako, S., Takemura, Y., Schmitt, L.M.: Minimizing wind resistance of vehicles with a parallel genetic algorithm. In: Madaan, A., Kikuchi, S., Bhalla, S. (eds.) DNIS 2014. LNCS, vol. 8381, pp. 214–231. Springer, Heidelberg (2014)
6. Ooi, C.H., Tan, P.: Genetic algorithms applied to multi-class prediction for the analysis of gene expression data. Bioinformatics **19**, 37–44 (2003)
7. Renner, G.B., Ekart, A.: Genetic algorithms in computer aided design. Comput.-Aided Des. **35**, 709–726 (2003)
8. Rocha, M., Neves, J.: Preventing premature convergence to local optima in genetic algorithms via random offspring generation. In: Imam, I., Kodratoff, Y., El-Dessouki, A., Ali, M. (eds.) IEA/AIE 1999. LNCS (LNAI), vol. 1611, pp. 127–136. Springer, Heidelberg (1999)

9. Lin, G., Huang, C., Zhan, S., Lu, X., Lu, Y.: Ranking based selection genetic algorithm for capacity flow assignments. In: Cai, Z., Tong, H., Kang, Z., Liu, Y. (eds.) ISICA 2010. CCIS, vol. 107, pp. 97–107. Springer, Heidelberg (2010)

10. Goldberg, D.: Genetic Algorithms in Search and Optimization. Addison-Wesley, Boston (1989)

11. Surjanovic, S., Bingham, D.: Virtual Library of Simulation Experiments: Test Functions and Datasets. Simon Fraser University. http://www.sfu.ca/~ssurjano/index.html. Accessed 30 Mar 2014

A New Heuristic Based on the Cuckoo Search for Cryptanalysis of Substitution Ciphers

Ashish Jain[1]([✉]) and Narendra S. Chaudhari[1,2]

[1] Discipline of Computer Science and Engineering,
Indian Institute of Technology Indore, Indore, India
phd11120101@iiti.ac.in
[2] Visvesvaraya National Institute of Technology Nagpur, Nagpur, India
nsc0183@gmail.com

Abstract. The efficient utilization of one of the latest search heuristic, namely, cuckoo search for automated cryptanalysis (or attack) of substitution ciphers is addressed. A previously proposed genetic algorithm based attack of the simple substitution cipher is enhanced. A comparison is obtained between various attack algorithms that are based on cuckoo search, genetic algorithm, enhanced genetic algorithm, tabu search and scatter search where cuckoo search shows the best performance. It is worth pointing out that the proposed cuckoo search algorithm provides a valid and efficient option for solving similar permutation problems.

Keywords: Metaheuristics · Cuckoo search · Genetic algorithm · Tabu search · Scatter search · Cryptanalysis · Substitution ciphers

1 Introduction

The goal of the cryptanalyst is to systematically recover the original text (plaintext) and/or key by mounting an attack on the cipher. The attack may involve several ciphertexts and/or some plaintexts, intelligent mathematical computer algorithms and usually some luck. A ciphertext-only attack is one where the objective of the adversary (or cryptanalyst) is to recover the plaintext and/or deduce decryption key by observing only the "known ciphertext". This class of attack is most challenging and in this paper, we study such a cryptanalytic attack on a classical cipher (a most general form –mono-alphabetic (or simple) substitution cipher, hereinafter, substitution cipher). Although classical ciphers, namely, substitution and transposition ciphers were used hundreds of years ago, a particular interest in the study of these ciphers has been retained due to the fact that most of the modern cryptosystems still utilize the functions of these ciphers as their basic building blocks. For instance, the Advanced Encryption Standard (AES) used all over the world by finance community is based on the principle of the substitution-permutation network. Due to such important facts, the classical ciphers are normally considered as first choice when investigating new attack strategies such as the one discussed in this paper. In literature, several heuristics have been applied successfully for cryptanalysis of classical ciphers. For example,

© Springer International Publishing Switzerland 2015
S. Arik et al. (Eds.): ICONIP 2015, Part II, LNCS 9490, pp. 206–215, 2015.
DOI: 10.1007/978-3-319-26535-3_24

[1–8], where the main goal of researchers was to develop efficient cryptanalytic algorithms by utilizing the search (or optimization) heuristics.

1.1   Simple Substitution Cipher

In this type of cipher a key can be represented as a permutation of the plaintext alphabet (for example, if the plaintext alphabet consists of 26 English alphabets and the space character, then a permutation of these 27 alphabet characters usually forms a key). During the encryption process each letter of the plaintext message is replaced by the corresponding key element (i.e. ciphertext alphabet) that forms a ciphertext of length equal to the length of the plaintext message. The original plaintext message from the ciphertext is then recovered by intended recipient using decryption process. For details about this cipher we refer the interested reader to [4].

1.2   Our Contributions

It is surprising that in the last one decade no heuristic attack more efficient than the tabu search attack has been reported for breaking the substitution cipher. In this paper, we overcome the limitation by developing a new attack algorithm by utilizing one of the latest search heuristic, namely, cuckoo search. Moreover, for optimizing the cryptanalysis process we fine-tune some of the parameters of the cuckoo search. Additionally, two related and efficient attacks of the substitution cipher proposed by Clark [4] and are based on the genetic algorithm and tabu search are implemented. Furthermore, the genetic algorithm attack proposed by Clark [4] is enhanced by incorporating a new adaptive mutation operator. Although the scatter search based attack of the substitution cipher proposed by Garici and Drias [5] is not efficient (with respect to time complexity), it is being considered to obtain a fair comparison between previously proposed evolutionary heuristic based attacks and those presented in this paper.

2   Cryptanalysis of Substitution Ciphers

The main weakness of substitution ciphers is that their character frequency statistics (or n-grams) are not changed by the encryption algorithm. In other words, for every grouping of characters in the plaintext there is a distinct and corresponding grouping of characters in the ciphertext [4]. Therefore, most attacks of substitution ciphers attempt to match the n-grams of the encrypted text with those of known language (for example, English). In this paper, we use this fact as the basis to attack the substitution cipher and at the same time automate the search for the required n-gram frequencies by developing efficient algorithms based on the optimization heuristics.

2.1 Fitness Function or Cost

During the optimization process, each candidate key (or individual solution) is used to decrypt the known ciphertext and at the same time the n-gram statistics of the decrypted text is compared to the language statistics. In general, Eq. (1) can be used for comparison of these statistics.

$$Cost_k = \alpha \left(\sum_{i \in \mathscr{A}} |\mathcal{K}_i^u - \mathcal{D}_i^u| \right) + \beta \left(\sum_{i \in \mathscr{A}} |\mathcal{K}_i^b - \mathcal{D}_i^b| \right) + \gamma \left(\sum_{i \in \mathscr{A}} |\mathcal{K}_i^t - \mathcal{D}_i^t| \right), \quad (1)$$

where $\mathscr{A}$ denotes the language alphabet (e.g. in English language: A, B, ..., Z, _), and $\mathcal{K}$ and $\mathcal{D}$ denote statistics of the known language and decrypted text, respectively. The indices u, b and t denote the unigram, bigram and trigram statistics, respectively. Different weights in the range (0.0–1.0) with step size 0.1 can be assigned to α, β and γ. However, the following constraint must be satisfied to keep the number of combinations of α, β and γ workable.

$$\alpha + \beta + \gamma = 1.0. \quad (2)$$

Generally, in order to do more accurate evaluation of candidate keys, the n-grams should be higher in number. Also as per literature, the inclusion of trigram statistics in the fitness function are generally the most effective basis for cryptanalysis of classical ciphers [4]. However, for cryptanalysis of the substitution cipher the fitness function used in literature (for example, [1,4]) is purely based only on the bigrams. The reason is that the complexity associated with computation of trigram statistics is high (order of N^3, where N is the key size), while the benefit over the bigrams is minimal. Moreover, even though small amount of ciphertexts are known, the attack on the substitution cipher is more effective using fitness function which is based on the bigrams only than one which utilizes trigrams alone [4]. Due to these facts, the following fitness function which is purely based on the bigrams is used in this paper to attack the substitution cipher.

$$Cost_k = \sum_{i \in \mathscr{A}} |\mathcal{K}_i^b - \mathcal{D}_i^b|. \quad (3)$$

2.2 Cryptanalysis Using Genetic Algorithms

In this section, we enhance the genetic algorithm attack that was proposed by Clark (particularly, mutation operator). Note, the crossover (or mating) operator employed in Algorithm 1 is identical to that proposed by Clark. Due to lack of space, we do not present the crossover operator, but we refer readers to [4].

Mutation-I. This mutation operator is utilized in most of the previously reported attacks on the classical cipher. It is based on the simple way of perturbing a cipher key where two randomly selected elements of a key are swapped. In addition to this mutation operator, we use a similar mutation operator but an adaptive one which can be described as follows.

Algorithm 1. GA and Enhanced GA Attack (GA/EGA Attack)

1: **Initialization:** Randomly generate the initial population (call this population, $Current_{pop}$) containing m individuals, where individuals are keys of the substitution cipher. Evaluate the fitness of each individuals using Eq. (3).

2: **repeat**

3:    Select $m/2$ pairs of individuals from the $Current_{pop}$. These pairs are referred to as parents of the new generation. Here we use a tournament selection method that choose a best individual (for each pairs of individuals) among five individuals, where the selection of these five individuals should be random.

4:    Each pair of parents are then mated that produces two offspring. These m offspring form the new population abbreviated as New_{pop}.

5:    **In case of GA-Attack: Apply Mutation-I, while in case of EGA-Attack: Apply Mutation-I and Mutation-II to each of the individual of New_{pop}.**

6:    Evaluate the cost of each of the individuals exist in the New_{pop} using Eq. (3).

7:    Merge $Current_{pop}$ with New_{pop} and in parallel remove duplicates. The merging process give a list of sorted (least cost to most cost) solutions, where the size of list will be between m and $2m$. The m best individuals from the merged list then become the new $Current_{pop}$.

8: **until** (Maximum-Iterations)

9: Output the best solution from the $Current_{pop}$.

Mutation-II. Interchange three randomly chosen elements of the key. However, this mutation operator is adapted when evolution starts to languish, i.e. no improvement in the solution is observed after some number of iterations. We limit this number to seven based on the experiments. The main benefit of the adaptive mutation operator is that it improves the rate of convergence and the success probability (in terms of full key recovery) of the genetic algorithm attack.

2.3 Cryptanalysis Using Cuckoo Search Combined with Lévy Flights

In 2009, Yang and Deb [9,10] have proposed a new nature-inspired population based metaheuristic known as cuckoo search via Lévy flights. "This metaheuristic is formed by inspiration from the obligate brood parasitic behavior of few cuckoo species in combination with Lévy flight behavior of some birds and fruit flies [9]". For simplicity, the standard cuckoo search method is described in Algorithm 2.

To generate a new nest (or a solution vector, see Algorithm 2, step 3) $x_j(t+1)$ from an existing nest $x_j(t)$, for, a cuckoo j; Lévy flight is performed as [9,11]:

$$x_j(t + 1) = xj(t) + \mu. \tag{4}$$

The above equation is essentially a stochastic equation for a random walk. In general, a random walk is a Markov chain whose next state/location depends only on the current location (the first term) and the transition probability (the second term) [9]. In the second term, $\mu > 0$ is a step-size scaling factor which should be associated to the scales of the problem of interest [12]. The term l is the step-size which is drawn from a probability distribution with a power law tail (also known as Lévy stable distribution) [13]. It is worth mentioning that the cuckoo search via random walk using Lévy flights is more efficient than other popular swarm intelligence techniques (e.g. PSO) in exploring the search space as its step-size distribution is pseudo-random [12]. Nonetheless, it is not a trivial task to produce pseudo-random steps that accurately obey the Lévy stable distribution. However, from the implementation aspects, Mantegna's [13] algorithm is one of the most efficient and yet straightforward method that generates a stochastic variable [12]. Note that the stochastic variable has probability density arbitrarily close to Lévy stable distribution characterized by an arbitrarily chosen control parameter $(0.3 \leq \lambda \leq 1.99)$ [13]. Finally, the step-size l using Mantegna's algorithm can be calculated as [13]:

$$l = \frac{u}{|v|^{1/\lambda}}, \tag{5}$$

where u and v are two Gaussian stochastic variables with a zero mean and standard deviations of $\sigma_u(\lambda)$ and $\sigma_v(\lambda)$, respectively, where $\sigma_u(\lambda)$ and $\sigma_v(\lambda)$ can be given by Eq. (6). Here, $\Gamma(z) = \int_0^\infty t^{z-1} e^{-z} dt$ [13]. For detailed description on the standard cuckoo search interested reader can refer [9,11,12].

$$\sigma_u(\lambda) = \left[\frac{\Gamma(1+\lambda) \sin(\pi\lambda/2)}{\Gamma((1+\lambda)/2)\lambda 2^{(\lambda-1)/2}} \right]^{1/\lambda} = 0.696575 \text{ and } \sigma_v(\lambda) = 1 (\text{if: } \lambda = 1.5),$$
$$\tag{6}$$

Cuckoo Search Attack. The cuckoo search attack algorithm for cryptanalysis of the substitution cipher is intuitive (see Algorithm 3). Therefore, here we discuss only the mapping of main components of the cuckoo search (i.e. nest, egg and Lévy flights) to the problem under consideration. In most applications of the cuckoo search, often it is assumed that each nest has one egg. However, in case of cryptanalysis of the substitution cipher, we consider each nest has N distinct eggs/elements (i.e. N distinct characters of a key: $n_1, n_2, ..., n_n$, where $N=27$ (i.e. A —Z and the space character). For simplicity, consider each egg has a unique number $(\in [1, N])$ associated with it. That is there must be a unique

---

Algorithm 2. Standard Cuckoo Search [9]

1: Generate the initial population of m host nests $\mathbf{x}_i, i = 1, 2, ..., m$.
2: **repeat**
3:    Get a cuckoo randomly (say, $\mathbf{x}_j$), and generate a new solution by Lévy flights
4:    Compute its cost, let it be f_j
5:    Randomly choose a nest among m host nests, say $\mathbf{x}_k$
6:    **if** $(f_j ¡ f_k)$ (Let the problem has minimization objective) **then**
7:       Replace $\mathbf{x}_k$ by the new solution $\mathbf{x}_j$
8:    **end if**
9:    Abandon a fraction (p_a) of worst nests/solutions and construct new ones
10:   Keep the best solutions; rank the solutions and find the current best
11: **until** (Termination condition is satisfied)
12: Post-process results

identity of each elements in the nest/solution. In other words, a key cannot have two similar characters. In this regard, the difficulty is how to preserve distinctness property of the key elements, since we can clearly observe that the Eq. (4) will destroy the distinctness property of the elements of the current solution during updation. Note that the importance of Eq. (4) is that it builds the new solution from an existing solution via Lévy flights which is an efficient approach, since the step-size is heavy tailed and any large step is possible. That is the existing solution has better chance to get changed to a better quality solution in lesser computations. Hence, we do not change this equation, rather we utilize its efficiency for improving existing solutions using the current best solution and new solutions generated by Eq. (4) (see Algorithm 3, step 7–11).

Algorithm 3. Cuckoo Search Attack (CS Attack)

1: **Initialization:** Randomly generate the initial population of m host nests. Call this population, POP_{nests}, where host nests are keys of the substitution cipher.

2: **repeat**

3:    Evaluate the cost of each nest of the POP_{nests} using Eq. (3). Select the nest form the POP_{nests} that has the lowest cost. Call this nest $BEST_{nest}$.

4:    Choose a nest randomly (say, POP^i_{nests}), and generate a new nest/solution (abbreviated as NEW_{nest}) by the Lévy flights as: $NEW_{nest}=POP^i_{nests}+\mu\ l$, where l is computed using Eq. (5) with μ=0.01 and λ=1.5.

5:    Evaluate the cost of NEW_{nest} using Eq. (3). If the cost of NEW_{nest} is better (lower) than $BEST_{nest}$ then it become the new $BEST_{nest}$. If it is then go to step 4, otherwise, update the ith existing solution (i.e. POP^i_{nests}) by using NEW_{nest} and $BEST_{nest}$ as follows.

6:    Generate a random number in the range [1,N], call this number n.

7:    **repeat**

8:       Find the next egg/element in the NEW_{nest} that has the identity less than n. Call the position of this element P_1. (Note: in case of first iteration, read first instead of next).

9:       Examine the identity of the element which is located at position P_1 in the $BEST_{nest}$. Call this identity n_1.

10:      Let P_2 is the position in POP^i_{nests} where the element of identity n_1 is located. Swap those elements of the POP^i_{nests} that are located at positions P_1 and P_2. As obvious this operation will store the element of identity n_1 at position P_1, since P_1 is the best position of the same identity element in the $BEST_{nest}$. Here we emphasized that this swapping operation is performed in order to preserve the element that have already placed in POP^i_{nests} to its best position.

11:   **until** (NEW_{nest} list has been traversed completely)

12:   If $BEST_{nest}$ and updated POP^i_{nests} are identical then swap the eggs that are located at two different positions in POP^i_{nests}; positions are determined randomly.

13:   Abandon a fraction (p_a) of worst nests. Create new nests as replacement of abandon nests. The new nests are created again from the best nest by swapping its two randomly chosen elements. In the experiment, we fixed p_a=0.01.

14: **until** (Maximum-Iterations)

15: Output the best solution from the POP_{nests}

2.4   Cryptanalysis Using Tabu Search

The main focus of this optimization algorithm [4,14] is to provide a heuristic for searching a good solution to the problem under consideration without becoming trapped in a local optima. This algorithm maintains a tabu list as a short term memory list where at each iteration the current key is added to the tabu list, and the key remains 'tabu' for a fixed number of iterations. An intuitive level heuristic for cryptanalysis of the substitution cipher is shown in Algorithm 4. Here, we mention that the choice of N_{poss} parameter must be less than $N(N\text{-}1)$, since this is the maximum number of distinct keys which can be created from the best key of the tabu by swapping two elements [4], where N is the key size (e.g. 27). For a detailed description of this heuristic we refer readers to [3,4].

Algorithm 4. Tabu Search Attack (TS Attack) [4]

1: **Initialization:** Randomly initialize the tabu list containing m keys of the substitution cipher. Set N_{poss}, the size of the possibilities list.
2: **repeat**
3:    Select the key form the tabu list that has the lowest cost. Call this key $TABU_{best}$.
4:    **for** $i = 1,2,...,\ N_{poss}$ **do**
5:        Choose n_1, $n_2 \in [1, N]$, such that $n_1 \neq n_2$.
6:        Construct a new key K_{new} by swapping the elements n_1 and n_2 of $TABU_{best}$.
7:        Ensure that K_{new} is not already in the possibilities list or the tabu list. If it is return to Step 5.
8:        Add K_{new} to the possibilities list and determine its cost.
9:    **end for**
10:    From the possibilities list find the key of lowest cost. Call this key $POSS_{best}$.
11:    From the tabu list find the key of highest cost. Call this key $TABU_{worst}$.
12:    **while** the cost of $POSS_{best}$ is less than the cost of $TABU_{worst}$ **do**
13:        Replace $Tabu_{worst}$ with $POSS_{best}$ and find the new $POSS_{best}$.
14:        Find the new $TABU_{worst}$.
15:    **end while**
16: **until** (Maximum-Iterations)
17: Output the best solution from the tabu list.

---

3   Parameters, Experimental Setup and Results

The inputs to all the above presented algorithms are: known ciphertext, its length and bigram statistics of the language (which are assumed to be known). The output of each algorithm is either full or partial substitution cipher key. Note that in order to recover the message that is readable, it is not essential to recover every element of the key, i.e. considerable partial key recovery is also significant to understand the message. The parameters such as m (i.e. population/tabu-list size), Maximum-Iterations (i.e. maximum number of iterations) and N_{poss} (i.e. size of possibilities list in case of tabu search) were fine-tuned by a combination of several experiments. Note that the fine-tuning was performed separately for each of the algorithms in order to optimize the cryptanalysis process.

In some scenarios guidelines are helpful, for example in Algorithm 3, $\mu=0.01$ and $\lambda=1.5$ are taken that has been reported in [10] as general choice. Furthermore, in order to obtain a clear comparison between above algorithms, the guidelines reported in [4] followed, i.e. three criteria were used: amount of known ciphertext available for the attack, number of keys examined prior to determination of correct solution and the time needed to find the correct solution.

Now, we discuss the experimental setup and their details. We tested each of the algorithms on 100 different messages. For each message, each of the cryptanalytic algorithms was run 3 times (it comes to a total of 300 times) and the best of the 3 was recorded. Afterwards, 100 best recorded results corresponding to each algorithm were then averaged. The above described process of recording and averaging is repeated for known ciphertext of size 100 to 800 with step size of 100. The average results of each of the algorithms are then plotted in Fig. 1. From Fig. 1, we can clearly observe that each of the attack algorithms perform well and comparably. Nevertheless, the best way to identify the proper efficiency of the approximation algorithms is; to compare them on the basis of two factors: state space searched (i.e. number of keys examined before evolving the best result) and complexity of the attack (i.e. performance time). For this purpose, we tested each of the attack algorithms on the 100 different known ciphertext of length 1000 characters and then recorded the amount of keys examined and the time taken by the attack. The average results are shown in Table 1. Note that the scatter search method has not been considered in Table 1 (and not re-implemented in this paper), since Garici and Drias (investigators of this algorithm) have concluded that the scatter search method takes 75 % more time than the genetic algorithms (please see Sect. 4.2 in [5]), while the quality of the results is only 15 % better. From Table 1, we can clearly see that the mean performance time of enhanced genetic algorithm is comparatively lesser than the genetic algorithm proposed by Clark, while the average number of keys examined is approximately equal. From Table 1, we can note that the tabu search is more efficient in both respects than genetic algorithms. However, it is also clear from the table that the cuckoo search outperforms tabu search in both respects. Most importantly, the overall performance of cuckoo search is much better among all the attack algorithms (including mean and standard deviation of the key elements correctly found).

Table 1. A comparison based on: mean performance time (T_{mean}), average number of keys examined before the best solution found (M_{avg}) and number of key elements correctly found–mean $(\bar{X})$ & standard deviation (S).

Method →	CS Attack	TS Attack	EGA Attack	GA Attack
T_{mean} (in sec.)	**0.137**	0.241	0.397	0.416
M_{avg}	**1823**	3306	3707	3695
$\bar{X}$	**25.4**	24.05	24.3	24.1
S	**1.04**	1.52	1.23	2.04

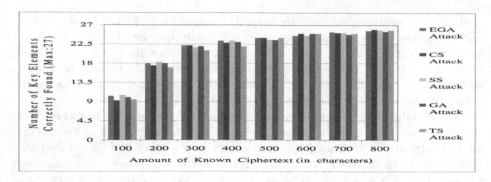

Fig. 1. A comparison based on the amount of known ciphertext. (In case of scatter search (SS) attack, the results were taken from Fig. 3 of [5] which may be not exact.)

4   Conclusion

This paper presents various attacks on the substitution cipher where genetic algorithms, tabu search and cuckoo search have been utilized. We examined that the adaptive mutation operator along with appropriate selection procedure improves the performance of previously proposed genetic algorithm attack. It is worth pointing out that the cuckoo search attack has shown the best performance among all attacks. Most importantly, we needed to fine-tune lesser number of parameters for cuckoo search than genetic algorithms and tabu search. This study indicates that the developed attack algorithm (which utilizes the cuckoo search) is able to produce results that are clearly better than previous attack algorithms of the substitution cipher, and therefore it can be used as a valid and efficient alternative for solving this kind of permutation problems.

References

1. Forsyth, W.S., Safavi-Naini, R.: Automated cryptanalysis of substitution ciphers. Cryptologia **17**(4), 407–418 (1993)
2. Spillman, R., Janssen, M., Nelson, B., Kepner, M.: Use of a genetic algorithm in the cryptanalysis of simple substitution ciphers. Cryptologia **17**(1), 31–44 (1993)
3. Clark, A.: Modern optimisation algorithms for cryptanalysis. In: Proceedings of the 1994 Second Australian and New Zealand Conference on Intelligent Information Systems, pp. 258–262. IEEE (1994)
4. Clark, A.J.: Optimisation heuristics for cryptology. Ph.D. thesis (1998)
5. Garici, M.A., Drias, H.: Cryptanalysis of substitution ciphers using scatter search. In: Mira, J., Álvarez, J.R. (eds.) IWINAC 2005. LNCS, vol. 3562, pp. 31–40. Springer, Heidelberg (2005)
6. Song, J., Yang, F., Wang, M., Zhang, H.: Cryptanalysis of transposition cipher using simulated annealing genetic algorithm. In: Kang, L., Cai, Z., Yan, X., Liu, Y. (eds.) ISICA 2008. LNCS, vol. 5370, pp. 795–802. Springer, Heidelberg (2008)
7. Cowan, M.J.: Breaking short playfair ciphers with the simulated annealing algorithm. Cryptologia **32**(1), 71–83 (2008)

8. Boryczka, U., Dworak, K.: Genetic transformation techniques in cryptanalysis. In: Nguyen, N.T., Attachoo, B., Trawiński, B., Somboonviwat, K. (eds.) ACIIDS 2014, Part II. LNCS, vol. 8398, pp. 147–156. Springer, Heidelberg (2014)
9. Yang, X.S., Deb, S.: Cuckoo search via lévy flights. In: World Congress on Nature & Biologically Inspired Computing, 2009, NaBIC 2009, pp. 210–214. IEEE (2009)
10. Yang, X.S., Deb, S.: Engineering optimisation by cuckoo search. Int. J. Math. Model. Numer. Optimisation **1**(4), 330–343 (2010)
11. Yang, X.S., Cui, Z., Xiao, R., Gandomi, A.H., Karamanoglu, M.: Swarm Intelligence and Bio-Inspired Computation: Theory and Applications. Elsevier, Waltham (2013)
12. Yang, X.S.: Nature-Inspired Optimization Algorithms. Elsevier, Amsterdam (2014)
13. Mantegna, R.N.: Fast, accurate algorithm for numerical simulation of levy stable stochastic processes. Phys. Rev. E **49**(5), 4677–83 (1994)
14. Glover, F.: Tabu search: a tutorial. Interfaces **20**(4), 74–94 (1990)

Spectral Clustering Trough Topological Learning for Large Datasets

Nicoleta Rogovschi[1], Nistor Grozavu[2]([✉]), and Lazhar Labiod[1]

[1] LIPADE, Paris Descartes University, 45, rue des Saints Pères, 75006 Paris, France
{nicoleta.rogovschi,lazhar.labiod}@parisdescartes.fr
[2] LIPN CNRS UMR 7030, Paris 13 University, 99, av. J-B Clement,
93430 Villetaneuse, France
nistor.grozavu@lipn.univ-paris13.fr

Abstract. This paper introduces a new approach for clustering large datasets based on spectral clustering and topological unsupervised learning. Spectral clustering method needs to construct an adjacency matrix and calculate the eigen-decomposition of the corresponding Laplacian matrix [4] which are computational expensive and is not easy to apply on large-scale data sets. Contrarily, the topological learning (i.e. SOM method) allows a projection of the dataset in low dimensional spaces that make it easy to use for very large datasets. The prototypes matrix weighted by the neighbourhood function will be used in this work to reduce the computational time of the clustering algorithm and to add the topological information to the final clustering result. We illustrate the power of this method with several real datasets. The results show a good quality of clustering results and a higher speed.

1 Introduction

Data mining, or knowledge discovery in databases (KDD), an evolving area in information technology, has received much interest in recent studies. The aim of data mining is to extract knowledge from data [16]. The data size can be measured in two dimensions, the size of features and the size of observations. Both dimensions can take very high values, which can cause problems during the exploration and analysis of the dataset. Models and tools are therefore required to process data for an improved understanding. In this work we will focus on clustering very large datasets based on both spectral and topological clustering.

Spectral clustering [9,12] has been well studied in the literature. It is easy to efficiently implement and often outperforms traditional clustering methods such as K-Means. There are numerous proposed different spectral clustering algorithms, such as Ratio Cut [6], K- way Ratio Cut [3], Normalized Cut [12,14] and Min-Max Cut [5]. The spectral clustering methods based on the graph partitioning theory focus on finding the best cuts of a graph that optimize certain predefined criterion functions. The optimization of the criterion functions usually leads to the computation of singular vectors or eigenvectors of certain graph

© Springer International Publishing Switzerland 2015
S. Arik et al. (Eds.): ICONIP 2015, Part II, LNCS 9490, pp. 216–223, 2015.
DOI: 10.1007/978-3-319-26535-3_25

affinity matrices. One of the disadvantages of this type of methods is the computing of the similarity matrix which is very expensive computationally.

In this paper we have focused on models that are based on spectral clustering and topological unsupervised learning i.e. self-organizing maps (SOM, [15]) in order to reduce the data size and to take advantage from the prototypes and neighbourhood information. SOM models are often used for visualization and unsupervised topological clustering, this technique allowing projection in low dimensional spaces that are generally two-dimensional. It is one of the most used connectionism methods among the unsupervised learning methods. This network has given rise to numerous practical applications in order to visualize and perform the information reduction.

At the end of this topographic learning, the "similar" data will be collected in clusters, which correspond to the sets of similar patterns. These clusters can be represented by more concise information than the brutal listing of their patterns, such as their gravity centre or different statistical moments. As expected, this information is easier to manipulate than the original data points.

The rest of this paper is organized as follows: we present the principle of Topological and Spectral clustering in Sects. 2 and 3. Our proposed approach is presented in Sect. 4. In Sect. 5 we present different results and, finally the paper ends with a conclusion and some future works.

2 Unsupervised Topological Learning

In this work we are interested in topological unsupervised learning, as basic model we will use Kohonen's Self-Organizing Maps [11].

The basic model proposed by Kohonen consists of a discrete set C of cells called "map". This map has a discrete topology defined by an undirected graph, which usually is a regular grid in two dimensions. For each pair of cells (j,k) on the map, the distance $\delta(j,k)$ is defined as the length of the shortest chain linking cells j and k on the grid. For each cell j this distance defines a neighbour cell; in order to control the neighbourhood area, we introduce a kernel positive function $\mathcal{K}$ ($\mathcal{K} \geq 0$ and $\lim_{|y|\to\infty} \mathcal{K}(y) = 0$). We define the mutual influence of two cells j and k by $\mathcal{K}_{j,k}$. In practice, as for traditional topological maps we use a smooth function to control the size of the neighbourhood as $\mathcal{K}_{j,k} = \exp(\frac{-\delta(j,k)}{T})$. Using this kernel function, T becomes a parameter of the model. As in the Kohonen algorithm, we decrease T from an initial value T_{max} to a final value T_{min}.

Let $\Re^d$ be the Euclidean data space and $E = \{\mathbf{x}_i; i = 1,\ldots,N\}$ a set of observations, where each observation $\mathbf{x}_i = (x_i^1, x_i^2, ..., x_i^d)$ is a vector in $\Re^d$. For each cell j of the grid (map), we associate a referent vector (prototype) $\mathbf{w}_i = (w_i^1, w_i^2, ..., w_i^d)$ which characterizes one cluster associated to cell i. We denote by $\mathcal{W} = \{\mathbf{w}_j, \mathbf{w}_j \in \Re^d\}_{j=1}^{|\mathcal{W}|}$ the set of the referent vectors. The set of parameter $\mathcal{W}$ has to be estimated iteratively by minimizing the classical cost function defined as follows:

$$R(\chi, \mathcal{W}) = \sum_{i=1}^{N} \sum_{j=1}^{|\mathcal{W}|} \mathcal{K}_{j,\chi(\mathbf{x}_i)} \|\mathbf{x}_i - \mathbf{w}_j\|^2 \tag{1}$$

where χ assigns each observation $\mathbf{x}_i$ to a single cell in the map $\mathcal{C}$. This cost function can be minimized using both stochastic and batch techniques [15]. The minimization of $R(\chi, \mathcal{W})$ is done by iteratively repeating the following three steps until stabilization. After the initialization step of prototype set $\mathcal{W}$, at each training step $(t+1)$, an observation $\mathbf{x}_i$ is randomly chosen from the input data set and the following operations are repeated:

– Each observation $(\mathbf{x}_i)$ is assigned to the closest prototype $\mathbf{w}_j$ using the assignment function defined as follows:

$$\chi(\mathbf{x}_i) = \arg\min_{i \leq j \leq |w|} \left(\|\mathbf{x}_i - \mathbf{w}_j\|^2 \right)$$

– The prototype vectors are updated using the gradient stochastic expression:

$$\mathbf{w}_j(t+1) = \mathbf{w}_j(t) + \epsilon(t)\mathcal{K}_{j,\chi(\mathbf{x}_i)}\left(\mathbf{x}_i - \mathbf{w}_j(t)\right)$$

At the end of the learning process the algorithm provides a prototypes matrix (topological map), a neighbourhood matrix and the affectations of data to each cell (best matching unit). This information will be used in our approach in order to improve the computational time of the spectral clustering and to give more information to the obtained clusters.

3   Proposed Method

Given a set of data points $x_1, x_2, ..., x_n$ $in\mathbb{R}^m$, spectral clustering first constructs an undirected graph $G = (V, E)$ represented by its adjacency matrix $W = (w_{ij})_{i,j=1}^n$, where $w_{ij} \geq 0$ denotes the similarity (affinity) between x_i and x_j. The degree matrix D is a diagonal matrix whose entries are column (or row, since W is symmetric) sums of W, $D_{ii} = \sum_j W_{ji}$. Let $L = D - W$, which is called Laplacian graph. Spectral clustering then use the top k eigenvectors of L (or, the normalized Laplacian $D^{-1/2}LD^{-1/2}$) corresponding to the k smallest eigenvalues as the low dimensional representations of the original data. Finally, k-means method is applied to obtain the clusters.

Spectral clustering method needs to construct an adjacency matrix and calculate the eigen-decomposition of the corresponding Laplacian matrix [4]. Both of these two steps are computational expensive. Then, it is not easy to apply spectral clustering on large-scale data sets.

The proposed method firstly performs SOM on the data set with a large number of p cells. Then, the traditional spectral clustering is applied on the p cells centres weighted by the neighbourhood function to give a topological view of the data distribution in clusters. The data point is assigned to the cluster as its nearest center.

The proposed approach (Algorithm 1) consists in two steps: (i) Compute the map using the SOM algorithm (presented in Sect. 2) and (ii) use the spectral clustering on the W prototypes matrix. This method is computationally simple and depends on the number of cells of the map and the number of clusters.

Algorithme 1. Spectral Clustering trough Topological Learning

---

Input: n data points $x_1, x_2, ..., x_n \in \mathbb{R}^m$; Cluster number k ;
Output: k clusters;

1. Compute p prototype points ($W \in R^{p \times m}$) and neighbourhood matrix $H \in R^{p \times p}$ using SOM
2. Construct the affinity matrix $A \in R^{p \times p}$ defined by $A_{ij} = exp(-||w_i - w_j||^2/2\sigma^2)$ if $i \neq j$ and $A_{ii} = 0$
3. Construct the affinity matrix $S = A * H$
4. Define D to be the diagonal matrix whose (i,i) element is the sum of A's i-th row, and construct the matrix $L = D^{-1/2}SD^{-1/2}$
5. Find $u_1, u_2, ..., u_k$, the k largest eigenvectors of L (chosen to be orthogonal to each other in the case of repeated eigenvalues), and form the matrix $U = [u_1 u_2 ... u_k] \in R^{p \times k}$ by stacking the eigenvectors in columns
6. Form the matrix Y from U by renormalizing each of U's rows to have unit length : $Y_{ij} = U_{ij}/(\sum_j U_{ij}^2)^{1/2}$
7. Cluster each row of Y into k clusters via k-means algorithm
8. Assign the prototypes w_i to cluster j if and only if row i of the matrix Y was assigned to cluster j.

---

4 Experimental Protocol

To evaluate the proposed method, we used several datasets of different size and complexity. For each dataset and experience we computed the external clustering validation indexes as the real class information is available for all of them, and we compared also the computational time between the classical spectral clustering and the proposed method.

4.1 Data Sets

We performed several experiments on four known problems from the UCI Repository of machine learning databases [2].

- Waveform dataset: This dataset consists of 5000 observations divided into three classes. The original dataset included 40 features, 19 of which were attributed to noise, with mean 0 and variance 1. Each class was generated from a combination of 2 of 3 "base" waves.

Dataset	nb. observations	nb. features	nb. classes
Waveform	5000	40	3
WDBC	569	32	2
Madelon	2000	500	3
SpamBase	4601	57	2

- Wisconsin Diagnostic Breast Cancer (WDBC): This dataset includes 569 observations with 32 features (ID, diagnosis, 30 real-valued input features). Each observation is labelled as benign (357) or malignant (212). Features are computed from a digitized image of a fine needle aspirate (FNA) of a breast mass. They describe characteristics of the cell nuclei present in the image.
- Madelon dataset: MADELON is an artificial dataset, with continuous input features. It formed part of the NIPS 2003 feature selection challenge. This dataset is a two-class classification problem which contains data points grouped into 32 clusters placed on the vertices of a five dimensional hypercube and randomly labelled +1 or −1. The five dimensions constitute five informative features. Fifteen linear combinations of these features were added to form a set of 20 (redundant) informative features. Based on these 20 features, the examples can be separated into the two classes (corresponding to the +/− 1 labels).
- SpamBase dataset: The SpamBase dataset is composed of 4601 observations described by 57 features. Every feature describes an e-mail and its category: spam or not-spam. Most of the attributes indicate whether a particular word or character occurs frequently in the e-mail. The run-length attributes (55–57) measure the length of sequences of consecutive capital letters.

4.2   Experimental Results

To evaluate the quality of clustering, we adopt the approach of comparing the results to a "ground truth". We use the clustering accuracy, the Rand index and Jaccard index for measuring the clustering results. This is a common approach in the general area of data clustering. In general, the result of clustering is usually assessed on the basis of some external knowledge about how clusters should be structured. This may imply evaluating separation, density, connectedness, and so on. The only way to assess the usefulness of a clustering result is indirect validation, whereby clusters are applied to the solution of a problem and the correctness is evaluated against objective external knowledge. This procedure is defined by [8] as "validating clustering by extrinsic classification", and has been followed in many other studies [1,7,10]. We feel that this approach is reasonable one if we don't want to judge clustering results by some cluster validity index, which is nothing but a bias toward some preferred cluster property (e.g., compact, or well separated, or connected). Thus, to adopt this approach we need labelled data sets, where the external (extrinsic) knowledge is the class information provided by labels. In this work we consider the purity (accuracy) index, Rand index and Jaccard index for validation.

Let C be the set of class labels and Ω - the clustering result, we denotes by a the number of pairs of points with the same label in C and assigned to the same cluster in Ω, b denotes the number of pairs with the same label, but in different clusters and c denotes the number of pairs in the same cluster, but with different class labels. The index produces a result in the range $[0, 1]$, where a value of 1 indicates that C and Ω are identical.

The purity of a cluster is the percentage of data belonging to the majority class. Each cluster is assigned to the class (real label) which is most frequent in the cluster, and then the purity of this assignment represents the number of correctly assigned objects and dividing by the total number of objects N:

$$\text{purity}(\Omega, \mathbb{C}) = \frac{1}{N} \sum_k \max_j |\omega_k \cap c_j|$$

The Rand index [13] or Rand measure is a measure of the similarity between two data clusters:

$$R = \frac{a+b}{a+b+c+d}; \tag{2}$$

The Rand index has a value between 0 and 1, with 0 indicating that the two data clusters do not agree on any pair of points and 1 indicating that the data clusters are exactly the same.

In the Jaccard index which has been commonly applied to assess the similarity between different partitions of the same dataset, the level of agreement between a set of class labels C and a clustering result Ω is determined by the number of pairs of points assigned to the same cluster in both partitions:

$$J(C, \Omega) = \frac{a}{a+b+c} \tag{3}$$

The Table 1 summarize the obtained results for the four datasets by applying the classical k-means, the spectral clustering and the proposed method. We can note that our method outperforms the classical k-means and sometimes the spectral clustering, but the goal is to decrease the computational time while keeping the same clustering performance as spectral clustering.

Table 1. Experimental results for k-means, Spectral Clustering and proposed approach

Dataset	Method	Acuracy	Rand	Jaccard	Comp. time (sec.)
Waveform	k-means	51.46	0.6216	0.5854	–
	spectral	37.5200	0.4233	0.4064	1592.4
	proposed method	54.267	0.6531	0.6068	**1.1272**
SpamBase	k-means	63.5949	0.5369	0.5086	–
	spectral	68.6807	0.5908	0.5217	1264.1
	proposed method	67.8353	0.5615	0.5064	**8.5481**
WDBC	k-means	83.413	0.7104	0.5499	–
	spectral	83.4798	0.7237	0.5799	2.441
	proposed method	85.0615	0.7475	0.60615	**1.1088**
Madelon	k-means	50.25	0.4998	0.3329	–
	spectral	51.05	0.5082	0.3717	87.5218
	proposed method	51.12	0.5107	0.3794	**15.7147**

The Fig. 1 represents the computational time for both: the proposed method (red plot) and the classical spectral clustering approach (blue plot). The data size (number of samples) is represented in axe X, and the computational time is for the axe Y. If the dataset is small (less than 2000 samples) the computational

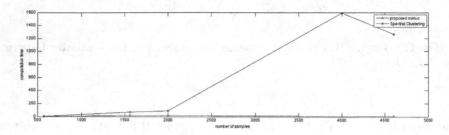

Fig. 1. Computational time for the proposed method and classical Spectral Clustering (Color figure online).

time is improved but not significantly, usually our method works twice faster. If the dataset is larger, i.e. containing 4000 – 5000 samples, the proposed method improves significantly the computational time: from 1592.4 seconds to 1.1272 seconds for the waveform dataset.

5 Conclusions

In this paper we proposed an approach for clustering very large datasets based on the Self-Organizing Maps and classical spectral clustering. The proposed method uses the prototypes and the neighbourhood matrix in order to decrease the high computational time due to the high complexity of the graph construction ($O(n^2)$) and the eigen-decomposition ($O(n^3)$). The obtained results show that our method is much faster compared do classical spectral clustering by maintaining the same clustering results or even improving them. Another aspect of the proposed method is the topology of the map that is not used here for clustering analysis and visualization but will be discussed in future works.

References

1. Andreopoulos, B., An, A., Wang, X.: Bi-level clustering of mixed categorical and numerical biomedical data. Int. J. Data Min. Bioinform. **1**(1), 19–56 (2006)
2. Asuncion, A., Newman, D.: UCI Machine Learning Repository (2007). http://www.ics.uci.edu/~mlearn/MLRepository.html
3. Chan, P.K., Schlag, M.D.F., Zien, J.Y.: Spectral K-way ratio-cut partitioning and clustering. IEEE Trans. Comput. Aaided Des. Integr. Circ.Syst. **13**(9), 1088–1096 (1994). http://ieeexplore.ieee.org/xpl/abs_free.jsp?arNumber=310898
4. Chung, F.R.K.: Spectral Graph Theory. American Mathematical Society, Providence (1997)

5. Ding, C.H.Q., He, X., Zha, H., Gu, M., Simon, H.: A min-max cut algorithm for graph partitioning and data clustering (2001)
6. Hagen, L., Member, S., Kahng, A.B.: New spectral methods for ratio cut partition and clustering. IEEE Trans. Comput. Aided Des., 1074–1085 (1992)
7. Jain, A.K., Murty, M.N., Flynn, P.J.: Data clustering: a review. ACM Comput. Surv. **31**(3), 264–323 (1999)
8. Jain, A.K., Dubes, R.C.: Algorithms for Clustering Data. Prentice-Hall Inc., Upper Saddle River (1988)
9. Jordan, M.I., Bach, F.R., Bach, F.R.: Learning spectral clustering. In: Advances in Neural Information Processing Systems 16. MIT Press (2003)
10. Khan, S.S., Kant, S.: Computation of initial modes for k-modes clustering algorithm using evidence accumulation. In: IJCAI, pp. 2784–2789 (2007)
11. Kohonen, T.: Self-Organizing Maps. Springer, Berlin (1995)
12. Ng, A.Y., Jordan, M.I., Weiss, Y.: On spectral clustering: analysis and an algorithm. In: Advances in Neural Information Processing Systems, pp. 849–856. MIT Press (2001)
13. Rand, W.: Objective criteria for the evaluation of clustering methods. J. Am. Stat. Assoc. **66**, 846–850 (1971)
14. Shi, J., Malik, J.: Normalized cuts and image segmentation. IEEE Trans. Pattern Anal. Mach. Intell. **22**(8), 888–905 (2000). http://dx.doi.org/10.1109/34.868688
15. Kohonen, T.: Self-Organizing Maps. Springer, Berlin (2001)
16. Wang, Q., Ye, Y., Huang, J.Z.: Fuzzy K-means with variable weighting in high dimensional data analysis. In: International Conference on Web-Age Information Management, pp. 365–372 (2008)

A Recommendation System Based on Unsupervised Topological Learning

Issam Falih[✉], Nistor Grozavu, Rushed Kanawati, and Younès Bennani

LIPN - UMR CNRS 7030, SPC, Université Paris 13, 99 Av. J.-B. Clément,
93430 Villetaneuse, France
{issam.falih,nistor.grozavu,rushed.kanawati,
younes.bennani}@lipn.univ-paris13.fr

Abstract. Recommendation systems provide the facility to understand a person's taste and find new, desirable content for them based on aggregation between their likes and rating of different items. In this paper, we propose a recommendation system that predict the note given by a user to an item. This recommendation system is mainly based on unsupervised topological learning. The proposed approach has been validated on MovieLens dataset and the obtained results have show very promising performances.

1 Introduction

Recommendation systems represents information filtering systems which aim to predict the "rating" (i.e. preference) that a user would give to an item (e.g. music file, book or any other product) or social element (e.g. people or groups) that he has not yet considered. Recommendation systems recommend those items predicted to better match user preferences, thereby reducing the user's cognitive and information overload. Collaborative filtering is the most widely used technique in recommender systems. To predict the rating value for an item by an user, the goal of the collaborative filtering methods is to find users which have similar preferences or behaviours using clustering methods for example.

In this work we propose an recommendation system, based on detection of different clusters for both items and users. As the first step for the proposed recommendation system we uses the topological unsupervised learning which allows to obtain more information i.e. topological structure, clustering and data distribution,... At the end of the topological learning, the "similar" data will be collected in clusters, which correspond to the sets of similar patterns. These clusters can be represented by a more concise information than the brutal listing of their patterns, such as their gravity center or different statistical moments. As expected, this information is easier to manipulate than the original data points. We apply different unsupervised topological learning approaches: community detection based clustering [5], the topological co-clustering [15] and the Generative Topographic Mapping [6].

The remainder of this paper is organized as follows. Next in Sect. 2 we present our recommendation system. In Sect. 3, we describe unsupervised topological

© Springer International Publishing Switzerland 2015
S. Arik et al. (Eds.): ICONIP 2015, Part II, LNCS 9490, pp. 224–232, 2015.
DOI: 10.1007/978-3-319-26535-3_26

learning used. In Sect. 4, we present and discuss the result found in Movielens database. Finally we conclude in Sect. 5.

2 Proposed Recommendation System

Recommender systems typically provide a list of recommended items that a user may be interested in. These systems help users to decide on appropriate items, and help the task of finding preferred items in a collection. The most state of art methods has been proposed to improve the accuracy of predicting rating values. To this end, test data is represented as a user-item rating matrix $R \in R^{u_n \times i_n}$, where u_n denotes the number of users, and i_n the number of items. $R_{u,i}$ is rating value of item i by user u. There are lots of missing values in the user-item rating-matrix R and the sparsity of R is often larger than 99 % in commercial systems [16]. Table 1 illustrates a example of rating-matrix concerning four users (denoted as $u1$ to $u4$) and six items (denoted as $i1$ to $i6$). Each user rates some items as to express the interest in each of the items. The ratings are usually on a numerical five-star scale, where one and two stars represent negative ratings, three stars represent ambivalence, while four and five stars represent positive ratings. The goal of a recommendation system algorithm is to predict the missing ratings in the matrix, and to recommend an item to a user if the predicted rating for the item is high. In the following we denote by $\widehat{R}$ the matrix of the predicted ratings.

Table 1. Example of user-item ratings matrix

	i_1	i_2	i_3	i_4	i_5	i_6
u_1	2	-	-	-	-	4
u_2	-	-	4	-	-	-
u_3	-	5	-	4	3	-
u_4	3	-	1	-	-	2

The proposed recommendation approach is composed from two steps :

1. Unsupervised topological learning to identify item and user clusters.
2. Apply the Algorithm 1 using clusters found in step 1.

Different unsupervised learning can be used as community detection based clustering, Generative Topographic Mapping and topological co-clustering. The second step of the proposed approach is presented in the Algorithm 1 which allows the prediction of the user's rates for the corresponding items.

First, for a given user (resp. item) we identify clusters in which the user (resp. item) belongs C_u (resp. C_i). Thereafter the predicted score will be the aggregation between all rates found by intersection between the clusters to which the user belongs and the cluster which the item belongs to. Different aggregation functions can be used i.e. the mean, the median or the mode.

Algorithm 1. Predicting the rate of an item given by a user

Require: $\mathcal{R}, \mathcal{P_U}, \mathcal{P_I}, u_k, i_{k'}$

- $\mathcal{R}$
- $\mathcal{P_U}$ users partitions
- $\mathcal{P_I}$ items partitions
- u_k user
- $i_{k'}$ item

Ensure: $\widehat{R}_{u_k, i'_k}$
1: $\mathcal{S} \leftarrow \emptyset$ { set of rates }
2: $\mathcal{C_U} \leftarrow \{C_u \in \mathcal{P_U} : u_k \in C_u\}$
3: $\mathcal{C_I} \leftarrow \{C_i \in \mathcal{P_I} : i_{k'} \in C_i\}$
4: **for** $u \in \mathcal{C_U}$ **do**
5:   **for** $i \in \mathcal{C_I}$ **do**
6:     **if** $exist(\mathcal{R}[u,i])$ **then**
7:       $\mathcal{S}.append(\mathcal{R}[u,i])$
8:     **end if**
9:   **end for**
10: **end for**
11: $\widehat{R}_{u_k, i'_k} = \mathbf{median}(\mathcal{S})$

---

3   Unsupervised Topological Learning Approaches

3.1   Community Detection

The community detection techniques can be used to treat the problem of recommendation system. Indeed, the community detection can be seen as a generalization of the principle of collaborative filtering which we can recommend to someone the good items assessed by members of community which belongs to. Items can also be grouped into communities as the reasons for their purchases allowing a customer to recommend similar products to what he liked in the past. Multiplex networks are also used to model interactions like in rating sites where consumers can assess on a scale of values (five-stars in general) proposed products (hotels, restaurants, movies, etc.). Each evaluation level corresponds to a layer in the multiplex network.

Few works have addressed the problem of community detection in multiplex networks. Two main trends can be distinguished:

1. *Applying monoplex approaches*: the basic idea is to transform the problem into a problem of community detection in simple networks [2,18].
2. Extending existing algorithms to deal directly with multiplex networks [1,11].

Applying Monoplex Approaches. One first naive approach consists on aggregating layers of a multiplex network in one layer [18]. Classical community detection algorithms can then be applied as Louvain and Wlaktrap...

In general, the layer aggregation approach (LA) consists on transforming a multiplex network into a weighted monoplex graph $G =< V, E, W >$ where W is a weight matrix. One simple aggregation function is the binary weighting: two nodes u, v are linked in the aggregated simple graph if there is at least one layer in the multiplex where these nodes are linked. Formally we have:

$$w_{ij} = \begin{cases} 1 & \text{if } \exists 1 \leq i \leq \alpha : (i,j) \in E_i \\ 0 & \text{otherwise} \end{cases} \tag{1}$$

Another family of approaches consists on applying a community detection algorithm to each layer then apply an consensus clustering approach in order to combine all obtained partitions [17].

Extending Monoplex Approaches to the Multiplex Case. Few studies have addressed the problem of simultaneous exploration of all layers of a multiplex network for the detection of communities. [19] is among the first studies that have tried to extend existing approaches to multiplex setting. The leading role that modularity and its optimization have played in the context of community detection in simple graphs has naturally motivated works to generalize the modularity to the case of multiplex networks. A generalized modularity function is proposed in [12] call Genlouvain algorithm. This is given as:

$$Q_{multiplex}(P) = \frac{1}{2\mu} \sum_{c \in P} \sum_{\substack{i,j \in c \\ k,l:1 \to \alpha}} \left(A_{ij}^{[k]} - \lambda_k \frac{d_i^{[k]} d_j^{[k]}}{2m^{[k]}} \right) \tag{2}$$

Where $\mu = \sum\limits_{k:1 \to \alpha} m^{[k]}$ is a normalization factor, and λ_k is a resolution factor as introduced [13] in order to cope with the modularity resolution problem.

Approaches based on optimizing the multiplex modularity are likely to have the same drawbacks of those optimizing the original modularity function for monoplex approaches [7]. This motivates exploring other approaches for community detection. Seed-centric approaches constitute an interesting option. Actually these approaches are mainly based on local computations that makes these suitable to apply to large-scale networks [9].

In [5], author propose an extension of a seed-centric approach [20], to cope with multiplex networks called muxLicod algorithm. Next, we present the general outlines of muxlicod algorithm:

1. Seed computation: The seed set computation is done in two steps. First all nodes having a degree centrality higher than of the centrality of their direct neighbours are selected as seeds. Let $\mathcal{L}$ be the set of identified leaders. Leaders in $\mathcal{L}$ sharing a high percentage of neighbours are merged into a same community.
2. Seed local community computation: this is also done in two steps. First, each node in the network (a leader or a follower) computes its membership degree

to each community in $\mathcal{C}$. A ranked list of communities can then be obtained, for each node, where communities with highest *membership degree* are ranked first. Next, each node will adjust its community membership preference list by merging this with preference lists of its direct neighbours in the network. Different strategies borrowed form the social choice theory can applied here to merge the different preference lists [4]. This step is iterated until stabilization of obtained ranked lists at each node.

3. Community computation: Finally, each node will be assigned to top ranked communities in its final obtained membership preference list.

3.2 Generative Topographic Mapping

Before applying the GTM algorithm to cluster the dataset, we use the singular value decomposition (SVD) in order to eliminate the sparsity and to reduce the dimensionality of the data. The GTM was proposed by Bishop et al. [3] as a probabilistic counterpart to the Self-organizing maps (SOM) [10]. GTM is defined as a mapping from a low dimensional latent space onto the observed data space. The mapping is carried through by a set of basis functions generating a constrained mixture density distribution. It is defined as a generalized linear regression model:

$$y = y(z, W) = W\Phi(z) \tag{3}$$

where y is a prototype vector in the D-dimensional data space, Φ is a matrix consisting of M basis functions $(\phi_1(z), \ldots, \phi_M(z))$, introducing the non-linearity, W is a $D \times M$ matrix of adaptive weights w_{dm} that defines the mapping, and z is a point in latent space. The standard definition of GTM considers spherically symmetric Gaussian as basis functions

Let $\mathcal{D} = (x_1, \ldots, x_N)$ be the data set of N data points. A probability distribution of a data point $x_n \in \Re^D$ is then defined as an isotropic Gaussian noise distribution with a single common inverse variance β:

$$p(x_n|z, W, \beta) = \mathcal{N}(y(z, W), \beta) = \left(\frac{\beta}{2\pi}\right)^{D/2} \exp\left\{-\frac{\beta}{2}\|x_n - y(z, W)\|^2\right\} \tag{4}$$

The distribution in x-space, for a given value of W, is then obtained by integration over the z-distribution and this integral can be approximated defining $p(z)$ as a set of K equally weighted delta functions on a regular grid:

$$p(x|W, \beta) = \frac{1}{K} \sum_{i=1}^{K} p(x|z_i, W, \beta) \tag{5}$$

For the data set $\mathcal{D}$, we can determine the parameter matrix W, and the inverse variance β, using maximum likelihood. In practice it is convenient to maximize the log likelihood, given by:

$$\mathcal{L}(W, \beta) = \ln \prod_{n=1}^{N} p(x_n|W, \beta) = \sum_{n=1}^{N} \ln \left\{\frac{1}{K} \sum_{i=1}^{K} p(x_n|z_i, W, \beta)\right\}$$

Next, the EM algorithm is used for the maximization of (6) which can be regarded as a missing-data problem in which the identity i of the component which generated each data point x_n is unknown.

3.3 Topological Co-Clustering

In this section we introduce the used topological spectral co-clustering (TSC) proposed in [15] which consists of two parts:

1. Spectral Decomposition of the Data Matrix: This step consists of a spectral decomposition of a weighted data matrix using an SVD and the construction of a new data matrix $\mathbf{D}$ adapted to the problem of co-clustering.
2. Co-clustering of original data A by applying the Self-Organizing Maps (SOM) algorithm on $\mathbf{D}$.

The algorithm begins by computing the first $(g-1)$ eigenvectors ignoring the trivial ones and is based on the following objective function as described in [15]:

$$\mathcal{J}^T(\varphi, W) = \sum_{i=1}^{N+M} \sum_{k=1}^{|\mathcal{W}|} \mathcal{K}^T_{(\delta(\varphi(i), \ell))} ||\mathbf{d}_i - \mathbf{w}_k||^2 \tag{6}$$

where φ the assignment function is defined by

$$\varphi(i) = arg \min_k ||\mathbf{d}_i - \mathbf{w}_k||^2 \tag{7}$$

The algorithm embed the input data into the Euclidean space by eigen-decomposing a suitable affinity matrix and then cluster $\mathbf{D}$ using a geometric clustering algorithm.

Let A be a N by M data matrix, R be a $N \times g$ index matrix and C be a $M \times g$ index matrix. The matrices R and C take these forms $R = (R_1|R_2|\ldots|R_g)$ and $C = (C_1|C_2|\ldots|C_g)$ where a column R_k or C_k is defined as follows: $r_{ik} = 1$ if row i belongs to cluster R_k and $r_{ik} = 0$ otherwise, and in the same manner $c_{jk} = 1$ if column j belongs to cluster C_k and $c_{jk} = 0$ otherwise. This method is well adapted for sparse data as the SVD is integrated in the model.

4   Experiments

To evaluate the proposed approach we used the MovieLens dataset. This dataset was collected through the MovieLens web site (movielens.umn.edu) during the seven-month period from September 19th, 1997 through April 22nd, 1998. This data has been cleaned up - users who had less than 20 ratings or did not have complete demographic information were removed from the data set. MovieLens data set consists of:

- 100,000 ratings (1–5) from 943 users on 1682 movies.
- Each user has rated at least 20 movies.
- The data sets are 80 %/20 % splits into training and test data.

4.1 Evaluation Criteria

There are two types of metrics to evaluate a recommendation system:

- Absolute Error (MAE) and Root Mean Square Error (RMSE) to evaluate the prediction accuracy.
- Precision, Recall and F-measure to measure the recommendation quality.

Mean Absolute Error (MAE) is defined as the average of the absolute error which is the difference between the predicted rating $\widehat{\mathcal{R}_{ij}}$ and actual rating $\mathcal{R}_{ij}$ [14]. Usually, a prediction algorithm tries to minimize the MAE.

$$MAE = \frac{1}{N} \sum_{k=1}^{N} |\mathcal{R}_{ij} - \tilde{\mathcal{R}}_{ij}| \tag{8}$$

Root Mean Square Error (RMSE) is similar to MAE and is biased to provide more weights to larger errors [8].

$$RMSE = \sqrt{\frac{1}{N} \sum_{k=1}^{N} (\mathcal{R}_{ij} - \tilde{\mathcal{R}}_{ij})^2} \tag{9}$$

where N is the number of ratings.

4.2 Experimental Results

The rating dataset is split into training set and test set, where the training set is used for model fitting and parameter tuning, and the test set serves for evaluating the recommendation system. We compared the proposed method using community detection algorithm, GTM and topological co-clustering. In the case of community detection based clustering we applied both tends of approach in particular approaches that uses monoplex algorithms (layer aggregation and partition aggregation) and approaches that extend monoplex algorithm to the multiplex case. We use Louvain and Walktrap algorithm for layer aggregation (LA) and partition aggregation (PA) approach. For the GTM and topological co-clustering, we fixed a 12×12 map size. The community detection based clustering found the number of clusters automatically: 10 clusters for muxlicod approach, 3 clusters for LA louvain, 4 clusters for LA walktrap, 3 clusters for PA louvain and PA walktrap. The aggregation function used for these results is the median as it gives better results compared to mean and the mode on this dataset.

The obtained results show that the recommendation system using layer aggregation approach algorithm outperforms other recommendation system in terms of prediction accuracy and in terms of recommendation quality. However, the goal here is not to improve these measures as the results are similar, but to add a new information to the recommendation results which is the clustering topology i.e. the neighbourhood information which will help the expert to analyse the results. This information is contained in the topology of the graph for the community detection based systems and in the maps for GTM and topological co-clustering (Table 2).

Table 2. Results of different methods on MovieLens dataset

	MAE	RMSE	Precision	Recall	F1-measure
GTM	0.9441	1.2549	0.2185	0.2207	0.2195
Topological co-clustering	0.9293	1.2562	0.25587	0.2094	0.2303
muxlicod	0.9635	1.2773	0.2274	0.2134	0.2202
LA louvain	0.8352	1.1509	**0.3113**	**0.2521**	**0.2779**
LA walktrap	**0.8216**	**1.1155**	0.2642	0.2233	0.2420
PA louvain	0.8713	1.1917	0.2532	0.2032	0.2245
PA walktrap	0.8801	1.2023	0.2705	0.2011	0.2283

5   Conclusion

In this paper, we focus on predicting the rate for a item given by user and we proposed a new recommendation system based on unsupervised topological learning. We test three main topological approaches: community detection based clustering, topological co-clustering and generative topographic maps. The use of consensus clustering methods is under consideration to combine partitions obtained by each unsupervised learning algorithm. An extension to integrate attributes of items and users in multiplexes networks is in perspective.

Acknowledgements. The authors would like to acknowledge the support of FUI HERMES for providing financial support.

References

1. Amelio, A., Pizzuti, C.: A cooperative evolutionary approach to learn communities in multilayer networks. In: Bartz-Beielstein, T., Branke, J., Filipič, B., Smith, J. (eds.) PPSN 2014. LNCS, vol. 8672, pp. 222–232. Springer, Heidelberg (2014)
2. Berlingerio, M., Pinelli, F., Calabrese, F.: Abacus: frequent pattern mining-based community discovery in multidimensional networks. Data Min. Knowl. Discov. **27**(3), 294–320 (2013)
3. Bishop, C.M., Svensén, M., Williams, C.K.I.: GTM: the generative topographic mapping. Neural Comput. **10**(1), 215–234 (1998)
4. Dwork, C., Kumar, R., Naor, M., Sivakumar, D.: Rank aggregation methods for the web. In: Proceedings of the 10th International Conference on World Wide Web, pp. 613–622. ACM (2001)
5. Falih, I., Hmimida, M., Kanawati, R.: Une approche centrée graine pour la détection de communautés dans les réseaux multiplexes. Revue des Nouvelles Technologies de l'Information Extraction et Gestion des Connaissances, RNTI-E-28, pp. 377–382 (2015)
6. Ghassany, M., Grozavu, N., Bennani, Y.: Collaborative multi-view clustering. In: The 2013 International Joint Conference on Neural Networks (IJCNN), pp. 1–8, August 2013

7. Good, B.H., de Montjoye, Y.A., Clauset, A.: The performance of modularity maximization in practical contexts. Phys. Rev. E **81**, 046106 (2010)
8. Jannach, D., Zanker, M., Felfernig, A., Friedrich, G.: Recommender Systems - An Introduction. Cambridge University Press, Cambridge (2010)
9. Kanawati, R.: Seed-centric approaches for community detection in complex networks. In: Meiselwitz, G. (ed.) SCSM 2014. LNCS, vol. 8531, pp. 197–208. Springer, Heidelberg (2014)
10. Kohonen, T.: Self-organizing Maps. Springer, Berlin (1995)
11. Lambiotte, R.: Multi-scale modularity in complex networks. In: WiOpt, pp. 546–553. IEEE (2010)
12. Mucha, P.J., Richardson, T., Macon, K., Porter, M.A., Onnela, J.P.: Community structure in time-dependent, multiscale, and multiplex networks. Science **328**(5980), 876–878 (2010)
13. Reichardt, J., Bornholdt, S.: Statistical mechanics of community detection. Phys. Rev. E 74(1) (2006)
14. Resnick, P., Iacovou, N., Suchak, M., Bergstrom, P., Riedl, J.: Grouplens: an open architecture for collaborative filtering of netnews. In: Proceedings of the 1994 ACM Conference on Computer Supported Cooperative Work, CSCW 1994, pp. 175–186. ACM (1994)
15. Rogovschi, N., Labiod, L., Nadif, M.: A spectral algorithm for topographical co-clustering. In: The 2012 International Joint Conference on Neural Networks (IJCNN), Brisbane, Australia, 10–15 June 2012, pp. 1–6 (2012)
16. Sarwar, B., Karypis, G., Konstan, J., Riedl, J.: Item-based collaborative filtering recommendation algorithms. In: Proceedings of the 10th International Conference on World Wide Web, pp. 285–295. ACM (2001)
17. Strehl, A., Ghosh, J., Cardie, C.: Cluster ensembles - a knowledge reuse framework for combining multiple partitions. J. Mach. Learn. Res. **3**, 583–617 (2002)
18. Suthers, D.D., Fusco, J., Schank, P.K., Chu, K.H., Schlager, M.S.: Discovery of community structures in a heterogeneous professional online network. In: HICSS, pp. 3262–3271. IEEE (2013)
19. Tang, L., Liu, H.: Community Detection and Mining in Social Media. Synthesis Lectures on Data Mining and Knowledge Discovery. Morgan and Claypool Publishers, San Francisco (2010)
20. Yakoubi, Z., Kanawati, R.: LICOD: a leader-driven algorithm for community detection in complex networks. Vietnam J. Comput. Sci. **1**(4), 241–256 (2014)

Online Training of an Opto-Electronic Reservoir Computer

Piotr Antonik[1](✉), François Duport[2], Anteo Smerieri[2], Michiel Hermans[2], Marc Haelterman[2], and Serge Massar[1]

[1] Laboratoire d'Information Quantique, Université Libre de Bruxelles, 50 Avenue F. D. Roosevelt, CP 225, 1050 Brussels, Belgium
pantonik@ulb.ac.be
[2] Service OPERA-Photonique, Université Libre de Bruxelles, 50 Avenue F. D. Roosevelt, CP 194/5, 1050 Brussels, Belgium

Abstract. Reservoir Computing is a bio-inspired computing paradigm for processing time dependent signals. Its analog implementations equal and sometimes outperform other digital algorithms on a series of benchmark tasks. Their performance can be increased by switching from offline to online training method. Here we present the first online trained opto-electronic reservoir computer. The system is tested on a channel equalisation task and the algorithm is executed by an FPGA chip. We report performances close to previous implementations and demonstrate the benefits of online training on a non-stationary task that could not be easily solved using offline methods.

1 Introduction

Reservoir Computing (RC) is a set of methods for designing and training artificial recurrent neural networks [9,12]. A typical reservoir is a randomly connected fixed network, with random coupling coefficients between the input signal and the nodes. This reduces the training process to solving a system of linear equations and yields performances equal, or even greater than other algorithms [7,11]. For instance the RC algorithm has been applied successfully to phoneme recognition [19], and won an international competition on prediction of future evolution of financial time series [1].

Reservoir Computing is well suited for analog implementations. The opto-electronic implementations [10,13,15] were the first fast enough for real time data processing. Since 2012, all-optical reservoir computers have been reported using a variety of approaches [5,6,20,21].

The performance of a reservoir computer greatly relies on the training technique. Up to now offline learning methods have been used [3,5,6,10,15,20]. In these approaches one first acquires a long sequence of training data, and then uses it to compute the readout weights. An alternative approach is to use online training in which the readout weights are progressively adapted in real time. To this end a variety of algorithms can be used, such as gradient descent or recursive least squares. Online training is well suited to solve non-stationary problems, as the weights can

© Springer International Publishing Switzerland 2015
S. Arik et al. (Eds.): ICONIP 2015, Part II, LNCS 9490, pp. 233–240, 2015.
DOI: 10.1007/978-3-319-26535-3_27

be adjusted while the task changes, and can also be applied to systems with complex nonlinear output, such as analog readout layers, proposed in [18].

Online training requires that the readout weights updates are computed in real time, as the reservoir is running. This cannot be achieved with a regular CPU and thus requires dedicated electronics. Field Programmable Gate Arrays (FPGAs) are particularly suited for such applications. Several RC implementations using an FPGA chip have been reported [2,8,17], but no interconnection between an FPGA and a physical reservoir computer has been reported so far.

In the present work we report the first online-trained experimental reservoir computer. It consists of an opto-electronic reservoir computer of the type reported in [10,15], combined with an FPGA board that generates the input sequence, collects the reservoir states and computes optimal readout weights using a simple gradient descent algorithm [4]. We evaluate the performance of our implementation on an example of a real-world task: the equalisation of a nonlinear communication channel [14]. We consider situations where the communication channel changes in time, and show that the online training can deal with this time-varying problem, whereas it would be difficult or impossible to adapt offline training methods to such a task. Furthermore, the use of an FPGA chip allows our system to be trained and tested over an arbitrarily long input sequence and significantly reduces the experimental runtime. The results (measured by the symbol error rate after equalization) are comparable to those reported previously [15].

2 Basic Principles

2.1 Reservoir Computing

A typical reservoir computer is depicted in Fig. 1. The nonlinear function used here is $f_{NL} = \sin(x)$, as in [10,15]. To simplify the interconnection matrix a_{ij}, we exploit the ring topology [16], so that only the first neighbour nodes are connected. The evolution equations thus become:

$$x_0(n + 1) = \sin\left(\alpha x_N(n - 1) + \beta b_0 u(n)\right), \tag{1a}$$

$$x_i(n + 1) = \sin\left(\alpha x_{i-1}(n) + \beta b_i u(n)\right), \tag{1b}$$

with $i = 1, \ldots, N - 1$, where α and β parameters are used to adjust the feedback and the input signals, respectively, and b_i is the input mask, drawn from a uniform distribution over the the interval $[-1, +1]$, as in [6,15,16].

2.2 Channel Equalisation Task

The performance of our implementation is evaluated on the channel equalisation task [9,14,16]. To demonstrate the benefits of the online learning, we investigated two variants of this task, a switching and a drifting channel, where some parameters of the channel vary during the experiment.

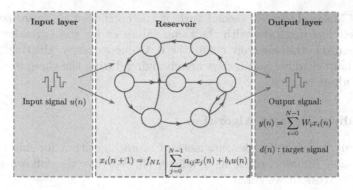

Fig. 1. Schematic representation of a typical reservoir computer. It contains a large number N of internal variables $x_i(n)$ evolving in discrete time $n \in \mathbb{Z}$, f_{NL} is a nonlinear function, $u(n)$ is an external signal that is injected into the system, a_{ij} and b_i are time-independent coefficients, drawn from a random distribution with zero mean and w_i are the readout weights, trained either offline (using standard linear regression methods), or online, as described in Sect. 2.3, in order to minimise the error between the output signal $y(n)$ and some target signal $d(n)$.

Constant Channel. An input signal $d(n) \in \{-3, -1, 1, 3\}$ is transmitted through a noisy nonlinear wireless communication channel, producing an output signal $u(n)$. The reservoir computer then has to recover the clean signal $d(n)$ from the noisy distorted signal $u(n)$. The channel is modelled by the following equations:

$$q(n) = 0.08d(n+2) - 0.120d(n+1) + d(n) + 0.18d(n-1)$$
$$- 0.10d(n-2) + 0.091d(n-3) - 0.05d(n-4) + 0.04d(n-5)$$
$$+ 0.03d(n-6) + 0.010d(n-7), \tag{2}$$

$$u(n) = p_1 q(n) + 0.036q^2(n) - 0.011q^3(n) + \nu(n), \tag{3}$$

where the first equation stands for the symbol interference, the second represents the nonlinear distortion and $\nu(n)$ is the noise, drawn from a uniform distribution over the interval $[-1, +1]$. The $p_1 \in [0, 1]$ parameter is used to tweak the nonlinearity of the channel. It is set to 1 for the constant channel, but varies in other cases. The performance of the reservoir computer is measured in terms of wrongly reconstructed symbols, given by the Symbol Error Rate (SER).

Switching Channel. In wireless communications, the properties of a channel are highly influenced by the environment. For better equalisation performance, it is crucial to be able to detect significant channel variations and adjust the RC readout weights in real time. We consider here the case of a "switching" channel, where p_1 is regularly switched between three predefined values $p_1 \in \{0.70, 0.85, 1.00\}$. These variations occur at very slow rates, much slower than the time required to train the reservoir computer.

Drifting Channel. In the second variant the coefficient p_1 varies faster than the equaliser learning rate, with the same values as for the switching channel. In this situation, the reservoir computer is trained over a "drifting" channel, that is, a channel whose parameters are changing during the computation of the optimal readout weights.

2.3 Gradient Descent Algorithm

The gradient, or steepest, descent method is an algorithm for finding a local minimum of a function using its gradient [4]. We use the following rule for updating the readout weights:

$$w_i(n+1) = w_i(n) + \lambda \left(d(n) - y(n)\right) x_i(n), \tag{4}$$

where λ is the step size, used to control the learning rate. It starts with a high value $\lambda(0) = \lambda_0$, and then gradually decreases to zero $\lambda(n+1) = \gamma\lambda(n)$ at rate $\gamma < 1$.

3 Experimental Setup

Our experimental setup is depicted in Fig. 2. It contains three distinctive components: the optoelectronic reservoir, the FPGA board implementing the input and the readout layers and the computer used to control the experiment.

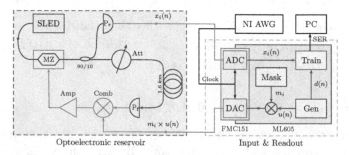

Fig. 2. Schematic representation of the experimental setup.

Optoelectronic Reservoir. The optoelectronic reservoir is based on the same scheme as in [15]. The reservoir states are encoded into the intensity of the incoherent light signal, produced by a superluminiscent diode (SLED). The Mach Zehnder intensity modulator (MZ) implements the nonlinear function. A fraction of the signal is extracted from the loop by the 90/10 beam splitter and sent to the readout photodiode (P_r), the resulting voltage signal is sent to the FPGA. The optical attenuator (Att) is used to set the feedback gain α of the system (see Eq. (1)). The fibre spool consists of about 1.6 km single mode fibre, giving a round trip time of 8.4056 μs. The resistive combiner (Comb) sums the electrical feedback signal, produced by the feedback photodiode (P_f), with the input signal from the FPGA to drive the Mach Zehnder modulator, with an additional amplification stage of +27 dB to span the entire V_π interval of the modulator.

FPGA Board. For our implementation, we use the Xilinx ML605 board, powered by the Virtex 6 XC6VLX240T FPGA chip. The board is paired with a 4DSP FMC151 daughter card, containing one two-channel ADC (Analog-to-Digital converter) and one two-channel DAC (Digital-to-Analog converter). The synchronisation of the FPGA board with the reservoir delay loop is achieved by using an external clock signal, generated by a NI PXI-5422 AWG card. Its maximal output frequency of 80 MHz limits the reservoir size of our implementation to 40 neurons.

The chip uses two Galois linear feedback shift registers with a total period of about 10^9 to generate pseudorandom symbols $d(n)$, which are multiplied by the input mask and sent to the optoelectronic reservoir through the DAC. The resulting reservoir states are collected and digitised by the ADC, and used to compute readout weights and the output signal. The latter is compared to the input signal $d(n)$ and the resulting SER is transmitted to the computer.

Personal Computer. The experiment is fully automated and controlled by a Matlab script, running on a computer. It is designed to run the experiment multiple times over a set of predefined values of parameters of interest and select the combination that yields the best results. For statistical purposes, each set of parameters is tested several times with different random input masks.

We scanned the following parameters: the input gain β, the decay rate γ, the channel signal-to-noise ratio and the feedback attenuation α. The other parameters were fixed during the experiments. Particularly, we used $\lambda_0 - 0.4$ and $\gamma = 0.999$ for the gradient descent algorithm.

4 Results

Constant Channel. Figure 3(a) presents the performance of our reservoir computer for different signal-to-noise ratios of the wireless channel (green squares). Each value is an average of the lowest rates obtained with different random input masks, and the error bars show the variations of the performance with these masks. We compare our results to those reported in [15], obtained with the same optoelectronic reservoir, trained offline (blue dots). For the lowest noise level (32 dB), our implementation produces an error rate of 4.14×10^{-4}, which is comparable to the 1.3×10^{-4} SER of the optoelectronic setup. Note that the latter value is a rough estimation of the reservoir's performance, due to the limited length of the input sequence (6 k symbols, see [15] for details), while in our case, the SER is estimated precisely over 100 k input symbols.

These results demonstrate that the online training algorithm works well, and is able to train an experimental reservoir computer. The obtained SERs are however somewhat worse than those reported in [15], sometimes with up to an order of magnitude. This may be due to the use of a smaller reservoir, with $N = 40$.

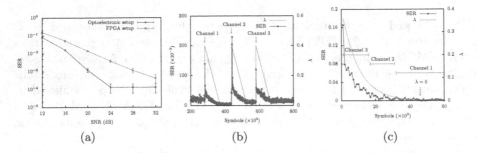

Fig. 3. (a) Comparison of experimental results for nonlinear channel equalisation. (b) Experimental results for switching channel task. After each switch, the SER goes up, the learning rate λ is reset to λ_0, and then decreases while the SER gradually goes back to a low value. (c) Simulation results in the case of the drifting channel. Even though the channel is changing during training, the online training method is able to reach a low SER, that is evaluated during the last 15 k points.

Switching Channel. Figure 3(b) shows the error rate produced by our experiment in case of a switching noiseless communication channel. The parameter p_1 is programmed to switch every 150 k symbols. Every switch is followed by a steep increase of the SER, as the reservoir computer is no longer optimised for the channel it is equalising. The performance degradation is detected by the algorithm, causing the learning rate λ to be reset to the initial value λ_0. The readout weights are then retrained to new optimal values, and the SER rate correspondingly goes down.

Drifting Channel. The reservoir computer is trained over a channel that gradually drifts from channel 3 (with $p_1 = 0.70$) to channel 1 (with $p_1 = 1.0$), and then tested over channel 1. Specifically, the system is presented with a sequence of 60 k input symbols: the first 15 k inputs are fed through channel 3, the next 15 k come out of channel 2, while the last 30 k are provided by channel 1. The reservoir is trained over the first 45 k inputs, while the channel is changing, and tested over the last 15 k inputs.

For simplicity, we simulate our implementation, which makes the reservoir states easily accessible. Figure 3(c) shows the simulation results. The SER gradually decreases during training. The average SER on the test sequence, when $\lambda = 0$, is 1.41×10^{-3}. For comparison, the same inputs were used for the offline training algorithm. We obtained a 1.77×10^{-2} on the test set, that is, a performance ten times worse. Thus online training seems particularly adapted to the drifting channel, as it naturally takes into account the changing nature of the channel, whereas offline training is not able to cope with this feature.

5 Conclusion

We reported here the first physical reservoir computer with online training implemented on an FPGA chip. We demonstrated how online training could deal with a time-dependent channel. These results could not, or only with difficulty, be obtained with traditional online training methods. The results reported here thus significantly broaden the range of potential applications of experimental reservoir computers. In addition, interfacing physical reservoir computers with an FPGA chip should allow the investigation of many other questions, including the robust training of analog output layers, and using controlled feedback to enrich the dynamics of the system. The present work is thus the first step in what we hope should be a very fruitful line of investigation.

Acknowledgements. We acknowledge financial support by Interuniversity Attraction Poles program of the Belgian Science Policy Office, under grant IAP P7-35 photonics@be and by the Fonds de la Recherche Scientifique FRS-FNRS.

References

1. The 2006/07 forecasting competition for neural networks & computational intelligence (2006). http://www.neural-forecasting-competition.com/NN3/. Accessed 21 February 2014
2. Antonik, P., Smerieri, A., Duport, F., Haelterman, M., Massar, S.: FPGA implementation of reservoir computing with online learning. In: Benelearn 2015: The 24th Belgian-Dutch Conference on Machine Learning (2015)
3. Appeltant, L., Soriano, M.C., Van der Sande, G., Danckaert, J., Massar, S., Dambre, J., Schrauwen, B., Mirasso, C.R., Fischer, I.: Information processing using a single dynamical node as complex system. Nat. Commun. **2**, 468 (2011)
4. Arfken, G.B.: Mathematical Methods for Physicists. Academic Press, Orlando (1985)
5. Brunner, D., Soriano, M.C., Mirasso, C.R., Fischer, I.: Parallel photonic information processing at gigabyte per second data rates using transient states. Nat. Commun. **4**, 1364 (2012)
6. Duport, F., Schneider, B., Smerieri, A., Haelterman, M., Massar, S.: All-optical reservoir computing. Opt. Express **20**, 22783–22795 (2012)
7. Hammer, B., Schrauwen, B., Steil, J.J.: Recent advances in efficient learning of recurrent networks. In: Proceedings of the European Symposium on Artificial Neural Networks, pp. 213–216, Bruges (Belgium), April 2009
8. Haynes, N.D., Soriano, M.C., Rosin, D.P., Fischer, I., Gauthier, D.J.: Reservoir computing with a single time-delay autonomous Boolean node, November 2014. arXiv preprint arXiv:1411.1398
9. Jaeger, H., Haas, H.: Harnessing nonlinearity: Predicting chaotic systems and saving energy in wireless communication. Science **304**, 78–80 (2004)
10. Larger, L., Soriano, M., Brunner, D., Appeltant, L., Gutiérrez, J.M., Pesquera, L., Mirasso, C.R., Fischer, I.: Photonic information processing beyond Turing: an optoelectronic implementation of reservoir computing. Opt. Express **20**, 3241–3249 (2012)

11. Lukoševičius, M., Jaeger, H.: Reservoir computing approaches to recurrent neural network training. Comp. Sci. Rev. **3**, 127–149 (2009)
12. Maass, W., Natschläger, T., Markram, H.: Real-time computing without stable states: A new framework for neural computation based on perturbations. Neural comput. **14**, 2531–2560 (2002)
13. Martinenghi, R., Rybalko, S., Jacquot, M., Chembo, Y.K., Larger, L.: Photonic nonlinear transient computing with multiple-delay wavelength dynamics. Phys. Rev. Lett. **108**, 244101 (2012)
14. Mathews, V.J., Lee, J.: Adaptive algorithms for bilinear filtering. In: SPIE's 1994 International Symposium on Optics, Imaging, and Instrumentation, pp. 317–327. International Society for Optics and Photonics (1994)
15. Paquot, Y., Duport, F., Smerieri, A., Dambre, J., Schrauwen, B., Haelterman, M., Massar, S.: Optoelectronic reservoir computing. Sci. Rep. **2**, 287 (2012)
16. Rodan, A., Tino, P.: Minimum complexity echo state network. IEEE Trans. Neural Netw. **22**, 131–144 (2011)
17. Schrauwen, B., D'Haene, M., Verstraeten, D., Campenhout, J.V.: Compact hardware liquid state machines on FPGA for real-time speech recognition. Neural Netw. **21**, 511–523 (2008)
18. Smerieri, A., Duport, F., Paquot, Y., Schrauwen, B., Haelterman, M., Massar, S.: Analog readout for optical reservoir computers. In: Advances in Neural Information Processing Systems 25, pp. 944–952. Curran Associates Inc. (2012)
19. Triefenbach, F., Jalalvand, A., Schrauwen, B., Martens, J.P.: Phoneme recognition with large hierarchical reservoirs. Adv. Neural Inf. Process. Syst. **23**, 2307–2315 (2010)
20. Vandoorne, K., Mechet, P., Van Vaerenbergh, T., Fiers, M., Morthier, G., Verstraeten, D., Schrauwen, B., Dambre, J., Bienstman, P.: Experimental demonstration of reservoir computing on a silicon photonics chip. Nat. Commun. **5**, 3541 (2014)
21. Vinckier, Q., Duport, F., Smerieri, A., Vandoorne, K., Bienstman, P., Haelterman, M., Massar, S.: High-performance photonic reservoir computer based on a coherently driven passive cavity. Optica **2**(5), 438–446 (2015)

Spike Train Pattern Discovery Using Interval Structure Alignment

Taro Tezuka[✉]

University of Tsukuba, Tsukuba, Japan
tezuka@slis.tsukuba.ac.jp

Abstract. A method of finding frequently occurring patterns in spike trains is proposed. Due to stochastic fluctuation, patterns do not appear in exactly the same way. Therefore spike train subsequences are clustered using a similarity measure based on a positive definite kernel for alignments of local interval structures. By applying the method to synthetic data, spike trains generated from different neuron types are classified correctly. For synthetic and real data, patterns that repeatedly appear were successfully extracted.

Keywords: Spike trains · Spiking patterns · Neuron types · Positive definite kernels

1 Introduction

Spike trains recorded from the brain are often considered as the most fundamental element of neural information processing. One critical research question is how information is coded by spike trains. At the lowest level, firing rates and spike timings are the main carrier of information. Researchers also seek for larger temporal structures that code information, where a certain number of spikes constitute an informational unit. This could metaphorically be expressed as an "alphabet" used by neural information transfer [2]. If such units were newly discovered, it would give a tremendous impact to neuroscience. Notable examples of larger temporal structures in spike trains are bursts and oscillations. These structures are easily found in real data, but more complex patterns are more difficult to be found by visual observation only. The use of pattern mining algorithms for conducting such an investigation, however, is surprisingly rare. General methods for temporal pattern mining are sometimes applied [8,10,13], but since many of them were originally developed for other data mining tasks, they are not fit to the data structure of spike trains. In order to propose a new pattern mining method for spike trains, one must define what a pattern is in this case. One way to define it is as a set of frequently occurring mutually similar short spike trains. This requires an adequate similarity measure between spike trains. To this end we define a new similarity measure based on alignments of spike intervals. One novel point of our method is that it aligns intervals between

© Springer International Publishing Switzerland 2015
S. Arik et al. (Eds.): ICONIP 2015, Part II, LNCS 9490, pp. 241–249, 2015.
DOI: 10.1007/978-3-319-26535-3_28

spikes, rather than spike timings themselves. Also, in order to cope with possible skipping of intervals, it allows repetition of intervals. The best alignment is found using dynamic programming. After local matches were extracted from pairs of sequences, they are extended to multiple sequences.

2   Related Work

There have been various proposals for finding significant patterns from spike trains. Torre et al. evaluated spike patterns extracted using a method of frequent item set mining [14]. Unnikrishnan et al. [16] proposed to apply the Frequent Episode Discovery framework [8], which is used in temporal data mining, on spike trains. Muiño et al. pointed out that there is stochastic fluctuation in timings of spikes, and proposed fuzzy modeling of influence from one neuron to another [12]. Using this model, they analyzed synchrony among neurons. These methods were based on traditional tasks employed in data mining, for example market basket analysis. There are also methods aimed at extracting patterns of spikes among multiple neurons. Lee and Wilson proposed a method that uses relative time order of spikes [9]. In case of multineuron spike trains, one of the most well-known examples of recurrently occurring patterns is a synfire chain [1,6]. Gerstein et al. proposed a method to detect synfire chains by searching for repetitions of synchronous activities [4]. It is also important to verify if the extracted patterns are actually statistically significant, based on sound statistic arguments. Grün investigated the statistical significance of extracted patterns, to judge whether discovered patterns are meaningful or not [5]. Toups et al. proposed to a measure of significance of spike train patterns in order to detect them [15].

3   Method

3.1   Interval Structure

A spike train can be expressed as a sequence of real numbers $(\tau_1, ..., \tau_m)$ where τ_i is the time of i-th spike and τ_0 is the starting time of recording. It can be converted to an *interval sequence* $(t_1, .., t_m)$ by setting $t_i = \tau_i - \tau_{i-1}$ for $i \in [1, m]$ and $\tau_0 = 0$. We use a short sequence of intervals as a local structure. Let w represent the number of intervals that will be used in the local interval structure. w is the *window size*. For each $i \in [1, m - w + 1]$, let x_i indicate a w-dimensional real vector constructed by sliding a window of size w intervals over an interval sequence. The value of w is determined either by physiological consideration or by optimization based on the training data. We will call $x_i = (t_i, t_{i+1}, ..., t_{i+w})^T$ an *interval structure*. Note that from a spike train with m spikes, $m - w + 1$ interval structures are obtained.

In this paper, we only consider the case where the interval structure can be represented by a vector. A further extension of this approach is to add more structure to it, for example a graph where each interval is a node and adjacencies between intervals are represented by edges. In that case, a kernel that respects such a structure could be more appropriate. For example, using a graph kernel is a possible option, because sequential order of the intervals can be represented as a graph. Here after, we use "symbol" and "interval structure" interchangeably. Since interval structures are finite dimensional vectors, we define a distance between two intervals structures x_i and y_j using the Euclidean distance, $d(x_i, y_j) = \|x_i - y_j\|_2$. A positive definite kernel is defined by $k_I(x_i, y_j) = \exp(-d(x_i, y_j)^2)$. From this symbol-wise kernel, we would like to define a similarity measure for two interval sequences. The simplest way is to use the product of interval-wise kernel values, i.e. $K_{\text{simp}}(x, y) = \prod_{i=1}^m k_I(x_i, y_i)$. This measure, however, cannot be applied unless the lengths of two sequences are equal. Also, in K_{simp}, intervals that appear at different locations in sequences are never matched. In reality, there are stochastic addition and deletion of spikes. It is therefore better to consider matching intervals in nearby locations as well. The propose method therefore modify the original sequences in different ways and check if that would make a better alignment.

3.2   Interval Structure Alignment Kernel

Let S be a set of symbols, $z \in S^\nu$ a sequence of symbols, and $a \in \mathbb{N}^\nu$ a sequence of positive integers. Define $p = \sum_{i=1}^\nu a_i$. Let $\epsilon^a \in \mathbb{Z}^p$ represent a sequence of positive integers where each integer $i \in [1, \nu]$ is repeated a_i times. ϵ^a can be used to define a lengthened sequence $s \in S^p$ from z by setting $s_\ell = z_{\epsilon_\ell^a}$ for $\ell = 1, ..., p$. In other words, s is a sequence of symbols created by repeating z_i for a_i times.

Let $\Omega_{\nu,p}$ be a set of all such ϵ^a. Alignment π between two sequences $x \in S^m$ and $y \in S^n$ is defined as a pair $(\pi^{(1)}, \pi^{(2)}) \in \Omega_{m,p} \times \Omega_{n,p}$ that fulfills the following three conditions: (1) $\pi_\ell^{(1)} - \pi_{\ell-1}^{(1)} \in \{0, 1\}$, (2) $\pi_\ell^{(2)} - \pi_{\ell-1}^{(2)} \in \{0, 1\}$, and (3) $(\pi_\ell^{(1)} - \pi_{\ell-1}^{(1)}) + (\pi_\ell^{(2)} - \pi_{\ell-1}^{(2)}) \subset \{1, 2\}$. The conditions (1) and (2) mean that symbols can be repeated but cannot be skipped. The condition (3) means that there is no simultaneous repetition of $\pi^{(1)}$ and $\pi^{(2)}$. This restriction is necessary to prohibit alignments that are infinitely long. Using π, two sequences are aligned so that $x_{\pi_\ell^{(1)}}$ is matched to $y_{\pi_\ell^{(2)}}$. Let $A(x, y)$ be the set of all alignments of x and y. In what follows, we use the absolute value sign to express the number of elements in a sequence. One similarity measure between symbol sequences is the global alignment kernel defined by Cuturi et al. [3]. It is defined as the sum of kernels for all possible alignments, $K_{GA}(x, y) = \sum_{\pi \in A(x,y)} \prod_{\ell=1}^{|\pi|} k_I(x_{\pi_\ell^{(1)}}, y_{\pi_\ell^{(2)}})$. All of the possible alignments appear in the sum, hence the name "global". The *interval structure alignment kernel* proposed in this paper is a global alignment kernel which uses the interval structures as symbols. It finds macro-scale patterns consisting of micro-scale patterns, where the latter is defined as the interval structure.

3.3   Pattern Retrieval and Multiple Alignment

Let x and y be two symbol sequences, i.e. sequences of interval structures. In order to find local match between x and y, we make a substring $u(\zeta, v)$ of

x, starting at and arbitrary offset ζ and length υ. In other words, $u(\zeta, \upsilon) = (x_\zeta, x_{\zeta+1}, ..., x_{\zeta+\upsilon-1})$. We call $u(\zeta, \upsilon)$ a *template*. For each value of offset ζ and length υ within a certain range, $u(\zeta, \upsilon)$ is matched to y using dynamic programming. In other words, templates taken from x are aligned to different subsequences of y. Local matches having similarity measures higher than the threshold are extracted as pairwise alignments. By finding local matches to multiple sequences, the system searches for frequently appearing patterns. In the process, pairwise local match is recursively applied, and a set of matched subsequences with high scores are stored as candidates for multiple alignments. Note that original subsequence $u(\zeta, \upsilon)$ are stored, instead of lengthened sequence $\{u_{\pi_\ell^{(1)}}\}_{\ell \in [1, p]}$, whose symbols are repeated. This is because if we use the lengthened sequence, in the next step it will be further modified, and modifications accumulates, departing from the original sequences. The process of multiple alignment is conducted as follows. Let $x^{(j)}$ be a symbol sequence. By changing offset ζ and pattern length υ, a set of subsequences $\{u^{(j)}(\zeta, \upsilon)\}$ is obtained. A multiple alignment $\mathfrak{m}$ is a set of subsequences that are similar to each other, i.e. $\{u^{(1)}, u^{(2)}, ..., u^{(\varsigma)}\}$.

4   Evaluation

We start with evaluating the classification accuracy when the interval structure kernel was used as a similarity measure between spike trains. If the similarity measure can classify patterns well, it means that the measure is capturing the qualitative features of the spike trains. We can then expect it to be a promising way for finding patterns, since patterns are to be grouped together based on its qualitative features. Mathematically, there is an infinite number of similarity measures, but what we want is the one that allows classification of spike trains into patterns. A similarity measure that we want is the one that can distinguish qualitative differences among spike trains, despite stochastic fluctuation. We therefore evaluated the classification accuracy using synthetic data, since in this case correct class labels are available. We use a neuron model that can emulate various neuron types. These types are defined based on already known patterns in spike trains, for example bursts.

4.1   Classification of Synthetic Data

Many researchers classify neurons into types, based on the patterns of spike trains they generate for different inputs. These types include regular spiking, intrinsically bursting, chattering, and fast spiking neurons. In order to generate synthetic spike trains belonging different neuron types, we used the Izhikevich neuron [7]. The Izhikevich neuron is one of the most widely used model for generating synthetic spike train data based on a system of ordinary differential equations. Izhikevich pointed out that by changing four parameters, it can emulate 20 types of neurons, each having characteristic spike train patterns [7]. The Izhikevich neuron is a deterministic system, but it generates different spike

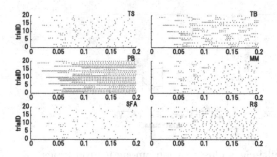

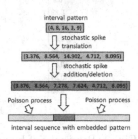

Fig. 1. Spike trains generated by different neuron types

Fig. 2. Embedding of a pattern to a spike train

patterns depending on the input current (injection). In order to generate various samples, we added stochastic variation to injections. Each injection is a step function with stochastically varying offset and amplitude. It is specified by the time point of discontinuity t' and the final amplitude i', as in $i(t) = i'H(t - t')$ where H is the Heaviside step function. We sampled offsets from the uniform distribution over $[0, cT]$, where T is the total time length of recording and c is some set constant. In the experiments, we set $T = 1000$ ms and $c = 1/2$. The amplitude of the input current is sampled from the uniform distribution over $[I_{min}, I_{max}]$. We used $I_{min} = 10$ and $I_{max} = 20$. Figure 1 illustrates example spike trains generated from different neuron types with different injections. Within each neuron types, exact timings of spikes varies. Note that to each neuron type, same randomly generated currents are injected in this visualization. We evaluated the interval structure alignment kernel by the precision of classification, using k-nearest neighbor classification (k-NN). Each neuron type corresponds to a class.

In k-NN classification, a test spike train is assigned to a class that constitutes the majority of the classes of k spike trains that are closest to it. In our case,

Table 1. Precisions of binary and multi-class classification

Method	Binary classification		Multi-class classification		Pattern discovery
	Avg. prec.	Std. dev.	Avg. prec.	Std. dev.	Success rate
Chance level	0.5000		0.1250		
Interval structure alignment, w = 1	0.8907	0.0040	0.6937	0.0353	0.0603
Interval structure alignment, w = 2	0.8933	0.0029	0.7461	0.0385	0.0597
Interval structure alignment, w = 3	0.8857	0.0049	0.6937	0.0384	0.0353
Interval structure alignment, w = 4	0.8787	0.0046	0.5609	0.1173	0.0303
Victor-Purpura distance (VP)	0.8466	0.0064	0.5031	0.0686	0.0187
van Rossum distance (VR)	0.8367	0.0044	0.4414	0.0663	0.0107

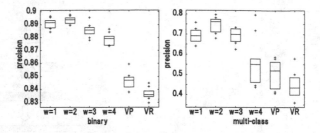

Fig. 3. Precisions of binary and multi-class classification among neuron types

closeness is defined either by a kernel or a distance. The test data is assigned to one of the most frequently appearing classes. If there is a tie, i.e. more than one spike trains had the same value of similarity to the test data, it is assigned to a randomly chosen class among the most similar k spike trains. In our experiments, we used $k = 3$. We evaluated the precision of k-NN classification using leave-one-out analysis, which is a special case of cross-validation. From whole set Z, spike trains are taken out one at a time. Let the taken-out spike train be z. For each $z' \in Z$, similarity measure $m(z, z')$ is calculated. The resulting set $\{m(z, z')|z' \in Z\}$ is sorted in descending order. z is assigned the label (class) of the top k most similar spike trains. We compared the interval structure alignment kernel (ISA) with two most commonly used spike train similarity measures, the Victor-Purpura distance (VP) [18] and the van Rossum distance (VR) [17]. The left graph of Fig. 3 indicates the precisions of binary classifications of neuron types using the interval structure alignment (ISA), the Victor-Purpura distance (VP), and the van Rossum distance (VR) are tested. It shows the average precision, 25 and 75 percentiles using the boxes. For the interval structure alignments, different windows sizes are tested, as indicated by w. It shows the average precision for all 20×19 combinations of 20 neuron types. 8 samples are generated for each neuron type. The Victor-Purpura distance has a parameter q that represents the inverse of the time scale. We tested $8 \times [1, .., 6]$ for q and found that $q = 16$ gives the best result for this data, so the figure shows the result for this parameter setting. Similarly, the van-Rossum distance has a parameter θ_{vR} which represents the time scale. We tested $5 \times 2^{[0,...,7]} \times 10^{-3}$ for θ_{vR} and found that $\theta_{vR} = 0.16$ gives the best result, so the figure shows the result for this setting.

The right graph of Fig. 3 indicates the classification accuracy of spike trains into 8 classes. The time length of each spike train is 200 ms. We used the first 8 neuron types of the Izhikevich neuron as classes. The chance level of estimating the correct class is therefore near $1/8$. For each class we generated 16 samples. We conducted 10 trials and evaluated the average and standard deviation of the precisions. Table 1 illustrates the average precisions and the standard deviation for the binary and multi-class classification task. The window size is represented by w. For both cases, the maximum precision is achieved when the window size is 2.

4.2 Pattern Discovery from Synthetic Data

Since correct patterns are not available for real data, we generate synthetic data and embed patterns into them. We then see if these patterns can correctly be retrieved using our algorithm. We generated spike trains with a given pattern embedded into samples of Poisson processes. The method is illustrated in Fig. 2. Since neural spike trains are a sequence of event times, they are often modeled by the Poisson process. The homogeneous Poisson process is specified by a single parameter λ, which indicates its firing frequency. Let $(\tau_1, ..., \tau_m)$ be a spike train pattern. The average appearance rate of this pattern is defined as $r = (\tau_m - \tau_1)/m$. A spike train embedding this pattern n times is generated by alternating between a pattern part and a non-pattern part. The non-pattern part is generated by Poisson process with appearance rate r. In this experiment we used $r = 0.7$. The time length of a non-pattern part is determined stochastically by sampling from a uniform distribution over $[0, 2(\tau_m - \tau_1)]$. The expected length of a non-pattern pattern part is equal to that of a pattern-part. A whole spike train starts with a non-pattern part, followed by a pattern part, then alternating between the two patterns, and finally ends with a non-pattern part.

In order to model stochastic fluctuation in spike trains, the pattern part is fluctuated using three basic stochastic operations, namely deletion, addition and translation of spikes. Specifically, each spike is deleted with probability γ_e. Spikes are added by superposing a Poisson process with appearance rate r_e. These spikes are overlaid to the original pattern. Finally, spikes are translated following the Gaussian distribution with standard deviation σ_e. In our experiment, we used $\gamma_e = 0.1, r_e = 0.1$, and $\sigma_e = 1$. We compared different methods using the efficacy of retrieving patterns embedded in the above defined way. We used a sequence $(4, 8, 16, 3, 9)$ as a pattern. To each spike train, a single pattern is embedded. The ratio of pattern parts to the whole sequence is set to 0.02, which means that the whole sequence is 50 times in the number of intervals than the pattern. We generated 3000 spike trains in this way. For each method, the same set of spike trains were presented. Pattern match is considered to be successful if the embedded pattern is contained in the top 5 patterns with highest similarity.

The last column of Table 1 indicate the ratio of successful pattern discovery using different methods. The interval structure alignment with the window size $w = 1$ has the highest success rate. For the interval structure alignment kernel, we used the parameters proposed for the global alignment kernel[1], namely $\sigma_p = $ median$\|x - y\|\sqrt{(\text{median}(\|x\|))}$. For the Victor-Purpura distance, we set the parameter to $8 \times [1, .., 6]$. For the van Rossum distance, we set the parameter to $5 \times 2^{[0,...,7]} \times 10^{-3}$. The interval structure alignment gave better result than the Victor-Purpura distance and the van Rossum distance. For this data, it works best when the window size is small.

4.3 Pattern Discovery from Real Data

For evaluating with real data, we used spike train data provided by CRCNS (Collaborative Research in Computational Neuroscience) data sharing[2]. CRCNS

---

[1] http://www.iip.ist.i.kyoto-u.ac.jp/member/cuturi/GA.html.
[2] http://crcns.org.

Table 2. Examples of extracted alignments for pairwise and multiple alignments

seq. 1	21	48	27	20	59	36	26	18	31	25
seq. 2	20	48	26	22	55	40	25	17	31	26
seq. 1	17	32	25	18	25	57	57	20	48	26
seq. 2	26	32	32	24	33	46	20	20	51	24
seq. 1	77	23	37	59	54	24	26	53	32	21
seq. 2	132	37	42	62	39	39	32	32	38	17
seq. 1	99	51	165	189	164	143	112	432	67	60
seq. 2	115	45	45	45	158	207	136	356	27	71

seq. 1	17	21	48	27	20	59	36	26	18	31
seq. 2	57	20	48	26	22	55	40	25	17	31
seq. 3	56	20	49	25	22	58	36	26	18	31
seq. 4	20	20	51	24	21	53	41	26	17	31
seq. 1	40	51	20	22	18	41	58	21	33	59
seq. 2	118	37	35	21	39	43	85	26	66	70
seq. 3	181	91	54	22	22	22	74	60	115	35
seq. 4	86	23	37	20	20	73	29	84	18	18

is a collection of neuroscience data taken at various laboratories. We used data from category "hc2". It consists of recordings of three male Long-Evans rats, with a 4-shank or 8-shank silicon probe in layer CA1 of the right dorsal hippocampus [11]. Using a spike sorting method, namely KlustaKwik, the recorded spikes were clustered into artifacts, noise/not-clusterable units, and isolated units, based on their shapes. Table 2 indicates some of the example alignments extracted from hippocampus data, 'ec013.527' and 'ec013.528'. The left table is for pairwise alignments and the right table is for multiple alignments.

5   Conclusion

We proposed a new similarity measure on spike trains based on the interval alignment, and applied it to spike train classification and pattern discovery. For synthetic data, the method was successful in classifying different neuron types and retrieving embedded patterns from a long sequence. Our method performed better than other existing distances on spike trains. For real spike train data, patterns that repeatedly appear in spike trains were extracted. In future work, we plan to evaluate the significance of the extracted patterns from physiologically perspective. This should be carried out in collaboration with physiologists. We hope that this work pave the way to further methods of pattern discovery for neural spike trains.

Acknowledgements. This work was supported in part by JSPS KAKENHI Grant Numbers 21700121, 25280110, and 25540159.

References

1. Abeles, M.: Local Cortical Circuits: An Electrophysiological Study. Springer, Heidelberg (1982)
2. Baram, Y.: Global attractor alphabet of neural firing modes. J. Neurophysiol. **110**, 907–915 (2013)
3. Cuturi, M.: Fast global alignment kernels. In: Proceedings of the 28th International Conference in Machine Learning, pp. 929–936 (2011)
4. Gerstein, G.L., Williams, E.R., Diesmann, M., Grün, S., Trengove, C.: Detecting synfire chains in parallel spike data. J. Neurosci. Methods **206**, 54–64 (2012)

5. Grün, S., Diesmann, M., Aertsen, A.: Unitary events in multiple single-neuron spiking activity: I. detection and significance. Neural Comput. **14**, 43–80 (2001)
6. Ikegaya, Y., Aaron, G., Cossart, R., Aronov, D., Lampl, I., Ferster, D., Yuste, R.: Synfire chains and cortical songs: temporal modules of cortical activity. Science **304**, 559–564 (2004)
7. Izhikevich, E.M.: Which model to use for cortical spiking neurons? IEEE Trans. Neural Netw. **15**(5), 1063–1070 (2004)
8. Laxman, S., Sastry, P.S., Unnikrishnan, K.P.: A fast algorithm for finding frequent episodes in event streams. In: Proceedings of the 13th ACM SIGKDD International Conference on Knowledge Discovery and Data Mining, pp. 410–419 (2007)
9. Lee, A.K., Wilson, M.A.: A combinatorial method for analyzing sequential firing patterns involving an arbitrary number of neurons based on relative time order. J. Neurophysiol. **92**, 2555–2573 (2004)
10. Mannila, H., Toivonen, H., Verkamo, A.I.: Discovery of frequent episodes in event sequences. Data Min. Knowl. Discov. **1**(3), 259–289 (1997)
11. Mizuseki, K., Sirota, A., Pastalkova, E., Buzsáki, G.: Theta oscillations provide temporal windows for local circuit computation in the entorhinal-hippocampal loop. Neuron **64**, 267–280 (2009)
12. Picado Muiño, D., Castro León, I., Borgelt, C.: Fuzzy frequent pattern mining in spike trains. In: Hollmén, J., Klawonn, F., Tucker, A. (eds.) IDA 2012. LNCS, vol. 7619, pp. 289–300. Springer, Heidelberg (2012)
13. Srikanth, R., Agrawal, R.: Mining sequential patterns: generalizations and performance improvements. In: Apers, P., Bouzeghoub, M., Gardarin, G. (eds.) Proceedings of the 5th International Conference on Extending Database Technology. Lecture Notes in Computer Science, vol. 1057, pp. 3–17. Springer, Heidelberg (1996)
14. Torre, E., Picado-Muino, D., Denker, M., Borgelt, C., Grün, S.: Statistical evaluation of synchronous spike patterns extracted by frequent item set mining. Front. Comput. Neurosci. **22** (2013)
15. Toups, J.V., Tiesinga, P.H.E.: Methods for finding and validating neural spike patterns. Neurocomputing **69**, 1362–1365 (2006)
16. Unnikrishnan, K.P., Patnaik, D., Sastry, P.S.: Discovering patterns in multineuronal spike trains using the frequent episode method. Computing Research Repository (2007)
17. van Rossum, M.C.W.: A novel spike distance. Neural Comput. **13**, 751–763 (2001)
18. Victor, J.D., Purpura, K.P.: Nature and precision of temporal coding in visual cortex: a metric-space analysis. J. Neurophysiol. **76**, 1310–1326 (1996)

Distant Supervision for Relation Extraction via Group Selection

Yang Xiang[✉], Xiaolong Wang, Yaoyun Zhang, Yang Qin, and Shixi Fan

Intelligence Computing Research Center, Harbin Institute of Technology Shenzhen Graduate
School, Shenzhen Guangdong, China
xiangyang.hitsz@gmail.com, wangxl@insun.hit.edu.cn,
yaoyun.zhang@uth.tmc.edu, yang.qin@hitsz.edu.cn,
shixifan@hit.edu.cn

Abstract. Distant supervision (DS) aligns relations between name entities from a knowledge base (KB) with free text and automatically annotates the training corpus with relation mentions. One big challenge of DS is that the heuristically generated relation labels usually tend to be noisy, when a pair of entity has multiple and/or incomplete relations in a KB. This paper proposes two ranking-based methods to reduce noise and select effective training data for multi-instance multi-label learning (MIML), one of the most popular learning paradigms for distantly supervised relation extraction. Through the proposed methods, training groups that are of low quality are excluded from the training data according to different ranking strategies. Experimental evaluation on the KBP dataset using state-of-the-art MIML algorithms in this community demonstrated that the proposed methods improved the performance significantly.

Keywords: Distant supervision · Relation extraction · Multi-instance multi-label learning · Data selection

1 Introduction

Distant supervision (also known as weak supervision) aligns relations between entities from a KB with free text and automatically annotates the training corpus with relation mentions, so as to build relation extractors. It has attracted much attention recently due to its natural advantage in reducing laborious efforts and cost on data annotation [1–6]. However, when a pair of entities has multiple relations, the heuristically generated relation labels usually contain a lot of noise (see Fig. 1 in [7]), resulting that many traditional machine learning algorithms perform poor when directly training on them. The challenge mainly lies in that we are not able to decide in advance what the latent relation each sentence can really express. In addition, it is not guaranteed that either the KB or the free text is complete, so that noisy instances (false positives (FPs) or false negatives (FNs)) are inevitable.

© Springer International Publishing Switzerland 2015
S. Arik et al. (Eds.): ICONIP 2015, Part II, LNCS 9490, pp. 250–258, 2015.
DOI: 10.1007/978-3-319-26535-3_29

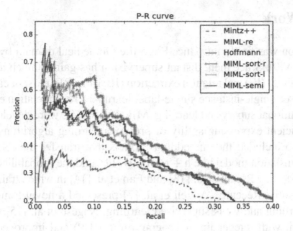

Fig. 1. P/R curves for MIML-semi and our methods.

MIML is one of the most popular learning paradigms for DS [3, 6, 8] extended from multi-instance learning (MIL) [1, 5]. Different from traditional machine learning algorithms, MIML can learn weak classifiers based on the multiple-instance multiple-label data structure (for a pair of entity). This paper propose two ranking-based methods and build corresponding models on top of a popular MIML algorithm –MIML-re [6], so as to select effective training groups and reduce noise before the training steps. The two methods are named as *MIML-sort-l* and *MIML-sort-r* respectively, defining different criteria to evaluate how effective a data group is (the quality for a group). Concretely, MIML-sort-l defines how close the instance-level labels are to the group-level labels, and MIML-sort-r evaluates the correctness of the group-level labels through computing pair-wise rankings.

In the community of distantly supervised relation extraction, Min et al.'s MIML-semi is the most closely related approaches to this work but there are major differences [8]: (1) They only considered the missing information in the KB but not in the text; (2) They explicitly modeled the incorrectness of group-level labels through another latent layer while we implicitly model it by means of ranking scores; (3) They changed the group-level labels before each training step while we remove the groups whose labels are likely to be wrong.

Experiments are setup on one of the landmark datasets–the KBP dataset [6]. We compare our approaches with one of the state-of-the-art models-MIML-semi and also report the results generated by three other baselines including MIML-re. The results show that our methods outperform all the previous work, with MIML-sort-r improves them significantly. In addition, the results on the whole testing dataset also go beyond previous work prominently, with an optimal enhancement of F_1 from 27.3 % to 29.4 %. Our contributions can be summarized as follows:

- The first to use the group-level information to select effective training data;
- Two effective ranking-based methods are proposed and validated;
- We show what types of data are removed after applying the methods.

2   Related Work

Distant supervision was firstly introduced into the biomedical domain by mapping databases to PubMED [9]. Since then, distant supervision has gained much attention in both information extraction (IE) and relation extraction (RE). For RE, most of earlier researches including [2, 4] used single-instance single-label learning to train relation extractors which are more like traditional supervised learning. MIL and MIML were developed afterwards due to the insufficient expressing ability of previous learning algorithms [1, 3, 5, 6, 8]. Related work also included the embedding models that transfer the relation extraction problem into a translation model like h + r ≈ t [10–12], and the probability matrix factorization (PMF) models by Riedel et al. [13] and Fan et al. [14] in which training and testing are carried out jointly. Recently, Angeli et al. [7] presented a novel framework by integrating active learning and weak supervised learning. Nagesh et al. [15] solved the label assigning problem with integer linear programming (ILP) and improved the baselines. Among previous researches, some were mainly aiming at reducing noise. For example, Intxaurrondo et al. [16] proposed several heuristic strategies to remove useless mentions but on a very simple learning framework. Xu et al. [17] employed a passage retrieval model to expand the training data. Min et al. [8] added another layer to MIML-re to model the true labels. Other related work also include [18, 19].

3   MIML via Group Selection

We built our models on top of MIML-re, one of the most popular distantly supervised learning paradigms for relation extraction. MIML-re places multiple sentences that contain the same entity pair into a group (multi instance), and enables these group members to share all existing relation labels for this pair (multi label). A two-level classifier is constructed to learn the latent sentential labels using all the positive training groups (with DS generated labels) and a subsample of the negative groups (without labels). In this work, we selected a subset of the data for training at the beginning of each training epoch. For a clearer description, we use the term *sort* to denote the ranking of groups, and the term *rank* to denote the ranking of instances.

3.1   MIML-Sort

Following MIML-re, MIML-sort defines all the possible relation labels from the KB as positive labels (P_i) for a group, while $Ł\backslash P_i$ as negative labels (N_i), where $Ł$ stands for the possible relation labels for the key entity (the first entity in the pair). The joint probability of the whole dataset D is defined as:

$$p(D) = \prod_{i=1}^{n} p(y_i|x_i, w_y, w_z) = \prod_{i=1}^{n} p(y_i, z_i|x_i, w_y, w_z) \tag{1}$$

$$p(y_i|x_i, w_y, w_z) = \prod_{m=1}^{M_i} p(z_i^m|w_z) \times \prod_{r \in P_i \cup N_i} p(y_i^r, |z_i, w_y^r) \tag{2}$$

The parameters of the model include the instance-level classifier w_z and the group-level classifier w_y. w_z is a multi-class classifier which maps each instance to one of the relation labels. w_y is a series of binary classifiers which transmits the information from the relation labels predicted by w_z to the labels for a group. We also use the hard EM algorithm to estimate the model parameters. More details can be found in [6].

We perform the sorting operation after each E-step in a training epoch when all instances have been assigned labels. Here we denote the score of each group as $f(x_i)$ (x_i stands for the ith group) which is further defined based on different strategies, and the training dataset is updated by selecting the top θ groups according to $f(x_i)$. We also use *Noisy-or* to infer the relations to avoid data sparsity and gain better results.

3.2 Sorting by Group Labels

As mentioned in the previous section, each training group is constituted by: a set of instances containing the same entity pair (E_i), a set of positive candidate relation labels that all instances share (P_i), and a set of negative relation labels that this entity pair can't express (N_i). We assign a score for each group by summing up the sub-scores for all positive labels and negative labels for the group.

K_l is defined as the key instance for relation l - the instance that has the maximum predicting score for a certain label in the group.

$$K_l = \arg\max_m p(z_m = l | w_z, x_i^m) \tag{3}$$

For each l in P_i, the sub-score is the predicting confidence to this label, while for each l' in N_i, the sub-score is the confidence for not predicting to the label. For each label in P_i and N_i, the sub-scores are selected from the key instances correspondingly. The final score for group i is defined as:

$$f(x_i) = \frac{1}{Z_P} \log \sum_{r \in P_i} p(K_r) + \frac{1}{Z_N} \log \sum_{\bar{r} \in N_i} [1 - p(K_{\bar{r}})] \tag{4}$$

where Z_p and Z_N are normalized factors which are set to be $|P_i|$ and $|N_i|$. This sorting strategy is inspired by the original MIML-re model in which the label assignment for an instance is partly determined by the group level generative probabilities. Intuitively, the higher score a group has, the more possible it contains enough instances to support the group-level positive labels and object to the negative labels. This method can implicitly model the missing of the text.

3.3 Sorting by Ranking

We define the term *ranking* according to the relative order of the positive and negative labels in a group. More specifically, if the predicting score for K_l is larger than that for $K_{l'}$, we say l has a higher rank than l'. We define:

R_l – the number of key instances that has a higher predicting score than that for label l within a group.

L_l – the ranking loss for a certain label l in a group which is related to R_l.

We use the above ranking loss to model the case when at least one non-positive label ranks higher than a positive label. More formally, we define:

$$R_l = \sum_{l' \notin P_i} I[p(K_{l'}) > p(K_l)] \tag{5}$$

where $I[\cdot]$ is an indicator function which returns 1 if the inner argument is true and 0 otherwise. And

$$f(x_i) = -\frac{1}{Z} \sum_{l \in P_i} L_l = -\frac{1}{Z_1} \sum_{l \in P_i} \frac{1}{Z_2} \sum_{t=1}^{R_l} 1/t \tag{6}$$

where Z_1 and Z_2 are the normalized factors which are finally set to be $|P_i|$ and $|U_i|$, respectively. Here, $|U_i|$ denotes the number of non-positive labels that has at least one instances in the group. The series $\sum_{t=1}^{R_l} 1/t$ is incremental, so that any positive label l that has a larger R would result in a larger loss, indicating that this label is inclined to be negative and vice versa. This strategy is inspired by [20, 21] where they define a similar ranking loss to train multiclass classifiers based on pair-wise structures. This sorting method aims at removing the groups that are more likely to have wrong generated labels. Intuitively, if a group has a low score, this group is likely to be less effective. This method can implicitly reflect the missing of both the KB and the text.

4   Experiment

4.1   Dataset

We test the models on the KBP dataset, one of the landmark datasets in this community created by Surdeanu et al. [6]. The free text provided contains approximately 1.5 million documents from a variety of sources, including newswire, blogs and telephone conversation transcripts. The KB is a snapshot of the English version Wikipedia. After the DS annotation and a subsample of 5 % as negative (following MIML-re), we finally got 524,777 groups including 950,102 instances. We used the above groups as the training set. The testing set contains 200 queries (one query refers to a key entity), which include 23 thousand instances. Following previous work, 40 out of 200 testing queries are made into the developing set and the rest 160 into the testing set.

4.2   Experimental Setup

Methods Implemented. We implemented MIML-semi [8], one of the state-of-the-art works in this field and compared it with our models. Following previous research, we also compared our models with the classical baseline MIML-re and two other landmark algorithms: (1) MultiR (denoted as Hoffmann in the experiment) [3], it is a multi-instance learning algorithm which supports overlapping relations. (2) Mintz++, Surdeanu et al.'s implementation of the Mintz model [4, 6], comparable to MIML-re in some cases.

Evaluation Metrics.

$$P/R\ curve + Max\ F_1\ \&\ Avg\ F_1 + Precision, Recall\ \&\ Final\ F_1$$

Similar to previous work, we report the Precision/Recall curve (P/R curve) to test the models' stability on the testing data. Generally speaking, if a curve is located on top of another, the corresponding method is better. Max F_1 means the maximum F_1 point on the P/R curve, and Avg F_1 represents the F_1 on the curve by averaging all points. These three metrics are evaluated to see the final performance on all the testing data. And F_1 is usually the main performance measure. In addition, the time complexity of the methods can be partly reflected from the parameters T and θ.

Following MIML-re, we also iterated up to 8 EM epochs (another one epoch for initialization) and used a 3-fold bagging strategy to avoid over-fitting. We fixed the negative samples and removed other random settings for the sake of fair comparisons in the experiment. For the sorting methods proposed, we tuned the parameters θ and T (number of epochs) on the developing set and used the optimal value on the testing set. Finally, we got $(\theta, T) = (0.98, 6)$ for MIML-sort-l and $(0.98, 2)$ for MIML-sort-r.

From Fig. 1 we see, MIML-sort-r significantly goes beyond other curves and extends the upper bound of recall to 0.4. MIML-sort-l performs comparatively with MIML-semi and obviously better than other baselines. More improvements can be read out from Table 1. We see that the best Final F_1, Max F_1 and Avg F_1 are all generated by our methods. Particularly, MIML-sort-l has the best F_1 on the whole testing set (Final F_1), reaching a state-of-the-art of 29.41. Max F_1 and Avg F_1 reflect that the curve of MIML-sort-r obtains the maximum point and best stability. From the parameters we get that our methods converge faster than previous work, especially MIML-sort-r (T = 2). Since the selecting operations are towards the data groups generated by MIML-re, we can see from the figure and table that removing the groups with low scores really help to improve the performance.

Table 1. Results of the proposed methods and baselines.

	Prec.	Rec.	Final F_1	Max F_1	Avg F_1	Parameters
Hoffmann	**30.65**	19.79	23.97	24.05	15.40	-
Mintz++	26.24	24.83	24.97	25.51	21.61	-
MIML-re	30.56	24.68	27.30	28.25	22.75	$T=8$
MIML-semi	13.38	**42.88**	20.39	28.28	23.27	$T=8$
MIML-sort-l	27.00	32.29	**29.41**	29.55	23.33	$T=6, \theta=98\%$
MIML-sort-r	20.95	39.93	27.48	**31.29**	**26.87**	$T=2, \theta=98\%$

What we are also interested in is that: what types of groups are removed through the sorting procedure. We analyze the filtered groups and draw Table 2. In the table, the first row stands for the size (number of instances) of the removed groups (i.e. S = 1 means the percentage of the singleton groups). Each field for a certain method is constituated by two rows, in which the first means the percentage of the corresponding removed groups, and the second shows the accumulation of the current and the previous (i.e. 72.09 = 46.10 + 25.99). It is easy to discover that the groups that contain smaller number of instances are more likely to be removed.

Table 2. The statistics for the removed data.

	S=1	S=2	S=3	S=4	S>=5
MIML-sort-l	46.10%	25.99%	11.94%	6.73%	9.24%
(sum)	(46.10%)	(72.09%)	(84.03%)	(90.76%)	(100%)
MIML-sort-r	75.53%	12.48%	5.07%	2.44%	4.48%
(sum)	(75.53) %	(88.01%)	(93.08%)	(95.52%)	(100%)

5 Conclusion

In this paper, we propose two ranking-based methods to select effective training groups for distantly supervised relation extraction. Experiments compared with four baselines validate the efficiency of the proposed methods. Particularly, MIML-sort-r boosts the F_1 significantly on the testing set by 2.1 %. We also believe that the proposed ranking-based methods can be integrated in other learning frameworks. From the results we notice, there is still large space for improvements, demanding more efficient and robust methods.

Acknowledgement. This work was supported in part by National 863 Program of China (2015AA015405), NSFCs (National Natural Science Foundation of China) (61402128, 61473101, 61173075 and 61272383).

References

1. Bunescu, R.C., Mooney, R.J.: Learning to extract relations from the web using minimal supervision. In: Proceedings of the 45th Annual Meeting of the Association for Computational Linguistics, pp. 576–583 (2007)
2. Bellare, K., McCallum, A.: Learning extractors from unlabeled text using relevant databases. In: Proceedings of the 6th International Workshop on Information Extraction on the Web, pp. 10–15 (2007)
3. Hoffmann, R., Zhang, C., Ling, X., Zettlemoyer L., Weld, D.S.: Knowledge based weak supervision for information extraction of overlapping relations. In: Proceedings of the 49th Annual Meeting of the Association for Computational Linguistics, pp. 541–550 (2011)

4. Mintz, M., Bills, S., Snow, R., Jurafsky, D.: Distant supervision for relation extraction without labeled data. In: Proceedings of the 47th Annual Meeting of the Association for Computational Linguistics and the 4th International Joint Conference on Natural Language Processing of the AFNLP.2, pp. 1003–1011(2009)
5. Riedel, S., Yao, L., McCallum, A.: Modeling relations and their mentions without labeled text. In: Proceedings of the European Conference on Machine Learning and Knowledge Discovery in Databases, pp. 148–163 (2010)
6. Surdeanu, M., Tibshirani, J., Nallapati, R., Manning, C.D.: Multi-instance multi-label learning for relation extraction. In: Proceedings of the 2012 Joint Conference on Empirical Methods in Natural Language Processing and Computational Natural Language Learning, pp. 455–465 (2012)
7. Angeli, G., Tibshirani, J., Wu, J.Y., Manning, C.D.: Combining distant and partial supervision for relation extraction. In: Proceedings of the 2014 Conference on Empirical Methods in Natural Language Processing, pp. 1556–1567 (2014)
8. Min, B., Grishman, R., Wan, L., Wang, C., Gondek, D.: Distant supervision for relation extraction with an incomplete knowledge base. In: Proceedings of the Conference of the North American Chapter of the Association for Computational Linguistics, pp. 777–782 (2013)
9. Craven, M., Kumlien, J.: Constructing biological knowledge bases by extracting information from text sources. In: Proceedings of the 7th International Conference on Intelligent Systems for Molecular Biology, pp. 77–86 (1999)
10. Bordes, A., Usunier, N., Garcia-Duran, A., Weston, J., Yakhnenko, O.: Translating embeddings for modeling multi-relational data. In: Proceedings of the 26th Advances in Neural Information Processing Systems, pp. 2787–2795(2013)
11. Wang, Z., Zhang, J., Feng, J., Chen, Z.: Knowledge graph embedding by translating on hyperplanes. In: Proceedings of the 29th AAAI Conference on Artificial Intelligence, pp. 1112–1119 (2014)
12. Lin, Y., Liu, Z., Sun, M., Liu, Y., Zhu, X.: Learning entity and relation embeddings for knowledge graph completion. In: Proceedings of the 30th AAAI Conference on Artificial Intelligence, pp. 2181–2187 (2015)
13. Riedel, S., Yao, L., McCallum, A., Marlin, B.M.: Relation extraction with matrix factorization and universal schemas. In: Proceedings of the 2013 Conference of the North American Chapter of the Association for Computational Linguistics: Human Language Technologies, pp. 74–84 (2013)
14. Fan, M., Zhao, D., Zhou, Q., et al.: Distant supervision for relation extraction with matrix completion. In: Proceedings of the 52nd Annual Meeting of the Association for Computational Linguistics, vol. 1, pp. 839–849 (2014)
15. Nagesh, A., Haffari, G., Ramakrishna, G.: Noisy or-based model for relation extraction using distant supervision. In: Proceedings of the 2014 Conference on Empirical Methods in Natural Language Processing, pp. 1937–1941 (2014)
16. Intxaurrondo, A., Surdeanu, M., de Lacalle, O.L., Agirre, E.: Removing noisy mentions for distant supervision. Procesamiento del Lenguaje Natural **51**, 41–48 (2013)
17. Xu, W., Hoffmann, R., Zhao, L., Grishman, R.: Filling knowledge base gaps for distant supervision of relation extraction. In: Proceedings of the 51st Annual Meeting of the Association for Computational Linguistics, pp. 665–670 (2013)
18. Takamatsu, S., Sato, I., Nakagawa, H.: Reducing wrong labels in distant supervision for relation extraction. In: Proceedings of the 50th Annual Meeting of the Association for Computational Linguistics, pp. 721–729 (2012)
19. Ritter, A., Zettlemoyer, L., Etzioni, O.: Modeling missing data in distant supervision for information extraction. Trans. Assoc. Comput. Linguist. **1**, 367–378 (2013)

20. Usunier, N., Buffoni, D., Gallinari, P.: Ranking with ordered weighted pairwise classification. In: Proceedings of the 26th International Conference on Machine Learning, pp. 1057–1064 (2009)
21. Weston, J., Bengio, S., Usunier, N.: Wsabie: scaling up to large vocabulary image annotation. In: Proceedings of the 22nd International Joint Conference on Artificial Intelligence, vol. 11, pp. 2764–2770 (2011)

SpikeComp: An Evolving Spiking Neural Network with Adaptive Compact Structure for Pattern Classification

Jinling Wang[1(✉)], Ammar Belatreche[1], Liam P. Maguire[1], and T. Martin McGinnity[1,2]

[1] Intelligent Systems Research Centre (ISRC), Faculty of Computing and Engineering,
University of Ulster, Magee Campus, Northland Road, Derry, BT48 7JL, UK
Wang-J1@email.ulster.ac.uk,
{A.Belatreche,LP.Maguire}@ulster.ac.uk
[2] School of Science and Technology, Nottingham Trent University, Nottingham, UK
TM.McGinnity@ulster.ac.uk

Abstract. This paper presents a new supervised learning algorithm (Spike-Comp) with an adaptive compact structure for Spiking Neural Networks (SNNs). SpikeComp consists of two layers of spiking neurons: an encoding layer which temporally encodes real valued features into spatio-temporal spike patterns, and an output layer of dynamically grown neurons which perform spatio-temporal pattern classification. The weights between the neurons in the encoding layer and the new added neuron in the output layer are initialised based on the precise spiking times in the encoding layer. New strategies are proposed to either add a new neuron, or update the network parameters when a new sample is presented to the network. The proposed learning algorithm was demonstrated on several benchmark classification datasets and the obtained results show that SpikeComp can perform pattern classification with a comparable performance and a much compact network structure compared with other existing SNN training algorithm.

Keywords: Spiking neurons · Supervised learning · Adaptive structure · Classification

1 Introduction

Artificial neural networks (ANNs) mimic the brain's ability to solve complex tasks, such as face recognition and object identification. The frequency of the spikes is used to encode information. However, existing studies have shown that biological neurons use precise spiking times to encode information [1]. Spiking neural networks (SNNs) use computational neural models which capture the firing dynamics of biological neurons [2]. SNNs offer a more biologically plausible model compared with their classical counterparts and are capable of handling spatio-temporal data. SpikeProp is an adaptation of the classical backpropagation algorithm which minimize the error between the predicted and desired spiking times. It can achieve comparable performance on several benchmark datasets to rate-coded networks [3]. However, it is slow when used in an online setting.

© Springer International Publishing Switzerland 2015
S. Arik et al. (Eds.): ICONIP 2015, Part II, LNCS 9490, pp. 259–267, 2015.
DOI: 10.1007/978-3-319-26535-3_30

Belatreche et al. [4] proposed a derivative-free supervised learning algorithm where an evolutionary strategy (ES) is used to minimise the error between the output firing times and the corresponding desired firing times. This algorithm achieves a better performance than SpikeProp. However, an ES-based iterative process makes the training procedure extremely time-consuming and is not suitable for online learning. It is important to note that fixed network architectures were employed in the above-mentioned approaches. For maximum adaptability, which is needed in a changing environment or when dealing with streams of data, it is desirable that both the network structure and its weights adapt to new data.

In [5] an evolving approach for SNNs (eSNN) has been presented. The learning rule is based on the rank order of the spiking times. The eSNN has successfully been used for on-line spatio-and spectro-temporal pattern recognition. We have developed an RBF-like evolving leaning algorithm for SNN in [6], the centre of each added hidden RBF spiking neuron is represented by its time to first spike, the adaptive SNN is able to classify spike-based spatio-temporal inputs after just one presentation of the training set. Recently, a Self Regulating Evolving SNN has been proposed for handling classification problems in [7]. The evolving algorithm utilizes the association between the spiking outputs of differing classes during training. These associations take a vital role in pattern classification tasks.

This paper presents a new supervised learning algorithm (called SpikeComp) with an adaptive compact structure for Spiking Neural Networks (SNNs). Extended from [6], SpikeComp assumes that each added output neuron has a centre which is represented by its time to first spike; there is an accommodation field around the centre. New strategies are proposed for adding a new neuron, updating the network weights, or just updating the neurons centre when a new sample is presented to the network. These strategies depend on whether the spiking time of a sample lies inside the accommodation field of a neuron centre and whether the winner output neurons are associated or not with the same class as the incoming input sample. SpikeComp has been validated on several benchmark datasets; the results show that using a much compact network, SpikeComp can achieve comparable performance using compared with other machine learning algorithms and the existing eSNN.

The remainder of this paper is structured as follows: Sect. 2 describes the employed neural model, the employed temporal encoding scheme, the SNN structure design and the proposed learning strategies. Section 3 presents experimental results for training SNNs on selected benchmark datasets from the UCI Machine Learning Repository. Finally Sect. 4 concludes the paper.

2   Learning and Structural Adaptation

2.1   Spiking Neural Model

The proposed SpikeComp employs the Integrate-and-Fire (IF) neurons in the output layer. The postsynaptic potential (PSP) of an output neuron i at time t relies on the spike times received from neurons in the encoding layer and can be described as:

$$PSP(i, t) = \sum_{j \in [1,N]} W_{ji} \exp\left(-\frac{t_j}{\tau}\right) \tag{1}$$

Where $j \in [1, N]$ represents the jth incoming connection, and N is the total number of incoming connections between the encoding layer and the output neuron i; t_j represents the precise spiking time of the jth encoding neuron; τ is a time constant and determines the range for which synaptic strengthening occurs; W_{ji} is the synaptic weight associated with the synaptic connection between output neuron i and encoding neuron j. If $PSP(i, t)$ is greater than the firing threshold of neuron i, denoted by, $PSP_{th}(i)$, then an output spike is produced at neuron i in the output layer, and the simulation for the current input sample is terminated and the PSP of firing output neuron is reset.

2.2 Information Encoding

Gaussian Receptive Field population encoding scheme, proposed by Bohte et al. [3], is used to encode continuous input variables into spike times. An example is shown in Fig. 1 where a real-valued feature of 5.1 (illustrated by a vertical dashed red line) is converted into spike times using eight Gaussian receptive fields. The resulting eight response values (y) are 0.0228, 0.4578, 0.9692, 0.2163, 0.0051, 0, 0 and 0 respectively. These values are then mapped linearly to spike times (see Eq. 2):

$$t = -9 * y + 9 \tag{2}$$

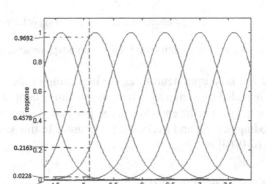

Fig. 1. Encoding of a real valued feature of 5.1 using 8 Gaussian receptive fields neurons.

If the resulting spike time is equal to 9 ms, this neuron is treated as 'silent', and is represented by a value of '−1'. The resulting spiking times are therefore represented by the following series of spiking times: 8.79 ms, 4.88 ms, 0.28 ms, 7.05 ms, 8.95, -1(silent), -1(silent) and -1(silent).

Time-to-first spike decoding is employed at the output layer where an input sample is considered to be correctly classified if the first spike is produced at an output neuron whose class label matches the class label of the current input sample; otherwise an input sample is considered to be incorrectly classified.

2.3   Network Topology

Figure 2 presents the proposed network topology that consists of a layer of encoding neurons and a layer of output neurons. Each neuron in the encoding layer provides a spike time which is fed to the next layer, and is fully connected to the neurons in the output layer. The number of neurons in this layer is determined by the dimensionality of the dataset (m) and the number of Gaussian receptive fields q, i.e., given by $m * q$. The set of spiking times represented by these neurons in the encoding layer is in the range of the time coding interval $[0,T_{ref}]$ and $T_{ref} = 9$ ms.

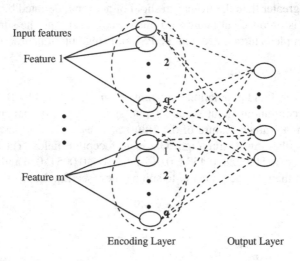

Fig. 2. A two layer feedforward SNN with adaptive structure.

The output layer has no output neurons at the beginning of the training process. A new output neuron is added dynamically when an incoming sample is received and is fully connected to the neurons in the encoding layer. Only one synaptic connection exists between every encoding neuron and every output neuron. In this study the simulation interval per sample is $[0\ 10]$ ms.

2.4   Learning and Structure Adaptation

As a new sample is presented to the network, SpikeComp finds the two winner neurons which have the same class label (i) and differing class label (k) as the incoming sample, and their first spiking times t_{cc} and t_{mc}, respectively (as illustrated in Fig. 3). Every added output neuron has a centre which is set based on the time of the first spike of the added output neuron ($T_c(i)$, i represents the ith added output neuron). In addition, we use two radius parameters, R_{cc} and R_{mc}, which represent accommodation boundaries when the neuron is a winner and has the same class label (R_{cc}) or a differing class label (R_{mc}) as the new incoming input sample, respectively. When added to the neuron centre value, i.e. $T_c + R_{cc} = Th_{cc}$, or $T_c + R_{mc} = Th_{mc}$, the resulting sums determine the neuron adding thresholds (Th_{cc} and Th_{mc}) when the output neuron has the same class label or different class label as the incoming sample, respectively. Based on the associations among the first spiking times (t_{cc} and t_{mc}) of the two winner neurons and the neuron adding thresholds (Th_{cc} and Th_{mc}), SpikeComp adds a new neuron (neuron addition strategy), update the network weights (weights update strategy) or just update the centre (update centre strategy).

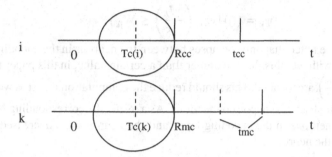

Fig. 3. Description of neuron addition with regard to classification or misclassification.

When a new output neuron (i) is created and its weights W_{ji} are initialised using Eq. 3;

$$w_{ji} = \exp\left(-\frac{t_j}{\tau}\right) \tag{3}$$

The maximum postsynaptic potential for this added output neuron ($PSP_{max}(i)$) is set when the training sample is propagated into the network, and is calculated using Eq. 1 when all potential input spikes have been used. The firing threshold of this added output neuron $PSP_{th}(i)$ for the postsynaptic potential in the network is determined by Eq. 4 as follows:

$$PSP_{th}(i) \leftarrow con * (PSP_{max}(i)) \tag{4}$$

Where $con \in [0, 1]$ and is set to 0.5 in this paper. The class label of this new added output neuron is set the same as that of this new sample.

Output Neuron Adding Strategy. SpikeComp adds a new output neuron during the training stage under three situations: (a) if an input sample is presented from a class that is different from those previously learned so an output neuron is added to cater for the new class. (b) As shown in Fig. 3, the output neuron i fires first ($t_{cc} < t_{mc}$, solid line for t_{mc} in neuron k), the spiking time (t_{cc}) of neuron i is greater than the adding threshold of neuron i, i.e., $t_{cc} < t_{mc}$ & $t_{cc} > Th_{cc}(i)$ or the output neuron k fires first ($t_{cc} > t_{mc}$, dashed line for t_{mc} in neuron k), the spiking time (t_{mc}) of neuron k is greater than the adding threshold of neuron k, i.e., $t_{mc} < t_{cc}$ & $t_{mc} > Th_{mc}(k)$. In these situations, a new output neuron is added. (c) if the winner output neuron (k) fires first, and its spiking time (t_{mc}) is less than the adding threshold of neuron k, i.e., $t_{mc} < t_{cc}$ & $t_{mc} < Th_{mc}(k)$, Hence, a new neuron needs to be added to represent this new incoming sample. The weights of this winner neuron k are updated so that $t_{mc} > Th_{mc}(k)$ using Eqs. (5) and (6). Where η is the learning rate and is set to 0.025.

$$W_{jk} = W_{jk} + \eta * \left(\exp \left(-\frac{t_j}{\tau} \right) - W_{jk} \right) \quad j\varepsilon [1N] \,\&\, \forall \,(j) \in \psi_k \tag{5}$$

$$\psi_k = \{(j) \mid \exp \left(-\frac{t_j}{\tau} \right) \leq W_{jk}\} \tag{6}$$

It should be noted that only synapses between the neurons in the encoding layer and the neuron k with weights that are higher than a certain value, in this paper this value is set to $\exp \left(-\frac{t}{\tau} \right)$, are updated. This should reduce the computational cost as weight update is limited to a subset of synaptic connections. As a result, the corresponding high weights between the neurons in the encoding layer and the neuron k are decreased, leading to late firing of the neuron k.

Weights Update Strategy. If the criteria for a new output neuron addition are not met, then relevant weights are considered to be updated. (a) if the spiking time difference between the neuron i and the neuron k is sufficiently large, i.e., $M/2 < (t_{mc} - t_{cc})/t_{cc} < M$. Furthermore, the neuron i fires first and the spiking time (t_{cc}) of neuron i is less than the adding threshold of neuron i, i.e., $t_{cc} < t_{mc} \,\&\, t_{cc} < Th_{cc}(i)$, M represents the separation constant, determines when a neuron should be added or just weights are updated, then only the weights for the winner neuron (i) are updated based on Eqs. (7) and (8). It should be noted that only synapses between the neurons in the encoding layer and the winner neuron i with weights that are lower than a certain value (set to $\exp \left(-\frac{t}{\tau} \right)$ in this paper) are updated. As a result, the corresponding low weights between the neurons in the encoding layer and the neuron i are increased, leading to earlier spiking of the neuron i.

$$W_{ji} = W_{ji} + \eta * \left(\exp \left(-\frac{t_j}{\tau} \right) - W_{ji} \right) \quad j\varepsilon [1N] \,\&\, \forall \,(j) \in \psi_i \tag{7}$$

$$\psi_i = \{(j) \mid W_{ji} \leq \exp \left(-\frac{t_j}{\tau} \right)\} \tag{8}$$

or (b). In this case where $|t_{mc} - t_{cc}|/t_{cc} <= M$, the weights between the neurons in the encoding layer and the neuron i, and neuron k are updated to avoid misclassification. The weights between the neurons in the encoding layer and the neuron i are updated using (7) and (8), the weights between the neurons in the encoding layer and the neuron k are updated using (5) and (6). As a result, the weights connecting to neuron i are strengthened, while those connecting to neuron k are weakened. This will increase the difference between the first spike times of the winner neuron i and the winner competing neuron k. It should be noted that neuron addition strategy and weights update strategy are inspired from [7].

Neuron Centre Update Strategy. If the learning strategies for neuron addition and weights update mentioned above are not satisfied, the centre of neuron i needs to be updated using Eq. (9). The total number of samples used to train the neuron i, represented by N_i^{sample}, is then incremented. So the neuron adding threshold Th_{cc} and Th_{mc} for the neuron i are also updated to accommodate the current input sample.

$$T_c(i) \leftarrow \frac{T_c(i) * N_i^{sample} + t_{cc}}{N_i^{sample} + 1}. \qquad (9)$$

3 Performance Evaluation of SpikeComp

In this section, the performance of the proposed SpikeComp is evaluated on the IRIS, Wisconsin Breast Cancer (*WBC*) and Ionosphere (*Ion*) benchmark datasets from the UCI machine learning repository. The datasets were each divided into two sets as outlined in Table 1, where the first *Tr* samples were used for training and the remaining *Ts* samples were used for testing. *T* represents the total number of samples, *c* is the number of classes. Table 1 also lists the values of the parameters *con*, τ, *M*, R_{cc} and R_{mc} that are related to the network parameter and structure learning for each dataset.

Table 1. Description of different datasets and their parameter values.

Dataset	T/Tr/Ts	m	c	q	N	con	τ	M	R_{cc}/R_{mc}
IRIS	150/30/120	4	3	10	40	0.5	25	0.1	2.5/2.0
WBC	683/40/643	9	2	8	72	0.5	30	0.2	2.5/2.5
Ion	351/100/251	33	2	7	231	0.5	30	0.1	2.5/2.5

The dynamic changes of the number of output neurons against the number of epochs during training for the IRIS, WBC and Ionosphere datasets are illustrated in Fig. 4. From Fig. 4 it could be seen that the network has grown a total of seven, fourteen and sixty-two neurons for IRIS, WBC and Ionosphere datasets, respectively.

Table 2 lists the classification accuracies obtained for each dataset and compares the classification performance of these datasets using SpikeComp algorithm with the Rank Order based eSNN method in [5] and the popular classical methods, namely KNN, SVM and MLP. *Acc_Tr/Acc_Ts* represents the training/testing classification accuracy; *Num_O* represents the total number of neurons in the output layer after training. From Table 2, it can be observed that the results of SpikeComp are comparable to the Rank Order based eSNN method [5] and other machine learning algorithms. In addition, these results clearly show that SpikeComp adds much less output neurons than the Rank Order based eSNN method.

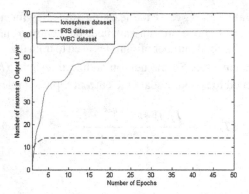

Fig. 4. Evolution of the output layer structure against the number of epochs during training for the IRIS, WBC and Ionosphere datasets.

Table 2. Comparison of the classification accuracy of SpikeComp with other approaches,

	SpikeComp/ (Num_o)	eSNNs/(Num_o)	MLP	KNN (k = 3)	SVM
Dataset	Acc_TR/Acc_TS (%)				
IRIS	100/**97.5**/(7)	100/95.0/(84)	100/92.7	96.0/92.0	100/96.0
WBC	97.5/96.7/(14)	99.6/**98.7**/(280)	98.0/88.5	97.3/98.5	96.6/98.5
Ion	96.0/91.2/(62)	81.6/74.4/(213)	100/78.3	88.6/**93.2**	100/90.3

4   Conclusion

In this paper, an Evolving Spiking Neural classifier, called SpikeComp, has been presented. SpikeComp employs a two layers feedforward network topology, and the Gaussian receptive fields population encoding is used to temporally encode real valued feature vectors into spatio-temporal spike patterns in the encoding layer. Output neurons that process spatio-temporal inputs from the encoding layer, are dynamically grown as new spatio-temporal spiking patterns are presented to the spiking neural network based on differing learning strategies. Evaluated on several benchmark tasks, the classification performance of SpikeComp has been compared with eSNN and other machine learning methods. The results clearly indicate that SpikeComp achieves comparable performance with much smaller network structure, but some datasets need more training epochs than other evolving SNNs methods. Future work will demonstrate these learning strategies on other neural models and a wider range of datasets.

References

1. Rullen, R., Van Guyonneau, R., Thorpe, S.J.: Spike times make sense. Trends Neurosci. **28**(1), 1–4 (2005)
2. Maass, W.: Networks of spiking neurons: the third generation of neural network models. Neural Netw. **10**(9), 1659–1671 (1997)
3. Bohte, S.M., Kok, J.N., Poutre, H.L.: Error-backprogation in temporally encoded networks of spiking neurons. Neurocomputing **48**, 17–37 (2002)
4. Belatreche, A., Maguire, L.P., McGinnity, T.M., Wu, Q.: Evolutionary design of spiking neural networks. J. New Math. Nat, Comput. **2**(3), 237–253 (2006)
5. Kasabov, N., Dhoble, K., Nuntalid, N., Indiveri, G.: Dynamic evolving spiking neural networks for on-line spatio-and spectro-temporal pattern recognition. Neural Netw. **41**, 188–201 (2013)
6. Wang, J., Belatreche, A., Maguire, L.P., McGinnity, T.M.: An online supervised learning method for spiking neural networks with adaptive structure. Neurocomputing **144**, 526–536 (2014)
7. Dora, S., Subramanian, K., Suresh, S., Sundararajan, N.: Development of a self regulating evolving spiking neural network for classification problem. Neurocomputing (2015, in press)

Formation of Momentum and Learning Rate Profile for Online Training and Testing of HMLP with ALRPE

Zuraidi Saad[1,2(✉)], Mohd Yusoff Mashor[2], and Wan Khairunizam Wan Ahmad[2]

[1] Faculty of Electrical Engineering, Universiti Teknologi MARA, 13500 Permatang Pauh, Penang, Malaysia
zuraidi570@ppinang.uitm.edu.my
[2] Electronic & Biomedical Intelligent Systems (EBItS) Research Group, School of Mechatronic Engineering, Universiti Malaysia Perlis, Pauh Putra Campus, 02600 Arau, Perlis, Malaysia

Abstract. The technique of momentum and learning rate formation for children pedagogy was adopted to improve the performance of learning algorithm in training the hybrid multilayered perceptron network (HMLP) using modified recursive prediction error (MRPE) algorithm. An Adaptive Learning Recursive Prediction Error Algorithm (ALRPE) is proposed as a second version of MRPE with a guidance of a new profile of learning and momentum rate. An online model was used to forecast speed, revolution and fuel balanced in a Proton Gen2 car tank. The car measured the injected fuel from fuel injection sensor and became an input to the HMLP model to forecast the speed, revolution and fuel balanced in tank. These forecasted variables were also measured from the car sensors. To date, there is a restricted study on the effect of the profile of learning and momentum rate to the performance of HMLP. This study proposes a new profile of momentum and learning rate to improve the performance of the nonlinear modelling using HMLP. Previous conventional profile was developed only based on its general algorithm. The effect of the profile of momentum and learning rate to the generalisation of the HMLP using MRPE network was lack of discussion and thus, motivates this research using the proposed technique. Experimental results showed that the proposed ALRPE profile of momentum and learning rate can improved the performance of nonlinear HMLP model in the range of 0.002 dB to 3.15 dB of mean square error (MSE) in the model validation.

1 Introduction

Most children pedagogy approaches allow them for more exploration starting from their first day in school even as the teacher continuously stimulate their learning potential and eventually they become matured and knowledgeable in the end of their schooling [1, 2]. These attempts of inspiration were adopted to train the neural network model. This idea induces more exploration at the beginning of learning and force the neural network to establish convergence before the end of the learning process. The selection of learning is of critical importance in finding the true global minimum of error distance. A high learning rate and a low momentum usually were selected as an early predetermined value. Then the learning rate will decrease and

© Springer International Publishing Switzerland 2015
S. Arik et al. (Eds.): ICONIP 2015, Part II, LNCS 9490, pp. 268–275, 2015.
DOI: 10.1007/978-3-319-26535-3_31

momentum will increase over the course. The modified recursive prediction error (MRPE) training algorithm with too small learning rate will make agonizingly slow progress. Too large a learning rate will proceed much faster, but may simply produced oscillations between relatively poor solutions. Empirical evidence in previous study [3, 4], showed that the use of momentum in MRPE algorithm can be helpful in speeding the convergence and avoiding local minima. The use of momentum is to stabilize the weight changes and suppresses cross-stitching, that is, it cancels side-to-side oscillations across the error valey. A learning rate of one with zero momentum means children are fully allowed exploring without a guidance of a teacher. This situation is only permitted as a predetermined of momentum and learning rate or when there is no problem with local minima or non-convergence is found. The online learning may have a better chance of finding a global minimum than the offline learning technique [5].

Previous studies were based on offline modelling technique of speed and fuel flow [3, 4]. Mashor [5] suggested that the HMLP network is suitable to be used for the online nonlinear modelling and forecasting of streamflow which is to reduce hydrological risks due to streamflow, such as flood. This model is adopted in this study by making simple modification of the HMLP input structure for the purposes of highly dynamic data forecasting.

2   Theoretical Background

The realization of artificial neural network as a system identification can be divided into two major part: (i) The selection of input output and its neural network structure (modeling of an unknown system), (ii) The development of learning algorithm to update the tunable weights contained by the neural network structure.

2.1   Hybrid Multilayered Perceptron Network (HMLP)

HMLP network is produced from MLP network with some additional linear input connections is shown in Fig. 1. These linear connections can be viewed as a linear model, which is in parallel to the standard MLP network; hence, the projected network is called a hybrid multilayered perceptron network. In the present study, theoretical description of the network will be limited to one hidden layer only [6]. The output of HMLP network with one hidden layer can be expressed by Eq. (1)

$$\hat{y}_k(t) = \sum_{j=1}^{n_h} w_{jk}^2 F\left(\sum_{i=1}^{n_i} w_{ij}^1 v_i^0(t) + b_j^1 \right) + \sum_{i=0}^{n_i} w_{ik}^l v_i^0(t) \quad for \ 1 \leq k \leq m \tag{1}$$

where w_{ij}^1 denotes the weights that connect the input and the hidden layers; b_j^1 and v_i^0 represents the threshold in hidden nodes and input supplied to the network; w_{jk}^2 denotes the weights that connect the hidden and output layer; w_{ik}^l is the weights connection between input and output layer; n_i and n_h are the number of input nodes and hidden nodes; m represents the number of output nodes while $F(\bullet)$ is an activation function which is normally selected as sigmoidal function. The weights w_{jk}^2, w_{ij}^1, w_{ik}^l and b_j^1 are unknown, and should be selected carefully in order to achieve minimum prediction error, defined as below:

$$\varepsilon_k(t) = y_k(t) - \hat{y}_k(t) \tag{2}$$

where $y_k(t)$ and $\hat{y}_k(t)$ are the actual and forecasted output. In this study, HMLP network is trained using Modified Recursive Prediction Error (MRPE) algorithm. MRPE algorithm is modified from Recursive Prediction Error (RPE) algorithm by varying the momentum and learning rate [6] compared to constant values applied in Chen et al. [7].

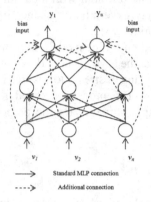

Fig. 1. Hybrid Multilayered Perceptron Network

2.2 Modified Recursive Prediction Error Algorithm (MRPE)

In this study the MRPE proposed by Mashor [6] is used for the online update of the tunable weights. Firstly, Chen et al. [7] modified the RPE algorithm to minimize the cost function given by:

$$J(\hat{\Theta}) = \frac{1}{2N} \sum_{t=1}^{N} \varepsilon^T(t, \hat{\Theta}) \Lambda^{-1} \varepsilon(t, \hat{\Theta}) \tag{3}$$

The modified RPE algorithm is used to update the estimated parameter $\hat{\Theta}$ (consists of ws and bs), recursively using the Gauss-Newton algorithm:

$$\hat{\Theta}(t) = \hat{\Theta}(t-1) + P(t)\Delta(t) \tag{4}$$

and

$$\Delta(t) = \alpha_m(t)\Delta(t-1) + \alpha_g(t)\psi(t)\varepsilon(t) \tag{5}$$

where $\varepsilon(t)$ and Λ are the prediction error and $m \times m$ symmetric positive definite matrix respectively. Meanwhile, m is the number output nodes of the network; $\alpha_m(t)$ and $\alpha_g(t)$

can be assigned within the range of 0 and 1, and the typical values of $\alpha_m(t)$ and $\alpha_g(t)$ are varied to improve further the convergence rate of the MRPE algorithm by Mashor [6] according to:

$$\alpha_m(t) = \alpha_m(t-1) + a \tag{6}$$

and

$$\alpha_g(t) = \alpha_g(0)(1 - \alpha_m(t)) \tag{7}$$

where a is a small constant (typically a = 0.01); $\alpha_m(0)$ and $\alpha_g(0)$ are the initial values of $\alpha_m(t)$ and $\alpha_g(t)$ that have the typical values of 0 and 0.5 respectively. $\psi(t)$ is the gradient of one step ahead predicted output with respect to the network parameters:

$$\psi(t, \hat{\Theta}) = \left[d\hat{y}(t, \hat{\Theta}) \Big/ d\hat{\Theta} \right] \tag{8}$$

The detailed and full explanation of MRPE for a one-hidden-layer HMLP network can be implemented as in [6].

2.3 Approach of Pedagogy in an Adaptive Learning Recursive Prediction Error Algorithm (ALRPE)

The philosophy of pedagogy stated that "all children are gifted, some just open their packages later than others". Just feed them with the breakfast of champions and know their learning style [1, 2]. An analogy to this concept, it is presumed that all learning algorithm are capable to achieve their maximum performance, the difference is just in term of speed of them to achieve the maximum convergence rate and the type of data being used to feed them. The more rich data being used to feed them, the less time is needed for the momentum to amplify the learning rate causing a faster convergence.

In this study the Adaptive Learning Recursive Prediction Error Algorithm (ALRPE) is proposed as a second version of MRPE with a guidance of a profile of α_g (learning rate) and α_m (momentum rate). The philosophy behind the formation of original profile is based only on the empirical evidence of the good generalization and analysis of the MSE during online training and testing of the proposed model. Generally, a learning rate of 1 with zero momentum means children are fully allows exploring without a guidance of a teacher. After momentum rate is begun to develop, the learning rate is begin to reduce. This concept means that the supervision of teacher will affect the children to focus more on the learning.

There is no proper study to select which profile of α_m and α_g are the best for modelling and forecasting. Formula of α_m and α_g as in Eqs. (6) and (7) was originally derived by Mashor [6] to train the HMLP network. The original shape of α_m and α_g are show as Fig. 2(a). The initial parameter of α_m and α_g are set as $\alpha_m(0) = 0$ and $\alpha_g(0) = 0.5$. The formation speed of momentum will depend on the value of small constant a. Constant a is set as 0.001 to form a slow momentum graph that will go across the formation of learning rate α_g and reach its maximum value of momentum rate. Learning rate is starting to form its profile from its initial value 0.5 and drastically shrink until it reaches a value of nearly zero to become a constant.

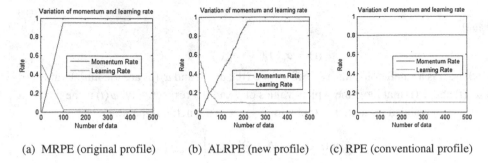

(a) MRPE (original profile)   (b) ALRPE (new profile)   (c) RPE (conventional profile)

Fig. 2. The profile of momentum rate and learning rate (α_m and α_g).

The adaptive learning formerly used by Jesus and Hagan [8, 9] just to adapt learning rate for backpropagation learning algorithm. In this study, adaptive learning was adopted for both learning rate and momentum rate. Finally a new formula for α_m and α_g was proposed as Eqs. (9)–(12) to improve the speed of convergence and improve the convergence rate of RPE learning algorithm for HMLP model. The Adaptive Learning RPE learning algorithm can be implemented as follows. If the value of $\varepsilon_k(t)$ is more than $\varepsilon_k(t - 1)$ then α_m and α_g will be update by using Eqs. (9) and (10). As well if the value of $\varepsilon_k(t)$ is less than $\varepsilon_k(t - 1)$, then α_m and α_g will be update by using Eqs. (11) and (12).

$$\alpha_m(t) = \alpha_m(t - 1) + \alpha_{m\_inc} \tag{9}$$

$$\alpha_g(t) = \alpha_g(t)(\alpha_{g\_dec}) \tag{10}$$

$$\alpha_m(t) = \alpha_m(t - 1) - \alpha_{m\_dec} \tag{11}$$

$$\alpha_g(t) = \alpha_g(t)(\alpha_{g\_inc}) \tag{12}$$

The new profile of α_m and α_g show as Fig. 2(b) is proposed with a zig zag formation of both momentum rate and learning rate that will become a constant after a number of training data. The initial parameter of α_m and α_g are set as $\alpha_m(0) = 0$ and $\alpha_g(0) = 0.5$. The formation speed of momentum will depend on the value of constant α_{m_inc} and α_{m_dec}. Constant value of α_{m_inc} and α_{m_dec} is set as 0.01 and 0.002 respectively to form a fast momentum graph that will go across the formation of learning rate α_g and reach its maximum value of momentum rate to become a constant momentum for the rest of training data. Learning rate (α_g) is starting to form its profile from its initial value 0.5. Constant value of α_{g_inc} and α_{g_dec} is set as 1.01 and 0.95 respectively to shrink down α_g with a form of zig zag until it cross the value of momentum (α_m) and reach its minimum value of learning rate to became a constant.

3   Results and Discussion

The performance of ALRPE, MRPE and RPE learning algorithm using HMLP network are compared. The performance comparison for speed, rpm, fuel balance and heat exchanger forecasting are carried out by using the same conditions mentioned in the previous section, such as number of training set and testing set, respectively. All the models in this study were simulated and compared by the following configuration base on the network node analysis:

$v_{speed}(t) = [u(t) u(t-1) u(t-2) u(t-3) y(t-1) y(t-2) y(t-3) e(t-1) e(t-2)]$ with bias input;

$v_{rpm}(t) = [u(t) u(t-1) u(t-2) u(t-3) y(t-1) y(t-2) y(t-3) e(t-1) e(t-2)]$ with bias input;

$v_{heat\ exchanger}(t) = [u(t-1) u(t-2) y(t-1) y(t-4) e(t-3) e(t-4) e(t-5)]$ with bias input;

with bias input 1, number of hidden nodes $n_h = 5$ and $P(t) = 1000I$. v is the input configuration of the one step ahead model. u and y are the input output of the predicted signal. y_{nn} is the output of HMLP network at one step ahead model.

The performance results of two different models (a) HMLP with ALRPE, (b) HMLP with MRPE and (c) HMLP with RPE are presented in Figs. 3, 4 and 5. These figures show that MSE calculated using testing data sets and indicated that the model parameters converge rapidly when the input is enriched with data from the proposed profile. The MSE converges to an acceptable value after about 770 data samples, suggesting that the model only requires 770 data samples to be trained properly. Generally a good predictions and a good MSE suggest that the model is unbiased and adequate to represent the identified system.

3.1   Modelling and Forecasting of Speed

The MSE comparison for speed with two different learning algorithms is shown in Fig. 3. From the results, it can be seen that HMLP with ALRPE model give a better performance than HMLP with MRPE and RPE model. The HMLP with ALRPE, MRPE and RPE model have converged rapidly with final MSE of 6.303, 6.305 and 6.580 respectively. ALPE learning algorithm stops to learn after 400 of training data while MRPE and RPE still continue to learn until 550 of training data and perform a better good final MSE.

3.2   Modelling and Forecasting of Revolution

The MSE comparison of revolution with two different learning algorithms is shown in Fig. 4. From the results, it can be seen that HMLP with ALRPE model give a better performance than HMLP with MRPE and RPE model. After 525 of training data the HMLP with ALRPE model has converge rapidly with the final MSE of 48.09. Difference to HMLP with MRPE and RPE model, it also needs 525 of training data to converge rapidly with the final MSE of 49.59 and 48.33 respectively.

3.3 Modelling and Forecasting of Heat Exchanger (Benchmark with Standard Data)

The MSE comparisons of heat exchanger with three different learning algorithms profile are shown in Fig. 6. From the results, it can be seen that HMLP with ALRPE model give a better performance than HMLP with MRPE and RPE profile. Before 400 of training data the HMLP with ALRPE model has converge rapidly with the best final MSE of -20.51. In contrast to HMLP with MRPE, it needs more than 350 of training data to converge rapidly with the lower final MSE of -17.36. Parallel to HMLP with RPE, the lowerst of final MSE of -15.95 has been achieve but with more than 300 training of data.

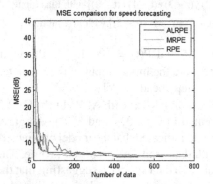
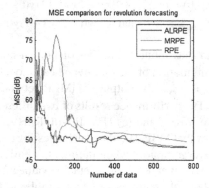

Fig. 3. MSE comparison for speed.   **Fig. 4.** MSE comparison for revolution.

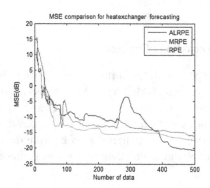

Fig. 5. MSE comparison for heat exchanger

4 Conclusion

This paper has proposed a new method for fine-tuning the new version of MRPE namely as ALRPE learning algorithm. The method uses an approach of pedagogy to develop the ALRPE as a second version of MRPE with a guidance of a profile for α_g (learning rate) and α_m (momentum rate). The hiphotesis behind this approach are proven by the empirical evidence of the good generalization and correct performance of the proposed model. Hypothesis of most learning algorithm are capable to achieve their maximum performance are proven as follows: The generalization performance of the new ALRPE learning algorithm is generally better in term of speed to converge and have a better MSE for most of data that has been test in this study. Hypothesis of the more rich data being used to feed them, the less time is needed to make them converge are proven as follows: The richest data in this study are speed and heat exchanger data. From the generalization performance it can be seen that time taken to reach a convergence level are less than 300 data that has been train. The less reach data are belonging to revolution. The result shows that less reach data type that need more than 500 data to be train to reach the convergence level. The overall result in modelling and forecasting show that the new profile of α_m and α_g use by ALRPE give a better MSE results compare to the original MRPE profile and conventional RPE profile. In general, the new ALRPE profile can continuesly stimulate the learning capability of the model to minimize the MSE and perform a better generalization performance than original MRPE and conventional RPE.

References

1. Robinson, A., Clinkenbeard, P.R.: History of giftedness: perspectives from the past presage modern scholarship. In: Pfeiffer, S.I. (ed.) Handbook of Giftedness in Children, pp. 13–31. Springer, New York (2008)
2. Nakao, H., Andrews, K.: Ready to teach or ready to learn: a critique of the natural pedagogy theory. Rev. Philos. Psychol. 5, 465–483 (2014)
3. Saad, Z., Osman, M.K., Mashor, M.Y.: Modelling and forecasting of car speed using hybrid multilayered perceptron network. J. Contemp. Eng. Sci. 7, 603–610 (2014). Hikari Ltd
4. Saad, Z., Mashor, M.Y.: Model structure selection for speed forecasting with nonlinear autoregressive with an exogenous input. In: The 4th International Conference on Intelligent Systems, Modelling and Simulation, Bangkok (2013)
5. Mashor, M.Y.: On-line nonlinear modelling and forecasting of streamflow using neural network. Int. J. Comput. Internet Manage. 17, 44–45 (2009)
6. Mashor, M.Y.: Hybrid multilayered perceptron network. Int. J. Syst. Sci. 31, 771–785 (2000)
7. Chen, S., Cowan, C.F.N., Billings, S.A., Grant, P.M.: A parallel recursive prediction error algorithm for training layered neural network. Int. J. Control 51, 1215–1228 (1990)
8. Demuth, H.B., Beale, M.: Users' Guide for the Neural Network Toolbox for MATLAB ver. 4.0. The Mathworks, Natick (2000)
9. Jesus, O.D., Hagan, M.: Backpropagation algorithms for a broad class of dynamic networks. IEEE Trans. Neural Netw. 18, 14–27 (2007)

Using Growing Neural Gas in Prototype Generation for Nearest Neighbor Classifiers

Jussara Dias, Marcos G. Quiles$^{(\boxtimes)}$, and Ana Carolina Lorena

Instituto de Ciência e Tecnologia, Universidade Federal de São Paulo,
São José Dos Campos, São Paulo, Brazil
{quiles,aclorena}@unifesp.br

Abstract. Instance-based learning algorithms, such as nearest neighbor (NN) classifiers, require storing all training instances and consulting them when making predictions. One alternative to overcome these costs is to reduce the learning dataset by a pre-processing step. This work deals with prototype generation, where new data points are generated from the original dataset. Reduction can be achieved by retaining less instances in the most representative areas of the dataset, which are represented by prototypes. Here Growing Neural Gas Networks are employed for generating the prototype instances. Experimentally, NN classifiers using the reduced datasets were able to maintain close accuracy to that of NN classifiers using the whole dataset.

1 Introduction

Various real problems require distinguishing data points into one of a number of categories or classes. Among the existing techniques for solving classification problems, one can cite instance-based learning algorithms, such as nearest neighbor classifiers (NN) [1]. These Machine Learning (ML) algorithms store a training dataset containing known instances and make new predictions based on a similarity criterion. For instance, the k-NN algorithm finds the k examples more similar to the new example and assigns their majority label to it.

NNs are among the top-ten algorithms in Data Mining [2], but they have drawbacks related to the scale of the dataset stored for classification. The more examples, the higher is the computational cost associated with the use of the NN algorithm, both in terms of storage and processing time. Furthermore, the presence of noisy examples, which are common in real datasets, can largely impair the predictive results of NN classifiers, specially for $k = 1$ [3,4]. In the face of these problems, there are some alternatives in the literature for pre-processing the learning dataset [3,5]. Two popular strategies are prototype selection (PS) [3] and prototype generation (PG) [4]. Both approaches reduce the learning dataset by retaining the most representative instances in it, which are thereby named prototypes. Nevertheless, while PS algorithms retain part of the original data points, PG methods can generate new examples if needed, which can be interesting for representing regions of lower density in the original dataset [4].

© Springer International Publishing Switzerland 2015
S. Arik et al. (Eds.): ICONIP 2015, Part II, LNCS 9490, pp. 276–283, 2015.
DOI: 10.1007/978-3-319-26535-3_32

In this paper, we use Growing Neural Gas (GNG) Networks [6] for PG. These Artificial Neural Network (ANN) models attempt to capture the topological data structure incrementally. Nodes are successively inserted, changed and removed from the GNG in order to fit the data points. We add some class supervision in this process, making the algorithm semi-supervised. The final structure of the GNG, that is, its nodes weights, define the prototypes. The main advantage of GNG for PG relies on its ability to capture automatically the topological structure of the dataset [6]. Furthermore, it is incremental and can easily accommodate new data points when needed.

This paper is organized as follows: Sect. 2 describes the PG problem and some related techniques. Section 3 presents the GNG for PG. Section 4 presents the experiments for evaluating the effectiveness of GNG in PG. Section 5 concludes this paper.

2 Prototype Generation

Prototype Generation (PG) is one alternative for reducing a dataset into a subset of representative instances [3]. In this process, new artificial data points can be generated. Such examples are usually positioned in places that better represent the original dataset. According to a taxonomy presented in [4], there are four main families of generation mechanisms:

1. **Class relabeling:** the labels of the learning examples are modified if they are likely to be either from other class or noisy;
2. **Centroid based:** generates prototypes by merging sets of similar examples;
3. **Space splitting:** heuristics are employed for partitioning the input space into regions. For each region one or more representatives are defined;
4. **Positioning adjustment:** the position of the prototypes are corrected according to an optimization process guided by the original data.

We chose algorithms from each of the previous families among those presented in [4] to compare to the GNG technique, which can itself be categorized as a positioning adjustment technique.

Generalized Editing using Nearest Neighbor (GENN) is based on an editing rule [7]. This rule, named kk'-NN, removes an example if k' out of $k-1$ of its nearest neighbors plus it are not from a same majority class. Otherwise, the example is relabeled according to the majority class of its neighbors.

Generalized Modified Changes Algorithm (GMCA) [8] uses a hierarchical clustering algorithm that only merges clusters of the same class label whether the resulting cluster is consistent. Cluster consistency is measured geometrically and asserts the prototype is able to correctly classify all cluster members.

Reduction by Space Partitioning 3 (RSP3) [9] employs an heuristic for finding data partitions until all subsets obtained are class homogeneous. From each partition a representative centroid is extracted.

The Learnig Vector Quantization (LVQ) [10] algorithm is a supervised technique which performs a Vector Quantization procedure.

The prototypes obtained are approximated or quantized versions of the input data. At each step, the set of prototypes, whose number is defined a priori, are adjusted in order to approximate the input examples, through a competitive learning procedure.

3   Growing Neural Gas Networks for PG

The Growing Neural Gas (GNG) ANN employs a competitive learning algorithm for adjusting the position of a set of neurons according to the structure of a dataset distributed in the Euclidean space [6]. Given a dataset T composed of m pairs $(\mathbf{x}_i, y_i)$, where each $\mathbf{x}_i$ is an example described by d features $(\mathbf{x}_i \in \Re^d)$ that has an associate label $y_i \in \{c_1, \ldots, c_k\}$, the objective of this work is to employ the GNG to obtain a new reduced dataset T' with $m' < m$ examples that better approximate the original entries. Each neuron of the GNG will also be a point in the d-dimensional space, with coordinates given by a weight vector $\mathbf{w}_j \in \Re^d$, which is optimized to represent some region of the original dataset. The final set of neurons will correspond to the set of prototypes output by the GNG in PG. Here, we made modifications into the original formulation of GNG, which considers the labels of the examples in some steps, as shown in Algorithm 1.

First T' is fed with one neuron per class, with coordinates given by the centroids of each of the classes (step 2). The GNG neurons have a local error attribute associated, which accounts for how well they are representing the original data. The farther a neuron is from the input data; the higher is its error. Initially the error values are null. In our proposal, the neurons have an associated label vector too. It contains k integer positions, which are all initialized as null. Each of the n_i initial neurons will have their label position correspondent to class c_i set to 1, indicating that they belong to this class (step 3).

Pairs of neurons of the GNG can be connected based on topological information. Neighbors are united based on the competitive learning rules, so that an induced Delaunay triangulation graph is obtained [11]. Each edge has an age attribute, used to control the removal of old edges in order to keep the topology updated. The edges have a null age when they are first created. All connections with a threshold age are eliminated from the GNG, as well as resulting isolated neurons. This allows the network to disregard regions that are not representative of the input data topology and corresponds to a shrinking step (step 19). In our proposal each initial neuron is connected to its nearest neighbor in T' (step 6).

When an example $\mathbf{x}_j$ is presented to the network (step 8), a winning neuron $s_1 \in T'$ will be determined (step 9) and will have its attributes updated. The class position of s_1 correspondent to y_j is incremented (step 10), representing a vote on the class of the example that activated it. The error of s_1 is updated by adding the squared distance between the input signal and its weight vector (step 11). Next s_1 and those neurons connected to it have their weight vectors moved towards $\mathbf{x}_j$ by fractions ϵ_b and ϵ_n, respectively (step 12). All edges connecting to s_1 have their ages incremented (step 13) and s_1 and the second winning neuron s_2 are connected or have their connection age reset.

Algorithm 1. GNG adapted for PG.

1: **for all** (classes c_i) **do**
2:     Insert a neuron n_i in T', with $\mathbf{w}_i$ given by the centroid of c_i, and $error_{n_i} = 0$;
3:     Set the c_i-th position of n_i's label vector to 1;
4: **end for**
5: **for all** (neurons n_i) **do**
6:     Connect n_i to its nearest neighbor in T';
7: **end for**
8: **loop**
9:     **repeat**
10:         Choose a random training example $\mathbf{x}_j$ from T to be presented to the network;
11:         Determine the winning neuron s_1 and the second winning neuron s_2:

$$s_1 = \arg\min_l \|\mathbf{x}_j - \mathbf{w}_l\|, \forall l \in T' \tag{1}$$

$$s_2 = \arg\min_l \|\mathbf{x}_j - \mathbf{w}_l\|, \forall l \in T' - s_1 \tag{2}$$

12:         Increment the class position equivalent to y_j in s_1;
13:         Update the error attribute of s_1 as:

$$error_{s_1} = error_{s_1} + \|\mathbf{w}_{s_1} - \mathbf{x}_j\|^2 \tag{3}$$

14:         Move s_1 and its direct topological neighbors towards the entry $\mathbf{x}_j$:

$$\mathbf{w}_{s_1} = \mathbf{w}_{s_1} + \epsilon_b \left(\mathbf{x}_j - \mathbf{w}_{s_1}\right) \tag{4}$$

$$\mathbf{w}_n = \mathbf{w}_n + \epsilon_n \left(\mathbf{x}_j - \mathbf{w}_n\right) \tag{5}$$

15:         Increment the age of all edges connected to s_1;
16:         **if** (s_1 and s_2 are connected by an edge) **then**
17:             Reset the age of this edge;
18:         **else**
19:             Create an edge between s_1 and s_2;
20:         **end if**
21:         Remove all edges with age higher than a_{max} and isolated nodes;
22:     **until** (all examples in T have been presented)
23:     Determine the neuron n_q with highest accumulated error;
24:     Determine the neuron n_f which has the largest error among n_q's neighbors;
25:     **if** (n_q and n_f belong to different classes) **then**
26:         Insert a new unit n_r halfway between n_q and n_f $\{\mathbf{w}_r = 0.5 * (\mathbf{w}_q + \mathbf{w}_f)\}$;
27:         Set the positions equivalent to the majority labels of n_q and n_r in n_r to 1;
28:         Insert edges between n_r and n_q and n_r and n_f;
29:         Remove the edge between n_q and n_f;
30:         Decrease the error attributes of n_q and n_f by a fraction α;
31:         Set $error_{n_r} = error_{n_q}$;
32:     **end if**
33:     **for all** (neurons n_i) **do**
34:         Determine the winning class of n_i and reset its label vector accordingly;
35:         Decrease error values $\{error_{n_j} = \beta * error_{n_j}\}$;
36:     **end for**
37:     **if** (there were no node insertions for λ consecutive iterations) **then**
38:         **return** T' {stops GNG training};
39:     **end if**
40: **end loop**

Other important step is the network growing, which consists of finding regions suitable for inserting new nodes (steps 20 to 29). This insertion is performed after each training epoch, that is, when all training examples in T have been presented to the network, in the neighborhood of neurons that show a high accumulated error (steps 23 and 24). Another modification introduced in this work is to grow the network only if these neighbor neurons have different labels (step 22). This allows to model more suitably conflicting regions between the classes. The class of a neuron is determined by its label vector position with the majority of the votes. The new neuron is positioned halfway between the determined neighbors (step 23) and has a vote for both of their classes (step 24). Next the connections between these neurons are adjusted (steps 25 and 26), as well as their errors.

When a training epoch is completed, the label vectors of all neurons in T' are reset so that the winning class position assumes the value 1, while all other positions are set to null. If there are no node insertions in the network for λ consecutive epochs, the GNG training is halted, and T' is output (step 38).

4   Experiments

We selected ten datasets from the Keel repository [12] to assess the results of the GNG in PG. Since the GNG formulation only handles numeric data, all of them have only numerical features. Table 1 presents a summary of the datasets used in this work, containing their numbers of classes, of examples and of features.

Table 1. Datasets employed in the experiments.

Dataset	♯Classes	♯Examples	♯Features
Bupa	2	345	6
Cleveland	5	297	13
Ecoli	8	336	7
Iris	3	150	4
Monk	2	432	6
NewThyroid	3	215	5
Pima	2	768	8
Vehicle	4	846	18
WDBC	2	569	30
Wine	3	178	13

The datasets were divided according to the 10-fold stratified cross-validation methodology. They are also normalized into the [0, 1] range using the Keel tool [12]. Each dataset training partition is then subject to PG. Its reduced version is fed into a 1-NN classifier. The 1-NN classifier accuracy is next evaluated on the corresponding test partition. This is done for all dataset folds, giving as a result an average accuracy and standard-deviation. Since it is a stochastic

Table 2. Accuracy Rates and Reduction Rates (in parenthesis)

Datasets	PG techniques					
	GNG	LVQ3	GMCA	GENN	RSP3	1NN
Bupa	*0.56 ± 0.07*	0.57 ± 0.08	0.59 ± 0.06	0.61 ± 0.08	0.57 ± 0.08	**0.61 ± 0.06**
	(52.65 ± 10.64)	**(98 ± 0.0)**	(66.67 ± 0.31)	*(32.83 ± 1.96)*	(48.38 ± 7.11)	–
Cleveland	**0.54 ± 0.05**	**0.54 ± 0.07**	*0.53 ± 0.06*	*0.53 ± 0.04*	**0.54 ± 0.06**	*0.53 ± 0.06*
	(97.00 ± 0.99)	(97.00 ± 0.21)	(67.06 ± 0.45)	*(37.29 ± 1.43)*	**(80.95 ± 0.0)**	–
Ecoli	0.71 ± 0.05	*0.68 ± 0.08*	0.78 ± 0.06	**0.82 ± 0.06**	0.81 ± 0.07	0.80 ± 0.07
	(94.80 ± 1.50)	**(96.35 ± 0.0)**	(65.56 ± 0.0)	*(14.23 ± 1.60)*	(74.0 ± 0.81)	–
Iris	**0.96 ± 0.04**	*0.93 ± 0.07*	0.94 ± 0.06	0.94 ± 0.04	0.94 ± 0.04	*0.93 ± 0.05*
	(93.18 ± 1.43)	**(97.03 ± 0.0)**	(83.70 ± 2.45)	*(4.14 ± 1.29)*	(87.85 ± 1.96)	–
Monks	*0.63 ± 0.08*	0.78 ± 0.06	**0.79 ± 0.06**	0.78 ± 0.05	0.70 ± 0.06	0.77 ± 0.05
	(82,42.03 ± 3.99)	**(98.19 ± 0.0)**	(68.29 ± 0.71)	*(17.98 ± 1.63)*	(50.25 ± 7.11)	–
New tyroid	*0.89 ± 0.08*	0.94 ± 0.05	0.94 ± 0.04	0.96 ± 0.03	0.95 ± 0.04	**0.97 ± 0.02**
	(97.46 ± 0.54)	(97.40 ± 0.0)	(82.17 ± 2.35)	*(3.10 ± 1.03)*	(86.21 ± 0.41)	–
Pima	*0.69 ± 0.04*	**0.73 ± 0.04**	*0.69 ± 0.04*	0.72 ± 0.03	0.70 ± 0.03	*0.69 ± 0.04*
	(98.04 ± 0.67)	**(98.11 ± 1.42)**	(67.43 ± 0.31)	*(20.98 ± 1.74)*	(68.16 ± 2.84)	–
Vehicle	0.58 ± 0.04	*0.55 ± 0.06*	0.68 ± 0.04	0.68 ± 0.05	0.67 ± 0.04	**0.70 ± 0.05**
	(92.55 ± 1.45)	**(98.22 ± 0.0)**	(72.42 ± 1.03)	*(30.68 ± 0.96)*	(63.23 ± 0.0)	–
WDBC	*0.92 ± 0.03*	0.93 ± 0.03	0.95 ± 0.03	**0.95 ± 0.02**	**0.95 ± 0.02**	0.95 ± 0.01
	(98.78 ± 0.46)	(98.24 ± 0.0)	(73.95 ± 1.38)	*(14.51 ± 2.93)*	(87.34 ± 0.0)	–
Wine	0.95 ± 0.05	0.95 ± 0.03	**0.96 ± 0.05**	0.95 ± 0.04	*0.93 ± 0.04*	0.95 ± 0.04
	(91.37 ± 2.86)	**(97.25 ± 0.03)**	(77.37 ± 1.78)	*(3.62 ± 1.54)*	(84.93 ± 2.41)	–

algorithm, GNG was also run 5 times with different random seeds and their results are averaged. 1-NN classifiers were obtained for the original datasets too, without reduction, as a baseline. The PG techniques described in Sect. 2 are also employed as implemented in the Keel software [12] (for LVQ we use the LVQ3 version). Their suggested default parameter values are used. Most of the GNG parameters are based on values recommended in [11] ($\epsilon_b = 0.05$, $\epsilon_n = 0.0006$, $\beta = 0.005$, $\alpha = 0.5$, $\lambda = 50$, $a_{max} = 88$).

Table 2 presents the accuracy rates (average ± standard deviation in 10-fold cross validation) achieved by the 1-NN classifiers in the experiments. The 1NN column refers to the results achieved when employing the original datasets. The best accuracy results in each dataset are highlighted in bold, while the worst results are highlighted in italics. Table 2 also shows, in parenthesis, the reductions in the number of examples achieved by each of the PG techniques. Higher reductions are highlighted in bold, while smaller reductions are in italics.

Comparing the accuracy values from Table 2 with the Friedman statistical test at 95 % of confidence level [13], no differences of overall performance are detected. Therefore, all techniques showed accuracy values comparable to each other when considering a multiple comparison scenario. 1-NN classifiers using GENN datasets achieved results closer to those of 1-NN employing all instances. Nonetheless, it is important to notice that GENN reductions are quite modest. All other techniques obtained the best and worst accuracy results in at least one of the datasets. It is worth noting that the LVQ3 and the GNG techniques were able to reduce sharply the datasets. But the GNG has the advantage of automatically finding a suitable number of instances, while this is an input parameter of LVQ3.

5   Conclusion

This paper presented a study on the use of GNG neural networks in prototype generation for the nearest neighbor classifier. Throughout, the basic GNG training algorithm for clustering was adapted to PG by taking into account the labels of the examples in some of its steps.

The results on ten benchmark classification datasets indicate that GNG is able to reduce the training dataset while keeping the accuracy results of 1-NN classifiers in general close to those calculated using all data. Nevertheless, the accuracy rates were in some cases impaired. This happened mostly for datasets with a high overlapping between the classes and shall be investigated further. In particular, the criterion adopted for inserting new nodes in the GNG, which privileges conflicting regions between the classes, might be causing it.

Future work should also include a sensibility analysis to the parameter values employed in the GNG. In our experiments, we limited the analysis to common parameter values adopted in the literature. Analysis on more datasets are also required. Finally, we intend to perform a sensibility analysis of the PG techniqhes to the noise level of the datasets. If a dataset has a high noise level, probably the efficiency of PG techniques, specially those based on positioning adjustment, will be impaired. In this case, an editing rule can be applied to remove suspicious cases first.

Acknowledgments. The authors would like to thank the Brazilian National Research Council (CNPq) and the So Paulo Research Foundation (FAPESP), Proc. 2011/18496-7 and 2012/22608-8, for financial support.

References

1. Aha, D.W., Kibler, D., Albert, M.K.: Instance-based learning algorithms. Mach. Learn. **6**(1), 37–66 (1991)
2. Wu, X., Kumar, V., Quinlan, J.R., Ghosh, J., Yang, Q., Motoda, H., McLachlan, G.J., Ng, A., Liu, B., Yu, P.S., Zhou, Z.-H., Steinbach, M., Hand, D.J., Steinberg, D.: Top 10 algorithms in data mining. Knowl. Inf. Syst. **14**(1), 1–37 (2008)
3. Garcia, S., Derrac, J., Cano, J.R., Herrera, F.: Prototype selection for nearest neighbor classification: taxonomy and empirical study. IEEE Trans. Pattern Anal. Mach. Intell. **34**(3), 417–435 (2012)
4. Triguero, I., Derrac, J., Garcia, S., Herrera, F.: A taxonomy and experimental study on prototype generation for nearest neighbor classification. IEEE Trans. Syst. Man, Cybern. Part C **42**(1), 86–100 (2012)
5. Liu, H.: Instance selection and construction for data mining. Springer, USA (2010)
6. Fritzke, B.: Advances in Neural Information Processing Systems 7. A growing neural gas network learns topologies, pp. 625–632. MIT Press, Cambridge (1995)
7. Koplowitz, J., Brown, T.A.: On the relation of performance to editing in nearest neighbor rules. Pattern Recogn. **13**(3), 251–255 (1981)
8. Mollineda, R.A., Ferri, F.J., Vidal, E.: A merge-based condensing strategy for multiple prototype classifiers. IEEE Trans. Syst. Man Cybern. Part B Cybern. **32**(5), 662–668 (2002)

9. Sánchez, J.S.: High training set size reduction by space partitioning and prototype abstraction. Pattern Recogn. **37**(7), 1561–1564 (2004)
10. Kohonen, T.: The self-organizing map. Proc. IEEE. **78**(9), 1464–1480 (1990)
11. Holmström, J.: Growing neural gas: Experiments with GNG, GNG with Utility and Supervised GNG. Uppsala Masters Thesis in Computer Science, Uppsala University Department of Information Technology (2002)
12. Alcalá, J., Fernández, A., Luengo, J., Derrac, J., García, S., Sánchez, L., Herrera, F.: Keel data-mining software tool: Data set repository, integration of algorithms and experimental analysis framework. J. Multiple-Valued Logic Soft Comput. **17**(2–3), 255–287 (2011)
13. Demšar, J.: Statistical comparisons of classifiers over multiple data sets. J. Mach. Learn. Res. **7**, 1–30 (2006)

Orthogonal Extreme Learning Machine Based P300 Visual Event-Related BCI

Yakup Kutlu[1(✉)], Apdullah Yayik[2], Esen Yildirim[1],
and Serdar Yildirim[1]

[1] İskenderun Technical University, Hatay, Turkey
{ykutlu, eyildirim, serdar}@mku.edu.tr
[2] Turkish Land Forces, Hatay, Turkey
ayayik@kkk.tsk.tr

Abstract. Brain Computer Interface (BCI) is a type of human-computer relationship research that directly translates electrical activity of brain into commands that can rule equipment and create novel communication channel for muscular disabled patients. In this study, in order to overcome shortcoming of Singular Value Decomposition in Extreme Learning Machine, iteratively optimized neuron numbered QR Decomposition technique with different approaches are proposed. QR Decomposition Extreme Learning Machine technique based P300 event-related potential BCI application that achieves almost % 100 classification accuracy with milliseconds is presented. QR decomposition based ELM and novel feature extraction method named Multi Order Difference Plot (MoDP) techniques are milestones of proposed BCI system.

Keywords: Brain computer interface · P300 · QR decomposition · MoDP method · Iteratively optimized neuron number

1 Introduction

Recently, different control systems based brain signal has been advanced and called as Brain Computer Interface (BCI) systems. Unlike traditional control systems that require motor ability, BCI system is controlled by human main control unit. Also, brain control is an interdisciplinary research that covers computer engineering, biology, neuroscience, physics and physiology. Neuro-scientist claim that even if peripheral nerves and muscles are harmed, brain can carry on its functional tasks. The tasks can be read from EEG signal because EEG signal demonstrates evidence associated with human's thinking or commands [1]. EEG channels are ports that are controlled by main brain unit. In middle of 20th century, scientist averaged the EEG and extracted the time-located brain response called event-related potential (ERP). Mainly there are two types of ERP; exogenous and endogenous component. While exogenous component is evoked by physical (light, sound …etc.) activity, endogenous component is evoked by mental activity. P300 (as an exogenous component) potential can be evoked approximately 300 ms after the visual stimulus [2]. P300 potential is widely used in BCI systems; spelling [3], robotic arm control [4], visual object detection [5]. Detection of P300 wave must be stable and quickly for online BCI system. For that reason

© Springer International Publishing Switzerland 2015
S. Arik et al. (Eds.): ICONIP 2015, Part II, LNCS 9490, pp. 284–291, 2015.
DOI: 10.1007/978-3-319-26535-3_33

extracting time-domain features is the most logical way to go. But, researchers mostly used frequency-domain features that are extracted using wavelet transform [6], fourier transform [7]. Contrary, in this study time-domain features named Multi Order Difference Plot that detects consecutive increase/decrease in different degrees. Also, neural networks have been good-studied and widely used with gradient descent algorithm in past decades. Researchers clearly agree with learning speed of neural network is so slowly than desired. Reasons behind here are; (1) conjugate gradient algorithm is used for learning, (2) Network parameters (weights and bias) are changed iteratively to minimize error. (3) Gradient descent algorithms may converge to local minimums. Many methods have been proposed to beat these problems like, second-order optimization methods and global optimization methods In 2006, extreme learning machine (ELM) is proposed by Huang [8] for single layer feed forward networks (SLFNs) to obtain high generalization performance in extremely rapid speed. Unlike traditional learning algorithms ELM randomly chooses the input weights and hidden neuron biases and determines output weights according to Moore Pensore generalized inverse conditions [9]. Different types of ELM are advanced to improve generalized performance; OP-ELM [10], OS-ELM, Evolutionary ELM. In this study QR decomposition approaches (HouseHolder-hh and Modified Gram Schmidt-mgs) based ELM and original ELM algorithm is applied exogenous ERP P300 potential component of EEG signal classification to detect visual objects. The remaining paper is organized as follows; in Sect. 2 experimental setup and database information are performed. Feature extraction and classification algorithms are described in Sect. 3. Performance results are given in Sect. 4. Conclusions are performed in Sect. 5.

2 Materials and Methods

2.1 Database Description

In this study P300 visual evoked potential database described in Hoffmann's article [5] is used. The images in Fig. 1 are flashed in random sequences. In each flash only one image was flashed. Each flash of image was performed 100 ms and during following 300 ms none of the images was flashed. So inter stimulus interval (ISI) was 400 ms. 32 electrodes are placed using 10–20 interval system and EEG was recorded at 2048 Hz sample rate. Biosemi Active Two device is used as EEG amplifier. In this study 4 channel EEG data (P7, P3, P4 and P8) is used. 4 disables and 5 healthy subject's EEG signal are recorded. Details about subjects are described in [5].

Fig. 1. The images used for evoking P300.

Four recording sessions are implemented for each subject. First and second are performed in one day, third and fourth are performed in another one day in two weeks. In each session 6 runs are consisted. The protocol used in each runs is described as follows; (i) Subjects were asked to count how many times a prescribed image was flashed in silence. (ii) On the screen images in Fig. 1 is displayed and a warning tone was given, (iii) 4000 ms after the warning tone, a random sequence of flashes was started and the EEG signal was recorded. The arrangement of flashes was block-randomized and the number of blocks was chosen randomly between 20 and 25, (iv) After each run subjects were asked what their counting result was. This was done in order to detect performance of the subjects.

2.2   Preprocessing

All of the recorded data is normalized between −1 and +1. Single trial extraction, First 100 ms of ISI (2048*0.1 ∼ 205) is extracted from signals to evoke P300 potentials. Following 300 ms is not used due to being non-P300 potentials. Each session has six runs and each run has at least 21 flashes. So 4*6*21*2048*0.1 data is extracted from subject for only one EEG channel. Overlapping between P300 and non-P300 potential EEG signal is not preferred [5].

3   Machine Learning and Classification

3.1   Multi-order Difference Plot (MoDP)

Multi Order Difference Plot (MoDP) is a novel time domain feature extraction system which is inspired from Poincare [11] and Second Order Difference Plot [12] techniques and firstly used in this study. MoDP consists of two stages, scattering of consecutive difference values with different degrees and analytically determining values in specified regions.

Scattering of Consecutive Difference Values with Different Degrees. For $D = [R_1, R_2 \ldots R_N] \in \Re^N$, is data matrix where R vectors are rows and N is state number of D. R vectors are divided into 2 vectors named x and y according following algorithm, where $x_{di} = [x_{1i}, x_{2i} \ldots]$, $y_{di} = [y_{1i}, y_{2i} \ldots]$ and d is degree of difference.

When $d = 1$, $x_{1i} = R_{i+1} \; y_{1i} = R_i$
When $d = 2$, $x_{2i} = x_{1(i+1)} - x_{1i} \; y_{2i} = y_{1(i+1)} - y_{1i}$
When $d = 3$, $x_{3i} = x_{2(i+1)} - x_{2i} \; y_{3i} = y_{2(i+1)} - y_{2i}$
When $d = 4$, $x_{4i} = x_{3(i+1)} - x_{3i} \; y_{4i} = y_{3(i+1)} - y_{3i}$

So, (x_{di}, y_{di}) data pairs in MoDP can be described as; For $d > 1$ and $d \in Z^+$,

$$\begin{cases} d = 1 \longrightarrow x_{di} = R_{i+1}, y_{di} = R_i \\ d > 1 \longrightarrow x_{(d-1)(i+1)} - x_{(d-1)i}, y_{(d-1)(i+1)} - y_{(d-1)i} \end{cases} \tag{1}$$

The statistical condition of consecutive differences with different degrees can be observed via MoDP. In this study 4 degrees of EEG domain signal is processed. More degrees can be used to compare their accuracies. Four feature vectors named 1st, 2nd, 3rd and 4th DP are extracted where DP means difference plot (As seen in Table 1).

Table 1. Classifiers' accuracy results

Classifier	Feature	Accuracy (%)	Neuron	Time (s)
ELM	1st DP	91,631	78	0,266
	2nd DP	97,894	24	0,016
	3rd DP	97,894	100	0,027
	4th DP	95,263	20	0,014
hhQRELM	1st DP	91,131	47	0,014
	2nd DP	97,368	35	0,012
	3rd DP	**98,421**	**64**	**0,031**
	4th DP	95,263	57	0,009
mgsQRELM	1st DP	92,684	47	0,017
	2nd DP	97,368	35	0,009
	3rd DP	98,421	64	0,023
	4th DP	95,263	57	0,018
MLP	1st DP	73,500	30-20-20-20	0,554
	2nd DP	68,710	30-20-20-21	0,512
	3rd DP	68,763	30-20-20-22	0,536
	4th DP	79,500	30-20-20-23	0,561
SVM	1st DP	87,484		
	2nd DP	92,146		
	3rd DP	92,670		
	4th DP	93,193		

Analytically Determining Values in Specified Regions. (x_{di}, y_{di}) data pairs determined in first stage are normalized between $[-1, 1]$ to fit easily into Cartesian Plane and easily determine values in specified regions described below. In Cartesian Plane 16 quadrant areas are specified using 4 circles (whose diameters are 0.25, 0.5, 0.75 and 1 respectively), and values in these areas are determined analytically detail (details are described in [12]). Statically meanings of these quadrants are; Detecting increase in x_{di} data with 4 different levels, detecting decrease in x_{di} data with 4 different levels, detecting increase in y_{di} data with 4 different levels and detecting decrease in y_{di} data with 4 different levels. Diameters of circles can be changed to detect variation of (x_{di}, y_{di}) described above more sensitively. In this study 4 diameter are preferred.

3.2 Extreme Learning Machine (ELM)

ELM is single layer neural network that has random nodes with random and fixed weights and learning capacity using Moore Pensore pseudoinverse conditions.

Learning method for SLFNs called extreme learning machine (ELM) can be summarized as follows: Given a training set $M = \{(x_i, t_i) | x_i \in R^n, t_i \in \Re^m, i = 1, \ldots N\}$, activation function $g(x)$ and hidden node number $N^{\approx}$; *Step 1*: Randomly assign input weight w_i and bias b_i, i, *Step 2*: Calculate the hidden layer output matrix H, *Step 3*: Calculate the output weight $\beta = H^+ T$ where H^+ is Moore Pensore inverse matrix of H [8]. Moore Pensore pseudoinverse conditions are given in [9].Here, if hidden layer output matrix H is non-square (that means $N^{\approx} < N$), determinant of $\det(H) = 0$ and $rank(H) < N^{\approx}$ (not-full rank), in linear algebra H matrix is assessed as Singular Matrix that cannot be directly inverted [13]. Using least square solution of matrix inverse problem can be given f as $\beta = (H^T H)^{-1} H^T T$. Now, here $H^T H$ square matrix must be inverted but in real world application $H^T H$ may be singular because of deficient (not full) rank. In this case Huang [8] has solved this issue using Singular Value Decomposition (SVD) approach to reach Moore Pensore pseudoinverse instead of risky least square solution. SVD can support both not-full rank and non-square matrix pseudoinverse.

3.3   Orthogonalization

Square matrix Q satisfying, $QQ^T = I$ is called orthogonal matrix, where I is identity matrix. So it holds $Q^T = Q^{-1}$. For a hidden layer output matrix H there exist an orthogonal matrix Q and an upper triangular (non-singular) matrix R, such that QR decomposition; $H = QR$. Orthogonal matrix has high advantages for solving linear equation. Here, inverse of Q easily determined using transpose, and inverse of R can be determined because of low conditioning. In Eq. (6) $\beta = (H^T H)^{-1} H^T T$, when we put H matrix in place in $\beta = (H^T H)^{-1} H^T T$ one can solve $H\beta = T$ linear equation problem with using inverse of R and transpose of Q. In literature, there are three main methods for matrix QR ortogonalization [13] (i) Gram-Schmidt Method, (ii) Household Reflection, (iii) Givens Transform. In this paper these two of these ortogonalization methods are used for determining output layer weights and learning algorithms are named as Modified Gram-Schmidt Extreme Learning (mgsQRELM) and Household Reflection Extreme Learning (hhQRELM) (Givens Transform is no preferred because of speed problem). Classification results are given in Table 1.

3.4   Iterative Neuron Number Optimization

Neuron number optimization process in Multilayer Neural Network (MLP) is high complex because multiple layers. MLP neuron numbers are always preferred using empirical techniques [14]. Fortunately, in ELM this process can be done much more simply. In this study, neuron number of single layer neural network is iteratively changed between 1 and attribute number and achieved classification results while changing are given for ELM, hhQRELM and mgsQRELM in Fig. 2.

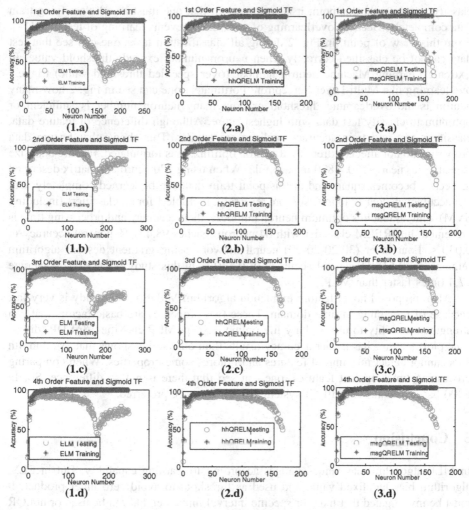

Fig. 2. ELM (1), hhQRELM (2) and mgsQRELM (3) classifiers with 1st DP (a), 2nd DP (b), 3rd DP (c) and 4th DP (d) features, train and test accuracies with iterative neuron number.

4   Results

In this study, in order to overcome shortcoming of Singular Value Decomposition in original Extreme Learning Machine, QR Decomposition based techniques with different approaches are proposed and iteratively neuron number optimization process is applied. Also, a novel feature extraction method Multi Order Difference Plot (MoDP) that includes scattering of consecutive difference values with different *degrees* of time domain signal is presented. In Fig. 2 ELM, hhQRELM and mgsQRELM classifiers train and test accuracies with iterative neuron number are shown. In classification problems feature data is divided into train and test (validate) data. Then firstly it is tried to classify train data correctly. If train data cannot be classified correctly, one can never classify test data. At this point it is so important that one must avoid classifying train data completely. Because overlearning of train data prevents learning of test data [14]. From this view of point, in Fig. 2 during all classification tasks one can see that test data cannot be classified correctly when neuron number exceeds threshold value. In Extreme Learning Machine neuron number (over specified threshold value) triggers overlearning like Multi-Layer Perceptron. Fortunately one can see in Fig. 2 how many neuron is the overlearning threshold or how many neuron number is sufficient or optimum to classify test data with highest score. Although difference of feature data, classifiers test and train accuracy inclines are similar. This shows that feature data properties do not affects extremely if neuron optimizing is formulated. Feature data size is nearly has nearly 170×820 in each fold. When using 170 neuron H matrix described in Sect. 2 becomes square and at this point train data can be learned completely. But test accuracy decreases to lowest score reversely. In Table 1 for all classifiers (including SVM) with all features optimum neuron numbers, their accuracy and processing time is presented. hhQRELM classifier with 64 neuron reaches 98,421 % general accuracy in 0.030 s. Four layer (30-20-20-20) neural network using gradient descent algorithm remains at 72,618 % general accuracy in 0,540 s. In this study hhQRELMis average 17,4 times faster than MLP.

Also, proposed novel feature extraction algorithm MoDP in this study is very fast because of remaining in time domain. These two properties are basic factors that can strengthen mobility to use in daily life online. Except MLP classifier, 3rd DP (difference plot) features has higher accuracy results than others, this shows that 1st, 2nd 4th DPs cannot detect meaningful features and ignores some properties. When comparing processing times against features, one can see that there is not so difference. So, In P300 based visual evoked BCI system 3rd DP must be preferred.

5   Conclusion

In MLP algorithm bias value changes according to learning capability. But in ELM algorithm bias has fixed value and used as threshold to avoid zero inner product. It must be investigated that fixed or specific interval input weights can be used or not.QR decomposition is used to overpower singular matrix inverse problem Other matrix decompositions that provides triangular and orthogonal matrices (that can simplify linear equations) may be tried also. Neuron number of SLFN structured ELM can be optimized or selected using detecting and removing which neuron causes ill-conditioned hidden layer output matrix.

References

1. Mesulam, M.: Principles of Behavioral and Cognitive Neurology. Oxford University Press, Oxford (2000)
2. Sutton, S.: Evoked-potential correlates of stimulus uncertainty. Science **150**, 1187–1188 (1965)

3. Rivet, B., Souloumiac, A., Gibert, G., Attina, V., Speller, P.: "P300 Speller" brain-computer interface: enhancement of P300 evoked potential by spatial filters. In: 16th European Signal Processing Conference (EUSIPCO-2008), pp. 1–5 (2008)
4. Potentials, A.B.: Robot control using anticipatory brain potentials. Autom. Control. Meas. Electron. Comput. Commun. **52**, 20–30 (2011)
5. Hoffmann, U., Vesin, J., Ebrahimi, T., Diserens, K.: An efficient P300-based brain-computer interface for disabled subjects. J. Neurosci. Methods **167**, 115–125 (2007)
6. Beverina, F., Palmas, G., Silvoni, S., Piccione, F., Giove, S.: User adaptive BCIs: SSVEP and P300 based interfaces. PsychNology J. **2003**(1), 331–354 (2003)
7. Abootalebi, V., Hassan, M., Ali, M.: A new approach for EEG feature extraction in P300-based lie. Comput. Methods Programs Biomed. **9**(4), 48–57 (2008)
8. Huang, G., Zhu, Q., Siew, C.: Extreme learning machine: theory and applications. Neurocomputing **70**, 489–501 (2006)
9. Regression, A.A.: The Moore-Penrose Pseudoinverse. Academic Press, New York (1972)
10. Miche, Y., Sorjamaa, A., Bas, P., Simula, O., Jutten, C., Lendasse, A.: OP-ELM: optimally pruned extreme learning machine. IEEE Trans. Neural Netw. **21**, 158–162 (2010)
11. Yayik, A., Kutlu, Y.: Diagnosis of congestive heart failure using poincare map plot. In: 2012 20th Signal Processing and Communications Applications Conference (SIU), pp. 1–4 (2012)
12. Yayik, A., Kutlu, Y., Yildirim, E.: Epileptic state detection : pre-ictal, inter-ictal, post-ictal. In: International Conference on Advanced Technology and Sciences (2014)
13. Demmel, J.W.: Applied Numerical Linear Algebra. Siam (1997)
14. Duda, R.O., Hart, P.E., Stork, D.G.: Pattern Classification, 2nd edn. Wiley, New York (2000)

A Neural Network Based Approach for Semantic Service Annotation

Supannada Chotipant[1(✉)], Farookh Khadeer Hussain[1],
Hai Dong[2], and Omar Khadeer Hussain[3]

[1] Decision Support and e-Service Intelligence Lab (DeSI), Quantum Computation and Intelligent Systems (QCIS) and School of Software, University of Technology Sydney (UTS), Sydney, NSW, Australia
Supannada.Chotipant@student.uts.edu.au,
Farookh.Hussain@uts.edu.au
[2] School of Computer Science and Information Technology, RMIT University, Melbourne, Australia
Hai.Dong@rmit.edu.au
[3] School of Business, University of New South Wales Canberra (UNSW Canberra), Australian Defence Force Academy, Canberra, ACT, Australia
O.Hussain@adfa.edu.au

Abstract. Nowadays, a large number of business owners provide advertising for their services on the web. Semantically annotating those services, which assists machines to understand their purpose, is a significant factor for improving the performance of automated service retrieval, selection, and composition. Unfortunately, most of the existing research into semantic service annotation focuses on annotating web services, not on business service information. Moreover, all are semi-automated approaches that require service providers to select proper annotations. As a result, those approaches are unsuitable for annotating very large numbers of services that have accrued or been updated over time. This paper outlines our proposal for a Neural Network (NN)-based approach to annotate business services. Its aim is to link a given service to a relevant service concept. In this case, we treat the task as a service classification problem. We apply a feed-forward neural network and a radial basis function network to determine relevance scores between service information and service concepts. A service is then linked to a service concept if its relevance score reaches the threshold. To evaluate the performance of this approach, it is compared with the ECBR algorithm. The experimental results demonstrate that the NN-based approach performs significantly better than the ECBR approach.

Keywords: Semantic service annotation · Service classification · Feed-forward neural network

1 Introduction

Understanding the meaning of services leads machines to automatically and efficiently retrieve and select composite service information for service consumers. Normally, the

© Springer International Publishing Switzerland 2015
S. Arik et al. (Eds.): ICONIP 2015, Part II, LNCS 9490, pp. 292–300, 2015.
DOI: 10.1007/978-3-319-26535-3_34

meanings of services are described by service providers. Although manual service annotations are accurate, they are time-consuming for service providers and therefore seldom updated. As a result, an automated semantic approach for an annotating service is desirable.

A business service is a service that advertises on a service directory web site, e.g. Yellow Pages. It consists of actual service components such as a service provider, their address, and a service description. Based on a thorough review of existing literature, current research has not yet developed approaches for the annotation of non-web services. Moreover, business services are ambiguous because they are described in human language. Soft computing techniques are well suited to the complex annotation process, but have not yet been investigated for semantic service annotation.

This paper proposes an NN-based approach to semantically annotate business services. We define the task as a classification problem -- that is, given a domain-specific service ontology, a service is classified and annotated to a relevant service concept. Our research applies a feed-forward neural network in order to identify the relevant concept from terms in the descriptions of all service concepts. Given a service S, its provider name and description are separated into terms. Term occurrences of S are input into the neural network, and the output is the relevant service concept.

This paper is organized as follows. Section 2 presents related work in the area of semantic service annotation. We then introduce the semantic service annotation problem in Sect. 3. Section 4 presents the proposed NN-based service annotation approach. An alternative automated annotation approach, the ECBR algorithm, is described in Sect. 5. The experiments and results are provided in Sect. 6, and the work is concluded in Sect. 7.

2   Related Work

Much research about web service annotation [1–8] has been published during the last decade. All focus on semi-automated approaches for annotating services. Given a service or services, the approaches suggest annotations to service providers and they select the most appropriate annotations.

To annotate web services, Stavropoulos et al. [1] developed a graphic tool that uses common words to match active ontology concepts to elements of the services. Likewise [1], Meyer et al. [2] propose a term-based technique for a light-weight community-based approach to semantic service annotation. This approach asks a user to categorize services with keywords. Inspired by the folksonomy annotation technique, users can tag documents and share them in a system.

In contrast, the publications [3–8] consider ontology-based annotation approaches. Reference [3] proposes the METEOR-S Framework. Given a web service document, the system suggests a list of matched ontologies for adding semantic information to that web service. The main idea is to convert both the web service and ontology into a SchemaGraph, and then match them by considering text-based and structure-based similarities. Inspired by METEOR-S, [4] attempts to match an XML schema, not a WSDL, with an ontology.

In addition, [5, 6] apply the idea of mapping web services to service ontologies. They then propose various ontology alignment methods to match the converted service ontology to relevant domain ontology. Given an input service and a domain ontology, [5] matches parameter concepts of the input service with concepts in the ontology by considering text and type similarities. Jiang et al. [6] includes an ontology in the web service description. The framework converts a web service description in the form of a WSDL document to an OWL-S file to annotate web services.

Based on a thorough review of the existing research, we conclude that although several semantic service annotation methods have been proposed, most researchers have only focused on annotations for web services. However, there is a great deal of service information on the web that is not in the form of WSDL, so semantic annotation approaches for online service information are required in order to improve the efficiency of service crawling and service retrieval. Current research has not developed approaches for automated service annotation of non-web services. To address this shortcoming, we have developed an NN-based method to solve the semantic annotation problem for online service information.

3   Semantic Service Annotation Problem

Given a service ontology, a service S is annotated to the relevant service concept, which reflects the purpose of S. For example, we assume that a service exists that is provided by "Virgin Blue Airline" in QLD Australia with the description "Low Fares, Great Service". Given a transport service ontology, service S may be annotated to the service concept "Airline Agent".

To solve this problem, we need to define services and a service ontology so we apply the representation from Hai's work [7]. Information about the actual services is represented by the Service Description Entity (SDE) metadata. SDE metadata consists of a service description and the provider's name, address, and contact number. For example, we assign "Low Fares, Great Service.", "Virgin Blue Airline", "131 Barry Pde Fortitude Valley QLD, 4006 Australia", and "Phone: 13 6789" as the service description, provider name, address, and contact number, respectively, of service S.

The structure of the service ontology is divided into two main layers: an abstract service layer and an actual service layer. Each service concept consists of two properties; the concept name and concept descriptions. For example, the concept named "Airline Agent" has two descriptions, "Airline Agent" and "Flight Agent".

Based on defined service information and the service ontology, semantically annotating services is achieved by linking SDE metadata to the relevant actual service concepts. To solve this problem, we need to find the relevance score between the SDE metadata and various actual service concepts. We then annotate the SDE to the concept that provides the highest relevance score. This task is challenging because human language often makes descriptions of services ambiguous. The length of the description may be too short to calculate the relevance score. For example, the service description; "Low Fares, Great Service", may not assist machines to increase the relevance score between the service and the concept "Airline Agent".

4 Neural Network Based Approach for Semantic Service Annotation

We propose a Neural Network (NN)-based approach for semantic service annotation. The proposed approach is divided into three modules. Given SDE metadata S and a service ontology O, the service-term extraction module will extract single terms from S. The service provider name and service description of S are stemmed, and the stop words are removed. The module separates the processed provider name and service description into terms. Those terms are input into a service classification module in order to find the relevance score between SDE S and each service concept C in the service ontology O. The SDE metadata S will be annotated to the service concept C if their relevance value is the highest value. The relevant concept C is linked to SDE S by the service-concept connection module.

4.1 Service Classification

The aim of this module is to calculate the relevance score between given SDE metadata and each service concept of domain-specific service ontology. We define this task as a classification problem. That is, if SDE metadata S is classified as service concept C, it means that S should be relevant to C.

We apply a two-layered feed-forward neural network in order to classify SDE metadata. The structure of the neural network is shown in Fig. 1.

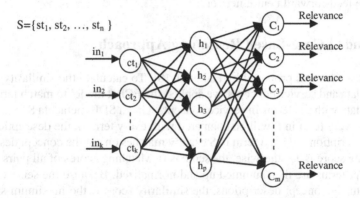

Fig. 1. The structure of feed-forward neural network for service classification

The input nodes of the neural network refer to concept terms for all service concepts. As mentioned in Sect. 3, each service concept consists of the concept name and concept description. The concept terms of service concept 'C' are terms that appear in the concept descriptions of 'C'. For example, the concept terms of the concept "Airline Agent", whose descriptions are "Airline Agent" and "Flight Agent", are "airline", "flight", and "agent". Thus, the number of input nodes is equal to the number of concept terms from all service concepts.

Given the service terms of the SDE from the previous step, the input value of each input node is calculated from the occurrences of the concept term in the SDE. We also use synonyms of the concept term to define the input value. Given SDE S, the input value of the input node i; $in_i(S)$, is calculated as follows

$$in_i(S) = \sum_{j=1}^{n} f\left(st_j, ct_i\right)$$

(1)

$$f(x,y) = \begin{cases} 1, & \textit{if x is equal to y} \\ 0.5, & \textit{if x is a synonym of y} \\ 0, & \textit{otherwise} \end{cases}$$

(2)

where SDE S consists of n service terms, st_j is the j-th service term of S, and ct_i is the concept term of input node i.

The input values of the neural network present similarities between terms in the SDE metadata and terms in the service concept description. That is, the SDE metadata S should be classified as a service concept C, if terms in S are similar to terms in C.

The output nodes of the neural network refer to the actual service concepts of a domain-specific service ontology. Therefore, the number of output nodes depends on the number of actual service concepts in the service ontology. The output value in each output node is the relevance score between the input SDE metadata and the service concept in that node. In addition, we also apply a radial basis function network to annotate SDE metadata. In this case, we define the input and output nodes as the same as those in the feed-forward neural network.

5   Extended Case-Based Reasoning Approach

The ECBR model was proposed by Hai et al. [8]. To calculate the similarity between SDE metadata and service concepts, we applied the ECBR model to match terms in the SDE metadata with the terms in service concepts. Given SDE metadata S and a service concept C, every term in S will be compared with every term in the descriptions of C. For each description of C, if a term in S exactly matches one in the concept description, the matching score is 1; otherwise, the score is 0. Matching scores of all pairs of meta-data-concept terms are then summed up and normalized. Because the service concept may have many concept descriptions, the similarity score is the maximum similarity score between metadata and each concept description. Then, S is annotated to C if the similarity score is more than the Relevance Threshold (RT).

We improve on the ECBR algorithm in [8] by adding one more condition. That is, if a service term is a synonym of a concept term, we calculate a matching score of 0.5.

6   Experiments and Results

This section compares the performance of the proposed NN-based annotation approach with the ECBR algorithm. We apply precision, recall, and a number of annotated samples to evaluate their performance.

6.1 Experimental Environment

The dataset from [8] was used. The data set was divided into two groups: the domain-specific service ontology, and the SDE metadata. This experiment applied the transport service ontology that consisted of 261 actual service concepts. We applied service metadata crawled from the transport category of the Australian Yellow Pages website. In total, 2948 samples of service metadata were found.

We applied the Neural Network Toolbox in Matlab to model a two-layer feed-forward neural network and a radial basis function network. The number of input nodes depended on the number of concept terms, which were extracted from descriptions of actual service concepts. In total, 257 input nodes existed in this experiment. The number of output nodes depended on the number of actual service concepts in the ontology. Based on the transport service ontology, 261 output nodes existed.

We used a supervised learning approach to annotate services. Input data and target data needed to be applied to simulate the neural network. We divided the 2948 service metadata samples into three sub data sets - 2064 samples (70 %) were used for training and 442 samples (15 %) were used for each of validation and testing.

6.2 Experimental Results

To evaluate the ECBR approach for semantic service annotation, we tested performance by setting the Relevance Threshold (RT) from 0.1 to 0.9 with an increment of 0.1. The results from Table 1 show that the increment of RT tended to increase the precision and recall values of the ECBR approach from 26.29 % in precision and 14.16 % in recall to 40.81 % in precision and 21.07 % in recall, with a setting for RT as 0.1 and 0.9, respectively. By contrast, the number of annotated samples declined from 96.38 % to 43.21 %.

Table 1. The performance values of the ECBR annotation approach.

Relevance threshold	Precision (%)	Recall (%)	Annotated samples (%)
0.1	26.29	14.16	96.38
0.2	26.49	14.28	94.80
0.3	27.23	14.75	91.40
0.4	26.77	14.04	89.59
0.5	38.26	19.60	52.04
0.6	38.26	19.60	52.04
0.7	40.48	20.71	47.51
0.8	40.72	21.01	43.89
0.9	40.84	21.07	43.21

To evaluate the NN-based approach, we tested the performance of the feed forward-based method by setting the number of hidden neurons from 10 to 90 with an increment of 10, and set the spread value of radial basis-based method from 0.1 to 0.9 with an increment of 0.1. The performance of both service annotation approaches are shown in Tables 2 and 3 respectively.

Table 2. The performance values of the feed-forward neural network based approach.

Neurons	Precision (%)	Recall (%)	Annotated samples (%)
10	23.08	19.85	100.00
20	37.56	24.85	100.00
30	38.24	26.47	100.00
40	38.46	27.00	100.00
50	38.91	29.69	100.00
60	44.57	29.54	100.00
70	56.56	37.46	100.00
80	42.99	28.00	100.00
90	63.35	41.57	100.00

Table 3. The performance values of the radial basis function network based approach.

Spread	Precision (%)	Recall (%)	Annotated samples (%)
0.1	60.41	38.65	100.00
0.2	61.54	39.22	100.00
0.3	61.31	39.67	100.00
0.4	61.31	39.67	100.00
0.5	61.31	39.67	100.00
0.6	62.90	41.23	100.00
0.7	60.63	40.03	100.00
0.8	53.17	35.20	100.00
0.9	48.64	32.52	100.00

The results in Table 2 show that the increment of hidden neurons assisted the feed-forward approach to increase the precision and recall values for service annotation. In the same way, in Table 3, the increment of spread value of radial basis function network tends to increase the performance of service annotation. The best performance of both NN-based methods was quite similar. That is, the precision and recall values were around 63 % and 41 % respectively, while the best performance of ECBR was about 41 % in precision and 21 % in recall. Moreover, the NN-based approaches were able to annotate all samples, while ECBR was only able to annotate 43.21 % because the service terms of unannotated samples did not match the concept terms of each service concept.

The experimental results demonstrate that the performance of the NN-based approach outperforms the ECBR approach. However, it seems that the recall value of NN-based approaches, which are less than 50 %, may be too low to apply to a real-world annotation task. This is because that the proposed approaches are single-label classification tasks. Unfortunately, in the real world, service metadata may relate to several service concepts.

7 Conclusion

This paper presents a Neural Network-based approach for semantic service annotation. The aim of this approach is to link given service metadata to a relevant service concept in a domain-specific service ontology. Service metadata is classified as a service class and the service metadata description and provider name are separated into service terms. The occurrence of those service terms are then input into a two-layer feed-forward neural network and a radial basis function network in order to find relevance scores for each service concept. The service metadata is then connected to the service concept whose relevance score is highest and is greater than the relevance threshold. The results demonstrate that the performance of the proposed NN-based approach is much higher than the ECBR algorithm. However, the recall values are less than 50 % and need to be improved for real-world annotation tasks.

References

1. Stavropoulos, T.G., Vrakas, D., Vlahavas, I.: Iridescent: a tool for rapid semantic annotation of web service descriptions. In: Proceedings of the 3rd International Conference on Web Intelligence, Mining and Semantics. ACM (2013)
2. Meyer, H., Weske, M.: Light-weight semantic service annotations through tagging. In: Dan, A., Lamersdorf, W. (eds.) ICSOC 2006. LNCS, vol. 4294, pp. 465–470. Springer, Heidelberg (2006)
3. Patil, A.A., et al.: Meteor-s web service annotation framework. In: Proceedings of the 13th International Conference on World Wide Web. ACM (2004)
4. Zhang, D., Li, J., Xu, B.: Web service annotation using ontology mapping. In: IEEE International Workshop on Service-Oriented System Engineering, pp. 235–242 (2005)
5. Wang, H., et al.: Constructing service network via classification and annotation. In: 2010 Fifth IEEE International Symposium on Service Oriented System Engineering (SOSE). IEEE (2010)
6. Jiang, B., Luo, Z.: A new algorithm for semantic web service matching. J. Softw. **8**, 351–356 (2013)

7. Dong, H., Hussain, F.K., Chang, E.: A service search engine for the industrial digital ecosystems. IEEE Trans. Industr. Electron. **58**(6), 2183–2196 (2011)
8. Dong, H., Hussain, F.K., Chang, E.: A transport service ontology-based focused crawler. In: Fourth International Conference on Semantics, Knowledge and Grid, pp. 49–56 (2008)

Neural Network-Aided Adaptive UKF
for Integrated Underwater Navigation

Meng Wu[1] and Ying Weng[2(✉)]

[1] School of Remote Sensing and Information Engineering,
Wuhan University, Wuhan, China
wumenghust@163.com
[2] School of Computer Science, Bangor University, Bangor, UK
y.weng@bangor.ac.uk

Abstract. Gravity aided navigation and geomagnetism aided navigation are equally important methods in the field of underwater navigation. However, the former is affected by terrain fluctuations, and the latter is sensitive to time-varying noise. Considering the characteristics that the gravity gradient vector can avoid the influence of time-varying noise and is less sensitive to terrain fluctuations, we propose to integrate the gravity gradient vector and geomagnetic vector together to achieve the merits of each aided navigation method. The gravity gradient vector and geomagnetic vector are used as measurement information from both local neural network-aided adaptive UKF filters, and then an information fusion algorithm based on weighted least squares estimation is used to combine the estimated values from each local filter to form an optimal estimated state value. Finally, the optimal estimated value is used to update the output values from each local neural network –aided adaptive UKF filter. Experimental results prove the feasibility of this integrated navigation method.

Keywords: Neural network · Adaptive UKF · Information fusion · Weighted least square estimation · Gravity gradient vector · Geomagnetic vector · Underwater navigation

1 Introduction

Mono-geomagnetic aided navigation and mono-gravity gradient aided navigation have both been used for underwater navigation. Guo et al. proposed a geomagnetic navigation algorithm based on the Sage-Husa adaptive Kalman filter to adjust the measurement noise matrix adaptively to eliminate the influence of magnetic storms [1]. Xiong et al. proposed a gravity gradient aided navigation algorithm based on the extended Kalman filter to correct the inertial navigation system's accumulated error [2]. Compared with geomagnetic aided navigation, gravity gradient aided navigation is insensitive to accelerations from various time-varying disturbances. Zheng et al. used a combined gravity- and geomagnetism-aided navigation method to improve the positioning success rate [3]. However, gravity aided navigation is affected by terrain fluctuations, and geomagnetic aided navigation is sensitive to time-varying noise. Considering various kinds of terrain fluctuations and time-varying noise, especially

© Springer International Publishing Switzerland 2015
S. Arik et al. (Eds.): ICONIP 2015, Part II, LNCS 9490, pp. 301–310, 2015.
DOI: 10.1007/978-3-319-26535-3_35

because the variations in gravity or geomagnetism are insufficient in some areas, it is better to combine the gravity gradient and geomagnetic information together to eliminate terrain fluctuations and time-varying disturbances. Based on such considerations, gravity gradient aided navigation and geomagnetism aided navigation are used as two local systems in an integrated navigation system. Then a kind of weighted least squares estimation algorithm is used to fuse information from each local system. Finally, the estimated optimal position and velocity values are used as feedback to update the estimated values of two local neural network-aided adaptive UKFs.

This paper is organized as follows. The adaptive UKF for neural network training is described in Sect. 2. The state model and measurement model of the integrated navigation system are introduced in Sect. 3. The information fusion algorithm is proposed in Sect. 4. Experimental results are discussed in Sect. 5, and conclusions are summarized in Sect. 6.

2 Adaptive UKF for Neural Network Training

As shown in Fig. 1, a multilayered Neural Network has K inputs $x_i(i = 1, 2, \ldots, K)$ and M outputs $y_j(j = 1, 2, \ldots, M)$ through the connecting weights w and the mapping function $g(\bullet)$.

To apply Adaptive UKF to Neural Network training, the first step is to organize all the inputs, outputs, and network weights as state vectors. The Adaptive UKF estimates the weights of Neural Network, which in turn are used to modify the state estimate predictions of the filter as observations are processed. The training can then be described as a state estimate problem with the following dynamic and observation equations [4]

$$\begin{cases} w_k = w_{k-1} \\ d_k = y_k + v_k = g(w_k, x_k) + v_k \end{cases} \qquad (1)$$

where d and y represent the desired and actual outputs, respectively; v denotes the random observation noise, and is assumed as zero-mean Gaussian and white with the covariance matrix $\mathbf{R}$.

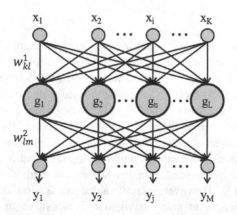

Fig. 1. Multilayered neural network structure.

For the sake of brevity, herein we present the Adaptive UKF for Neural Network training in the following steps

Step (1) Initialization

$$\left\{ \begin{array}{l} \widehat{w}_0 = E[w_0] \\ P_0 = E[(w_0 - \widehat{w}_0)(w_0 - \widehat{w}_0)^T] \end{array} \right. \tag{2}$$

Step (2) Calculation of sigma points with corresponding weights

$$\left\{ \begin{array}{ll} (\chi_t^a)_0 = \widehat{x}_t^a & \\ (\chi_t^a)_i = \widehat{x}_t^a + (\sqrt{(L+\lambda)P_t^a})_i & i = 1, \ldots, L \\ (\chi_t^a)_i = \widehat{x}_t^a - (\sqrt{(L+\lambda)P_t^a})_{i-L} & i = L+1, \ldots, 2L \\ W_0^{(m)} = \lambda/(L+\lambda) & \\ W_0^{(c)} = \lambda/(L+\lambda) + (1 - \alpha^2 + \beta) & \\ W_i^{(m)} = W_i^{(c)} = 1/\{2(L+\lambda)\} & i = 1, \ldots, 2L \end{array} \right. \tag{3}$$

Step (3) Time update

$$\left\{ \begin{array}{l} (\chi_{t+1|t}^a)_i = \Phi[(\chi_t^x)_i, (\chi_t^w)_i] \\ \widehat{x}_{t+1|t}^a = \sum_{i=0}^{2L} W_i^{(m)}(\chi_{t+1|t}^a)_i \\ P_{t+1|t}^a = \sum_{i=0}^{2L} W_i^{(c)}[(\chi_{t+1|t}^a)_i - \widehat{x}_{t+1|t}^a][(\chi_{t+1|t}^a)_i - \widehat{x}_{t+1|t}^a]^T \\ (y_{t+1|t})_i = H[(\chi_{t+1|t}^x)_i] \\ \widehat{y}_{t+1|t} = \sum_{i=0}^{2L} W_i^{(m)}(y_{t+1|t})_i \end{array} \right. \tag{4}$$

Step (4) Measurement update

$$\left\{ \begin{array}{l} P_{y_{t+1|t} \, y_{t+1|t}}^a = \sigma_k \sum_{i=0}^{2L} W_i^{(c)}[(y_{t+1|t})_i - \widehat{y}_{t+1|t}][(y_{t+1|t})_i - \widehat{y}_{t+1|t}]^T \\ P_{x_{t+1|t} \, y_{t+1|t}}^a = \sigma_k \sum_{i=0}^{2L} W_i^{(c)}[(\chi_{t+1|t}^x)_i - \widehat{x}_{t+1|t}^a][(y_{t+1|t})_i - \widehat{y}_{t+1|t}]^T \\ V_k'' = (z_{t+1|t})_i - \widehat{z}_{t+1|t} \\ \sigma_k = \left\{ \begin{array}{ll} \sigma_k = 1 & tr(V_k'' V_k''^T) \leq tr\left(P_{x_{t+1|t} \, y_{t+1|t}}^a\right) \\ \sigma_k = \dfrac{tr\left(P_{x_{t+1|t} \, y_{t+1|t}}^a\right)}{tr(V_k'' V_k''^T)} & tr(V_k'' V_k''^T) > tr\left(P_{x_{t+1|t} \, y_{t+1|t}}^a\right) \end{array} \right. \\ \kappa_{t+1} = P_{x_{t+1|t} \, y_{t+1|t}}^a \left(P_{y_{t+1|t} \, y_{t+1|t}}^a\right)^{-1} \\ \widehat{x}_{t+1}^a = \widehat{x}_{t+1|t}^a + \kappa_{t+1}(y_{t+1} - \widehat{y}_{t+1|t}) \\ P_{t+1}^a = P_{t+1|t}^a - \kappa_{t+1} P_{y_{t+1|t} \, y_{t+1|t}}^a \kappa_{t+1}^T \end{array} \right. \tag{5}$$

where the parameter σ_k is adopted to improve the robustness of UKF towards a large initial state error and time-vary noises from measurement sensors, and keep the measurement covariance matrix to be positive definite. Furthermore, using such parameter also makes UKF achieve stronger robustness under influences from outliers.

3   State Model and Measurement Model

The process and observation models of the autonomous underwater vehicle system exhibit significant nonlinearities. When the underwater vehicle is in a state of motion, the values of gravity gradient and of geomagnetism are measured at discrete points simultaneously using sensors fixed on underwater vehicles. An approach to model the integrated navigation system is

$$\begin{cases} x_{t+1} = \Phi(t)x_t + n_t \\ \mathbf{z}_{t+1} = H(x_{t+1}) + v_{t+1} \end{cases} \tag{6a, b}$$

Equation (6a, b) describes the propagation of system state in time, where $x_t = [\delta r_x(t), \delta r_y(t), \delta r_z(t), \delta v_x(t), \delta v_y(t), \delta v_z(t)]^T$ is the state of the system at time step t; δ is defined as the error between the actual state and the reference state which is presented by INS; (r_x, r_y, r_z) are the coordinates of the underwater vehicle's position in the Cartesian reference frame in x, y and z directions, respectively; v_x, v_y and v_z are the velocity components of the vehicle in x, y and z directions, respectively; $\Phi(t)$ represents the state transition matrix, and n_t denotes the process noise of dynamic system. H is the measurement model of the integrated navigation system, v_{t+1} is the measurement error of the dynamic system, $\mathbf{z}$ is the output measurement value of the system. Let e_t denote the error between the true model and the priori known mathematical model $\widehat{\Phi}(x_t)$ and $e_t = \Phi(x_t) - \widehat{\Phi}(x_t)$. The Neural Network is used to approximate the error, and we adjust the weights of the Neural Network, when $e_t - g(x_t, w_t) \rightarrow 0$, the error e_t is well approximated. Then we have

$$x_{t+1} = \widehat{\Phi}(x_t) + g(x_t, w_t)$$
$$w_{t+1} = w_t \tag{7}$$

$$\Phi(x_t) = \begin{bmatrix} 1 & 0 & 0 & \Delta t & 0 & 0 \\ 0 & 1 & 0 & 0 & \Delta t & 0 \\ 0 & 0 & 1 & 0 & 0 & \Delta t \\ 0 & 0 & 0 & 1 & 0 & 0 \\ 0 & 0 & 0 & 0 & 1 & 0 \\ 0 & 0 & 0 & 0 & 0 & 1 \end{bmatrix} \times x_t \tag{8}$$

If we represent the augmented state vector as $x_t^a = [x_t^T w_t^T]^T$, then the Eq. (7) can be redefined as

$$x_{t+1}^a = \begin{bmatrix} x_{t+1} \\ w_{t+1} \end{bmatrix} = \begin{bmatrix} \hat{\Phi}(x_t) + g(x_t, w_t) \\ w_t \end{bmatrix} \tag{9}$$

The Eq. (6a, b) is the measurement equation, so we have

$$\begin{cases} \mathbf{z\_geo} = \delta geo = geo_{actual} - geo_{INS} \\ \mathbf{z\_gg} = \delta gg = gg_{actual} - gg_{INS} \end{cases} \tag{10}$$

where geo represents geomagnetism, gg represents gravity gradient, and δ is the error between the actual geophysical signal (the gravity gradient vector and the geomagnetic vector) measured values and the referenced geophysical signals, which are given according to the INS state and the underwater gravity gradient/geomagnetism maps.

4 Information Fusion Algorithm Based on Gravity Gradient and Geomagnetism Measurement Information

As the state equation and measurement equation are nonlinear, a Neural Network-aided Adaptive UKF is used in each local filter. Equation (10) is taken to form two local NN-aided adaptive UKF filters for gravity gradient and geomagnetism measurement information separately. Using the weighted least squares estimation algorithm, a master filter fuses the local optimal state estimates $\mu_i(k)$ ($i = 1, 2$) from the two local filters to obtain the entire optimal state estimate. A block diagram in Fig. 2 illustrates the information fusion algorithm of the integrated navigation system based on the gravity gradient and geomagnetism sub-filters.

After the positioning results from the gravity gradient local filter and the geomagnetism local filter have been obtained, denoted by p_1 and p_2 respectively, the weighted least squares estimation is utilized to combine the gravity gradient aided navigation sub-system and the geomagnetism aided navigation sub-system. They are affected by various errors, including measurement error, database error, and algorithm error. Such errors can be treated as Gaussian noise. Then the weighted least squares estimation method for the combined gravity gradient and geomagnetism measurement information can be constructed as

$$WP = [w_1, w_2][p_1, p_2]^T \tag{11}$$

Where w_1 and w_2 are the weights of the gravity gradient position result and the geomagnetism position result, respectively. In other words, the mathematical expectation of the position result after fusion is the weighted expectation of the gravity gradient local filter and the geomagnetism local filter individually. The accuracy of the integrated navigation method is

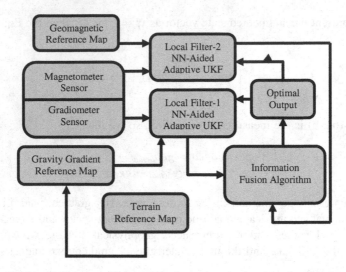

Fig. 2. The block diagram of information fusion on the integrated navigation system.

$$\begin{cases} \delta_T = \sum \sqrt{w_i^2 \delta_i^2} \\ \sum w_i = 1 \end{cases} \quad i = 1, 2 \tag{12}$$

Where δ_i is the residual error of each local adaptive UKF filter. Thus a kind of conditional Lagrange multiplier extreme value equation F can be written as

$$\begin{cases} F = \sum w_i^2 \delta_i^2 + \lambda(\sum w_i - 1) \\ \frac{\partial F}{\partial w_i} = 0 \end{cases} \quad i = 1, 2 \tag{13}$$

According to Eq. (13), the weights of gravity gradient and geomagnetism filters are obtained as follows

$$w_i = \frac{1}{\delta_i^2 \sum \frac{1}{\delta_i^2}}, \qquad i = 1, 2 \tag{14}$$

Then the final optimal state can be described as

$$\begin{cases} \hat{\mu}_g(k) = \sum w_i \mu_i(k) \\ \hat{p}_g(k) = \sum w_i p_i(k) \end{cases} \quad i = 1, 2 \tag{15}$$

Where $\hat{\mu}_g(k)$ is the optimal state of local UKFs, and $\hat{P}_g(k)$ is the optimal covariance matrix of state.

5   Experimental Results

This section presents the simulation conditions and shows the experimental results. The initial position and velocity errors are assumed zero. The parameters related to the local UKFs are shown in Table 1, where Q is the covariance of state process noise in the geomagnetic navigation system and Gravity Gradient Navigation system, q_1 is the position error, q_2 is the velocity error, R_1 is the covariance of measurement noise in the gravity gradient navigation system, and R_2 is the covariance of measurement noise in the geomagnetic navigation system. Disturbances and noise are assumed to be kinds of white noise. The underwater geomagnetic reference map is shown in Fig. 3, and the computed gravity gradient map from the DEM of the terrain-reference map (the method details in [5]) is illustrated in Fig. 4.

Table 1. Parameters of the integrated navigation simulation

Parameters	Values
Q	$\mathbf{Q} = diag[q_1^2, q_1^2, q_1^2, q_2^2, q_2^2]$
	$q_1 = 5\text{m}, \quad q_2 = 0.2\text{m/s}$
R_1	$\mathbf{R}_1 = diag[(0.2)^2, (0.1)^2, (0.4)^2, (0.3)^2, (0.1)^2, (0.2)^2]E$
R_2	$\mathbf{R}_2 = diag[(0.5)^2, (0.2)^2, (0.3)^2]nT$

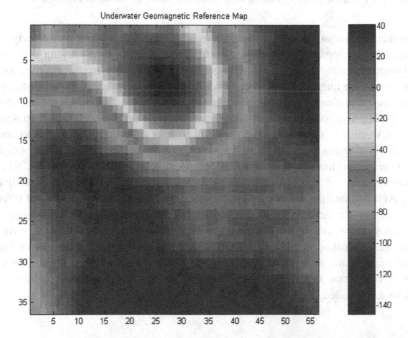

Fig. 3. Underwater geomagnetic reference map.

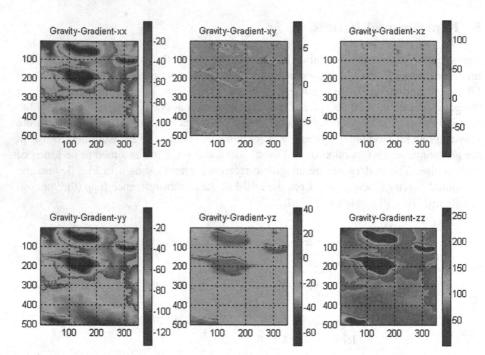

Fig. 4. Computed gravity gradient map from the DEM of the terrain-reference map.

Considering various kinds of measurement noise which have particularly great influence on the integrated navigation system, some kinds of time-varying noise make the mono-geomagnetic and mono-gravity gradient navigation systems become unstable and the covariance matrix of state lose its positive definite character. In addition, a large time-varying disturbance introduces singularities into the two kinds of navigation system. To improve robustness of the integrated navigation, the weighted least squares estimation is utilized to adapt to the residual error of each local UKF and to restrict various kinds of sensor noise and time-varying disturbances. In particular, the weighted least squares algorithm is a good method for effectively updating each sub-UKF, and the optimal states are robust to the uncertainties of a complex integrated navigation system. At the same time, by using a kind of modified parameter σ_k in the local filters, it makes the gravity gradient and geomagnetism sub-systems more robust towards time-varying noise and terrain fluctuations. Figure 5 shows clearly that the weighted least squares estimation combining with NN-Aided adaptive UKF is the most effective algorithm among the three algorithms examined. By achieving the optimal states, the integrated navigation system has improved its robustness and reduced certain disturbances to a controlled level.

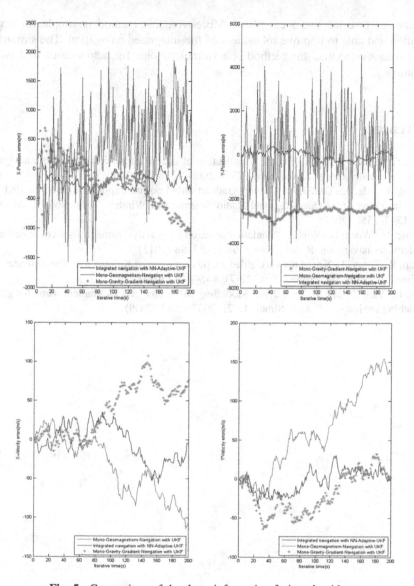

Fig. 5. Comparison of the three information fusion algorithms.

6   Conclusions

This paper presents an integrated underwater navigation, which effectively combines gravity gradient aided navigation and geomagnetic aided navigation together. The proposed weighted least squares estimation and Neural Network-Aided adaptive UKF method exhibits better performance than the MMAE-UKF method and the adaptive-UKF method. A feasible explanation is that the proposed method takes full advantage of the two types of measurement and uses the weighted least squares algorithm to fuse the two local filters. Moreover, this method is simple for implementation and able to improve robustness of the integrated navigation. The simulation results demonstrate that this method is a suitable choice for autonomous underwater navigation.

References

1. Guo, C., Hu, Z., Zhao, X., Cai, H.: Application of adaptive filter in geomagnetic navigation under magnetic storms. J. Astronaut. **8**(31), 1000–1328 (2010)
2. Xiong, L., Ma, J., Tian, J.W.: Gravity gradient aided position approach based on EKF and NN. In: Cross Strait Quad-Regional Radio Science and Wireless Technology Conference, pp. 1347–1350 (2011)
3. Zheng, H., Wang, H., Wu, L.: Simulation research on gravity-geomagnetism combined aided underwater navigation. R. Inst. Navig. **66**(1), 83–98 (2013)
4. Zhan, R., Wang, J.: Neural network-aided adaptive unscented Kalman filter for nonlinear state estimation. IEEE Sig. Process. Lett. **13**(7), 445–448 (2006)
5. Wu, L., Yao, Z., Ma, J., Tian, J.W.: Modelling full-tensor gravity gradient maps for gravity matching navigation. J. Syst. Simul. **12**(2), 7037–7041 (2009)

A Stochastic Approximation Algorithm for Quantile Estimation

Ajin George Joseph$^{(\boxtimes)}$ and Shalabh Bhatnagar

Department of Computer Science and Automation,
Indian Institute of Science, Bangalore, India
{ajin,shalabh}@csa.iisc.ernet.in

Abstract. In this paper, we present two new stochastic approximation algorithms for the problem of quantile estimation. The algorithms uses the characterization of the quantile provided in terms of an optimization problem in [1]. The algorithms take the shape of a stochastic gradient descent which minimizes the optimization problem. Asymptotic convergence of the algorithms to the true quantile is proven using the ODE method. The theoretical results are also supplemented through empirical evidence. The algorithms are shown to provide significant improvement in terms of memory requirement and accuracy.

1 Introduction and Preliminaries

Quantiles form a class of performance measures. Given a probability space $(\Omega, \mathcal{F}, \mathbb{P})$ and a bounded Lebesgue measurable function $H : \mathbb{R}^d \to \mathbb{R}$ with $H \in [H_l, H_u]$ and H_l, H_u both finite, the ρth-quantile (denoted as $\gamma(\rho)$) where $\rho \in [0, 1]$ is defined as

$$\gamma(\rho) = \sup_{y \in \mathcal{Y}} \{ \mathbb{P}[H(X) < y] < \rho \}, \quad \text{with } \mathcal{Y} = H(\mathbb{R}^d). \tag{1}$$

If the distribution function of X is continuous, then $\gamma(\rho) = F_X^{-1}(\rho)$ where F_X is the cumulative distribution function (CDF) of X. The problem of quantile estimation is to efficiently estimate the true quantile when the distribution is unknown but only samples chosen using the distribution are available.

The importance of this problem is evident from the application it finds in diverse fields ranging from medical domain to environmental modelling to humanities and social sciences. Examples of the application of quantile estimation in manufacturing context and hypothesis testing are provided in [4]. A wide range of examples is also provided in [3].

The most common approach in quantile estimation is to use Monte-Carlo (MC) method where a collection $C_H = \{H(X_1), H(X_2), \ldots, H(X_N)\}$ of N samples are chosen using the distribution defined by $\mathbb{P}$ and the ρth-quantile estimate $\bar{\gamma}_{\text{MC}}(\rho) = C_H^s[\lceil pN \rceil]$ where C_H^s is the sorted form of C_H and $\lceil . \rceil$ is the integer ceiling function *i.e.*, for any $x \in \mathbb{R}$, $\lceil x \rceil$ is the least integer upper bound of x.

© Springer International Publishing Switzerland 2015
S. Arik et al. (Eds.): ICONIP 2015, Part II, LNCS 9490, pp. 311–319, 2015.
DOI: 10.1007/978-3-319-26535-3_36

This estimate is asymptotically normal [8]. Thus

$$\sqrt{n}(\bar{\gamma}_{MC}(\rho) - \gamma(\rho)) \xrightarrow{n \to \infty} \mathcal{N}(0, \sigma^2_{MC}) \text{ where } \sigma^2_{MC} = \frac{\rho(1 - \rho)}{f_X(\gamma(\rho))} \tag{2}$$

with $f_X(.)$ being the probability density function (PDF) of X defined by $\mathbb{P}$ i.e. $f_X(x) = \frac{d}{dx}\mathbb{P}(X \leq x)$.

The problem with the MC approach is the high variance of the estimates for the tail quantiles ($\rho < 0.1$ or $\rho > 0.9$) since $f_X \approx 0$ in this window, which will cause the estimates obtained from small number of samples to be far away from the true value. Various variance reduction techniques like Controlled Stratification [8], Latin Hypercube Sampling [7] and Importance Sampling [5] can be used to bring down the deviation.

In this paper, we provide a stochastic approximation algorithm which converges to the true ρ^{th}-quantile asymptotically. This is the first time a stochastic approximation algorithm is being proposed for the quantile estimation problem. The evolutionary nature of the algorithm eliminates the need to store the N samples (which requires $N * d$ space where d is the dimension) and sorting computation (which requires $O(NlogN)$ time), both of which are needed in the Monte-Carlo algorithm.

2   Our Proposed Algorithm

In [1], the quantile problem is reformulated as an optimization problem

$$\gamma(\rho) = \min_{w \in \mathcal{W}} \mathbb{E}_x \left[\psi(H(X), w; \rho) \right], \tag{3}$$

where $\psi(H(x), w; \rho) = \rho(H(x) - w)I_{\{H(x) > w\}} + (1 - \rho)(w - H(x))I_{\{H(x) \leq w\}}$ and $\mathbb{E}_x[.]$ is the expectation w.r.t. $\mathbb{P}$. I is the indicator function defined as

$$I_A(x) = \begin{cases} 1 \text{ if } x \in A \\ 0 \text{ if } x \notin A. \end{cases} \tag{4}$$

This can be easily verified by assigning the gradient of $\psi(H(X), w; \rho)$ to 0.The gradient operator can be taken inside the expectation as the function ψ is bounded.

$$\begin{aligned} \nabla_w \mathbb{E} \left[\psi(H(X), w; \rho) \right] &= \mathbb{E} \left[\nabla \psi(H(X), w; \rho) \right] \\ &= \mathbb{E} \left[-\rho I_{\{H(z) > w\}} + (1 - \rho)I_{\{H(z) \leq w\}} \right] \\ &= -\rho \mathbb{P}\{H(z) > w\} + (1 - \rho)\mathbb{P}\{H(z) \leq w\} \\ &= -\rho(\mathbb{P}\{H(z) > w\} + \mathbb{P}\{H(z) \leq w\}) + \mathbb{P}\{H(z) \leq w\} \\ &= -\rho + F_H(w), \end{aligned}$$

where F_H is the cumulative distribution function (CDF) of $H(X)$, i.e., $F_H(x) = P\{H(X) \leq x\}$. Equating the gradient to 0 we obtain

$$\begin{aligned} \nabla_w \mathbb{E} \left[\psi(H(X), w; \rho) \right] &= 0. \\ \Rightarrow -\rho + F_H(w) &= 0. \\ \Rightarrow F_H(w) &= \rho. \end{aligned} \tag{5}$$

2.1   Algorithm 1

The solution to the optimization problem (3) can be obtained via the stochastic approximation recursion

$$\gamma_{k+1} = \gamma_k - \alpha_{k+1} \left(-\rho I_{\{H(z_{k+1}) > \gamma_k\}} + (1 - \rho) I_{\{H(z_{k+1}) \leq \gamma_k\}} \right), \qquad (6)$$

where z_k are *i.i.d.* and $z_k \sim \mathbb{P}$ and α_k is the learning rate with $\Sigma_k \alpha_k = \infty$ and $\Sigma_k \alpha_k^2 < \infty$. The step sizes control the rate of convergence of the iterates. We can also use constant values as the step size. In that case, the convergence to the true value is not guaranteed, but to a small enough neighbourhood of the true value can be obtained, depending on the value of the step-size parameter.

The algorithm for quantile estimation is given in below.

Algorithm 1. Quantile Estimation

Data: $\alpha_k \in (0, 1)$, $\rho \in [0, 1]$, $T \in \mathbb{Z}_+$
Initialization: $\gamma_0 = 0$, $k = 0$;
while $k < T$ **do**
$\quad\mid\quad z_{k+1} \sim \mathbb{P}$.
$\quad\mid\quad \gamma_{k+1} = \gamma_k - \alpha_{k+1} \left(-\rho I_{\{H(z_{k+1}) > \gamma_k\}} + (1 - \rho) I_{\{H(z_{k+1}) \leq \gamma_k\}} \right)$.
$\quad\mid\quad k := k + 1$.
return γ_T.

---

3   Proof of Convergence

We shall assume for convenience that $\sup_k |\gamma_k| < \infty$ *w.p.* 1. Note that this is not straightforward and, in practice, one needs to show that this holds, *i.e.*, the iterates remain uniformly bounded. In most practical scenarios, one imposes this requirement by projecting the updates to a compact and convex set. The ODE analysis in such a case follows roughly along the same lines as below. For more see [2].

Define the filtration $\{\mathcal{F}_k\}$ where $\mathcal{F}_k = \sigma(\gamma_0, \gamma_i, 1 \leq i \leq k)$.

$\gamma_{k+1} = \gamma_k - \alpha_{k+1} \nabla \psi(H(z_{k+1}), \gamma_k)$.

$\gamma_{k+1} = \gamma_k + \alpha_{k+1} \left(-\nabla \psi(H(z_{k+1}), \gamma_k) + \mathbb{E}\left[\nabla \psi(H(z_{k+1}), \gamma_k)\right] - \mathbb{E}\left[\nabla \psi(H(z_{k+1}), \gamma_k)\right] \right)$.

$\gamma_{k+1} = \gamma_k + \alpha_{k+1} \left(\mathbb{M}_{k+1} + h(\gamma_k) \right)$,

where $\mathbb{M}_{k+1} = \mathbb{E}[\nabla \psi(H(z), \gamma_k)] - \nabla \psi(H(z_{k+1}), \gamma_k)$ and $h(\gamma) = -\mathbb{E}[\nabla \psi(H(z), \gamma)]$.

Lemma 1. $h(.)$ *is Lipschitz continuous.*

Proof. For any $\gamma_1, \gamma_2 \in \mathbb{R}$, $\gamma_2 > \gamma_1$,

$$\left| h(\gamma_1) - h(\gamma_2) \right| = \left| \mathbb{E}\left[\nabla \psi(H(z), \gamma_1)\right] - \mathbb{E}\left[\nabla \psi(H(z), \gamma_2)\right] \right|$$

$$= \left| \mathbb{E}[-\rho I_{\{H(z) \geq \gamma_1\}} + (1 - \rho) I_{\{H(z) < \gamma_1\}}] - \mathbb{E}[-\rho I_{\{H(z) \geq \gamma_2\}} + (1 - \rho) I_{\{H(z) < \gamma_2\}}] \right|$$

$$= \left| \mathbb{P}(H(z) < \gamma_1) - \rho - \mathbb{P}(H(z) < \gamma_2) + \rho \right|$$
$$= \mathbb{P}(\gamma_1 < H(X) < \gamma_2)$$
$$= \int_{\gamma_1}^{\gamma_2} f_H(x)dx$$
$$= L(\gamma_2 - \gamma_1), L > 0.$$

The last inequality above is obtained using the Mean Value Theorem for Integrals. Hence the claim follows. ∎

Lemma 2. $\mathbb{M}_k$, $k \geq 0$ *is a martingale difference noise sequence.*

Proof. Note that $\mathbb{M}_k$ is $\mathcal{F}_k$-measurable $\forall k$ and is also integrable. Now we have almost surely

$$\mathbb{E}[\mathbb{M}_{k+1}|\mathcal{F}_k] = \mathbb{E}[\mathbb{E}[\nabla\psi(H(z),\gamma_k)] - \nabla\psi(H(z_{k+1}),\gamma_k)|\mathcal{F}_k]$$
$$= \mathbb{E}[\mathbb{E}[\nabla\psi(H(z),\gamma_k)]|\mathcal{F}_k] - \mathbb{E}[\nabla\psi(H(z_{k+1}),\gamma_k)|\mathcal{F}_k]$$
$$= \mathbb{E}[\mathbb{E}[\nabla\psi(H(z),\gamma_k)]|\mathcal{F}_k] - \mathbb{E}[-\rho I_{\{H(z_{k+1})\geq\gamma_k\}} + (1-\rho)I_{\{H(z_{k+1})<\gamma_k\}}|\mathcal{F}_k]$$
$$= \mathbb{E}[\nabla\psi(H(z),\gamma_k)] - \mathbb{E}[-\rho I_{\{H(z)\geq\gamma_k\}} + (1-\rho)I_{\{H(z)<\gamma_k\}}]$$
$$= \mathbb{E}[\nabla\psi(H(z),\gamma_k)] - \mathbb{E}[\nabla\psi(H(z),\gamma_k)]$$
$$= 0.$$

The third equality above holds because z_k, $k \geq 0$ are *i.i.d.*. Hence the claim follows. ∎

Lemma 3. $\mathbb{E}[\|\mathbb{M}_{k+1}\|^2|\mathcal{F}_k] \leq K(1 + \|\gamma_k\|^2)$, $K > 0$.

Proof. It is easy to see that $\nabla\psi(H(z_{k+1}),\gamma_k)$ has finite first and second order moments. So the claim follows. ∎

By Lemma 1, Chap. 2 in [6] and by Lemmas 1–3 above, we can claim that asymptotic behaviour of the stochastic process $\{\gamma_k\}$ is similar to the deterministic system

$$\dot{\gamma} = \mathbb{E}[\nabla\psi(H(z),\gamma)]. \tag{7}$$

The following theorem (Theorem 2) in this paper shows convergence of the iterates γ_k. This is achieved using Theorem 2, Chap. 2 in [6]. For that we need to prove the global asymptotic stability of the ODE (7). This is achieved using the Lyapunov method.

Theorem 1. $\lim_{k\to\infty} \gamma_k = \gamma^*$ *a.s., where* γ^* *satisfies* $\mathbb{E}[\nabla\psi(H(z),\gamma^*)] = 0$.

Proof. Define $V(x) = (x - F_H^{-1}(\rho))^\top (x - F_H^{-1}(\rho))$. It is easy to verify that $V(x) > 0$, $\forall x \in \mathbb{R}^d \backslash \{F_H^{-1}(\rho)\}$. Further $V(F_H^{-1}(\rho)) = 0$ and $V(x) \to \infty$ as $\|x\| \to \infty$. So $V(x)$ is a Lyapunov function. Also note that, by using the monotonicity of $F_H(x)$ we can show that $< \nabla V(x), h(x) > = (x - F_H^{-1}(\rho))(\rho - F_H(x)) \leq 0$. So the ODE defined in (7) is globally asymptotically stable at $F_H^{-1}(\rho)$. Thus by Theorem 2 of Chap. 2 in [6], the iterates γ_k converges to $F_H^{-1}(\rho) = \gamma(\rho)$. And by (5), $\gamma(\rho)$ satisfies $\mathbb{E}[\nabla\psi(H(z),\gamma(\rho))] = 0$. ∎

3.1 Algorithm 2

Algorithm 1 has the disadvantage of having slow rate of convergence. In this section, we provide a new version of the algorithm which has better convergence rate. The new algorithm uses double sampling where two independent sample z_{k+1} and z'_{k+1} are used to evaluate γ_{k+1}. The algorithm is given below. Note that Π is the truncation operator where $\Pi(\gamma) \in [H_l, H_u]$.

Algorithm 2. Quantile Estimation

Data: $\alpha_k \in (0,1)$, $\rho \in [0,1]$, $T \in \mathbb{Z}_+$
Initialization: $\gamma_0 = 0$, $k = 0$;
while $k < T$ do

$\quad z_{k+1}, z'_{k+1} \sim \mathbb{P}.$

$\quad \gamma_{k+1} = \Pi\Big(\gamma_k -$

$\quad\quad \alpha_{k+1}\left(|H(z'_{k+1})| + 1\right)\left(-\rho I_{\{H(z_{k+1})>\gamma_k\}} + (1-\rho)I_{\{H(z_{k+1})\leq\gamma_k\}}\right)\Big).$

$\quad k := k + 1.$
return $\gamma_T.$

Theorem 2. *For a given $\rho \in (0,1)$, $\lim_{k\to\infty} \gamma_k = \gamma(\rho)$ almost surely.*

Proof. Using the same proof technique as in Theorem 1, we can show that γ_k converges and the limit of γ_k satisfies

$$\mathbb{E}\left[(|H(z')| + 1)(\nabla\psi(H(z), \gamma))\right] = 0.$$
$$\Rightarrow \mathbb{E}\left[(|H(z')| + 1)\right] \mathbb{E}\left[(\nabla\psi(H(z), \gamma))\right] = 0$$
$$\Rightarrow \mathbb{E}\left[(\nabla\psi(H(z), \gamma))\right] = 0,$$

since $\mathbb{E}\left[(|H(z')| + 1)\right] \neq 0$. ∎

4 Experimental Results

The algorithm was tested on two functions $H_1(x) = \frac{\prod_{i=1}^{i=d} x_i}{1+\sum_{i=1}^{i=d} x_i^2}$, where $x \in \mathbb{R}^d$, $d = 100$ and $H_2(x) = \frac{\prod_{i=1}^{i=d} x_i}{1+\sum_{i=1}^{i=d} |x_i|}$, where $x \in \mathbb{R}^d$, $d = 1000$ and $\rho = 0.5$. Here, H_1 is continuous, differentiable function on a 100-dimensional space while H_2 is continuous, non-differentiable function on a 1000-dimensional space. The distribution chosen is the d-dimensional multivariate standard Gaussian distribution $\mathcal{N}_d(0_{d\times1}, I_{d\times d})$. The true 0.5−quantile in both cases is 0 since H_1 and H_2 are odd functions and the distribution is symmetric. Figure 1 shows the evolutionary path of the algorithm.

Experiments were also conducted to estimate the quantile of various distributions using both Algorithms 1 and 2. Normal, Chi Square, Weibull and Cauchy

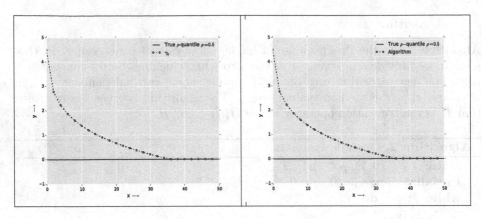

Fig. 1. Evolutionary track of the iterates $\{\gamma_k\}$ for $\rho = 0.5$. (a) $H_1(x) = \frac{\prod_{i=1}^{i=d} x_i}{1+\sum_{i=1}^{i=d} x_i^2}$ and $d = 100$. (b) $H_2(x) = \frac{\prod_{i=1}^{i=d} x_i}{1+\sum_{i=1}^{i=d} |x_i|}$, $d = 1000$, $f_X = \mathcal{N}_d(0_{d\times 1}, \mathrm{I}_{d\times d})$.

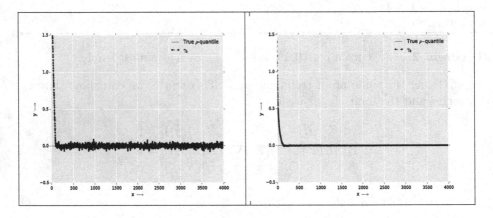

Fig. 2. Comparison of the evolutionary track of the iterates $\{\gamma_k\}$ between constant step sizes and decreasing ones. $H(x) = x^5 + x + \sin(x)$ and $\rho = 0.5$. (a) $\alpha_k = 0.001, \forall k$ (b) $\alpha_k = \frac{1}{k^{0.95}}$. The distribution used is the von Mises distribution.

are the distributions on which tests were conducted. Table 1 compares the estimates of the various quantiles of the above distributions with the true values. Note that to estimate the quantile of a distribution, $H(x) = x$.

Tests were also conducted on the von Mises distribution. The PDF of von Mises distribution is given by

$$f_X(x) = \frac{e^{\kappa \cos(x-\mu)}}{2\pi I_0(\kappa)}, \text{ where } I_0 \text{ is the Bessel function.}$$

In this case, $H(x) = x^5 + x + \sin(x)$ and estimates were obtained both by using constant step sizes and decreasing step sizes. The trajectory of γ_k is shown in Fig. 2. It shows that in constant step size scenario, convergence to the true

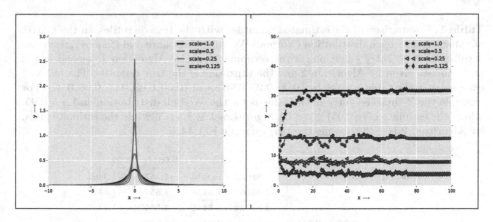

Fig. 3. Robustness Evaluation: Performance of the algorithm on various scale parameter of the Cauchy distribution. As the scale parameter approaches 0, the distribution becomes more and more narrow. Even in the narrowest case, the algorithm shows convergence at a very good rate.

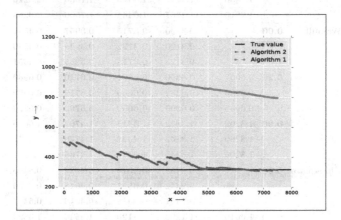

Fig. 4. Comparison of the rate of convergence of Algorithms 1 and 2. The initial point for both algorithm is chosen to be 1000.0. The Algorithm 2 is shown to converge faster to the true value.

quantile is not obtained but only to a certain neighbourhood of the true value can be achieved.

Tests were conducted to measure the robustness of the algorithm by checking the tail quantiles of Cauchy distribution for various values of its scale parameter. The PDF of the Cauchy distribution is given by

$$f_X(x) = \frac{1}{\pi c \left[1 + \left(\frac{x-x_0}{c}\right)^2\right]}$$ where c and x_0 are scale and location parameters.

Table 1. Comparison of the estimated quantiles with the true quantiles. In the matrix of values against each distribution (Normal, Weibull, Chisquare and Cauchy), there are 3 values in each entry - first one is the estimate given by Algorithm 1, second one is the estimate given by Algorithm 2 and the third one is the true quantile. The index of each entry in the table is given by $\{0.00, 0.20, 0.90\}$ and $\{0.00, 0.03, 0.05, 0.06, 0.09\}$. For example the '*' marked entry in the table is for the Weibull distribution and $\rho = 0.95$, where the estimate of $\gamma(0.95)$ given by Algorithm 1 is 1.4415, while the estimate given by Algorithm 2 is 1.4422, while the true quantile is 1.4415.

		0.00	0.03	0.05	0.06	0.09
Normal	0.00	—	−1.8807	−1.6430	−1.5573	−1.3405
		—	−1.8800	−1.6438	−1.5548	−1.3400
		—	−1.8807	−1.6448	−1.5547	−1.3407
	0.20	−0.8416	−0.7389	−0.6750	−0.6443	−0.5535
		−0.8411	−0.7383	−0.6749	−0.6423	−0.5590
		−0.8416	−0.7388	−0.6744	−0.6433	−0.5534
	0.90	1.2815	1.4717	1.6446	1.7503	2.3263
		1.2836	1.4727	1.6443	1.7500	2.3220
		1.2815	1.4757	1.6444	1.7506	2.3263
Weibull	0.00	—	0.3130	0.3715	0.3957	0.4541
		—	0.3120	0.3722	0.3948	0.4540
		—	0.3123	0.3715	0.3955	0.4551
	0.20	0.6067	0.6390	0.6601	0.6710	0.6989
		0.6090	0.6382	0.6600	0.6712	0.6999
		0.6065	0.6390	0.6600	0.6703	0.6990
	0.90	1.3202	1.3842	1.4415*	1.4763	1.6615
		1.3200	1.3842	1.4422*	1.4770	1.6641
		1.3205	1.3845	1.4415*	1.4765	1.6637
Chi square	0.00	—	0.2455	0.3510	0.4022	0.5405
		—	0.2442	0.3527	0.4018	0.5400
		—	0.2451	0.3518	0.4012	0.5401
	0.20	1.0010	1.1290	7.8120	1.2540	1.3820
		1.0014	1.1294	7.8145	1.2530	1.3814
		1.0050	1.1293	7.8147	1.2543	1.3810
	0.90	6.2560	7.0630	7.8120	8.3020	11.3500
		6.2500	7.0640	7.8145	8.3120	11.3350
		6.2510	7.0600	7.8150	8.3120	11.3450
Cauchy	0.00	—	−10.5774	−6.3168	−5.2481	−3.4498
		—	−10.5667	−6.3184	−5.2429	−3.4520
		—	−10.5780	−6.3138	−5.2421	−3.4420
	0.20	−1.3798	−1.1367	1.0000	−0.9320	−0.7759
		−1.3799	−1.1340	−1.0010	−0.9384	−0.7729
		−1.3763	−1.1343	−1.0010	−0.9390	−0.7756
	0.90	3.0777	4.4737	6.3100	7.9116	31.5203
		3.0693	4.4865	6.3100	7.9208	31.6561
		3.0776	4.4737	6.3138	7.9158	31.8205

In the experiment, the location parameter is set to 0. The results obtained are shown in Fig. 3. The results obtained show that the rate of convergence almost remains constant for various values of the scale parameter.

We also compared the rate of convergence of both Algorithms 1 and 2. In this case, $H(x) = 10x$, $\rho = 0.99$ and the distribution is the standard Cauchy, $i.e.$, loc=0 and scale=1. The results obtained are shown in Fig. 4. The empirical results verify the claim that Algorithm 2 is showing better rate of convergence.

5   Conclusions

In this paper, we provided two stochastic approximation algorithms for the problem of quantile estimation. The algorithms show good performance both in terms of accuracy and memory requirements. The evolutionary nature of the algorithms eliminates the need for both the storage of the samples (which requires $N * d$ space where d is the dimension) and sorting computation (which requires $O(NlogN)$ time), both of which are needed in the Monte-Carlo algorithm. The stochastic approximation nature of the algorithms helps to control the learning rate which provides an extra dimension when compared with the Monte-Carlo methods. We also provide in this paper the proof of concept and the proof of convergence to the true value. Results on the rate of convergence and lock-in probability to a neighbourhood of the true value remain to be obtained. The empirical evidence provided shows the robustness of the algorithms where the algorithms are shown to perform well even when the distribution is heavy tailed or when $\rho < 0.1$ or $\rho > 0.9$.

References

1. Hu, J., Hu, P., Chang, H.S.: A stochastic approximation framework for a class of randomized optimization algorithms. IEEE Trans. Autom. Control 57(1), 165–178 (2012)
2. Kushner, H.J., Clark, D.S.: Stochastic Approximation Methods for Constrained and Unconstrained Systems, vol. 26. Springer Science & Business Media, New York (2012)
3. Yu, K., Zudi, L., Stander, J.: Quantile regression: applications and current research areas. J. Royal Stat. Soc. Ser. D (Stat.) 52(3), 331–350 (2003)
4. Chen, E.J., Kelton, W.D.: Simulation-based estimation of quantiles. In: Simulation Conference Proceedings, 1999 Winter, vol. 1. IEEE (1999)
5. Glynn, P.W.: Importance sampling for Monte Carlo estimation of quantiles. In: Mathematical Methods in Stochastic Simulation and Experimental Design: Proceedings of the 2nd St. Petersburg Workshop on Simulation (1996)
6. Borkar, V.S.: Stochastic Approximation: A Dynamical Systems Viewpoint. Cambridge University Press, Cambridge (2008)
7. Avramidis, A.N., Wilson, J.R.: Correlation-induction techniques for estimating quantiles in simulation experiments. Oper. Res. 46(4), 574–591 (1998)
8. Cannamela, C., Garnier, J., Iooss, B.: Controlled stratification for quantile estimation. Ann. Appl. Stat. 2, 1554–1580 (2008)

Reinforcement Learning in Continuous Spaces by Using Learning Fuzzy Classifier Systems

Gang Chen[1,2]($\boxtimes$), Colin Douch[1,2], Mengjie Zhang[1,2], and Shaoning Pang[1,2]

[1] Victoria University of Wellington, Wellington, New Zealand
[2] Unitec Institute of Technology, Mount Albert, New Zealand
{aaron.chen,mengjie.zhang}@ecs.vuw.ac.nz,
douchcoli@myvuw.ac.nz, ppang@unitec.ac.nz

Abstract. Aimed at achieving multi-step reinforcement learning in continuous spaces, many Learning Classifier Systems have been developed recently to learn fuzzy logic rules. Among these systems, accuracy-based Michigan learning fuzzy classifier systems are gaining increasing research attention. However, in order to learn effectively, existing accuracy-based systems often require the action space to be discrete. Without this restriction, only single-step learning may be supported. In this paper, we will develop a new accuracy-based learning fuzzy classifier system that can perform multi-step reinforcement learning in completely continuous domains. To achieve this goal, a special fuzzy logic system will be introduced in this paper where the output action from the system is modelled through a continuous probability distribution. A natural gradient learning technique will be further exploited to fine-tune the action outputs of individual fuzzy rules. The effectiveness of our learning system has been verified on several benchmark problems.

Keywords: Fuzzy system · Reinforcement learning · Learning classifier system

1 Introduction

Learning classifier systems (LCSs) are evolutionary machine learning technologies. They seek to solve machine learning problems by evolving a group of IF-THEN rules [8]. Recently, Michigan LCSs such as the XCS algorithm have been extensively studied for their remarkable ability of multi-step reinforcement learning [5,6]. Despite of its proven effectiveness, XCS cannot be applied directly to a learning environment with continuous states and actions. To address this limitation, many learning fuzzy classifier systems (LFCS) have been proposed to learn fuzzy logic rules [4].

In the literature, accuracy-based LFCSs are gaining increasing research attention since they are widely shown to evolve accurate and optimally general rules [3,4]. For example, in [4], the authors seek to tackle single-step *function approximation problems* by using an accuracy-based LFCS derived from XCS. Accuracy-based LFCSs have also been popularly used to perform pattern recognition and

© Springer International Publishing Switzerland 2015
S. Arik et al. (Eds.): ICONIP 2015, Part II, LNCS 9490, pp. 320–328, 2015.
DOI: 10.1007/978-3-319-26535-3_37

classification tasks, which are also single-step learning problems [2]. In addition to single-step problems, multi-step learning problems was explicitly handled by some LFCSs such as the fuzzy Q-learning system proposed in [7].

Although closely related to our research in this paper, the fuzzy Q-learning system only supports discrete action outputs. In fact, according our knowledge, few accuracy-based LFCSs have been successfully developed to address multi-step reinforcement learning problems that require continuous actions. Moreover, to achieve effective learning, an LFCS needs to flexibly adjust its action output based on the environmental feedback. Such adjustment should be performed at a *fine-grained level* by constantly updating the action output from each fuzzy rule without changing the learned fuzzy rule set. As far as we know, no accuracy-based LFCSs have ever adopted such a mechanism for flexible learning.

In an attempt to tackle these technical challenges, a new LFCS will be developed in this paper to explicitly support multi-step reinforcement learning in completely continuous spaces. To improve learning flexibility, we will utilize a fuzzy logic system where the output action from the system is modelled through a *continuous probability distribution* and will be further fine-tuned effectively by using a recently developed natural gradient learning technique [9]. Our experiment results on several benchmark problems clearly show that the newly developed LFCS can effectively handle both single-step and multi-step reinforcement learning problems in continuous spaces.

2   Reinforcement Learning in Continuous Spaces

Reinforcement learning is a general problem for multi-step decision making [11]. The learning environment can be understood as a *Markov Decision Process* (MDP) in discrete time with continuous states $s \in \mathbb{S} \subseteq \mathbb{R}^n$ and continuous actions $a \in \mathbb{A} \subseteq \mathbb{R}$. Each time when a learning agent performs an action a in state s, the environment will provide a feedback in the form of a scalar reward $r(s, a) \in \mathbb{R}$. Starting from state s at time $t = 0$, the long-term payoff is defined as below

$$V(s) = E\left\{ \sum_{t=0}^{\infty} \gamma^t r_t \, | s_0 = s \right\} \tag{1}$$

where $\gamma \in (0, 1)$ is the *discount factor*. r_t refers to reward received at time t. To solve a reinforcement learning problem, we have shown in [5,6] that it is often beneficial for an agent to learn *stochastic policies*. Particularly, in any state s_t at time t, a stochastic policy $\pi(s_t, a_t)$ describes a *continuous probability distribution*, from which the actual action a_t to be performed by the agent will be sampled. Following this idea, the *state-action value function* Q can be recursively defined as below.

$$Q^\pi(s_t, a_t) = E(r_t) + \gamma \cdot \int_{a \in \mathbb{A}} Q^\pi(s_{t+1}, a) \cdot \pi(s_{t+1}, a) \, da \tag{2}$$

The *long-term payoff* obtainable in any state s therefore is

$$V^\pi(s) = \int_{a \in \mathbb{A}} Q^\pi(s, a) \cdot \pi(s, a) \, da \tag{3}$$

Starting from $t = 0$, the learning objective is hence to identify the optimal policy π^*:

$$\pi^* = \arg\max_\pi V^\pi(s_0). \tag{4}$$

3   A Fuzzy Logic System for Learning

In this paper, stochastic policies are represented by a group of fuzzy rules. Similar with [4], the *disjunctive normal form* (DNF) fuzzy rules with the structure shown below will be utilized in this paper.

$$\textbf{IF } s^1 \textit{ is } \tilde{S}_1 \textit{ and } \dots \textit{ and } s^n \textit{ is } \tilde{S}_n \textbf{ THEN } a \textit{ is } \tilde{A} \tag{5}$$

The state input to a learning agent is modelled as an n-dimensional vector $s = \{s^1, \dots, s^n\} \in \mathbb{S}$. According to (5), the IF part of the a fuzzy rule is made up of multiple elementary conditions. Each condition involves one dimension s_i of s, which takes as its value a *linguistic term* $\tilde{S}_i$ described through a separate *membership function*.

For any fuzzy rule R, let's denote its degree of matching the state input s as $R(s)$. If $R(s) > 0$, rule R will recommend an action $\tilde{A}_R$ as specified in its THEN part. In order to model stochastic policies, $\tilde{A}_R$ in this paper represents a *normal probability distribution* with a given *mean* $\mu_{\tilde{A}_R}$ and *standard deviation* $\sigma_{\tilde{A}_R}$. It is therefore called a *stochastic action*.

We associate three *policy parameters* with each fuzzy rule. For rule R, they are denoted respectively as θ_R, μ_R, and σ_R. These policy parameters enable us to fine-tune the action output of every fuzzy rule. Specifically, at any time t, the collection of fuzzy rules that match the state input s_t forms the match set $[M]_{s_t}$. The *combined stochastic action* from $[M]_{s_t}$ is denoted as $\tilde{A}_{[M]_{s_t}}$. Its mean and standard deviation can be obtained as below.

$$\mu_{\tilde{A}_{[M]_{s_t}}} = \frac{\displaystyle\sum_{R \in [M]_{s_t}} e^{\theta_R} \cdot R(s_t) \cdot (\mu_{\tilde{A}_R} + \mu_R)}{\displaystyle\sum_{R \in [M]_{s_t}} e^{\theta_R} \cdot R(s_t)} \tag{6}$$

$$\sigma_{\tilde{A}_{[M]_{s_t}}} = \frac{\displaystyle\sum_{R \in [M]_{s_t}} e^{\theta_R} \cdot R(s_t) \cdot (\sigma_{\tilde{A}_R} \times e^{\sigma_R})}{\displaystyle\sum_{R \in [M]_{s_t}} e^{\theta_R} \cdot R(s_t)} \tag{7}$$

As evidenced in (6) and (7), policy parameters make the mean and standard deviation of the combined stochastic action $\tilde{A}_{[M]_{s_t}}$ easily adjustable. Once $\tilde{A}_{[M]_{s_t}}$ is determined, the actual action to be performed in state s_t at time t will be obtained by taking a random sample from $\tilde{A}_{[M]_{s_t}}$. It should be clear now that our fuzzy logic system introduced in this section essentially defines a stochastic policy. Specifically, we have $\pi(s_t, a_t) = \tilde{A}_{[M]_{s_t}}(a_t)$.

4 Natural Fuzzy-XCS

In this section, a new LFCS called *Natural Fuzzy-XCS* will be developed based on XCS. Particularly, we will extend two existing learning components in XCS, i.e. the reinforcement component and the discovery component. Meanwhile, a new policy parameter learning component will also be introduced to facilitate fine-grained learning of policy parameters.

4.1 The Reinforcement Component

The reinforcement component is responsible for updating the prediction parameter and other non-policy parameters of each fuzzy rule in the *action set*. At time t, the action set $[A]_t$ can be created from the following definition:

$$[A]_t = \left\{ R \middle| R \in [M]_{s_t} \;\&\&\; \tilde{A}_R(a_t) > 0 \right\} \tag{8}$$

Once the action set is identified, the next step is to perform reinforcement update of the prediction parameter p_R of every rule $R \in [A]_t$. Since the updating rule used in XCS is no longer applicable, we have developed a new updating rule given below

$$p_R(t+1) \leftarrow (1-\beta) \cdot p_R(t) + \beta \cdot \tag{9}$$

$$\left(r_t + \gamma \int_{a \in \mathbb{A}} \tilde{Q}(s_{t+1}, a) \cdot \tilde{A}_{[M]_{s_{t+1}}}(a)\, da \right)$$

$$= (1-\beta) \cdot p_R(t) + \beta \left(r_t + \gamma \tilde{V}(s_{t+1}) \right)$$

where $\tilde{V}(s_{t+1})$ refers to the learned payoff in state s_{t+1}. Given W random actions, i.e. $\{\hat{a}^1, \ldots, \hat{a}^W\}$, sampled independently from $\tilde{A}_{[M]_{s_{t+1}}}$, $\tilde{V}(s_{t+1})$ is estimated as

$$\tilde{V}(s_{t+1}) \approx \frac{1}{W} \sum_{i=1}^{W} \tilde{Q}(s_{t+1}, \hat{a}^i) \tag{10}$$

where the state-action value function $\tilde{Q}$ in (10) can be estimated as

$$Q(s_{t+1}, a) \approx \tilde{Q}(s_{t+1}, a) = \frac{\displaystyle\sum_{R \in [M]_{s_{t+1}}^a} p_R \times F_R}{\displaystyle\sum_{R \in [M]_{s_{t+1}}^a} F_R} \tag{11}$$

F_R in (11) refers to the fitness of rule R. $[M]_{s_{t+1}}^a$ is defined as

$$[M]_{s_{t+1}}^a = \left\{ R \middle| R \in [M]_{s_{t+1}} \;\&\&\; \tilde{A}_R(a) > 0 \right\}$$

With (9), (10), and (11), a complete process has now been established for learning the prediction parameters.

4.2   The Discovery Component

In the literature, the niche GA is shown to promote learning of accurate and general rules in XCS [3]. Following this idea, the discovery component in Natural Fuzzy-XCS is also applied to the action set $[A]_t$ at any time t. Before activating the niche GA, the average time period since the last GA application in the action set is calculated. If the result is greater than the threshold θ_{GA}, roulette-wheel selection will be performed so that two parent rules will be chosen randomly from the action set $[A]_t$ according to their fitness levels. Subsequently, two offspring rules will be created by applying the single-point crossover operator with probability χ_c and then applying the mutation operator with probability μ_m.

4.3   The Policy Parameter Learning Component

To develop a new component for learning the policy parameters introduced in Sect. 3, we need to understand the potential impact of these parameters on the learning performance. Peters and Schaal showed in [9] that the learning performance J can be determined as

$$J = \int_{\boldsymbol{s} \in \mathbb{S}} d^\pi(\boldsymbol{s}) \left(\int_{a \in \mathbb{A}} \tilde{A}_{[M]_{\boldsymbol{s}}}(a) E(r(\boldsymbol{s}, a)) \mathrm{d}a \right) \mathrm{d}\boldsymbol{s} \tag{12}$$

where the *stationary probability* d^π in (12) is defined as

$$d^\pi(\boldsymbol{s}) = Pr(\boldsymbol{s}_0 = \boldsymbol{s}) + \sum_{i=1}^{\infty} \gamma^r Pr^\pi(\boldsymbol{s}_t = \boldsymbol{s}) \tag{13}$$

Because of $\tilde{A}_{[M]_{\boldsymbol{s}}}$ in (12), J can be considered as a function of the policy parameters θ_R, μ_R, and σ_R of every fuzzy rule R in Natural Fuzzy-XCS.

In order to maximize J, we will utilize a *natural gradient learning technique*. Natural gradient is defined in a *Riemannian parametric space* [1]. Since each fuzzy rule has three different policy parameters, for each type of parameter, a separate parametric space can be constructed. For the Riemannian space that contains parameter θ_R of every rule R, the distance between any two neighboring points in the space, i.e. $\boldsymbol{\theta}_t$ and $\boldsymbol{\theta}_t + \boldsymbol{\Delta}$, is

$$\|\boldsymbol{\Delta}\| = \sqrt{\boldsymbol{\Delta}^T \cdot G(\boldsymbol{\theta}_t) \cdot \boldsymbol{\Delta}} \tag{14}$$

where $G(\boldsymbol{\theta}_t)$ is the *Riemannian metric tensor*. Based on this distance measure, the steepest ascent direction of J in the Riemannian space of $\boldsymbol{\theta}_t$ will be called the natural gradient of J, denoted as $\tilde{\nabla}_{\boldsymbol{\theta}_t} J$. Amari showed in [1] that the natural gradient can be determined as

$$\tilde{\nabla}_{\boldsymbol{\theta}_t} J = G(\boldsymbol{\theta}_t)^{-1} \cdot \nabla_{\boldsymbol{\theta}_t} J. \tag{15}$$

According to [1], the *Fisher information matrix* as defined below can be utilized as the Riemannian metric tensor.

$$G(\boldsymbol{\theta}_t) = \int_{\boldsymbol{s} \in \mathbb{S}} d^\pi(\boldsymbol{s}) \int_{a \in \mathbb{A}} \pi \cdot \nabla_{\boldsymbol{\theta}_t} \log \tilde{A}_{M_{\boldsymbol{s}}} \cdot \nabla_{\boldsymbol{\theta}_t} \log \tilde{A}_{M_{\boldsymbol{s}}}^T \, \mathrm{d}a \, \mathrm{d}\boldsymbol{s} \tag{16}$$

In this paper, we adopt this approach. Consequently, by following the derivation process presented in [5], an efficient rule for learning θ_t can be obtained, as shown below

$$\theta_{t+1} \leftarrow \theta_t + \lambda \cdot \delta_t \cdot \nabla_{\theta_t} \log \tilde{A}_{M_{s_t}} \qquad (17)$$

where δ_t estimates the *advantage function* $A(s_t, a_t)$ through

$$A(s_t, a_t) = Q(s_t, a_t) - V(s) \approx \delta_t = r_t + \gamma \cdot \tilde{V}(s_{t+1}) - \tilde{V}(s_t) \qquad (18)$$

To implement (17), a simple numerical approximation of $\nabla_{\theta_t} \log \tilde{A}_{M_{s_t}}$ as illustrated below will be used.

$$\frac{\partial \log \tilde{A}_{M_{s_t}}}{\partial \theta_R} \approx \frac{\log \tilde{A}_{[M]_{s_t}}(a, \theta_R + \Delta) - \log \tilde{A}_{[M]_{s_t}}(a, \theta_R)}{\Delta} \qquad (19)$$

In our implementation of Natural Fuzzy-XCS, Δ in (19) is set to a very small number, i.e. 0.0001. Up to now, we have described the complete process for learning policy parameter θ_R of every rule R. The same method will also be used to learn the other two policy parameters, i.e. μ_R and σ_R. The details will be omitted.

5   Experimental Results

Many benchmark problems have been utilized to examine the usefulness of Natural Fuzzy-XCS. Due to space limitation, we will present experiment results on two benchmark problems in this paper, i.e. the single-step Frog problem and the multi-step Cart-Pole problem. To facilitate a comparison study, we have implemented and tested a neuro-fuzzy network (FACLN) specifically designed for reinforcement learning in continuous spaces [12] and the Fuzzy Q-learning system [7]. Because the Fuzzy Q-learning system only supports discrete actions, it cannot be tested on the Frog problem.

5.1   Experiments on the Frog Problem

The frog problem is a single-step continuous action problem. In this problem, an agent behaves as a frog and tries to learn the best-sized jump in order to catch a fly. Let D, $0 \leq D \leq 1$, represent the original distance between the frog (i.e. the agent) and the fly. Meanwhile, $s = 1 - D$ is used to denote the one-dimensional continuous sensory input to the agent. Subsequently, the goal is for the agent to learn to choose the best action a^* so as to maximize its payoff, which is further defined below

$$r(s, a) = \begin{cases} s + a, & \text{if } s + a \leq 1 \\ 2 - (s + a), & \text{otherwise} \end{cases} \qquad (20)$$

Our experiment is conducted by performing 30 independent trials in which a learn system is run for 5000 learning problems. The learned policy was evaluated

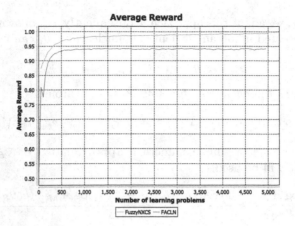

Fig. 1. The average performance (or reward) per learning problem for Natural Fuzzy-XCS and FACLN in the Frog problem.

after every 50 learning problems, and the average performance is presented in Fig. 1.

As can be seen in Fig. 1, after 5000 learning problems, Natural Fuzzy-XCS and FACLN eventually achieved average performance levels of 0.995 and 0.942 respectively. A Levene test between the two groups of experiment results presented in Fig. 1 gives a p-value of 0.00015, suggesting that we can discard the null hypothesis that both groups have equal variance. Therefore, after running a Welch t-test (assuming unequal variance) in between the performance results of Natural Fuzzy-XCS and FACLN, a p-value of 7.66×10^{-10} is obtained. In view of this, we believe that Natural Fuzzy-XCS can effective handle single-step learning problems and may achieve significantly higher performance than FACLN.

5.2   Experiments on the Cart-Pole Problem

The cart-pole (or inverted pendulum) is a multi-step continuous action problem. In this problem, a rigid pole is hinged onto a cart which can move in a single dimension along a track of fixed size. The goal of an learning agent is to learn the forces F to apply to the cart in order to balance the pole. At any time t, the agent has access to four input values: x_t, the position of the center of the cart; $\dot{x}_t$, the linear velocity of the cart; θ_t, the current radius angle of the pole; and $\dot{\theta}_t$, the current angular velocity of the pole. Given these inputs, the system applies a force F_t to the cart in the range $[-10, 10]$. The mechanics of the system using this force are defined below.

$$\ddot{x} = \frac{(F + ML\dot{\theta}^2 \sin\theta) - ML\ddot{\theta}\cos\theta}{m + M}$$

$$\ddot{\theta} = \frac{g\sin\theta - \cos\theta(F + ML\dot{\theta}^2 \sin\theta)}{\frac{4}{3}L - \frac{m\cos\theta^2}{m+M}}$$

In our implementation of the cart-pole problem, $M = 0.1\,$kg is the mass of the pole; $L = 0.5\,$m is half the length of the pole; $m = 1\,$kg is the mass of the cart; and $g = 9.81\frac{m}{s^2}$ is the gravitational acceleration. To compare the learning performance in the cart-pole problem, the average radius angles of the pole achieved by Natural Fuzzy-XCS, FACLN and Fuzzy Q-Learning has been presented in Fig. 2.

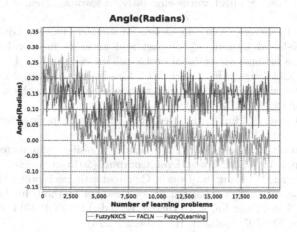

Fig. 2. The average performance over time for Natural Fuzzy-XCS, FACLN, and Fuzzy Q-Learning in the cart-pole problem.

Our aim in Fig. 2 is to bring the radius angle of the pole as close to 0 as possible. To understand the performance difference, an ANOVA test has been performed over the experiment results, giving a p-value of 0.00987. Tukey's post-hoc analysis further confirms that Natural Fuzzy-XCS can better balance the pole than both FACLN and Fuzzy Q-Learning. Similar results have also been observed in several other multi-step benchmark learning problems, including the puddle world problem and the mountain car problem [10]. Based on the promising results, we believe that Natural Fuzzy-XCS may potentially be useful for many practical reinforcement learning tasks.

6 Conclusions

Aimed at enabling multi-step reinforcement learning in completely continuous spaces, a new accuracy-based learning fuzzy classifier systems named Natural Fuzzy-XCS has been successfully developed in this paper. To achieve fine-grained learning, we have proposed to use a fuzzy logic system where the output action from the system is modelled through a continuous probability distribution. The learning performance is further improved by constantly updating the action output of each fuzzy rule with the help of a natural gradient learning technique.

Natural Fuzzy-XCS has been tested on several benchmark problems with successful results. Our experiments also showed that Natural Fuzzy-XCS can perform better than some recently developed learning systems.

References

1. Amari, S.: Natural gradient works efficiently in learning. Neural Comput. **10**(2), 251–276 (1998)
2. Berlanga, F.J., Rivera, A.J., del Jesus, M.J., Herrera, F.: Gp-coach: genetic programming-based learning of compact and accurate fuzzy rule-based classification systems for high-dimensional problems. Inf. Sci. **180**(8), 1183–1200 (2010)
3. Butz, M.V., Goldberg, D.E., Lanzi, P.L.: Gradient descent methods in learning classifier systems: improving XCS performance in multistep problems. IEEE Trans. Evol. Comput. **9**, 452–473 (2005)
4. Casillas, J., Carse, B., Bull, L.: Fuzzy-XCS: a michigan genetic fuzzy system. IEEE Trans. Fuzzy Syst. **15**, 536–550 (2007)
5. Chen, G., Douch, C., Zhang, M.: Using learning classifier systems to learn stochastic decision policies. IEEE Trans. Evol. Comput. (2015, to appear)
6. Chen, G., Zhang, M., Pang, S., Douch, C.: Stochastic decision making in learning classifier systems through a natural policy gradient method. In: Loo, C.K., Yap, K.S., Wong, K.W., Beng Jin, A.T., Huang, K. (eds.) ICONIP 2014, Part III. LNCS, vol. 8836, pp. 300–307. Springer, Heidelberg (2014)
7. Gu, D., Hu, H.: Accuracy based fuzzy q-learning for robot behaviours. In: Proceedings of 2004 IEEE International Conference on Fuzzy Systems (2004)
8. Holland, J.H.: Adaptation in Natural and Artificial Systems. University of Michigan Press, Ann Arbor (1975)
9. Peters, J., Schaal, S.: Natural actor-critic. Neurocomputing **71**, 1180–1190 (2008)
10. Sutton, R.: Generalization in reinforcement learning: successful examples using sparse coarse coding. Adv. Neural Inf. Process. Syst. **8**, 1038–1044 (1996)
11. Sutton, R.S., Barto, A.G.: Reinforcement Learning: An Introduction. MIT Press, Cambridge (1998)
12. Wang, X.S., Cheng, Y.H., Yi, J.Q.: A fuzzy actor-critic reinforcement learning network. Inf. Sci. **177**, 3764–3781 (2007)

A Malware Classification Method Based on Generic Malware Information

Jiyeon Choi[1,2], HeeSeok Kim[1,2], Jangwon Choi[1], and Jungsuk Song[1,2](✉)

[1] Korea Institute of Science and Technology Information, Daejeon, Korea
[2] Korea University of Science and Technology, Daejeon, Korea
{ji935,hs,jwchoi,song}@kisti.re.kr

Abstract. Since attackers easily have been making malware using dedicated malware generation tools, the number of malware is increasing rapidly. However, it is hard to analyze all malwares because of rise in high-volume of malwares. For this reason, many researchers have proposed the malware classification methods for classifying new and well-known types of malwares in order to focus on analyzing new malwares. The existing methods mostly try to find out good features which are used as a criterion of calculating a similarity between malwares for improving a classification accuracy. So, these methods extract the features including malicious behavior information by performing static and dynamic analysis, but analyzing many malwares itself spends too much time and efforts. In this paper, we propose a malware classification method for finding new types from large scale malwares using generic malware information. Proposed method can be used for a pre-step so as to help the existing methods reduce the spending time in analysis and classification for malwares. It improve the classificaion accuracy of malwares by using an imphash and proved a classification accuracy based on the imphash is more than 99 % while maintaining a low false positive rate.

Keywords: Malware · Malware Classification · Imphash

1 Introduction

With the rapid growth of the Internet, the number of cyber attacks whose purpose is threatening crucial computer systems is continually increasing and evolving. One of the serious security threats in the cyber space is a malware, i.e., malicious software, designed for damaging the target softwares or systems. Since attackers can easily create malwares using dedicated malware generation tools, according to a recent report, 387 malwares are made every minute [1]. Because of this, numerous antivirus companies are spending a lot of time analyzing a large number of malwares. However, it is hard to analyze all malwares due to an increase of high-volume malwares.

In order to solve this problem, many researchers have previously proposed the malware classification methods for finding new types from the large scale malwares. [2–9]. In general, the existing methods extract features including malicious

© Springer International Publishing Switzerland 2015
S. Arik et al. (Eds.): ICONIP 2015, Part II, LNCS 9490, pp. 329–336, 2015.
DOI: 10.1007/978-3-319-26535-3_38

behavior information before performing the malware classification, because the extracted features are then used for a criterion of computing a similarity between malwares. So, most of the malware classification methods carry out static and dynamic analysis for features extraction. Moreover, they mostly make an effort to find good features for improving a classification accuracy. But the existing methods have a weakness that analyzing many malwares, one by one, is required to spend too much time and efforts. To cope with this problem, our research focuses on finding new types of malware and helps the existing approaches reduce the spending time in analysis and classification for malwares.

In this paper, we propose a malware classification method which effectively identifies whether malware is a new type or not, by using generic malware information without detailed analysis. Generic malware information includes hash, antivirus detection results, and values about import address table (IAT) of malwares, these information also is simply obtained from web services. Especially in the proposed method, we use an imphash that is a hash of an import address table as the features for malware classification. The imphash can be an important value made by IAT including specific order of functions within the source file. In addition, our method classifies the large malicious softwares obtained from honeypot system which is a trap set to attract hackers for finding malwares used in real attack. In order to verify the effectiveness of the imphash used in the proposed method, we used 245 malwares collected from the honeypot of S&T-SEC Korea for experiment. We proved that the imphash accuracy is 99 % while maintaining a low false positive rate.

The rest of this paper is organized as follows. Section 2 briefly describes related works on malware classification. Section 3 introduces the proposed malware classification method in detail. Evaluation of the proposed method is shown in Sect. 4. Finally, we conclude and suggest for future research in Sect. 5.

2   Related Work

Many researchers have made an effort to find new types of malware for preventing a unknown attack. So, they have proposed a lot of methods for the malware classification [2–7]. Most of approaches to classify malwares have focused on seeking good features used as a criterion of calculating a similarity between malwares. For example, Kong et al. proposed a new framework for automated malware classification using function call graph. So, their framework extracted structural information based on function call graph of each malware program and evaluated the similarity between two malwares [8]. Besides, Park et al. presented a new malware classification method based on behavior graph that is extracted by capturing the system calls when malware is executed [9]. Then, they computed the similarity between behavior graph of different malwares.

These approaches have a limitation; in order to classify malwares using features extracted by malware analysis, analyzing high-volume malwares which is increasing steadily itself spends too much time and efforts. Therefore, we propose a malware classification method for finding new types from large scale malwares based on generic malware information, so as to help the existing approaches reduce the spending time in analysis and classification for malwares.

3   Proposed Method

3.1   Overall Architecture

This section presents a malware classification method to find the malwares of new type. Figure 1 shows a framework of the proposed method. In order to classify high-volume malwares into unknown or not, the proposed method consists of three steps: Hash value check, Antivirus check and Imphash check. Firstly, the purpose of the Hash value check step is to filter out already known malwares in high-volume malwares. If their hash values are same after comparing a hash about input to a hash of existing malwares, we treat the input is already analyzed as malicious software. The Antivirus check step secondly finds new types not analyzed from antivirus vendors. So, it confirms the antivirus check results of malwares which were decided as unknown types in the previous step and decides whether malwares are normal or not depending on the results detected by antivirus vendors. Finally, the Imphash check step classifies malwares that are received from the previous step using their imphash so that we find a similar group of the sample. In this step, we select malware groups or make new groups after comparing the imphash between malwares. With the help of these 3 steps, we reduce large volume of malwares to small scale of malwares which are assumed to be new types.

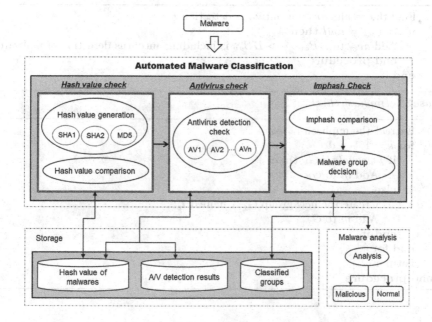

Fig. 1. Overall framework of the proposed method

3.2 Design of the Architecture

In this section, we describe the detailed procedure about the architecture for the proposed method. Algorithm 1 shows the process of our method. The proposed method consists of three phases as mentioned in the Sect. 3.1. Classifying malwares with Algorithm 1 is as follows:

Algorithm 1. Algorithm for malware classification

Input: malware m, Malware groups $G_j (0 \leq j \leq the\ number\ of\ group\ in\ DB_i)$
Output: Updated malware groups G_{j+1}

Phase 1 − Hash value check

1: **procedure** MALWARE CLASSIFICATION
2:     Extract the hash value h_m of input malware m
3:     **if** $h_m \in DB_h$ **then**
4:         **end procedure**
5:     **else**
6:         Add h_m to DB_h ▷ DB_h is database including hash values of malwares
7:     **end if**

Phase 2 − Antivirus check

8:     Find the results av_m for antivirus detection h_m
9:     **if** $AV_{check} \neq null$ **then**
10:         Add av_m to DB_{av} ▷ DB_{av} is including antivirus detection of malwares
11:         **end procedure**
12:     **end if**

Phase 3 − Imphash check

13:     Extract the imphash i_m of m
14:     **for** $k = 1 \rightarrow j$ **do**
15:         **if** $i_m\ equals\ to\ G_k$ **then**
16:             Add m to G_k
17:         **else**
18:             Create the new group G_{j+1} named i_m
19:             Add m to G_{j+1}
20:         **end if**
21:     **end for**
22:     **return** G_{j+1}
23: **end procedure**

---

1. Hash value check: This phase filters out already known malwares from high-volume malwares. For this phase, it extracts hash values about a input malware. These hash are values of MD5, SHA1 and SHA256. It can be also extracted by simply searching in the web services such as the virustotal and malwares.com. As shown in the Fig. 1, the proposed method stores information for the hash values of malwares, antivirus detection results and classified groups in databases. So, after extracting the hash values, it checks an existence of the extracted hash values in the database. And then, if the same value is existed, the process is terminated. Otherwise, the hash value of input malwares is added to the database.

2. Antivirus check: The Antivirus check is carried out in two steps. First, it searches antivirus detection results for the input of malware by using hash values gotten from the database including hash values of malwares. Checking the antivirus detection results can use API served from malwares.com or virustotal in order to receive automatically the detection results. Second step is to confirm the search results of the input. If it is detected as malware from at least one product, it assumes that the malware is already known malicious software. So it adds the antivirus detection results to the database of A/V detection results in Fig. 1 and stop the procedure. However, if not, it proceeds to the next phase for a group selection.

3. Imphash check: In the final phase, The Imphash check phase classifies the input which is supposed to unknown malware from the previous phases. In this paper, it use an imphash as the features for malware classification.

– An imphash is as follows: Mandiant, information security company, created a hash called an imphash (for import hash) in order to track threat groups. The imphash is a value obtained by hashing an import address table (IAT), which is filled with function imported from other files. The Imphash can be an important and unique value to identify similar malwares. Because when the IAT is generated and built by the compiler's linker, it includes particular order of primary functions within the source file. Figure 2 shows the example about comparison of source code.
  Each source code of case 1 and 2 looks very similar, but it's order is slightly different. When some order in the code of two cases is just changed, their

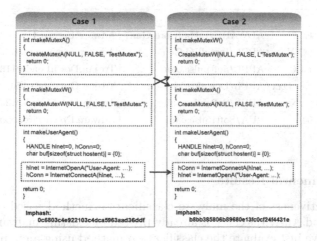

Fig. 2. Example of the source code

imphash are created as completely different value. If imphash values of some files are identical, it assumes that they are included in same group. In addition, not only websites like virustotal, malwares.com but also tool as viper, binary analysis tool serve the imphash by uploading files.

In the Imphash check phase, it uses the imphash as a name for each group. So, this phase first extracts the imphash of input by uploading malware to the websites and then performs to compare the imphash with names of other groups. If the imphash equals to any group, it can select that group. Otherwise, it makes a new group named imphash of the input and stores this information in the database of the Classified groups.

4   Evaluation

4.1   Experimental Environment

In order to verify the effectiveness about the classification by imphash used in the proposed method, we prepared 351 malwares collected from a honeypot of S&T-SEC Korea. For the experiment, we classified malwares using the analysis results by three antivirus venors: AVG, Avira and DrWeb. In general, malware names assigned by them have a form like hierarchy structure, i.e., a prefix, actual name of malware and suffix. If the malware names in antivirus products are all identical, we can regard them as a same group while malwares with different names in antivirus products can be regarded as separate groups. This is the reason why we adopt these malware names of antivirus products for the evaluation dataset. Table 1 shows the results for the number of malwares in each group and malware names by three products. We used the 245 classified malwares of 6 groups in the Table 1 for the experiment.

Table 1. The number of elements of malware groups

Group	AVG	Avira	DrWeb	$\#(M)$
G1	Generic.EC7	ADWARE/InstallCore.Gen7	Adware.Eorezo.414	75
G2	InstallIQ.CW	APPL/InstallIQ.Gen4	Adware.Downware.9715	61
G3	Generic_r.PQ	APPL/Softpulse.oanr	Trojan.DownLoader11.25503	48
G4	Downloader.Dxx	APPL/Downloader.Gen	Trojan.OutBrowse.54	25
G5	InstallIQ.CN	APPL/Softpulse.oanr	Trojan.DownLoader12.5732	22
G6	Generic.040	PUA/SoftPulse.oanr	Adware.Downware.xxxx	14

4.2   Experimental Results

Since an effectiveness of the malware classification method based on the Hash value check and Antivirus check is already proved by existing malware classification studies, we just evaluate the classification method using an imphash in this section. Firstly, we extracted the imphash for 245 malwares in the Table 1. As a result of the imphash extraction, Table 2 shows a representation of the imphash of each group.

Table 2. The imphash of malware groups

Group	Imphash
G1	884310b1928934402ea6fec1dbd3cf5e
G2	5755157383f7a8880e67d13d7c58f726
G3	0d6515bc121719663c20a7477edf2c90
G4	7fa974366048f9c551ef45714595665e
G5	4f40fa997203943468668e92edfd6750
G6	8c2086e82cad1736a562bb87ccb7b8df

Figure 3 shows the classification result of the malwares. Secondly, we classified malwares by applying our method to 245 malwares, in order to identify the accuracy for the proposed classification method.

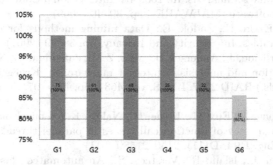

Fig. 3. Accuracy of the proposed classification method

As a result, we observed that 75 malwares of group 1 was classified (100 %) correctly because their imphash is equally '884310b1928934402ea6fec1dbd3cf5e'. Furthermore, we recognized that the group 2 to 5 also appeared the classification accuracy of 100 %. However, we can see that the accuracy of group 6 was 86 % and lower than other groups. In case of this group, 2 malwares which have other imphash value were confirmed as programs packed by a packer. This means that IAT is typically changed when packing the malwares. So, we can conclude that unless the malware is packed, the proposed method has 99 % accuracy and it is very effective to classify malwares.

5   Conclusion

In this paper, we have proposed a malware classification method using generic malware information. The main purpose of our classification method is to find new types of malware and concurrently reduce the spending time in classification perfomed by exsisting methods. The proposed method consists of three steps: the Hash value check, Antivirus check and imphash check. Especially, in the imphash check step, we classify malwares using the imphash as features used in malware classification. The experimental results showed that an accuracy of classification by imphash is 99 %.

However, the proposed method has a weakness for the packed malwares whose import address table changed by various types of packers. Because of this, in the future work, we need to focus on devising an enhanced classification method for finding out new types of malwares including packed malwares.

References

1. McAfee Labs Threats Report: February 2015. http://www.mcafee.com/us/resources/reports/rp-quarterly-threat-q4-2014.pdf
2. Leder, F., Steinbock, B., Martini, P.: Classification and detection of metamorphic malware using value set analysis. In: 2009 4th International Conference on Malicious and Unwanted Software (MALWARE), pp. 39–46 (2009)
3. Schultz, M.G., Eskin, E., Zadok, E.: Data mining methods for detection of new malicious executables. In: Security and Privacy, pp. 38–49 (2001)
4. Bailey, M., Oberheide, J., Andersen, J., Mao, Z.M., Jahanian, F., Nazario, J.: Automated classification and analysis of internet malware. In: Kruegel, C., Lippmann, R., Clark, A. (eds.) RAID 2007. LNCS, vol. 4637, pp. 178–197. Springer, Heidelberg (2007)
5. Nakazato, J., Song, J., Eto, M., Inoue, D., Nakao, K.: A novel malware clustering method using frequency of function call traces in parallel threads. Inst. Electron. Inf. Commun. Eng. E94–D(11), 2150–2158 (2011)
6. Tian, R., Batten, L., Islam, R., Versteeg, S.: An automated classification system based on the strings of trojan and virus families. In: 4th International Conference on Malicious and Unwanted Software, 2009. IEEE, pp. 23–30 (2009)
7. Iwamoto, K., Wasaki, K.: Malware classification based on extracted API sequences using static analysis. In: 12th Proceedings of the Asian Internet Engineering Conference, pp. 31–38 (2012)
8. Kong, D., Yan, G.: Discriminant malware distance learning on structural information for automated malware classification. In: ACM SIGMETRICS/International Conference on Measurement and Modeling of Computer Systems, pp. 347–348 (2013)
9. Park, Y., Reeves, D., Mulukutla, V., Sundaravel, B.: Fast malware classification by automated behavioral graph matching. In: 6th Annual Workshop on Cyber Security and Information Intelligence Research, Article no. 45 (2010)

Secure Multi-party Computation Based Privacy Preserving Extreme Learning Machine Algorithm Over Vertically Distributed Data

Ferhat Özgür Çatak[✉]

TÜBİTAK BİLGEM, Cyber Security Institute, Gebze, Kocaeli, Turkey
ozgur.catak@tubitak.gov.tr

Abstract. Especially in the Big Data era, the usage of different classification methods is increasing day by day. The success of these classification methods depends on the effectiveness of learning methods. Extreme learning machine (ELM) classification algorithm is a relatively new learning method built on feed-forward neural-network. ELM classification algorithm is a simple and fast method that can create a model from high-dimensional data sets. Traditional ELM learning algorithm implicitly assumes complete access to whole data set. This is a major privacy concern in most of cases. Sharing of private data (i.e. medical records) is prevented because of security concerns. In this research, we propose an efficient and secure privacy-preserving learning algorithm for ELM classification over data that is vertically partitioned among several parties. The new learning method preserves the privacy on numerical attributes, builds a classification model without sharing private data without disclosing the data of each party to others.

Keywords: Extreme learning machine · Privacy preserving data analysis · Secure multi-party computation

1 Introduction

The main purpose of machine learning can be expressed as to find the patterns and summarize the data form of high-dimensional data sets. The classification algorithms [1,2] is one of the most widely used method of machine learning in real-life problems. Data sets used in real-life problems are high dimensional as a result, analysis of them is a complicated process.

Extreme Learning Machine (ELM) was proposed by [3] based on generalized Single-hidden Layer Feed-forward Networks (SLFNs). Main characteristics of ELM are small training time compared to traditional gradient-based learning methods, high generalization property on predicting unseen examples with multi-class labels and parameter free with randomly generated hidden nodes.

Background knowledge attack uses quasi-identifier attributes of a dataset and reduce the possible values of output sensitive information. A well-known example

© Springer International Publishing Switzerland 2015
S. Arik et al. (Eds.): ICONIP 2015, Part II, LNCS 9490, pp. 337–345, 2015.
DOI: 10.1007/978-3-319-26535-3_39

about background knowledge attack is the personal health information about of Massachusetts governor William Weld using a anonymized data set [4]. In order to overcome these types of attacks, various anonymization methods have been developed like k-anonymity [4], l-diversity [5], t-closeness [6].

Although the anonymization methods are applied to data sets to protect sensitive data, though, the sensitive data is still accessed by an attacker in various ways [7]. Also, data anonymization methods are not applicable in some cases. In another scenario, consider the situation when two or more hospitals wants to analyze patient data [8] through collaborative processes that require using each other's databases. In such cases, it is necessary to find a secure training method that can run jointly on private union databases, without revealing or pooling their sensitive data. Privacy-preserving ELM learning systems are one of the methods that the only information learned by the different parties is the output model of learning method.

In this research, we propose a privacy-preserving ELM training model that constructs the global ELM classification model from the distributed data sets in multiple parties. The training data set is vertically partitioned among the parties, and the final distributed model is constructed at an independent party to securely predict the correct label for the new input data.

The content of this paper is as follows: Related work is reviewed in Sect. 2. In Sect. 3, ELM, secure multi-party computation and the secure addition are explained. In Sect. 4, our new privacy-preserving ELM for vertically partitioned data is proposed. Section 5 emprically shows the timing results of our method with different public data sets.

2   Related Works

In this section, we review the existing works that have been developed for different machine learning methods. The major differences between our learning model and existing work are highlighted.

Recently, there has been significant contributions in privacy-preserving machine learning. Secretans et al. [9] presents a probabilistic neural network (PNN) model. The PNN is an approximation of the theoretically optimal classifier, known as the Bayesian optimal classifier. There are at least three parties involved in the computation of the secure matrix summation to add the partial class conditional probability vectors together. Aggarwal et al. [10] developed condensation based learning method. They show that an anonymized data closely matches the characteristics of the original data. Samet et al. [11] present new privacy-preserving protocols for both the back-propagation and ELM algorithms among several parties. The protocols are presented for perceptron learning algorithm and applied only single layer models. Oliveria et al. [12] proposed methods distort confidential numerical features to protect privacy for clustering analysis. Guang et al. [13] proposed a privacy-preserving back-propagation algorithm for horizontally partitioned databases for multi-party case. They use secure sum in their protocols. Yu et al. [14] proposed a privacy-preserving solution for support

vector machine classification. Their approach constructs the global SVM classification model from the data distributed at multiple parties, without disclosing
the data of each party to others.

3 Preliminaries

In this section, we introduce preliminary knowledge of ELM, secure multi-party
computation and secure addition briefly.

3.1 Extreme Learning Machine

ELM was originally proposed for the single-hidden layer feed-forward neural
networks [3, 15, 16]. Then, ELM was extended to the generalized single-hidden
layer feed-forward networks where the hidden layer may not be neuron like [17,
18]. Main advantages of ELM classification algorithm is that ELM can be trained
hundred times faster than traditional neural network or support vector machine
algorithm since its input weights and hidden node biases are randomly created
and output layer weights can be analytically calculated by using a least-squares
method [19, 20]. The most noticeable feature of ELM is that its hidden layer
parameters are selected randomly.

Given a set of training data $\mathcal{D} = \{(\mathbf{x}_i, y_i) \mid i = 1, ..., n\}, \mathbf{x}_i \in \mathbb{R}^p, y_i \in \{1, 2, ..., K\}\}$ sampled independently and identically distributed (i.i.d.) from
some unknown distribution. The goal of a neural network is to learn a func-
tion $f : \mathcal{X} \to \mathcal{Y}$ where $\mathcal{X}$ is instances and $\mathcal{Y}$ is the set of all possible labels. The
output label of an single hidden-layer feed-forward neural networks (SLFNs)
with N hidden nodes can be described as

$$f_N(\mathbf{x}) = \sum_{i=1}^{N} \beta_i G(\mathbf{a}_i, b_i, \mathbf{x}), \ \mathbf{x} \in \mathbb{R}^n, \ \mathbf{a}_i \in \mathbb{R}^n \qquad (1)$$

where $\mathbf{a}_i$ and b_i are the learning parameters of hidden nodes and β_i is the weight
connecting the ith hidden node to the output node. The output function of ELM
for generalized SLFNs can be identified by

$$f_N(\mathbf{x}) = \sum_{i=1}^{N} \beta_i G(\mathbf{a}_i, b_i, \mathbf{x}) = \beta \times h(\mathbf{x}) \qquad (2)$$

For the binary classification applications, the decision function of ELM becomes

$$f_N(\mathbf{x}) = sign\left(\sum_{i=1}^{N} \beta_i G(\mathbf{a}_i, b_i, \mathbf{x})\right) = sign\left(\beta \times h(\mathbf{x})\right) \qquad (3)$$

Equation 2 can be written in another form as

$$\mathbf{H}\beta = \mathbf{T} \qquad (4)$$

H and T are respectively hidden layer matrix and output matrix.

$$\beta = \mathbf{H}^{\dagger}\mathbf{T} \tag{5}$$

$\mathbf{H}^{\dagger}$ is the *Moore-Penrose generalized inverse of matrix* $\mathbf{H}$. Hidden layer matrix can be described as

$$H(\tilde{a}, \tilde{b}, \tilde{x}) = \begin{bmatrix} G(a_1, b_1, x_1) & \cdots & G(a_L, b_L, x_1) \\ \vdots & \ddots & \vdots \\ G(a_1, b_1, x_N) & \cdots & G(a_L, b_L, x_N) \end{bmatrix}_{N \times L} \tag{6}$$

where $\tilde{a} = a_1, ..., a_L$, $\tilde{b} = b_1, ..., b_L$, $\tilde{x} = x_1, ..., x_N$. Output matrix can be described as

$$T = [t_1 \dots t_N]^T \tag{7}$$

The hidden nodes of SLFNs can be randomly generated. They can be independent of the training data.

3.2   Secure Multi Party Computation

In vertically partitioned data, each party holds different attributes of same data set. Let's have n input instances, $\mathcal{D} = \{(\mathbf{x}_i, y_i) \mid \mathbf{x}_i \in \mathbb{R}^p, y_i \in \mathbb{R}\}_{i=1}^n$. The partition strategy is shown in Fig. 1.

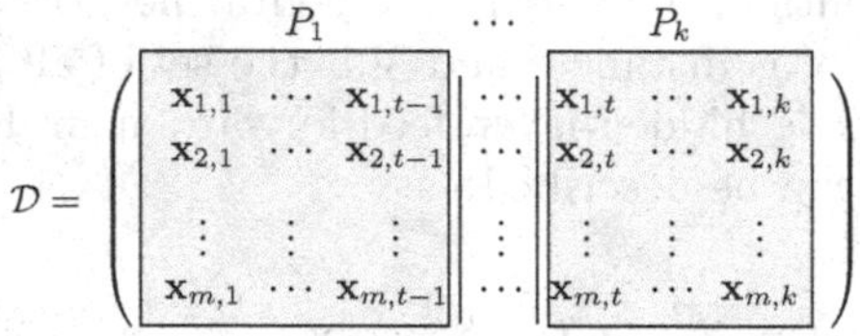

$$\mathcal{D} = \begin{pmatrix} \begin{bmatrix} \mathbf{x}_{1,1} & \cdots & \mathbf{x}_{1,t-1} \\ \mathbf{x}_{2,1} & \cdots & \mathbf{x}_{2,t-1} \\ \vdots & \vdots & \vdots \\ \mathbf{x}_{m,1} & \cdots & \mathbf{x}_{m,t-1} \end{bmatrix} & \cdots & \begin{bmatrix} \mathbf{x}_{1,t} & \cdots & \mathbf{x}_{1,k} \\ \mathbf{x}_{2,t} & \cdots & \mathbf{x}_{2,k} \\ \vdots & \vdots & \vdots \\ \mathbf{x}_{m,t} & \cdots & \mathbf{x}_{m,k} \end{bmatrix} \end{pmatrix}$$

Fig. 1. Vertically partitioned data set $\mathcal{D}$.

Secure Multi-party Addition. In secure multi-party addition (SMA), each party, P_i, has a private local value, x_i. At the end of the computation, we obtain the sum, $x = \sum_{i=0}^{k-1}$. For this works, we applied the Yu et al. [21] secure addition procedure. Their approach is a generalization of the existing works [22] that uses secure communication and trusted party. Canonical order based, $P_0, \cdots, P_{k-1}$, protocol is applied. The SMA method is show in Algorithm 1. This protocol calculates the required sum in secure manner.

4   Privacy-Preserving ELM Over Vertically Partitioned Data

The data set that one wants to find a classifier for consists of m instances in n-dimensional space is shown with $\mathcal{D} \in \mathbb{R}^{m \times n}$. Each instances of the data set

Algorithm 1. Secure multi-party addition

1: **procedure** SMA(**P**)
2:     $P_0 : R \leftarrow rand(\mathcal{F})$ ▷ P_0 randomly chooses a number R
3:     $V \leftarrow R + x_0 \mod \mathcal{F}$
4:     P_0 sends V to node P_1
5:     **for** $i = 1, \cdots, k - 1$ **do**
6:         P_i receives $V = R + \sum_{j=0}^{i-1} x_j \mod \mathcal{F}$
7:         P_i computes $V = \left(R + \sum_{j=1}^{i} x_j \mod \mathcal{F}\right) = ((x_i + V) \mod \mathcal{F})$
8:         P_i sends V to node P_{i+1}
9:     **end for**
10:    $P_0 : V \leftarrow (V - R) = (V - R \mod \mathcal{F})$ ▷ Actual addition result
11: **end procedure**

has values for n features. Matrix $\mathcal{D}$ is vertically partitioned into i parties of $P_0, P_1, \cdots, P_i$ and each features of instances owned by a party that is private shown as Fig. 1 illustrates. As shown in Eq. 6, at second stage of ELM learning, hidden layer output matrix, **H** is calculated using randomly assigned hidden node parameters **w**, **b** and β. ELM calculates the matrix **H** and output weight vector, β, is obtained by multiplying **H** and **T**. Each member of **H** is computed with an activation function g such that $G(\mathbf{w}_i, \mathbf{x}_i, b_i) = g(\mathbf{x}_i \cdot \mathbf{w}_i + b_i)$ for sigmoid or $G(\mathbf{w}_i, \mathbf{x}_i, b_i) = g(b_i - ||\mathbf{x}_i - \mathbf{w}_i||)$ for radial based functions. An $(i, j)^{th}$ element of **H** is

$$G(\mathbf{w}_i, \mathbf{x}_i, b_i) = sign(\mathbf{x}_i \cdot \mathbf{w}_i + b_i) \tag{8}$$

where $\mathbf{x}_i$ is the ith instances of data set and $\mathbf{w}_i$ is hidden node input weight of ith instance and $\mathbf{x}_i, \mathbf{w}_i \in \mathbb{R}^n$.

Let $\mathbf{x}_i^1, \cdots \mathbf{x}_i^k$ be vertically partitioned vectors of input instance $\mathbf{x}_i$ and $\mathbf{w}_j^1, \cdots \mathbf{w}_j^k$ be vertically partitioned vectors of jth hidden node input weight $\mathbf{w}_j, b_j^0, \cdots, b_j^k$ be jth node input bias over k different parties. Then the output of jth input node with ith instance of input data set using k sites is

$$sign(\mathbf{x}_i \cdot \mathbf{w}_j + b_j) = sign\left((\mathbf{x}_i^0 \cdot \mathbf{w}_j^0 + b_j^0) + \cdots + (\mathbf{x}_j^k \cdot \mathbf{w}_j^k + b_j^k)\right) \tag{9}$$

From Eq. 9, calculation of hidden layer output matrix, **H** can be decomposed into k different parties using secure sum of matrices, such that

$$\mathbf{H} = sign\left(\mathbf{T}_1 + \cdots + \mathbf{T}_k\right) \tag{10}$$

where

$$\mathbf{T}_i = \begin{bmatrix} \left(\mathbf{x}_1^i \cdot \mathbf{w}_1^i + b_1^i\right) & \cdots & \left(\mathbf{x}_1^i \cdot \mathbf{w}_L^i + b_L^i\right) \\ \vdots & \ddots & \vdots \\ \left(\mathbf{x}_N^i \cdot \mathbf{w}_1^i + b_1^i\right) & \cdots & \left(\mathbf{x}_L^i \cdot \mathbf{w}_N^i + b_N^i\right) \end{bmatrix}_{N \times L} \tag{11}$$

Privacy Preserving ELM Algorithm. Let $\mathcal{D} \in \mathbb{R}^{M \times N}$, and the number of input layer size be L, the number of parties be k, then the our training model becomes:

1. Master party creates weight matrix, $\mathbf{W} \in \mathbb{R}^{L \times N}$.
2. Master party distributes partition $\mathbf{W}$ with same feature size for each parties.
3. Party P_0 creates a random matrix, $\mathbf{R} = \begin{bmatrix} rand_{1,1}(\mathcal{F}) & \cdots & rand_{1,L}(\mathcal{F}) \\ \vdots & \ddots & \vdots \\ rand_{N,1}(\mathcal{F}) & \cdots & rand_{N,L}(\mathcal{F}) \end{bmatrix}_{N \times L}$.
4. Party P_0 creates perturbated output, $\mathbf{V} = \mathbf{R} + \begin{bmatrix} (\mathbf{x}_1^0 \cdot \mathbf{w}_1^0 + b_1^0) & \cdots & (\mathbf{x}_1^0 \cdot \mathbf{w}_L^0 + b_L^0) \\ \vdots & \ddots & \vdots \\ (\mathbf{x}_N^0 \cdot \mathbf{w}_1^0 + b_1^0) & \cdots & (\mathbf{x}_N^0 \cdot \mathbf{w}_L^0 + b_L^0) \end{bmatrix}$.
5. for $i = 1, \cdots, k-1$

   – P_i computes $\mathbf{V} = \mathbf{V} + \begin{bmatrix} (\mathbf{x}_1^i \cdot \mathbf{w}_1^i + b_1^i) & \cdots & (\mathbf{x}_1^i \cdot \mathbf{w}_L^i + b_L^i) \\ \vdots & \ddots & \vdots \\ (\mathbf{x}_N^i \cdot \mathbf{w}_1^i + b_1^i) & \cdots & (\mathbf{x}_L^i \cdot \mathbf{w}_N^i + b_N^i) \end{bmatrix}$

   – P_i sends $\mathbf{V}$ to P_{i+1}.
6. P_0 subtracts random matrix, $\mathbf{R}$, from the received matrix $\mathbf{V}$. $\mathbf{H} = (\mathbf{V} - \mathbf{R})$ mod $\mathcal{F}$.
7. Hidden layer node weight vector, β, is calculated. $\beta = \mathbf{H}^\dagger \cdot \mathbf{T}$.

5 Experiments

In this section, we perform experiments on real-world data sets from the public available data set repositories. Public data sets are used to evaluate the proposed learning method. Classification models of each data set are compared for accuracy results without using secure multi-party computation.

Experimental Setup: In this section, our approach is applied to six different data sets to verify model affectivity and efficiency. The data sets are summarized in Table 1, including australian, colon-cancer, diabetes, duke, heart, ionosphere.

For each data set in Table 1, we vary number of party size, k from 2 to number of feature, n, of the data set. For instance, when our party size is three, $k = 3$, and attribute size fourteen, $n = 14$, then the first two party have 5 attributes, and last party has 4 attributes.

Simulation Results: The accuracy of secure multi-party computation based ELM is exactly same for the traditional ELM training algorithm. Figure 2 shows

Table 1. Description of the testing data sets used in the experiments.

Data set	#Train	#Classes	#Attributes
Australian [23]	690	2	14
Colon-cancer [24]	62	2	2,000
Diabetes [25]	768	2	8
Duke breast cancer [26]	44	2	7,129
Heart [27]	270	2	13
Ionosphere [28]	351	2	34

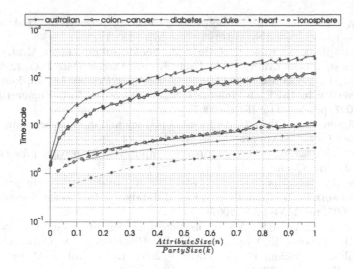

Fig. 2. Vertically partitioned data set $\mathcal{D}$.

results of our simulations. As shown in figure, time scale becomes its steady state position when number of parties, k, moves closer to number of attributes, k.

6 Conclusion and Future Works

ELM learning algorithm, a new method compared to other classification algorithms. ELM outperforms traditional Single Layer Feed-forward Neural-networks and Support Vector Machines for big data [29]. The ELM is applied in many fields. Almost, in all fields that ELM is applied (i.e. medical records, business, government), privacy is a major concern.

A new privacy-preserving learning model is proposed for ELM in vertically partitioned data in multi-party partitioning without sharing the data of each site to the others. In order to save the privacy of input data set, master party divides weight vector, and each party calculates the activation function result with its data and weight vector. Extending the privacy-preserving ELM to horizontally distributed data set is a future work for this approach.

References

1. Anderson, J.R., Michalski, R.S., Carbonell, J.G., Mitchell, T.M.: Machine Learning: An Artificial Intelligence Approach, vol. 2. Morgan Kaufmann, San Mateo (1986)
2. Ramakrishnan, R., Gehrke, J.: Database Management Systems. Osborne/McGraw-Hill, Berkeley (2000)
3. Huang, G.B., Zhu, Q.Y., Siew, C.K.: Extreme learning machine: theory and applications. Neurocomputing **70**(13), 489–501 (2006)

4. Sweeney, L.: k-anonymity: A model for protecting privacy. Int. J. Uncertainty Fuzziness Knowl. Based Syst. **10**(05), 557–570 (2002)
5. Machanavajjhala, A., Kifer, D., Gehrke, J., Venkitasubramaniam, M.: L-diversity: privacy beyond k-anonymity. ACM Trans. Knowl. Discov. Data **1**(1), 42–93 (2007)
6. Li, N., Li, T., Venkatasubramanian, S.: t-closeness: Privacy beyond k-anonymity and l-diversity. In: IEEE 23rd International Conference on Data Engineering, 2007, ICDE 2007, pp. 106–115. IEEE (2007)
7. Ji, Z., Lipton, Z.C., Elkan, C.: Differential privacy and machine learning: a survey and review (2014). arXiv preprint arXiv:1412.7584
8. Lindell, Y., Pinkas, B.: Secure multiparty computation for privacy-preserving data mining. J. Priv. Confidentiality **1**(1), 5 (2009)
9. Secretan, J., Georgiopoulos, M., Castro, J.: A privacy preserving probabilistic neural network for horizontally partitioned databases. In: Neural Networks, 2007, IJCNN 2007, pp. 1554–1559 (2007)
10. Aggarwal, C.C., Yu, P.S.: A condensation approach to privacy preserving data mining. In: Bertino, E., Christodoulakis, S., Plexousakis, D., Christophides, V., Koubarakis, M., Böhm, K. (eds.) EDBT 2004. LNCS, vol. 2992, pp. 183–199. Springer, Heidelberg (2004)
11. Samet, S., Miri, A.: Privacy-preserving back-propagation and extreme learning machine algorithms. Data Knowl. Eng. **79–80**, 40–61 (2012)
12. Oliveira, S.R., Zaiane, O.R.: Privacy preserving clustering by data transformation. J. Inf. Data Manag. **1**(1), 37 (2010)
13. Guang, L., Ya-Dong, W., Xiao-Hong, S.: A privacy preserving neural network learning algorithm for horizontally partitioned databases. Inform. Technol. J. **9**, 1–10 (2009)
14. Yu, H., Jiang, X., Vaidya, J.: Privacy-preserving svm using nonlinear kernels on horizontally partitioned data. In: Proceedings of the 2006 ACM Symposium on Applied Computing, SAC 2006, pp. 603–610. ACM, New York (2006)
15. bin Huang, G., yu Zhu, Q., kheong Siew, C.: Extreme learning machine: a new learning scheme of feedforward neural networks. In: Proceedings International Joint Conference Neural Networks, pp. 985–990 (2006)
16. Huang, G.B., Chen, L., Siew, C.K.: Universal approximation using incremental constructive feedforward networks with random hidden nodes. IEEE Trans. Neural Netw. **17**(4), 879–892 (2006)
17. Huang, G.B., Chen, L.: Convex incremental extreme learning machine. Neurocomputing **70**(1618), 3056–3062 (2007)
18. Huang, G.B., Chen, L.: Enhanced random search based incremental extreme learning machine. Neurocomputing **71**(1618), 3460–3468 (2008)
19. Tang, J., Deng, C., Huang, G.B., Zhao, B.: Compressed-domain ship detection on spaceborne optical image using deep neural network and extreme learning machine. IEEE Trans. Geosci. Remote Sens. **53**(3), 1174–1185 (2015)
20. Huang, G.B., Li, M.B., Chen, L., Siew, C.K.: Incremental extreme learning machine with fully complex hidden nodes. Neurocomputing **71**(46), 576–583 (2008)
21. Yu, H., Vaidya, J., Jiang, X.: Privacy-preserving SVM classification on vertically partitioned data. In: Ng, W.-K., Kitsuregawa, M., Li, J., Chang, K. (eds.) PAKDD 2006. LNCS (LNAI), vol. 3918, pp. 647–656. Springer, Heidelberg (2006)
22. Sweeney, L., Shamos, M.: Multiparty computation for randomly ordering players and making random selections. Technical report, Carnegie Mellon University (2004)
23. Quinlan, J.R.: Simplifying decision trees. Int. J. Man Mach. Stud. **27**(3), 221–234 (1987)

24. Alon, U., Barkai, N., Notterman, D.A., Gish, K., Ybarra, S., Mack, D., Levine, A.J.: Broad patterns of gene expression revealed by clustering analysis of tumor and normal colon tissues probed by oligonucleotide arrays. Proc. Natl. Acad. Sci. **96**(12), 6745–6750 (1999)
25. Smith, J.W., Everhart, J., Dickson, W., Knowler, W., Johannes, R.: Using the adap learning algorithm to forecast the onset of diabetes mellitus. In: Proceedings of the Annual Symposium on Computer Application in Medical Care, American Medical Informatics Association, pp. 261–265 (1988)
26. West, M., Blanchette, C., Dressman, H., Huang, E., Ishida, S., Spang, R., Zuzan, H., Olson, J.A., Marks, J.R., Nevins, J.R.: Predicting the clinical status of human breast cancer by using gene expression profiles. Proc. Natl. Acad. Sci. **98**(20), 11462–11467 (2001)
27. UCI: Statlog (heart) data set (2015). https://archive.ics.uci.edu/ml/datasets/Statlog+(Heart)
28. Sigillito, V.G., Wing, S.P., Hutton, L.V., Baker, K.B.: Classification of radar returns from the ionosphere using neural networks. Johns Hopkins APL Tech. Digest **10**, 262–266 (1989)
29. Cambria, E., Huang, G.B., et al.: Extreme learning machines [trends controversies]. Intell. Syst. IEEE **28**(6), 30–59 (2013)

A Hybrid Model of Fuzzy ARTMAP
and the Genetic Algorithm
for Data Classification

Manjeevan Seera[1]([✉]), Wei Shiung Liew[2], and Chu Kiong Loo[2]

[1] Faculty of Engineering, Computing and Science,
Swinburne University of Technology, Kuching, Sarawak, Malaysia
mseera@gmail.com
[2] Faculty of Computer Science and Information Technology, University of Malaya,
Kuala Lumpur, Malaysia
liew.wei.shiung@gmail.com, ckloo.um@um.edu.my

Abstract. A framework for optimizing Fuzzy ARTMAP (FAM) neural networks using Genetic Algorithms (GAs) is proposed in this paper. A number of variables were identified for optimization, which include the presentation order of training data during the learning step, the feature subset selection of the training data, and the internal parameters of the FAM such as baseline vigilance and match tracking. A single configuration of all three variables were encoded as a chromosome string and evaluated by creating and training the FAM according to the variables. The fitness of the chromosome is determined by the final classification accuracy of the FAM. Evaluation on benchmark data sets are conducted with the results compared with literature. Experimental results indicate the effectiveness of the proposed framework in undertaking data classification tasks.

1 Introduction

The fuzzy ARTMAP (FAM) neural network is able to learn a pattern classification task relatively quickly while incorporating new information into its existing knowledge base without the need to retrain the previously learned information. Classification accuracy of FAM is dependent on two things: parameter settings and order of the training data. Getting the best possible combination in order to achieve the highest possible accuracy is essentially an optimization problem.

Genetic Algorithms (GAs) are typically used for their capability to rapidly converge to optimum solutions. Every potential solution is encoded as a chromosome, which consists of all the parameters to be optimized. Every chromosome is then assessed using a fitness function to determine its effectiveness. Over multiple iterations, the competitive eliminations will discard chromosomes with sub-par combinations. This leaves only the fittest survivors, with the best combination of parameter settings. The genetic reproduction then creates variants of the survivors for another round of evaluation and eliminations.

© Springer International Publishing Switzerland 2015
S. Arik et al. (Eds.): ICONIP 2015, Part II, LNCS 9490, pp. 346–353, 2015.
DOI: 10.1007/978-3-319-26535-3_40

This paper is organized as follow. Related works are presented in Sect. 2. The optimization methodology is detailed in Sect. 3. Experiments using benchmark data is analyzed and discussed in Sect. 4. Concluding remarks are finally given in Sect. 5.

2   Related Works

The related works are presented in this section, with a review on multi-objective and hybrid GAs. In optimizing a neural network structure for dynamic system modeling, the multi-objective (MO) GA is detailed [1]. The goal was to meet the objectives with good accuracy, while having a minimum model structure. Results from simulations indicate the algorithm is able to correctly identify the examples [1]. A MO hybrid GA (MO-HGA) using variable neighborhood descent algorithm for local search is detailed in [2]. MO-HGA is applied for scheduling problem of the thin-film transistor-liquid crystal display module. The proposed approach acquired good results, when compared with conventional approaches [2].

In designing electrical distribution networks, a MO GA approach is used [3]. The set objectives are system failure index and monetary cost index. Information acquired from the Pareto-optimal solution is helpful in the decision-making process of the distribution network evolution planning [3]. A MO method is proposed for sizing of distributed generation resources to existing distribution networks [4]. The GA-based method allows the planner to decide the best compromise between network upgrading, cost of power loss, energy required, and energy not supplied. The proposed method is proven based on a number of given examples [4].

A hybrid Taguchi-GA model is proposed in [5] to solve global numerical optimization problems with continuous variables. The model is used to solve a number of benchmark problems with large dimensions with large numbers of local minima. Results from the experiments show the proposed model is much more robust than algorithms in literature [5]. In improving search efficiency, a stochastic GA is presented in [6]. The search space is dynamically divided to various sections by employing a novel stochastic coding strategy. Experiments on various complexities and test functions indicate the model is able to acquire near-optimal solution in most cases [6].

3   Optimization Methodology

In this section, the proposed methodology using GAs to optimize a number of variables is detailed. These variables were identified to have an effect on performance of FAM in learning and classifying the data. The overall process flow is shown in Fig. 1.

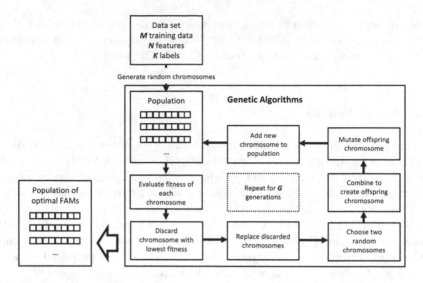

Fig. 1. Flowchart of the methodology for optimizing Fuzzy ARTMAP parameters

3.1 Fuzzy ARTMAP

The FAM learns using supervised learning. A single training pattern is represented as a vector of numerical features, which is given a group label. In the supervised learning process, a number of labelled patterns are given to FAM for training. This creates an association between the patterns and the labels, by adjusting the weights of the nodes in the mapping fields within the FAM. However when FAM receives an unkown pattern as an input for classification, FAM selects the label with the strongest association. More details on FAM can be found in [7].

FAM however is not without its downsides. One of the issues is known as category proliferation. This issue has been tackled in a number of ways, which includes a modified neural network architecture to mitigate or bypass entirely the category proliferation problem. As an example, the Gaussian ARTMAP [8] exhibits noise resistance traits, while the Biased ARTMAP [9] modifies the supervised learning process.

The category proliferation issue in FAM can be tackled by manipulating the order of training patterns that are being presented during training. A min-max clustering method is used by Dagher et al. [10] to identify the training order that would result in a FAM with the best classification accuracy. In finding the exact sequence of training data, especially if the data set is large, is a time-consuming process. As such, the GA was proposed for a more efficient searching process.

3.2 FAM Optimization Using Genetic Algorithms

GAs imitates the principles of natural evolution in seeking for optimum solutions, in a given problem. A popular application for GAs is the multi-parameter optimization, in which each parameter is represented as a single gene. A chromosome consisting of a number of genes incorporates a specific combination of parameter settings. Effectiveness of the chromosome is tested by implementing parameters defined in the genes, as in context of the problem. When provided with a population of chromosomes, high-fitness solutions are kept to generate new variants for the next round of competitive eliminations, while low-fitness solutions are discarded. Finally at the end of the optimization sequence, the population contains the best-performing combination of parameter settings for high-accuracy FAMs.

The process of the optimization parameters being encoded as a single chromosome string is shown in Fig. 2. The training data sequence is encoded as a string of progressive integers from 1 to N, where N is the number of instances from the data set to be used for training the FAM. Then, feature selection is encoded as a binary string to denote inclusion or exclusion of a particular feature during FAM training. The final FAM parameters then consists of *(i)* baseline vigilance, *(ii)* learning rate, *(iii)* ARTMAP parameters: baseline vigilance, choice parameter, match tracking parameter, and learning rate.

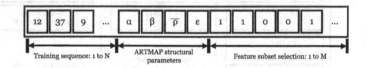

Fig. 2. Structure of a single chromosome, encoding parameters for optimizing performance of a single FAM neural network

Steps of optimizing FAM is listed as follows. Initially, a population of randomly generated chromosomes was created. A single chromosome was parsed to determine the three main parameters: sequence of training data to be presented, features to include/exclude from training, and the ARTMAP parameters. The initial FAM was constructed using the ARTMAP parameters. A copy of the data set was rearranged and filtered according to the training sequence and excluded features. The FAM was then trained using the rearranged data subset using the ten-fold cross-validation method. The data subset was divided equally into ten sub-subsets. Nine sub-subsets were used for training the FAM while the remaining was used for testing. This was performed ten times, each time using a different test set. The overall effectiveness or fitness of the chromosome was determined by the average classification accuracy of the testing data. Of all the chromosomes in the population, the ones with the lowest fitness were discarded. For each chromosome discarded this way, two chromosomes remaining in the population were selected at random to create an offspring chromosome. The reproduction and mutation process are as shown in Fig. 3.

− The training sequences of both chromosomes were compared for any common sequences, which would be passed down to the offspring. The remaining were randomized sequences.

- The feature selection segments in binary were compared. Features that were selected or deselected in both parent chromosomes were likewise selected or deselected in the offspring. Otherwise, the selection/deselection of a feature in the offspring was performed at random.
- The ARTMAP parameters in the offspring were computed by averaging the parameters of both parent chromosomes.
- After successfully creating the offspring chromosome, mutation was introduced according to the pre-determined genetic mutation parameter. For example, a genetic mutation rate of 0.10 in a chromosome with 100 elements would mutate a total of ten elements. For the training sequence, mutation involved swapping the positions of two elements in the sequence. For feature selection, mutation involved bit-flipping 0 s to 1 s and vice versa. Mutating the ARTMAP parameters involved adding or subtracting a small fraction.

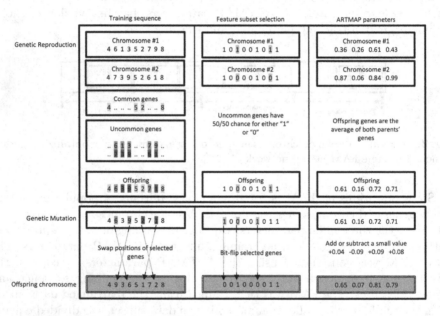

Fig. 3. Reproduction by combining the common traits of two parent chromosomes. Mutation was introduced to prevent stagnation.

This process is repeated for a set number of generations or until a stopping criterion is achieved. The final population of chromosomes would acquire good classification accuracy by keeping and propagating desirable genetic traits. Using this step, the next GA is then executed to select the combination of individual FAMs to be assembled into a classifier ensemble.

4 Experimental Results

Benchmarking is performed in this section to test the performance of the methodology in optimizing FAMs. This is done using selected data sets, i.e. Glass, Ionosphere, and Wine from the UCI Machine Learning Repository [11]. Details of instances, features, and classes are given in Table 1.

Table 1. Details of UCI data sets

Data set	Instances	Features	Classes
Diabetes	768	8	2
Glass	214	9	6
Ionosphere	351	34	2
Iris	150	4	3
Wine	178	13	3

Table 2 presents the results of the experiment. Each FAM was generated using the parameters (i.e. vigilance, learning rate) and then trained using the given data set. The data set is ordered and feature-selected according to the chromosome. Training and testing were performed using the 10-fold cross validation method with the average testing accuracy reported. Accuracy is defined as the ratio of classifications that are assigned to the correct class, over total number of classifications.

The classification performance of the GA-optimized FAMs were compared to that of FAMs generated with completely randomized variables. To ensure fairness, for each generation of optimization performed by the GA, the population of randomized FAMs was also filtered to remove the lowest-performing FAMs to be replaced by new randomly-generated FAMs. Results of FAM, FAM-GA and comparison with literature are shown in Table 2.

Table 2. Accuracy comparison methods in literature with FAM and FAM-GA

Ref	Classifier	Diabetes	Glass	Ionosphere	Iris	Wine
[12]	C4.5, RF	0.7681	0.7290	0.9342	-	-
[13]	kNN, SVM	-	0.7090	0.8821	0.9627	0.9764
[14]	FAM	0.7075	0.8913	-	0.9858	0.9838
[15]	kNN, SVM	-	-	0.8999	0.9521	0.9817
	FAM	0.7156	0.6342	0.7034	0.9737	0.9005
	FAM-GA	0.7617	0.7771	0.9340	0.9867	0.9847
		±0.0266	±0.0175	±0.0247	±0.0038	±0.0046

It can be seen that results of FAM-GA acquired good results in all three data sets, having the highest accuracy in two of the three data sets, as in Table 2. Results of standalone FAM was the lowest compared to other methods. C4.5 and Random Forest (RF) was used in [12], k-Nearest Neighbors and Support Vector Machines was used in [13] and [15]. In the Glass data set, [12] had the second highest accuracy after FAM-GA. The accuracy of FAM-GA however was 0.002 lower than in [12] for the Ionosphere data set. For the Wine data set, FAM-GA had the highest accuracy, followed by [15].

5 Conclusions

Optimization of FAM neural networks using the GAs have been proposed in this paper. The presentation order of training data during the learning step, feature subset selection of the training data, and the internal parameters (baseline vigilance and match tracking) of the FAM were used in the optimization process. These three variables are encoded as a chromose string and evaluated by creating and training the FAM according to the variables. Fitness of the chromosome is determined by the final classification accuracy of the FAM. The proposed framework was evaluated using benchmark data sets from UCI. Results showed good performances and effectiveness of the proposed framework for data classification, as compared with other methods in literature. Future work will look into implementing a method for subsampling larger data sets, in order to improve the algorithm speed.

Acknowledgments. This research is supported by Collaborative Research in Engineering, Science Technology (CREST) Grant P05C2-14 and University of Malaya Grant UM.C/625/1/HIR/MOHE/FCSIT/10.

References

1. Loghmanian, S.M.R., Jamaluddin, H., Ahmad, R., Yusof, R., Khalid, M.: Structure optimization of neural network for dynamic system modeling using multi-objective genetic algorithm. Neural Comput. Appl. **21**(6), 1281–1295 (2012)
2. Chou, C.W., Chien, C.F., Gen, M.: A multiobjective hybrid genetic algorithm for TFT-LCD module assembly scheduling. IEEE Trans. Autom. Sci. Eng. **11**(3), 692–705 (2014)
3. Carrano, E.G., Soares, L.A., Takahashi, R.H., Saldanha, R.R., Neto, O.M.: Electric distribution network multiobjective design using a problem-specific genetic algorithm. IEEE Trans. Power Delivery **21**(2), 995–1005 (2006)
4. Celli, G., Ghiani, E., Mocci, S., Pilo, F.: A multiobjective evolutionary algorithm for the sizing and siting of distributed generation. IEEE Trans. Power Syst. **20**(2), 750–757 (2005)
5. Tsai, J.T., Liu, T.K., Chou, J.H.: Hybrid Taguchi-genetic algorithm for global numerical optimization. IEEE Trans. Evol. Comput. **8**(4), 365–377 (2004)
6. Tu, Z., Lu, Y.: A robust stochastic genetic algorithm (StGA) for global numerical optimization. IEEE Trans. Evol. Comput. **8**(5), 456–470 (2004)

 7. Carpenter, G.A., Grossberg, S., Marzukon, N., Reynolds, J.H., Rosen, D.B.: Fuzzy ARTMAP: a neural network architecture for incremental supervised learning of analog multidimensional maps. IEEE Trans. Neural Netw. **3**(5), 698–713 (1992)
 8. Williamson, J.R.: Gaussian ARTMAP: a neural network for fast incremental learning of noisy multidimensional maps. Neural Netw. **9**(5), 881–897 (1996)
 9. Carpenter, G.A., Gaddam, S.C.: Biased ART: a neural architecture that shifts attention toward previously disregarded features following an incorrect prediction. Neural Netw. **23**(3), 435–451 (2010)
10. Dagher, I., Georgiopoulos, M., Heileman, G.L., Bebis, G.: An ordering algorithm for pattern presentation in fuzzy ARTMAP that tends to improve generalization performance. IEEE Trans. Neural Netw. **10**(4), 768–778 (1999)
11. Frank, A., Asuncion, A.: UCI machine learning repository (2011). http://archive.ics.uci.edu/ml. Accessed March 2015
12. Ordóñez, F.J., Ledezma, A., Sanchis, A.: Genetic approach for optimizing ensembles of classifiers. In: FLAIRS Conference, pp. 89–94 (2008)
13. Woloszynski, T., Kurzynski, M., Podsiadlo, P., Stachowiak, G.W.: A measure of competence based on random classification for dynamic ensemble selection. Inf. Fusion **13**(3), 207–213 (2012)
14. Yaghini, M., Shadmani, M.A.: GOFAM: a hybrid neural network classifier combining fuzzy ARTMAP and genetic algorithm. Artif. Intell. Rev. **39**(3), 183–193 (2013)
15. Wei, H., Lin, X., Xu, X., Li, L., Zhang, W., Wang, X.: A novel ensemble classifier based on multiple diverse classification methods. In: 2014 11th International Conference on Fuzzy Systems and Knowledge Discovery, pp. 301–305 (2014)

Is DeCAF Good Enough for Accurate Image Classification?

Yajuan Cai[1], Guoqiang Zhong[1]([⊠]), Yuchen Zheng[1], Kaizhu Huang[2],
and Junyu Dong[1]

[1] Department of Computer Science and Technology,
Ocean University of China, 266100 Qingdao, China
cyj-ouc@hotmail.com, {gqzhong,dongjunyu}@ouc.edu.cn
zhengyuchen@live.shop.edu.cn
[2] Department of Electrical and Electronic Engineering,
Xi'an Jiaotong-Liverpool University, SIP, 215123 Suzhou, China
Kaizhu.Huang@xjtlu.edu.cn

Abstract. In recent years, deep learning has attracted much interest for addressing complex AI tasks. However, most of the deep learning models need to be trained for a long time in order to obtain good results. To overcome this problem, the deep convolutional activation feature (DeCAF) was proposed, which is directly extracted from the activation of a well trained deep convolutional neural network. Nevertheless, the dimensionality of DeCAF is simply fixed to a constant number. In this case, one may ask whether DeCAF is good enough for image classification applications and whether we can further improve its performance? To answer these two questions, we propose a new model called RS-DeCAF based on "reducing" and "stretching" the dimensionality of DeCAF. In the implementation of RS-DeCAF, we reduce the dimensionality of DeCAF using dimensionality reduction methods, such as PCA, and meanwhile increase the dimensionality by stretching the weight matrix between successive layers. RS-DeCAF is aimed to discover the effective representations of data for classification tasks. As there is no back propagation is needed for network training, RS-DeCAF is very efficient and can be easily applied to large scale problems. Extensive experiments on image classification show that RS-DeCAF not only slightly improves DeCAF, but dramatically outperforms previous "stretching" and other state-of-the-art approaches. Hence, RS-DeCAF can be considered as an effective substitute for previous DeCAF and "stretching" approaches.

Keywords: Image classification · Feature learning · Deep convolutional neural network · DeCAF · Stretching

1 Introduction

In recent years, the development of feature learning algorithms, new weight initialization methods and inexpensive GPUs have brought back deep neural

© Springer International Publishing Switzerland 2015
S. Arik et al. (Eds.): ICONIP 2015, Part II, LNCS 9490, pp. 354–363, 2015.
DOI: 10.1007/978-3-319-26535-3_41

networks into the main stream for solving complex AI tasks. Deep neural networks have been successfully applied to many tasks, such as handwritten digits, face, object recognition and more others [1, 5–8, 22, 23]. Particularly, a number of convolutional neural networks (CNNs) based models have been proposed for large-scale visual recognition applications [10, 11]. With the availability of large-scale sources of labeled training data [12], and effective regularization method ("dropout") [1, 24], they have shown superior performance on a large-scale object recognition and detection challenge [15]. However, despite the attractive qualities of CNNs-based models, they are prohibitively expensive to be applied to large-scale problems in order to obtain good results. In fact, this is also a common problem for most of the deep learning models.

To overcome this problem, Donahue et al. proposed the deep convolutional activation feature (DeCAF) [2], which is extracted from the activation of a deep convolutional network that is well trained on a large-scale object recognition task. DeCAF can be adapted to novel generic tasks that may significantly differ from the originally training task. Experimental results in [2] have showed that DeCAF outperforms the state-of-the-art approaches on several vision challenges. However, considering that DeCAF is extracted from one fully-connected hidden layer of the trained deep convolutional network, its dimensionality is always fixed to a constant. One may ask whether DeCAF is good enough for image classification applications and whether we can further improve its performance? To answer these two questions, in this paper, we propose a novel model called RS-DeCAF based on "reducing" and "stretching" the dimensionality of DeCAF. Here, the "reducing" and "stretching" operations are aimed to discover the effective representations of data for accurate classification tasks. As no back propagation is needed for network fine-tuning, the learning of RS-DeCAF is very efficient. This guarantees that it can be easily applied to large scale classifications tasks. We evaluated RS-DeCAF on several image classification challenges, and observed that it not only slightly improves DeCAF, but dramatically outperforms previous "stretching" and other state-of-the-art approaches.

In the following section, we briefly introduce related work. In Sect. 3, we describe our proposed model, RS-DeCAF, in detail. Section 4 shows experimental settings and results with comparison against DeCAF and other state-of-the-art approaches. Section 5 concludes this paper with remarks and future work.

2   Related Work

Due to the great success of deep convolutional neural networks (CNNs) based models in many challenging applications, here we only focus on CNNs based models. CNNs are of a specific topology of artificial neural network, which is inspired by biological visual cortex and tailored for computer vision tasks by LeCun et al. [10]. In recent years, CNNs based models, in particular the landmark "AlexNet" model, have won many challenging competitions on large-scale data sets [1, 13, 14]. Among others, the model most closely related to our work is called DeCAF [2]. DeCAF is the feature extracted from the activation of a well trained deep convolutional neural network. It can be easily repurposed to novel generic tasks. However, the dimensionality of DeCAF is primarily fixed to a constant number. This may hamper its application to various visual recognition tasks. In this paper, we propose a novel model called RS-DeCAF based on "reducing" and "stretching" the dimensionality of DeCAF. Here, the "reducing" and "'stretching" operations are aimed to discover the effective representations of data for recognition tasks.

In the literature, many papers have been proposed for dimensionality reduction or called feature learning algorithm, such as principal component analysis (PCA) [17] and linear discriminant analysis (LDA) [18]. In general, from the perspective of the mapping function between the original and latent spaces, dimensionality reduction methods can be classified into two categories: linear and nonlinear. For example, PCA and LDA are both linear dimensionality reduction methods, whilst their kernel extensions, kernel PCA [25] and generalized discriminant analysis (GDA) [26], are nonlinear methods. For simplicity, in this paper, we only apply PCA for dimensionality reduction. Other methods can also be considered instead.

Kernel tricks are widely used to map the observed data into a high dimensional space, even that with infinite dimensionality [28,29]. In [9], Pandey and Dukkipati proposed an approach to stretch the weight matrix of a neural network. Particularly, the weight matrix can be stretched to infinity with the arccosine kernels [31]. In this paper, we stretch the weight matrix or projection matrix of PCA learned based on DeCAF, to seek the proper representations of data for accurate image classification. Experimental results show that the proposed model, RS-DeCAF, dramatically improves the method of [9].

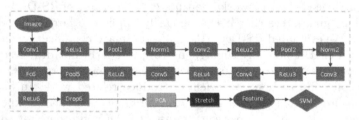

Fig. 1. The architecture of the RS-DeCAF model. Here, we use the activation output of the 6th hidden layer (Fc6) for further dimensionality reducing and stretching.

3   The Proposed Model

In this section, we introduce the RS-DeCAF model, in detail. Figure 1 shows the architecture of RS-DeCAF. Here, the framework of RS-DeCAF is based on the pre-trained model provided by Caffe [3]. However, different from DeCAF, RS-DeCAF reduces the dimensionality of DeCAF using PCA and stretches the projection matrix of PCA to map the features into a higher dimensional space. It is expected that RS-DeCAF can perform better than DeCAF in this space. As the experiments shown in [2] that DeCAF extracted from the activation output of the 6th hidden layer (Fc6) generally performs well, here we only consider the activation output of Fc6.

As mentioned above, when implementing RS-DeCAF, we reduce the dimensionality of the activation output using PCA. To further improve the performance of RS-DeCAF, we stretch the projection matrix of PCA to map the features into a higher dimensional space using the "stretching" technique [9].

Suppose that the projection matrix of PCA is $\mathbf{A} \in R^{4096 \times d}$, and let $\mathbf{W} \in \mathbb{R}^{d \times L}$, $L > d$, be a matrix whose entries are sampled from the standard normal distribution $\mathcal{N}(0, 1)$. The stretched matrix $\mathbf{A}_s$ can be calculated as

$$\mathbf{A}_s = \frac{1}{\sqrt{L}} (\mathbf{A} \times \mathbf{W}). \tag{1}$$

For each input instance $\mathbf{x}$, its corresponding output representation can be computed as

$$h_l(\mathbf{x}) = (\mathbf{x}^T \mathbf{A}_s)_+, \tag{2}$$

where the $(f)_+ = \max(0, f)$ is the ReLu function.

Assume $\mathbf{x}, \mathbf{y} \in \Re^{4096 \times 1}$. According to [9], we have

$$\lim_{L \to \infty} \mathbf{h_x}^T \mathbf{h_y} = \frac{1}{2\pi} \|\mathbf{x}^T \mathbf{A}\| \|\mathbf{y}^T \mathbf{A}\| (\sin \theta_\mathbf{A} + (\pi - \theta_\mathbf{A}) \cos \theta_\mathbf{A}), \tag{3}$$

where

$$\theta_\mathbf{A} = \cos^{-1} \frac{\mathbf{x}^T \mathbf{A} \mathbf{A}^T \mathbf{y}}{\|\mathbf{A}^T \mathbf{x}\| \|\mathbf{A}^T \mathbf{y}\|}. \tag{4}$$

If we replace $\mathbf{\Sigma} = \frac{1}{M} \mathbf{A} \mathbf{A}^T$, we can define

$$k(\mathbf{x}, \mathbf{y}) = \lim_{L \to \infty} \mathbf{h_x^T h_y} = \frac{M}{2\pi} \sqrt{\mathbf{x}^T \mathbf{\Sigma} \mathbf{x}} \sqrt{\mathbf{y}^T \mathbf{\Sigma} \mathbf{y}} (\sin \theta_\mathbf{\Sigma} + (\pi - \theta_\mathbf{\Sigma}) \cos \theta_\mathbf{\Sigma}), \tag{5}$$

where $\theta_\mathbf{\Sigma} = \cos^{-1} \frac{\mathbf{x}^T \mathbf{\Sigma} \mathbf{y}}{\sqrt{\mathbf{x}^T \mathbf{\Sigma} \mathbf{x}} \sqrt{\mathbf{y}^T \mathbf{\Sigma} \mathbf{y}}}$. Hence, based on Eq. (3), we can map the features to a space even with infinite dimension.

The entire algorithm of RS-DeCAF is informally described in Algorithm 1.

Algorithm 1. RS-DeCAF

Require:

A set of training instances $\mathbf{X} = (\mathbf{x}_1, \cdots, \mathbf{x}_i, \cdots, \mathbf{x}_n)$.

Steps:

- Extract the activation output of the 6th hidden layer from the pre-trained model for each instance $\mathbf{x}_i$.
- Reduce the dimensionality of the activation outputs using PCA and obtain the projection matrix $\mathbf{A}$.
- Compute the mapped features according to Eq. (2) with finite L or Eq. (3) with the arc-cosine kernel ($L = \infty$).

---

4   Experiments

To evaluate the RS-DeCAF model, we conducted extensive experiments on three challenging data sets: Caltech101 [19], Caltech-UCSD birds [20] and STL10 [21]. We set DeCAF as the baseline [2], and meanwhile, compared RS-DeCAF to previous "stretching" and other state-of-the-art approaches. For RS-DeCAF, we extracted the features from the 6th hidden layer (Fc6) of the pre-trained "AlexNet" model [3]. Similar to DeCAF, the dimensionality of these features is 4096. Then we projected these features to a PCA subspace of dimensionality 2048. To test the effect of the stretching technique, we applied the stretching technique on both the original 4096D and 2048D PCA representations of data. For concreteness, to map the 4096D representations of data into a higher dimensional space, we applied the stretching operation on the weight matrix between Fc6 and Fc7, while to map the 2048D representations of data, we applied the stretching operation on the projection matrix of PCA. For the dimensionality of the higher dimensional space after stretching, $L = \{4096, 8192, 16384, 32768, 45000, \ldots, \infty\}$ were tested until stretching cannot improve the classification accuracy. For both DeCAF and RS-DeCAF, we report the classification results obtained using support vector machines (SVMs) with the parameter cross-validated over the training data.

Following [1], in both training and test procedures, each RGB image was pre-processed by resizing the smallest dimension to 256, cropping the center 256×256 region, subtracting the per-pixel mean (across all images) and then using 10 different sub-crops of size 227×227 (corners + center with(out) horizontal flips). Figure 2 shows an image in the Caltech-USCB birds data set and its corresponding 10 patches.

Fig. 2. Visualization of the patches of a bird image in the Caltech-USCB birds data set.

4.1   Results on the Caltech101 Data Set

The Caltech101 dataset [19] includes 101 categories of object images. Following [2], we selected 30 images per class for training and the rest for test. The parameters were selected by 6-fold cross validation on a 25 train/5 validation subsplit in each category. Table 1 shows the results obtained by DeCAF and RS-DeCAF. From Table 1, we can observe three interesting points:

- DeCAF learns fairly discriminative representations of data. Previous stretching approaches on the weight matrix cannot help to improve DeCAF any more.

Table 1. Classification accuracies on the Caltech-101 data set. In the second column, we show the accuracy obtained by DeCAF (4096D) and its stretching variants. In the third column, we show that obtained by RS-DeCAF. The best result is highlighted with boldface.

Dim (L)	DeCAF+ (%)	RS-DeCAF (%)
2048	-	83.49
4096	86.10	84.52
8192	83.22	85.56
16384	83.36	85.72
32768	83.38	**86.21**
45000	83.64	85.90
65536	83.30	85.79
∞	3.08	0.88

Table 2. Classification accuracies obtained by RS-DeCAF, previous stretching approach and the state-of-the-art on the Caltech-101 data set.

Method	Accuracy(%)
Model of [9]	74.3
Model of [16]	**86.5**
RS-DeCAF	86.21

- Stretching improves the accuracy of RS-DeCAF, but its effect turns worse from a point as $L \to +\infty$.
- RS-DeCAF slightly improves DeCAF. The best results obtained by RS-DeCAF is about 0.1 % more than that of DeCAF.

Table 2 shows the results obtained by RS-DeCAF, a previous stretching [9] and a state-of-the-art approach [16]. The model used in [16] was an 8 layered convolutional deep belief networks (DBN) with 5 convolution layers and 3 pooling layers [16]. We can see that, RS-DeCAF performs close to the state-of-the-art, but with much less training time. Meantime, RS-DeCAF performs dramatically better than the previous stretching method, since the activation outputs were extracted from a well trained deep convolutional network [3].

4.2 Results on the Caltech-UCSD Birds Data Set

The Caltech-UCSD Birds data set [20] includes image photos of 200 bird species with 6,033 images. Following [2], we preprocess each image to feed them into the Caffe pipeline [3]. Table 3 shows the comparison results between DeCAF and RS-DeCAF. We can easily see that, on this data set, RS-DeCAF (35.31 %) outperforms DeCAF (34.49 %) with about 1 % accuracy.

Table 3. Classification accuracies on the Caltech-USCB birds data set.

Dim (L)	DeCAF+ (%)	RS-DeCAF (%)
2048	-	32.08
4096	34.49	33.95
8192	26.38	34.69
16384	27.27	**35.31**
32768	27.50	34.95
45000	27.47	34.85
∞	0.26	0.26

4.3   Results on the STL10 Data Set

The STL10 data set [21] includes 10 classes of images with 500 training and 800 test samples per class. Table 4 shows the comparison results between DeCAF and RS-DeCAF. We can see from these results that RS-DeCAF slightly improves DeCAF. Table 5 shows the results obtained by RS-DeCAF and the models of model of [9,27,30], respectively. As we can see, RS-DeCAF not only improves previous stretching method, but also the state-of-the-art.

Table 4. Classification accuracies on the STL10 data set.

Dim (L)	DeCAF+ (%)	RS-DeCAF (%)
2048	-	85.75
4096	90.54	90.15
8192	89.49	90.39
16384	89.56	**90.66**
32768	89.73	90.54
45000	89.68	90.49
∞	10.89	16.79

4.4   Discussion

In the literature, many deep learning approaches have been proposed [7,11,23]. To obtain good results, these deep models generally need to be trained for a long time. However, following the idea of DeCAF, RS-DeCAF has no need to use back propagation for network fine-tuning. Even dimensionality reduction and the "stretching" operations are applied, the computation is still efficient. One may argue that SVMs with non-linear kernel are not suitable for the testing of large scale problems. In this case, one solution is to use other classifiers, such as "softmax"; the other one is to adopt efficient training algorithms of SVMs, such as that introduced in [4]. Hence, RS-DeCAF can be easily applied to large scale problems.

Table 5. Classification accuracies obtained by RS-DeCAF, previous stretching approach and the state-of-the-art on the STL10 data set.

Method	Accuracy(%)
Model of [30]	70.1
Model of [27]	72.8
Model of [9]	90.54
Our model	**90.66**

5 Conclusion

In this paper, we propose a novel model called RS-DeCAF based on "reducing" and "stretching" the dimensionality of DeCAF. Extensive experiments on some image classification challenges show that RS-DeCAF not only outperforms DeCAF, but dramatically improves previous "stretching" and other stat-of-the-art approaches. Hence, one can achieve much better accuracy with RS-DeCAF rather than training deep neural networks. And more, RS-DeCAF can be easily applied to large scale problems. For the future work, to learn more compact representations of the activation outputs, we plan to test other dimensionality reduction methods instead of PCA in the implementation of RS-DeCAF; we would like to fine-tune the deep architecture of RS-DeCAF to further improve the classification accuracy.

Acknowledgments. This work was supported by the National Basic Research Program of China (No. 2012CB316301), the National Natural Science Foundation of China (No. 61271405, No. 61403353, No. 61473236 and No. 61401413), the Ph.D. Program Foundation Of Ministry Of Education Of China (No. 20120132110018), the Open Project Program of the National Laboratory of Pattern Recognition (NLPR) and the Fundamental Research Funds for the Central Universities of China.

References

1. Krizhevsky, A., Sutskever, I., Hinton, G.E.: Imagenet classification with deep convolutional neural networks. In: Neural Information Processing Systems, pp. 1097–1105 (2012)
2. Donahue, J., Jia, Y., Vinyals, O., Hoffman, J., Zhang, N., Tzen, E., Darrel, T.: DeCAF: a deep convolutional activation feature for generic visual recognition. In: The Internalational Compouter Machine Learning, pp. 647–655 (2014)
3. Jia, Y., Shelhamer, E., Donahue, J., Karayev, S., Long, J., Girshick, R., Darrell, T.: Caffe: convolutional architecture for fast feature embedding. In: Proceedings of the ACM International Conference on Multimedia, pp. 675–678. ACM (2013). http://caffe.berkeleyvision.org
4. Tsang, I., Kowk, J., Cheung, P.-M.: Core vector machines: fast SVM training on very large data sets. J. Mach. Learn. Res. 6(2), 363–392 (2005)
5. Lin, M., Chen, Q., Yan, S.: Network in network, J. CoRR (2013). abs/1312.4400

6. Sun, Y., Chen, Y., Wang, X., Tang, X.: Deep learning face representation by joint identification-verification. In: Neural Information Processing Systems, pp. 1988–1996 (2014)
7. Zheng, Y., Zhong, G., Liu, J., Cai, X., Dong, J.: Visual texture perception with feature learning models and deep architectures. In: Li, S., Liu, C., Wang, Y. (eds.) CCPR 2014, Part I. CCIS, vol. 483, pp. 401–410. Springer, Heidelberg (2014)
8. Wang, N., Yeung, D.Y.: Ensemble-based tracking: aggregating crowdsourced structured time series data. In: Proceedings of the 31st International Conference on Machine Learning (ICML-14), pp. 1107–1115 (2014)
9. Pandey, G., Dukkipati, A.: Learning by stretching deep networks. In: Proceedings of the 31st International Conference on Machine Learning (ICML-14), pp. 1719–1727 (2014)
10. LeCun, Y., Boser, B., Denker, J.S., Henderson, D., Howard, R.E., Hubbard, W., Jackel, L.D.: Backpropagation applied to handwritten zip code recognition. J. Neural Comput. $1(4)$, 541–551 (1989)
11. LeCun, Y., Bottou, L., Bengio, Y., Haffner, P.: Gradient-based learning applied to document recognition. J. Proc. IEEE $86(11)$, 2278–2324 (1998)
12. Deng, J., Dong, W., Socher, R., Li, L.J., Li, K., Fei-Fei, L.: ImageNet: a large-scale hierarchical image database. In: Computer Vision and Pattern Recognition, pp. 248–255 (2009)
13. Sermanet, P., Eigen, D., Zhang, X., Mathieu, M., Fergus, R., LeCun, Y.: Overfeat: integrated recognition, localization and detection using convolutional networks (2013). arXiv preprint arXiv:1312.6229
14. Szegedy, C., Liu, W., Jia, Y., Sermanet, P., Reed, S., Anguelov, D., Rabinovich, A.: Going deeper with convolutions (2014)
15. Russakovsky, O., Deng, J., Su, H., Krause, J., Satheesh, S., Ma, S., Fei-Fei, L.: Imagenet large scale visual recognition challenge (2014)
16. Zeiler, M.D., Fergus, R.: Visualizing and understanding convolutional networks. In: Fleet, D., Pajdla, T., Schiele, B., Tuytelaars, T. (eds.) ECCV 2014, Part I. LNCS, vol. 8689, pp. 818–833. Springer, Heidelberg (2014)
17. Jolliffe, I.: Principal Component Analysis. Wiley, New York (2002)
18. Fisher, R.A.: The use of multiple measurements in taxonomic problems. J. Ann. Eugenics $7(2)$, 179–188 (1936)
19. Fei-Fei, L., Fergus, R., Perona, P.: One-shot learning of object categories. J. IEEE Trans. Pattern Anal. Mach. Intell. $28(4)$, 594–611 (2006)
20. Welinder, P., Branson, S., Mita, T., Wah, C., Schroff, F., Belongie, S., Perona, P.: Caltech-UCSD Birds 200 (2010)
21. Coates, A., Ng, A.Y., Lee, H.: An analysis of single-layer networks in unsupervised feature learning. Int. Conf. Artif. Intell. Stat. 15, 215–223 (2011)
22. Hinton, G., Osindero, S., Teh, Y.W.: A fast learning algorithm for deep belief nets. J. Neural Comput. $18(7)$, 1527–1554 (2006)
23. Hinton, G.E., Salakhutdinov, R.R.: Reducing the dimensionality of data with neural networks. J. Sci. $313(5786)$, 504–507 (2006)
24. Hinton, G.E., Srivastava, N., Krizhevsky, A., Sutskever, I., Salakhutdinov, R.R.: Improving neural networks by preventing co-adaptation of feature detectors (2012). arXiv preprint arXiv:1207.0580
25. Scholkopf, B., Smola, A., Mller, K.R.: Nonlinear component analysis as a kernel eigenvalue problem. J. Neural Comput. $10(5)$, 1299–1319 (1998)
26. Baudat, G., Anouar, F.: Generalized discriminant analysis using a kernel approach. J. Neural Comput. $12(10)$, 2385–2404 (2000)

27. Dosovitskiy, A., Springenberg, J.T., Riedmiller, M., Brox, T.: Discriminative unsupervised feature learning with convolutional neural networks. In: Neural Information Processing Systems, pp. 766–774 (2014)
28. Scholkopf, B., Smola, A.J.: Learning with Kernels: Support Vector Machines, Regularization, Optimization, and Beyond. MIT press, Cambridge (2002)
29. Vapnik, V.N., Vapnik, V.: Statistical Learning Theory, vol. 1. Wiley, New York (1998)
30. Swersky, K., Snoek, J., Adams, R.P.: Multi-task bayesian optimization. In: Neural Information Processing Systems, pp. 2004–2012 (2013)
31. Cho, Y., Saul, L.K.: Large-margin classification in infinite neural networks. J. Neural Comput. 22(10), 2678–2697 (2010)

Optimal Feature Subset Selection for Neuron Spike Sorting Using the Genetic Algorithm

Burhan Khan$^{(\boxtimes)}$, Asim Bhatti, Michael Johnstone, Samer Hanoun, Douglas Creighton, and Saeid Nahavandi

Centre for Intelligent Systems Research (CISR),
Deakin University, Geelong, Australia
{burhan.khan,asim.bhatti,michael.johnstone,
samer.hanoun,douglas.creighton,
saeid.nahavandi}@deakin.edu.au

Abstract. It is crucial for a neuron spike sorting algorithm to cluster data from different neurons efficiently. In this study, the search capability of the Genetic Algorithm (GA) is exploited for identifying the optimal feature subset for neuron spike sorting with a clustering algorithm. Two important objectives of the optimization process are considered: to reduce the number of features and increase the clustering performance. Specifically, we employ a binary GA with the silhouette evaluation criterion as the fitness function for neuron spike sorting using the Super-Paramagnetic Clustering (SPC) algorithm. The clustering results of SPC with and without the GA-based feature selector are evaluated using benchmark synthetic neuron spike data sets. The outcome indicates the usefulness of the GA in identifying a smaller feature set with improved clustering performance.

Keywords: Genetic algorithm · Super-Paramagnetic clustering · Neuron spike sorting · Features selection · Optimization

1 Introduction

The human brain is a complex organ consisting of millions of neuron cells. These neuron cells form a complex network of communication that allows humans to perceive information and perform various actions. The communication between two neuron cells is realized by a phenomenon called action potential [1]. Action potential is basically the flow of current from one cell to another due to the difference in voltage, which is known as neuron spikes in neuroscience [2]. In this regard, one of the key challenges faced by neuroscientists is to analyse the enormous number of spikes data generated from the brain. This analysis includes spike sorting, i.e. the process to identify and cluster the acquired spikes according to the most relevant neurons that generate the spikes. In essence, the spike sorting process involves three major steps, *viz.* spike detection, features extraction, and spike clustering [3, 4].

As shown in Fig. 1, feature extraction is an important step to identify the most relevant and distinctive representation of a spike data set. The extracted features play a key role in producing a better clustering outcome. Feature extraction also acts as a

© Springer International Publishing Switzerland 2015
S. Arik et al. (Eds.): ICONIP 2015, Part II, LNCS 9490, pp. 364–370, 2015.
DOI: 10.1007/978-3-319-26535-3_42

dimension reduction scheme for tackling problems with a large number of features [5]. As stated in [6], a subset of extracted features is able to improve efficacy of classification algorithms in a complex, high-dimensional data space. As such, this research focuses on improving the neuron spike sorting process by identifying a set of optimal features for the spike clustering algorithm.

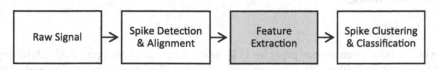

Fig. 1. Spike sorting using WaveClus

A comprehensive performance overview for spike sorting algorithms is provided in [4]. Using the Adjusted Mutual Information (AMI) [7] as the clustering performance measure, three spike sorting programs were compared, i.e. WaveClus, KlustaKwik, and OSort. According to the findings in [2], WaveClus performed well as compared with the other two when the algorithms were tested for different parameters, noise levels, and number of neuron settings. As such, we use WaveClus [8] as the underlying program for spike clustering in this study.

WaveClus uses an unsupervised spike detection and sorting algorithm. Specifically, it employs wavelet transform for feature extraction and a super-paramagnetic clustering (SPC) algorithm. In this paper, we propose to exploit the search capability of the GA to identify the optimal number of wavelet features that can provide the best performance in neuron spike sorting using SPC. The clustering performance of SPC is compared with and without using the GA. The outcome shows a significant improvement when the GA is employed for feature selection, especially when more difficult data sets are used for evaluated.

The rest of this paper is organized as follows. In Sect. 2, the related work is presented. Section 3 provides the background of the proposed method. Section 4 describes the details of the proposed method. Section 5 presents the results and discussion. Conclusions and suggestions for further work are given in Sect. 6.

2 Related Work

The use of GA-based techniques for feature selection is not new. Ekbal et al. [9] used NSGA-II for named entity recognition (NER) in two languages. Using recall and precision as two competing objectives and binary string representation of features in the optimization algorithm, improved results as compared with those from the existing NER methods were reported. More recently, Le and Tran [10] proposed a similar method in NER applications by using the GA to search for an optimal subset of features with improved computation time. Huang et al. [11] used the GA at two levels for feature selection, in an attempt to improve predictive accuracy and classification accuracy. At an initial global level, the most relevant features were selected for the classifier. Then, the selected features were filtered at the local level to eliminate redundant features. Another related work was presented by Tan et al. [12]. The proposed method used different techniques to extract features and then employed the GA to search for the most appropriate feature subset out of the extracted pool of features.

3 Methodology

3.1 WaveClus

WaveClus is an open source program [13]. It implements an unsupervised (clustering) spike sorting algorithm [8] to detect and extract spikes from various neurons. The wavelet transformation method is used for feature extraction, while the SPC algorithm is employed to cluster the spikes based on the extracted features. There are many parameters used in the SPC algorithm. However, one of the most important parameters is "temperature", which directly effects the number of clusters produced. A lower temperature setting generates fewer number of clusters, and vice versa. Figure 1 illustrates the main steps for spike sorting using WaveClus.

3.2 The Genetic Algorithm

The GA is inspired by the process of natural selection. It belongs to the category of stochastic search optimization techniques [14]. The GA is famous for its characteristic of parallel search owing to the use of multiple chromosomes in a population; therefore producing multiple feasible solutions at the same time. In this study, we exploit the search capability of the GA to identify an optimal subset of wavelet features from the pool of extracted features from neuron signals. Three key GA steps are selection, crossover, and mutation. We use tournament selection, single-point crossover, and binary mutation in this study. Equation (1) shows a single-objective optimization problem, where $f(x)$ is the objective function to be minimized and k is the number of decision variables.

$$min f(x) \ \forall \ x = 1, 2, 3, \ldots, k. \tag{1}$$

4 The Proposed Method

SPC uses wavelet coefficients for neuron spike clustering [8]. We use a string comprising binary bits to indicate the selection of the respective wavelet coefficients (input features). The string is used as the decision variables encoded as an individual chromosome in the GA population. Table 1 illustrates an example of selecting (represented by 1) or masking (represented by 0) a specific wavelet coefficient in a chromosome with binary bits. The wavelet coefficients corresponding to 1's are passed to SPC for clustering purposes.

Table 1. GA chromosome representation

WC$_1$	WC$_2$	WC$_3$	WC$_4$	WC$_5$	WC$_6$	WC$_7$	WC$_8$	WC$_9$	WC$_{10}$
1	0	1	1	1	0	0	1	0	1

4.1 Fitness Computation

The GA requires fitness calculation so that it is able to determine the quality of the solutions. Two equally important objectives considered in this study are improving the clustering performance and reducing the number of input features. Both objectives can be combined in a way that the clustering performance index is evaluated by the number of features. As such, we employ a single-objective GA with the silhouette index [15] for evaluation of the clustering performance as an objective function for evaluation of the clustering performance.

The silhouette index examines each individual data sample within a spike data set to determine how well the respective data sample is clustered within its own cluster and relative to other clusters. As shown in Eq. (2), a_i is the average difference (Euclidean distance) between data sample i and all other data samples belonging to the same cluster. Similarly, b_i is the minimum average difference between data sample i and those belonging to other remaining clusters.

$$sm_i = \frac{b_i - a_i}{max(a_i, b_i)} \Rightarrow -1 \leq sm_i \leq 1 \tag{2}$$

The silhouette index of a data sample lies between -1 and 1. A score close to 1 indicates that the data samples in a cluster are well clustered, and vice versa. Equation (3) illustrates the average silhouette index sa_i for all data samples, which serves as the fitness value for an individual chromosome in the GA.

$$sa_i = \frac{1}{n} \sum_{i=1}^{n} sm_i. \tag{3}$$

4.2 Algorithm

The inputs to WaveClus comprise spike data (RawSpikes). Filtered spike data (Spikes) from noise are obtained by applying the thresholding and alignment techniques in the *DetectSpikes* procedure. Filtered spike data are then used for feature extraction (WaveCoeffs) in the *WaveletFeatureExtraction* procedure. A randomly generated population (a set of chromosmes) is created by the *InitilaisePopulation* procedure. Features (WaveCoeffs) are selected/masked using the binary bits from the selected chromosome in the *FeaturesSubset* function, and a new subset of features (SFeatures) is returned. This new subset of features (SFeatures) is subject to clustering in the *ClusteringSPC* procedure, which returns the detected spike clusters (SC) and detected spike times (ST). To determine the fitness of a chromosome, we use the *SilhouetteEvaluation* procedure as the fitness function. All chromosomes are provided to the *GA* procedure once every individual chromosme is assigned a fitness value (Fit). Furthermore, the functions highlighted bold in the algorithm below are used from the previous work [8].

Algorithm

```
INPUT: Raw Spike Data
OUTPUT: Features & Clustering Performance Measure
```

```
RawSpikes ← LoadRawData()
Spikes ← DetectSpikes(RawSpikes)
WaveCoeffs ← WaveletFeatureExtraction(Spikes)
Chromosomes ← InitializePopulation()
for i = 1:Generations do
  for j = 1:length(Chromosomes) do
    SFeatures ← FeaturesSubset(Chromosomes, WaveCoeffs)
    [ST,SC] ← ClusteringSPC(SFeatures,Spikes)
    Fit ← SilhouetteEvaluation (ST,SC)
    if Fit < BestAccuracy then
      BestAccuracy = Fit
      BestFeatures = Chromosomes(j)
    end if
    OffspringChromosomes ← GA(Chromosomes)
    Chromosomes = [Chromosomes;OffspringChromosomes]
    Chromosomes = Sort(Chromosomes)
    Chromosomes = Chromosomes/2;
  end for
end for
```
---

5 Results and Discussion

In this study, we used a number of synthetic data sets in [11] for evaluation purposes. There are a variety of data sets with different difficulty levels based on the number of data and noise, as shown in Table 2. In order to compare our results, we conducted two simulation runs of the spike sorting algorithm: with and without the GA for feature selection. In the GA, we used 50 generations, a population size of 20, and 10 features represented by 10 individual binary bits in each chromosome of the population. All other parameters used were set to the default setting in WaveClus for the spike sorting algorithm.

Table 2 summarizes the overall performance. The synthetic data samples were generated by three sources. By using the GA-based method, we successfully obtained 3 clusters for all data sets with reduced number of features and higher clustering performance. Moreover, the performance of the spike sorting algorithm without the GA decreased with increasing difficulty level of the data sets. By using the GA for optimal feature subset selection, improvement in performance corresponding to increasing

Table 2. A summary of the data sets used for evaluation

Dataset	Algorithm	GA Features Representation	Features	Silhouette Index
C_Easy1_noise01_short	SPC+GA	1011111010	7	0.616
	SPC	1111111111	10	0.546
C_Easy1_noise01	SPC+GA	1100110101	6	0.582
	SPC	1111111111	10	-0.245
C_Difficult1_noise01	SPC+GA	1001101011	6	0.295
	SPC	1111111111	10	-0.528
C_Difficult2_noise005	SPC+GA	1110001101	6	0.436
	SPC	1111111111	10	-0.077
C_Difficult2_noise015	SPC+GA	0011101100	5	0.464
	SPC	1111111111	10	-0.388

difficulty level of the data sets can be observed. In other words, an increase in clustering quality with a compact set of features can be obtained using the GA for feature selection.

Our approach have shown significant improvement in producing better quality clusters when applied to difficult datasets. As shown in Table 2, Datasets: C_Difficult1_noise01, C_Difficult2_noise005, and C_Difficult2_noise015 produced negative values for silhouette index with plain SPC, representing poor quality of clusters.

6 Conclusions

In this research, we have investigated the clustering performance through optimal feature subset selection using the GA for an unsupervised spike sorting algorithm. Based on WaveClus, we have demonstrated the usefulness of the GA for neuron spike sorting. The outcome of our approach clearly shows better silhouette index values compared to the outcome of running plain SPC. Our approach successfully enables SPC to produce better quality clusters using less number of features when applied to difficult datasets. In future work, we will investigate the GA-based method for other problems and compare the results obtained from various feature extraction methods with different clustering techniques.

References

1. Rey, H.G., Pedreira, C., Quian Quiroga, R.: Past, present and future of spike sorting techniques. Brain Res. Bull. (2015). ISSN: 0361-9230. http://dx.doi.org/10.1016/j.brainresbull.2015.04.007, http://www.sciencedirect.com/science/article/pii/S0361923015000684

2. Lewicki, M.S.: A review of methods for spike sorting: the detection and classification of neural action potentials. Netw. Comput. Neural Syst. **9**, 53–78 (1998)
3. Haggag, S., Mohamed, S., Bhatti, A., Haggag, H., Nahavandi, S.: Neural spike representation using Cepstrum. In: 2014 9th International Conference on System of Systems Engineering (SOSE), pp. 97–100 (2014)
4. Wild, J., Prekopcsak, Z., Sieger, T., Novak, D., Jech, R.: Performance comparison of extracellular spike sorting algorithms for single-channel recordings. J. Neurosci. Methods **203**, 369–376 (2012)
5. Ahmed, S., Zhang, M., Peng, L.: Feature selection and classification of high dimensional mass spectrometry data: a genetic programming approach. In: Vanneschi, L., Bush, W.S., Giacobini, M. (eds.) EvoBIO 2013. LNCS, vol. 7833, pp. 43–55. Springer, Heidelberg (2013)
6. Abdel-Aal, R.E.: GMDH-based feature ranking and selection for improved classification of medical data. J. Biomed. Inform. **38**, 456–468 (2005)
7. Vinh, N.X., Epps, J., Bailey, J.: Information theoretic measures for clusterings comparison: variants, properties, normalization and correction for chance. J. Mach. Learn. Res. **11**, 2837–2854 (2010)
8. Quiroga, R.Q., Nadasdy, Z., Ben-Shaul, Y.: Unsupervised spike detection and sorting with wavelets and superparamagnetic clustering. Neural Comput. **16**, 1661–1687 (2004)
9. Ekbal, A., Saha, S., Garbe, C.S.: Feature selection using multiobjective optimization for named entity recognition. In: 2010 20th International Conference on Pattern Recognition (ICPR), pp. 1937–1940 (2010)
10. Le, H.T., Tran, L.V.: Automatic feature selection for named entity recognition using genetic algorithm. Presented at the Proceedings of the Fourth Symposium on Information and Communication Technology, Danang, Vietnam (2013)
11. Huang, J., Cai, Y., Xu, X.: A hybrid genetic algorithm for feature selection wrapper based on mutual information. Pattern Recogn. Lett. **28**, 1825–1844 (2007)
12. Tan, F., Fu, X., Zhang, Y., Bourgeois, A.: A genetic algorithm-based method for feature subset selection. Soft. Comput. **12**, 111–120 (2008)
13. Quiroga, R.Q.: Wave_clus: Unsupervised spike detection and sorting. https://vis.caltech.edu/~rodri/Wave_clus/Wave_clus_home.htm
14. Goldberg, D.E.: Genetic Algorithms. Pearson Education, New York (2006)
15. Rousseeuw, P.J.: Silhouettes: a graphical aid to the interpretation and validation of cluster analysis. J. Comput. Appl. Math. **20**, 53–65 (1987)

Continuous Summarization for Microblog Streams Based on Clustering

Qunhui Wu[✉], Jianghua Lv, and Shilong Ma

School of Computer Science and Engineering, Beihang University, Beijing,
People's Republic of China
wuqunhui@buaa.edu.cn, jhlv@nlsde.buaa.edu.cn

Abstract. With rapid growth of information found on microblog services, dynamic summarization of evolving information has become an important task. However, the existing work on continuous microblog stream summarization cannot effectively work due to the enormous noises and redundancies. We tackle this problem using a two-step process, first by clustering online microblog streams and maintaining cluster feature vectors. Then, the dynamic summaries of arbitrary time durations are generated from the microblog cluster features. This helps users to better find the worthy interpretations of the online microblog streams. We make use of features to calculate the importance of similar sentences in each cluster for these two steps. Our approach integrates these cluster information with an unsupervised topic evolvement detection model, and illustrate that latent topics to capture the feature dependencies summaries with better performance. Finally, the experimental results on real microblogs demonstrate that our summarization framework can significantly improve the performance and make it comparable to the state-of-the-art summarization approaches.

Keywords: Microblog · Topic detection model · Summarization · Clustering

1 Introduction

Microblog services have explosively increased its popularity in recent decades. Users convey their opinions and share their thoughts on Microblog platforms, such as Twitter, Weibo and Tumblr. To help users to obtain the interesting contents through the tremendous number of microblogs (or tweets) efficiently, continuous microblog summarization has been attracted more and more attention.

The importance of continuous summarization has become the focal point to be studied and developed by the research community. For example, in real-world application, the specific natural hazard topic evolves in time, and the topic-related tweet stream needs to be deal without delay. Although the most existing studies on multi-document summarization [9, 10], e.g., clustering-based summarization methods usually group the tweet cluster vectors acquired using the similarity calculation to centroid [15]. However, updating periodically summaries using current multi-document summarization approaches are still mainly based on processing previous documents repeat, which can

© Springer International Publishing Switzerland 2015
S. Arik et al. (Eds.): ICONIP 2015, Part II, LNCS 9490, pp. 371–379, 2015.
DOI: 10.1007/978-3-319-26535-3_43

be a time-consuming operation. In this paper, we tackle above mentioned problems by making use of search engines to collect microblog messages dynamically which are topic-relevant, use clustering model to maintain topic delegates automatically, and organize summary coverage and newness for the time duration.

The key contributions of this work conclude: (1) we overcome the lack of contextual information for microblog messages. (2) We propose a continuous microblog stream summarization model, and detect the topic evolvement for arbitrary time span. (3) Extensive experiments on real datasets, which improve the quality and efficiency.

This paper is organized as follows. Section 2 covers related work. Section 3 is devoted to provide concepts in this work. In Sect. 4, we introduce continuous summarization model for microblog stream in detail. In Sect. 5, we report our experiment settings and results. Section 6 concludes the paper.

2 Related Works

While document summarization is a widely studied topic, the analysis of the impact caused by microblog stream aspects has only started in the last decade. There are researches close to our study: i.e., data stream clustering, microblog summarization.

Data Stream Clustering. In literature, data stream clustering task has received considerable attention, and most of methods build hierarchical tree to fully complete online clustering [11]. However, with the rapid growth of microblog over the internet, there is a great necessity to generate the evolvement clusters over different time duration. In particular, Aggarwal [2] has proposed to group microblogs into duration-based clusters by extending pyramidal time frame [1]. These approaches depend on the availability of generating large number clusters to a large extent. To break the current impasse between efficiency and storage utilization, we tap on additional cluster feature vectors from automatically acquired messages.

Microblog Summarization and Mining. It is understandable that many of methodologies were applied to microblog summarization [4]. Shou et al. proposed the Sumblr algorithm to summarize multiple tweet streams [13], which compressed tweets into Tweet Cluster Vectors and used a TCV-Rank summarization for generating online summaries. In [17], Wang and Li introduced an incremental hierarchical clustering framework which updates summaries as soon as new microblog message arrives in real time. Sharifi et al. [12] leveraged on trending phrases by taking advantage of the link structure among words, which achieved great improvements in the quality of summarizations. Other techniques, e.g., contextual topic model [16] and canonical differential evolution [3] have also been applied to this application.

Most of these methods focus on existing content of microblogs, a key difference between our works and above mentioned is our novel use search engine result to overcome the shortage of sufficient contextual information inherent to microblogs. In addition, we generate the evolvement summaries during the arbitrary time durations.

3 Preliminary

Before microblog message presentation, a pre-processing step is performed to alleviate the problem associated with the informal language which is often used in messages. Microblog services emphasize that the messages should be imposed with character limit. Users often utilize the abbreviation to maximize the information. For example, users may just use an abbreviated word "*app*" instead of "*application*", here the approach [18] is employed to normalize these messages. We introduce a lexicon which consists of 727 abbreviations for unique words [4].

In this paper, a normalized microblog is represented as a feature vector, while the significance of each feature is the weight of a word with the same semantics. We utilize the improved TF-IDF score and entropy value to calculate the value of each feature. For a limited time, generating summaries and providing solutions with stable performance become the emphasis of our study.

As a result, we define a microblog m_i as a tuple: (mv_i, mt_i), where mv_i is the feature vector which denotes as the textual of message, mt_i is the posted timestamp. It is necessary to calculate the statistic results during clustering process. Let the Microblog Feature Cluster Vector for cluster c be defined as: $MFCV_c = (mv_{sum}, mv_w, mt, n, sm_{set})$.

Where, c insists of microblogs $m_1, m_2, \dots m_n$; $mv_{sum} = \sum_{i=1}^{n} mv_i / \|mv\|$ is the sum of message feature vectors; $mv_w = \sum_{i=1}^{n} w_i \cdot mv_i$ is the sum of weighted message feature vectors (e.g., $w_i \cdot mv_i = \sum_{j=1}^{k} w_i^{(j)} \cdot mv_i^{(j)}$, all the $mv_i^{(j)}$ with the same semantics); $mt = \max(mt_1, mt_2, \dots mt_n)$ is the latest posted timestamp in cluster c; n is the number of messages in cluster c; sm_{set} is demoted as the closest k messages to the cluster centroid, the cosine similarity is used as the distance criteria of measurement.

In this paper, $MFCV_c$ is an extension of the tweet textual vector in [8], and $MFCV_c$ is deigned to update the evolvement summaries over time.

4 The Continuous Summarization by Clustering Stream Model

4.1 Microblog Stream Clustering

We use an incremental clustering method to build the feature vector hierarchy of the document inputs. Then the set of $MFCV_c s$ are incremental updated.

We query the Google search engine with the specific topic, the top 20 messages matched results are performed with pre-processing. Namely, the removal of stop words, normalization of these messages [18], word segmentation, and discovery of feature vector. Applications of natural language processing [14], word segmentation [7] and semantic word features extraction [8] are utilized in this process. Assuming that a microblog m arrives at time mt, N clusters are obtained at this time. m is assigned to one of the clusters, by calculating the distance between the feature vector of m(i.e., mv) and cluster centroids, here the centroid is denoted as $cv = mv_w / n$. Through computing the cosine similarity between cv and m, if the maximum of similarity $S(cv_c, m)$ is determined, c is the closet to m. Here $S(cv_c, m) = \frac{cv_c \cdot m}{\|cv_c\| \cdot \|m\|}$.

Our algorithm performs one of four possible operations: insert, create, merge: and delete. Case (1) and (2): In such case, an improved Minimum Bounding Similarity (iMBS) is described as follows. If the maximum of the similarity $S\left(cv_c, m\right)$ is smaller than iMBS, microblog m is generated a new cluster, otherwise, we add m to its closest cluster. The iMBS is defined as $\alpha \cdot \overline{S\left(cv_c, m_i\right)}\left(0 < \alpha < 1, m_i \in c\right)$.

Where α is a bounding factor, and $\overline{S\left(cv_c, m_i\right)}$ is the average cosine similarity between cv_c and m_i included in cluster c. α can be calculated by Eq. 1.

$$\alpha = \sum\nolimits_{i=1}^{n}\left(1 - w_{i0}\right)\left(1 - w_i\right) w_i^{n-1} \Big/ n \tag{1}$$

Where, w_{i0} is the initial weight of mv_i, w_i is the current weight of mv_i.

$$\overline{S\left(cv_c, m_i\right)} = mv_{sum} \cdot mv_w \big/ n \cdot \|mv_w\| \tag{2}$$

Case (3): If merge two clusters a and b into a new cluster c, the microblog obtaining the higher similarity with topic is selected as the representative message.

$$mr_c = \arg\max_{mr_a, mr_b}\left(S\left(topic, mr_a\right), S\left(topic, mr_b\right)\right) \tag{3}$$

Where, $mr_c = \arg\max\left(sm_{set}\right)$, mr_c is the most representative microblog of cluster c, mr_a and mr_b have similar meaning, $topic$ is the specific topic that we are concerned.

Furthermore, sm_{set} is demoted as the closest k messages to the cluster centroid, we use the Eq. 4 to compute the representative microblogs for cluster c.

$$sm_{set} = top\_k\left((1 - \beta) S\left(topic, m_i\right) + \frac{\beta}{n}\sum_{i \neq j} S\left(m_i, m_j\right)\right) \tag{4}$$

Where, $S()$ is the similarity function between pairs of microblog feature vectors, $S\left(m_i, m_j\right) = mv_i \cdot mv_j \big/ \|mv_i\| \cdot \|mv_j\|$. β is a parameter, and is set to 0.6 empirically.

Case (4): When the sub-topic is more attractive, the topic is discussed rarely, we employ the standard normal distribution of the posted timestamp to determine the expired cluster and remove it [13].

4.2   The CLR & Summ Model

To generate summaries of the microblog messages during arbitrary time duration, we propose a novel way to make use of external resourced to improve the quality of the obtained summaries. In general, suppose the start time of time duration is mt_s and the ending time is mt_e, the summaries between mt_s and mt_e are described as the summaries between $CS\left(mt_s\right)$ and $CS\left(mt_e\right)$. Where, $CS\left(mt_s\right)$ and $CS\left(mt_e\right)$ are the cluster sets at mt_s and mt_e, respectively. The corresponding $MFCV$ set of this subtraction is represented as $D\left(MFCV\right)$. All the messages in sm_{set} set of $D\left(MFCV\right)$ is denoted as $sm_{set}\left(D\right)$, the representative microblogs mr_{set} are messages with the highest similarity, we design a model to form summaries from selected representative microblogs.

Algorithm 1: CLR&Summ

   Input: the microblogs $sm_{set}(D)$,

   Output: the summaries $summ$,

1   $summ = \varnothing$;

2   **for** each microblog $m_i \in sm_{set}(D)$ **do**

3   build graph with node m ; calculate the similarity score of the node;

4   generate mr_{set} by calculating the centrality scores for $sm_{set}(D)$;

5   **for** each microblog $m_j \in mr_{set}$ **do**

6   calculate the similarity score by using the extension of LexRank and TextRank;

7   select microblog message m_{max} with the highest score;

8   $summ = summ \cup m_{max}$

9   **if** $len(summ) < 150$ **then**

10       $mr_{set} = mr_{set} - m_{max}$;

11   repeat step 5-8;

We draw our continuous microblog summarizations to implement graph-based approaches. In this representation, each microblog $m \in sm_{set}(D)$ is represented as a node in a graph. The similarity scores of the nodes are utilized to weight the edges linking up pairs of the nodes. Next, the extension of LexRank [5] and LexpageRank [6] are applied to calculate the centrality scores for microblogs $sm_{set}(D)$ [13]. The final summaries are selected the highest scoring microblogs until 150 words.

5   Experiments

We assembled real-world dataset from a popular microblog platform, Sina Weibo. We selected the dataset by conducting earthquake filtering form August to November 2014. By combining the main topic with earthquake, we got a query tracking that was then used to collect social media data from the Sina Weibo APIs. The real-world dataset consisted of earthquake topic and its sub-topics observed at different focuses approximately 32,000 messages. We implement the experiment to compare it against the following widely used summarization methods as the baseline systems, including: (1) Random, (2) ClusterSum [10] and (3) LexRank [6]. In the evaluation, we compare the results by different methods using the popular ROUGE toolkit for evaluation. We report ROUGE-1, ROUGE-2 and ROUGE-SU to compare our proposed method with the baseline methods. Experiments are implemented on a PC with CPU Inter(R) Core(TM) i7-2600, frequency 2.9 GHz, memory 8.00 GB. The Operation System is Microsoft Windows 7 Enterprise Edition. The development software is JRE 1.6.0 14.

5.1   Illustrative Example

The dataset is sampled for the illustrative example by Earthquake Management from August 20 to September 5 2014. The collection contains approximately 3,500 messages, which have been categorized into 3 phases based on the status: (1) damage during the growth of earthquake, (2) carry out rescue activities, (3) the recovery after earthquake. The approximate numbers of documents obtained at the end of the three phases are 1340, 760, and 1400, respectively. The summaries are listed in Table 1.

Table 1. Summaries of selected part for earthquake

1. The earthquake of XX Country was happened at XX time, the magnitude is XX, XX people die, XX people injure. 2. Traffic and road were damaged in a large extent, building collapse, the direct economic loss reached XX. XX people have been acquired salvation.
1. The malignance sub-event of this earthquake was happened at XX time, the number of casualties was reached XX people, the epicentral intensity level was XX level, the main economic losses was XX Counties. 2. The relevant department should be completed the further recues, guarantee the sufficient medical treatments. 3. Many Country Public Schools closed through XX days ago, transit bus, rail station service and international airport remain closed.
1. The medical treatments are sufficient to help all the people diastasis trouble, the foods and clothes are sufficient for victims. 2. The post-quake reconstruction is developing actively, the transit bus, rail station service and international airport will recover in near future. 3. Recently, the government receives many society donations, including fund, clothes, equipment....

5.2   Experimental Results

The evaluation experiments we will describe looks at how well our CLR&Summ does in microblogs summarization. Figure 1 presents the results for ROUGE-1, ROUGE-2 and ROUGE-SU scores. From the comparison results, we have the following observations: (1) Random has the worst performance. (2) ClusterSum and LexRank achieve better summary quality than Random. (3) CLR&Summ outperforms all the baseline methods in term of summarization quality. A major problem of the baseline methods: these methods rely on non-continuous stream to produce quality summaries. Unfortunately, the continuous stream leads to very frequent and expensive computations for them. In contrast, CLR&Summ has a good performance of summarization, because the nature of our method can better capture the event change and update.

We progressively increase the amount of data size available to different methods, at incremental steps of 5000. The effect this has on scalability results is plotted in Fig. 2. We see that our method outperforms the others significantly. CLR&Summ is faster than LexRank by nearly an order of magnitude at the data sizes 10000 and 15000.

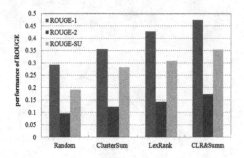

Fig. 1. Average summarization performance using ROUGE-1, ROUGE-2, ROUGE-SU

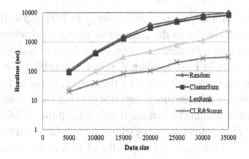

Fig. 2. Scalability with different data size

In following experiments, we vary only one parameter and fix others and only show the ROUGE-1 results in the following results. Now we examine the parameter k used in sm_{set}, which is demoted as the closest k messages to the cluster centroid. We gradually adjust the value of k from 10 to 50, and Fig. 3 demonstrates the influence of k. As shown in Fig. 4, the summary quality improves when k increases from 10 to 35. When $k \geq 35$, the improvement is not prominent, and more storage overhead will be incurred with a larger k. Thus, we choose $k = 35$ in our research.

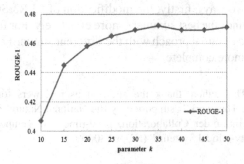

Fig. 3. Weight parameter k tuning using ROUGE-1

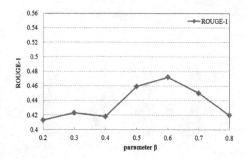

Fig. 4. Weight parameter β tuning using ROUGE-1

In the meanwhile, another parameter in sm_{set} is β $(0 < β < 0)$. When β $= 1$, the microblog message most similar to the topic is selected. And when β $= 0$, the centroid microblog is selected as the representative sentence. We gradually adjust the value of β from 0.2 to 0.8. As shown in Fig. 4, when β ≤ 0.4, microblogs related to other topics may be classified into the current clusters, so the performance of ROUGE-1 is with the low quality. When β $= 0.6$, CLR&Summ has the good performance, while β ≥ 0.6 lead to summarization quality reduction. An ideal value for β mc is 0.6.

Our work currently proposes the use of dynamic clustering to compress microblogs. CLR&Summ is helpful in improving the quality of generated summaries.

6   Conclusion

This paper discusses work on continuous microblog stream summarization to create updatable summaries on the related topic. The proposed approach CLR&Summ adopts a microblog stream clustering algorithm to understand messages and compress them into $MFCV_cs$ on-line. Then, a summary of contextual contents with arbitrary time duration is created, which organizes key sentences to detect topic evolvement dynamically. We showed that the resulting summarization model based on the proposed clustering approach is efficiency and effectiveness. In further work, we will further improve our approach mainly in two ways: firstly, the modification of CLR&Summ algorithm will be developed so that finds the best summary more effectively. For further work, we plan to develop multi-topic of our approach with the different similarity measures, and evaluate its organization more complete.

Acknowledgments. The authors thank the anonymous reviewers for their insightful and constructive comments. This work is supported by the National Natural Science Foundation of China "Research on High-order Collaboration, Real-time and Temporal Characteristics in Automatic Test of Safety-critical Systems" (NO. 61300007).

References

1. Aggarwal, C.C., Han, J., Wang, J., Yu, P.S.: A framework for clustering evolving data streams. In: VLDB, pp. 81–92 (2003)

2. Aggarwal, C.C., Yu, P.S.: On clustering massive text and categorical data streams. Knowl. Inf. Syst. **24**(2), 171–196 (2010)
3. Alguliev, R.M., Aliguliyev, R.M., Isazade, N.R.: DESAMC + DocSum: differential evolution with self-adaptive mutation and crossover parameters for multi-document summarization. Knowl.-Based Syst. **36**, 21–38 (2012)
4. Chen, Y., Zhang, X.M., Li, Z.J., Ng, J.P.: Search engine reinforced semi-supervised classification and graph-based summarization of microblogs. Neurocomputing **152**, 274–286 (2015)
5. Erkan, G., Radev, D.R.: LexRank: graph-based lexical centrality as salience in text summarization. J. Artif. Intell. Res. **22**, 457–479 (2004)
6. Erkan, G., Radev, D.R.: Lexpagerank: prestige in multi-document text summarization. In: EMNLP, pp. 365–371 (2004)
7. George, T.: Optimizing word segmentation tasks using ant colony metaheuristics. Literary Linguist. Comput. **29**(2), 234–254 (2014)
8. Han, X.P., Zhao, J.: Named entity disambiguation by leveraging wikipedia semantic knowledge. In: CIKM, pp. 215–224 (2009)
9. Harabagiu, S.M., Hickl, A.: Relevance modeling for microblog summarization. In: ICWSM (2011)
10. Inouye, D., Kalita, J.K.: Comparing twitter summarization algorithms for multiple post summaries. In: SocialCom, pp. 298–306 (2011)
11. Peng, T., Liu, L.: A novel incremental conceptual hierarchical text clustering method using CFu-tree. Appl. Soft Comput. **27**, 269–278 (2015)
12. Sharifi, B., Hutton, M.A., Kalita, J.: Summarizing microblogs automatically. In: HLT-NAACL, pp. 685–688 (2010)
13. Shou, L.D., Wang, Z.H., Chen, K., Chen, G.: Sumblr: continuous summarization of evolving tweet streams. In: SIGIR, pp. 533–542 (2013)
14. Verma, S., Vieweg, S., Corvey, W.J., Palen, L., Martin, J.H., Palmer, M., Schram, A., Anderson, K.M.: Natural language processing to the rescue? Extracting situational awareness tweets during mass emergency. In: ICWSM, pp. 49–57 (2011)
15. Wan, X.J., Yang, J.W.: Multi-document summarization using cluster-based link analysis. In: SIGIR, pp. 299–306 (2008)
16. Yang, G.B., Wen, D.W., Kinshuk, Chen, N.S., Sutinen, E.: A novel contextual topic model for multi-document summarization. Expert Syst. Appl. **42**(3), 1340–1352 (2015)
17. Wang, D.D., Li, T.: Document update summarization using incremental hierarchical clustering. In: CIKM, pp. 279–288 (2010)
18. Zhang, C., Baldwin, T., Ho, H., Kimelfeld, B., Li, Y.: Adaptive parser-centric text normalization. In: ACL, pp. 1159–1168 (2013)

Versatile English Learning System Using Webpages as Learning Materials

Yuki Oikawa, Kozo Mizutani, and Masayuki Arai[✉]

Department of Human Information Systems, Teikyo University, Tokyo, Japan
{mizutani,arai}@ics.teikyo-u.ac.jp

Abstract. In Japan, there is a strong demand for English language instruction not only from students, but also from working adults. However, the nation has few systems that have functions for improving English skills (listening, speaking, reading, and writing), are simple to operate, and inexpensive. To solve the problem, we propose a versatile English learning system that uses webpages as learning materials. The proposed system has following characteristics: it creates Japanese versions of webpages translated from English, it can determine a user's English skill, it is capable of speaking and listening interactions, it does not require an installation fee, and it is easy to operate.

Keywords: English · Learning system · Using webpage · Multi-functionality

1 Introduction

In the face of the increasing globalization of today, Japan's Ministry of Education, Culture, Sports, Science and Technology (MEXT) has mandated changes to the nation's high school English language curriculum, with the goal of giving students, by the time of their high school graduation, English skills in the four areas of listening, speaking, reading, and writing that can be expected to be useful throughout their entire lives [1].

Additionally, today's working adults are also strongly motivated to improve their English skills because the number of overseas local subsidiaries of Japanese companies has increased about four times the level of 30 years ago, from 4,000 in 1979 to 17,000 in 2007, and currently more than 70 % of all Japanese companies consider their employees' English skill levels to be more important than they did 5 years ago [2].

Thus, it is clear that the demand for English instruction is rising for working adults as well as students. However, the nation has few systems that have functions for improving English skills (listening, speaking, reading, and writing) that are simple to use and inexpensive. To solve this problem, we propose a versatile English learning system that uses webpages as learning materials.

The paper consists of six sections. In Sect. 2, we provide an outline of the system and describe its requirements; in Sect. 3, the system configuration and procedures are discussed. In Sect. 4, we describe system functions and implementation methods. In Sect. 5, we review related systems. Finally, in Sect. 6, we provide our conclusions.

© Springer International Publishing Switzerland 2015
S. Arik et al. (Eds.): ICONIP 2015, Part II, LNCS 9490, pp. 380–389, 2015.
DOI: 10.1007/978-3-319-26535-3_44

2   Requirements and System Outline

When formulating the requirements for our system, we surveyed students and faculty members of our university. Table 1 shows the results. Most students and faculty members said they wanted a system with functions that would help them to learn the four basic English skills (listening, speaking, reading, and writing), that they wanted to use materials that were appropriate to their skill levels, and that they wanted to utilize the system without the necessity of a troublesome installation process. Some people added that they wanted to use their favorite English webpages as learning materials.

Table 1. System and English learning requirement

English learning requirements
(1) Utilize the user's favorite webpages for learning
(2) Improve reading comprehension by sentence or paragraph units
(3) Review words previously studied
(4) Learn unfamiliar words
(5) Improve speaking skill
(6) Improve listening skill
(7) Improve writing skill
(8) Use teaching materials that match user levels
Learning systems requirements
(9) Does not require installation
(10) Simple operation

Figure 1 describes how to execute our system. First, users access their favorite webpage, then click the bookmark "English Learning" button. This allows them to obtain the learning materials shown in the right side of the figure.

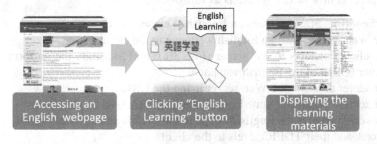

Fig. 1. System execution

Figure 2 shows a view of the client, which consists of three panels, as shown in Fig. 2(a), (b), (c). The system can utilize the users' favorite webpage as shown in Fig. 2(a), which corresponds to requirement 1 of Table 1. It is also capable of displaying a Japanese translation of an English page, as shown in Fig. 2(b), which meets requirement 2 of Table 1. Additionally, users can select unlearned or previously learned words, as shown in Fig. 2(c), which meets requirements 3 and 4 in Table 1. Furthermore, as also shown in Fig. 2(c), users can study speaking and listening (thus meeting requirements 5 and 6 in Table 1) and select their Test of English for International Communication (TOEIC) [3] level as shown in Fig. 2(c), thus matching the requirement 8 of Table 1.

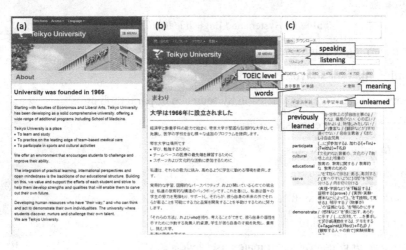

Fig. 2. System overview

3   System Configuration and Procedures

Figure 3 depicts the configuration of the system and its procedures. First, a client computer (Fig. 3(a)) accesses an English Web server (Fig. 3(b)) on the Internet and obtains English learning materials based on the webpage received from our developed server (Fig. 3(c)). One of the materials is a Japanese translation of an English webpage generated by the English-Japanese translation server (Fig. 3(d)).

The process flow of the system is as follows:

(1) A client (Fig. 3(a)) accesses an English Web server (Fig. 3(b)).
(2) The Web server sends an English webpage to the client.
(3) The client connects to our server (Fig. 3(c)) and transmits the uniform resource locater (URL) of the English webpage.
(4) Our server accesses the Web server using the URL.
(5) Our server obtains the same English webpage received by the client.
(6) Our server divides the English webpage into words and send the words, their definitions, and their TOEIC levels to the client.

(7) The client then connects an English–Japanese translation server (Fig. 3(d)) and sends it the URL of the English webpage.

(8) The client receives the Japanese translation of the English webpage from the translation server.

We used the hypertext markup language (HTML), cascading style sheets (CSS), and the JavaScript, Java Servlet, JQuery, and Web Speech application programming interfaces (APIs) to construct the system.

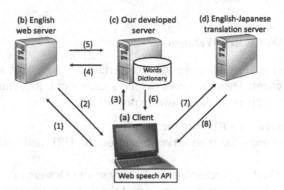

Fig. 3. System configuration and procedures

4 Implementation Methods

4.1 Obtaining the Webpage URL

To implement the functions for translating English webpages and unlearned or previously learned words on the pages, the JavaScript bookmarklet [4] shown in Fig. 4 is installed in the clients.

```
Javascript:(
  function(){
    var url = location.href;
    var redirect_url = "http://example.com/?url=" + encodeURIComponent(url);
    document.location = redirect_url;
  }
)()
```

Fig. 4. JavaScript bookmarklet for obtaining URL

4.2 Generating English–Japanese Word Dictionary

Our developed server utilizes an English–Japanese word dictionary (EJDIC) [5] and an English word frequency list [6] that are generated in advance. Each dictionary record consists of an English word, its grammatical category, definitions, example sentences, and TOEIC level. Table 2 shows the frequency to TOEIC level conversion table.

Table 2. Frequency order to TOEIC level conversion table

Frequency order	TOEIC level
Top 2000 words	340
3600	470
5000	600
8000	730
10000	860

4.3   Displaying Words in a Webpage

Our server obtains an English webpage, converted the page, and sends the words dictionary to the client. Then the client displays each word, its definition, and other information in the webpage as the following procedure:

(1) The server receives a URL from the client.
(2) The server connects the Web server based on the URL and obtains the English webpage.
(3) The server converts uppercase letters of every word to lowercase except for HTML tags, and sends the converted webpage and the words dictionary.
(4) The client extracts every word in the webpage using the JavaScript split method.
(5) The client finds each word in the dictionary and gets its meaning, example sentences, and TOEIC level.
(6) In situations where users select "previously learned words," the client selects the words and other information that are beneath the user's TOEIC level. In contrast, if users selects "unlearned words," the client selects the words and other information that are above the user's TOEIC level to the client.

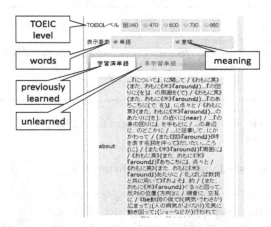

Fig. 5. Example situation where user selects 340 for TOEIC level, words, meaning, and previously learned

Figure 5 depicts an example for situations where the user selects "340" for the TOEIC level, "words," "meaning," and "previously-learned," while Fig. 6 shows an example of a situation where the user selects "600" for the TOEIC level, "words," and "unlearned."

Fig. 6. Example situation where user selects 600 for TOEIC level, words, and unlearned

4.4 Displaying Translated Japanese Page

In order to display a Japanese page (as shown in Fig. 2(b)) translated from the English webpage (as shown in Fig. 2(a)), we utilize an outside service such as EXCITE [7] to perform the translation. Figure 7 shows an example of URLs for using the service. The URL in the upper part of Fig. 7 is for accessing an English Web site, and the bottom is for obtaining the Japanese translation of the English page.

> URL for the English Web server
> http://www.teikyo-u.ac.jp/english/about/
>
> URL for the English-Japanese Web server
> http://www.excite-webtl.jp/world/english/web/?wb_url=
> http://www.teikyo-u.ac.jp/english/about/&wb_lp=ENJA

Fig. 7. Example of URLs for accessing the English Web server and the English–Japanese translation server

4.5 Speaking

We implement the speaking function via the Web Speech API in client computers. Figure 8 shows part of the JavaScript codes used for this function. It begins by generating an instance of webkitSpeechRecognition shown in Fig. 8(1). To listen continuously to the user's voice, the parameter shown in Fig. 8(2) is set to true. If the system obtains a candidate word temporarily, the word can be displayed. Therefore, the parameter shown in Fig. 8(3) must be set to true. The code in Fig. 8(4) means the English language is used. Finally, the system displays the word recognition result as shown in Fig. 8(5).

The system can depict a word as shown in Fig. 9 in situations where it recognizes the voice inputted from a microphone.

```
//(1) generating an instance
var recognition = new webkitSpeechRecognition();

//(2) cotinuing voice recognition
recognition.continuous = true;

//(3) indicationg an interim word
recognition.interimResults = true;
            :

//starting voice recognition
function startButton(event) {
            :
    //(4) setting language
    recogniton.lang="en-US";
    recognition.start();
};

// go to the right column
```

```
//extracting a word from the voice
recognition.onresult = function(event) {
            :
    if(typeof(event.results) == 'undefined') {
        recognition.onend = null;
        recognition.stop();
        return;
    }
            :

//(5) extracting a candidate word
for(var i = event.resultIndex; i < event.results.length;
++i) {
    if (event.results[i].isFinal) {
        final_transcript += event.results[i][0].transcript;
    }
    else{
        interim_transcript += event.results[i][0].transcript;
    }
}
            :
}
```

Fig. 8. Part of JavaScript for implementing speaking function using Web Speech API

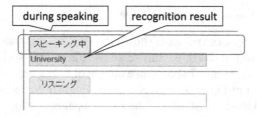

Fig. 9. Example of executing the speaking function

4.6 Listening

The listening function is also implemented via the Web Speech API. Figure 10 shows a part of JavaScript for implementing this function. The code shown in Fig. 10(1) generates an instance of SpeechSynthesisUtterance. Figure 10(2) is the text attribute, and it can set the text for speaking sentences or words. The maximum length of the text is 32,767 letters. The language attribute shown in Fig. 10(3) gives the system the ability to specify the text language. For US English, we set "en-US." The volume attribute shown in Fig. 10(4) can be used to adjust the volume of the speech. A float value between

0.0 and 1.0 should be specified. The rate attribute shown in Fig. 10(5), which is a float value between 0.0 and 10.0, defines the speed at which the text should be spoken. The pitch attribute, which is a float value between 0.0 and 2.0., controls the frequency level at which the text is spoken.

```
//(1) generating an instance
var msg = new SpeechSynthesisUtterance();

//(2) text attribute
msg.text = "" ;

//(3) lang attribute
msg.lang = "en-US";

//(4) volume attribute (float value 0.0-1.0)
msg.volume = 1;

//(5) rate attribute (float value 0.0-10.0)
msg.rate = 1;

//(6) pitch attribute (float value 0.0-2.0)
msg.pitch = 2;
```

Fig. 10. Example for executing the speaking function using Web Speech API

When a user clicks the "previously learned words" or "unlearned words" tabs, window.getSelection() is called and obtains the word corresponding to the tab. Figure 11 shows the process that occurs when a user selects "university." As can be seen in the figure, the system displays the word in the "listening" textbox. Then the user can hear the pronunciation of the word when he or she clicks the "listening" button.

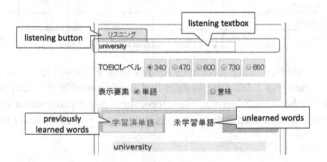

Fig. 11. Example for executing the listening function

5   Related Systems and Research

In this section, we review some of the more popular English learning systems designed for Japanese users that are similar to our proposed system. For example, the P-Study System [8] employs four selectable modes and a typing mode for studying English words and idiomatic phrases. It also includes questions for junior high school students, the 2nd Grade of the Society for Testing English Proficiency (STEP) Test in Practical English Proficiency, and TOEIC. Additionally, users can learn English words and sentences via listening exercises and can confirm pronunciation for English words using the voice replay function.

Another system, Reader's Digest English20 [9], permits the use of movies and dramas and learners study English by listening to the characters talks. Additionally, AmiVoice® CALL -pronunciation- [10] and Hatsuon Kentei® [11] can be used for improving English speaking skills. These systems can record learners' pronunciation, analyze voices, compare them to the correct pronunciation, indicate incorrect articulation, and suggest improvements. For tablet and smartphone users, MyET [12] is available. This system allows users to recognize their weaknesses by listening to their own voices. Table 3 shows comparisons between the various English learning systems for Japanese learners, including our proposed system. As can be seen in the table, the unique characteristics of your system include its use of webpages and the fact that no installation is required.

Table 3. English learning systems comparison

	P-Study System	Reader's Digest English20	AmiVoice ® CALL - pronunciation-	Hatsuon Kentei ®	MyET	Our system
Word Level Learning	✓					✓
Sentence Level Learning	✓					✓
Listening	✓	✓				✓
Speaking		✓	✓	✓	✓	✓
Writing	✓	✓				
Using teaching materials in accordance with the user's level	✓					✓
Using favorite web pages						✓
Without installation						✓
Simple operation	✓	✓	✓	✓	✓	✓

Sano and Ino studied using webpages for learning English [13]. The goal of their research was to develop a learning system that could provide free English materials adapted to the needs of each learner. Therefore, their paper proposed a way to extract learning materials from a corpus consisting of webpages and evaluated the difficulty of the material.

6 Conclusion

In this paper, we described an English learning system that uses webpages as learning materials. In the sections above, we discussed the outline, characteristics, and implementation methods for the proposed system. As part of our future studies, we intend to implement a function that will permit the system to recommend appropriate learning materials based on users' skill levels and will evaluate this system via practical use in actual classes.

Acknowledgement. The authors would like to thank the members of the Arai Laboratory, Department of Human Information Systems, School of Science and Engineering, Teikyo University for their useful advice and assistance during the system evaluation.

References

1. The Ministry of Education, Culture, Sports, Science and Technology Japan. http://www.mext. go.jp/b_menu/shingi/chousa/shotou/102/houkoku/attach/1352463.htm. Accessed 28 February 2015
2. Koike, I., Takada, S., Matsui, J., Terauchi, H.: The Institute for International Business Communication, "English Skills: What Do Companies really need," Asahi Press (2010, in Japanese)
3. TOEIC. http://www.toeic.or.jp. Accessed 28 February 2015
4. Negrino, T., Smith, D.: Visual QuickStart Guide: JavaScript, 9th edn. Peachpit Press, San Francisco (2014)
5. Ejdic. http://kujirahand.com/web-tools/EJDictFreeDL.php.Accessed 28 February 2015
6. Wiktionary:Frequencylists. http://en.wiktionary.org/wiki/Wiktionary:Frequency_lists Accessed 28 February 2015
7. EXCITE translation service. http://www.excite.co.jp/world/english/web/. Accessed 28 February 2015
8. P-Study System. http://www.takke.jp/. Accessed 2 March 2015
9. Reader's Digest English20. http://www.sourcenext.com/product/pc/edu/pc_edu_000654/. Accessed 2 March 2015
10. AmiVoice® CALL –pronunciation. http://www.advanced-media.co.jp/products/ami voicecallpronunciation.html. Accessed 2 March 2015
11. Hatsuon Kentei®. http://www.prontest.co.jp/wp/new-bar/. Accessed 2 March 2015
12. MyET. http://www.alc.co.jp/speaking/myet/. Accessed 2 March 2015
13. Sano, H., Ino, M.: Measurement of difficulty on English Grammar and automatic analysis. IPSJ Trans. CE **117**, 1–12 (2000). (in Japanese)

A Fast Algorithm for Local Rank Distance: Application to Arabic Native Language Identification

Radu Tudor Ionescu[✉]

Faculty of Mathematics and Computer Science,
University of Bucharest, 14 Academiei Street, Bucharest, Romania
raducu.ionescu@gmail.com

Abstract. A novel distance measure for strings, termed Local Rank Distance (LRD), was recently introduced. LRD is inspired from rank distance, but it is designed to conform to more general principles, while being more adapted for specific data types, such as DNA strings or text. More precisely, LRD measures the local displacement of character n-grams among two strings. Local Rank Distance has already demonstrated promising results in computational biology and native language identification, but the algorithm used to compute LRD is computationally expensive. In this paper, an efficient algorithm for LRD is proposed. The main efficiency improvement is to build a positional inverted index for the character n-grams in one of the compared strings. Then, for each n-gram in the other string, a binary search is used to find the position of the nearest matching n-gram in the positional inverted index. The proposed algorithm is more than two orders of magnitude faster than the original algorithm. An application of the described algorithm is also exhibited in this paper. Indeed, state of the art results are presented for Arabic native language identification from text documents.

Keywords: Local Rank Distance · Rank distance · String kernel · String similarity · Character n-grams · Native language identification · Fast algorithm

1 Introduction

Researchers have developed a wide variety of methods for string data, that can be applied with success in different fields such as computational biology, natural language processing, information retrieval and so on. Such methods range from clustering techniques used to analyze the phylogenetic trees of different organisms, to kernel methods used to identify authorship or native language from text. Several state of the art methods for solving such tasks are based on analyzing and comparing the strings at the character level. These methods are often based on a similarity or distance function between strings. Popular choices for recent techniques are the Hamming distance [25], edit distance [22], Kendall-tau distance [20], rank distance [3], string kernels [18,21], among others.

© Springer International Publishing Switzerland 2015
S. Arik et al. (Eds.): ICONIP 2015, Part II, LNCS 9490, pp. 390–400, 2015.
DOI: 10.1007/978-3-319-26535-3_45

Rank distance [6] is a low computational complexity measure of similarity with various applications in computational biology [3,8] and natural language processing [7]. Despite of having such broad applications, rank distance might sometimes be not fully adequate for specific data types, such as DNA strings. This probably happens because rank distance is a rather trivial extension for strings of the Spearman's Footrule which was developed to measure the distance between permutations. A recently introduced distance measure, termed Local Rank Distance (LRD) [10], comes from the idea of better adapting rank distance to string data, in order to comprise a better similarity (or dissimilarity) between strings, such as DNA sequences or text documents. In fact, LRD borrows some concepts from Local Patch Dissimilarity (LPD) [4], which is an extension of rank distance for images. Rank distance needs to annotate strings with indexes in order to eliminate duplicate characters. Unlike rank distance, LRD does not require for strings to be annotated, just as LPD does not require for image pixels to be annotated. Furthermore, LRD is based on character n-grams instead of single characters, in the same way as LPD is based on patches instead of single pixels. Local Rank Distance has already shown promising results in computational biology [5,10] and native language identification [11,19]. Having such broad applications, LRD has proved to be a general approach to measure string similarity. However, the brute force search algorithm for computing LRD presented in [10] is computationally expensive. In this paper, a fast algorithm for computing LRD is proposed. The main efficiency improvement brought to the LRD algorithm is to store n-gram positions in a hash table for one of the compared strings. More precisely, the hash table constructed from a string will contain an array (or set) for each n-gram the occurs in the respective string. The array will contain all the positions of a particular n-gram in the string. This hash table is actually a positional inverted index structure that is very popular in information retrieval. When LRD is computed, it is no longer necessary to do an extensive search within a window of fixed size to find equal n-grams between the two strings, as described in [10]. Instead, for each n-gram in one of the strings, binary search is employed to find the position of the nearest identical n-gram in the positional inverted index built from the other string. An empirical evaluation is conducted to quantify the amount of improvement in terms of time. The results indicate that the efficient algorithm is more than two orders of magnitude faster than the brute force search algorithm. Furthermore, an application of the fast algorithm is also presented in this paper. More precisely, a kernel based on LRD is used for Arabic native language identification. Compared with the state of the art method described in [15], the approach based on LRD is up to 10 % better in terms of accuracy.

The paper is organized as follows. A state of the art of string similarity measures and approaches that work at the character level is given in Sect. 2. The efficient LRD algorithm is described in Sect. 3. The experiments on native language identification and the time evaluation are presented in Sect. 4. The final remarks are given in Sect. 5.

2 Related Work

Natural language processing (NLP) is a vast domain that studies machine translation, document classification by topic, authorship identification, sentiment analysis, among others. Using words is natural in such text analysis and classification tasks. Perhaps surprisingly, recent results have proved that methods handling the text at character level can also be very effective for specific text analysis tasks [9,14,18,21]. In [14] string kernels were used for document categorization with very good results. String kernels were also successfully used in authorship identification [18,21] and native language identification [19].

Methods working at the word level or above very often restrict their feature space according to theoretical or empirical principles. For example, they select only features that reflect various types of spelling errors or only some type of words such as function words. These features prove to be very effective for specific tasks, but other, potentially good features, depending on the particular task, may exist. String kernels embed the texts in a very large feature space, given by all the substrings of length n, and leave it to the learning algorithm to select important features for the specific task, by highly weighting these features. A kernel based on Local Rank Distance was successfully used for English native language identification [11,19]. In this paper, a similar approach is used for Arabic native language identification.

The first application of string kernel ideas came in the field of text categorization, with the work of [14], followed by applications in bioinformatics [12]. In the recent years, character n-grams have found increasingly new applications in both of these fields. In computational biology, character n-grams are more commonly referred to as k-mers or polymers. There are recent studies that use k-mers for the phylogenetic analysis of organisms [13], or for sequence alignment [17]. Analyzing DNA at the substring level is also more suited from a biological point of view, because DNA substrings may contain meaningful information. For example, genes are encoded by a number close to 100 base pairs, or codons that encode the twenty standard amino acids are formed of 3-mers.

3 Efficient Local Rank Distance Algorithm

Local Rank Distance [10] is a recently introduced distance measure that emerged from the idea of redesigning rank distance [6] in order to capture a better similarity between strings, such as DNA sequences or text documents. More precisely, LRD borrows some of the concepts used to develop Local Patch Dissimilarity [4], which is an extension of rank distance to images. More details about LRD and the brute force search algorithm to compute it are given in [10].

Performing an extensive search within a window of fixed size to find the spatial displacement of equal n-grams between two strings is computationally expensive. The efficient algorithm aims to improve the computational time of LRD by avoiding this costly search step. Given two strings x and y, the efficient algorithm to compute $\Delta_{LRD}(x, y)$ works as described next. First of all, a hash table is used to store all the n-grams and their positions in y. More precisely, each n-gram that occurs at least once in y will correspond to a key in the hash table. The value associated to a specific key (n-gram) is a set that will contain all the positions of the respective n-gram in the string y. The hash table h can also be described as a positional inverted index structure. The next step is to consider every n-gram in x and to look it up in the hash table h. Let i denote the position of the currently considered n-gram in x. If the n-gram is found in h, the next step is to carry out a binary search in the set stored at $h(x[i : i+n])$, in order to find the nearest position j that minimizes $|i - j|$. The spatial offset $|i - j|$ is added to the distance sum only if $|i - j|$ is less than the maximal offset m, otherwise, m is added. If the n-gram $x[i : i + n]$ is not found in h, m is added to the distance sum. The final sum obtained by this algorithm is the $\Delta_{left}(x, y)$ partial sum from Definition 1 of [10]. It can be easily observed that $\Delta_{right}(x, y) = \Delta_{left}(y, x)$. Therefore, by switching the roles of x and y in the algorithm described so far, the $\Delta_{right}(x, y)$ partial sum can be obtained.

The efficient algorithm to compute either one of the partial sums is formally described in Algorithm 1. The $[i]$ operator is used in the algorithm to identify the i-th element of a set, such as the set of positions stored in h for a certain n-gram. The algorithm can be divided into two stages. In the first stage (steps 11–15), the positional inverted index table h is computed from the input string y. In the second stage (steps 16–37), a binary search algorithm is employed for each n-gram that x and y have in common, in order to find the position in y that minimizes the spatial offset from the position in x of each common n-gram. In order to compute LRD, the algorithm has to be executed twice, once to compute $\Delta_{left}(x, y)$ and once to compute $\Delta_{right}(x, y)$. The analysis of the computational complexity of Algorithm 1 is straightforward. Let $l_1 = |x|$ and $l_2 = |y|$ denote the lengths of the two strings. Searching and storing keys in the hash table can be done in $O(1)$ time. The binary search takes $O(\log n)$. The first stage of the algorithm can be computed in $O(l_2)$ time. The second phase is more computationally expensive, requiring $O(l_1 \cdot \log l_2)$ time. Consequently, the overall complexity of Algorithm 1 is $O(l_1 \cdot \log l_2)$. Since Algorithm 1 has to be executed twice to compute each partial sum, LRD can be computed in $O(l_1 \cdot \log l_2 + l_2 \cdot \log l_1)$. Unlike the brute force search LRD algorithm [10], the computational time of the efficient LRD algorithm is no longer limited by the maximal offset m of LRD. This is a clear advantage of this faster implementation.

In practice, the input parameters of Algorithm 1 should be carefully adjusted with respect to the kind of strings (texts or DNA sequences) and the approached task. For example, setting $n = 10$ to use 10-mers for DNA strings of 100 or 200 bases is probably not reasonable, since finding similar 10-mers in such short DNA strings is rare, if not almost impossible. But 3 to 5-mers are probably more suitable for such short DNA strings. In the case of authorship identification, it is worthwhile using 2 to 5-grams that correspond to function words which have been found to provide a good indication of the author style [18]. Thus, good knowledge of the domain and the approached task is very useful, but the parameters can be easily tuned through a validation procedure, as shown in

---

Algorithm 1. Efficient LRD Algorithm

```
 1  Input:
 2  x, y – the input strings;
 3  n – the length of the n-grams to be compared;
 4  m – the maximum spatial offset;

 5  Notations:
 6  h – the positional inverted index table;
 7  o – the minimum offset between matched n-grams;
 8  beg, mid, end – indexes used in the binary search algorithm;
 9  ⌈r⌉ – the rounding function that returns the nearest integer value to r;

10  Computation:
11  for i ∈ {1, ..., |y| − n + 1} do
12      if h(y[i : i + n]) ≠ ∅ then
13          ⌊ h(y[i : i + n]) ← h(y[i : i + n]) ∪ {i};

14      else
15          ⌊ h(y[i : i + n]) ← {i};

16  Δ ← 0;
17  for i ∈ {1, ..., |x| − n + 1} do
18      if h(x[i : i + n]) ≠ ∅ then
19          o ← m;
20          beg ← 1;
21          end ← |h(x[i : i + n])|;
22          while end − beg > 1 do
23              mid ← ⌊(beg + end)/2⌉;
24              if i = h(x[i : i + n])[mid] then
25                  ⌈ beg ← mid;
26                  ⌊ end ← mid;

27              else
28                  if i < h(x[i : i + n])[mid] then
29                      ⌊ end ← mid;

30                  else
31                      ⌊ beg ← mid;

32          for j ∈ {beg, ..., end} do
33              if |i − h(x[i : i + n])[j]| < o then
34                  ⌊ o ← |i − h(x[i : i + n])[j]|;

35          Δ ← Δ + o;

36      else
37          ⌊ Δ ← Δ + m;

38  Output:
39  Δ - the Δ_left partial sum between x and y.
```

Sect. 4.3. An interesting note on improving the efficiency when computing LRD for multiple strings is to build the hash table for each string only once. For this, the two stages of Algorithm 1 need to be completely separated. Once the hash table for a certain string y has been built, it can be reused to compute all partial sums $\Delta_{left}(x, y), \forall x \in X − \{y\}$, where X is the set of strings. Furthermore, it is sufficient to keep only one hash table in memory at a time, considerably reducing the memory footprint. Nevertheless, the algorithm can also be easily executed in a parallel computing environment.

4   Arabic Native Language Identification

The goal of automatic native language identification (NLI) is to determine the native language of a language learner, based on a piece of writing in a foreign language. Most research has focused on identifying the native language of English language learners [24], though there have been some efforts recently to identify the native language of writing in other languages, such as Chinese [16] or Arabic [15]. This section presents results based on LRD for Arabic native language identification and compares them with the results presented in [15].

4.1   Data Set

The second version of the Arabic Learner Corpus was recently introduced in [1]. The corpus includes essays by Arabic learners studying in Saudi Arabia. There are 66 different mother tongue (L1) representations in the corpus, but the majority of these have less than 10 essays. To compare the LRD approach with the state of the art method, the subset of ALC used in [15] is also used in this paper. The subset is obtained by considering only the top 7 native languages by number of essays. The ALC subset used in the experiments contains 329 essays in total, which are not evenly distributed per native language. Indeed, there are 76 essays written by Chinese speakers, 35 by English speakers, 44 by French speakers, 36 by Fulani speakers, 46 by Malay speakers, 64 by Urdu speakers, and 28 by Yoruba speakers. LRD treats the essays simply as strings. Because LRD works at the character level, there is no need to split the texts into words, or to do any NLP-specific pre-processing. The only editing done to the texts was the replacing of sequences of consecutive space characters (space, tab, new line, and so on) with a single space character. This normalization was needed in order to demote the artificial increase or decrease of the similarity between texts as a result of different spacing.

4.2   Learning Method

By normalizing LRD, it becomes a dissimilarity measure. LRD can also be used as a kernel, since kernel methods are based on similarity. The classical way to transform a distance or dissimilarity measure into a similarity measure is by using the RBF kernel [23]:

$$k_n^{LRD}(x, y) = e^{-\frac{\Delta_{LRD}(x, y)}{2\sigma^2}},$$

where x and y are two strings and n is the n-grams length. The parameter σ is usually chosen so that values of $k(x, y)$ are well scaled. In the above equation, Δ_{LRD} is already normalized to a value in the $[0, 1]$ interval to ensure a fair comparison of strings of different length, and so, σ is set to 1 in the experiments.

Kernel-based learning algorithms work by embedding the data into a Hilbert feature space, and searching for linear relations in that space. The embedding is performed implicitly, that is by specifying the inner product between each pair of points rather than by giving their coordinates explicitly. More precisely, a kernel matrix that contains the pairwise similarities between every pair of training samples is used in the learning stage to assign a vector of weights to the training samples. Various kernel classifiers differ in the way they learn to separate the samples. In the case of binary classification problems, Kernel Ridge Regression (KRR) selects the vector of weights that simultaneously has small empirical error and small norm in the Reproducing Kernel Hilbert Space generated by the kernel function. The KRR classifier is chosen for the NLI experiments.

4.3 Parameter Tuning for LRD Kernel

A set of preliminary experiments were performed to determine the range of n-grams that gives the most accurate results. A 10-fold cross-validation procedure was conducted on half of the ALC subset in order to evaluate the n-grams in the range 2–7. The best performance was obtained using 4-grams, which is correlated to the average word length in the ALC corpus (4.36). In general, a blended spectrum of n-grams produces better results, and therefore, the combination of 3–5 n-grams (centered around 4-grams) was chosen for the rest of the experiments on the ALC corpus. Preliminary results on the Arabic corpus indicate that using the range of 3–5 n-grams gives a roughly 2 % improvement over using n-grams of length 4 alone. Other ranges of n-grams were not evaluated to prevent any overfitting. Remarkably, other authors [2,19] also report better results on English corpora by using n-grams with the length in a range, rather than using n-grams of fixed length. Finally, the maximum offset parameter m involved in the computation of LRD was chosen so that it generates a search window size close to the average number of letters per document from the ALC set. There are 232792 letters and 329 documents in the ALC corpus, so there are roughly 708 characters per documents. Thus, m was set to 708.

4.4 Results and Discussion

The authors of [15] are the first (and only) to show NLI results on Arabic data. In their evaluation, the 10-fold CV procedure was used to evaluate an SVM model based on several types of features including Context-Free Grammar (CFG) rules, function words, and part-of-speech n-grams. To directly compare the system based on LRD with the approach described in [15], the same folds should have been used. Since the folds are not publicly available, two solutions were adopted to fairly compare the two NLI approaches. First of all, the system based on LRD is evaluated by repeating the 10-fold CV procedure for 20 times and averaging the resulted accuracy rates. The folds are randomly selected at each trial. This helps to reduce the amount of accuracy variation introduced by using a different partition of the data set for the 10-fold CV procedure. To give an idea of the amount of variation in each trial, the standard deviations for the computed average accuracy rates are also reported. Second of all, the leave-one-out (LOO) cross-validation procedure was also adopted because it involves a predefined

Table 1. Accuracy rates of the KRR based on the LRD kernel compared with a state of the art approach on ALC subset. The accuracy rates are reported using two cross-validation procedures, one based on 10 folds and one based on LOO. The 10-fold CV procedure was repeated for 20 times and the results were averaged to reduce the accuracy variation introduced by randomly selecting the folds. The standard deviations for the computed average accuracy rates are also given. The best accuracy rate is highlighted in bold.

Method	10-fold CV	LOO CV
SVM and combined features [15]	41.0 %	41.6 %
KRR and k_{3-5}^{LRD}	50.1 % ± 1.2	51.4 %

partitioning of the data set. Furthermore, the LOO procedure can easily be performed on a small data set, such as the subset of 329 samples of the ALC. Thus, the LOO procedure is more suitable for this NLI experiment, since it is straight forward to compare newly developed systems with the previous state of art systems. The authors of [15] were asked to re-evaluate their system using the LOO CV procedure and they kindly provided additional results which are included in Table 1.

Table 1 presents the results of the classification system based on LRD in contrast to the results of the SVM model based on several combined features. The results clearly indicate the advantage of using an approach that works at the character level. When the 10-fold CV procedure is used, the average accuracy rate of the system based on LRD is 50.1 %. The kernel based on LRD is considerably better than the model described in [15]. An important remark is that the standard deviations computed over the 20 trials is 1.2 %, which means the amount of accuracy variation is too small to have an influence on the overall conclusion. Furthermore, the results obtained using the 10-fold CV procedure are consistent with the results obtained using the leave-one-out CV procedure. Both models reach better accuracy rates when the LOO procedure is used, most likely because there are more samples available for training. When the LOO CV procedure is used, the system based on LRD attains an accuracy rate that is almost 10 % higher than the accuracy rate of the state of the art SVM model.

The topic distribution per native language is unknown for the ALC subset, so the authors of [15] removed any lexical information from their model to prevent any topic bias. However, the authors of [1] state in their paper presenting the second version of the Arabic Learner Corpus that: "We decided to choose two common genres in the learner corpora, narrative and discussion. For the written part, learners had to write narrative or discussion essays under two specific titles which were likely to suit both native and non-native learners of Arabic, entitled *a vacation trip* for the narrative and *my study interest* for the discussion type.", which means that there are precisely two topics in the corpus. Since there are only two topics and 7 native languages in the ALC subset used in this experiment, there is no chance that the accuracy level obtained by the system based on the LRD kernel is due any kind of topic bias.

Table 2. Running times of the brute force search LRD algorithm versus the efficient algorithm based on the hash table. The reported times are measured in seconds.

LRD Algorithm	3-grams	4-grams	5-grams	Total
Brute force search [10]	688.0	1096.0	1367.0	3151.0
Hash table	8.6	7.5	7.0	23.1

4.5   Time Evaluation

Table 2 shows the time taken to compute the kernel matrices of 329×329 components corresponding to the 329 documents of the ALC subset. There are three kernel matrices, each one corresponding to a different n-gram length. The accuracy rates reported in Table 1 are based on the sum of all these matrices. Thus, the total time required to compute the kernel matrix corresponding to k_{3-5}^{LRD} is reported on the last column. The reported times were measured on a computer with Intel Core i7 2.3 GHz processor and 8 GB of RAM using a single Core. The time of the brute force search algorithm is directly proportional to the n-gram length. This can be explained by the fact that it is more likely to find an occurrence of a shorter n-gram using brute force. On the other hand, searching for longer n-grams can even result in going through the entire search space without finding any occurrences. For the very same reason, the efficient algorithm based on the positional inverted index table runs increasingly faster for longer n-grams. Longer n-grams will produce a rather more sparse hash table, in the sense that number of occurrences of each n-gram decreases as the n-gram length increases. Nevertheless, the algorithm based on the hash table structure is considerably faster than the brute force search algorithm for all three kernel matrices. More precisely, the efficient algorithm is more than two orders of magnitude faster for the kernels based on 4-grams and 5-grams, respectively. Moreover, It is almost two orders of magnitude faster for the kernel based on 3-grams. In conclusion, the overall picture shows that the proposed algorithm is significantly faster than the brute force search algorithm proposed in [10].

5   Conclusion and Further Work

This work proposed an efficient algorithm for Local Rank Distance. The main efficiency improvement of the algorithm was to build a positional inverted index for the character n-grams in one of the compared strings. For each n-gram in the other string, a binary search was employed to find the position of the nearest matching n-gram in the positional inverted index. The empirical evaluation showed that the proposed algorithm is more than two orders of magnitude faster than the brute force search algorithm. The efficient algorithm was used for the task of Arabic native language identification. A kernel classification system based on LRD obtained state of the art accuracy rates on the NLI task. In future work, more applications of LRD will be investigated, ranging from sentiment analysis to phylogenetic analysis.

References

1. Alfaifi, A., Atwell, E., Hedaya, I.: Arabic learner corpus (ALC) v2: a new written and spoken corpus of Arabic learners. In: Proceedings of the Learner Corpus Studies in Asia and the World, May 2014

2. Bykh, S., Meurers, D.: Native language identification using recurring n-grams - investigating abstraction and domain dependence. In: Proceedings of COLING 2012, pp. 425–440, December 2012

3. Dinu, L.P., Ionescu, R.T.: An efficient rank based approach for closest string and closest substring. PLoS ONE **7**(6), e37576 (2012)

4. Dinu, L.P., Ionescu, R.-T., Popescu, M.: Local patch dissimilarity for images. In: Huang, T., Zeng, Z., Li, C., Leung, C.S. (eds.) ICONIP 2012, Part I. LNCS, vol. 7663, pp. 117–126. Springer, Heidelberg (2012)

5. Dinu, L.P., Ionescu, R.T., Tomescu, A.I.: A rank-based sequence aligner with applications in phylogenetic analysis. PLoS ONE **9**(8), e104006 (2014)

6. Dinu, L.P., Manea, F.: An efficient approach for the rank aggregation problem. Theor. Comput. Sci. **359**(1–3), 455–461 (2006)

7. Dinu, L.P., Popescu, M., Dinu, A.: Authorship identification of romanian texts with controversial paternity. In: Proceedings of LREC (2008)

8. Dinu, L.P., Sgarro, A.: A low-complexity distance for DNA strings. Fundam. Informaticae **73**(3), 361–372 (2006)

9. Grozea, C., Gehl, C., Popescu, M.: ENCOPLOT: pairwise sequence matching in linear time applied to plagiarism detection. In: 3rd Pan Workshop on Uncovering Plagiarism, Authorship and Social Software Misuse, p. 10 (2009)

10. Ionescu, R.T.: Local rank distance. In: Proceedings of SYNASC, pp. 219–226 (2013)

11. Ionescu, R.T., Popescu, M., Cahill, A.: Can characters reveal your native language? a language-independent approach to native language identification. In: Proceedings of EMNLP, pp. 1363–1373, October 2014

12. Leslie, C.S., Eskin, E., Noble, W.S.: The spectrum kernel: a string kernel for SVM protein classification. In: Proceedings of Pacific Symposium on Biocomputing, pp. 566–575 (2002)

13. Li, M., Chen, X., Li, X., Ma, B., Vitanyi, P.M.B.: The similarity metric. IEEE Trans. Inf. Theor. **50**(12), 3250–3264 (2004)

14. Lodhi, H., Saunders, C., Shawe-Taylor, J., Cristianini, N., Watkins, C.J.C.H.: Text classification using string kernels. J. Mach. Learn. Res. **2**, 419–444 (2002)

15. Malmasi, S., Dras, M.: Arabic native language identification. In: Proceedings of the EMNLP 2014 Workshop on Arabic Natural Language Processing (ANLP), pp. 180–186, October 2014

16. Malmasi, S., Dras, M.: Chinese native language identification. In: Proceedings of EACL, vol. 2, pp. 95–99 (2014)

17. Melsted, P., Pritchard, J.: Efficient counting of k-mers in DNA sequences using a bloom filter. BMC Bioinformatics **12**(1), 333 (2011)

18. Popescu, M., Grozea, C.: Kernel methods and string kernels for authorship analysis. In: CLEF (Online Working Notes/Labs/Workshop), September 2012

19. Popescu, M., Ionescu, R.T.: The story of the characters, the DNA and the native language. In: Proceedings of the Eighth Workshop on Innovative Use of NLP for Building Educational Applications, pp. 270–278, June 2013

20. Popov, Y.V.: Multiple genome rearrangement by swaps and by element duplications. Theor. Comput. Sci. **385**(1–3), 115–126 (2007)

21. Sanderson, C., Guenter, S.: Short text authorship attribution via sequence kernels, markov chains and author unmasking: an investigation. In: Proceedings of EMNLP, pp. 482–491, July 2006
22. Shapira, D., Storer, J.A.: Large edit distance with multiple block operations. In: Nascimento, M.A., de Moura, E.S., Oliveira, A.L. (eds.) SPIRE 2003. LNCS, vol. 2857, pp. 369–377. Springer, Heidelberg (2003)
23. Shawe-Taylor, J., Cristianini, N.: Kernel Methods for Pattern Analysis. Cambridge University Press, New York (2004)
24. Tetreault, J., Blanchard, D., Cahill, A.: A report on the first native language identification shared task. In: Proceedings of the Eighth Workshop on Innovative Use of NLP for Building Educational Applications, pp. 48–57, June 2013
25. Vezzi, F., Fabbro, C.D., Tomescu, A.I., Policriti, A.: rNA: a fast and accurate short reads numerical aligner. Bioinformatics **28**(1), 123–124 (2012)

Human Perception-Based Washout Filtering Using Genetic Algorithm

Houshyar Asadi[✉], Shady Mohamed, Kyle Nelson, Saeid Nahavandi,
and Delpak Rahim Zadeh

Centre for Intelligent Systems Research, Deakin University, Geelong, Australia
{hasadi,shady.mohamed,kyle.nelson,
saeid.nahavandi,d.rahimzadeh}@deakin.edu.au

Abstract. The Motion Cueing Algorithm (MCA) transforms longitudinal and rotational motions into simulator movement, aiming to regenerate high fidelity motion within the simulators physical limitations. Classical washout filters are widely used in commercial simulators because of their relative simplicity and reasonable performance. The main drawback of classical washout filters is the inappropriate empirical parameter tuning method that is based on trial-and-error, and is effected by programmers' experience. This is the most important obstacle to exploiting the platform efficiently. Consequently, the conservative motion produces false cue motions. Lack of consideration for human perception error is another deficiency of classical washout filters and also there is difficulty in understanding the effect of classical washout filter parameters on generated motion cues. The aim of this study is to present an effortless optimization method for adjusting the classical MCA parameters, based on the Genetic Algorithm (GA) for a vehicle simulator in order to minimize human sensation error between the real and simulator driver while exploiting the platform within its physical limitations. The vestibular sensation error between the real and simulator driver as well as motion limitations have been taken into account during optimization. The proposed optimized MCA based on GA is implemented in MATLAB/Simulink. The results show the superiority of the proposed MCA as it improved the human sensation, maximized reference signal shape following and exploited the platform more efficiently within the motion constraints.

Keywords: Washout filter · Motion cueing algorithm (MCA) · Genetic algorithm (GA) · Human sensation

1 Introduction

The Motion Cueing Algorithm (MCA) is the process of extracting translational and rotational motion commands from a real vehicle and transforming them into the simulator platforms, aiming to regenerate realistic motion and the same sensation within the simulator's physical limitations.

© Springer International Publishing Switzerland 2015
S. Arik et al. (Eds.): ICONIP 2015, Part II, LNCS 9490, pp. 401–411, 2015.
DOI: 10.1007/978-3-319-26535-3_46

The design of MCAs in driving simulators is a complex task and depends on simulator architecture as well as the kind of maneuver that we want to regenerate [1, 2]. The first type of MCA are known as classical washout filters. Conrad and Schmidt [3, 4] have provided the basic setup for the classical MCA which was generally developed in [5]. It is currently the basic solution widely used in various simulators because of its relative simplicity. There are additional advantages to using classic washout algorithms such as short processing time, fast computation and stable performance [6, 7]. The second type of MCA is adaptive MCA which was devised by Parrish et al. [8] and then developed at the University of Toronto [5]. Optimal MCA was provided by Sivan et al. [9] and it has been developed further by Telban et al. [10]. This was to meet the drawbacks of the classical MCA, and take human perception into account [11].

In the research conducted by Nehaoua et al. [12–14], the classical MCA has better performance in terms of motion fidelity and human sensation compared with the adaptive and optimal MCAs for small driving simulators. However, while the classical washout filter has some advantages, it produces notable false cues because of its filters' characteristics as they are tuned empirically based on programmers experience. The trial-and-error adjusting of classical washout filters is a time-consuming procedure and requires expert knowledge. This type of tuning is the most important factor restricting full motion platform utilization. This is the main cause behind the generation of wrong cues and makes the classical MCA inappropriate for all circumstances [6, 11, 15]. Furthermore, the classical washout filtering technique does not consider human sensation error in its algorithm. A study on optimization methods for tuning the MCA parameters, based on nonlinear filtering and genetic algorithm was conducted to remove backlash produced in high-pass filters [16]. Due to the drawbacks of classical washout filters, there is a need to present an optimization method for tuning classical washout filters.

The aim of this paper is to present an optimized classical washout filter for a driving simulator based on Genetic Algorithm (GA). This is an effective method to find the most appropriate washout filter parameters that is able to decrease human sensation error between real and simulator driver within the physical constraints. This is achieved by taking the vestibular system model and platform physical constraints into account. The GA, a popular optimization strategy that can effectively solve optimization problems, is applied in this classical washout filter. Using this method can remove manual trial-and-error effort for adjusting washout filters.

This paper is organized as follows. Section 2 introduces the classical washout filter. The applied vestibular system and related mathematical models are provided in Sect. 3. The proposed optimized MCA based on GA is provided in Sect. 4. The results of the proposed method are presented in Sect. 5.

2   Classical Washout Filter

The structure of the classical washout filter consists of translational, tilt-coordination, and rotational motion channels. The classical washout filter composed of empirically tuned high and low-pass filters for generating the translational and the rotational motions. A crossover path including tilt-coordination is the main part of this structure.

This is required to present the steady state or gravity alignment cues in the translational direction. Figure 1 illustrates the structure of the classic algorithm.

High-pass filters are applied in the classical structure to longitudinal and angular motion input from a vehicle dynamic model. This is used to extract its transient component. Due to constraints of driving simulators, it is not possible to reproduce low frequency rotational and translational motions, as these kind of motions can move the platform to its physical limitations. The translational channel filter, which has a 3rd order high-pass filter [1], is given as follows.

$$\frac{s^2}{s^2 + 2\zeta_h\omega_h s + \omega_h^2} \cdot \frac{s}{s + \omega_b} \tag{1}$$

where ω_h and ω_b are the translational high-pass cut-off frequency and, ζ_h is the translational high-pass damping ratio.

The rotational channel filter, which has a 1st order high-pass filter [17], is given as

$$\frac{s}{s + \omega_{h\theta}} \tag{2}$$

where $\omega_{h\theta}$ is the high-pass cut-off frequency in the rotational channel.

Through a low-pass filter in tilt-coordination the sustained component of the acceleration signal can be extracted. A low-pass filter can extract the sustained acceleration signal by tilt coordination that uses gravity as an illusory sustained acceleration. This tilt coordination must produce motions below the driver's perception threshold. The low-pass filter in the tilt-coordination channel [17] is given as

$$\frac{\omega_l^2}{s^2 + 2\zeta_l\omega_l s + \omega_l^2} \tag{3}$$

where ω_l and ζ_l are the second-order low-pass cut-off frequency and damping ratio in the tilt-coordination channel respectively.

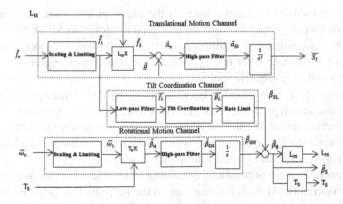

Fig. 1. Classical washout filter structure.

3   Human Vestibular System

The human vestibular system includes semicircular canals and otolith organs and is located in the inner ear. The vestibular semicircular canals and otoliths are the main sensors in the human body to perceive rotational and linear motion. The sensed specific force through otolith is achieved by deducting acceleration of gravity $\vec{g}$ from linear acceleration $\vec{a}$ as given

$$\vec{sF} = \vec{a} - \vec{g} \tag{4}$$

The transfer function of specific force sensation model that can link sensed specific force to the real input specific force (mathematical model for otolith.) is provided [18] as below

$$\frac{\hat{f}}{f} = \frac{K(1 + \tau_a s)}{(\tau_L s + 1)(\tau_s s + 1)} \tag{5}$$

where τ_a, τ_s, τ_L in the abovementioned equation are the neural processing term lead operator, short time constant and long time constant of the otolith respectively.

According to [19], the rotational sensation transfer function model that can link sensed angular velocity to the actual angular velocity input is presented as

$$\frac{\hat{\omega}}{\omega} = \frac{\tau_a s}{(\tau_a s + 1)} \cdot \frac{\tau_l s}{(\tau_l s + 1)(\tau_2 s + 1)} \tag{6}$$

where τ_l, τ_2, τ_a are the long time constant, short time constant and adaptation time constant of semicircular canals respectively.

4   Proposed Linear Motion Cueing Algorithm Based on GA

To overcome and solve the disadvantages of the classical MCA such as inappropriate tuning, lack of human perception and conservative motions, an MCA based on GA is proposed. This is designed to eliminate the lack of human perception and motion constraints in the washout filter tuning, which is a time-consuming empirical process based on the worst case scenario.

GAs are one of the most popular EAs, as shown by their wide application [20]. A GA is a heuristic search procedure that generates a solution through a processes similar to natural selection and evolution, in order to optimize nonlinear systems with many variables. GAs are stochastic algorithms with less chance of falling into a local minima, while it determines the best optimal solution for the problem [21].

In GA, generations are populations consisting of individual chromosomes that have evolutionary operators. At each iteration and in the reproduction process, the chromosomes with better fitness are given more chance of surviving to the next generations.

It employs heuristics such as evaluation, selection, crossover, and mutation which are the basic GA operators to develop better solutions.

Selection is the step of a GA in which chromosome genomes are selected from a population for future breeding through crossover and mutation. Therefore, the selection operator shows how the GA selects parents for the next generation. In order to generate better offspring, cross over and mutation are applied in the GA inspired by natural evolution processes.

Crossover operators show the random sexual generation of a child, or offspring from two parents or combination of parent pairs from selected reproductive individuals to produce offspring. The offspring have partial information from their parents. Mutation is a GA operator applied to maintain genetic diversity from one generation in a GA population to the next generation. Mutation randomly changes one or more gene values in a small fraction of chromosomes from its initial state. The mutation operator may change the solution completely from the previous solution. The individual chromosomes are evaluated and ranked. Evaluation operations are used on each chromosome through a fitness function. GA repeats the process for new produced generation in a loop until a predefined stopping, a maximum number of allowed steps with no improvement, or terminating criteria met [22].

Designing of MCA based on GA tuning can decrease the sensed false cues as the human sensation and all motion limitations are considered in this method. Tuning efficiently means eliminating poor usage of the platform and exploiting platform workspace effectively in order to reduce human sensation error.

4.1 Genetic Algorithm Based Washout Filter Tuning

In classical washout filters, the cut-off frequency and damping ratio of the high-pass and low-pass filter are tuned off-line empirically through a trial-and-error process. The GA is used in the proposed method to generate the optimal washout filter parameters as it is an effectual method of searching in a wide and complex solution space. It can present optimal solutions much faster compared with the random trial-and-error method. Moreover, GA is a more effective approach and avoids local minima compared to other methods. A GA chromosome at iteration i, consisting of six genes is represented by

$$X_d^i = [x_{d,1}^i, x_{d,2}^i, \ldots x_{d,6}^i] \tag{7}$$

where d is the number of the chromosomes in the population. The chromosome genes $x_{d,1}^i, x_{d,2}^i, \ldots x_{d,6}^i$ denote ω_h, ω_l, ζ_h, ζ_l, $\omega_{h\theta}$, ω_b which are the classical high and low-pass filter parameters. Figure 2 indicates the flow chart for applying GA in the proposed MCA.

The aim of GA in the proposed optimal washout filter based on GA method is to minimize the fitness function including translational and rotational human sensation error and motion displacement. This is designed to decrease human sensation error, track the sensed reference signal more effectively, and deliver the optimized balance between the human sensation error and workspace usage within the physical constraints.

Human sensation error, which is the difference between driver sensation and the simulated driver sensation caused by simulator motion, should be minimized to eliminate any simulator sickness.

GA start tuning washout filter parameters to minimize human sensation error and platform motion fitness function. For evaluating the total fitness value, a fitness function is defined and given as below

$$J\left(X_d^i\right) = J_{d,sensation\ error}^i + J_{d,limitation}^i \tag{8}$$

where $J_{d,sensation\ error}^i$ is the total human sensation error fitness function and $J_{d,limitation}^i$ is the motion displacement fitness function in translational and rotational directions.

The fitness function of human sensation error in the translational and rotational directions is assessed with the Integral of Squared Error (ISE) performance index as given below

$$J_{d,sensation}^i = \int (k_{e,f}\hat{e}_f^2 + k_{e,\omega}\,\hat{e}_\omega^2)dt \tag{9}$$

where $k_{e,f}$ and $k_{e,\omega}$ represent weights for sensed specific force and angular velocity error between the real and simulator driver. These penalty weights are determined based on our design requirements to affect the fitness function and consequently the human sensation and generated motion. The specific force and angular velocity sensation errors between the real and simulator driver are provided as

$$\hat{e}_f(t) = \hat{f}_{v(ref)}(t) - \hat{f}_s(t) \tag{10}$$

$$\hat{e}_\omega(t) = \hat{\omega}_{v(ref)}(t) - \hat{\omega}_s(t) \tag{11}$$

In addition, in the next step in order to constrain the simulator motion, it is necessary for the washout filters to limit the translational and rotational motions. Translational acceleration a and rotational velocity $\dot{\theta}$ have been also taken into account in the proposed fitness function. Furthermore, the translational and rotational displacement (x_d and θ) must be within the physical limitation ranges. The integral of displacement x_d and velocity v_{HP} should be considered in the fitness function. It is a necessary term in the fitness function as realistic motion is required to be regenerated within the platform motion limitations.

The total displacement fitness function in the translational and rotational directions is given as

$$\begin{aligned}J_{d,limitation}^i &= \int k_{x_d}x_d^2dt + \int k_{return}\left(\int x_d dt\right)^2 dt + \int k_{v_{HP}}v_{HP}^2dt + \int k_\theta\theta^2dt \\ &\quad + \int k_a a^2dt + \int k_{\dot{\theta}}\dot{\theta}^2dt\end{aligned} \tag{12}$$

where $k_{x_d}, k_{return}, k_{v_{HP}}, k_\theta, k_a, k_{\dot{\theta}}$ are the penalty weights of platform displacements and motions.

There are some advantages in using optimal algorithms over classical washout algorithms such as having a fitness function to optimize parameters wisely. In addition, it takes the human sensation error into account during parameter tuning while the previous method was not able to do so.

The main aim of using the proposed GA-based washout filter tuning method is to minimize the sensation error and false cue motions while improving the generated motion in following the reference signal precisely within the work space physical constraints. GA is used to present an optimized balance between sensed errors and simulator work space usage. To effectively achieve this aim, GA minimizes the total fitness function.

The most favorable filter parameters [17] according to driver comments are considered in this research. In addition, they are used as initial values in the proposed washout filter tuning based on GA method. This will ensure that even in the worst case GA solution, the obtained parameters will be near to the prior parameters. Furthermore, in this method during the iterations the effects of the rate limiter in the tilt coordination channel are considered in the offline tuning, while the threshold effect on the produced signal has been ignored in the previous algorithms.

The parameters are adjusted for minimizing the total fitness function in MATLAB/ Simulink and the results of MCA based on GA are presented and evaluated in the result and discussion section below.

5   Results and Discussion

In the proposed washout filter based on GA, we employed standard parameters for the GA, such as: crossover rate 0.8; mutation rate 0.2; and elite reproduction 5 % of the population size. A population size of 200 is chosen as being appropriate for the generation of precise genetic maps [23]. Higher population sizes can deliver desirable results owing to the larger search space. If the population size was chosen to be bigger than this, there would be no guarantee of further decreasing the fitness value, while the computational time would be higher. Figure 3 shows the convergence of the GA for the population size of 200. This figure shows the GA convergence result with respect to iterations for finding the best fitness function value. For this population, convergence is obtained with best fitness value at iteration 586, where GA stopped searching.

A driving scenario consisting of acceleration, deceleration and braking is applied in the longitudinal direction to the proposed and classical methods, as shown in Fig. 4.

Figure 5 shows the comparison of longitudinal sensed specific force for the reference (real driver), the proposed method based on GA and classical washout filter. According to this comparison, the sensed specific force of the proposed method based on GA has followed the sensed reference signal more accurately compared with the classical washout filter since the human vestibular sensation and platform motions have been taken into account in tuning of the washout filter parameters. The classical sensed specific result has some false cues which shows the inconsistency of this method.

The cross correlation coefficient values of these generated signals have been evaluated and compared. The cross correlation coefficient shows the dependency of the two signals.

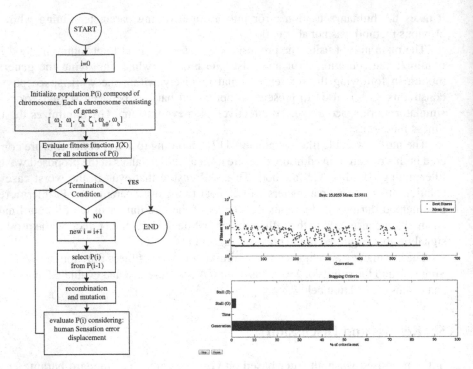

Fig. 2. Flow chart of genetic optimization for the proposed MCA parameters

Fig. 3. Genetic algorithm convergence result for finding the total minimum fitness function value

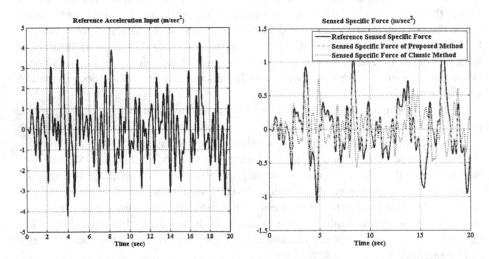

Fig. 4. Translational reference acceleration input

Fig. 5. Comparison of the sensed specific force for the proposed and classic MCA

It can show the shape following criterion between the sensed result and sensed reference signal. The cross correlation coefficients between the sensed specific force and sensed reference signal in the classical method was 10 % which was improved to 99 % in the proposed method based on GA. This shows that the proposed method based on GA outperforms the previous method in terms of tracking the reference sensation as the capability of tracking the reference signal is improved significantly.

The sensed specific force error between the simulator and real driver for the proposed method and classic method are shown in Fig. 6. According to this figure, the sensed human sensation error is decreased significantly and it has been suppressed under human threshold limit. The Root Mean Square (RMS) value of human sensation for classical washout filter is 0.44 m/s² and this is reduced to 0.08 m/s² by using the proposed method through tuning the parameters based on GA.

Figure 7 compares the displacement for the proposed method based on GA and classic method. This figure shows the proposed method based on GA exploits 21 % of the platform workspace whereas the classical MCA exploited only 9 %. It should be noted that the platform surge limitation is 1.5 m. The proposed MCA eliminate conservativeness in exploiting the platform workspace and illustrates that the optimized proposed method based on GA can exploit the platform workspace more efficiently compared to the classical method. Therefore, the proposed method decreased human sensation error and increased the cross correlation coefficient which leads to following the reference signal more effectively. Simultaneously, the displacement usage capability is improved and the workspace can be exploited more efficiently. This shows the optimized balance between the human sensation error and exploiting the workspace within the physical constraints.

According to the evaluated result, the new optimized washout filter based on genetic tuning improved the human sensation caused by generated simulator motion while the

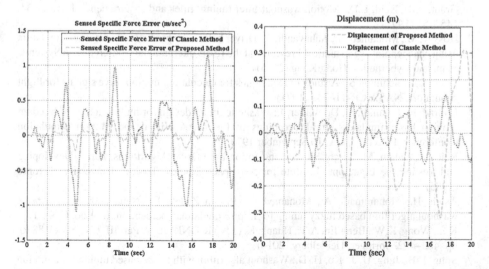

Fig. 6. Comparison of the sensed specific force error for the proposed and classic MCA

Fig. 7. Comparison of displacement for the proposed and classic MCA

platform workspace is exploited more efficiently aiming to follow the reference signal. This shows successful optimization of the GA-based classical washout filter.

6 Conclusion

The Motion Cueing Algorithm (MCA) aims to transform longitudinal and rotational motions into simulator platform motion, for reproducing high fidelity motion within the simulator's constraints. Classical washout filters which are the most popular MCAs have some drawbacks, including the unsuitable empirical worst case parameter tuning based on designer's knowledge and experience; lack of human perception in the algorithm; and poor and conservative usage of platform workspace. This causes false cue motions in some situations consisting of acceleration, deceleration and braking. To solve these problems, classical washout filter parameters are tuned by GA optimization method which considers human perception and motion limitations. The GA has tuned parameters of a conventional classical washout filter including two high-pass filter and one low-pass filter in the translational, rotational and tilt coordination channels, in order to improve produced motion cues and motion fidelity.

Based on the obtained results, applying this method minimized the sensed artifacts and incorrect motion cues, and decreased human sensation error between the real and simulator driver. Through the proposed method, a simulator is able to exploit platform limitations more efficiently in order to deliver the most realistic motions. Therefore, the results have shown that the proposed method out-performs classical washout filter in terms of performance, sensation error and workspace usage.

References

1. Grant, P.R., Reid, L.D.: Motion washout filter tuning: rules and requirements. J. Aircr. **34**, 145–151 (1997)
2. Hossny, M., Mohamed, S., Nahavandi, S.: Driver behaviour prediction for motion simulators using changepoint segmentation. Presented at the IEEE International Conference on Systems, Man, and Cybernetics, Hong Kong (2015)
3. Conrad, B., Schmidt, S.: A study of techniques for calculating motion drive signals for flight simulators. NASA CR-114345 (1971)
4. Conrad, B., Schmidt, S., Douvillier, J.: Washout circuit design for multi-degrees of freedom moving base simulators. In: Proceedings of the AiAA Visual and Motion Simulation Conference, Palo Alto, CA, 10 September 1973
5. Reid, L., Nahon, M.: Flight Simulation Motion-base Drive Algorithms: Part 1-Developing and Testing the Equations. Institute for Aerospace Studies, Toronto University, Toronto (1985)
6. Asadi, H., Mohammadi, A., Mohamed, S., Rahim Zadeh, D., Nahavandi, S.: Adaptive washout algorithm based fuzzy tuning for improving human perception. In: Loo, C.K., Yap, K.S., Wong, K.W., Beng Jin, A.T., Huang, K. (eds.) ICONIP 2014, Part III. LNCS, vol. 8836, pp. 483–492. Springer, Heidelberg (2014)
7. Song, J.-B., Jung, U.-J., Ko, H.-D.: Washout algorithm with fuzzy-based tuning for a motion simulator. KSME Int. J. **17**, 221–229 (2003)

8. Parrish, R.V., Dieudonne, J.E., Martin Jr., D.J.: Coordinated adaptive washout for motion simulators. J. Aircr. **12**, 44–50 (1975)
9. Sivan, R., Ish-Shalom, J., Huang, J.-K.: An optimal control approach to the design of moving flight simulators. IEEE Trans. Syst. Man Cybern. **12**, 818–827 (1982)
10. Telban, R.J., Cardullo, F.M.: Motion cueing algorithm development: human-centered linear and nonlinear approaches. NASA TechReport CR-2005-213747 (2005)
11. Asadi, H., Mohamed, S., Nahavandi, S.: Incorporating human perception with the motion washout filter using fuzzy logic control. IEEE/ASME Trans. Mechatron. **PP**, 1–9 (2015)
12. Nehaoua, L., Amouri, A., Arioui, H.: Classic and adaptive washout comparison for a low cost driving simulator. In: Proceedings of the 2005 IEEE International Symposium on Intelligent Control, Mediterrean Conference on Control and Automation, pp. 586–591 (2005)
13. Nehaoua, L., Arioui, H., Mohellebi, H., Espie, S.: Restitution movement for a low cost driving simulator. In: American Control Conference 2006, p. 6 (2006)
14. Nehaoua, L., Mohellebi, H., Amouri, A., Arioui, H., Espié, S., Kheddar, A.: Design and control of a small-clearance driving simulator. IEEE Trans. Veh. Technol. **57**, 736–746 (2008)
15. Asadi, H., Mohammadi, A., Mohamed, S., Nahavandi, S.: Adaptive translational cueing motion algorithm using fuzzy based tilt coordination. In: Loo, C.K., Yap, K.S., Wong, K.W., Beng Jin, A.T., Huang, K. (eds.) ICONIP 2014, Part III. LNCS, vol. 8836, pp. 474–482. Springer, Heidelberg (2014)
16. Asadi, H., Mohamed, S., Rahim Zadeh, D., Nahavandi, S.: Optimisation of nonlinear motion cueing algorithm based on genetic algorithm. Veh. Syst. Dyn. 1–20 (2015)
17. Nahon, M.A., Reid, L.D.: Simulator motion-drive algorithms-a designer's perspective. J. Guidance Control Dyn. **13**, 356–362 (1990)
18. Meiry, J., Young, L.: A revised dynamic otolith model (1968)
19. Zacharias, G.L.: Motion cue models for pilot-vehicle analysis. DTIC Document (1978)
20. Yang, X.-S.: Chapter 5 - genetic algorithms. In: Yang, X.-S. (ed.) Nature-Inspired Optimization Algorithms, pp. 77–87. Elsevier, Oxford (2014)
21. Kantardzic, M.: Data Mining: Concepts, Models, Methods, and Algorithms. Wiley, New York (2011)
22. Gosselin, L., Tye-Gingras, M., Mathieu-Potvin, F.: Review of utilization of genetic algorithms in heat transfer problems. Int. J. Heat Mass Transf. **52**, 2169–2188 (2009)
23. Ferreira, A., da Silva, M.F., Cruz, C.D.: Estimating the effects of population size and type on the accuracy of genetic maps. Genet. Mol. Biol. **29**, 187–192 (2006)

Conjugate Gradient Algorithms
for Complex-Valued Neural Networks

Călin-Adrian Popa[✉]

Department of Computer and Software Engineering,
Polytechnic University Timişoara, Blvd. V. Pârvan, No. 2,
300223 Timişoara, Romania
calin.popa@cs.upt.ro

Abstract. In this paper, conjugate gradient algorithms for complex-valued feedforward neural networks are proposed. Since these algorithms yielded better training results for the real-valued case, an extension to the complex-valued case is a natural option to enhance the performance of the complex backpropagation algorithm. The full deduction of the classical variants of the conjugate gradient algorithm is presented, and the resulting training methods are exemplified on synthetic and real-world applications. The experimental results show a significant improvement over the complex gradient descent algorithm.

Keywords: Complex-valued neural networks · Conjugate gradient algorithm · Time series prediction

1 Introduction

Over the last few years, the domain of complex-valued neural networks has received an increasing interest. Some popular applications include antenna design, estimation of direction of arrival and beamforming, radar imaging, communications signal processing, image processing, and many others (for an extensive presentation of the applications of complex-valued neural networks, see [13]).

Several methods have been proposed to increase the efficiency of learning in complex-valued neural networks. These methods include different network architectures and different learning algorithms, some of which are specially designed for this type of networks, and others extended from the real-valued case. One method that has been proven to be very efficient in the real-valued case is the conjugate gradient learning algorithm. First proposed, among others, by [6,15], this algorithm has become today one of the most known and used methods to train feedforward neural networks. Having this idea in mind, it seems natural to extend this learning algorithm to complex-valued neural networks, also.

In this paper, we present the full deduction of the most known and used variants of the conjugate gradient algorithm. We test the proposed conjugate gradient algorithms on both synthetic and real-world applications. The synthetic applications include a fully complex function approximation problem and

© Springer International Publishing Switzerland 2015
S. Arik et al. (Eds.): ICONIP 2015, Part II, LNCS 9490, pp. 412–422, 2015.
DOI: 10.1007/978-3-319-26535-3_47

a split complex function approximation problem. The real-world application is a nonlinear time series prediction problem.

The remainder of this paper is organized as follows: Sect. 2 gives a thorough explanation of line search algorithms in general, and linear and nonlinear conjugate gradient algorithms in particular for optimization of an error function defined on the complex plane. The experimental results of the three applications of the proposed algorithms are shown and discussed in Sect. 3, along with a detailed explanation of the nature of each problem. Section 4 is dedicated to presenting the conclusions of the study.

2   Conjugate Gradient Algorithms

Conjugate gradient methods belong to the larger class of line search algorithms. For minimizing the error function of a neural network, a series of steps through the weight space are necessary to find the weight for which the minimum of the function is attained. Each step is determined by the search direction and a real number telling us how far in that direction we should move. In the classical gradient descent, the search direction is that of the negative gradient and the real number is the learning rate parameter. In the general case, we can consider some particular search direction, and then determine the minimum of the error function in that direction, thus yielding the real number that tells us how far in that direction we should move. This represents the line search algorithm, and constitutes the basis for a family of methods that perform better than the classical gradient descent. Our deduction of conjugate gradient algorithms follows mainly the one presented in [4], which too follows that in [15].

Let's assume that we have a complex-valued neural network with an error function denoted by E, and an $2N$-dimensional weight vector denoted by $\mathbf{w} = (w_1^R, w_1^I, \ldots, w_N^R, w_N^I)^T$. We must find the weight vector denoted by $\mathbf{w}^*$ that minimizes the function $E(\mathbf{w})$. Suppose we are iterating through the weight space to find the value of $\mathbf{w}^*$ or a very good approximation of it. Further, let's assume that at step k in the iteration, we want the search direction to be $\mathbf{p}_k$, where $\mathbf{p}_k$ is obviously an $2N$-dimensional vector. In this case, the next value for the weight vector is given by $\mathbf{w}_{k+1} = \mathbf{w}_k + \lambda_k \mathbf{p}_k$, where the parameter λ_k is a real number telling us how far in the direction of $\mathbf{p}_k$ we want to go, which means that λ_k should be chosen to minimize $E(\lambda) = E(\mathbf{w}_k + \lambda \mathbf{p}_k)$.

This is a real-valued function in one real variable, so its minimum is attained when $\frac{\partial E(\lambda)}{\partial \lambda} = \frac{\partial E(\mathbf{w}_k + \lambda \mathbf{p}_k)}{\partial \lambda} = 0$. By the chain rule, we can write that

$$
\begin{aligned}
\frac{\partial E(\mathbf{w}_k + \lambda \mathbf{p}_k)}{\partial \lambda} &= \sum_{i=1}^{N} \frac{\partial E(\mathbf{w}_k + \lambda \mathbf{p}_k)}{\partial (w_i^{k,R} + \lambda p_i^{k,R})} \frac{\partial (w_i^{k,R} + \lambda p_i^{k,R})}{\partial \lambda} \\
&+ \sum_{i=1}^{N} \frac{\partial E(\mathbf{w}_k + \lambda \mathbf{p}_k)}{\partial (w_i^{k,I} + \lambda p_i^{k,I})} \frac{\partial (w_i^{k,I} + \lambda p_i^{k,I})}{\partial \lambda} \\
&= \sum_{i=1}^{N} \frac{\partial E(\mathbf{w}_{k+1})}{\partial w_i^{k+1,R}} p_i^{k,R} + \frac{\partial E(\mathbf{w}_{k+1})}{\partial w_i^{k+1,I}} p_i^{k,I} \\
&= \langle \nabla E(\mathbf{w}_{k+1}), \mathbf{p}_k \rangle,
\end{aligned}
\tag{1}
$$

where $\langle \mathbf{a}, \mathbf{b} \rangle$ is the Euclidean scalar product in $\mathbb{R}^{2N} \simeq \mathbb{C}^N$, given by $\langle \mathbf{a}, \mathbf{b} \rangle = \left(\sum_{i=1}^{N} a_i \overline{b_i} \right)^R = \sum_{i=1}^{N} a_i^R b_i^R + a_i^I b_i^I$, for all $\mathbf{a}, \mathbf{b} \in \mathbb{R}^{2N} \simeq \mathbb{C}^N$, and by a^R and a^I we denoted the real and imaginary part of the complex number a, and by $\overline{a}$ the conjugate of the complex number a.

So, if we denote $\mathbf{g} := \nabla E$, then (1) can be written in the form

$$\langle \mathbf{g}(\mathbf{w}_{k+1}), \mathbf{p}_k \rangle = 0, \tag{2}$$

which shows that the vectors $\mathbf{g}(\mathbf{w}_{k+1})$ and $\mathbf{p}_k$ are orthogonal in $\mathbb{R}^{2N}$.

The next search direction $\mathbf{p}_{k+1}$ is chosen so that the component of the gradient parallel to the previous search direction $\mathbf{p}_k$ remains zero. As a consequence, we have that $\langle \mathbf{g}(\mathbf{w}_{k+1} + \lambda \mathbf{p}_{k+1}), \mathbf{p}_k \rangle = 0$. By the Taylor series expansion to the first order, we have that $\mathbf{g}(\mathbf{w}_{k+1} + \lambda \mathbf{p}_{k+1}) = \mathbf{g}(\mathbf{w}_{k+1}) + \nabla \mathbf{g}(\mathbf{w}_{k+1})^T \lambda \mathbf{p}_{k+1}$, and then, if we take (2) into account, we obtain that $\lambda \langle \nabla \mathbf{g}(\mathbf{w}_{k+1})^T \mathbf{p}_{k+1}, \mathbf{p}_k \rangle = 0$, which is equivalent to $\langle \mathbf{H}^T(\mathbf{w}_{k+1}) \mathbf{p}_{k+1}, \mathbf{p}_k \rangle = 0$, or, further to

$$\langle \mathbf{p}_{k+1}, \mathbf{H}(\mathbf{w}_{k+1}) \mathbf{p}_k \rangle = 0, \tag{3}$$

where we denoted by $\mathbf{H}(\mathbf{w}_{k+1})$ the Hessian of the error function $E(\mathbf{w})$, because $\mathbf{g} = \nabla E$, and thus $\nabla \mathbf{g}$ is the Hessian.

The search directions that satisfy Eq. (3) are said to be conjugate directions. The conjugate gradient algorithm builds the search directions $\mathbf{p}_k$, such that each new direction is conjugate to all the previous ones.

Next, we will explain the linear conjugate gradient algorithm. For this, we consider an error function of the form

$$E(\mathbf{w}) = E_0 + \mathbf{b}^T \mathbf{w} + \frac{1}{2} \mathbf{w}^T \mathbf{H} \mathbf{w}, \tag{4}$$

where $\mathbf{b}$ and $\mathbf{H}$ are constants, and the matrix $\mathbf{H}$ is assumed to be positive definite. The gradient of this function is given by

$$\mathbf{g}(\mathbf{w}) = \mathbf{b} + \mathbf{H} \mathbf{w}. \tag{5}$$

The weight vector $\mathbf{w}^*$ that minimizes the function $E(\mathbf{w})$ satisfies the equation

$$\mathbf{b} + \mathbf{H} \mathbf{w}^* = 0. \tag{6}$$

As we saw earlier from Eq. (3), a set of $2N$ nonzero vectors $\{\mathbf{p}_1, \mathbf{p}_2, \ldots, \mathbf{p}_{2N}\} \subset \mathbb{R}^{2N}$ is said to be conjugate with respect to the positive definite matrix $\mathbf{H}$ if and only if

$$\langle \mathbf{p}_i, \mathbf{H} \mathbf{p}_j \rangle = 0, \forall i \neq j. \tag{7}$$

It is easy to show that, in these conditions, the set of $2N$ vectors is also linearly independent, which means that they form a basis in $\mathbb{R}^{2N}$. If we start from the initial point $\mathbf{w}_1$ and want to find the value of $\mathbf{w}^*$ that minimizes the error function given in (4), taking into account the above remarks, we can write that

$$\mathbf{w}^* - \mathbf{w}_1 = \sum_{i=1}^{2N} \alpha_i \mathbf{p}_i. \tag{8}$$

Now, if we set

$$\mathbf{w}_k = \mathbf{w}_1 + \sum_{i=1}^{k-1} \alpha_i \mathbf{p}_i, \qquad (9)$$

then (8) can be written in the iterative form

$$\mathbf{w}_{k+1} = \mathbf{w}_k + \alpha_k \mathbf{p}_k, \qquad (10)$$

which means that the value of $\mathbf{w}^*$ can be determined in at most $2N$ steps for the error function (4), using the above iteration. We still have to determine the real parameters α_k that tell us how much we should go in any of the $2N$ conjugate directions $\mathbf{p}_k$.

For this, we will multiply Eq. (8) by $\mathbf{H}$ to the left, and take the Euclidean $\mathbb{R}^{2N}$ scalar product with $\mathbf{p}_k$. Taking into account Eq. (6), we obtain that $-\langle \mathbf{p}_k, \mathbf{b} + \mathbf{Hw}_1 \rangle = \sum_{i=1}^{2N} \alpha_i \langle \mathbf{p}_k, \mathbf{Hp}_i \rangle$. But, because the directions $\mathbf{p}_k$ are conjugate with respect to matrix $\mathbf{H}$, we have from (7) that $\langle \mathbf{p}_k, \mathbf{Hp}_i \rangle = 0, \forall i \neq k$, so the above equation yields the following value for α_k:

$$\alpha_k = -\frac{\langle \mathbf{p}_k, \mathbf{b} + \mathbf{Hw}_1 \rangle}{\langle \mathbf{p}_k, \mathbf{Hp}_k \rangle}. \qquad (11)$$

Now, if we multiply Eq. (9) by $\mathbf{H}$ to the left, and take the Euclidean $\mathbb{R}^{2N}$ scalar product with $\mathbf{p}_k$, we have that: $\langle \mathbf{p}_k, \mathbf{Hw}_k \rangle = \langle \mathbf{p}_k, \mathbf{Hw}_1 \rangle + \sum_{i=1}^{k-1} \alpha_i \langle \mathbf{p}_k, \mathbf{Hp}_i \rangle$, or, taking into account that $\langle \mathbf{p}_k, \mathbf{Hp}_i \rangle = 0, \forall i \neq k$, we get that $\langle \mathbf{p}_k, \mathbf{Hw}_k \rangle = \langle \mathbf{p}_k, \mathbf{Hw}_1 \rangle$, and so the relation (11) for calculating α_k becomes:

$$\alpha_k = -\frac{\langle \mathbf{p}_k, \mathbf{b} + \mathbf{Hw}_k \rangle}{\langle \mathbf{p}_k, \mathbf{Hp}_k \rangle} = -\frac{\langle \mathbf{p}_k, \mathbf{g}_k \rangle}{\langle \mathbf{p}_k, \mathbf{Hp}_k \rangle}, \qquad (12)$$

where $\mathbf{g}_k := \mathbf{g}(\mathbf{w}_k) = \mathbf{b} + \mathbf{Hw}_k$, as relation (5) shows.

Finally, we need to construct the mutually conjugate directions $\mathbf{p}_k$. For this, the first direction is initialized by the negative gradient of the error function at the initial point $\mathbf{w}_1$, i.e. $\mathbf{p}_1 = -\mathbf{g}_1$. We have the following update rule for the conjugate directions:

$$\mathbf{p}_{k+1} = -\mathbf{g}_{k+1} + \beta_k \mathbf{p}_k. \qquad (13)$$

Taking the Euclidean $\mathbb{R}^{2N}$ scalar product with $\mathbf{Hp}_k$, and imposing the conjugacy condition $\langle \mathbf{p}_{k+1}, \mathbf{Hp}_k \rangle = 0$, we obtain that

$$\beta_k = \frac{\langle \mathbf{g}_{k+1}, \mathbf{Hp}_k \rangle}{\langle \mathbf{p}_k, \mathbf{Hp}_k \rangle}. \qquad (14)$$

It can be easily shown by induction that repeated application of the relations (13) and (14), yields a set of mutually conjugate directions with respect to the positive definite matrix $\mathbf{H}$.

So far, we have dealt with a quadratic error function that has a positive definite Hessian matrix $\mathbf{H}$. But, in practical applications, the error function may be far from quadratic, and so the expressions for calculating α_k and β_k that we deduced above, may not be as accurate as in the quadratic case. Furthermore, these expressions need the explicit calculation of the Hessian matrix $\mathbf{H}$ for each step of the algorithm, because the Hessian is constant only in the case of the quadratic error function. This calculation is computationally intensive and should be avoided. In what follows, we will deduce expressions for α_k and β_k that do not need the explicit calculation of the Hessian matrix, and don't even assume that the Hessian is positive definite.

First of all, let's consider the expression for α_k, given in (12). Because of the iterative relation (10), we can replace the explicit calculation of α_k with an inexact line search that minimizes $E(\mathbf{w}_{k+1}) = E(\mathbf{w}_k + \alpha_k \mathbf{p}_k)$, i.e. a line minimization along the search direction $\mathbf{p}_k$, starting at the point $\mathbf{w}_k$. In our experiments, we used the golden section search, which is guaranteed to have linear convergence, see [16].

Now, let's turn our attention to β_k. From (5), we have that

$$\mathbf{g}_{k+1} - \mathbf{g}_k = \mathbf{H}(\mathbf{w}_{k+1} - \mathbf{w}_k) = \alpha_k \mathbf{H} \mathbf{p}_k, \tag{15}$$

and so the expression (14) becomes:

$$\beta_k = \frac{\langle \mathbf{g}_{k+1}, \mathbf{g}_{k+1} - \mathbf{g}_k \rangle}{\langle \mathbf{p}_k, \mathbf{g}_{k+1} - \mathbf{g}_k \rangle}. \tag{16}$$

This is known as the *Hestenes-Stiefel* update expression, see [12].

To process this even further, we must take the Euclidean $\mathbb{R}^{2N}$ scalar product of Eq. (15) with $\mathbf{p}_k$ and take into account the expression (12) for α_k to give

$$\langle \mathbf{p}_k, \mathbf{g}_{k+1} \rangle = 0. \tag{17}$$

Taking the scalar product of (13) with $\mathbf{g}_{k+1}$ and using the above equation, yields

$$\langle \mathbf{p}_{k+1}, \mathbf{g}_{k+1} \rangle = -\langle \mathbf{g}_{k+1}, \mathbf{g}_{k+1} \rangle. \tag{18}$$

Now, inserting relations (17) and (18) into expression (16), we obtain the *Polak-Ribiere* update expression (see [18]):

$$\beta_k = \frac{\langle \mathbf{g}_{k+1}, \mathbf{g}_{k+1} - \mathbf{g}_k \rangle}{\langle \mathbf{g}_k, \mathbf{g}_k \rangle}. \tag{19}$$

Next, we will take the Euclidean $\mathbb{R}^{2N}$ scalar product of Eq. (15) with $\mathbf{p}_i$, $i < k$, to obtain

$$\langle \mathbf{p}_i, \mathbf{g}_{k+1} - \mathbf{g}_k \rangle = \alpha_k \langle \mathbf{p}_i, \mathbf{H} \mathbf{p}_k \rangle = 0, \tag{20}$$

where we took into account the fact that $\mathbf{p}_i$ and $\mathbf{p}_k$ are conjugate directions, for $i < k$. Equations (20) and (17) can be used to prove by induction that we have

$$\langle \mathbf{p}_i, \mathbf{g}_k \rangle = 0, \forall i < k \leq 2N. \tag{21}$$

From the expression for $\mathbf{p}_{k+1}$ given in Eq. (13), we deduce that $\mathbf{p}_k$ is given by a linear combination of all the previous gradient vectors, namely we can write

$$\mathbf{p}_i = -\mathbf{g}_i + \sum_{j=1}^{k-1} \gamma_j \mathbf{g}_j. \tag{22}$$

Taking the scalar product of Eq. (22) with $\mathbf{p}_i$, $i < k$, and using Eq. (21), we obtain the following relation: $\langle \mathbf{g}_i, \mathbf{g}_k \rangle = \sum_{j=1}^{k-1} \gamma_j \langle \mathbf{g}_j, \mathbf{g}_k \rangle, \forall i < k \leq 2N$. Starting from the fact that $\mathbf{p}_1 = -\mathbf{g}_1$ and Eq. (21), we obtain inductively that

$$\langle \mathbf{g}_i, \mathbf{g}_k \rangle = 0, \forall i < k \leq 2N. \tag{23}$$

In particular, we have that $\langle \mathbf{g}_k, \mathbf{g}_{k+1} \rangle = 0$, and so expression (19) becomes: $\beta_k = \frac{\langle \mathbf{g}_{k+1}, \mathbf{g}_{k+1} \rangle}{\langle \mathbf{g}_k, \mathbf{g}_k \rangle}$. This expression is known as the *Fletcher-Reeves* update formula, see [20].

Lastly, using (23), namely that $\langle \mathbf{g}_k, \mathbf{g}_{k+1} \rangle = 0$, in Eq. (16), we obtain the following update formula: $\beta_k = \frac{\langle \mathbf{g}_{k+1}, \mathbf{g}_{k+1} \rangle}{\langle \mathbf{p}_k, \mathbf{g}_{k+1} - \mathbf{g}_k \rangle}$, which is known as the *Dai-Yuan* update expression, see [7].

Two variations of the Hestenes-Stiefel and Polak-Ribiere update formulas exist, which have been proven in [8] to converge even if inexact line search is used to determine the value of α_k. If we denote by β_k^{HS}, and respectively by β_k^{PR}, the values of the updates in the case of the two algorithms, the expressions for the updates that guarantee convergence are $\beta_k^{HS+} = \max\{0, \beta_k^{HS}\}$, and respectively $\beta_k^{PR+} = \max\{0, \beta_k^{PR}\}$.

If the error function E is quadratic, then the conjugate gradient algorithm is guaranteed to find its minimum in at most $2N$ steps. But, in general, the error function may be far from quadratic, and so the algorithm will need more than $2N$ steps to approach the minimum. Over these steps, the conjugacy of the search directions tends to deteriorate, so it is a common practice to restart the algorithm with the negative gradient $\mathbf{p}_k = -\mathbf{g}_k$, after $2N$ steps.

A more sophisticated restart algorithm, proposed by Powell in [19], following an idea by Beale in [3], is to restart if there is little orthogonality left between the current gradient and the previous gradient. To test this, we verify that the following inequality holds: $|\langle \mathbf{g}_k, \mathbf{g}_{k+1} \rangle| \geq 0.2 \langle \mathbf{g}_{k+1}, \mathbf{g}_{k+1} \rangle$. The update rule (13) for the search direction $\mathbf{p}_k$ is also changed to $\mathbf{p}_{k+1} = -\mathbf{g}_{k+1} + \beta_k \mathbf{p}_k + \gamma_k \mathbf{p}_t$, where $\beta_k = \frac{\langle \mathbf{g}_{k+1}, \mathbf{g}_{k+1} - \mathbf{g}_k \rangle}{\langle \mathbf{p}_k, \mathbf{g}_{k+1} - \mathbf{g}_k \rangle}$, $\gamma_k = \frac{\langle \mathbf{g}_{k+1}, \mathbf{g}_{t+1} - \mathbf{g}_t \rangle}{\langle \mathbf{p}_t, \mathbf{g}_{t+1} - \mathbf{g}_t \rangle}$, and $\mathbf{p}_t$ is an arbitrary downhill restarting direction. These expressions are deducted in the same way as the ones above. The conjugate gradient algorithm with these characteristics is called the *Powell-Beale* algorithm.

In order to apply the conjugate gradient algorithms to complex-valued feedforward neural networks, we only need to calculate the gradient $\mathbf{g}_k$ of the error function E at different steps, which we do by using the backpropagation algorithm.

3 Experimental Results

3.1 Fully Complex Synthetic Function

The synthetic fully complex function we will test the proposed algorithms on is the simple quadratic function $f_1(z_1, z_2) = \frac{1}{6}\left(z_1^2 + z_2^2\right)$. The most important property of fully complex functions is that they treat complex numbers as a whole, and not the real and imaginary parts separately, as the split complex functions do. This function was used to test the efficiency of different architectures and learning algorithms, for example in [1,22,23].

To train the networks, we randomly generated 3000 training samples, such that each sample had the inputs z_1, z_2 inside the disk centered at the origin and with radius 2.5. For testing, we generated 1000 testing samples having the same characteristics. All the networks had one hidden layer comprised of 15 neurons and were trained for 5000 epochs. The activation function for the hidden layer was the fully complex hyperbolic tangent function $G(z) = \tanh z = \frac{e^z - e^{-z}}{e^z + e^{-z}}$, and the activation function for the output layer was the identity.

In our experiments, we trained complex-valued feedforward neural networks using the classical gradient descent algorithm (abbreviated GD), the conjugate gradient algorithm with Hestenes-Stiefel updates (CGHS), the conjugate gradient algorithm with Polak-Ribiere updates (CGPR), the conjugate gradient algorithm with Fletcher-Reeves updates (CGFR), the conjugate gradient algorithm with Dai-Yuan updates (CGDY), the variations of the Hestenes-Stiefel (CGHS+) and the Polak-Ribiere (CGPR+) algorithms, and the conjugate gradient algorithm with Powell-Beale restarts (CGPB).

Table 1. Experimental results for the function f_1

Algorithm	Training MSE	Testing MSE
GD	**5.23e-5 ± 9.29e-6**	**5.84e-5 ± 1.16e-5**
CGHS	**1.21e-7 ± 2.96e-8**	**1.38e-7 ± 3.70e-8**
CGPR	1.61e-5 ± 5.14e-6	1.80e-5 ± 6.19e-6
CGFR	1.51e-7 ± 3.95e-8	1.78e-7 ± 4.98e-8
CGDY	1.45e-7 ± 4.02e-8	1.69e-7 ± 4.66e-8
CGHS+	1.38e-7 ± 2.01e-8	1.63e-7 ± 2.74e-8
CGPR+	1.73e-5 ± 4.18e-6	1.89e-5 ± 4.75e-6
CGPB	1.78e-7 ± 4.98e-8	2.21e-7 ± 5.78e-8
FC-RBF [21]	3.61e-6	9.00e-6
FC-RBF with KMC [21]	2.01e-6	1.87e-6
Mc-FCRBF [22]	2.50e-5	2.56e-6
CMRAN [21]	4.60e-3	4.90e-3

Training was repeated 50 times for each algorithm, and the resulted mean and standard deviation of the mean squared error (MSE) are given in Table 1. The table also gives the MSE of other algorithms used to learn this function, together with the references in which they first appeared.

3.2 Split Complex Synthetic Function

We now test the proposed algorithms on a split complex function. The function, also used in [2,5,14], is $f_2(x + iy) = \sin x \cosh y + i \cos x \sinh y$.

The training set had 3000 samples and the test set had 1000 samples randomly generated from the unit disk. The neural networks had 15 neurons on a single hidden layer. The activation functions were split hyperbolic tangent for the hidden layer: $G(x + iy) = \tanh x + i \tanh y = \frac{e^x - e^{-x}}{e^x + e^{-x}} + i \frac{e^y - e^{-y}}{e^y + e^{-y}}$, and the identity function for the output layer. The mean and standard deviation of the mean squared error (MSE) over 50 runs are presented in Table 2.

Table 2. Experimental results for the function f_2

Algorithm	Training MSE	Testing MSE
GD	**7.29e-4 ± 1.71e-4**	**7.72e-4 ± 1.78e-4**
CGHS	8.83e-6 ± 2.57e-6	9.82e-6 ± 2.83e-6
CGPR	1.88e-4 ± 4.14e-5	1.88e-4 ± 4.14e-5
CGFR	8.02e-6 ± 1.85e-6	8.83e-6 ± 2.15e-6
CGDY	5.97e-6 ± 1.28e-6	6.52e-6 ± 1.43e-6
CGHS+	8.48e-6 ± 2.57e-6	9.42e-6 ± 2.92e-6
CGPR+	1.87e-4 ± 4.63e-5	2.03e-4 ± 5.00e-5
CGPB	**5.57e-6 ± 2.30e-6**	**6.13e-6 ± 2.53e-6**

3.3 Nonlinear Time Series Prediction

The last experiment deals with the prediction of complex-valued nonlinear signals. It involves passing the output of the autoregressive filter given by $y(k) = 1.79y(k - 1) - 1.85y(k - 2) + 1.27y(k - 3) - 0.41y(k - 4) + n(k)$, through the nonlinearity given by $z(k) = \frac{z(k-1)}{1+z^2(k-1)} + y^3(k)$, which was proposed in [17], and then used in [9–11].

The complex-valued noise $n(k)$ was chosen so that the variance of the signal as a whole is 1, taking into account the fact that $\sigma^2 = (\sigma^R)^2 + (\sigma^I)^2$. The tap input of the filter was 4, and so the networks had 4 inputs, 4 hidden neurons and one output neuron. They were trained for 5000 epochs with 5000 training samples.

After running each algorithm 50 times, the results are given in Table 3. In the table, we presented a measure of performance called *prediction gain*, defined by $R_p = 10 \log_{10} \frac{\sigma_x^2}{\sigma_e^2}$, where σ_x^2 represents the variance of the input signal and σ_e^2 represents the variance of the prediction error. The prediction gain is given in dB. It is obvious that, because of the way it is defined, a bigger prediction gain means better performance. The table also presents the performances of some classical algorithms and network architectures found in the literature.

Table 3. Experimental results for nonlinear time series prediction

Algorithm	Prediction gain
GD	**3.95 ± 4.67e-1**
CGHS	8.35 ± 8.34e-4
CGPR	8.31 ± 2.59e-2
CGFR	8.34 ± 7.49e-4
CGDY	8.35 ± 1.09e-3
CGHS+	8.35 ± 6.61e-4
CGPR+	6.45 ± 3.90e-1
CGPB	**8.35 ± 4.11e-4**
CLMS [24]	1.87
CNGD [11]	2.50
CRTRL [9]	3.76

4   Conclusions

The full deduction of the most known variants of the conjugate gradient algorithm for training complex-valued neural networks was presented. The seven variants of the conjugate gradient algorithm with different updates and restart procedures were applied for training networks used to solve three well-known synthetic and real-world problems.

The experimental results shown that all the conjugate gradient algorithms performed better on the proposed problems than the classical gradient descent algorithm, in some cases as much as an order or two orders of magnitude better in terms of training or testing mean square error. Conjugate gradient with Powell-Beale restarts and conjugate gradient with Dai-Yuan updates generally had the best performances among all the other conjugate gradient variants. These algorithms were generally followed by conjugate gradient with Hestenes-Stiefel updates and its variation that keeps the update positive at all times. Conjugate gradient with Fletcher-Reeves followed, and last, having a performance sometimes an order of magnitude worse, were the conjugate gradient with Polak-Ribiere and its variation. But still, even with such bad performance, these last two algorithms performed better than the gradient descent.

As a conclusion, it can be said that conjugate gradient algorithms represent an efficient method of training feedforward complex-valued neural networks, as it was shown by their performance in solving heterogeneous synthetic and real-world problems.

References

1. Amin, M., Savitha, R., Amin, M., Murase, K.: Complex-valued functional link network design by orthogonal least squares method for function approximation problems. In: International Joint Conference on Neural Networks (IJCNN), pp. 1489–1496. IEEE, July 2011
2. Arena, P., Fortuna, L., Re, R., Xibilia, M.: Multilayer perceptrons to approximate complex valued functions. Int. J. Neural Syst. **6**(4), 435–446 (1995)
3. Beale, E.: A derivation of conjugate gradients. In: Lootsma, F.A. (ed.) Numerical Methods for Nonlinear Optimization, pp. 39–43. Academic Press, London (1972)
4. Bishop, C.: Neural Networks for Pattern Recognition. Oxford University Press Inc., New York (1995)
5. Buchholz, S., Sommer, G.: On clifford neurons and clifford multi-layer perceptrons. Neural Netw. **21**(7), 925–935 (2008)
6. Charalambous, C.: Conjugate gradient algorithm for efficient training of artificial neural networks. IEEE Proc. G Circ. Devices Syst. **139**(3), 301–310 (1992)
7. Dai, Y., Yuan, Y.: A nonlinear conjugate gradient method with a strong global convergence property. SIAM J. Optim. **10**(1), 177–182 (1999)
8. Gilbert, J., Nocedal, J.: Global convergence properties of conjugate gradient methods for optimization. SIAM J. Optim. **2**(1), 21–42 (1992)
9. Goh, S., Mandic, D.: A complex-valued rtrl algorithm for recurrent neural networks. Neural Comput. **16**(12), 2699–2713 (2004)
10. Goh, S., Mandic, D.: Nonlinear adaptive prediction of complex-valued signals by complex-valued prnn. IEEE Trans. Signal Process. **53**(5), 1827–1836 (2005)
11. Goh, S., Mandic, D.: Stochastic gradient-adaptive complex-valued nonlinear neural adaptive filters with a gradient-adaptive step size. IEEE Trans. Neural Netw. **18**(5), 1511–1516 (2007)
12. Hestenes, M., Stiefel, E.: Methods of conjugate gradients for solving linear systems. J. Res. Natl Bur. Stan. **49**(6), 409–436 (1952)
13. Hirose, A.: Complex-Valued Neural Networks: Advances and Applications. Wiley, New York (2013)
14. Huang, G.B., Li, M.B., Chen, L., Siew, C.K.: Incremental extreme learning machine with fully complex hidden nodes. Neurocomputing **71**(4–6), 576–583 (2008)
15. Johansson, E., Dowla, F., Goodman, D.: Backpropagation learning for multilayer feed-forward neural networks using the conjugate gradient method. Int. J. Neural Syst. **2**(4), 291–301 (1991)
16. Luenberger, D., Ye, Y.: Linear and Nonlinear Programming. International Series in Operations Research & Management Science, vol. 116. Springer, US (2008)
17. Narendra, K., Parthasarathy, K.: Identification and control of dynamical systems using neural networks. IEEE Trans. Neural Netw. **1**(1), 4–27 (1990)
18. Polak, E., Ribiere, G.: Note sur la convergence de méthodes de directions conjuguées. Rev. Fr. d'Informatique et de Rech. Opérationnelle **3**(16), 35–43 (1969)
19. Powell, M.: Restart procedures for the conjugate gradient method. Math. Program. **12**(1), 241–254 (1977)
20. Reeves, C., Fletcher, R.: Function minimization by conjugate gradients. Comput. J. **7**(2), 149–154 (1964)
21. Savitha, R., Suresh, S., Sundararajan, N.: A fully complex-valued radial basis function network and its learning algorithm. Int. J. Neural Syst. **19**(4), 253–267 (2009)
22. Savitha, R., Suresh, S., Sundararajan, N.: A self-regulated learning in fully complex-valued radial basis function networks. In: International Joint Conference on Neural Networks (IJCNN), pp. 1–8. IEEE, July 2010

23. Savitha, R., Suresh, S., Sundararajan, N.: A meta-cognitive learning algorithm for a fully complex-valued relaxation network. Neural Netw. **32**, 209–218 (2012)
24. Widrow, B., McCool, J., Ball, M.: The complex lms algorithm. Proc. IEEE **63**(4), 719–720 (1975)

XaIBO: An Extension of aIB for Trajectory Clustering with Outlier

Yuejun Guo[1], Qing Xu[1(✉)], Sheng Liang[1], Yang Fan[1], and Mateu Sbert[2]

[1] School of Computer Science and Technology, Tianjin University, Tianjin, China
{guoyuejun13,qingxu.itcn,tjulsg,Michellehexing}@gmail.com
qingxu@tju.edu.cn
[2] Graphics and Imaging Laboratory, Universitat de Girona, Girona, Spain
mateu@ima.udg.edu

Abstract. Clustering plays an important role for trajectory analysis. The agglomerative Information Bottleneck (aIB) approach is effective for successfully managing an optimum number of clusters without the need of an explicit measure of trajectory distance, which is usually very difficult to be defined. In this paper, we propose to utilize a statistically representation of the trajectory shape to perform the aIB based trajectory clustering. In addition, an extension of aIB is proposed to cope with the clustering on trajectories with outliers (for brevity, we call this extended version of aIB as XaIBO) and in this case, XaIBO can be widely used in practice for dealing with complex trajectory data. Extensive experiments on synthetic, simulated and real trajectory data have shown that XaIBO achieves the trajectory clustering very well.

Keywords: XaIBO · aIB · Trajectory clustering · Outlier

1 Introduction

Clustering is vital important for extracting valuable structures and analyzing potential knowledge in plenty of trajectory data [6]. Lots of methods of trajectory clustering have been, so far, advanced and shown to perform well and, we refer readers to a recent overview for detailed information [9]. However, two difficult problems for trajectory clustering have been preventing progress of its wide usage in real applications. One issue is to determine an optimum number of clusters for complex trajectory data [6]. Another challenge is to define a good measure on trajectory similarity/difference [3], which is proper for any trajectory data.

Notably, a very latest approach has been developed to adaptively achieve an optimal number of the clusters without needing an explicit trajectory distance [3], on the basis of making use of the agglomerative Information Bottleneck (aIB) [14] technique from information theory literature [12]. The point of using aIB for

This work has been funded by Natural Science Foundation of China (61471261, 61179067, U1333110), and by grants TIN2013-47276-C6-1-R from Spanish Government and 2014-SGR-1232 from Catalan Government (Spain).

© Springer International Publishing Switzerland 2015
S. Arik et al. (Eds.): ICONIP 2015, Part II, LNCS 9490, pp. 423–431, 2015.
DOI: 10.1007/978-3-319-26535-3_48

achieving a clustered structure of one random variable X is to define and utilize a relevant variable Y for X [14]. And, this Y can be easily obtained, for example, by using a simply partitioned version of X [3,15]. In this paper, we propose an alternative utilization of aIB for the clustering of motion trajectory X, in which a probabilistic distribution of the tangential angles happening to the sample positions of the corresponding moving object is treated as the relevant random variable Y for X. More importantly, we have observed that, in the framework of aIB, outliers occurring in the trajectory data could degrade the clustering performance. In this paper, we additionally propose a new trajectory clustering scheme based on aIB, called XaIBO, to fuse the identification of the possible outlier trajectories and the avoidance of their mistaken use into aIB, achieving largely better clustering.

The rest of this paper is organized as follows: Sect. 2 covers the related work. Section 3 fully describes our proposed XaIBO based trajectory clustering approach. Experimental results are presented and discussed in Sect. 4. Finally, Sect. 5 provides the conclusion and our future work.

2   Related Work

According to the IB principle [12], to obtain a clustered representation of one random variable X, denoted with $\widehat{X}$, is done by maximizing the mutual information with respect to another relevant variable Y, $I(\widehat{X}; Y)$, and meanwhile by compacting the information shared between $\widehat{X}$ and X, $I(\widehat{X}; X)$,

$$\max\{I(\widehat{X}; Y) - \beta I(\widehat{X}; X)\}, \tag{1}$$

where β is the Lagrange multiplier, controlling the compromise between data compaction and information preservation. In practice, the solution is usually achieved with a greedy algorithm based on a bottom-up merging strategy by minimizing the loss of mutual information, $I(X; Y) - I(\widehat{X}; Y)$, called agglomerative Information Bottleneck (aIB) [14].

The aIB algorithm was first proposed, for finding a locally optimal clustering, by Slonim and Tishby in 1999 [14]. So far, aIB has been widely applied in a variety of application domains, such as document processing [14], image retrieval [2], galaxy data engineering [13] and neural information processing [11]. Some literatures represent the data to be compacted by a Gaussian Mixture Model (GMM) to obtain the relevant variable [2], and some others partition the dataset to achieve this [3,15].

3   Method

3.1   Trajectory Clustering Based on aIB

With a given set of trajectory data $T = \{t\}$, in general, a feature space $F = \{f\}$ of T can be always constructed and obviously, T correlated with F. For the sake of this paper, both T and F should be random variables. Here, the shape of a trajectory is made use of to form the feature space. Typically and actually, the shape of a motion trajectory is probabilistically featured by a set of the tangential angles at the sample positions of the moving object (for the purpose of simple description, here we call these positions as the sample points), in order to deal with some inevitable noise embedded in the given trajectory data [3]. The normalized set of angle histogram is utilized as the relevant random variable F for T. As a result, we create an information channel between T and F, and three basic elements of the information channel $T \to F$ are as follows:

- Input random variable T representing a trajectory distribution. In this paper, an uniform probability distribution is simply used for the trajectory data. Apparently the probability of T is $p(t) = 1/n_T$, here $n_T = |T|$.
- Conditional probability $p(f|t)$. Given a trajectory t with n angles, the Kernel Density Estimation (KDE) technique [3] is taken to build up a Probability Density Function (PDF) using the angles of this trajectory

$$p(f|t) = \sum_\theta p(\theta) = \sum_\theta \frac{1}{n} \sum_{i=1}^n \{|h|^{-\frac{1}{2}} K[h^{-\frac{1}{2}}(\theta - \theta_i)]\} \tag{2}$$

where θ is the angle range of f. $h = s_\theta^2 n^{-\frac{2}{5}}$ is the bandwidth and s_θ^2 is the variance of angles. $K(\cdot)$ is the Gaussian kernel function.
- Output probability $p(f)$. This probability is obtained by

$$p(f) = \sum_{t \in T} p(t)p(f|t). \tag{3}$$

Based on taking advantage of aIB, a compacted T, represented by $\widehat{T}$, is iteratively obtained with the minimal loss of the mutual information $I(\widehat{T}; F)$. In each iteration of aIB, the two clusters $\hat{t}_a$ and $\hat{t}_b$ are considered as candidates if the merging of them results in the lowest information loss [14]

$$\begin{aligned}
\Delta I(\hat{t}_a, \hat{t}_b) &= I(\widehat{T}; F) - I(\widehat{T}^*; F) \\
&= \sum_{f, i=a,b} p(\hat{t}_i, f) \log \frac{p(\hat{t}_i, f)}{p(\hat{t}_i)p(y)} - \sum_y p(\hat{t}^*, f) \log \frac{p(\hat{t}^*, f)}{p(\hat{t}^*)p(f)} \\
&= p(\hat{t}^*)(H(\sum_{i=a,b} \pi_i p(\hat{t}_i)) - \sum_{i=a,b} \pi_i H(p(\hat{t}_i))) \\
&= p(\hat{t}^*)JSD(\pi_a, \pi_b; p(f|\hat{t}_a), p(f|\hat{t}_b))
\end{aligned} \tag{4}$$

where $\widehat{T}$ and $\widehat{T}^*$ are the clusters before and after merging $\hat{t}_a$ and $\hat{t}_b$, respectively. $H(\cdot)$ is the Shannon entropy and, $JSD(\cdot)$ is the *Jensen-Shannon divergence* [14]. After merging, a new cluster, denoted by $\hat{t}^*$, is $\hat{t}_a \cup \hat{t}_b$, and the corresponding conditional probability is

$$p(f|\hat{t}^*) = \frac{p(\hat{t}_a)}{p(\hat{t}^*)} p(f|\hat{t}_a) + \frac{p(\hat{t}_b)}{p(\hat{t}^*)} p(f|\hat{t}_b) \tag{5}$$

where

$$p(\hat{t}^*) = p(\hat{t}_a) + p(\hat{t}_b). \tag{6}$$

Notably, in the procedure of aIB, some clusters may have many trajectories due to merging, and the marginal and conditional probabilities of them are large (see Formulae (6) and (5)). In contrast, the others may only possess a few trajectories (for example, the cluster containing outliers just has a small number of trajectories), resulting in their small marginal and conditional probabilities. As a mater of fact, this kind of big probability difference can seriously degrade the clustering performance, due to a wrong choice of two clusters to be merged. This results from a quite small information loss by merging two clusters that have largely different trajectory numbers. For better explanation, we illustrate an example of a wrong choice of candidate clusters to be merged as follows.

Eg. 1. *As shown in Fig. 1, $\hat{t}_1$ and $\hat{t}_2$ are similar, and both of them are highly different with $\hat{t}_3$. Thus, $\hat{t}_1$ and $\hat{t}_2$ should be the candidate clusters for merging. Unfortunately, due to the large differences in both conditional and marginal probabilities, $\Delta I(\hat{t}_1, \hat{t}_2)$ is not the lowest value as shown in Table 1, causing a wrong merge of outlier $\hat{t}_3$ with $\hat{t}_1$.*

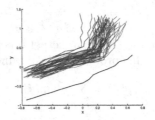

Fig. 1. An example of imbalanced clusters in an iteration of aIB. $\widehat{T} = \{\hat{t}_1, \hat{t}_2, \hat{t}_3\}$ ($\hat{t}_3$ is an outlier). The number of trajectories in $\hat{t}_1$ (red), $\hat{t}_2$ (blue) and $\hat{t}_3$ (black) are 28, 22 and 1, respectively. The rose mark "*" indicates the starting point of each trajectory (Color figure online).

In order to avoid some possible mistaken clustering induced by outlier trajectories, our XaIBO applies a proposed filter to obtain the appropriate candidate clusters for merging. Algorithm 1 clearly presents the procedure of XaIBO. Notice that the iterative clustering can be adaptively terminated by a ratio $\delta = \frac{I(\widehat{T};F)}{H(F)}$. Obviously, the maximal of mutual information $I(\widehat{T};F)$ obtained is the entropy $H(F)$. In the meantime, an optimal number of clusters are achieved. We have observed that $\delta = 75\%$ can perform very well.

3.2 Filtering Criterion

In each iteration, the distance between clusters is employed to filter out the mistaken merging with outlier. Firstly, All the information losses between each two clusters are sorted ascendingly. And then we propose the following filtering criterion to determine whether two clusters $\hat{t}_a$ and $\hat{t}_b$ in $\widehat{T}$ should be merged or not

Table 1. Conditional, marginal probabilities and information loss of $\widehat{T}$ in Fig. 1. $F = \{f_i\}, i = 1, 2, \ldots, 9$.

$p(f|\hat{t})$\ f $\hat{t}$	f_1	f_2	f_3	f_4	f_5	f_6	f_7	f_8	f_9	$p(\hat{t})$	ΔI (bit)
$\hat{t}_1$	$3.09e^{-1}$	$4.45e^{-1}$	$1.94e^{-1}$	$1.12e^{-2}$	$5.84e^{-5}$	$2.83e^{-8}$	$1.63e^{-12}$	$9.85e^{-18}$	$4.44e^{-24}$	$28/51$	$\Delta I(\hat{t}_1,\hat{t}_2)$=0.0650
$\hat{t}_2$	$2.73e^{-1}$	$3.81e^{-1}$	$2.55e^{-1}$	$3.70e^{-2}$	$1.12e^{-3}$	$9.62e^{-6}$	$2.16e^{-8}$	$9.65e^{-12}$	$7.51e^{-16}$	$22/51$	$\Delta I(\hat{t}_2,\hat{t}_3)$=0.0423
$\hat{t}_3$	$1.03e^{-211}$	$2.06e^{-123}$	$3.17e^{-59}$	$4.51e^{-19}$	$2.57e^{-2}$	$8.66e^{-1}$	$1.09e^{-1}$	$5.15e^{-13}$	$2.83e^{-47}$	$1/51$	$\Delta I(\hat{t}_1,\hat{t}_3)$=**0.0419**

Algorithm 1. XaIBO algorithm

Input : $N_T = |T|$: Number of trajectories
$N_F = |F|$: Number of elements in F
$p(f|t)$: Conditional probability distribution.
Output: $\widehat{T} = \{\hat{t}_1, \hat{t}_2, \ldots, \hat{t}_m\}$: Set of clusters
begin
  Create $\widehat{T} = T$
  foreach $i \in \{1, 2, \ldots, N_T\}$, $f \in F$ **do** $p(\hat{t}_i) = p(t_i)$, $p(f|\hat{t}_i) = p(f|t_i)$
  foreach $i, j \in \{1, 2, \ldots |\widehat{T}|\}, j > i$, $f \in F$ **do** Calculate $\Delta I(\hat{t}_i, \hat{t}_j)$ by Formula (4)
  while $\delta \geq 75\%$ **do**
    foreach $\hat{t}_i, \hat{t}_j \in \widehat{T}$, $i \neq j$ **do** Sort $\{\hat{t}_i, \hat{t}_j\}$ by $\Delta I(\hat{t}_i, \hat{t}_j)$ ascendingly
    for $i = 1$ **to** $\frac{|\widehat{T}| \times (|\widehat{T}|-1)}{2}$ **do**
      Filter the i-th possible candidate clusters $\{\hat{t}_a, \hat{t}_b\}$
      if $R(\hat{t}_a, \hat{t}_b) < 1$ *by Formula* (7) **then**
        $\hat{t}_a$ and $\hat{t}_b$ are candidate clusters
        Break
      end
    end
    Create $\hat{t}^* = \hat{t}_a \cup \hat{t}_b$
    Calculate $p(\hat{t}^*)$ by Formula (6)
    foreach $f \in F$ **do** Calculate $p(f|\hat{t}^*)$ by Formula (5)
    Update $\widehat{T} = \{\widehat{T} - \hat{t}_a \cup \hat{t}_b\} \cup \hat{t}^*$
    foreach $\hat{t} \in \{\widehat{T} - \hat{t}^*\}$ **do** Calculate $\Delta I(\hat{t}^*, \hat{t})$
  end
end

$$R(\hat{t}_a, \hat{t}_b) = \frac{\min \text{JSD}(t_i, t_j)}{D}, t_i \in \hat{t}_a, t_j \in \hat{t}_b \qquad (7)$$

where D is the diameter of a cluster, the maximum distance between two trajectories in this cluster. D takes different values depending on three different cases: if $|\hat{t}_a| \geq 2$ or $|\hat{t}_b| \geq 2$, D is the bigger one between the diameters of $\hat{t}_a$ and $\hat{t}_b$; if $|\hat{t}_a| = 1$ and $|\hat{t}_b| = 1$ and there exists a $\hat{t} \in \widehat{T} - \hat{t}_a \cup \hat{t}_b$ with more than two trajectories, then D is the biggest diameter for all the clusters in $\widehat{T}$; if $|\hat{t}_a| = 1$ and $|\hat{t}_a| = 1$ and any $\hat{t} \in \widehat{T} - \hat{t}_a \cup \hat{t}_b$ with at most two trajectories, then $D = \infty$. In this way, as for the clusters to be merged, if $R(\hat{t}_a, \hat{t}_b) > 1$, $\hat{t}_a$ and $\hat{t}_b$ must be very dissimilar in marginal and conditional probabilities, and one of them is outlier. As a result, $\hat{t}_a$ and $\hat{t}_b$ should be filtered out.

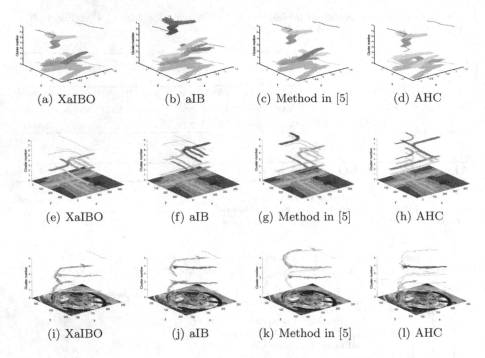

(a) XaIBO   (b) aIB   (c) Method in [5]   (d) AHC

(e) XaIBO   (f) aIB   (g) Method in [5]   (h) AHC

(i) XaIBO   (j) aIB   (k) Method in [5]   (l) AHC

Fig. 2. Clustering results by four approaches on synthetic (first row, due to space limitation, only "TS299" is listed), simulated (second row) and real datasets (third row), respectively. The complete original data are visualized with grey on horizontal plane. Different clusters are visualized with different colors (except grey) (Color figure online).

4   Experiments

We conduct a series of experiments with 20 synthetic[1] [10], a simulated and a real[2] [8] datasets to evaluate the performance of our XaIBO algorithm. Additionally, we compare XaIBO with aIB, method in [5] and agglomerative hierarchical clustering (AHC). The method in [5] preprocesses the trajectories by Discrete Fourier Transform (DFT), and then do clustering using Density-Based Spatial Clustering of Applications with Noise (DBSCAN), which is capable of detecting noise. For fair comparison, we also do DFT as the data preprocessing for AHC. All the parameters in [5] and AHC are tuned to obtain their best possible results.

As shown in Fig. 2(a,e,i), XaIBO can accurately separate the trajectories with different shapes. Apparently, aIB fails to detect outliers, blue and red trajectories, shown in Fig. 2(a). Actually aIB wrongly mixes these outliers with other clusters, and very similar trajectories, such as yellow and green and, brown and light blue trajectories (Fig. 2(b)), which should be merged, are mistakenly

---

[1] http://avires.dimi.uniud.it/papers/trclust.
[2] http://cvrr.ucsd.edu/bmorris/datasets/dataset_trajectory_analysis.html.

Table 2. Comparison results on Dunn index and F1-Measure of four approaches.

	XaIBO		aIB		Method in [5]		AHC	
	Dunn index	F1-Measure	Dunn index	F1-Measure	Dunn index	F1-Measure	Dunn index	F1-Measure
TS10	0.3118	1.0000	0.0004	0.8729	0.0633	0.9779	0.0050	0.2710
TS20	0.2651	1.0000	0.1487	0.9209	0.0628	0.8989	0.0332	0.0748
TS120	0.4326	1.0000	0.0883	0.8628	0.0498	0.6502	0.0532	0.4934
TS132	0.2918	1.0000	0.0380	0.8890	0.0889	0.9947	0.2918	1.0000
TS150	0.4134	1.0000	0.0647	0.8905	0.0580	0.9872	0.0549	0.4934
TS182	0.1338	0.9505	0.0543	0.6451	0.0886	0.1094	0.0384	0.0748
TS184	1.3637	1.0000	0.0210	0.9482	0.4902	0.9947	1.3637	1.0000
TS186	0.1343	1.0000	0.0826	0.8790	0.1038	0.9947	0.1343	1.0000
TS198	0.4192	1.0000	0.0071	0.8786	0.1245	0.9947	0.4192	1.0000
TS214	0.2883	1.0000	0.0831	0.8912	0.0426	0.9726	0.0294	0.0825
TS228	0.1191	0.9348	0.0723	0.8789	0.0673	0.9731	0.0529	0.1012
TS233	0.2079	1.0000	0.0632	0.8873	0.1491	0.9947	0.2079	1.0000
TS238	0.6026	1.0000	0.0633	0.8755	0.3559	0.5952	0.0259	0.0825
TS242	0.5409	1.0000	0.0723	0.9064	0.4170	0.9947	0.0810	0.5297
TS251	0.5149	1.0000	0.0498	0.9228	0.4350	0.9947	0.0413	0.4934
TS262	0.4738	1.0000	0.0555	0.6225	0.3545	0.9947	0.4738	1.0000
TS264	0.6542	1.0000	0.0811	0.9076	0.0474	0.9626	0.0252	0.0697
TS265	0.3808	1.0000	0.0647	0.9024	0.0150	0.7552	0.0148	0.2692
TS295	0.3139	0.9900	0.0711	0.8940	0.0340	0.7927	0.0491	0.7475
TS299	0.7898	1.0000	0.0929	0.9125	0.5048	0.9947	0.0724	0.7334
Simulated	0.6047	1.0000	0.0155	0.8695	0.0005	0.9291	0.0014	0.5395
Real	0.2105	1.0000	0.0833	0.8485	0.0082	0.5840	0.0076	0.5760
Average	**0.4304**	**0.9943**	0.0624	0.8685	0.1619	0.7525	0.1652	0.4295

separated. The same problem also happens to Fig. 2(f) and Fig. 2(j). As shown in Fig. 2(c), the method presented in [5] detects the two outliers in Fig. 2(a), and groups them to the same cluster though their shapes are very different. This is because the DBSCAN clustering method in [5] is sensitive to both distance measure and parameters selection. Obviously, AHC works unsatisfactorily. It clusters the dissimilar trajectories with yellow color (Fig. 2(d,h,l)) together. This is due to that AHC also resorts to the difficult distance measure and, a wrong merge of two clusters in its greedy procedure seriously degrades the performance.

In addition, we make use of two metrics, Dunn index [1] and F1-Measure [4], to measure the clustering quality quantitatively. Dunn index aims to measure the within-class similarity and inter-class dissimilarity through the distance of trajectory-to-trajectory and cluster-to-cluster distance, respectively. F1-Measure gives an overall score by combining precision and recall values. In this paper, we apply Jensen-Shannon divergence for the distance measure in Dunn index. For Dunn index and F1-Measure, the higher is the better. As demonstrated in Table 2, XaIBO obtains the highest scores for the two metrics, which is consistent with the visual clustering results in Fig. 2.

5 Conclusion

In this paper, we have presented an application of aIB for effective trajectory clustering. Considering that the global shape is important to characterize a trajectory, we take the trajectory shape as a relevant variable for the original data to be clustered. Trajectory shape is statistically represented by using Kernel Density Estimation (KDE), based on a collection of tangential angles at the sample points of the trajectory. In addition, we extend aIB to XaIBO, satisfactorily coping with outliers in the procedure of trajectory clustering and, outperforming the state-of-the-art methods.

As for the future work, *Information Bottleneck* technique will be tried to consider online based trajectory clustering and outlier detection. Additionally, *visual analytics* tool [7] will be exploited to improve on the clustering process.

References

1. Dang, X.H., Bailey, J.: Generation of alternative clusterings using the cami approach. In: SDM, vol. 10, pp. 118–129. SIAM (2010)
2. Goldberger, J., Gordon, S., Greenspan, H.: Unsupervised image-set clustering using an information theoretic framework. IEEE Trans. Image Process. **15**(2), 449–458 (2006)
3. Guo, Y., Xu, Q., Yang, Y., Liang, S., Liu, Y., Sbert, M.: Anomaly detection based on trajectory analysis using kernel density estimation and information bottleneck techniques. Technical report 108, University of Girona (2014)
4. Hromic, H., Prangnawarat, N., Hulpuș, I., Karnstedt, M., Hayes, C.: Graph-based methods for clustering topics of interest in twitter. In: Cimiano, P., Frasincar, F., Houben, G.-J., Schwabe, D. (eds.) ICWE 2015. LNCS, vol. 9114, pp. 701–704. Springer, Heidelberg (2015)
5. Annoni Jr., R., Forster, C.H.Q.: Analysis of aircraft trajectories using fourier descriptors and kernel density estimation. In: Proceedings of the 15th International IEEE Conference on Intelligent Transportation Systems, pp. 1441–1446 (2012)
6. Laxhammar, R., Falkman, G.: Online learning and sequential anomaly detection in trajectories. IEEE Trans. Pattern Anal. Mach. Intell. **36**(6), 1158–1173 (2014)
7. May, R., Hanrahan, P., Keim, D.A., Shneiderman, B., Card, S.: The state of visual analytics: views on what visual analytics is and where it is going. In: IEEE Symposium on Visual Analytics Science and Technology (VAST). pp. 257–259. IEEE, Salt Lake City, UT (2010)
8. Morris, B.T., Trivedi, M.M.: Trajectory learning for activity understanding: unsupervised, multilevel, and long-term adaptive approach. IEEE Trans. Pattern Anal. Mach. Intell. **33**(11), 2287–2301 (2011)
9. Morris, B.T., Trivedi, M.M.: Understanding vehicular traffic behavior from video: a survey of unsupervised approaches. J. Electron. Imaging **22**(4), 041113–041113 (2013)
10. Piciarelli, C., Micheloni, C., Foresti, G.L.: Trajectory-based anomalous event detection. IEEE Trans. Circuits Syst. Video Technol. **18**(11), 1544–1554 (2008)
11. Schneidman, E., Slonim, N., Tishby, N., deRuyter van Steveninck, R., Bialek, W.: Analyzing neural codes using the information bottleneck method. In: Advances in Neural Information Processing Systems 15 (2002)

12. Slonim, N.: The information bottleneck: Theory and applications. Ph.D. thesis, Hebrew University of Jerusalem (2002)
13. Slonim, N., Somerville, R., Tishby, N., Lahav, O.: Objective classification of galaxy spectra using the information bottleneck method. Mon. Not. R. Astron. Soc. **323**(2), 270–284 (2001)
14. Slonim, N., Tishby, N.: Agglomerative information bottleneck. In: Advances in Neural Information Processing Systems, vol. 12, pp. 617–623. Citeseer (1999)
15. Slonim, N., Tishby, N.: Document clustering using word clusters via the information bottleneck method. In: Proceedings of the 23rd Annual International ACM SIGIR Conference on Research and Development in Information Retrieval, pp. 208–215. ACM (2000)

A Model of Motor Impairment After Stroke for Predicting Muscle Activation Patterns

Yuki Ueyama[✉]

Department of Rehabilitation Engineering,
Research Institute of National Rehabilitation Center for Persons with Disabilities,
Tokorozawa, Japan
ueyama-yuki@rehab.go.jp

Abstract. Several studies have aimed to provide computational evidence of post-stroke interventions because how to optimize motor recovery has been unclear. Although muscle synergies may be the basic control modules on which the central nervous system relies to generate motion, previous computational evidence ignored muscle activity. This study proposes a model of motor impairment after stroke for predicting muscle activity patterns. This model can reproduce a peculiar muscle activation pattern observed in stroke patients. Moreover, we carried out a simulation of the motor recovery process by minimizing the output torque error. As a result, the muscle activation patterns could not be modified to the intact condition, because the recovery process might fall into a local minimum. Thus, we suggest that our model could reproduce muscle activities after stroke, and that muscle synergy cannot be recovered by 'conventional' processes of the post-stroke rehabilitation.

Keywords: Rehabilitation · Neural network · Motor cortex · Preferred direction · Muscle synergy · EMG

1 Introduction

The incidence of stroke was approximately 17 million cases in 2010 worldwide, and there were approximately 33 million people who had previously suffered a stroke and were still alive [1]. Although many survive a first stroke, they often have significant morbidity; loss of functional movement is a common consequence of stroke. However, how to optimize motor recovery has been unclear. Thus, various interventions to help retrieve motor function—such as constraint-induced movement therapy (CIMT), mental practice with motor imagery, and electromyographic (EMG) biofeedback—have been developed. In particular, a therapist's decision regarding the program used to optimize motor recovery could be determined using EMG data [2, 3], as many studies suggest that muscle synergies may be the basic control modules on which the central nervous system (CNS) relies to generate motion because the dynamic behavior of the musculoskeletal system seems to be captured by the structure of the synergies [4, 5]. Moreover, it has been suggested that the assessment of muscle synergies should be used to evaluate different therapy modalities in post-stroke rehabilitation [6]. However, although motor

© Springer International Publishing Switzerland 2015
S. Arik et al. (Eds.): ICONIP 2015, Part II, LNCS 9490, pp. 432–439, 2015.
DOI: 10.1007/978-3-319-26535-3_49

recovery has several distinct traits beyond motor learning, motor learning models were often adopted as the foundation for organizing therapy, with the aim of altering the underlying neural architecture and connectivity to optimize recovery [7]. However, little is known about what constitutes the best practice after a stroke.

A previous study proposed a computational model of motor impairment after a stroke in order to explain a cause of directional errors commonly observed in stroke patients [8]. The model was based on a specific feature of motor cortex neurons known as the preferred direction (PD) [9], and the population vector (PV) model [10]. The PD is the direction in which a neuron is activated to the greatest degree when the hand is moved in several directions. Because the model could reproduce impairment of motion in a certain direction, it has been used to evaluate the feasibility of rehabilitation protocols [11, 12]. However, the model ignored muscle activities, despite the fact that stroke induces the disruption of muscle synergies (i.e., the coordinated recruitment of groups of muscles with a specific activation profile).

In this study, we provide a model to reproduce muscle activation patterns impaired after a stroke. This model was constructed from a simple neural network model, and based on a distribution of neural PDs in joint-torque space. Using the model, we predicted muscle activation patterns under intact and post-stroke conditions. Moreover, we carried out a simulation of a conventional motor recovery process based on error-descent learning. As a result, some muscles could not be modified adequately to the intact condition, and the recovery process may fall into a local minimum. Thus, we suggest that 'conventional' rehabilitation may not be sufficient to recover muscle synergy during post-stroke rehabilitation.

2 Methods

2.1 Simulation Model

We used a two-link six-muscle arm model as the musculoskeletal model (Fig. 1A). The movement of the arm is restricted in the horizontal plane. Additionally, we modeled motor cortex neurons using a simple neural network model (Fig. 1B). The neural network model generates the output torque to transform the desired torque. The input layer receives the desired torque of the shoulder and elbow joints, and the first hidden layer transforms the inputs into neural activities, such as in the motor cortex neurons. The second hidden layer transforms the neural activities into muscle activities, such as the corticospinal tract, and the muscles generate joint torque in the output layer through the moment arms.

The number of neurons in the first hidden layer is set to 1,000, and the neural activities $r \in R^{1000}$ are given by

$$r = W_1 \cdot \tau_d, \tag{1}$$

where $\tau_d \in R^2$ and $W_1 \in R^{1000 \times 2}$ are the desired torque and synaptic weights, respectively. The muscle activities $u \in R^6$ are given by

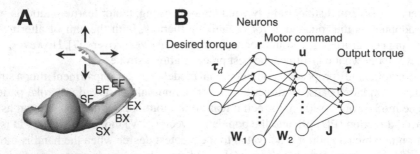

Fig. 1. Simulation model. (A) Two-link six-muscle arm model. The motion is restricted in the horizontal plane. SF, SX, BF, BX, EF, and EX denote arm muscles: SF: shoulder flexor, SX: shoulder extensor, BF: biauriculate flexor, BX: biarticulate extensor, EF: elbow flexor, EX: elbow extensor. (B) Neural network model generating joint toque. $\mathbf{W}_1$, $\mathbf{W}_2$ are the synaptic weights of the motor cortex neurons and connections between the motor cortex and spinal cord, respectively, and $\mathbf{J}$ is the moment arm.

$$\mathbf{u} = [\mathbf{W}_2(\mathbf{r} + \boldsymbol{\varepsilon})]_+, \tag{2}$$

where $\mathbf{W}_2 \in R^{6 \times 1000}$ is the synaptic weight matrix, and $[\cdot]_+$ is a function defined as

$$[\cdot]_+ \equiv \max\{\cdot, 0\},$$

where the function prohibits the muscles from generating negative activities not to allow induction of a pushing force. $\boldsymbol{\varepsilon} \in R^{1000}$ is the neural noise vector, which is dependent on the neural activity, and denoted by a Gaussian distribution $\boldsymbol{\nu}$

$$\boldsymbol{\varepsilon} = \mathbf{r} \cdot \boldsymbol{\nu}, \text{ with } \boldsymbol{\nu} \sim N(0, \sigma^2)$$

where $\sigma = 0.25$ in accordance with a previous report [13]. The output torque is given by

$$\boldsymbol{\tau} = \mathbf{J} \cdot \mathbf{u}, \tag{3}$$

where $\mathbf{J} \in R^{2 \times 6}$ is the moment arms. According to a previous study [14], the parameter was defined as

$$\mathbf{J} = \begin{bmatrix} 2.6 & -1.3 & 0 & 0 & 0.7 & -2.5 \\ 0 & 0 & 1.2 & -1.7 & 1.6 & -1.1 \end{bmatrix} \times 10^{-2}.$$

2.2 Optimization of the Neural Network Model

In this study, the synaptic weights of the input layer $\mathbf{W}_1$ were modeled in the motor cortex neurons. Other synaptic weights of the first hidden layer $\mathbf{W}_2$, which modeled connections between the motor cortex and spinal cord such as the corticospinal tract, were given by a uniform distribution within a closed interval [0 1]. Here, we adapted error-descent learning with a "slightly forgetting rule" to optimize the synaptic weight $\mathbf{W}_1$ [15]. The synaptic weight $\mathbf{W}_2$ and the moment arm $\mathbf{J}$ were not changed by this optimization process. The synaptic weight w_{ij} of $\mathbf{W}_1$ was updated with the following rule:

$$w_{ij} \leftarrow w_{ij} - \alpha \frac{\partial \left\| \tau - \tau_d \right\|^2}{\partial w_{ij}} - \beta w_{ij}, \qquad (4)$$

where α, β are the learning parameter and decay parameter, respectively, and set to $\alpha = 1.0$ and $\beta = 1.0 \times 10^{-4}$. The synaptic weight W_1 was initialized with a Gaussian distribution with zero mean and unit variance. The norm of the desired torque was set to 1, and the direction was selected randomly from eight directions: 0°, 45°, 90°, 135°, 180°, 225°, 270°, and 315°.

2.3 Motor Impairment After a Stroke

Stroke seems to affect movement only within a certain range of directions. A previous study proposed a model to simulate the directional control of reaching after a stroke as cell death in a PV model of movement control [8]. To model the effect, we removed neurons with PDs in the third quadrant of the joint-torque plane (i.e., [180° 270°]), similar to the previous model, because extension movements are impaired in most stroke patients. Then, we had the synaptic weights re-learned according to the optimization rule described by (4) to simulate the recovery process. The desired torque was provided randomly from eight directions, similar to the optimization process.

3   Results

3.1   Optimization Process

We optimized the neural network model to initialize the neural state as an intact condition, according to a previous study [15]. Then, the squared error was decreased across training trials (Fig. 2A), and the mean norm of the synaptic weight W_1 was also decreased and converged to an optimal state (Fig. 2B). After 40,000 trials, output torque was almost equal to the desired torque (Fig. 2C).

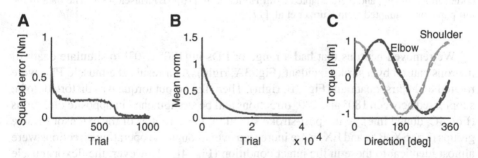

Fig. 2. Optimization of the neural network model. (A) Squared error of the joint torques across trials. (B) Mean norm of a synaptic weight of W_1. (C) Output torques across several directions after 40,000 trials. Solid and dashed lines indicate output and desired torque.

3.2 Effects of Motor Impairment

A primate study showed that the distribution of neural PDs in the primary motor cortex was biased to the second and fourth quadrants in joint-torque plane [16] (Fig. 3A, left). Our simulation could reproduce the same features, similar to a previous report [15] (Fig. 3A, center). It was subsequently reported that the muscle PDs were also skewed to the second and fourth quadrants [17] (Fig. 3B, left). The muscle PD is the direction in which maximal muscle force is generated. Our simulation could reproduce the feature of the muscle PDs skewed in the second and fourth quadrants (Fig. 3B, center).

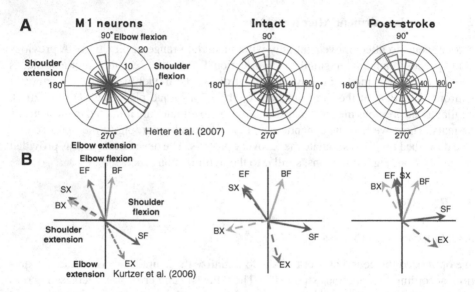

Fig. 3. Distribution of neural PDs and muscle PDs. In (A)–(B), the left, center, and right panels indicate the experimental measurement, intact condition, and post-stroke condition, respectively. (A) Distributions of neural PD. The data in the left panel were recorded from the primary motor cortex in a monkey, and were adapted from Herter et al. [16]. (B) Muscle PDs. The data in the left panel were adapted from Kurtzer et al. [17].

We removed neurons that had a range of PDs in [180° 270°] to simulate death of neurons caused by a stroke incident (Fig. 3A, right). As a result, the muscle PDs were rotated to the first quadrant (Fig. 3B, right). Then, the output torque was distorted, to be suppressed between 180° and 270° directions and be strengthened in opposite directions (Fig. 4C, dotted line). In the post-stroke situation, the activities of the extensor muscle groups (i.e., SX, BX, and EX) were increased even though the operating directions were almost identical to those in the intact condition (Fig. 4E). However, the flexor muscle groups (i.e., SF, BF, and EF) worked in deflected directions versus the intact condition. Such distorted muscle activation patterns have been observed in EMGs of stroke patients [7].

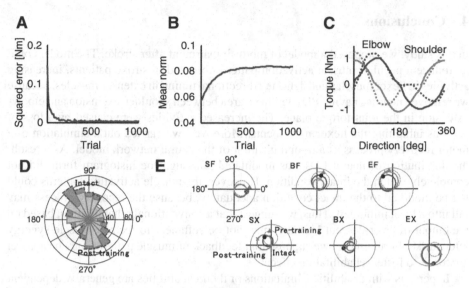

Fig. 4. Recovery process as a re-optimization of the neural network. (A) Squared error of the output torque across training trials. (B) Mean norm of a synaptic weight of $\mathbf{W}_1$. (C) Output torque for several desired directions after 40,000 trials. Solid and dotted lines indicate output torque in the post-training and pre-training conditions, respectively. (D) Distributions of neural PDs in the intact and post-training conditions. Solid red and dashed blue lines indicate histograms of the post-training and intact conditions, respectively. (E) Simulated muscle activation patterns in the joint-torque plane. Solid blue and red lines indicate the intact and post-stroke conditions, respectively. The dotted and solid red lines indicate the pre-training and post-training conditions. Markers '●', '▲', '×' indicate the muscle PDs in the intact, pre-training, and post-training conditions, respectively.

3.3   Recovery Process

We retrained the neural network model in the post-stroke condition to simulate rehabilitation effects on the neural state and muscle activities. We adopted the same learning rule as the optimization process for the recovery process. In the recovery process, the error was decreased across the trials (Fig. 4A), and the synaptic weights were increased (Fig. 4B). After 40,000 trials, output torque was modified from the distorted formation to fit the desired torque (Fig. 4C). Then, the distribution of neural PDs was also modulated, although the histogram form did not completely match the intact condition (Fig. 4D). Although SF and EX muscles in the post-training state were corrected from those of the post-stroke situation to essentially match the intact condition (Fig. 4E), other muscles could not be modified adequately to fit the intact condition because the recovery process may fall into a local minimum.

4   Conclusions

In this study, we proposed a model of motor impairment after stroke. The model could reproduce a peculiar muscle activation pattern observed in stroke patients, increasing activities in flexor muscles and deflected directional tuning in extensor muscles, because we removed neurons having PDs within an area between shoulder-extension and elbow-extension in the joint-torque plane. The increased activities were caused by a loss of neurons inhibiting the flexor movements. Moreover, we carried out a simulation of a motor recovery process as a re-optimization of the neural network model. As a result, the distribution of neural PDs was modulated, although the histogram form did not completely match the intact condition. However, the muscle activation patterns could not be modified to the intact condition adequately, because the recovery process may fall into a local minimum. Thus, we suggest that a conventional rehabilitation, based on the output of error-descent learning, may not be sufficient to recover muscle synergy after stroke. Furthermore, we suppose that feedback of muscle patterns may be a novel approach to foster rehabilitation effects.

In persons with disabilities, limitations of their capabilities are generally dependent on the differences between individual situations, such as the damaged brain region(s). Our proposed model may be used to evaluate and identify disabilities from EMG signals. Moreover, we suppose that ultimately rehabilitation may be optimized for individuals with disabilities in accordance with the evaluations.

Acknowledgement.   This work was supported by JSPS KAKENHI Grant Number 26702023.

References

1. Feigin, V.L., Forouzanfar, M.H., Krishnamurthi, R., Mensah, G.A., Connor, M., Bennett, D.A., Moran, A.E., Sacco, R.L., Anderson, L., Truelsen, T.: Global and regional burden of stroke during 1990–2010: findings from the global burden of disease study 2010. Lancet **383**, 245–255 (2014)
2. Johnson, M.J.: Recent trends in robot-assisted therapy environments to improve real-life functional performance after stroke. J. Neuroeng. Rehabil. **3**, 29 (2006)
3. Dipietro, L., Ferraro, M., Palazzolo, J.J., Krebs, H.I., Volpe, B.T., Hogan, N.: Customized interactive robotic treatment for stroke: EMG-triggered therapy. IEEE Trans. Neural Syst. Rehabil. Eng. **13**, 325–334 (2005)
4. d'Avella, A., Portone, A., Fernandez, L., Lacquaniti, F.: Control of fast-reaching movements by muscle synergy combinations. J. Neuroscience. **26**, 7791–7810 (2006)
5. Ueyama, Y.: Effects of cost structure in optimal control on biological arm movement: a simulation study. In: Lee, M., Hirose, A., Hou, Z.-G., Kil, R.M. (eds.) ICONIP 2013. LNCS, vol. 8226, pp. 241–248. Springer, Heidelberg (2013)
6. Cheung, V.C., Piron, L., Agostini, M., Silvoni, S., Turolla, A., Bizzi, E.: Stability of muscle synergies for voluntary actions after cortical stroke in humans. Pro. Nat. Acad. Sci. **106**, 19563–19568 (2009)
7. Cesqui, B., Tropea, P., Micera, S., Krebs, H.I.: EMG-based pattern recognition approach in post stroke robot-aided rehabilitation: a feasibility study. J. Neuroeng. Rehabil. **10**, 75 (2013)

8. Reinkensmeyer, D.J., Iobbi, M.G., Kahn, L.E., Kamper, D.G., Takahashi, C.D.: Modeling reaching impairment after stroke using a population vector model of movement control that incorporates neural firing-rate variability. Neural Comput. **15**, 2619–2642 (2003)
9. Georgopoulos, A.P., Schwartz, A.B., Kettner, R.E.: Neuronal population coding of movement direction. Science **233**, 1416–1419 (1986)
10. Schwartz, A.B., Kettner, R.E., Georgopoulos, A.P.: Primate motor cortex and free arm movements to visual targets in three-dimensional space. I. Relations between single cell discharge and direction of movement. J. Neurosci. **8**, 2913–2927 (1988)
11. Takiyama, K., Okada, M.: Recovery in stroke rehabilitation through the rotation of preferred directions induced by bimanual movements: a computational study. PLoS ONE **7**, e37594 (2012)
12. Han, C.E., Arbib, M.A., Schweighofer, N.: Stroke rehabilitation reaches a threshold. PLoS Comput. Biol. **4**, e1000133 (2008)
13. Matthews, P.: Relationship of firing intervals of human motor units to the trajectory of post-spike after-hyperpolarization and synaptic noise. J Physiol-London **492**, 597–628 (1996)
14. Ueyama, Y., Miyashita, E.: Optimal feedback control for predicting dynamic stiffness during arm movement. IEEE Trans. Ind. Electron. **61**, 1044–1052 (2014)
15. Hirashima, M., Nozaki, D.: Learning with slight forgetting optimizes sensorimotor transformation in redundant motor systems. PLoS Comput. Biol. **8**, e1002590 (2012)
16. Herter, T.M., Kurtzer, I., Cabel, D.W., Haunts, K.A., Scott, S.H.: Characterization of torque-related activity in primary motor cortex during a multijoint postural task. J. Neurophysiol. **97**, 2887–2899 (2007)
17. Kurtzer, I., Pruszynski, J.A., Herter, T.M., Scott, S.H.: Primate upper limb muscles exhibit activity patterns that differ from their anatomical action during a postural task. J. Neurophysiol. **95**, 493–504 (2006)

Cost Reduction in Thyroid Diagnosis: A Hybrid Model with SOM and C4.5 Decision Trees

Ahmet Cumhur Kinaci[✉] and Sait Can Yucebas

Computer Engineering Department,
Canakkale Onsekiz Mart University, Çanakkale, Turkey
{cumhur.kinaci,can}@comu.edu.tr

Abstract. The main objective of this paper is to introduce a hybrid model of Self Organizing Maps (SOM) and C4.5, to reduce the costs while maintaining an acceptable diagnostic performance. In this hybrid model, SOM is used first to form clusters and then C4.5 trees specific to each cluster is constructed. The proposed hybrid model is tested on multiclass Thyroid Data and compared to standalone C4.5 tree. Costs were reduced by 22 %–27 % and performance results vary between 88 % and 97 % in terms of accuracy and 90 %–97 % in terms of sensitivity. Cost and performance differences between the hybrid model and standalone C4.5 found to be statistically significant according to Wilcoxon signed-rank test.

Keywords: Clinical Decision Support Systems · Cost reduction · Clustering · Decision trees · SOM · C4.5

1 Introduction

Early Clinical Decision Support Systems (CDSS), mainly focus on diagnosis [6], had many advantages such as improving practitioner performance, increasing patient satisfaction, decreasing medical errors. And among these cost reduction stood as a byproduct. Later CDSS such as reminder systems, drug dosing, drug interaction, treatment planning and follow up systems [19] also cut the costs, by reducing medical errors and by preventing to prescribe unnecessary and/or duplicate tests and treatments.

Today reducing costs while aiding diagnosis is an emerging need for CDSS from healthcare perspective. This need arises from increasing healthcare expenditure costs which introduces major load on both nations and their citizens. According to WHOs Global Health Expenditure Database, in year 2013 the percentages of total health expenditures to gross domestic product in several countries are as follows: USA 17 %, Canada and Germany 11 %, Japan 10 %, UK 9 % and Turkey 6 % [22]. Although the need for cost reduction is obvious and is an important issue in healthcare, there are very few CDSS systems that directly reveal their cost effectiveness [11].

© Springer International Publishing Switzerland 2015
S. Arik et al. (Eds.): ICONIP 2015, Part II, LNCS 9490, pp. 440–448, 2015.
DOI: 10.1007/978-3-319-26535-3_50

The work [1] mainly focused on the correct prescription of antibiotics to patients with pneumonia. As a byproduct, cost effectiveness of CDSS was calculated in terms of average cost of prescribed antibiotics per patient. According to this study, average cost of prescribed antibiotics was decreased in CDSS group.

Another study compares the effectiveness of CDSS for patients with renal insufficiency [3]. The CDSS was used for drug order and dosing and the effectiveness of the system was measured in terms of length of stay, cost of hospital and pharmacies. Unfortunately there is no significant cost effectiveness was found.

Cost effectiveness of CDSS is studied on diagnose related groups (DRGs). In this work [9] CDSS decreased the charges by 10 % and decreased the costs by 12 % over given trials. In [15] CDSS reduces costs per patient by $37.64. Another study reports that the use of CDSS affects antibiotic treatment positively and thus cuts the costs of prescribed drugs [18]. There are other similar studies that try to measure the effectiveness of CDSS. These studies measure the effectiveness in terms of length of stay, cost reduction, mortality rate etc. [10,16,17].

According to our literature search and to the best of our knowledge there are very few CDSS that are directly designed for cost reduction. The most up to date study is the work by Chi, Street and Katz [4]. In this work a three phase algorithm, Optimal Decision Path Finder (ODPF), was designed. ODPF shows promising results in terms of cost reduction and diagnostic ability.

As the need is obvious and the examples in the domain are very few, we aim to build a CDSS that is designed mainly for cost reduction. Our methodology is to form a hybrid model of clustering and classification schemes. The model clusters the data with selected attributes by Self Organizing Maps (SOMs), so the most similar cases are grouped together, and then for each cluster a specific C4.5 is formed with the rest of the attributes. This scheme introduces smaller C4.5 trees for clusters and eliminates the clinical tests that are not necessary for diagnosing a condition in that group. By eliminating the tests in each cluster, a cost reduction is achieved. The model is compared in terms of cost reduction and diagnostic performance to standalone C4.5 trees constructed with all attributes. Although diagnostic performances are similar, standalone C4.5 is much deeper thus introducing more clinical test for almost all cases so it is not cost effective as the proposed model.

2 Materials and Methods

2.1 Materials

We tested our model with multiclass thyroid data [13]. The data consists of 2800 training instances and 972 test instances. There are 28 attributes and their values are either boolean or integer or real. The selected attributes are integer or real values. And these are age, TSH, T3, TT4 and T4U. FT attribute is discarded since it has no cost value in the dataset.

The data consists of three sub datasets of the same patient group with names binding protein, hypothyroid and hyperthyroid[1]. Each dataset has multiclass labels. Instead of building a specific model for each of the given conditions, we used a matching scheme to combine all these data. By this way our model be able to determine all three conditions. The number of classes, the number of the instances in classes, and the combination scheme is summarized in the Table 1.

Table 1. Thyroid class scheme

Binding protein	Hyperthyroid	Hypothyroid	Class	#Instances
Decreased	Negative	Compensated hypothyroid	1	1
Decreased	Negative	Negative	2	13
Increased	Negative	Compensated hypothyroid	3	6
Increased	Negative	Negative	4	141
Negative	Negative	Compensated hypothyroid	5	187
Negative	Negative	Negative	6	3205
Negative	Hyperthyroid	Negative	7	99
Negative	T3Toxic	Negative	8	10
Negative	Goiter	Negative	9	12
Negative	Negative	Primary hypothyroid	10	95
Negative	Negative	Secondary hypothyroid	11	2

The data is preprocessed by eliminating instances with missing values. The missing value rate is nearly 34 %. So the tuples with missing values were excluded from the dataset. In the last step of data preprocessing, attributes are normalized into [0–1] range by span normalization.

2.2   Methods

The purpose of this study is to introduce a model that will cut costs and maintain acceptable diagnosis performance. In order to meet these aims, we suggest using a hybrid model of clustering and classification.

When the data is sparse and not homogeneously distributed among multiple classes, standalone classification methods could fail to perform well [20,21]. In such cases optimization of methods by using statistical analysis or combination with other methods could be preferred. In our case, the dataset is not sparse and distribution of the instances among classes is not homogenous as shown in Table 1. In order to provide homogeneity, a clustering method is applied first as recommended in [5].

---

[1] http://archive.ics.uci.edu/ml/machine-learning-databases/thyroid-disease/
   File names: allbp.data, allhyper.data and allhypo.data.

As represented in Fig. 1, our proposed model consists of two phases. In first phase an artificial neural network clustering algorithm, Self Organizing Map (SOM), is used. SOM preferred because of its good representation power over clusters, robustness to noise and outliers [12] and its wide use in clinical decision making [2, 5, 14]. The second phase of the model is the classification of the clusters that are obtained from SOM. For classification widely used C4.5 is selected to handle continuous valued attributes and to provide better interpretation of the classification scheme.

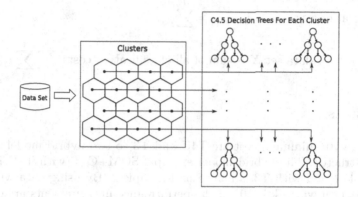

Fig. 1. Overall structure of SOM–C4.5 Hybrid Model

SOM clustering and C4.5 classification is applied by using an open source tool Orange [8]. For SOM, a hexagonal topology of 4 by 4 is used. As a distance measure Gaussian Neighborhood is preferred and initial radius is set to 3, final radius is set to 1. This topology was trained by 100 iterations. For C4.5, minimal examples in leaves is set to 2 and post prunning confidence level is set to 0.25.

Attribute Selection for Clustering. Since the main purpose of the proposed method is reducing the cost while keeping the performance same, for clustering the selected attributes have the minimum cost compared to other attributes. T4U and T3 attributes are medical tests which have the same cost and they are less expensive than the other tests. So for clustering this attributes are selected in addition to age attribute which has zero cost. As a result first example model clustering is based on age and T4U attributes and the second is based on age and T3. Cost of total clustering is simply multiplying single test price with data set size (age attribute has no cost). By selecting the cheapest test the total clustering cost will be the minimum.

Cost Calculation. A is the set of all attributes in data set X. C is the set of clusters c_m generated by SOM. SOM clustering cost is simply sum of input attributes costs multiplied by number of instances. For every cluster c_m there exists a clustered data subset $X_m \subset X$ and a decision tree t_j generated by using X_m. Every c_m has the same input attributes A^c and $A^c \subset A$. And each t_j is constructed with attributes in A^t and $A^t \subset A$. $A^c \cap A^t = \emptyset$ and $(A^c \cup A^t) \subseteq A$.

Total cost for a decision tree is calculated as follows:

- For a given $x_i \in X_m$ find all the nodes in all depths of decision tree t_j that belong to decision path.
- Find the set of unique attributes set $A^t_{m,i} \subset A^t$ of these nodes. For each $a_j \in A^t_{m,i}$ the cost is denoted as p_j.
- Classification cost for x_i is $\alpha_{m,i} = \sum_1^j p_j$.

- Total cost for data set X is sum of all classification costs: $\sum_1^m \sum_1^i \alpha_{m,i}$.

3   Results

Attributes with minimum cost are T4U and T3 so two hybrid model instances were constructed. The hybrid model example SOM+C4.5 with T4U is named as Example-1 and with T3 is named as Example-2. By using data with k-fold cross validation where k is 10, all the performance measurements are obtained. The difference between costs and performance comparison based on accuracy and sensitivity of C4.5 and Hybrid Model is given in the Table 2.

Table 2. Average performance values (with standard deviations in parenthesis) of 10 folds for each model example

	C4.5		SOM+C4.5 with T4U		SOM+C4.5 with T3	
	Training data	Test data	Training data	Test data	Training data	Test data
Cost per patient	59.21	59.22	46.11	45.82	43.32	43.16
	(0.30)	(0.32)	(0.92)	(1.11)	(1.06)	(1.24)
Accuracy	0.9619	0.9185	0.9722	0.8975	0.9783	0.8851
	(0.0028)	(0.0091)	(0.0033)	(0.0101)	(0.0029)	(0.0165)
Sensitivity	0.9698	0.9299	0.9730	0.9155	0.9777	0.9065
	(0.0023)	(0.0136)	(0.0036)	(0.0080)	(0.0034)	(0.0171)

As given in Table 2, our Hybrid Model cut the costs per patient by 22.12 % in Example-1 and 26.83 % in Example-2 when tested with training data. For test data, costs per patient is reduced by 22.62 % in Example-1 and 27.12 % in Example-2. In clinical performance perspective, hybrid model slightly performed better than C4.5 when training data is used. In terms of accuracy, performance increased by 1 % in Example-1 and 1.67 % in Example-2. Sensitivity increased by 0.32 % in Example-1 and 0.8 % in Example-2. When test data is used increase in performance was in favor of C4.5. Accuracy decreased by 2.28 % in Example-1 and decreased by 3.63 % in Example-2. Decrease in sensitivity for Example-1 and Example-2 is 1.5 % and 3.5 % respectively.

In order to test the performance differences between methods, Wilcoxon signed-rank test is used as recommended in [7]. General form of the null hypothesis is as follows:

H_0: Median difference between the pairs is zero.

In Tables 3, 4 and 5 all p values for paired Wilcoxon signed-rank tests are given. For the p-values which are less than the 0.05 significance level (marked with * at the tables) the null hypothesis is rejected.

Table 3. P values of Wilcoxon signed-rank test for Cost

		C4.5		SOM+C4.5 with T4U		SOM+C4.5 with T3	
		Training	Test	Training	Test	Training	Test
C4.5	Training	-	1.0	0.001953*	-	0.001953*	-
	Test	1.0	-	-	0.001953*	-	0.001953*
SOM + DT with T4U	Training	0.001953*	-	-	0.02734*	0.001953*	-
	Test	-	0.001953*	0.02734*	-	-	0.005859*
SOM + DT with T3	Training	0.001953*	-	0.001953*	-	-	0.7695
	Test	-	0.001953*	-	0.005859*	0.7695	-

From the cost efficiency perspective it is statistically significant (Table 3) that there is a difference between the proposed method and C4.5 algorithm. As given in Table 2 average diagnosis cost per patient is reduced by using proposed Hybrid Model.

Table 4. P values of Wilcoxon signed-rank test for Accuracy

		C4.5		SOM+DT with T4U		SOM+DT with T3	
		Training	Test	Training	Test	Training	Test
C4.5	Training	-	0.001953*	0.001953*	-	0.005857*	-
	Test	0.001953*	-	-	0.005793*	-	0.005889*
SOM + DT with T4U	Training	0.001953*	-	-	0.02734*	0.003906*	-
	Test	-	0.005793*	0.02734*	-	-	0.1934
SOM + DT with T3	Training	0.005857*	-	0.001953*	-	-	0.001953*
	Test	-	0.005889*	-	0.001953*	0.001953*	-

According to Wilcoxon signed-rank test results presented by Tables 3, 4 and 5, both the average costs and performance differences between the Hybrid Model and C4.5 are statistically significant. As these results indicate, the proposed Hybrid Model was able to cut the costs. When the same dataset is used both for training and testing, diagnostic performance of the hybrid model is slightly better than C4.5. But when a separate data set is used for testing, diagnostic performance of the hybrid model was slightly under the performance of C4.5

Table 5. P values of Wilcoxon signed-rank test for Sensitivity

		C4.5		SOM+DT with T4U		SOM+DT with T3	
		Training	Test	Training	Test	Training	Test
C4.5	Training	-	0.001953*	0.06445	-	0.001953*	-
	Test	0.001953*	-	-	0.009766*	-	0.005859*
SOM + DT with T4U	Training	0.06445	-	-	0.001953*	0.03711*	-
	Test	-	0.009766*	0.001953*	-	-	0.3223
SOM + DT with T3	Training	0.001953*	-	0.03711*	-	-	0.001953*
	Test	-	0.005859*	-	0.3223	0.001953*	-

4   Conclusion

The proposed Hybrid Model was able to meet the cost reduction aims. In Example-1 the average cost per patient is cut by 22.12 %–22.62 % and in Example-2 costs cut by 26.83 %–27.12 %. From clinical perspective the performance results differ according to training and testing data used. In worst case, for Example-1 accuracy decreased by 2.28 % and sensitivity decreased by 1.5 %, for Example-2 these indicators decreased by 3.63 % and 3.5 % when compared to stand alone C4.5.

These results show that, the Hybrid Model was very successful for cutting the average costs per patient in thyroid diagnosis as aimed. The clinical diagnosis performance of the model is also very high with accuracy over 88 % and sensitivity over 90 %. Depending on the dataset used, diagnostic performance of standalone C4.5 could be slightly better than the hybrid model. However, when the higher amount of cost savings and the general diagnostic performance is taken into consideration, the proposed Hybrid Model shows promising results.

The results seem to be promising in terms of cost reduction and clinical decision performance and this gives us courage to apply this Hybrid Model on different datasets.

References

1. Buising, K.L., Thursky, K.A., Black, J.F., MacGregor, L., Street, A.C., Kennedy, M.P., Brown, G.V.: Improving antibiotic prescribing for adults with community acquired pneumonia: does a computerised decision support system achieve more than academic detailing alone?-a time series analysis. BMC Med. Inform. Decis. Mak. **8**(1), 35 (2008)
2. Chen, D.R., Chang, R.F., Huang, Y.L.: Breast cancer diagnosis using self-organizing map for sonography. Ultrasound Med. Biol. **26**(3), 405–411 (2000)
3. Chertow, G.M., Lee, J., Kuperman, G.J., Burdick, E., Horsky, J., Seger, D.L., Lee, R., Mekala, A., Song, J., Komaroff, A.L., et al.: Guided medication dosing for inpatients with renal insufficiency. Jama **286**(22), 2839–2844 (2001)
4. Chi, C.L., Street, W.N., Katz, D.A.: A decision support system for cost-effective diagnosis. Artif. Intell. Med. **50**(3), 149–161 (2010)

5. Churilov, L., Bagirov, A., Schwartz, D., Smith, K., Dally, M.: Data mining with combined use of optimization techniques and self-organizing maps for improving risk grouping rules: application to prostate cancer patients. J. Manag. Inf. Syst. **21**(4), 85–100 (2005)
6. Coiera, E.: Guide to Health Informatics. Hodder Arnold (2003)
7. Demar, J.: Statistical comparisons of classifiers over multiple data sets. J. Mach. Learn. Res. **7**, 1–30 (2006)
8. Demšar, J., Curk, T., Erjavec, A., Gorup, Č., Hočevar, T., Milutinovič, M., Možina, M., Polajnar, M., Toplak, M., Starič, A., Štajdohar, M., Umek, L., Žagar, L., Žbontar, J., Žitnik, M., Zupan, B.: Orange: data mining toolbox in python. J. Mach. Learn. Res. **14**, 2349–2353 (2013). http://jmlr.org/papers/v14/demsar13a.html
9. Elkin, P.L., Liebow, M., Bauer, B.A., Chaliki, S., Wahner-Roedler, D., Bundrick, J., Lee, M., Brown, S.H., Froehling, D., Bailey, K., et al.: The introduction of a diagnostic decision support system (dxplain) into the workflow of a teaching hospital service can decrease the cost of service for diagnostically challenging diagnostic related groups (drgs). Int. J. Med. Inform. **79**(11), 772–777 (2010)
10. Evans, R.S., Pestotnik, S.L., Classen, D.C., Clemmer, T.P., Weaver, L.K., Orme Jr, J.F., Lloyd, J.F., Burke, J.P.: A computer-assisted management program for antibiotics and other antiinfective agents. N. Engl. J. Med. **338**(4), 232–238 (1998)
11. Fillmore, C.L., Bray, B.E., Kawamoto, K.: Systematic review of clinical decision support interventions with potential for inpatient cost reduction. BMC Med. Inform. Decis. Mak. **13**(1), 135 (2013)
12. Kaski, S.: Data exploration using self-organizing maps. In: Acta Polytechnica Scandinavica: Mathematics, Computing and Management in Engineering Series no. 82. Citeseer (1997)
13. Lichman, M.: UCI machine learning repository (2013). http://archive.ics.uci.edu/ml
14. Markey, M.K., Lo, J.Y., Tourassi, G.D., Floyd, C.E.: Self-organizing map for cluster analysis of a breast cancer database. Artif. Intell. Med. **27**(2), 113–127 (2003)
15. McGregor, J.C., Weekes, E., Forrest, G.N., Standiford, H.C., Perencevich, E.N., Furuno, J.P., Harris, A.D.: Impact of a computerized clinical decision support system on reducing inappropriate antimicrobial use: a randomized controlled trial. J. Am. Med. Inform. Assoc. **13**(4), 378–384 (2006)
16. Mekhjian, H.S., Kumar, R.R., Kuehn, L., Bentley, T.D., Teater, P., Thomas, A., Payne, B., Ahmad, A.: Immediate benefits realized following implementation of physician order entry at an academic medical center. J. Am. Med. Inform. Assoc. **9**(5), 529–539 (2002)
17. Mullett, C.J., Evans, R.S., Christenson, J.C., Dean, J.M.: Development and impact of a computerized pediatric antiinfective decision support program. Pediatrics **108**(4), e75 (2001)
18. Paul, M., Andreassen, S., Tacconelli, E., Nielsen, A.D., Almanasreh, N., Frank, U., Cauda, R., Leibovici, L., et al.: Improving empirical antibiotic treatment using treat, a computerized decision support system: cluster randomized trial. J. Antimicrob. Chemother. **58**(6), 1238–1245 (2006)
19. Sen, A., Banerjee, A., Sinha, A.P., Bansal, M.: Clinical decision support: converging toward an integrated architecture. J. Biomed. Inform. **45**(5), 1009–1017 (2012)
20. Shahshahani, B.M., Landgrebe, D.A.: The effect of unlabeled samples in reducing the small sample size problem and mitigating the hughes phenomenon. IEEE Trans. Geosci. Remote Sens. **32**(5), 1087–1095 (1994)

21. Tsoumakas, G., Katakis, I.: Multi-label classification: An overview. Dept. of Informatics, Aristotle University of Thessaloniki, Greece (2006)
22. WHO: Global health expenditure database (2013). http://apps.who.int/nha/database

Webcam Based Real-Time Robust Optical Mark Recognition

Huseyin Atasoy[(✉)], Esen Yildirim, Yakup Kutlu, and Kadir Tohma

Department of Computer Engineering, Iskenderun Technical University,
Hatay, Turkey
{hatasoy,eyildirim,ykutlu,ktohma}@mku.edu.tr

Abstract. This study proposes a robust, low cost, real-time optical mark recognition (OMR) system that uses a webcam and a small OMR form to read hand-marked data from plain paper. The system is designed to read data from any user-designed OMR form which can be customized for any purpose. It was implemented and tested on examination papers to read students' numbers and their examination results automatically. Results and numbers on 87 out of 88 papers were correctly identified. It was tested under different lighting conditions and with different mark colors. The results indicate that the system is robust and reliable.

Keywords: Optical mark recognition · Clustering · Mark detection

1 Introduction

Optical mark recognition is a generic name for techniques of data extraction from hand-filled forms. This technology provides a solution for reading and processing large number of forms such as questionnaires or multiple-choice tests. It is widely used, especially for grading students in schools.

There are many solutions that use devices specialized for holding large number of forms and reading them as they are being fed into the devices. But these solutions require scanner devices and a special kind of paper which is more expensive than a regular sheet and has a fixed layout. Therefore, cheaper solutions are needed for simpler purposes that do not need a whole OMR sheet. For instance to read student number from a regular sheet that can contain anything beside the OMR form.

Several studies have been proposed for optical mark recognition so far. Perez-Benedito et al. have tested and compared two softwares to decrease the time consumed for the process of student assessment, as a part of their educational innovation project [1]. Nguyen et al. have proposed a camera based multiple choice test grading method [2]. In another work, Xiu et al. have proposed a style-based ballot recognition approach and have compared their approach with the others [3]. Deng et al. have presented an OMR solution that uses scanners and supports plain sheets and low printing quality [4]. Kubo et al. have developed a web based integrated OMR system that needs scanner devices [5].

© Springer International Publishing Switzerland 2015
S. Arik et al. (Eds.): ICONIP 2015, Part II, LNCS 9490, pp. 449–456, 2015.
DOI: 10.1007/978-3-319-26535-3_51

In this paper we propose a webcam-based, real-time OMR system that can reads data from user-designed OMR forms. The system is focused on reading user-designed small forms rather than special OMR sheets and it can perform the task using a simple webcam, under bad lighting conditions, different mark colors and intensities.

2 Methods

The proposed system consists of two steps, namely, adaptive binarization and mark detection. In the first step, some well-known basic image processing techniques such as histogram equalization, erosion, dilation are applied to the image to obtain the binary form of the image without being affected by lighting conditions or image quality. In the second step, binary image is subjected to some processes to detect traces and to determine corners of the OMR form and locations of hand-drawn marks. The proposed system is shown in Fig. 1.

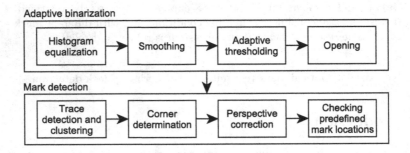

Fig. 1. Proposed mark recognition system

2.1 Adaptive Binarization

Binarization is a simple process of converting images from grayscale to binary format by using a pixel value threshold. A constant threshold is not applicable to images captured under different lighting conditions. For this reason, adaptive threshold determination algorithms such as Otsu's method [6] are preferred. But in optical mark recognition, marks have not always the same color or intensity and the system is desired to be invariant to lighting conditions, mark color and intensity. Therefore Otsu's method was used along with histogram equalization, smoothing and opening in the order shown in Fig. 1. Noise is reduced on these steps gradually.

Histogram Equalization: Histogram equalization enhances contrast in an image by stretching intensity range. The main idea of the histogram equalization is to build a lookup table (Eq. 2) for mapping intensity values to new values that make distribution of them wider and more uniform. Calculation of cumulative histogram, constructing a lookup table and mapping intensities are given in Eqs. 1, 2 and 3, respectively, where H is the histogram, H' is the cumulative histogram, L is the lookup table, M is the maximum intensity and N is the number of pixels in the image.

$$H'(i) = \sum_{j=0}^{i} H(j) \tag{1}$$

$$L(i) = \frac{M \times H'(i)}{N} \tag{2}$$

$$dst(x, y) = L(src(x, y)). \tag{3}$$

Smoothing: Beside increasing the image contrast, histogram equalization may increase noise contrast too. So it is important to reduce the noise after this process. Smoothing is one of the methods that reduce noise in an image. Since the system is wanted to run in real-time, the simplest and the fastest smoothing approach that is based on changing each pixel value to sum of pixel values in its NxN neighbourhood, was used [7]. Default value of N is set to 4 and user is allowed to change it from the user interface. Pixels' final values are not scaled after the summation to whiten the image.

Adaptive Thresholding: Grayscale images can simply be converted to binary images by using constant threshold values. But constant values are not useful for images that are captured under different environmental conditions. Otsu's method [6] is used to determine an adaptive threshold value in this study.

Otsu's method searches for the threshold that minimizes the within-class variance where classes are assumed to be background and foreground. Since the between-class variance is faster to calculate and it takes its maximum value when the within-class variance is minimum, the threshold that makes the between-class variance maximum is considered as the best threshold. The between-class variance is defined as

$$W_b W_f \left(\mu_b - \mu_f \right)^2 \tag{4}$$

where W_b and W_f are weights of background and foreground pixels (ratios of background and foreground pixels to total number of pixels) and μ represents the means.

Opening: If two or more marks are too close in the image, they may seem like a single mark. To reduce the possibility of such confusion, opening which is an erosion (shrinking) followed by a dilation (expanding) [8] is applied to the image. This also helps to eliminate small traces and this is the final step to remove the noise.

Results that were obtained after each step of the first step are shown in Fig. 2.

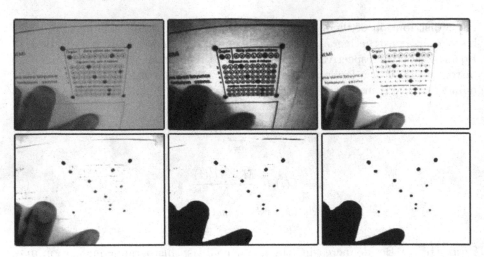

Fig. 2. Results of each step in step 1, from left to right, from top to bottom: original gray image, histogram equalization, smoothing, erosion, adaptive thresholding and opening.

2.2   Mark Detection

Trace Detection and Clustering: After obtaining the binary image, gravitational center of each element in the binary image is determined by calculating means of coordinates of pixels that belongs to the same element. These centers are considered as centers of mark candidates. Then Euclidean distances between the center points and a predefined clustering point are calculated. $M + 4$ points that has the shortest distance values are considered as marks where M is the number of marks to be drawn by hand and 4 is the number of points that will be used to determine the corners.

Blue circles with fixed radii are drawn on the points in the cluster to be used on the next steps. The other points that are not included in the cluster are marked with circles filled with red color.

Corner Determination: Corners of the form are assumed to be the nearest points to the corners of the image. Therefore, 4 points that has the shortest distances to the corner points of the image are selected and considered as the corners of the OMR form.

Perspective Correction: Since the corners may not form a rectangle depending on the view angle, a perspective transform [7] is applied to the corners of the form to correct the perspective. The area that contains the marks is scaled to a fixed size with a correct perspective in this manner.

Checking Predefined Mark Locations: Predefined locations where the user can draw marks are checked if they include blue pixels. Because if these areas are marked by user, blue dots are drawn on them on the "Trace Detection and Clustering" step. Locations that include blue pixels are considered as marked fields.

Figure 3 shows the image after applying the mark detection step. On the left the red point (in this example there is only 1 red point) are the traces which are not included in the cluster. The blue points are the points that are considered as marks and the blue points rounded by red circles represent the corners of the OMR form. The area rounded by a red rectangle is the view of the form after perspective correction (on the right bottom corner of Fig. 3a). On Fig. 3b, the red lines are the distances between gravitational centers of the elements in the binary image and the clustering point. In this example, $M = 8$ and the nearest $8 + 4$ points to the clustering point are selected. The blue lines represent the distances between the corners of the image and the closest points in the cluster to the corners. The red circle was drawn to show approximate location of the clustering center.

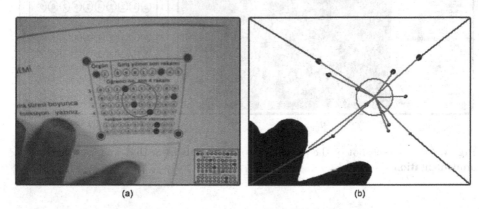

(a)   (b)

Fig. 3. (a) Image after the mark detection step. (b) The cluster and the corners on the same image (Color figure online).

3   Implementation

The proposed system was implemented as a customizable software in Visual Basic.NET and implementations of OpenCV [9] were used for basic image processing methods. The software is customized to read students' numbers and their examination results from a webcam in real-time and writes them into a file. It can be adapted for different purposes by changing its configurations using the provided interface (Fig. 4).

3.1   User-Designed OMR Forms

The implementation allows users to design their own OMR forms. But some requirements need to be satisfied while designing the form:

- Corners of the forms must be pre-marked to help the system to determine the corners.
- A clustering point must be predefined for the proposed mark determination method.
- The number of marks to be drawn by users must be fixed, optional fields are not allowed.

A screen-shot of the program and the form that was designed to test the implementation are given in Fig. 4.

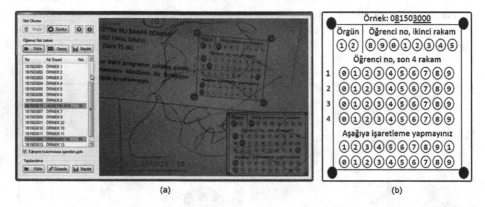

(a)   (b)

Fig. 4. (a) A screenshot of the program. (b) The OMR form that was used to test the implementation.

4   Experimental Results

The application was tested with 88 examination papers to read students' numbers and their examination results. A small OMR form which was designed specifically for this purpose was placed on top right of the first page of each examination paper and students were asked to fill out the forms with required information to construct the student number. Exam result was marked in the form by filling out two bubbles on the bottom of the form.

87 of the 88 forms were correctly read by the application with an average processing time of 23.8 ms which means that the theoretical processing speed is 42 fps (frames per second). But the webcam that was used in this experiment was not able to provide images more than 25 in a second. So the speed is limited to 20 fps and thus, possible excessive processor usage is also prevented. The form that could not be read was filled incorrectly (Fig. 5a).

A form was scratched and shown to two webcams that provide different image qualities under two different lighting and background to test robustness of the system under different conditions. They were successfully read. The results are shown in Fig. 5.

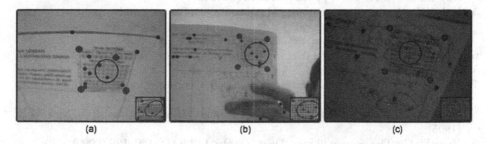

(a)   (b)   (c)

Fig. 5. (a) The form that was filled incorrectly and could not be read properly. Because two marks were drawn in the same row. (b) A scratched form that was read successfully over an unconstrained background. (c) The same scratched form that was read successfully again, under low lighting.

5   Conclusion

In this paper a webcam-based, fast and robust OMR system was presented. The system was implemented and tested on real examination papers to read student number and grade and to save them as an Excel file. All the forms, except one that was filled incorrectly, were read successfully in a very short average processing time.

The experiments show that the system is fast and robust enough to be used in real-time applications even with low-quality webcams. It can be adapted for different purposes using different OMR forms that are designed to be filled by hand with more specific data.

But it can not read forms that have optional fields because of the clustering method which was proposed to determine marks. In future works, different clustering methods can be investigated to make the system able to read OMR forms that has optional fields.

References

1. Perez-Benedito, J.L., Querol Aragon, E., Alriols, J.A., Medic, L.: Optical mark recognition in student continuous assessment. Tecnologias del Aprendizaje, IEEE Rev. Iberoamericana de **9**(4), 133–138 (2014)
2. Nguyen, T.D., Manh, Q.H., Minh, P.B., Thanh, L.N., Hoang, T.M.: Efficient and reliable camera based multiple-choice test grading system. In: 2011 International Conference on Advanced Technologies for Communications (ATC), pp. 268–271. IEEE (2011)

3. Xiu, P., Lopresti, D., Baird, H., Nagy, G., Smith, E.B.: Style-based ballot mark recognition. In: 10th International Conference on Document Analysis and Recognition, ICDAR 2009, pp. 216–220. IEEE (2009)
4. Deng, H., Wang, F., Liang, B.: A low-cost OMR solution for educational applications. In: International Symposium on Parallel and Distributed Processing with Applications ISPA 2008, pp. 967–970. IEEE (2008)
5. Kubo, H., Ohashi, H., Tamamura, M., Kowata, T.: Shared questionnaire system for school community management. In: International Symposium on Applications and the Internet Workshops, pp. 439–445. IEEE (2004)
6. Otsu, N.: A threshold selection method from gray-level histograms. Automatica 11(285–296), 23–27 (1975)
7. Forsyth, D.A., Ponce, J.: A Modern Approach. Computer Vision: A Modern Approach (2003)
8. Jain, R., Kasturi, R., Schunck, B.G.: Machine Vision. McGraw-Hill, New York (1995)
9. Bradski, G.: The opencv library. Doct. Dobbs J. 25(11), 120–126 (2000)

Webcam-Based Visual Gaze Estimation Under Desktop Environment

Shujian Yu[1]([⊠]), Weihua Ou[2], Xinge You[3,4], Xiubao Jiang[3], Yun Zhu[1],
Yi Mou[3], Weigang Guo[3], Yuanyan Tang[3,5], and C.L. Philip Chen[5]

[1] Department of Electrical and Computer Engineering, University of Florida,
Gainesville, FL, USA
yusjlcy9011@ufl.edu
[2] School of Mathematics and Computer Science, Guizhou Normal University,
Guiyang, Guizhou, China
[3] School of Electronic Information and Communications,
Huazhong University of Science and Technology, Wuhan, Hubei, China
[4] Research Institute of Huazhong University of Science and Technology in Shenzhen,
Shenzhen, Guangdong, China
[5] Faculty of Science and Technology, University of Macau, Macau, China

Abstract. Image-based visual gaze estimation has been widely used in various scientific and application-oriented disciplines. However, the high cost and tedious calibration procedure impede its generalization in real scenarios. In this paper, we develop a low cost yet effective webcam based visual gaze estimation system. Different from previous works, we aim at minimizing the system cost, and at the same time, making the system more flexible and feasible to users. More specifically, only a single ordinary webcam is used in our system. Meanwhile, we also proposed a novel calibration mechanism which takes account binocular feature vectors simultaneously, and uses only four visual target points. We compare our system with the state of the art webcam based visual gaze estimation methods. Experimental results demonstrate that our system can achieve satisfactory performance without the requirements of dedicated hardware or tedious calibration procedure.

Keywords: Visual gaze estimation · Ordinary webcam · Desktop environment · Low cost · Flexible · Binocular calibration

1 Introduction

Image-based visual gaze estimation has attracted many potential applications, spanning from human machine interactions (HMI) to attentive user interfaces [25]. As one of the most salient features within human face, eye movements and visual attention are implicitly acknowledged as the way we negotiate with the visual world [10,15,16]. Previous image-based visual gaze estimation systems usually employ two types of imaging techniques: *infrared imaging* and *visible imaging* [3]. Although the high-contrast infrared images can significantly improve

© Springer International Publishing Switzerland 2015
S. Arik et al. (Eds.): ICONIP 2015, Part II, LNCS 9490, pp. 457–466, 2015.
DOI: 10.1007/978-3-319-26535-3_52

the accuracy of eye feature detection, the high cost reduces the feasibility of widespread applications. Therefore, visible-image-based visual gaze estimation systems with low cost are preferred for general public use [10].

Traditional visible-image-based visual gaze estimation systems fall into two categories: feature-based systems and appearance-based systems [3,8]. Feature-based systems rely on the extraction of salient eye features, including iris center, iris contour or eye corners, to provide eye movement information. On the other hand, appearance-based systems employ the whole extracted eye regions to estimate visual gaze. Since the appearance-based systems usually require a large number of training samples to construct the mapping function (or classifier) during calibration, this paper focuses on developing a novel feature-based visual gaze estimation system with an ordinary webcam.

A pioneering work for feature-based visual gaze estimation was proposed in [29], where the authors combined a novel iris center and eye corner detector with a simple linear calibration mechanism to provide a computational efficient solution. Then, in [26], the authors introduced a state-of-the-art iris center locator and eye corner detector to realize the same function. Followed works in [26], they further incorporated head pose information to present a more sophisticated system in [25]. Meanwhile, [12] also designed a novel system which features two simple yet effective algorithms for iris detection and eye movement direction estimation, respectively. A most recent study is available in [3]. Many other works can also guarantee desirable performance under specific constraints at the expense of dedicated hardware, deliberate initialization, or a large number of training data [8,17].

In this paper, we develop a novel single-webcam-based visual gaze estimation system in a desktop environment. Our system features two main ingredients: firstly, an eye feature tracker which can quickly and accurately locate iris center and detect eye inner corner given a video frame; and secondly, an effective mapping function between extracted eye feature vector (connecting iris center and eye inner corner) and screen coordinate. To accomplish these two ingredients, we combine a state-of-the-art eye corner detector with an improved iris center locator to extract eye feature vectors. Meanwhile, we also develop a novel calibration mechanism based on binocular feature vectors to estimate the mapping function, during which only four visual target points are used.

The remainder of this paper is organized as follow: In Sects. 2 and 3 the technical details for the two main ingredients are elaborated step by step. Extensive experiments are conducted and discussed in Sect. 4 prior to the conclusions in Sect. 5.

2   Eye Feature Vector Extraction

This section discusses the reasons behind the choice and the theory of the used eye corner detector (only inner corner is used in our system, since it is more robust to facial expression or eye status [3]) and iris center locator for our system.

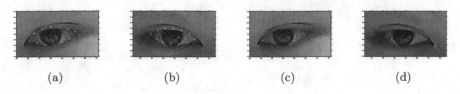

| (a) | (b) | (c) | (d) |

Fig. 1. Eye inner corner detection: (a)–(d): eye landmarks detected with SDM; (c)–(d): final eye inner corner localization with Gabor eye corner filter, the green crosses are the initial corner locations detected using SDM, while the red asterisks mark the final corner locations (Color figure online).

2.1 Eye Inner Corner Detection

A large diverse number of approaches, ranging from simple techniques based on the application of Harris corner detector to more sophisticated Active Shape Model (ASM) [5] or Active Appearance Model (AAM) [4] based approaches, have been proposed for eye inner corner detection in the last decades.

In this paper, we use the state-of-the-art Supervised Descent Method (SDM) [27] to detect and track eye inner corners across the whole video frames. As suggested in [18], 12 landmarks are used to model sclera contours, and the end points of two eyelid curves describe the locations of eye corners (Fig. 1(a)–(b)).

For SDM, assuming we are given a set of training images $\{d^i\}$ and the corresponding hand-labeled landmarks $\{x_*^i\}$. Then, the learning process for SDM is posed as the cascade regression problem in which the following function is optimized at each step:

$$\{R_k, b_k\} = \operatorname*{argmin}_{R_k, b_k} \sum_{d^i} \sum_{x_k^i} \|x_*^i - x_k^i - R_k \phi_k^i - b_k\|_2^2 \tag{1}$$

where x_k^i is the estimated landmarks at kth step, $\phi_k^i = h(d^i(x_k^i))$ is the updated feature vector for x_k^i (h is a non-linear feature extraction function), and R_k and b_k are the generic descent directions and bias term that should be learnt in the kth step. After several steps, the estimated shape x_k^i converges to the true shape x_*^i for all images d^i in training set, and the learning process is done. For more details on deformable object fitting with SDM and a detailed explanation to its solution, the reader is referred to [27].

It is worth noting that ASM, AAM, or SDM can only roughly yet robustly determine the initial locations of eye corners [2,10]. To further improve its accuracy, this paper introduces a Gabor eye corner filter [28] with size 7×7 to final locate the true eye corners, i.e., one need to convolve the Gabor representation of a small rectangle centered at the initial corner to the constructed Gabor eye corner filter, the candidate with the highest response is determined as the final corner location (Fig. 1(c)–(d)). Interested readers can refer to [28] for the construction of Gabor eye corner filters.

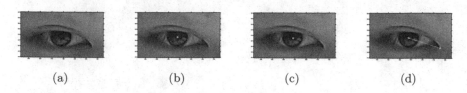

Fig. 2. Iris center localization: (a) is the raw eye region; (b) and (c) demonstrate the iris center localization results with methods proposed in [23,24], respectively; (d) is the eye feature vector.

2.2   Iris Center Localization

Different methods have been developed on robust iris center localization for visible-image, an overview to recent proposed iris segmentation and center localization methods is available in [21]. Note that, ordinary webcam images often suffer from low resolution and ambient noises, and accurate localization of pupil center is not feasible [3,26] (Fig. 2(a)), thus methods which require to locate pupil center prior to determine iris center fall out the scope of this paper.

Among the remaining methods, Daugman's integro-differential operator [6] and its enhanced versions [23] are well acknowledged for both commercial and academic purpose [8]. Symmetry Transform based method and its enhanced versions [1,14] as well as the method using isophote curvature [24] all demonstrate satisfactory performance in specific scenarios [10]. With respect to these methods, both integro-differential operator and isophote curvature can achieve accurate iris center localization in the desktop environment, without heavy constraints on noise artifacts or slightly pose changes (Fig. 2(b)–(c)). For the sake of feasibility and generalization, method proposed in [23] is adopted in this paper.

Traditional integro-differential operator aims at maximizing the following function within an iris image $I_e(x, y)$ to search a circular boundary described with center (x_0, y_0) and radius r:

$$\max_{r,x_0,y_0} \left| G_\sigma(r) * \frac{\partial}{\partial r} \oint_{r,x_0,y_0} \frac{I_e(x,y)}{2\pi r} ds \right| \tag{2}$$

where G_σ is a Gaussian smoothing kernel with variance σ^2. In (2), the operator applies a Gaussian kernel to search the optimal circular boundary that yields the maximum intensity difference. Different from the traditional integro-differential operator, to tackle the problem of heavy computation, [23] introduces *integro-differential-ring* and *integro-differential-constellation* to facilitate the searching process. Interested readers can refer to [23] for more details.

3   Calibration and Eye Gaze Estimation

In a real visual gaze estimation system, a typical calibration procedure presents the user a set of visual target points that he/she has to gaze at while the

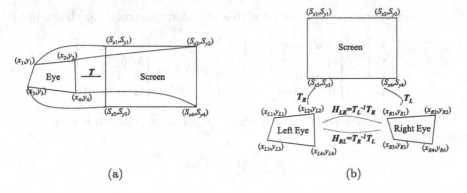

(a)   (b)

Fig. 3. Calibration settings: (a) is the basic paradigm for classical single-eye-based calibration, while (b) depicts the geometry for our proposed joint calibration based on binocular feature vectors.

corresponding measurements, i.e., the screen coordinate (s_x, s_y) and its corresponding eye feature vector $e = e_{iris} - e_{corner}$ (e_{iris} denotes the iris center, and e_{corner} denotes the eye inner corner, Fig. 2(d)), are taken. From these correspondences, the mapping function T, which maps $e = [x, y]^T$ to (s_x, s_y), can be estimated (Fig. 3(a)). Different methods have been proposed to estimate T. Among them, linear model based method [29] and polynomial model based method [19] are two of the most famous methods. An overview to recently proposed calibration methods with the required number of visual target points can be found in [11,13].

However, almost all the current calibration method suffer from two main issues. Firstly, too many visual target points (usually more than 9) are required to guarantee satisfactory performance [13], which makes it inconvenience and tedious [20]. Actually, apartment from the four corners on the screen, it is unrealistic to ask uses to look at visual target points in other positions without manually labeling or dedicated setups [11]. Secondly, almost all the previous works are based on single eye, and disregard useful information provided with another eye. However, as described in [22], binocular data can always improve the accuracy and robustness of eye gaze estimation. To overcome these two issues simultaneously, we developed a novel joint calibration mechanism with binocular feature vectors in this section.

3.1   Joint Calibration with Binocular Feature Vectors

For classical calibration method, a polynomial function is often used to describe the relationship between $e = [x, y]^T$ and (s_x, s_y):

$$\begin{cases} s_x = a_0 + a_1 x + a_2 y + a_3 xy \\ s_y = b_0 + b_1 x + b_2 y + b_3 xy \end{cases} \tag{3}$$

Equation (3) can be represented with a transform matrix T, as shown in (4):

$$\begin{bmatrix} s_x \\ s_y \\ 0 \\ 0 \end{bmatrix} = \begin{bmatrix} a_1 & a_2 & a_3 & a_0 \\ b_1 & b_2 & b_3 & b_0 \\ 0 & 0 & 0 & 0 \\ 0 & 0 & 0 & 0 \end{bmatrix} \times \begin{bmatrix} x \\ y \\ xy \\ 1 \end{bmatrix} \quad with \ T = \begin{bmatrix} a_1 & a_2 & a_3 & a_0 \\ b_1 & b_2 & b_3 & b_0 \\ 0 & 0 & 0 & 0 \\ 0 & 0 & 0 & 0 \end{bmatrix} \quad (4)$$

To calibrate the system, the following optimization function is used to estimate T:

$$\min_{T} \sum_{i=1}^{N} ||L_i - Tg_i||_2^2 \quad (5)$$

where N is the number of visual target points (4 in our system), $L = [s_x, s_y, 0, 0]^T$, $g = [x, y, xy, 1]^T$.

Normally, only single eye feature vector is used to estimate T. Thus, a question arises: how can we combine the information of both eyes (actually two eyes vectors) together to provide a more robust and accurate calibration mechanism. Since there is no training data assumed available beforehand, learning techniques are not feasible herein. Therefore, a novel calibration mechanism is proposed as follow:

$$\min_{T_L, T_R} \sum_{i=1}^{N} ||L_i - T_L g_i^L||_2^2 + ||L_i - T_R g_i^R||_2^2$$
$$+ 1/2 ||g_i^R - T_R^{-1} T_L g_i^L||_2^2 + 1/2 ||g_i^L - T_L^{-1} T_R g_i^R||_2^2 \quad (6)$$

where the superscripts (or subscripts) L and R represent left eye and right eye, respectively. The last two items in (6) assume the existence of a transform H that can transform the coordinates in left eye to the right eye, and vice visa (Fig. 3(b)). Note that, to avoid singularity, we add a small disturbance ηI to both T_L and T_R, and η is fixed to 10^{-7} throughout this paper.

3.2 Eye Gaze Estimation

Once the calibration procedure is performed, given the extracted two eye feature vectors in each video frame, we can obtain the gaze point (s_x, s_y) via (7), where $L = [s_x, s_y, 0, 0]^T$ as described in (5):

$$\min_{L} ||L - T_L g_i^L||_2^2 + ||L - T_R g_i^R||_2^2. \quad (7)$$

4   Experiments

This section evaluates the webcam based visual gaze estimation performance of our proposed system and several other methods via two groups of experiments. Four existing methods are re-implemented for comparison purpose: Zhu2002 [29], Valenti2009 [26], Valenti2012 [25], Cheung2015 [3]. Among them, Zhu2002 is the benchmark method for comparison, Valenti2009 almost shares the same purpose with our system, Valenti2012 and Cheung2015 are two of the state-of-the-art methods in this topic.

Note that, both Valenti2012 and Cheung2015 combine head pose estimation to provide a more reliable estimation results. However, head pose estimator is not considered in our system for three main reasons: firstly, the high computation requirements, complex head models as well as accurate initialization of head pose estimators contradict to the low cost and lightweight requirements of our system; secondly, as pointed out in [9], visual gaze estimation can only be estimated if the head is complete stable when the system employs only one camera and one light source; and thirdly, small mistakes in pose estimation will introduce additional errors in the final visual gaze estimation.

Nine subjects (4 males and 5 females), with different ages (22–45), participated the experiments. Subjects are asked to sit still in front of an ordinary camera at a distance of approximately 0.5 m. Three different ordinary cameras are used herein: (1) internal webcam embedded in a laptop (Satellite P55-At200 by Toshiba Inc.); (2) webcam (Logitech C920) externally connected to the same laptop; (3) internal webcam embedded in another laptop (Macbook Pro by Apple Inc.). During experiments, all the subjects were asked to hold his/her head stationary at the same position during calibration and the following eye gaze tracking process.

4.1   Visual Gaze Estimation to Artificial Targets

In the first group of experiment, all the subjects are asked to gaze at 9 equally distributed target points on the screen, with approximately 10 s for each points. We evaluate the visual gaze estimation with angular degree (A_{dg}) which is defined by $A_{dg} = \arctan(A_d/A_g)$, where A_d is the distance between the estimated gaze position and the real observed position, and A_g represents the distance between the subject and the screen. The smaller A_{dg}, the higher accuracy of visual gaze estimation. To provide a fair comparison, the number of visual target points during calibration is fixed to 4 to all the methods. Figure 4 demonstrate three representative visual gaze estimation results of our proposed method for camera 1, camera 2 and camera 3 respectively, where the red crosses denote the ground truth gaze targets, while the blue points denote the estimated gaze locations. As can be seen, our proposed system outperforms Zhu2002 [29], Valenti2009 [26], and Valenti2012 [25], except for Cheung2015 [3]. Nevertheless, different from Cheung2015 [3], our system does not require deliberate initialization and sophisticated mathematical model to estimate head pose. Table 1 reports the average accuracy of the nine subjects for three cameras.

4.2   Visual Gaze Estimation to Webpages

In the second group of experiment, all the subjects are asked to free-browse the webpages from upper left to lower right without any other specific goals. We then estimate their gaze directions for each webpage and check whether their

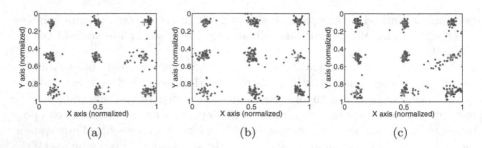

Fig. 4. Visual gaze points (blue) estimated with our proposed method for three different cameras, where the red crosses represent ground truth fixation targets (Color figure online).

Table 1. Accuracy (degree) comparison of different methods (mean)

Different methods	Ordinary camera 1	Ordinary camera 2	Ordinary camera 3
Zhu2002 [29]	5.70	5.18	5.47
Valenti2009 [26]	5.25	4.54	4.92
Valenti2012 [25]	3.89	3.57	3.52
Cheung2015 [3]	3.65	2.90	3.43
Ours	3.66	3.41	3.55

fixations fall within the corresponding image areas and follow specific patterns. 29 screenshots of webpages are collected herein, which can be categorized into *pictorial, text* and *mixed* according to the composition of text and pictures. The average gaze estimation results of nine subjects for three representative webpages are visualized with heat maps, as shown in Fig. 5. As can be seen, the fixations for *text* webpage approximately follow a F-shape viewing pattern, while fixations are tend to focus on image areas for *pictorial* webpage (similar phenomena have already been reported in [7]).

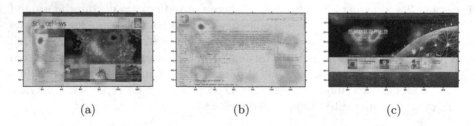

Fig. 5. Heat map visualization of (a) *pictorial* webpage; (b) *text* webpage and (c) *mixed* webpage.

5   Conclusion

This paper proposed a novel webcam based visual gaze estimation system under desktop environment. The new system features an improved iris center locator, a state-of-the-art eye corner detector, and a novel calibration mechanism based on binocular feature vectors. Compared with the benchmark methods, the main advantages of our proposed system are threefold: (1) it does not require any dedicated equipment; (2) it does not require training data for iris center localization or calibration, which makes it more flexible and feasible; and (3) the novel calibration mechanism based on binocular feature vectors can effectively combine the information of both eyes. Experimental results demonstrate that our system can provide reasonable accuracy even compared with the current more sophisticated system.

Acknowledgments. This work is supported partially by the National Natural Science Foundation of China (no.61402122) and the 2014 Ph.D. Recruitment Program of Guizhou Normal University.

References

1. Bai, L., Shen, L., Wang, Y.: A novel eye location algorithm based on radial symmetry transform. In: International Conference on Pattern Recognition, vol. 3, pp. 511–514. IEEE (2006)
2. Bengoechea, J.J., Cerrolaza, J.J., Villanueva, A., Cabeza, R.: Evaluation of accurate eye corner detection methods for gaze estimation. In: International Workshop, Pervasive Eye Tracking Mobile Eye-Based Interaction (2013)
3. Cheung, Y.M., Peng, Q.: Eye gaze tracking with a web camera in a desktop environment. IEEE Trans. Hum. Mach. Syst. **45**(4), 419–430 (2015)
4. Cootes, T.F., Edwards, G.J., Taylor, C.J.: Active appearance models. IEEE Trans. Pattern Anal. Mach. Intell. **23**(6), 681–685 (2001)
5. Cootes, T.F., Taylor, C.J., Cooper, D.H., Graham, J.: Active shape models-their training and application. Comput. Vis. Image Underst. **61**(1), 38–59 (1995)
6. Daugman, J.: How iris recognition works. IEEE Trans. Circ. Syst. Video Technol. **14**(1), 21–30 (2004)
7. Djamasbi, S., Siegel, M., Tullis, T.: Visual hierarchy and viewing behavior: an eye tracking study. In: Jacko, J.A. (ed.) Human-Computer Interaction, Part I, HCII 2011. LNCS, vol. 6761, pp. 331–340. Springer, Heidelberg (2011)
8. Duchowski, A.: Eye Tracking Methodology: Theory and Practice, vol. 373. Springer Science & Business Media, London (2007)
9. Guestrin, E.D., Eizenman, E.: General theory of remote gaze estimation using the pupil center and corneal reflections. IEEE Trans. Biomed. Eng. **53**(6), 1124–1133 (2006)
10. Hansen, D.W., Ji, Q.: In the eye of the beholder: a survey of models for eyes and gaze. IEEE Trans. Pattern Anal. Mach. Intell. **32**(3), 478–500 (2010)
11. Hansen, D.W., Pece, A.E.: Eye tracking in the wild. Comput. Vis. Image Underst. **98**(1), 155–181 (2005)
12. Ince, I.F., Kim, J.W.: A 2d eye gaze estimation system with low-resolution webcam images. EURASIP J. Adv. Sig. Process. **2011**(1), 1–11 (2011)

13. Lee, J.W., Heo, H., Park, K.R.: A novel gaze tracking method based on the generation of virtual calibration points. Sensors **13**(8), 10802–10822 (2013)
14. Loy, G., Zelinsky, A.: A fast radial symmetry transform for detecting points of interest. In: Heyden, A., Sparr, G., Nielsen, M., Johansen, P. (eds.) ECCV 2002, Part I. LNCS, vol. 2350, pp. 358–368. Springer, Heidelberg (2002)
15. Lv, Z., Feng, L., Li, H., Feng, S.: Hand-free motion interaction on google glass. In: SIGGRAPH Asia 2014 Mobile Graphics and Interactive Applications (2014)
16. Lv, Z., Halawani, A., Feng, S., Rhman, S.U., Li, H.: Touch-less interactive augmented reality game on vision based wearable device. Pers. Ubiquit. Comput. **19**(3–4), 551–567 (2015)
17. Majaranta, P.: Gaze Interaction and Applications of Eye Tracking: Advances in Assistive Technologies. IGI Global, Hershey (2011)
18. Alabort-i-Medina, J., Qu, B., Zafeiriou, S.: Statistically learned deformable eye models. In: Agapito, L., Bronstein, M.M., Rother, C. (eds.) ECCV 2014 Workshops. LNCS, vol. 8925, pp. 285–295. Springer, Heidelberg (2015)
19. Morimoto, C.H., Koons, D., Amir, A., Flickner, M.: Pupil detection and tracking using multiple light sources. Image Vis. Comput. **18**(4), 331–335 (2000)
20. Pfeuffer, K., Vidal, M., Turner, J., Bulling, A., Gellersen, H.: Pursuit calibration: making gaze calibration less tedious and more flexible. In: Proceedings of the 26th annual ACM symposium on User interface software and technology, pp. 261–270. ACM (2013)
21. Proenca, H.: Iris recognition: on the segmentation of degraded images acquired in the visible wavelength. IEEE Trans. Pattern Anal. Mach. Intell. **32**(8), 1502–1516 (2010)
22. Sesma-Sanchez, L., Villanueva, A., Cabeza, R.: Design issues of remote eye tracking systems with large range of movement. In: Proceedings of the Symposium on Eye Tracking Research and Applications, pp. 243–246. ACM (2014)
23. Tan, T., He, Z., Sun, Z.: Efficient and robust segmentation of noisy iris images for non-cooperative iris recognition. Image Vis. Comput. **28**(2), 223–230 (2010)
24. Valenti, R., Gevers, T.: Accurate eye center location and tracking using isophote curvature. In: IEEE Conference on Computer Vision and Pattern Recognition, pp. 1–8. IEEE (2008)
25. Valenti, R., Sebe, N., Gevers, T.: Combining head pose and eye location information for gaze estimation. IEEE Trans. Image Process. **21**(2), 802–815 (2012)
26. Valenti, R., Staiano, J., Sebe, N., Gevers, T.: Webcam-based visual gaze estimation. In: Foggia, P., Sansone, C., Vento, M. (eds.) ICIAP 2009. LNCS, vol. 5716, pp. 662–671. Springer, Heidelberg (2009)
27. Xiong, X., De la Torre, F.: Supervised descent method and its applications to face alignment. In: IEEE Conference on Computer Vision and Pattern Recognition (CVPR), pp. 532–539. IEEE (2013)
28. Zheng, Z., Yang, J., Yang, L.: A robust method for eye features extraction on color image. Pattern Recogn. Lett. **26**(14), 2252–2261 (2005)
29. Zhu, J., Yang, J.: Subpixel eye gaze tracking. In: Fifth IEEE International Conference on Automatic Face and Gesture Recognition, pp. 124–129. IEEE (2002)

Influence of Previous Choice and Outcome in a Two-Alternative Decision-Making Task

Manisha Chawla[✉] and Krishna P. Miyapuram

Center for Cognitive Science, Indian Institute of Technology Gandhinagar,
Ahmedbad 382424, India
{manisha.chawla,kprasad}@iitgn.ac.in

Abstract. In everyday life, we encounter many such situations in which we need to make decisions. A major gap in decision-making research so far is that decision-making paradigms are often limited to specific domains, such as perceptual, value-based, or social stimuli. We use two-alternative forced choice experiment as a unified paradigm to present different types of reinforcers – Numbers, shapes, money, and faces. We gave participants different probabilities of winning on two-cards. Although the probabilities that varied between different blocks of trials were unknown, our results confirmed participants follow probability matching behavior. Our results show that the context of previous trial's outcome (i.e. win or loss) modulated the staying and switching patterns and this effect was stronger for numbers choice task.

Keywords: 2AFC · Matching · Reinforcement learning model · Win-stay lose switch · Decision making

1 Introduction

The decision-making scenarios in day-to-day life are often repetitive in nature (e.g. purchases in a supermarket, recruitment of personnel, getting from point A to point B, etc.). An individual, during the process of making repeated choices, is considering a tradeoff between exploration (switching to a new choice) or exploitation (staying with the previous choice) based on the outcomes from previous decisions. Although, decisions on repeated choices in experimental setups seem stochastic and unpredictable, the cumulative rewards earned can be approximated by matching law (Herrnstein 1961). Matching law states that the ratio of choices made by the participants will be proportional to the ratio of relative reinforcement between the two choices. Another competing view in individual choice scenarios is that of a rational agent perspective, which suggests that people should choose an alternative that maximizes the expected utility. We do not yet understand sufficiently how valuation is performed in decision-making scenarios that lead to maximizing or matching behaviour.

This research was supported by grant from Cognitive Science research Initiative, Department of Science & Technology, India (SR/CSI/54/2012).

© Springer International Publishing Switzerland 2015
S. Arik et al. (Eds.): ICONIP 2015, Part II, LNCS 9490, pp. 467–474, 2015.
DOI: 10.1007/978-3-319-26535-3_53

Much of decision-making work has been focused on perceptual, value-based and social decisions domains which are often not useful to study decision-making phenomena in another domain. Value-based decision-making paradigms support the notion of a common currency of subjective value that is to be maximised. In perceptual decision making, participants are typically required to choose one option from a set of alternatives based on available sensory information. This information then must be interpreted and translated into behaviour. For e.g. in a moving-dot paradigm, participants are expected to identify the dominant direction amongst a set of randomly moving dots. Individuals do not necessarily always follow the maximum utility approach, particularly in social decision-making scenarios involving trust, fairness, altruism, cooperation, and reciprocation, where context plays a major role. The computation of subjective value is dynamically modulated by different contexts such as time and resources. Some studies have shown bias in subjective valuation due to change in contexts (e.g. Ariely and Norton 2008; Framing effect, Tversky and Kahneman 1981; Iyengar and Lepper 2000). Such contextual influences on decision-making can be a result of top-down factors, for example - the temporal history of previous choices, (i.e. a Markov Decision Process) or other factors such as the social setting (as often encountered in strategic decision making) and transient personal factors (e.g. risk attitudes, emotions, etc.). We use a two-alternative forced choice paradigm, where individuals will choose the two alternatives matching with the respective probabilities of obtaining a reward. We modify this setup by having a contextual dependence based on previous trials, i.e. the conditional probability of obtaining a reward given the choices made in previous n trials. When, the probabilities of winning given a certain choice are not known, participants need to estimate these based on previous trials.

2   Materials and Methods

In the current experimental work we investigate individuals' choices under different contexts. Experiment 1 varied the type of reinforcer keeping the probability of reinforcement constant at 50 % for each of the two alternatives. Experiment 2 varied the reinforcement probabilities between different blocks. We also look at outcomes from previous trial, taking previous trial choices and accuracy as context and studying their effects on choice modulations made in the current trial. The different probabilities of winning for the left and right cards were independently manipulated: 70 %–30 %, 70 %–50 %, 30 %–50 %, and 50 %–50 %. 10 student volunteers (3 Females, mean age = 26.13 years, range 19–31 years) participated in experiment 1. 9 student volunteers (5 Females, mean age = 24.6 years, range 19–31 years) participated in experiment 2. All participants were from IIT Gandhinagar, right handed, and had normal or corrected to normal vision. All participants gave informed consent. Stimulus presentation and response-recording were controlled by Psychtoolbox (Brainard 1997) software running within Matlab on a Windows PC.

The experiment consisted of a modified 2AFC setup where two cards were displayed facing down on the left and right side of the display screen with some content (perceptual information) hidden behind them. The spatial position of the cards was fixed with the two cards horizontally aligned towards the left and right side of the screen. In each trial, after a choice was made, the chosen card was flipped to reveal the information behind it. Participants had to choose one of the two cards (left/right) by responding with a key press corresponding to the location of the card. Assuming no prior information, the individual makes repeated choices between the left and right options, and thereby calculates an estimate of the underlying distribution of reward values by successive sampling. In experiment 1, a correct card (i.e. win) on each trial was equally likely for the left and right cards. In experiment 2, the reinforcement probabilities were varied (see subsection below). In both experiments, winning card for each trial was independent of the previous choice a participant made. The transition probabilities were held constant at 50 % throughout the experiment (i.e., if there is a win on the left card on the previous trial, then in the current trial it is equally likely that the left or the right card is a winning card). The non-chosen card on any particular trial, across all blocks remained as is, i.e. no counterfactual information is displayed. Pay off received by the participants was fixed at 10 points on each correct trial. Participants attempted to maximize the cumulative points earned as they were paid on the basis of the cumulative points earned over the course of the experiment. Stimulus presentation and response-recording were controlled by Psychtoolbox (Brainard 1997) software running within Matlab on a Windows PC.

Experiment 1. The experiment was divided in to four blocks (of 20 trials each repeated thrice) with different conditions corresponding to the category of outcomes in the binary choice task. The outcome on any trial after flipping of the chosen card belonged to one of the following categories; uniform distribution of numbers between 1 to 10, 5 different sizes of squares or circles, Happy/Sad, Male or Female faces and notes or coins of Rs. 1, 2, 5, and 10. In the numeric choice task, participants were asked to choose a card that had a number greater than 5 behind it (winning card). Choosing of a left or right card resulted in flipping of the chosen card and the information contained behind the card was displayed. If the chosen card was a winning card, on flipping, it displayed a number between 6 and 10 and was marked with a green border. If the chosen card was incorrect it was depicted by a red border and a number between 1 and 5 was displayed. In the monetary choice task, in order to win, the participants were asked to choose a card that had a coin behind it on each trial. If the chosen card was a winning card, on flipping, it displayed a coin that could have values of Re 1, Rs. 2, Rs. 5 or Rs. 10 and the card was marked with a green border. If the chosen card was incorrect it was depicted by a red border and a rupee note of values Re 1, Rs. 2, Rs. 5 or Rs. 10 was displayed. Note here, that the monetary value on the winning and losing card is the same (i.e. equal choice value on both cards) but the stimulus values (win on coins and loss on rupee notes) differ. In the face choice task block, participants won on a trial if they chose a card which had a happy male/female face behind it. The correct card was marked with a green border on flipping and displayed either a male happy or a female happy face. Choosing an incorrect card resulted in the flipped card displaying a female or a male sad face with a red border around the card. In the shape choice block, participants were asked to choose a card that had a circle behind it. Correct card when chosen displayed a circle that could have one of the 5 different predefined radii and a green border. An incorrect chosen card resulted in a display of red border and a square of one of the predefined 5 dimensions. Order of the four blocks was randomized for each participant.

Experiment 2. The experiment was divided into two blocks with four conditions each according to the variable reward reinforcement probabilities on each left/right card. Block 1 consisted of the following 4 conditions: Condition 1 had a probability of 70 % win and 30 % loss on the left card and 30 % win and 70 % loss on the right card. The win-loss reward probabilities were independent for both the left and the right cards, i.e. it may so happen that in a given trial both the left and right cards are the winning (or losing) cards. In Condition 2 probability of winning and losing on the left card was 70 %–30 % respectively and probability of winning and losing on the right card was 50 %–50 % respectively. Notice here that the probabilities of winning (or losing) on the left and the right cards do not sum to 100 this is because the underlying distribution for both the left and right cards was calculated separately across all conditions. Similarly in condition 3 the probability of win and loss on the left and right cards was 30 %–70 % and 50 %–50 % respectively. In condition 4, the winning reward probabilities were kept 50 % for both left and the right cards. Block 2 had exact same conditions as block 1 with the difference being that this time the reward probabilities were flipped between the left and right cards, i.e. if block 1 condition 2 had 70 %–30 % on the left card and 50 %–50 % on the right card then this order was reversed in block 2. So, block 2 condition 2 had 50 %–50 % probability on the left card and 70 %–30 % on the right card. This procedure was followed for all the conditions. This was done to remove any choice biases for the left and right sides. Apart from the different reward probabilities in each condition we kept the transition probabilities constant at 50 % throughout the experiment (i.e., if there is a win on the left card on the previous trial, then in the current trial it is equally likely that the left or the right card is a winning card) given the underlying distribution of the set reward probabilities is followed. Order of the conditions was randomized for each participant.

Each participant completed one session consisting of 2 blocks with 4 conditions each. Each block began with a set of instructions at the centre of the screen. Each condition consisted of 30 trials where the participants had to choose of the two cards displayed on the screen to maximize their cumulative points. The goal of the participants was to maximize their cumulative points over all conditions and trials as they were paid on the basis of the total points earned. The cards remained on the screen until a response was made, upto a maximum limit of 300 ms. After a choice (response) was made a feedback was provided to the participant about the correctness of their choice made in terms of total points and flipping of the chosen card. Chosen card was flipped and content behind the chosen card was displayed to the participant, i.e. a treasure box on the winning card and a red cross sign depicting loss over that particular trial. In addition, after every trial total points won so far was displayed at the bottom of the screen. After every condition the participants were given a self paced break.

3 Results

Experiment 1. We first calculated the probabilities of choosing left and right options in all the four choice tasks. Since, the probabilities of winning on each trial was set to 50 %, we would predict that the participants choose equally between the left and right options (Matching law). However, the experienced winning probability for each of the left and right cards will depend on the participant's actual choices made, i.e. these will not be the same as the predetermined probabilities. Hence, we calculated the correlation between experienced winning probabilities and choice made, which was found to be statistically significant ($r^2 = 0.97$).

One of the aims of the present experiment was to find the effects of context of previous trial on the current choice. To avoid any bias, we had also set the transition probabilities of winning on the left and right cards to be equal to 50 %. i.e. given that the left card (or equivalently the right card) would be the winning card in the current trial, the next trial would have equal probability for left and right cards to be the winning cards. Hence, we verified whether the choice made in the current trial would depend upon their choice in the immediately previous trial, i.e. we calculated whether the choices made by participants matched the transition probabilities. Given the participants had chosen a left (or right) card in the previous trial, we found equal probability for choosing left or right cards in the current trial. Next, we calculated whether the probabilities of choosing left or right card depended on the winning (or losing) in the previous trial. This also does not appear to be the case (Fig. 1).

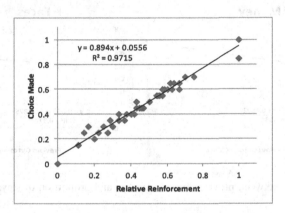

Fig. 1. Correlation between experienced winning probabilities and choice made in agreement with Matching Law.

According to Win-Stay and Lose-switch model, participants should stay with their previous choice based on previous trial's accuracy. We therefore checked whether the probability of stay or switch depended upon (a) previous choice and (b) previous trial's accuracy (i.e. win or lose). Repeated measures Anova with previous Accuracy (2 levels – win, lose), and parametric outcome value (5 or 4 levels) revealed significant main effect for the number task [$F(1,9) = 9$, $p < 0.05$] but not for other three types of tasks. In all tasks, the higher half values of outcome were designated as win and the lower half of the outcomes were designated as a loss trial. The outcome value was directly related to the actual number shown, the size of the shape, or the denomination of monetary value. However the face stimuli did not differ parametrically and only their valence was manipulated (happy Vs sad face).

We correlated the stay patterns with the previous choice's outcome in each of the four tasks. We found stronger correlation for the Number task ($r^2 = 0.67$) followed by shape choice task ($r^2 = 0.52$). Interestingly smaller effects of stay pattern based on previous outcome was observed for money stimuli ($r^2 = 0.53$) and face stimuli ($r^2 = 0.37$) (Fig. 2).

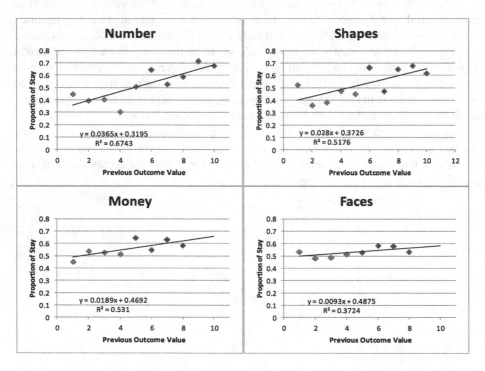

Fig. 2. Correlation between previous trial's outcome and proportion of staying with previous choice

Experiment 2. We first calculated the probabilities of choosing left and right options across all four conditions and blocks. We predicted that participant's choices would be proportional to the previously set reinforcement probabilities. But we found that apart from the 50 %–50 % condition proportion of participants' choices did not follow the set reinforcement probabilities. A 3-way RM ANOVA with Blocks (2levels) X Conditions (4 levels – 70 %–30 %, 70 %–50 %, 30 %–50 %, 50 %–50 %) X Trials (30 levels – for each condition) as within subject factors. We found no significant effects of Blocks [F (1, 8) = 1.20, p = 0.3] which confirmed that flipping of probabilities over the left and right side did not have any significant effect on the participants choices. We found significant effects across conditions [F (3, 24) = 5.29, p < 0.001]. Pairwise comparisons revealed significant differences between 70 %–50 % condition and other 3 conditions (Fig. 3).

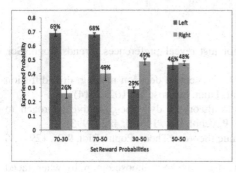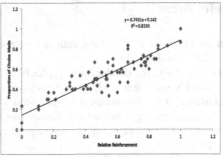

Fig. 3. Experienced reward reinforcement Probabilities and Correlation with choice made

We further calculated the correlation between experienced winning probabilities for every 30 trials and choice made for all participants, which was found to be statistically significant ($r^2 = 0.83$). Since, in different conditions we kept the probabilities independent on the left and the right cards we calculated the experience probabilities to check if the probabilities experienced are different from what we set. A 3-way RM ANOVA revealed significant effects of Blocks [$F(1,8) = 16.08$, $p < 0.05$], conditions [$F(3,24) = 3.42$, $p < 0.05$], block x conditions [$F(3,24) = 27$, $p < 0.001$] and all pairwise conditions. We further analyzed probability of stay after winning or losing on trials and found the probability of staying after a win was higher.

4   Discussion

Our research has used a 2-alternative forced choice paradigm to study the different types of reinforcers – numbers, shapes, money and faces. We further investigated the effects of Matching law – whether the left and right cards chosen by participants was proportional to their corresponding probabilities of reinforcement. Given, our experimental design, we set out to investigate the contextual influence of effect of previous trial on the decision-making strategies used by the participants. Our results showed that participants not only matched the probabilities, but also the transition probabilities. These results support and extend the matching law. Further previous trial's accuracy did not influence the choice made (i.e. choosing left or right card) but had an influence on the participants' staying and switching patterns. These results are in agreement with the Win-stay Lose-switch strategy. Assuming no prior information of winning probabilities, the individual is making repeated choices between the left and right options, and thereby calculating an estimate of the underlying distribution of reward values by successive sampling. Our results also showed that probability of stay after win is higher in support of the win-stay lose-switch model. In sum, our results show a correspondence between matching law, win-stay and lose-switch model of simple value based decision making.

The long-term goal of this research can be to have integrated computational models combining different approaches of reinforcement learning and Bayesian modelling. Of particular interest would be to see how many previous trials have an influence on the current choice.

References

Ariely, D., Norton, M.I.: How actions create–not just reveal–preferences. Trends Cogn. Sci. **12**(1), 13–16 (2008)

Bitzer, S., Park, H., Blankenburg, F., Kiebel, S.J.: Perceptual decision making: drift-diffusion model is equivalent to a Bayesian model. Front. Hum. Neurosci. **8**, 102 (2014)

Busemeyer, J.R., Townsend, J.T.: Decision field theory: a dynamic-cognitive approach to decision-making in an uncertain environment. Psychol. Rev. **100**(3), 432 (1993)

Charnov, E.L.: Optimal foraging, the marginal value theorem. Theor. Popul. Biol. **9**(2), 129–136 (1976)

Feng, S., Holmes, P., Rorie, A., Newsome, W.T.: Can monkeys choose optimally when faced with noisy stimuli and unequal rewards? PLoS Comput. Biol. **5**(2), e1000284 (2009)

Gao, J., Tortell, R., McClelland, J.L.: Dynamic integration of reward and stimulus information in perceptual decision-making. PLoS ONE **6**(3), e16749 (2011)

Herrnstein, R.J.: Relative and absolute strength of response as a function of frequency of reinforcement 1, 2. J. Exp. Anal. Behav. **4**(3), 267–272 (1961)

Iyengar, S.S., Lepper, M.R.: When choice is demotivating: Can one desire too much of a good thing? J. Pers. Soc. Psychol. **79**(6), 995 (2000)

Kahneman, D., Tversky, A.: Prospect theory: an analysis of decision under risk. Econ. J. Econ. Soc. **47**, 263–291 (1979)

Krajbich, I., Rangel, A.: Multialternative drift-diffusion model predicts the relationship between visual fixations and choice in value-based decisions. Proc. Natl. Acad. Sci. **108**(33), 13852–13857 (2011)

Loomes, G., Sugden, R.: Regret theory: an alternative theory of rational choice under uncertainty. Econ. J. **92**, 805–824 (1982)

Pleger, B., Blankenburg, F., Ruff, C.C., Driver, J., Dolan, R.J.: Reward facilitates tactile judgments and modulates hemodynamic responses in human primary somatosensory cortex. The Journal of Neuroscience **28**(33), 8161–8168 (2008)

Ratcliff, R.: A theory of memory retrieval. Psychol. Rev. **85**(2), 59 (1978)

Rorie, A.E., Gao, J., McClelland, J.L., Newsome, W.T.: Integration of sensory and reward information during perceptual decision-making in lateral intraparietal cortex (LIP) of the macaque monkey. PLoS ONE **5**(2), e9308 (2010)

Stephens, D.W., Krebs, J.R. Foraging Theory (1986)

Sugrue, L.P., Corrado, G.S., Newsome, W.T.: Choosing the greater of two goods: neural currencies for valuation and decision making. Nat. Rev. Neurosci. **6**(5), 363–375 (2005)

Tversky, A., Kahneman, D.: The framing of decisions and the psychology of choice. Science **211**(4481), 453–458 (1981)

Von Neumann, J., Morgenstern, O.: Theory of games and economic behavior. Bull. Amer. Math. Soc **51**(7), 498–504 (1944)

Using Modern Neural Networks to Predict the Decisions of Supreme Court of the United States with State-of-the-Art Accuracy

Ranti Dev Sharma[✉], Sudhanshu Mittal, Samarth Tripathi, and Shrinivas Acharya

Department of Computer Science, Indian Institute of Technology, Guwahati, India
{ranti.iitg,sudhanshumittal1992,samarthtripathi,shrinivas.iitg}@gmail.com

Abstract. Deep neural networks are a family of statistical learning models inspired by biological neural networks and are used to estimate functions that can depend on a large number of inputs and are generally unknown. In this paper we build upon the works of Katz, Bommarito and Blackman 2014, who use extremely randomized trees and feature engineering to help in predicting the behaviour of Supreme Court of United States. We explore Machine Learning techniques to achieve our goals including SVM and Neural Networks, but attain state-of-the-art accuracy with Deep Neural Networks trained using momentum methods and incorporating the Dropout technique. We explicitly use only data available prior to the decision and predict the decisions with 70.4 percent accuracy across 7,700 cases with nearly 70,000 justice votes. Our model is simple yet robust, uses far less feature vectors to train and still provides excellent accuracy, but most importantly deploys no feature engineering.

Keywords: Deep neural networks · Supreme court · Dropout technique

1 Introduction

The Supreme Court of the United States is the highest federal court in the United States. If the court agrees to a petition, each of the parties to the litigation will then submit written materials on the relevant issues and later provide oral argument before the Court. The Court ultimately votes whether to affirm or reverse the decision of the lower court unanimously. From a prediction standpoint, the main question is: Will the Court affirm or reverse the lower Courts judgment? (Katz, Bommarito and Blackman 2014). As this question is of cardinal importance for the people involved with the case and other academia, we seek to provide an unbiased, robust and fully predictive model to methodically predict the outcome, using the recent advances in artificially intelligent prediction models. In particular we examine modern neural networks with great attention as our experiments yield their best results with it. Neural networks is a machine that is designed to model the way our brain performs a particular task,

© Springer International Publishing Switzerland 2015
S. Arik et al. (Eds.): ICONIP 2015, Part II, LNCS 9490, pp. 475–483, 2015.
DOI: 10.1007/978-3-319-26535-3_54

where the key concepts of brain as a complex, non-linear and parallel computer are imitated (Simon Haykin 2004). Neural networks possess the ability to model and estimate complex functions depending on multitude of factors, and fit our current needs perfectly.

Predicting decisions of the Supreme Court has been studied before in detail by various scholars. We closely follow the research of Katz, Bommarito and Blackman 2014 where they use extremely random trees for classfying and predicting the overall Supreme court decision and the decisions of individual judges. In this paper we follow a similar approach but instead of using random trees for prediction we use modern machine learning techniques to emulate their results. Among other scholars who carried out the similar task of predicting Supreme court decision, most notable is the work of Ruger *et al.* 2004 who use statistical models to predict the results of judges.

Our study incorporates two of the major Machine Learning algorithms discovered and improved recently, for the task of decision prediction. We utilize Support Vector Machines(linear, non-linear, and polynomial kernel models) to predict using the shallow learning methods, and Modern Neural Networks for deep learning methods. For our neural networks model, we train the network with the dropout technique using both RMSprop and Momentum techniques to compare results. Our research concludes with detailed analysis of the accuracy of both approaches and in depth analysis of their shortcomings. In the process of our study we also provide for state-of-the-art accuracy in prediction of decisions of Supreme Court, faring better than Katz, Bommarito and Blackman's 2014 work. Our model yields 70.4 % accuracy on predicting the overall decisions of Supreme Court, using only data available prior to the case and utilizing far less features for the prediction. Our prime accuracy is unmatched in research academia, and substantially eclipses human experts accuracy reported by Ruger *et al.* 2004, while our model is simple, robust, fully predictive, scalable and could be easily extended as it involves no feature engineering.

2   Related Works

As we discussed, predicting decisions of Supreme Court of the US has been studied in detail by certain research academia, but mostly from a legal perspective and not from a pattern matching and machine learning perspective. Our work is greatly inspired from the previous study carried out by Katz, Bommarito and Blackman 2014, who use extremely randomized tree method to form a classification algorithm, similar to random forest, with novel feature engineering to obtain excellent results. Their model incorporates key properties of being robust, generalized and fully predictive. Our study is an attempt to achieve the same goals using various Machine learning approaches especially deep neural networks, advancing their research where it left of. Their feature engineering is extremely detailed and complex, and use upto fifty feature vectors, mostly derived from complex mathematical normalizations. Our work seeks to tackle this issue by using very generic and simple features, normalized using simple techniques, yet providing improved accuracy. They report a commendable accuracy of 69.7 % accuracy of the Supreme Court decision.

Guiemera and Sales-Pardon 2011 also endeavour to find the predictability of judges decisions, using stochastic block models for vote prediction. It relies on assumptions that justices and cases can be grouped into blocks, as per information about justices voting pattern. Then they proceed to use Bayesian approach to average over them. Ruger *et al.* 2004 and Martin *et al.* 2004 try to predict every case during 2002 term using statistical models and achieve upto 75 % accuracy, emulating human experts who correctly identified only 67.9 % cases. Their model used minimal data to construct a decision based classification tree but their concept and methodology cannot be extended or generalized without serious compromises.

Our approaches include Support Vector Machines for shallow learning, and neural networks for deep learning. Support vector machines (SVM) are a group of supervised learning methods that can be applied to classification or regression. Support vector machines represent an extension to nonlinear models of the generalized portrait algorithm developed by Vladimir Vapnik 1995. SVM are used extensively for pattern recognition, throughout the field of machine learning. Formally, a neural network is a massively parallel distributed processor made up of simple processing units which has a natural propensity for storing experiential knowledge and making it available for use. It resembles our brain in two key respects. Firstly, knowledge is acquired from the environment through a learning process. Secondly, interneuron connection strength, called synaptic weights, are used to stored the acquired information (Simon Haykin 2004). Our implementation of neural network uses python's Theano library (Bastien *et al.* 2012 and Bergstra *et al.* 2010). The approach was further refined by using RMSProp (Dauphin *et al.* 2015 and Tieleman and Hinton 2012) for optimization which divides the gradient by the running average of its recent magnitudes. The standard momentum method first computes the gradient at the current location and then takes a big jump in the direction of the updated accumulated gradient as described by Nesterov 1983. To further improve our algorithm we utilized the dropout technique presented by *et al.* 2014, which provide a simple way to prevent neural networks from overfitting.

3   Initial Dataset

We derive our initial dataset for training our model from the Supreme Court Database (SCDB) by Harold *et al.* 2014. The Database contains over two hundred pieces of information about each case decided by the Court between the 1946 and 2013 terms. The database has been widely used over research academia with consistent reliability and features up to two hundred and forty seven variables for each case including background variables, chronological variables, substantive variables, outcome variables, voting variables and opinion variables. As we seek to build upon the work of Katz 2014, we also define our prediction outcome variable based on the decision of the lower court. We consider that for a given

Table 1. Features from SCDB used as initial dataset

Docket Id	Natural court
Actual outcome	Origin of case
Origin of case state	Source of case
Source of case state	Law type
Lower court disposition	Lower court disposition direction
Lower court disagreemen	Issue
Issue area	Manner in which the court takes jurisdiction
Petitioner	Respondent
Reason for granting cert	Administrative action preceeding litigation

case whether the Supreme Court will affirm or reverse the decision of the lower court. We consider many features from SCDB however, unlike Katz 2014 we do not deploy extensive feature engineering. Our initial dataset consists of eighteen features from SCDB, enumerated in Table 1.

4   Prediction Using Deep Neural Networks

We use the eighteen variables previously mentioned of SCDB as the input features for training our model, and expect only the predicted outcome for the case as output. The various constant parameters used in the functions are discussed in detail and also provided in Table 2. Our deep neural network model is implemented on Pythons Theano library.

Out of the total 8386 unique dockets provided by the SCDB, we use the first 7500 as our initial training dataset, and remaining 886 (approximately 10 %) as our test dataset. We individually scale each of the features in our initial training dataset to have unit variance and a zero mean using scikit's scale library. This ensures stable convergence of the neural networks weights and biases. To remove any unwanted biases because of the chronological enumeration that SCDB follows in its data, we also randomly shuffle our initial dataset. To speed up the learning rates we experiment with and compare two learning techniques, namely RMSProp and Momentum-Method. RMSProp divides the learning rate for a weight by a running average of the magnitudes of recent gradients for that weight. We have used a learning rate of 0.001 and a gradient direction of 0.9 for our RMSProp learning. On the other hand, Momentum Method first computes the gradient at the current location and then takes a big jump in the direction of the updated accumulated gradient. For momentum technique we use the learning rate of 0.01 and momentum value of 0.9. We will discuss the results of using either approaches in the next section.

Neural networks generally use Activation functions which are used to transform the activation level of a unit (neuron) into an output signal. A nonlinear activation function allows Neural networks to compute nontrivial problems using

Table 2. Table defining constant values for parameters

Parameter	Variable	Value
Number of training cases	TrainingTotal	7500
Number of test cases	testTotal	886
Momentum learning rate	momLearn	0.01
Momentum gradient value	momValue	0.9
RMSProp learning rate	rmsLearn	0.001
RMSProp gradient direction value	rmsValue	0.9
RMSProp epsilon	rmsEpsilon	Pow(e,-6)
Dropout probability first input neural layer	dropInput	0.2
Dropout probability hidden neural layer	dropHidden	0.5
Total neural layers (inclusive of first, last and hidden)	layerTotal	10
Models input layer dimensionality	dimInput	16
Models intermediate hidden Layer dimensionality	dimHidden	10
Models output layer dimensionality	dimOutput	2
Total cases in each epoch	batchTotal	128
Epoch	epochTotal	500

only a small number of nodes. We deploy Rectified Linear Units (ReLU) as our non-linear activation function. Among various non-linear activation functions like sigmoid, tan hyperbolic etc. ReLU has also been argued to be more biologically plausible and practical (Glorut et al. 2011 and LeCun et al. 2012).

Deep learning usually involves neural layers learning on huge datasets, in the order of millions. However as dataset size is constrained in our problem, we use dropout technique to obtain good results. The dropout technique probabilistically drops a few nodes from the neural network while training to avoid as a prevention against overfitting. While using the dropout technique it is important to use large number of epochs. An epoch is a measure of the number of times all of the training vectors are used once to update the weights. For batch training all of the training samples pass through the learning algorithm simultaneously in one epoch before weights are updated.

We train our model in batches of 128, with an epoch of 500 to provide for good detailed learning process. For our input layer we use a dropout probability of 0.2, while for the successive hidden layers we use a dropout factor of 0.5. These probabilities have been empirically shown to yield best results which we also affirm in our own tests. Our model trains on a total of 10 layers - 9 ReLU hidden layers and an output Softmax layer. The input dimension size is sixteen (all variables in feature space except Docket Id and Outcome), which are used to map the Docket Id to the actual result of a court trial. For our input neural layer we keep the nodes as sixteen but for the rest of the hidden layers we reduce the weights dimensional space to ten, as this allows the model to be trained faster and remove unimportant features and noise from the initial dataset. The last hidden layer finally reduces the dimensionality to our two variables, the unique Docket id and the predicted outcome, using hot encoding. Our implementation of the neural model is represented by Algorithm 1.

Algorithm 1. Neural Model:

Require: Initial dataset $trainDataSet$, Weight matrix (2 dimensional matrix of size 2 * layerTotal) $dimenMatrix$, Dropout input layer Probability $dropInputProb$, Dropout Hidden Layer Probability $dropHiddenProb$

Ensure: Neural layers Matrix in Rectilinear units matrix (2 dimensional matrix of size 2 * layerTotal) $hiddenMatrix$

1: **procedure** NeuralModel   (
   $trainDataSet, dimenMatrix[layerTotal][2], dropInputProb, dropHiddenProb)$
2:    $hiddenMatrix = \emptyset$
3:    $dataset = dropout(trainDataSet, dropInputProb)$
4:    $interimOutput = dot(dataset, dimenMatrix[0][1])$
                                                  ▷ dot represents the dot product
5:    $nextLayerOutput = Relu(Intermediate_o utput)$
                                                  ▷ apply non linear activation for eg. ReLu
6:    $hiddenMatrix+ = nextLayerOutput$
7:    $inputPreviousLayer = nextLayerOutput$
8:    **for all** $layer in (1 : layerTotal)$ **do**
9:       $dataset = dropout(inputPreviousLayer, dropHiddenProb)$
10:      $interimOutput = dot(dataset, dimenMatrix[layer][1])$
11:      $nextLayerOutput = Relu(interimOutput)$
12:      $hiddenMatrix+ = nextLayerOutput$
13:      $inputPreviousLayer = nextLayerOutput$
14:   **end for**
15:   $outputProbability$ =
   $softmax(dot(inputPreviousLayer, dimenMatrix[layerTotal][1]))$
                                                  ▷ where softmax is the softmax function
16:   $hiddenMatrix+ = nextLayerOutput$
17:   **return** $hiddenMatrix$
18: **end procedure**

The neural networks works by minimizing the cross entropy or noise between the actual and predicted outcomes. We allow our theano model to downcast which means that the values passed as inputs when calling the function can be silently downcasted to fit the dtype of the corresponding variable, which may lose precision. Deep learning usually involves neural layers learning on huge datasets, in the order of millions. However as dataset is constrained, we use dropout technique, and obtain good results. While using the dropout technique it is important to use a high epoch. An epoch is a measure of the number of times all of the training vectors are used once to update the weights. For batch training all of the training samples pass through the learning algorithm simultaneously in one epoch before weights are updated. We train our model in batches of 128, with an epoch of 500 to provide for good detailed learning process.

We now compare our deep Learning algorithm with shallow learning algorithms like support vector machines (SVM) and convoluted neural networks. For SVMs we exhaustively choose linear, polynomial and non-linear kernel models.

5   Results

Of the 8487 total trial dockets available to us, court decision for only 8385 was known. We took 885 dockets as the test data while the rest 7500 were used to train our deep neural network.

Table 3. Contribution of various features in our model

Feature	Contribution	Feature	Contribution
dateDecision	0.05412011	Jurisdiction	0.05446927
Respondent	0.05412011	caseSourceState	0.05307263
lawType	0.05219972	caseOrigin	0.0523743
issueArea	0.05307263	Petitioner	0.05377095
naturalCourt	0.05272346	lcDisposition	0.05219972
adminAction	0.05097765	lcDisagreement	0.05307263
caseOriginState	0.05202514	certReason	0.05412011
caseSource	0.05272346	Issue	0.0523743
dateArgument	0.10038408	lcDispositionDirection	0.05219972

5.1   Contribution of Each of the Features Towards the Accuracy of Our Model

In order to evaluate the relative contribution of a feature F mentioned in Table 2 in the overall prediction of the result, we evaluate the accuracy by dropping feature F during the testing phase and then calculating the accuracy of the model. Clearly, the lower the accuracy achieved after dropping a feature, the more important it is to the model. Table 3 gives the relative importance of various features. We observe that Date Of Argument plays a major role in determining the outcome of the case.

5.2   Graphs Depicting Classification Error Rate and Accuracy for Models

This section depicts the path followed by our models that lead to their results. For Deep Neural Networks model the classification error's fall is depicted by Fig. 1. Figure 1 also portrays the rise of Deep Neural Networks model's accuracy with training data and epoch until the model starts overfitting and the accuracy deteriorates.

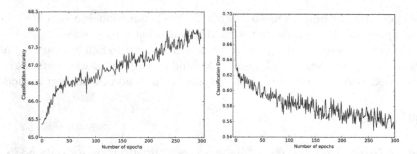

Fig. 1. Figure representing deep neural networks model's training accuracy and classification error

5.3   Accuracy of Various Models and Comparisons

We now present the comparative study of various models we used and the accuracy they generated. We also compare our accuracy with the results provided by Katz 2014, in Table 4. As it is clear from the table our results achieved from Modern Neural Network model with the accuracy of 70.4 % emulate the previous best results in research academia while still being fully predictive and generalized. Our best results were obtained using a Deep Neural Network with Momentum optimization, incorporating the Dropout technique and using Rectilinear Units for non-linearity.

Table 4. Comparison between accuracies of various models

Model	Accuracy
Katz 2014	69.7 %
SVM linear kernel	65.2 %
SVM polynomial kernel	65.5 %
SVM non-linear(rbf) kernel	66.7 %
Deep neural networks (with RMSProp)	69.7 %
Deep neural networks (with Momentum)	70.2 %

6   Conclusion

In this paper we set out to build upon the works of Katz, Bommarito and Blackman 2014 by predicting the behaviour of Supreme Court of United States, using various Machine Learning Models. In the course of our experiment the results achieved from Modern Neural Network model with the accuracy of 70.4 % emulate the previous best results in research academia while still being fully predictive and generalized. The model used Deep Neural Network with Momentum optimization, incorporating the Dropout technique and using Rectilinear Units for non-linearity. Neural networks possess the ability to model and estimate complex functions depending on multitude of factors, and hence provide state-of-the-art accuracy for our experiment, while keeping the model robust, simple and easily extendible as it deploys no feature engineering.

References

Katz, D.M., Bommarito, M.J., Blackman, J.: Predicting the Behavior of the Supreme Court of the United States: A General Approach. Available at SSRN 2463244

Haykin, S., Network, N.: A comprehensive foundation. Neural Networks **2**, 1–3 (2004)

Ruger, T.W., Kim, P.T., Martin, A.D., Quinn, K.M.: The supreme court forecasting project: legal and political science approaches to predicting supreme court decision-making. Columbia Law Rev. **104**(4), 1150–1210 (2004)

Guimera, R., Sales-Pardo, M.: Justice blocks and predictability of us supreme court votes. PloS one **6**(11), e27188 (2011)

Martin, A.D., Quinn, K.M., Ruger, T.W., Kim, P.T.: Competing approaches to predicting supreme court decision making. Perspect. Polit. **2**(04), 761767 (2004)

Bastien, F., Lamblin, P., Pascanu, R., Bergstra, J., Goodfellow, I., Bergeron, A., Bouchard, N., Warde-Farley, D., Bengio, Y.: Theano: new features and speed improvements. In: NIPS 2012 deep learning workshop (2012)

Bergstra, J., Breuleux, O., Bastien, F., Lamblin, P., Pascanu, R., Desjardins, G., Turian, J., Warde-Farley, D., Bengio, Y.: Theano: a CPU and GPU math expression compiler. In: Proceedings of the Python for Scientific Computing Conference (SciPy), 30 June–3 July, Austin, TX (2010)

Tieleman, T., Hinton. G.: Lecture 6.5-rmsprop: divide the gradient by a running average of its recent magnitude. In: COURSERA: Neural Networks for Machine Learning, vol. 4 (2012)

Dauphin, Y.N., de Vries, H., Chung, J., Bengio, Y.: RMSProp and equilibrated adaptive learning rates for non-convex optimization (2015). arXiv preprint arXiv:1502.04390

Srivastava, N., Hinton, G., Krizhevsky, A., Sutskever, I., Salakhutdinov, R.: Dropout: a simple way to prevent neural networks from overfitting. J. Mach. Learn. Res. **15**(1), 1929–1958 (2014)

Nesterov, Y.: A method of solving a convex programming problem with convergence rate O(1/sqr(k)). Sov. Math. Dokl. **27**, 372376 (1983)

Vapnik, V.N.: The Nature of Statistical Learning Theory. Springer, New York (1995)

Harold J.S., Epstein, L., Martin, A.D., Segal, J.A., Ruger, T.J., Benesh, S.C.: Supreme Court Database, Version 2014 Release 01. http://Supremecourtdatabase.org

Glorot, X., Bordes, A., Bengio, Y.: Deep sparse rectifier neural networks. In: International Conference on Artificial Intelligence and Statistics, pp. 315–323

LeCun, Y.A., Bottou, L., Orr, G.B., Müller, K.-R.: Efficient BackProp. In: Orr, G.B., Müller, K.-R. (eds.) NIPS-WS 1996. LNCS, vol. 1524, pp. 9–48. Springer, Heidelberg (1998)

Analysis of Mixed-Rule Cellular Automata Based on Simple Feature Quantities

Naoki Tada and Toshimichi Saito[✉]

Hosei University, Koganei, Tokyo 184–8584, Japan
tsaito@hosei.ac.jp

Abstract. This paper studies cellular automata with mixed rules (MCA) that can generate various spatiotemporal patterns. The dynamics is integrated into the digital return map on a set of lattice points. In order to analyze the dynamics, we present two simple feature quantities of steady state and transient phenomena. In elementary numerical experiments, plentifulness of the dynamics is confirmed.

Keywords: Cellular automata · Digital return maps · Feature quantities

1 Introduction

The elementary cellular automata (ECAs) are simple dynamical system in which time, space, and state are all discrete [1]. Governed by one rule, the ECA can generate a variety of spatio-temporal patterns. Plural patterns can co-exist and ECA can output either one depending on an initial condition. The ECAs have been studied not only as typical systems in fundamental study of nonlinear dynamics but also as basic system for various engineering applications. The applications include signal processing, information compressions, self-replications, and ciphers [2–5].

This paper studies the cellular automata with mixed rules (MCAs). The MCA is governed by combination of two rules and can generate rich phenomena. In order to visualize the dynamics, we introduce the digital return map (Dmap) on a set of lattice points. In order to analyze the dynamics, we introduce two simple feature quantities. The first quantity is the rate of co-existing PEPs. The MCA can have plural periodic patterns and exhibits either of them depending on the initial condition. It can characterize plentifulness of the steady states. The second quantity is the concentricity of initial conditions to the PEPs. This quantity can characterize deviation of the transient phenomena. Note that the feature quantities are applicable not only to MCA but also to various digital dynamical systems including digital spike maps and dynamic binary neural networks [6–8].

In order to perform basic numerical experiments of MCA on a fixed space, we select several rules of ECAs that give very simple dynamics: either no transient phenomena or only one fixed point (steady state). Combining two rules of the selected rules, we construct MCAs. Calculating feature quantities of the MCAs,

© Springer International Publishing Switzerland 2015
S. Arik et al. (Eds.): ICONIP 2015, Part II, LNCS 9490, pp. 484–491, 2015.
DOI: 10.1007/978-3-319-26535-3_55

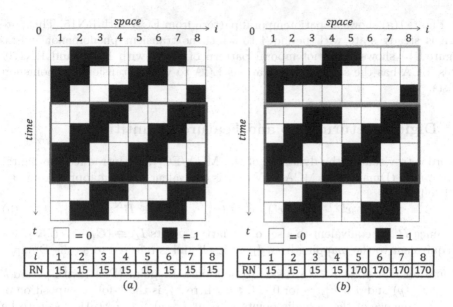

Fig. 1. Examples of spatiotemporal patterns. (a) ECA. (b) MCA.

we show that the MCAs can generate a variety of steady state and transient phenomena. Note that the feature quantities based analysis of MCA/ECA has not been discussed in existing works including [9].

2 Cellular Automata with Mixed Rules

We define the ECA and MCA on the ring of N cells. Let $x_i^t \in \{0,1\} \equiv B$ be the binary state of the i-th cell at discrete time t where $i = 1 \sim N$. The time evolution of x_i^t is governed by the Boolean function of x_i^t and its closest neighbors:

$$x_i^{t+1} = F_i(x_{i-1}^t, x_i^t, x_{i+1}^t), \ i = 1 \sim N \tag{1}$$

where $x_{-1}^t \equiv x_N^t$ and $x_{i+1}^t \equiv x_1^t$ on the ring. F_i is referred to as a rule. In the ECA, F_i does not depend on i ($F_i = F$) and the dynamics is defined by one rule

$$y_0 = F(0,0,0), \ y_1 = F(0,0,1), \ y_2 = F(0,1,0), \ y_3 = F(0,1,1)$$
$$y_4 = F(1,0,0), \ y_5 = F(1,0,1), \ y_6 = F(1,1,0), \ y_7 = F(1,1,1) \tag{2}$$

where $y_j \in B$ and $j = 0 \sim 7$. $(y_0, \cdots, y_7)$ is referred to as a rule table and is equivalent to an 8 bits binary number that is referred to as the rule number (RN) [1]. There exist 2^{2^3} rules for the ECA.

In the MCA, the rule depends on the cell i. However, it is hard to consider all the combinations of the rules. For simplicity, this paper considers combination of two rules: $F_i \in \{F_A, F_B\}$.

Figure 1(a) shows a spatiotemporal pattern from ECA with RN15. This pattern is steady state with period 4 to which no transient phenomenon exists. Figure 1(b) shows a spatiotemporal pattern of MCA with RN15 and RN170. This MCA has the same steady state as ECA to which transient phenomenon exists.

3   Digital Return Map and Feature Quantities

In order to visualize the dynamics of the MCA/ECA, we introduce the digital return map (Dmap). The MCA of N cells is equivalent to the mapping from B^N to itself:

$$x^{t+1} = F_D(x^t), \quad x^t \equiv (x_1^t, \cdots, x_N^t) \in B^N \tag{3}$$

Since B^N is equivalent to a set of 2^N lattice points $I_D \equiv \{C_1, \cdots, C_{2^N}\}$, F_D is equivalent to the mapping from I_D to itself and is referred to as the Dmap.

Definition 1: A point $p \in I_D$ is said to be a periodic point (PEP) with period k if $p = F_D^k(p)$ and $p \neq F_D^l(p)$ for $0 < l < k$ where F_D^k is the k-fold composition of F_D. A sequence of the periodic points $\{F_D(p), F_D^2(p), \cdots, F_D^k(p)\}$ is said to be a periodic orbit (PEO). A PEO with period 1 is referred to as a fixed points.

Since the number of lattice points is finite, the steady state is a PEO(PEP). Note that at least one PEO must exist.

Figure 2 shows examples of Dmaps for ECA and MCA. ECA of RN15 and ECA of RN170 have the same steady state (PEO with period 4) as the MCA of RN15 & 170, however, transient phenomena are different: the ECAs have no transient phenomenon whereas the MCA has it. ECA of RN60 and ECA of RN153 have only one fixed point, however, MCA of RN60 & 153 has various PEOs including PEO with period 4 as shown in Fig. 2(f). In order to consider the dynamics, we introduce two simple feature quantities.

Definition 2: Let N_p be the number of PEPs. The first feature quantity is the rate of PEPs:

$$\alpha = \frac{N_P}{2^N}, \quad \frac{1}{2^N} \leq \alpha \leq 1 \tag{4}$$

α can characterize plentifulness of the steady states.

Definition 3: Let N_p be the number of PEPs. Let M_i be the number of initial points falling into the i-th PEP. The second feature quantity is the concentricity of transition to the PEPs:

$$\beta = \sum_{i=1}^{N_P} \left(\frac{M_i}{2^N}\right)^2, \quad \frac{1}{2^N} \leq \beta \leq 1 \tag{5}$$

β can characterize deviation of the transient phenomena and is based on the results in [6,10]. In Fig. 3(a), the Dmap has one PEO with period 4 (consisting of the 1st to the 4th PEPs) and distributions of M_i ($i = 1 \sim 4, N = 3$) is uniform. In this case, we obtain $\alpha = \frac{4}{2^3}$ and $\beta = 4 \times (2/8)^2 = \frac{1}{4}$. In Fig. 3(b),

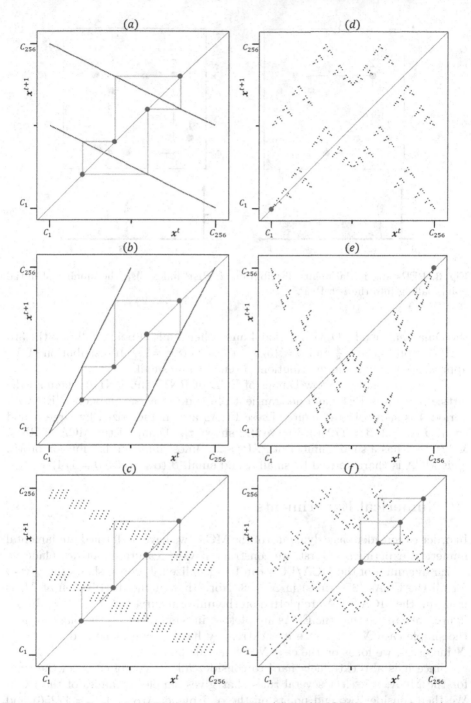

Fig. 2. Dmaps. Red point: PEP; blue orbit: PEO. (a) ECA of RN15, $\alpha = 1$, $\beta = 1/256$. (b) ECA of RN170, $\alpha = 1$, $\beta = 1/256$. (c) MCA (RN15 for $i = 1 \sim 4$, RN170 for $i = 5 \sim 8$). $\alpha = 1/64$, $\beta = 1/4$. (d) ECA of RN60, $\alpha = 1/256$, $\beta = 1$. (e) ECA of RN153, $\alpha = 1/256$, $\beta = 1$. (f) MCA (RN60 for $i = 1 \sim 4$, RN153 for $i = 5 \sim 8$). $\alpha = 1/4$, $\beta = 1/64$ (Color figure online).

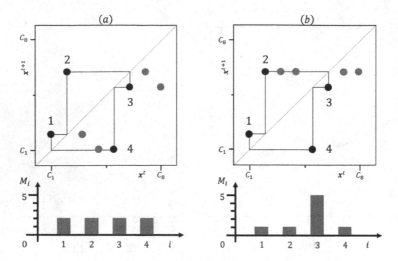

Fig. 3. PEPs and initial points distribution. i: PEP index. M_i: the number of initial points falling into the i-th PEP.

the Dmap has one PEO with period 4 and other 4 initial points fall into the 3rd PEP. We obtain $\alpha = \frac{4}{8}$ and $\beta = (5/8)^2 + 3 \times (1/8)^2 = \frac{28}{64}$. If distribution of M_i approaches the uniform distribution, β tends to be small.

Figure 2(a) and (b) show Dmaps of ECA of RN15 and RN170 where all the lattice points are PEPs and no transient phenomena exist. Only one PEO with period 4 is shown in the figure. These ECAs are characterized by large α and small β ($\alpha = 1, \beta = 1/256$). Figure 2(c) shows the Dmap of the MCA of RN15 & 170 that has a small number of PEPs and almost uniform distribution of M_i. This MCA is characterized by small α and small β ($\alpha = 1/64, \beta = 1/4$).

4   Numerical Experiments

In order to consider basic dynamics of the MCA, we have performed fundamental numerical experiments. First, we construct α-vs-β feature quantities plane on which dynamics of the MCA/ECA can be visualized. Figure 4 shows the α-vs-β for all the ECAs (256 rules) in $N = 8$. Note that, if the distribution of M_i is uniform, the MCA/ECA are plotted on the uniform curve $\alpha\beta = 1/2^N$ [8]. In the figure, we can see that the ECAs are plotted in wide range on the plane even in the simple case $N = 8$. Since it is extremely hard to consider the dynamics as N increases, we focus on the case $N = 8$ in this paper.

Since it is also extremely hard to consider all the combination of the rules for the MCA, we select several rules that gives simple dynamics of the ECA. We then consider two end points on the α-β plane: A($\alpha = 1, \beta = 1/256$) and B($\alpha = 1/256, \beta = 1$). The point A corresponds to the following 8 rules:

$$\text{RS-A} = \{15, 51, 85, 105, 150, 170, 204, 240\} \tag{6}$$

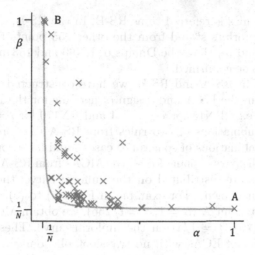

Fig. 4. Feature quantity plane for all the rules of ECAs ($N = 8$). The uniform curve in green (Color figure online).

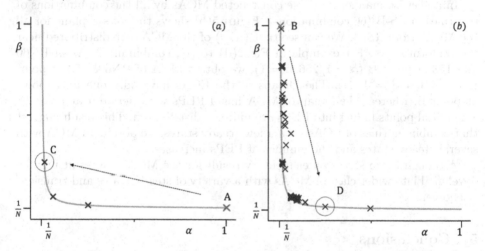

Fig. 5. Feature quantity plane for MCAs (blue crosses, $N = 8$) The uniform curve in green. (a) All the combinations of two rules in RS-A (at point A for ECA). The arrow corresponds to transition in Fig. 2(a) to (c). Point C: Fig. 2(c). (b) All the combinations of two rules in RS-B (at point B for ECA). The arrow corresponds to transition in Fig. 2(d) to (f). Point C: Fig. 2(f) (Color figure online).

This set of the rules is referred to as RS-A in which the RN15 and 170 in Fig. 2 are included. In the RS-A, the ECA has 256 PEPs (all the lattice points are PEPs) and has no transient phenomena. The point B corresponds to the following 12 rules:

$$RS\text{-}B = \{0, 8, 60, 64, 90, 102, 153, 165, 195, 239, 253, 255\} \tag{7}$$

This set of the rules is referred to as RS-B. In the RS-B, the ECA has only one fixed point and orbits stated from the other 255 points fall into the fixed point. Figure 2(d) and (e) show the Dmaps of RN60 and 153 in the RS-B where one fixed point can be confirmed.

Using the rules in RS-A and RS-B, we have constructed MCAs. First, we select two rules from the RS-A and assign either rule for the half of the spaces as shown in Fig. 2, e.g., RN15 for $i = 1 \sim 4$ and RN170 for $i = 5 \sim 8$. We have selected all the combinations of two rules from RS-A (28 combinations), have assigned all the combinations of spaces (70 cases), and have constructed MCAs. Figure 5(a) shows the α-vs-β plane for all the MCAs from RS-A. We can see that (α, β) of the MCAs are distributed on the uniform curve: the MCA can have various transient phenomena. For example in Fig. 2(a) to (c), combining ECAs of RN15 and 170 at point A ($\alpha = 1, \beta = 1/256$), we obtain MCA of RN15&170 at point C ($\alpha = 1/64, \beta = 1/4$) on the uniform curve. These results suggest that combining rules of ECAs with no transient phenomena, we obtain MCA with uniform distribution of transient phenomena and the stability of PEPs is reinforced.

In a likewise manner, we have constructed MCAs by all the combinations of two rules in RS-B (66 combinations). Figure 5(b) shows the α-vs-β plane for all the MCAs from RS-B. We can see that (α, β) of the MCAs are distributed near the uniform curve. For example in Fig. 2(d) to (f), combining ECAs of RN60 and 153 at point B ($\alpha = 1/256, \beta = 1$), we obtain MCA of RN60&153 at point D ($\alpha = 1/4, \beta = 1/64$). The Dmaps of the ECAs have only one fixed point at point B, whereas the Dmap of MCA has 4 PEPs with period 4 at point D. The initial points falling into PEPs are uniform distribution. This result suggest that combining rules of ECAs with a few steady states, we obtain an MCA with several steady states and the number of PEPs increases.

Although Fig. 5 shows an elementary result for the MCA, the result may be developed into wider class of MCAs with a variety of steady states and transient phenomena.

5   Conclusions

In order to consider basic dynamics of the MCAs, two basic feature quantities (α and β) are introduced in this paper. Constructing the α-vs-β plane for all the ECAs in $N = 8$, variety of dynamics is demonstrated. Extracting rules from two end points A and B in the plane of ECAs, we have constructed MCAs of two rules. Constructing the α-vs-β planes for the MCAs, we have confirmed that (α, β) is distributed on and near the uniform curve: MCAs from point A (no transient phenomena) can have various transient phenomena and MCAs from point B (only one fixed point) can have various PEPs.

Future problems include analysis of the dynamics of various MCAs, synthesis of desired MCAs, and engineering applications.

References

1. Chua, L.O.: A Nonlinear Dynamics Perspective of Wolfram's New Kind of Science, I, II. World Scientific, Singapore (2005)
2. Wada, W., Kuroiwa, J., Nara, S.: Completely reproducible description of digital sound data with cellular automata. Phys. Lett. A **306**, 110–115 (2002)
3. Seredynski, M., Bouvry, P.: Block cipher based on reversible cellular automata. In: Proceedings of IEEE/CEC, pp. 2138–2143 (2004)
4. Rosin, P.L.: Training cellular automata for image processing. IEEE Trans. Image Process. **15**(7), 2076–2087 (2006)
5. Lohn, J.D., Reggia, J.A.: Automatic discovery of self-replicating structures in cellular automata. IEEE Trans. Evol. Comput. **1**(3), 165–178 (1997)
6. Yamaoka, H., Horimoto, N., Saito, T.: Basic feature quantities of digital spike maps. In: Wermter, S., Weber, C., Duch, W., Honkela, T., Koprinkova-Hristova, P., Magg, S., Palm, G., Villa, A.E.P. (eds.) ICANN 2014. LNCS, vol. 8681, pp. 73–80. Springer, Heidelberg (2014)
7. Moriyasu, J., Saito, T.: A deep dynamic binary neural network and its application to matrix converters. In: Wermter, S., Weber, C., Duch, W., Honkela, T., Koprinkova-Hristova, P., Magg, S., Palm, G., Villa, A.E.P. (eds.) ICANN 2014. LNCS, vol. 8681, pp. 611–618. Springer, Heidelberg (2014)
8. Yamaoka, H., Saito, T.: Analysis of digital spike maps based on simple feature quantities. In: Proceedings of NOLTA (2015, to appear)
9. Sawayama, R., Saito, T.: Evolutionary learning and stability of mixed-rule cellular automata. In: Loo, C.K., Yap, K.S., Wong, K.W., Teoh, A., Huang, K. (eds.) ICONIP 2014, Part II. LNCS, vol. 8835, pp. 271–278. Springer, Heidelberg (2014)
10. Amari, S.: A method of statistical neurodynamics. Kybernetik **14**, 201–215 (1974)

Depth Map Upsampler Using Common Edge Detection

Soo-Yeon Shin[✉] and Jae-Won Suh

College of Electrical and Computer Engineering, Chungbuk National University,
Chungdae-ro 1, Seowon-gu, Cheongju, Chungbuk, Korea
{ssy8992,sjwon}@cbnu.ac.kr

Abstract. In this paper, we present an improved upsampling technique for depth map. The proposed edge preserving depth map upsampler is based on a common edge region of depth map, segmentated color image, and HSV color image. First, we interpolate the depth map by using bilinear interpolation to make a high resolution image. Next, if the interpolated pixel in a higher resolution image belongs to the common edge region, it should be cost calculated by using neighboring nine candidate pixels. The cost calculation function jointly uses spatial, color, and range weighting functions. Simulation results show that the proposed algorithm outperforms compared with four existing upsamplers.

Keywords: 3D · Depth image · Upsampler · Bilateral filtering

1 Introduction

In recent years, depth map cameras has been developed but its resolution is limited. For instance, an high definition (HD) sequence has a relatively large resolution of 1920 × 1080 but resolution of depth map is from 200 × 200 to 640 × 480. In addition, the depth map generates significant misalignments between the depth map and color image [1]. Several depth map upsampling methods have been proposed to solve the problem [2–7].

In general, upsampling methods can be categorized into two classes. One category is simple interpolation methods such as the nearest neighbor or bilinear interpolation [2]. Although interpolation methods are simple and fast, the result shows blurring effects or jagging effects around the edge region. These errors make serious misalignment when synthesize multi-view images.

The other category is based on bilateral filter [3]. The bilateral filter is one of the most widely used smoothing filters. The bilateral filter uses both a spatial filter kernel and a range filter kernel evaluated on the data values. Petschnigg *et al.* [4] have introduced the concept of a joint bilateral filter (JBF) to improve the quality of a non-flash image using its associated flash image. Kopf *et al.* [5] proposed depth upsampling method using a joint bilateral upsampling (JBU) filter. The bilateral filter is the spatial smoothing kernel by multiplying with a color similarity term, which try to yield an edge preserving interpolation.

© Springer International Publishing Switzerland 2015
S. Arik et al. (Eds.): ICONIP 2015, Part II, LNCS 9490, pp. 492–499, 2015.
DOI: 10.1007/978-3-319-26535-3_56

This approach takes advantage of high resolution color information to preserve edge region. Yang *et al.* [6] used joint bilateral filtering to interpolate the high-resolution depth values. Since filtering can often over smooth the interpolated depth values, especially along the depth discontinuity boundaries, they quantized the depth values into several discrete layers. However in case that depth edges are corresponding to homogenous color areas, it causes blurring artifacts at edge region. Kopf *et al.* [5] and Yang *et al.* [6] suffer from texture copying and edge blurring artifacts. Texture copying occurs in smooth areas with noisy depth data and textures in the color image, while edge blurring occurs in transition areas if different objects have similar color. In addition, fast edge-preserving depth map upsampling algorithm (FEDU) [7] has been presented to suppress visual artifacts fabricated from useless color data. This method successivly preserved edge region also show an improvement of computing time. However if some depth edge regions are not belong to common edge region, it makes blurring artifacts.

In this paper, we propose a depth map upsampler based on joint bilinear filtering method. Our algorithm essentially uses bilinear interpolation, although a bilinear interpolation method causes blurring effects at the edge region. Therefore, preserving the sharpness of the edge regions by cost calculation was attempted. First, the depth map was resized using bilinear interpolation. Second, the common edge region was detected by using a depth map, color image segmentation, and HSV space image. Third, the cost of each of the nine neighbor candidates was calculated by using joint bilateral weights. Finally, the pixel value was changed to the minimum cost pixel among nine candidates. The paper is organized as follows: Sect. 2 describes the proposed algorithm in detail. Next, experimental results are compared with other algorithms in Sect. 3. Finally, we conclude with an overall discussion in Sect. 4.

2   Proposed Depth Map Upsampler

2.1   Depth Upsampler

The proposed algorithm increases the resolution of the depth map to make it to fit the size of the color image. Our upsampling method is an edge-preserving depth map upsampler. In this section, our depth map algorithm framework is introduced. An overview of our depth map upsampling framework is provided in Fig. 1.

- Step 1: We upsample the low resolution depth map D_p^l, to be the same size as the color image resolution using bilinear interpolation.
- Step 2: We define the common edge regions E_C.
- Step 3: If an interpolated pixel p, belongs to E_C, nine candidates $q_1 \sim q_9$ are selected, and their costs are calcuated by bilateral filter weights, based on spatial weight function, color weight function, and range weight function.
- Step 4: We determine the minimum depth value q_x among nine candidates.
- Step 5: The upsampled depth map D_p^h at p is replaced by the depth value U_{qx}^h at q_x. If p is out of E_C, D_p^h is assigned by U_p^h directly.

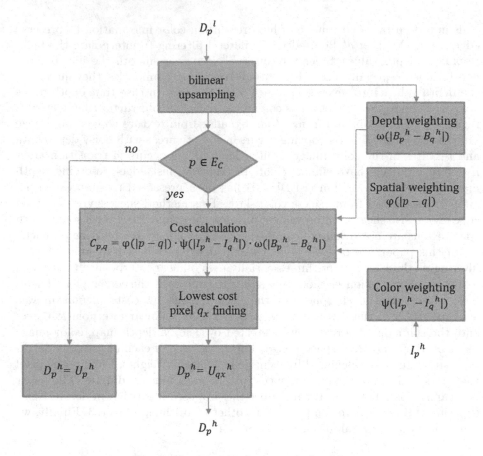

Fig. 1. Framework of proposed algorithm

2.2 Common Edge Detection

The common edge detection technique is suggested for detecting accurate edge region. Common edge regions are defined by the intersected edge areas of plural dilated edge maps generated from depth map and its color image. FEDU [7] uses RGB color space of a color image. However, RGB color space rarely detects correct edge because it is strongly influenced by intensity. Therefore we use HSV color space instead of the RGB color space. The HSV color space can provide a correct edge region based on clear color difference. Also we add an additional edge region, which detected from segmentated color image. An overview of the proposed common edge detection algorithm is provided in Fig. 2.

- Step 1: We down sample the color image I_p^h as the same size as the depth map D_p^l.
- Step 2: Segment color information for detecting accurate color edge.
- Step 3: We find edge regions from segmentated color image, HSV space of the color image and the depth map, respectively.

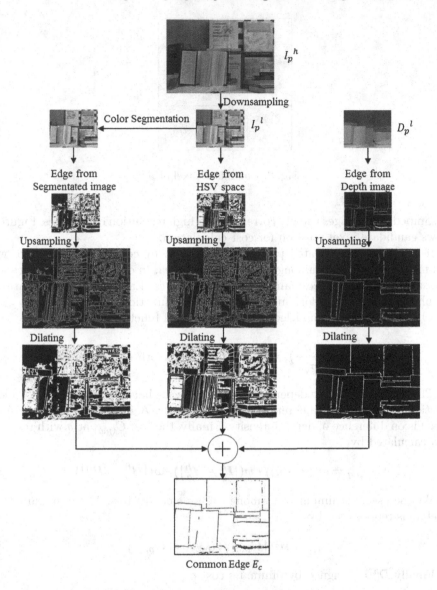

Fig. 2. Proposed common edge detection

- Step 4: We upsample each edge images and dilates upsampled edge images.
- Step 5: We find the common edge region E_C, which commonly exists at each dilated edge region.

2.3 Cost Calculation

After the common edge region is detected, we seek minimum cost among nine candidates. The cost calculation refers to color data under the assumption that

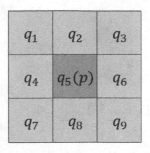

Fig. 3. Candidate pixel of p

uplsamped depth edges usually correspond to high resolution color edges. Figure 3 shows candidate pixel position for cost calculation.

If a bilinear interpolated pixel U_p^h is included by common edge region, we find most low cost pixel among the 3×3 pixel area. In cost calculation method, we seek the minimum cost among nine candidates. Each cost of candidate is calculated by spatial, color, and range weighting function φ, ψ, and ω.

The weightings are modeled by an exponential function like (1) [3].

$$\varphi(n) = exp(\frac{-n^2}{\sigma_\varphi}), \quad \text{where the } \sigma_\varphi \text{ is smooth parameter of } \varphi. \tag{1}$$

The spatial fuction φ depends on pixel distance between p and q. The color function ψ depends on difference of color intensity. Also, the range function ω depends on difference of depth intensity. Finally, the cost $C_{(p,q)}$ at q with respect to p calculated by

$$C_{p,q} = \varphi(|p - q|) \cdot \psi(|I_p^h - I_q^h|) \cdot \omega(|B_p^h - B_q^h|) \tag{2}$$

We then seek minimum cost among the nine candidates. The minimum cost pixel q_x is represented by

$$q_x : min\{C_{p,q_1}, C_{p,q_2} \ldots C_{p,q_9}\} \tag{3}$$

Finally D_p^h is assigned by minimum cost q_x.

$$D_p^h = B_q^h. \tag{4}$$

3   Experimental Results

The simulation results for depth map upsampling are compared: proposed algorithm, bilinear interpolation (BI) [2], bilinear interpolation with bilateral filter (BF) [3], bilinear interpolation with joint bilateral filter (JBF) [4] and fast edge preserving depth upsampler (FEDU) [7]. To evaluate the performance of the proposed upsampler, we tested with a synthetic image set having ground truth

Table 1. PSNR Comparison (UNIT:dB, Downsampling factor:4)

Images	BI [2]	BF [3]	JBF [4]	FEDU [7]	Proposed
Art	32.71	32.91	33.41	32.82	33.22
Barn	38.99	38.34	39.33	40.18	40.23
Flowerpots	29.02	29.15	29.40	29.06	28.04
Teddy	30.45	30.62	30.85	30.34	30.95
Sawtooth	36.94	36.58	38.94	37.12	40.79
Venus	41.64	40.83	42.80	41.74	44.27
Avg.PSNR	34.95	34.73	35.78	35.21	36.25

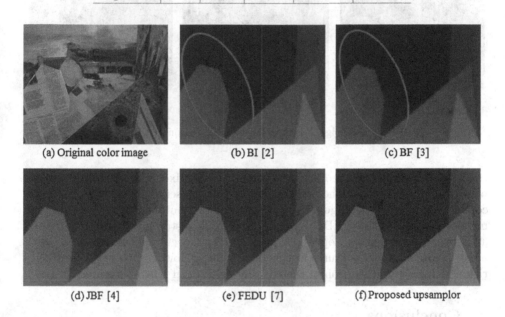

(a) Original color image   (b) BI [2]   (c) BF [3]

(d) JBF [4]   (e) FEDU [7]   (f) Proposed upsamplor

Fig. 4. Blurred edge region of upsampled depth maps (Color figure online)

depth data [6]. These test data are Art, Barn, Flowerpots, Teddy, Sawtooth, and Venus. Each ground truth depth map is downsampled by a factor 4 to generate input low-resolution depth maps. For objective evaluation, we compare PSNR values between the upsampled depth map and ground truth depth map. In the simulation, the cost function parameters in (4) were set by $\sigma_\varphi = 2$, $\sigma_\psi = 0.1$, $\sigma_\omega = 0.1$. Also, we use the canny edge detector for common edge detection. The parameters for the canny edge detector were set by $T_1 = 10$, $T_2 = 100$.

Table 1 shows the test results of the PSNR comparison. These results indicate that the proposed upsampler has higher PSNR, as much as about 1.29 dB, 1.21 dB, 0.46 dB, 1.04 dB than BI [2], BF [3], JBF [4], and FEDU [7] on average respectively. Figure 4 shows blurring effect has occurred in BI [2], BF [3] and JBF [4] methods. FEDU [7] reduces blurring effect at the edge region, but

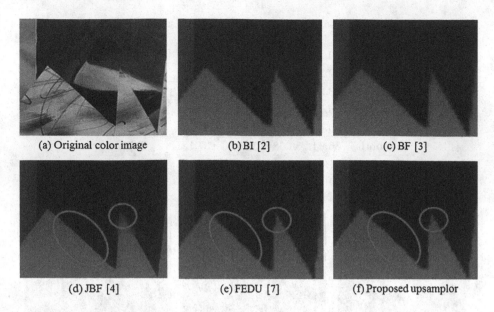

(a) Original color image   (b) BI [2]   (c) BF [3]

(d) JBF [4]   (e) FEDU [7]   (f) Proposed upsamplor

Fig. 5. Distortion of edge region

there are remaining blurred artifacts when object pixel value is similar to the background. In contrast, the proposed algorithm preserves sharpness of edge by correct edge detection. Figure 5 shows edge distortion occurred by upsampling methods. JBF [4] and FEDU [7] generated edge distortion in the highlighted area because of texture copying. These artifacts occur in smooth areas with noisy depth map and textures in the color image. However, the proposed algorithm reduces the distortion of the edge by additional cost operation.

4   Conclusions

In this paper, an upsampling method for depth maps was presented. The upsampler adapts the novel common edge detector and additional cost calculation. The performance of the common edge detector is increased by adding HSV color space information. In addition, the temporal common edge pixel is regenerated by cost calculation. Therefore, the blurring effects in edge region are reduced. As shown in the experimental results, the proposed algorithm produce high PSNR value compared with four of the other algorithms while keeping edge sharpness without edge distortion. When applied on depth map signals for 3D media, spatial consistency of the depth map image relates to virtual view synthesis, and therefore the improved depth map upsampler will improve the quality of multi-view 3D videos.

Acknowledgment. This research was supported by Basic Science Research Program through the National Research Foundation of Korea (NRF) funded by the Ministry of Education (2014R1A1A2057662)]

References

1. Merkle, P., Müller, K., Wiegand, T.: 3D video: acquisition, coding, and display. IEEE Trans. Consum. Electron. **56**(2), 7–42 (2010)
2. Gonzalez, R.C., Woods, R.E.: Digital Image Processing. PrenticeHall, New Jersey (2002)
3. Tomasi, C., Manduchi, R.: Bilateral filtering for gray and color images. In: Proceedings of the IEEE International Conference on Computer Vision, pp. 839–846 (1998)
4. Petschnigg, G., Agrawala, M., Hoppe, H., Szeliski, R., Cohen, M., Toyama, K.: Digital photography with flash and no-flash image pairs. ACM Trans. Graph. **23**(3), 664–672 (2004)
5. Kopf, J., Cohen, M.F., Lischinski, D., Uyttendaele, M.: Joint bilateral upsampling. ACM Trans. Graph. **26**(3), 1–6 (2007)
6. Yang, Q., Yang, R., Davis, J.: Spatial-depth super resolution for range images. In: Proceedings of In CVPR, pp. 1–8 (2007)
7. Kim, S.Y., Ho, Y.: Fast edge-preserving depth image upsampler. IEEE Trans. Consum. Electron. **21**(32), 1176–1190 (2012)
8. Scharstein, D., Szeliski, R.: A taxonomy and evaluation of dense two-frame stereo correspondence algorithms. Int. J. Comput. Vis. **7**(1), 7–42 (2002)

Application of Mathematical Morphology in Diesel Vibration Signals

Hongxia Pan[✉], An Dong, and Manliang Cao

School of Mechanical and Power Engineering, North University of China,
Xueyuan Road No. 3, Taiyuan 030051, China
panhx1015@163.com

Abstract. In this paper, according to the characteristics of the diesel engine vibration signal, the design for a class of adaptive generalized morphological filter is applied to the noise reduction of diesel engine vibration signal. After pre-processing the diesel engine vibration signal, then it is designed the shape, width and height of the structure elements. This paper, according to the characteristics of the diesel engine noise, chooses the semicircle structural elements and calculates the local maximum signal sequence and local minimum signal sequence so as to strike the height and width of the structural elements, then uses the gradient method to find the weight value adaptive. As a result, it will make noise reduction to achieve effectively, and the result has some superiorities compared with the traditional wavelet noise reduction.

Keywords: Morphological filtering (MF) · Structural elements · De-noise · Vibration signal

1 Introduction

Mathematical morphology has always been applied to digital image processing technology since it was proposed, and now in various aspects like language, electric power, ecg and eeg signals has applied successfully. Mathematical morphology in one-dimensional mechanical vibration signal is at the beginning stage. In recent years, mathematical morphology has got more and more research results in mechanical fault diagnosis of vibration signal processing. An Liansuo proposed on - off and closed - composite morphological filter which is used for the purification of rotor axis trajectory [1]. Zhang Lijun used form closed operation method to extract the impact of the gear fault feature [2]. Hu Aijun used combinatorial mathematics morphology filter to reduce the rotating machinery vibration signal noise [3]. Du Biqiang used morphology filter(MF) to process rotating machinery vibration signal [4]. Zhang Wenbin used adaptive morphological filtering method to purify the rotor axis trajectory [5]. Shen Lu used generalized morphological filter to reduce rotating machinery vibration signal noise. But, MF technology applied in diesel engine vibration signal is relatively rare. At present, the morphology applied in the vibration signal of engine has obtained certain research results [7, 8]. This paper, according to the characteristics of the diesel engine vibration signal, has explored the application of morphology in the diesel engine vibration signal de-noising.

© Springer International Publishing Switzerland 2015
S. Arik et al. (Eds.): ICONIP 2015, Part II, LNCS 9490, pp. 500–509, 2015.
DOI: 10.1007/978-3-319-26535-3_57

2   Basic Principle of Mathematical Morphology Filter

Mathematical morphology abandons the traditional view of numerical modeling, and uses the collection to depict and analyze the object, and uses *probe* (structural elements) to detect the target signal of each location, so as to obtain the geometry information and the relationship between them. This method does not need to analyze the signal frequency domain information, and its algorithm is simple with fast computing speed. The collected mechanical vibration signals in the experiments is generally a one-dimensional discrete time sequence, gray value is done in the morphological transform. Common morphological operators are corrosion, expansion, opening operation and closing operation and cascade combination between them. $F = (0, 1, \cdots, N - 1)$ whose discrete function is a vibration signal $f(n)$, $G = (0, 1, \cdots, M - 1)$ whose discrete function is structural elements $g(n)$, $N \gg M$. $f(n)$ and $g(n)$ whose corrosion and expansion operations are respectively defined as:

$$(f \Theta g)(n) = \min[f(n + m) - g(m)] \tag{1}$$

$$(f \oplus g)(n) = \max[f(n - m) + g(m)] \tag{2}$$

By (1) and (2) can lead to morphological open operation and closing operation, whose mathematical expressions are respectively defined as:

$$(f \circ g)(n) = (f \Theta g \oplus g)(n) \tag{3}$$

$$(f \bullet g)(n) = (f \oplus g \Theta g)(n) \tag{4}$$

From (1) to (4), min is minimum operation, max is maximum operation, Θ is corrosion operation, $\oplus$ is Expansion operation, $\bigcirc$ is open operation, $\bullet$ is close operation. Based on morphological open operation and close operation can also build three types of filter: alternately filter, hybrid filter and alternating hybrid filter [9].

Alternate filter has two kinds, form on-off and form off-on, whose mathematical expressions are respectively as follows:

$$[(f)oc(g)](n) = (f \circ g \bullet g)(n) \tag{5}$$

$$[(f)co(g)](n) = (f \bullet g \circ g)(n) \tag{6}$$

Hybrid filter is the arithmetic average of open operation and close operation, whose formula is defined as follows:

$$[(f)hf(g)](n) = (f \circ g + f \bullet g)(n)/2 \tag{7}$$

Alternating hybrid filter its formula is defined as follows:

$$[(f)ah(g)](n) = [(f)oc(g) + (f)co(g)](n)/2 \tag{8}$$

Form (5) to (8), *oc* is the form on-off alternate filtering operation, *co* is the form off-on alternate filtering operation, *hf* is the hybrid filter operation, ah is the alternating hybrid filter operation.

3   Choice of Structural Elements

Structural elements are the most important factors that influence the effect of morphological filtering. Structural elements in morphological transform is similar to a filter
window or referential template. Three elements of the structural elements: shape, width
(the width of the domain of definition) and height (amplitude). When the selected structural elements and the processing signals match well, the morphological filtering
performs well.

3.1   Shape of Structural Elements

In the morphological transform, there are many kinds of types of structural elements.
Common type of structural elements: flat shape (straight), triangle (upper triangle or
lower triangle), sinusoid, semicircle (the upper half of circle or the lower half of circle)
and so on. In general, the selection of structural elements is made according to the characteristics of the signal. Relevant researches showed that in the procedure of vibration
signal de-noising, structural elements of the flat shape (straight), semicircle and triangle
type can obtain good filtering effects. Flat-shape-type structural element can hold the
shape characteristics of the signal, semicircle-type structural element makes for reducing
the interference of random noise, and triangle-type structural element has a good effect
when processing the interference of pulse signal. There are some random noise when
diesel engine running. According to the characteristics of the signal, semicircle-type
structural elements is selected in this paper, of which formula is as shown in (9).

$$g(i) = H \times \sqrt{1 - (\frac{2i}{L-1})^2}, (i = -\frac{L-1}{2}, \cdots, \frac{L-1}{2}) \tag{9}$$

The width of the structural elements L takes for odd number commonly, as the peak
of such structural elements can be located at the origin. H is the height of the structural
elements.

3.2   Width of Structural Elements

The vibration signal of the diesel engine $X = \{x_i/i = 1, 2, \ldots, N\}$, in which N is the
amount of data of the original signal. Remove the trend item of original signals, and preprocess by low-pass filter and calculate the local maximum value and the local minimum
value of the signal sequence. The local maximum value sequence is denoted as
$PE = \{x_i|i = 1, 2, \ldots, N_{PE}\}$, and the local minimum value sequence is denoted as
$NE = \{x_i|i = 1, 2, \ldots, N_{NE}\}$, in which N_{PE} and N_{NE} is the number of local maximum and
small values respectively. Let the interval of the local maximum value
$D_p = \{d_p|d_{pi} = PE_{i+1} - PE_i, i = 1, 2, \ldots, N_{PE} - 1\}$, and the interval of the local minimum
value $D_N = \{d_N|d_{Ni} = NE_{i+1} - NE_i, i = 1, 2, \ldots, N_{NE} - 1\}$. According to the maximum
interval and minimum interval of the elements, the maximum and minimum values of
the width of the structural elements can be calculated:

$$K_{l\max} = \max(\lfloor \max(d_{Pi}) - 1)/2 \rfloor, \lfloor \max(d_{Ni}) - 1)/2 \rfloor)$$
$$K_{l\min} = \min(\lceil \min(d_{Pi}) - 1)/2 \rceil, \lceil \min(d_{Ni}) - 1)/2 \rceil).$$

In the formula, $\lfloor \bullet \rfloor$, $\lceil \bullet \rceil$ is the operation of rounding down and rounding up respectively.

Then, the sequence of the width of structure elements is:

$$K_l = \left\{ K_{l\min}, K_{l\min} + 1, \cdots, K_{l\max} - 1, K_{l\max} \right\}$$

The scale sequence of the width of the structural elements in the upper formula can be used to local feature of the signal, thus avoiding the blindness and the uncertainty in the selection of the width of the structure elements.

3.3 Height of Structural Elements

In general, the structural elements of the height can be selected according to the experience, it is appropriate that in the vibration signal the height takes usually 1 %–5 % of the height of the original signal (profile of main wave), but such a value is with a certain blindness in experience. Then, the local maximum and local minimum of the signal is defined, and can be used to determine the height of the structural element. Set the maximum value and minimum value of amplitudes of the local maximum value sequence for $P_{P\max}$, $P_{P\min}$, and similarly the maximum value and minimum value of amplitudes of local minimum value sequence for $P_{N\max}$, $P_{N\min}$. The difference in height between local maximum and local minimum of the signal $H_{Pe} = P_{P\max} - P_{P\min}$, $H_{Ne} = P_{N\max} - P_{N\min}$, the amplitude of local extremum of the signal $H_e = \max(H_{Pe}, H_{Ne})$. Let the height sequence of the structural elements H_l and the width of the scale K_l correspond, and define the height sequence H_l as:

$$H_l = \left\{ \alpha[H_e/(K_{l\max} - K_{l\min} + 1) + (j - 1)H_e/(K_{l\max} - K_{l\min} + 1)] \right\}$$
$$j = 1, 2, \cdots, K_{l\max} - K_{l\min} + 1 \tag{10}$$

In the formula, α is a high proportion factor, generally can be based on practical experience value.

4   Design of Generalized Morphological Filter

Corrosion arithmetic eliminates the positive pulse and enhances the negative pulse; expansion operation eliminates the negative pulse and enhances positive pulse; open operation eliminates the positive pulse and reserves the negative pulse; close operation eliminates the negative pulse but reserves the positive pulse. Form closed operation has extensibility, while form open operation has the contrary extensibility. In addition to eliminate the negative pulse at the same time also strength the positive pulse, the open operation cannot effectively filter all of the positive pulse, nor can form on-off operation efficiently filter out all the negative pulse. Therefore, the traditional morphological filters can't completely filter out noise. With the lack of a priory knowledge of signal, so it is difficult to obtain the signal characteristics. Diesel engine contains a variety of noise signal, the distribution of noise in the signal is often uneven. Therefore, in order to overcome this shortcoming here uses the different scales of structural elements of the diesel engine signal composite filtering. Through the signal of local maximum value and the local minimum value, it determines the scope of the structural elements of the width and height, according to the characteristics of the signal itself to select structure element. Then, construct three kinds of generalized morphological filter (GMF).

(a) Morphological corrosion and inflation mean filter:

$$G_1(f) = \frac{1}{4}(f\Theta g_1 + f \oplus g_1 + f\Theta g_2 + f \oplus g_2)(n) \tag{11}$$

(b) Morphology and closed mean filter:

$$G_2(f) = \frac{1}{2}(f\Theta g_1 \oplus g_2 + f \oplus g_1\Theta g_2)(n) \tag{12}$$

(c) Morphological on-off and off-on mean filter:

$$G_3(f) = \frac{1}{2}(f\circ g_1 \bullet g_2 + f \bullet g_1\circ g_2)(n) \tag{13}$$

Above the formulas, $f(n)$ is the input of discrete signal, $G1$, $G2$ whose structural element functions respectively are $g1, g2, g1 \subseteq g2, G_1 \subseteq G_2$.

This paper applies adaptive generalized morphological filter method to the noise reduction of diesel engine vibration signal. First, preprocess the diesel engine vibration signal. Diesel engine R6105AZLD with speed of 1200 r/min, in normal working condition of a cycle of the vibration signal is analyzed. For four-stroke diesel engine, the crankshaft turns two laps for each cycle; the sampling frequency is 48 kHz, a period of time is $T = 2 \times 60/1200 = 0.1$ s, each cycle analysis points reach to 4800 points. Eliminate trend item by the least squares method, and then filter part of the high frequency signal through a low-pass filter so as to preprocess the original signal.

The maximum interval and minimum interval of local maximum value and the local minimum values are as follows:

$$max\left(D_p\right) = 19, max\left(D_n\right) = 26, min\left(D_p\right) = 2, min\left(D_n\right) = 2$$

The width of the structural elements for scale sequence is $Kl = [1\ 2\ 3\ 4\ 5\ 6\ 7\ 8\ 9\ 10\ 11\ 12\ 13]$. The width of structural elements has the largest influence on filtering effect. In this paper, after the normal signal was tested a few times, when the width of the semi-circular structure elements was $L1 = 3, L2 = 5$, it achieved the best noise reduction effect. At the same time, it calculated the local extreme amplitude of signal is $He = 397.1551$, spacing is $Klmax - Klmin + 1=13$.

The height of the structure element has a little influence on the filter effect. It is appropriate to take the vibration signal height from 1 % to 5 %. In this paper, the main outline of the signal is between 50 ~ 150, and its structural elements height generally between $0.5–7.5$ is more appropriate. Therefore, the value of α is general according to the experience. This article α is 0.03, the width of the scale sequence for the corresponding height sequence is as follows:

H_l = [0.9165 1.8330 2.7495 3.6660 4.5825 5.4990 6.4155 7.3320 8.2486 9.1651 10.0816 10.9981 11.9146].

The widths of the two structural elements are L1 = 3, L2 = 5, corresponding heights are *H1 = 2.75, H2 = 4.58*.

The above three kinds of average filter in signal de-noising effect evaluation index [10] are shown in Table 1. The signal-to-noise ratio (SNR) formula and root mean square error (RMSE) formula are as follows:

$$SNR = 10\log \frac{\frac{1}{N}\sum_{k=1}^{N} x^2(k)}{\frac{1}{N}\sum_{k=1}^{N} [x(k) - y(k)]^2} \tag{14}$$

$$RMSE = \sqrt{\frac{1}{N}\sum_{k=1}^{N} [x(k) - y(k)]^2} \tag{15}$$

Table 1. Noise reduction effect of three kinds of filter

Evaluation index	Corrosion and inflation average filter	Open and close mean filter	On-off and off-open filter
SNR	14.3299	18.6264	16.6117
RMSE	15.4076	12.4265	13.7407

From Table 1, it can come to a conclusion that open and close mean filter has the best de-noise effect, whose signal-to-noise ratio is maximum and root mean square error is minimal. So this paper chosen open and close mean filter.

Because it is the mean filter, its weight coefficient as a fixed value 0.5; if such weight in the process of filter coefficient remain fixed, it will make filtering is not adaptive to achieve the best effect. In order to make the filter to achieve the optimal effect, this article used the Gradient method [11] (Gradient) to determine the optimal weight value.

The actual input of the original signal is $x(n) = s(n) + i(n)$, (n = 0,1, …, N − 1), in which $s(n)$ is the ideal signal, $i(n)$ is the noise signal. $e(n)$ is the error signal, which is the ideal signal minuses the output signal of filter, namely $e(n) = s(n) - y(n)$. The formulas of the generalized morphological opening and closing are as follows:

$$y_1(n) = G_o = (x \ominus g_1 \oplus g_2)(n) \tag{16}$$

$$y_2(n) = G_c = (x \oplus g_1 \ominus g_2)(n) \tag{17}$$

So the output signal of filter is:

$$y(n) = a_1(n)y_1(n) + a_2(n)y_2(n) = \sum_{i=1}^{2} a_i(n)y_i(n) \tag{18}$$

Mean square error of the output signal is:

$$E[e^2(n)] = E[|s(n) - y(n)|^2] = E\left[\left|s(n) - \sum_{i=1}^{2} a_i(n)y_i(n)\right|^2\right] \tag{19}$$

Here by gradient method to gradually modified weight coefficient, it makes the output signal in the minimum mean square error (LMS) condition so as to reach the ideal signal. Here takes a single sample error square of $e^2(n)$ to estimate the mean square error $E[e^2(n)]$, $e^2(n)$ to the weight coefficient of the gradient is defined as:

$$\nabla e^2(n) = \left[\frac{\partial[e^2(n)]}{\partial a_1(n)} \frac{\partial[e^2(n)]}{\partial a_2(n)}\right]^T = [T_1 \ T_2]^T \tag{20}$$

Based on the gradient method of adaptive morphological filtering iterative calculation process is as follows:

(a) To process the original signal by generalized morphological opened and generalized morphological closing operation, it will get $y_1(n)$ and $y_2(n)$, after calculated by k times, the k iteration filter output is:

$$y^k(n) = a_1^k(n)y_1(n) + a_2^k(n)y_2(n) = \sum_{i=1}^{2} a_i^k(n)y_i(n) \tag{21}$$

In which $a_i^k(i = 1, 2)$ is the weight coefficient of k iteration

(b) A single sample of error is:

$$[e(n)]^k = y^{k-1}(n) - y^k(n) \tag{22}$$

Here $y^{k-1}(n)$ replaces $s(n)$, so the gradient is:

$$[\nabla e^2(n)]^k == [T_1^k \ T_2^k]^T \tag{23}$$

(c) Direction vector and weight are as follows:

$$p^k = [\ p_1^k \ p_2^k \ ]^T \tag{24}$$

$$p_i^k = -T_i^k + \beta_i^k p_i^{k-1} \tag{25}$$

p_i^k is the *ith* component of direction vector of the k iteration, $i = 1, 2$.

(d) The coefficient is:

$$\beta_i^k = \frac{\left(T_i^k\right)^T T_i^k}{\left(T_i^{k-1}\right)^T T_i^{k-1}} \tag{26}$$

Weight coefficient is:

$$a_i^{k+1} = a_i^k + \mu p_i^k \tag{27}$$

In the formula, if μ, the step length is too large, it will cause a oscillation during the convergence then lead to instabilities of the system; and if μ is too small, the convergence rate of the system is affected. In this paper, μ is set to 0.05 through tests.

k ($k = 0, 1, 2, \cdots$) is the number of iterations. Initial weight coefficient $a_i = 0.5$. Set $\beta_i^0 = 0$, then $p_i^0 = -T_i^0$ ($i = 1, 2$). In the first round of iteration, set $[e(n)]^0 = x(n) - y^0(n)$. Use formula (21) to (27) to adaptive correct weight coefficient. The schematic diagram of adaptive generalized morphological filter based on gradient method is shown in Fig. 1.

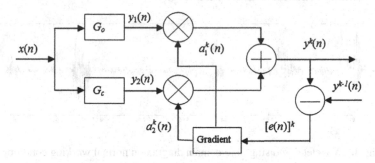

Fig. 1. Schematic diagram of adaptive generalized morphological filter based on gradient method

Adaptive adjust weight according to gradient method, when a1 = 0.5110 a2 = 0.489 is obtained, the minimum mean square error (MSE) is 154.40. The relationship between the weights A1 and the mean square error is shown in Fig. 2. Time domain diagram of adaptive morphological filtering in normal condition is shown in Fig. 3.

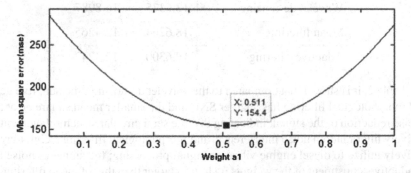

Fig. 2. Relationship between the weights A1 and MSE

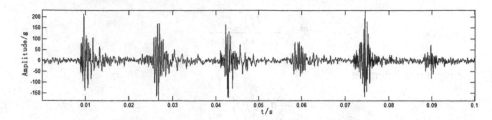

Fig. 3. Time domain diagram of adaptive MF in normal condition

In order to demonstrate the effect of adaptive morphological noise reduction, the wavelet de-noising is used in this paper to conduct a comparison. First, wavelet threshold function wbmpen is used to obtain adaptive threshold; then wavelet threshold de-noising function wdencmp is used to de-noise the pre-processed signal, in which the wavelet base is 'DB2', and 5 layer decomposition. The wavelet de-noising time-domain diagram in normal working condition is as shown in Fig. 4.

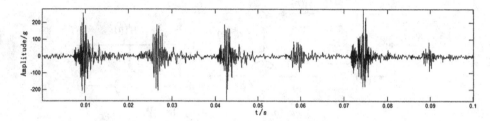

Fig. 4. Wavelet de-noising time-domain diagram in normal working condition

The comparison of the wavelet de-noising, morphological opening and closing mean filtering and adaptive morphological filtering is as shown in Table 2.

Table 2. Comparison of evaluation index of noise reduction method

Noise reduction method	SNR	RMSE
Wavelet de-noising	13.6375	16.8987
Mean filtering	18.6264	12.4265
Adaptive filtering	18.6300	12.4242

In Table 2, it is shown that compared to the wavelet de-noising, the adaptive generalized morphological filtering has a larger SNR and the smaller mean square error after the noise reduction of the signal, indicating that the semi circular structural element can effectively filter random noise interference, and morphological filtering technology can effectively utilize to diesel engine vibration signal processing; the signal to noise ratio after adaptive adjustment of the weights is slightly larger than that of mean filtering, and the root mean square error is slightly smaller, indicating that the adaption is adjusted to the best filtering effect.

5 Conclusion

(1) The corrosion and expansion mean filter, the opening and closing mean filter, and the open-close-open mean filter are designed. In contrast, the noise reduction effect of morphological opening and closing mean filter is optimal.

(2) In this paper, aiming at diesel engine vibration signal characteristics, a kind of adaptive generalized morphological filter is designed and used to reduce the noise of the diesel engine vibration signal. Compared with the traditional wavelet de-noising, the morphological filters designed in this paper made better noise reduction effect.

Acknowledgment. The work described in this paper has been supported by National Natural Science Foundation of China under the grant No. 51175480 and Natural Science Foundation of Shanxi under the grant No. 2012021014-2. The authors would like to express their gratitude for the support of this study.

References

1. An, L.S., Hu, A.J., Tang, G.J.: Purification of rotor center's orbit with mathematical morphology filters. Power Eng. **25**(4), 550–553 (2005)
2. Zhang, L.J., Yang, D.B., Xu, J.W.: Approach to extracting gear fault feature based on mathematical morphology filtering. Chi. J. Mech. Eng. **43**(2), 71–75 (2007)
3. Hu, A.J., Tang, G.J., An, L.S.: De-noising technique for vibration signal of rotating machinery based on mathematical morphology filter. Method Chi. J. Mech. Eng. **42**(4), 127–130 (2006)
4. Du, B.Q., Tan, G.J., Shi, J.J.: Design and analysis of morphological filter for vibration signals of a rotating machinery vibration signal form filter design and analysis. J. Vib. Shock **28**(9), 79–81 (2009)
5. Zhang, W.B.: Research on state trend prediction and fault diagnosis methods for turbo-generator unit. Doctoral Thesis, Hangzhou, Zhe Jiang University (2009)
6. Shen, L.: Study on Application of morphology in machinery fault diagnosis. Doctoral thesis, Hangzhou, Zhe Jiang University (2010)
7. Li, B., Zhang, P.L., Mi, S.S.: Mathematical Morphology Analysis and Intelligent Classification of Mechanical Fault Signal. National Defence Industry Press, Beijing (2011)
8. Li, B.L., Wei, M.X.: Cylinder pressure signal processing in engine based on morphological filter. Transducer. Micro Sys. Tech. **28**(7), 56–58 (2009)
9. Chen, P., Li, Q.M.: Design and analysis of mathematical morphology-based digital filters. Proc. CSEE. **25**(11), 60–65 (2005)
10. Tao, K., Zhu, J.J.: A comparative study on validity assessment of wavelet de-noising. J. Geodesy Geodyn. **32**(2), 128–133 (2012)
11. Jiang, Z., Deng, A.D., Cai, B.H.: Application of an adaptive generalized morphological filter based on the gradient method in rubbing acoustic emission signal de-noise. Proc. CSEE **31**(8), 87–92 (2011)

On the Discovery of Time Distance Constrained Temporal Association Rules

Heitor Murilo Gomes[1]([⊠]), Deborah Ribeiro de Carvalho[2], Lourdes Zubieta[3],
Jean Paul Barddal[1], and Andreia Malucelli[1]

[1] Programa de Pós-Graduação em Informática,
Pontifícia Universidade Católica do Paraná, Curitiba, Brazil
hmgomes@ppgia.pucpr.br
[2] Programa de Pós-Graduação em Tecnologia Aplicada em Saúde, Pontifícia
Universidade Católica do Paraná, Curitiba, Brazil
[3] Williams School of Business, Bishop's University, Sherbrooke, QC, Canada

Abstract. The increased use of data mining algorithms reflects the need for automatic extraction of knowledge from large volumes of data. This work presents a temporal data mining algorithm that discovers frequent Association Rules from timestamped data. These rules are named Cause-Effect Rules, each represented by a multiset of unordered events (Cause) followed by a singleton event (Effect). Also, a Cause-Effect Rule is valid within an specific constraint that defines the minimum and maximum time distance between its Cause and Effect. Our algorithm was tested on a data set from two hospital emergency departments in Sherbrooke, QC, Canada.

1 Introduction

The increased use of data mining algorithms reflects the need for automatic extraction of knowledge from large volumes of data. In many contexts, data contains timestamps or other sequential index. To extract knowledge from this type of data, the temporal data mining research area was established. Temporal data mining can be defined as the efficient discovery of frequent sequential patterns. Identifying such patterns provides the ability to predict a future event, with certain confidence, if a set of events is identified. Its applications cover different areas of knowledge ranging from banking and telecommunications to web browsing and adverse drug reactions. The data being processed in each of these areas varies greatly with regard to the number of examples, attributes, noise level, and others. Temporal data mining algorithms are designed to cope with specific requirements from the domain in which they are applied. For example, in healthcare analysis there is a vast amount of data arranged as transactions of events with timestamps. Specialists in healthcare analysis often want to find associations between events in patients' medical history over time in order to make evidence-based decisions. Informative patterns can be extracted to identify, for example, that "10 days before being hospitalized patients frequently

© Springer International Publishing Switzerland 2015
S. Arik et al. (Eds.): ICONIP 2015, Part II, LNCS 9490, pp. 510–519, 2015.
DOI: 10.1007/978-3-319-26535-3_58

took test A and B" rather than "hospitalized patients frequently took test A and B".

Sequential data mining algorithms attempt to find frequent occurrences of itemsets in a specific order. Introduced by [1], the itemset is a non-empty set of items, while a sequence [2] is an ordered non-empty list of itemsets. Previous works [4,7,8] identify sequential patterns that appear more often than a user-specified minimum support while maintaining their item occurrence order. Time intervals are commonly added by defining time distance constraints, limiting the timespan between events. More recently, there has been increasing interest in extracting Association Rules from temporal data [3].

In this paper we present a new temporal data mining algorithm, namely Cause-Effect Rules (CER). CER extends and combines definitions found in previous works to generate insightful, yet simple to understand patterns. CER generates Association Rules based on the assumption that an event (effect or consequent) E might have been caused by a set of other events (cause set or antecedent) C if, and only if, all events in C occurred before E within a timespan δ. In other words, not every event that happened before E could have influenced it. For example, if C is composed of patients' medical history and we are looking for a specific event like being hospitalized, it is likely that events that occurred years before the hospital admission may not have influenced E as much as events that happened a few days or weeks before. Obtaining rules in a similar format, although from sequential data, has been explored in previous works, like the MARBLES algorithm [3]. However, combining it with the well-known technique [6] of time distance constraint has not yet been explored in the literature and we show that it is very useful for analysing timestamped data, such as emergency department visits' data.

The rest of this paper is organized as follows. Section 2 briefly surveys temporal data mining methods. Section 3 formally describes the problem of discovering CERs, while Sect. 4 outlines CER implementation. In Sect. 5, we present the results and discuss around the experiments conducted on an Emergency Department dataset. Finally, Sect. 6 concludes the paper.

2   Related Work

In [2] the problem of mining sequential patterns was introduced along with three algorithms to solve it. Despite the difference between the three approaches, they aim for the same goal: discovering frequent sequences. A variation of rule support [1] was used to quantitatively assess the discovered sequences. This support measure reflects the fraction of transactions in which a sequence occurs. At a later date the same authors of [2] presented the GSP algorithm in [7], which performed better than the previous algorithms and permitted the user to specify constrains to the discovered sequences, such as the minimum and maximum gap between adjacent itemsets.

In [8] the SPADE algorithm for discovering frequent sequences was presented. It diverges from previous work as it does not inherit from Association Rules

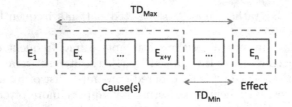

Fig. 1. Cause-Effect Sequence derived from a raw sequence of events.

discovery techniques, i.e., it does not explore the anti-monotonicity property, instead the authors use a Lattice-based approach along with efficient search methods to discover and count the occurrence of sequences.

In [4], authors present Chrono Assoc, a hybrid data mining method, designed to combine knowledge from an Association Rules discovery [1] process along with chronological order evaluation. The algorithm receives as input Association Rules and the dataset used to generate them, incrementing chronological support and confidence for each rule based on events that are consistent with the hypothesis "every antecedent must have happened before consequent".

The MARBLES [3] algorithm is capable of uncovering Association Rules between general episodes. A general episode is formed by a parallel episode that precede a sequential episode, i.e. an unordered set of events that occur before a set of strictly ordered events. Works before MARBLES used to focus on either sequential or parallel episodes [6], yet it is intuitive that rules combining both of them can be useful in many problems domains.

Even though these algorithms can uncover meaningful knowledge they do not fully explore important information included in some transactional data-bases. We claim that the next step beyond the informative rules obtained by an algorithm such as MARBLES, is the exploitation of time distance constraints according to the problem domain.

3   Cause-Effect Rules

A Cause-Effect Sequence (CES) is defined as a triplet (C, E, δ) where C is a cause multiset[1] of events, E is a singleton effect event, and $\delta = (TD_{Min}, TD_{Max})$ is a Time Distance Constraint (TDC) over which the sequence is valid. It is assumed that events in C happened close to each other in time, thus their order is deemed as irrelevant. δ delimits the minimum (TD_{Min}) and maximum (TD_{Max}) time distance between C and E. Concretely, δ is the building template for CES, following the intuitive notion that causes may have influenced a later effect if, and only if, they have happened neither too close, nor too far apart in time. Figure 1 presents how a CES is built given a raw event sequence.

---

[1] The multiset is a flexible representation that permits the Cause to contain repeated events.

The complete set of CES C_δ of the input database is a different representation of the original data restricted to a specific TDC δ. Therefore, not all CES represent frequent patterns, they are only a representation of the original data. A Cause-Effect Rule (CER) represents a CES that satisfies the user-given thresholds: Minimum Effect Frequency (E_{min}), Minimum Support S_{min}, and Minimum Confidence C_{min}. E_{min} is meant to filter out any candidate CER which effect is not frequent, while S_{min} and C_{min} are more specific to the rule frequency as a whole, and related to our Support and Confidence definitions (see Eqs. 2 and 3).

Ultimately, one CER is a CES that either occur or are contained in "a sufficiently large number of CES". A CES x is said to be contained within CES y if, and only if, they share the same E and δ, and all events in x cause appear in y cause. To assess the obtained CERs we use three measures. The first one is the Effect Frequency $E_f(e, \delta)$ (Eq. 1), which represents the ratio of occurrence of an effect for a given TDC δ.

$$E_f(e, \delta) = \frac{|\{x : x \in C_\delta, \text{Effect}(x) = e\}|}{|C_\delta|} \qquad (1)$$

The other two measures are inspired by the Support and Confidence definition are based on those presented originally on [1]. Although, these measures must be defined differently for CER. The Cause-Effect Support (S_{CE}) outlines the ratio of CES that are consistent with a given CER (Eq. 2). In other words, it is the ratio of occurrence of a given CER according to all C_δ.

$$S_{CE}(r, \delta) = \frac{|\{x : x \in C_\delta, r \subseteq x\}|}{|C_\delta|} \qquad (2)$$

Cause-Effect Confidence (C_{CE}) represents the ratio of CES that are consistent with a given CER, limited to those that have the same Effect as that CER (Eq. 3).

$$C_{CE}(r, \delta) = \frac{|\{x : x \in C_\delta, r \subseteq x\}|}{|\{x : x \in C_\delta, \text{Effect}(x) = \text{Effect}(r)\}|} \qquad (3)$$

Other works have been developed to uncover frequent sequences; in CER the only sense of order that matters is that all events in C must have happened at a constrained distance from the event in E. Nevertheless, we do treat differently the same event happening multiple times as this can yield different rules. For example, a situation in which a patient has visited the emergence department with the same problem four times in a row is different than a situation with a single visit.

4   CER 1.0 ALGORITHM

The proposed algorithm for Cause-Effect Rules mining, namely CER 1.0, performs multiple passes over a transformed version of the original database. This new database is represented in terms of Cause-Effect Sequences. A key component of the counting subroutine is the data structure, which represents the

CER candidates. We use an n-ary tree, which accommodates all the candidates associated with the same effect for a TDC. The algorithm is divided into three phases: Transformation, Initialization and Discovery, detailed in the following sections.

4.1 Transformation Phase

Some preparation steps are performed in the first pass over the database, including mapping the events description from string to integer representation. The most important and time-consuming step is the transformation of original input database to CESs. The CES form can be used to directly match CER candidates. It is straightforward to count the occurrences of individual events, e.g., causes and effects, during this phase, since the whole database is inspected. By doing that, the first candidates generation can be restricted to those that can actually yield a CER. At the end of this phase, the input dataset is transformed from a raw event data into a collection of CES, according to the time constraint chosen. The effect counter is stored for later use during the Effect Frequency calculation explained next.

4.2 Initialization Phase

In the Initialization phase, the first generation of candidate trees are created, such that for each combination of event e, and time constraint δ the corresponding candidate trees $M_\delta = \mu_1, \mu_2, \cdots, \mu_e$ are created. This phase also includes the calculation of E_f and the removal of CES whose E_f is less than E_{min}, shrinking the scope of the mining process. The counters used to calculate E_f are updated during the Transformation phase (see Sect. 4.1). Therefore, it is only necessary to iterate over every effect and calculate their frequency.

The initialization of the candidate trees requires combining frequent effects and causes that withstand filtering. If none were filtered out, the number of candidates of the first generation grows as much as the number of occurrences of its effect $N_e \times N_\delta$, although this is rare in practice. The number of trees is exactly the same as the number of Frequent Effects, since each tree represents candidates that share the same effect. The assertion behind which candidates can yield a CER is based on the anti-monotonicity (Apriori) property, which states that one set can only be frequent if each of its subsets is also frequent. Another interpretation, focused on the current work, is that a CES can only become a CER if its Effect and its Cause set are frequent. Besides, one Cause set can only be frequent if each of its subsets are also frequent.

4.3 Discovery Phase

The last phase incrementally expands the candidate trees and filters out candidates. There are two levels of filtering. The first filter is the Cause-Effect Support (S_{CE}) followed by the Cause-Effect Confidence (C_{CE}), both set by the user as

Table 1. Number of CERs found for different cause set sizes ($|C|$).

TD_{min}	TD_{max}	$|C| = 1$		$|C| = 2$		$|C| = 3$		$|C| = 4$	
		#	%	#	%	#	%	#	%
1 s	1 h	4	4.3	4	1.9	0	0	0	0
1 h	1 day	17	18.1	29	13.6	5	7.6	1	5
1 h	2 days	18	19.1	44	20.7	11	16.7	1	5
1 h	3 days	19	20.2	52	24.4	20	30.3	8	40
1 h	4 days	19	20.2	55	25.8	25	37.9	9	45

Table 2. CERs with effect = 24 (LWBS).

Frequent CERs	S_{CE} (%)	C_{CE} (%)
24 (1 h, 1 day, 24)	5.00	52.15
24 (1 h, 2 days, 24)	5.11	45.42
24 (1 h, 3 days, 24)	4.74	40.99
24 (1 h, 4 days, 24)	4.58	39.35
17 (1 h, 1 day, 24)	0.92	09.57
17 (1 h, 3 days, 24)	1.23	10.62
5 (1 h, 2 day, 24)	1.10	09.76
5 (1 h, 3 day, 24)	1.23	10.62
5 (1 h, 4 day, 24)	1.43	12.26

the minimum support (S_{min}) and Minimum Confidence (C_{min}), respectively. In order to calculate a candidate's S_{CE} and C_{CE} it is necessary to count how many CES contains it. A given CES is said to contain a candidate CER if their effects match and if every member of the candidate's cause multiset is present in the CES cause multiset. For example, consider candidate CER $x = (\{A, B, C\}, D, \delta_z)$ and CES $y = (\{B, C, D, A\}, D, \delta_z)$, then y contains x. The candidates' cause sets are represented recursively in the candidate trees, as follows:

Nodes. Each node contains a list of event identifiers. If it is a parent node, each of its events may or may not have a child node.

Leaves. Every event on the leaves list is the last cause of a candidate; thus the size of all leaves lists combined corresponds exactly to the number of candidates that exists for a tree.

Height. A tree with height h represents candidates up to generation $h + 1$, e.g., trees with height equals two have candidates on the third candidate (Gen_3) generation.

New generations, except the first one (Gen_1), are created by adding a level on all existing trees. The last cause of a candidate, i.e., the event at the leaf, will receive a child node if, and only if, the candidate has surpassed S_{min} and C_{min} thresholds. This effectively eliminates candidates that cannot become CER. Trees that do not have nodes on the current generation level are removed. For every generation, the set of CES is compared to each tree according to the effect and TDC that the tree represents. The comparison of a CES and a candidate tree starts at the root and advances to lower levels according to the CES events in the cause set. If a leaf node is reached, then a counter is incremented indicating that the CER candidate, represented by the inverse traverse of the tree (leaf to root), support has been increased.

5   Experiments

We tested the CER 1.0 algorithm on a dataset from two hospital emergency departments (ED) in Sherbrooke, QC. Canada. The dataset contained 138,107 low acuity visits to the EDs during a period of 3.5 years, from June 2006 to December 2009. Besides the visit's date and arrival time, each record include a 40-character patient identifier that allows to follow the person over time; other visit details include age, sex, postal code, diagnostic code(s) as written by a doctor, and whether the patient was admitted to the hospital. We were particularly interested in uncompleted visits because the patient left without being seen by a doctor (LWBS). These LWBS visits receive a particular code (24) and represent 13.9 % of all visits in the dataset. The rate of LWBS has been used as an indicator of quality of care, although there is controversy about its use and interpretation. LWBS behaviour has been associated with very long waiting times at the ED, lack of access to medical care, and higher risks of short term adverse events [5]. A sequence was defined as a list of visits by the same patient. The TDCs used to mine the dataset were small time intervals: between 1 s and 1 h, between 1 h and 1 day, 1 h and 2 days, 1 h and 3 days, and 1 h and 4 days, for $|C| \in [1; 5]$.

5.1   Results Before a LWBS Visit

In the first set of experiments we find out which diagnostic codes patients received previously to a LWBS visit; the "causes" are the previous diagnostic codes and the "effect" is an incomplete visit (code 24). We found many CERs having effect = 24 (LWBS), but not many with a high support. For each time interval, code 24 came out as the second or third most frequent one. Other frequent CERs had effect = 17 (Infections and parasitic diseases), 5 (Mental illness) and 16 (Ill-defined conditions). These are also the most frequent codes in the original database [9]. The most frequent rules for effect = 24 are shown in Table 2. For each TDC, the most common event prior to a LWBS visit is another LWBS visit, so patients who left the ED are more likely to come back in a very near future and leave again. The rules with the highest support were $24 \rightarrow 24$, with near 5 % support but with high confidence values (C_{CE}) ranging from 52 % for a repeat behaviour within a day, down to 39.35 % within 4 days. The originality of our results is the fact that patients come back and leave again within a short period of time, indicating their healthcare needs were unfulfilled.

5.2   Results Before a Hospital Admission via ED

In the second set of experiments we find out whether LWBS patients were hospitalized after an incomplete visit to the ED; the "effect" is being hospitalized and the "causes" are any previous diagnostic codes received by the patients. As we do not have data about all hospitalizations that occurred during the time intervals studied (only those admissions from the ED), we must be careful in the interpretation of the results.

Table 3. CERs of size 1 with effect = hospitalization.

TD_{min}	TD_{max}	Cause	Effect	E_F (%)	S_{CE} (%)	C_{CE} (%)	TD_{min}	TD_{max}	Cause	Effect	E_F (%)	S_{CE} (%)	C_{CE} (%)
1sec	1h	{24}	Hosp.	5.75	2.30	40.00	1 h	3 days	{17}	Hosp.	6.93	1.41	20.31
1sec	1 day	{17}	Hosp.	7.72	1.72	22.22	1 h	3 days	{16}	Hosp.	6.93	1.30	18.77
1sec	1 day	{16}	Hosp.	7.72	1.39	17.99	1 h	3 days	{8}	Hosp.	6.93	0.82	11.88
1sec	1 day	{24}	Hosp.	7.72	0.86	11.11	1 h	3 days	{Hosp.}	Hosp.	6.93	0.82	11.88
1 h	1 day	{17}	Hosp.	7.79	1.78	22.83	1 h	3 days	{9}	Hosp.	6.93	0.74	10.73
1 h	1 day	{16}	Hosp.	7.79	1.44	18.48	1 h	3 days	{24}	Hosp.	6.93	0.69	09.96
1 h	1 day	{24}	Hosp.	7.79	0.80	10.33	1 h	4 days	{17}	Hosp.	6.51	1.26	19.42
1 h	2 days	{17}	Hosp.	6.99	1.56	22.37	1 h	4 days	{16}	Hosp.	6.51	1.15	17.63
1 h	2 days	{16}	Hosp.	6.99	1.34	19.18	1 h	4 days	{Hosp.}	Hosp.	6.51	0.87	13.31
1 h	2 days	{9}	Hosp.	6.99	0.70	10.05	1 h	4 days	{8}	Hosp.	6.51	0.75	11.51
1 h	2 days	{24}	Hosp.	6.99	0.67	09.59	1 h	4 days	{9}	Hosp.	6.51	0.70	10.79
							1 h	4 days	{24}	Hosp.	6.51	0.63	09.71

Table 4. CERs with effect = LWBS visit (24).

TD_{min}	TD_{max}	Cause	Effect	E_f %	S_{CE} %	C_{CE} %
1 s	1h	{24}	24	09.20	4.60	50.00
1 s	1h	{5, 5}	24	09.20	1.15	12.50
1 s	1h	{16, 9, 5}	24	08.86	0.04	00.46
1h	1 day	{24}	24	08.84	4.61	52.15
1h	1 day	{5, 5}	24	08.84	0.17	01.91
1h	1 day	{16, 9, 5}	24	08.84	0.04	00.48
1h	2 days	{24}	24	10.46	4.75	45.43
1h	2 days	{5, 5}	24	10.46	0.38	03.66
1h	2 days	{16, 9, 5}	24	10.46	0.03	00.30
1h	3 days	{24}	24	10.76	4.41	40.99
1h	3 days	{5, 5}	24	10.76	0.45	04.20
1h	3 days	{16, 9, 5}	24	10.76	0.03	00.25
1h	4 days	{24}	24	10.89	4.28	39.35
1h	4 days	{5, 5}	24	10.89	0.56	05.16
1h	4 days	{16, 9, 5}	24	10.89	0.02	00.22

When considering only the most recent previous visit, we found LWBS was the most frequent "cause" only in the very short interval [1 s, 1 h]. For other periods of time and with 2 or more previous visits, LWBS was not the most frequent code. Other diagnostic codes like ill-defined conditions (16), diseases of the digestive system (9), respiratory problems (8) and a previous hospitalization were found more frequently than a LWBS visit, but we noticed the low frequence values E_f of all rules. Table 3 shows the detailed results for $|C| = 1$ only.

5.3 Discussion

Experiments conducted using the CER 1.0 algorithm revealed for the first time according to the authors, that patients who leave the emergency department before seeing a doctor (LWBS) are more likely to come back in a short period of time ranging up to 4 days, and leave again before seeing a doctor. This behaviour was found more frequently in men than in women. We also found that other common diagnostic codes previous to an LWBS visit are infections and parasitic diseases and mental illnesses, indicating that the healthcare needs of these patients were not fulfilled. A third result of our experiments showed no major risk of hospitalization after an incomplete LWBS visit. However, our study is not conclusive because of two reasons: (i) we do not know if these patients came back to the ED but were assigned a more urgent triage code, or (ii) they were admitted to the hospital shortly after an LWBS visit by other pathway than the ED.

6 Conclusion

In this paper we presented a new data mining algorithm for finding temporal rules constrained by user-defined Time Distance Constraints. To illustrate our method, we applied the algorithm to a dataset of low acuity emergency department visits from two hospitals in Sherbrooke, QC, Canada. Our results provide information on specific diagnosis associated with a higher likelihood of bouncing back and contributed to assess the risks of short term adverse effects like hospitalization. The results are noteworthy as the same rule structures held for different TDCs. Future work includes automatic discovery of TDCs, since these require domain knowledge and may be difficult to determine, and generating rules that combine static attributes and temporal events.

References

1. Agrawal, R., Srikant, R.: Fast algorithms for mining association rules. Proc. VLDB **1215**, 487–499 (1994)
2. Agrawal, R., Srikant, R.: Mining sequential patterns. In: Proceedings of the 11th ICDE, pp. 3–14. IEEE (1995)
3. Cule, B., Tatti, N., Goethals, B.: Marbles: mining association rules buried in long event sequences. Stat. Anal. DM: ASA Data Sci. J. **7**(2), 93–110 (2014)
4. Gomes, H.M., Carvalho, D.R.: A hybrid data mining method: exploring sequential indicators over association rules. Iberoamerican J. Appl. Comput. **1**(1), 40–60 (2011)
5. Lucas, J., Batt, R.J., Soremekun, O.A.: Setting wait times to achieve targeted left-without-being-seen rates. Am. J. Emerg. Med. **32**(4), 342–345 (2014)
6. Mannila, H., Toivonen, H., Inkeri Verkamo, A.: Discovery of frequent episodes in event sequences. Data Min. Knowl. Discov. **1**(3), 259–289 (1997)
7. Srikant, R., Agrawal, R.: Mining sequential patterns: generalizations and performance improvements. In: Apers, P.M.G., Bouzeghoub, M., Gardarin, G. (eds.) EDBT 1996. LNCS, vol. 1057, pp. 1–17. Springer, Heidelberg (1996)

8. Zaki, M.J.: Spade: an efficient algorithm for mining frequent sequences. Mach. Learn. **42**(1–2), 31–60 (2001)
9. Zubieta, L., Fernández-Peña, J.R.: A retrospective study on emergency visits in two hospitals in sherbrooke, Canada. J. Hosp. Adm. **3**(4), p61 (2014)

Software Clone Detection Using Clustering Approach

Bikash Joshi, Puskar Budhathoki, Wei Lee Woon, and Davor Svetinovic(✉)

Department of Electrical Engineering and Computer Science,
Institute Center for Smart and Sustainable Systems,
Masdar Institute of Science and Technology, P.O. Box 54224, Abu Dhabi, UAE
{bjoshi,pbudhathoki,wwoon,dsvetinovic}@masdar.ac.ae

Abstract. Code clones are highly similar or identical code segments. Identification of clones helps improve software quality through managed evolution, refactoring, complexity reduction, etc. In this study, we investigate Type 1 and Type 2 function clones using a data mining technique. First, we create a dataset by collecting metrics for all functions in a software system. Second, we apply DBSCAN clustering algorithm on the dataset so that each cluster can be analysed to detect Type 1 and Type 2 function clones. We evaluate our approach by analyzing an open source software Bitmessage. We calculate the precision value to show the effectiveness of our approach in detecting function clones. We show that our approach for functional clone detection is effective with high precision value and number of function clones detected.

Keywords: Clone detection · Data mining · Software metrics · Function clones

1 Introduction

Considerable amount of research has been conducted for clone detection in software systems [1–3]. Software clones arise from activities such as copy-paste and adaptation. Identification and evaluation of software clones is important because of its essential role in various software engineering tasks such as software maintenance, code quality analysis, plagiarism detection, code compaction (for example in mobile devices), bug detection, etc. [4]. It takes extra effort to enhance or adapt a system containing duplicate code fragments. Research has shown that 7–23 % of software systems contain cloned code [2]. Code clones usually arise from two kinds of similarities: textual similarity and functional similarity. Based on these similarities code clones can be classified as [3]:

- Type 1: Similar code segments which only differ in whitespace, layout and comments.
- Type 2: Similar code segments which differ in variables, literals, types, whitespace, layout, comments or function identifiers.

S. Arik et al. (Eds.): ICONIP 2015, Part II, LNCS 9490, pp. 520–527, 2015.
DOI: 10.1007/978-3-319-26535-3_59

- Type 3: Code segments with additional variations in changed, added or removed statements in addition to the variations in Type 1 and Type 2 clones.

Although a large number of clone detection studies has been published, very few of them have used a hybrid technique using data mining and metric based approach. In [2], a hybrid metric-based data mining approach for function clone detection has been proposed. Metric-based approaches use metric vectors calculated from different code fragments rather than using code for comparison. Distance between these metric vectors are used as the indicator for code similarity. It uses fractal clustering as data mining technique for clone detection. The evaluation of this approach is performed on open source software systems developed in C programming language. Four different subsets of metrics are used for evaluation of the proposed technique. Precision and recall are assessed for the software systems using the subsets of metrics.

Following the approach in [2], we propose a function clone detection technique using DBSCAN clustering algorithm. DBSCAN is a widely used [5,6] clustering algorithm proposed in [7]. As opposed to the other clustering algorithms like k-means, fuzzy c-means, DBSCAN does not require the number of clusters to be specified a priori. We assess the performance of the proposed technique by manual verification of the detected clones in the Bitmessage software. We also analyze the precision and number of detected clones to determine the suitable value of Epsilon for DBSCAN clustering.

In summary, this paper makes the following contributions:

- We adapt the approach presented in [2] using DBSCAN clustering algorithm on the Bitmessage software system.
- We evaluate the effectiveness of our approach by calculating the precision and the number of function clones.
- We analyze the precision and number of detected clones for different values of Epsilon to determine the suitable value of Epsilon.

2 Related Work

Existing research in software clone detection include categorization of different clones, experimental studies and evaluations, tracing of clones as they evolve along with software system enhancement, automatic clone detection system, refactoring clones and studying effects of cloning in software maintainability.

Basit et al. [8] formulate structural clone approach and use it to detect higher-level similarities using frequent closed item sets and clustering data mining technique. Clone Miner [8], Deckard [9] and semantic clone detector [10] are used to identify space and recorded clones. CCFinder [11] convert the code into token sequence based on lexical analyzer and rule based transformation is applied afterwards. Their usefulness is analyzed by detecting clones in commercial and public domain software systems.

Zhuo et al. [1] introduce metric space into code clone detection and measure the level of similarity using iterative approach of metric space distance to detect clones. The approach is applied in industrial projects providing electronic online reporting services to institutional investors and the result is more accurate when metric space distance is less.

Roy et al. [12] provide a comprehensive survey of textual, syntactic, lexical and semantic clone detection technique. It helps new potential users in selecting appropriate clone detection method for their needs.

Basit et al. [13] explore string algorithm for suitable data structures and algorithms for the development of Repeated Tokens Finder using efficient token based clone detection. Memory efficient suffix array linear time algorithm is developed to detect string matches with customizable tokenization mechanism.

Tree-based approach in software clone detection, as in [14], represents source code and clones as subtrees of abstract syntax trees; and structural characteristic vector measures the code similarity and describes code changes as tree editing scripts. Koschke et al. [15] present suffix trees to find clone in abstract syntax tree along with syntactic clone in linear time and space.

We used data mining techniques in our previous research on clustering goals [16], within the context of [17]. Besides use in refactoring, the ability to detect clones is useful in the early stages of development cycle too [18,19].

3   Proposed Method

The proposed software clone detection method consists of following three main steps:

- Feature Extraction: We analyze all functions of the software system and extract a set of metrics associated with each function.
- Clustering of Dataset: After preparing our metrics dataset for each function in the software system, we apply the DBSCAN clustering algorithm. This clustering algorithm partitions the dataset into clusters. Each cluster represents a group of functions with high similarity to each other in terms of used metrics.
- Function Clone Identification: We analyze each cluster and identify the function clones present in the software system.

Each of the three steps listed above is discussed in following subsections.

3.1   Feature Extraction

Suppose the software system consists of N functions (f_1, f_2, ..., f_N) and each function contains D software metrics (m_1, m_2, ..., m_D), then it can be represented as N X D dataset, where N corresponds to the number of functions and the dimensionality D, represents the metrics used to represent each function:

- Number of declarative statements
- Number of conditional statements
- Number of looping statements
- Number of parameters
- Number of called functions
- Number of return statements
- Number of lines of code (LOC).

3.2   Clustering of Dataset

In our proposed method, we use DBSCAN algorithm for the clustering of our dataset. The Density Based Spatial Clustering of Applications with Noise (DBSCAN) is a density based data clustering algorithm which helps to find the clusters of arbitrary shape based on the density distribution of the data [7].

DBSCAN is based on the concept of density reachability. A point P1 is considered as density reachable from point P2 if they are not separated by a distance of more than ϵ and if P2 is surrounded by sufficient number of points such that they can be considered as the part of the same cluster.

DBSCAN algorithm has two parameters: ϵ (Epsilon) and minimum number of points required to form a cluster (minPts). Epsilon denotes the distance between the two points and minPts denotes the number of points specified to form a cluster. Epsilon is necessary to find the ϵ neighborhood which means the number of points that are less than Epsilon distance than a particular data point. The algorithm chooses an arbitrary data point and finds its ϵ neighborhood. If the Epsilon neighborhood has sufficient number of data points, then a cluster is formed; otherwise, the point is considered as an outlier (noise). Since DBSCAN uses Epsilon and minPts for clustering, the algorithm may merge two clusters if they are sufficiently close to each other. Therefore, based on the values of Epsilon and minPts, the number of clusters formed may vary. The detailed description, analysis and evaluation of the algorithm are given in [7] where the authors show that DBSCAN effectively deals with large data sets and noise.

We have decided to use DBSCAN algorithm as it posses two main advantages over the other data clustering algorithms [20]:

- Minimum domain knowledge required: DBSCAN does not require much prior knowledge of the data. This algorithm has only two free parameters: Epsilon and minPts, which can be determined experimentally. As opposed to the other partition based clustering algorithms such as k-means, fuzzy c-means, DBSCAN does not require the determination of the number of clusters in advance.
- Not sensitive to noise: Most of the clustering algorithms are sensitive to the data noise, which makes them unsuitable for data having considerable amount of noise. DBSCAN has a notion for noise as it is based on density clustering. As such, DBSCAN can identify the noise points and consider them as not a part of a cluster.

3.3   Function Clone Identification

As a result of DBSCAN, the functions of the case study Bitmessage software system are partitioned into separate clusters of varying sizes. The functions present inside a cluster have a similar set of metrics. This property defines them to be potential function clones. As such, the functions present in one cluster are considered to be function clones for potential removal, maintenance, etc.

4 Results and Discussion

4.1 Dataset Preparation

The case study in this paper is focused on finding the function clones in Bitmessage code. Although Bitmessage is not a very large system, it is rich in function clones. We calculated each of the seven metrics for each function in the software system.

The version of the Bitmessage system that we analyzed contains 260 functions. Most of the clone detection techniques presented in the literature identify function clones which are more than five lines long. The reason is that the metrics extracted from short functions tend to be the same and hence are identified as clones by the clustering algorithm. We also observed that such reasoning applies to our case as well. Thus, we also considered functions which contain five or more lines of code only. So, after this pre-processing step we identified a dataset with 208 functions.

4.2 Results

Usually the performance of the clone detection techniques is assessed in terms of precision and recall. Precision and recall are be defined as:

$$Precision = \frac{TP}{TP + FP} * 100\,\%$$

$$Recall = \frac{TP}{TP + FN} * 100\,\%$$

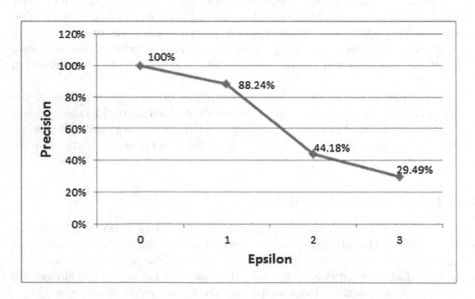

Fig. 1. Precision values for different values of Epsilon

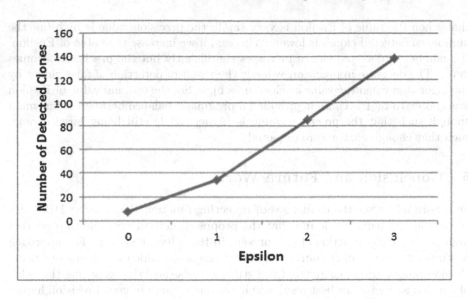

Fig. 2. Number of detected clones for different values of Epsilon

Where, True Positive (TP), False Positive (FP), True Negative (TN) and False Negative (FN) represent number of correctly identified clones, number of incorrectly identified clones, number of correctly rejected clones and number of incorrectly rejected clones, respectively.

Precision represents the percentage of correctly identified clones (TP) out of the total number of detected clones, and recall represents the percentage of correctly identified clones (TP) detected by our approach with respect to the actually present clones in the software system. So, recall is difficult to assess as it requires already existing standard data to compare with. In our case, it is the total number of accurate clones present in a software system. As there is no existing tool to identify function clones in software developed in Python, we could not evaluate our method in terms of recall. But, we manually verified our detected clones to find out how many of them are true clones. Thus, we were able to assess precision of the method.

An important consideration, while using DBSCAN clustering, is the choice of Epsilon. Epsilon determines the distance between two points considered for the clustering task. If the value of Epsilon is very high many functions will be grouped inside the same cluster causing many false detected clones, whereas if the value is Epsilon is very low, each of the functions will form their own clusters causing no clones detected. So, the value of Epsilon should be large enough to detect most of the actual clones as well as small enough to not contain false clones. Thus, to determine the optimal value of Epsilon, the performance in terms of precision and number of detected clones was calculated for a range of Epsilon values. The variation of precision and number of detected clones for different values of Epsilon is shown in Figs. 1 and 2. It can be seen from Figs. 1 and 2 that when the value of Epsilon is very small, the precision value is high but the number of detected clones is lower. Whereas, if we increase the value of Epsilon, the number of detected clones increases significantly but the precision becomes lower. The decrease in precision value is the result of detection of false clones by the algorithm when the value of Epsilon is high. So, the optimal value of Epsilon was chosen to be 1 because it provided a pragmatic balance between being small enough such that the precision value is higher, while still being large enough such that enough clones were detected.

5 Conclusion and Future Work

Our work addresses the challenges of detecting function clones using DBSCAN clustering approach. We found that the proposed method was able to retrieve a high percentage of function clones present in the software system. The approach was evaluated on an open source software Bitmessage, which was developed using Python programming language. Case study results show that choosing the value of Epsilon as 1 gave the best results as it is small enough to keep precision higher and large enough to detect sufficient number of clones. Our proposed function clone detection approach effectively determines the sufficient number of clones with higher precision value.

References

1. Li, Z., Sun, J.: An iterative, metric space based software clone detection approach. In: 2010 2nd International Conference on Software Engineering and Data Mining (SEDM), pp. 111–116 (2010)
2. Abd-El-Hafiz, S.K.: A metrics-based data mining approach for software clone detection. In: Computer Software and Applications Conference (COMPSAC), 2012 IEEE 36th Annual, pp. 35–41 (2012)
3. Saha, R.K., Roy, C.K., Schneider, K.A., Perry, D.E.: Understanding the evolution of type-3 clones: an exploratory study. In: Proceedings of the 10th Working Conference on Mining Software Repositories, MSR 2013, pp. 139–148. IEEE Press, Piscataway (2013)
4. Roy, C.K., Cordy, J.R., Koschke, R.: Comparison and evaluation of code clone detection techniques and tools: a qualitative approach. Sci. Comput. Program. **74**(7), 470–495 (2009)
5. Erman, J., Arlitt, M., Mahanti, A.: Traffic classification using clustering algorithms. In: Proceedings of the 2006 SIGCOMM Workshop on Mining Network Data, MineNet 2006, pp. 281–286. ACM, New York (2006)
6. Kriegel, H.-P., Pfeifle, M.: Density-based clustering of uncertain data. In: Proceedings of the Eleventh ACM SIGKDD International Conference on Knowledge Discovery in Data Mining, KDD 2005, pp. 672–677. ACM, New York (2005)
7. Ester, M., Kriegel, H.P., Sander, J., Xu, X.: A density-based algorithm for discovering clusters in large spatial databases with noise, pp. 226–231. AAAI Press (1996)
8. Basit, H.A., Jarzabek, S.: A data mining approach for detecting higher-level clones in software. IEEE Trans. Softw. Eng. **35**(4), 497–514 (2009)

9. Jiang, L., Misherghi, G., Su, Z., Glondu, S.: Deckard: scalable and accurate tree-based detection of code clones. In: Proceedings of the 29th International Conference on Software Engineering, ICSE 2007, pp. 96–105. IEEE Computer Society, Washington, DC (2007)

10. Gabel, M., Jiang, L., Su, Z.: Scalable detection of semantic clones. In: Proceedings of the 30th International Conference on Software engineering, pp. 321–330. ACM, New York (2008)

11. Kamiya, T., Kusumoto, S., Inoue, K.: Ccfinder: a multilinguistic token-based code clone detection system for large scale source code. IEEE Trans. Softw. Eng. **28**(7), 654–670 (2002)

12. Roy, C.K., Cordy, J.R.: Scenario-based comparison of clone detection techniques. In: The 16th IEEE International Conference on Program Comprehension, ICPC 2008, pp. 153–162 (2008)

13. Basit, H.A., Jarzabek, S.: Efficient token based clone detection with flexible tokenization. In: Proceedings of the the 6th Joint Meeting of the European Software Engineering Conference and the ACM SIGSOFT Symposium on the Foundations of Software Engineering, ESEC-FSE 2007, pp. 513–516. ACM, New York (2007)

14. Nguyen, H.A., Nguyen, T.T., Pham, N.H., Al-Kofahi, J., Nguyen, T.N.: Clone management for evolving software. IEEE Trans. Softw. Eng. **38**(5), 1008–1026 (2012)

15. Koschke, R., Falke, R., Frenzel, P.: Clone detection using abstract syntax suffix trees. In: 13th Working Conference on Reverse Engineering, WCRE 2006, pp. 253–262 (2006)

16. Casagrande, E., Woldeamlak, S., Woon, W.L., Zeineldin, H.H., Svetinovic, D.: Nlp-kaos for systems goal elicitation: smart metering system case study. IEEE Trans. Softw. Eng. **40**(10), 941–956 (2014)

17. Svetinovic, D.: Strategic requirements engineering for complex sustainable systems. Syst. Eng. **16**(2), 165–174 (2013)

18. Svetinovic, D., Berry, D.M., Day, N.A., Godfrey, M.W.: Unified use case statecharts: case studies. Requirements Eng. **12**(4), 245–264 (2007)

19. Svetinovic, D.: Architecture-level requirements specification. In: STRAW, pp. 14–19 (2003)

20. Mumtaz, K., Duraiswamy, K.: An analysis on density based clustering of multi dimensional spatial data. Indian J. Comput. Sci. Eng. **1**(1), 8–12 (2010)

Bilingual Lexicon Extraction with Forced Correlation from Comparable Corpora

Chunyue Zhang and Tiejun Zhao[✉]

School of Computer Science and Technology,
Harbin Institute of Technology, Harbin, China
cyzhang@mtlab.hit.edu.cn, tjzhao@hit.edu.cn

Abstract. Recently a simple linear transformation with word embedding has been found to be highly effective to extract a bilingual lexicon from comparable corpora. However, the pairs of bilingual word embedding for training this transformation are assumed to satisfy a linear relationship automatically which actually can't be guaranteed absolutely in practice. This paper proposes a simple solution based on canonical correlation analysis (CCA) which forces the bilingual word embedding for training the transformation to be maximally linearly correlated onto the projection subspaces. After projecting the original word embedding into the new correlation subspace in two languages, a better transformation matrix is again learned with the new projected word embeddings as before. The experimental results confirm that the proposed solution can achieve a significant improvement of 62 % in the precision at Top-1 over the baseline approach on the English-to-Chinese bilingual lexicon extraction task.

1 Introduction

Bilingual lexicons serve as an invaluable resource of knowledge in various natural language processing tasks, such as cross-lingual information retrieval and machine translation. Compiling such lexicons manually is often an expensive task, whereas the methods for mining the lexicons from parallel corpora are not applicable for language pairs and domains where such corpora is unavailable or missing. Therefor the automatic bilingual lexicon extraction (BLE) from comparable corpora [3,5,7,13], where documents are not direct translations but share a topic or domain, has attracted many researchers.

In recent years, **Distributed Representation** [1,4,14] for a word, which is often called **word embedding**, has been extensively studied. Word embedding projects discrete words to a dense low-dimensional and continuous vector space where co-occurred words are located close to each other. Inspired by the approximate linear relation in the bilingual scenario, a linear transformation [12] is learned to project semantically identical words from a language to another with the corresponding word embedding. Although its simplicity, the authors reported a high accuracy on a bilingual lexicon extraction task.

© Springer International Publishing Switzerland 2015
S. Arik et al. (Eds.): ICONIP 2015, Part II, LNCS 9490, pp. 528–535, 2015.
DOI: 10.1007/978-3-319-26535-3_60

In practice, the bilingual word embedding for words translating each other in two languages is not easy to be linearly mapped with comparable corpora. The reasons can be explained as follows:

- the word embeddings obtained in [12] are learned from the WMT11 parallel corpora other than comparable corpora which is often extremely asymmetrical.
- the performance achieved by the approach [12] is very high on the English to Spanish direction and very low on the English to Vietnamese direction. It shows the quality of linear relation of bilingual word embedding automatically learned is very diverse for different language pairs.
- all the word embeddings in a language are learned independently from some copora so that the relation between the corresponding word embeddings in two languages is not explicit to be guaranteed.

In this paper, we argue the monolingual word embeddings of two words translating each other as the different views of the same latent semantic signal. Inspired with this, we propose a solution by exploiting the canonical correlation analysis (CCA) to force the bilingual word embedding for training the transformation to be correlated linearly in the new latent semantic subspace. Specifically, in this work we:

- firstly learn the projection directions in the two language on the bilingual training word embedding pairs with CCA to maximize the correlation,
- then project the original word embeddings into their maximally correlated subspaces with the correlation directions learned above,
- finally, learn a new linear transform matrix from the new word embedding in two languages.

2   Background: Linear Transformation for BLE

Mikolov et al. proposed a skip-gram model [11] which aims at predicting the context words with a word in the central position. Mathematically, the training process maximizes the following likelihood function with a word sequence $w_1, w_2, \ldots, w_N$:

$$\frac{1}{N} \sum_{i=1}^{N} \sum_{-C \leq j \leq C} log P(w_{i+j}|w_i) \tag{1}$$

where C is the length of the context in concern, and the prediction probability is given by a softmax function:

$$P(w_{i+j}|w_i) = \frac{exp(c_{w_{i+j}}^T c_{w_i})}{\sum_w exp(c_w^T c_{w_i})} \tag{2}$$

where w is any word in the vocabulary, and c_w is a continuous vector which denotes the **word embedding** of word w.

Mikolov et al. [12] solved the bilingual lexicon extraction task with learning a linear transformation matrix from the source language to the target language by the linear regression. Let $\Sigma \in R^{n_1 \times d_1}$ and $\Omega \in R^{n_2 \times d_2}$ be word embeddings space of two different vocabularies in two language where rows represent words. During the training period we choose $\Sigma' \subset \Sigma$ where every word in Σ' is translated to one other word in $\Omega' \in \Omega$ and $\Sigma' \in R^{n \times d_1}$ and $\Omega' \in R^{n \times d_2}$.

The objective function is as follows:

$$\hat{W} = \underset{W \in R^{d_2 \times d_1}}{\mathrm{argmin}} \sum_{i=1}^{n} \|W x_i^\top - y_i^\top\|^2 \tag{3}$$

where $x_i \in \Sigma'$ is the word embedding of word i in the source language and $x_i^\top$ is its transpose, $y_i \in \Omega'$ is the word embedding of the translation of x_i, $\hat{W}$ is called the bilingual transformation matrix.

During the prediction period, given a new source word embedding x, the standard way to retrieve its translation word in the target language is to return the nearest neighbour (in terms of cosine similarity measure) of mapped $y^\top = \hat{W} x^\top$ from the set of word embedding of the target language, i.e. Ω.

3   Forced Correlation with CCA

As discussed at the Sect. 1, the pairs of bilingual word embedding for training the transformation are assumed to be linearly correlated which actually can't be guaranteed very well in practice. To counter this problem we use the canonical correlation analysis (CCA)[1] which is a way of measuring the linear relationship between two multidimensional variables. It finds two projection vectors, one for each variable, that are optimal with respect to correlations. The dimensionality of these new projected vectors is equal to or less than the smaller dimensionality of the two variables.

Let x and y be two corresponding vectors from Σ' and Ω' (defined in Sect. 2), and v and u be two projection directions. Then, the projected vectors are:

$$x' = x v^\top \qquad y' = y u^\top \tag{4}$$

and the correlation between the projected vectors can be written as:

$$\rho((x', y')) = \frac{E[x' y']}{\sqrt{E[x'^2] E[y'^2]}} \tag{5}$$

CCA maximizes ρ for the given set of vectors Σ' and Ω' and outputs two projection vectors v and u:

$$v, u = CCA(x, y) \tag{6}$$
$$= \underset{v,u}{\mathrm{argmax}}\, \rho(x v^\top, y u^\top)$$

---

[1] We use the MATLAB module for CCA: http://www.mathworks.com/help/stats/canoncorr.html.

Using these two projection vectors we can project the entire vocabulary of the two language Σ and Ω using Eq. 4. Summarizing:

$$V, U = CCA(\Sigma', \Omega') \tag{7}$$

$$\Sigma^* = \Sigma V \quad \Omega^* = \Omega U \tag{8}$$

where, $V \in R^{d_1 \times d}$, $U \in R^{d_2 \times d}$ contain the projection vectors and $d = min$ $\{rank(V), rank(U)\}$. Thus, the resulting vectors can't be longer than the original vectors.

Since CCA gives us correlations and corresponding projection vectors d dimensions which can be large, we perform experiments by taking projections of the original word vectors across only the top k correlated dimensions. This is trivial to implement as the projection vectors V, U in Eq. 7 are already sorted in descending order of correlation. Therefor in,

$$\Sigma_k^* = \Sigma V_k \quad \Omega^* = \Omega U_k \tag{9}$$

where V_k and U_k are the column truncated matrices, and Σ_k^* and Ω^* are now the new word embedding space, each word embedding in which is a k-dimensions vector projected along the top k correlated dimensions. After obtaining the Σ_k^* and Ω^*, we can learn a new bilingual transformation matrix as introduces in Sect. 2 and retrieve the new translations for a given word.

4   Experiment and Results

4.1   Experimental Settings

In this paper, we carry on the bilingual lexicon extraction task in the English-to-Chinese direction. For the comparable corpora, we use the English Gigaword Corpus (LDC2009T13) and the Chinese Gigaword Corpus (LDC2009T27). In order to align these two comparable corpora better, we select the part of the two corpus published by Xinhua News Agency which contains news articles from January 1995 to December 2008.

In the corpus preprocessing step, we performed these operations:

- For the English corpus, we tokenize the text using the scripts from www. statmt.org
- For the Chinese corpus, we segment the text using the Stanford Chinese Word Segmenter[2]
- Duplicate sentences are removed
- Numeric values are rewritten as a single token according to the heuristic rules.

Details of every corpus are reported in Table 1.

From every language, we use the same setup to train the skip-model. We use the word2vec toolkit[3] to learn a 200-dimensional word embedding. We just

---

[2] http://nlp.stanford.edu/software/segmenter.shtml.
[3] https://code.google.com/p/word2vec.

Table 1. The size of the monolingual corpora for English and Chinese. The vocabularies consist of the words that occurred at least ten times in the corpus.

	English	Chinese
Training tokens	326 M	346 M
Vocabulary size	136 K	205 K

Table 2. The statistics of the train set, dev set and test set for BLE. The average number of the translations for a English word is more than 4.

	Train set	Dev set	Test set
Entries	64692	2062	10576
Words	14413	500	2500
Avg	4.48	4.12	4.23

consider the words occurred at least 10 times, and set a context windows of 3 words to either side of the center. Other hyper-parameters follows the default software setup.

To obtain a bilingual lexicon between English-to-Chinese to train the bilingual transformation matrix, we use an in-house dictionary which consists of 55668 English words and 137420 Chinese words. We filter this dictionary with the vocabulary of monolingual corpora for English and Chinese. From the filtered dictionary, we randomly select 2500 (500) different English words and its translation as test set (dev set), and the left dictionary as the training set. Details of the train set and test set are listed in Table 2.

4.2   Results

In order to decide how many the top k correlated dimensions are the most suitable for our task, we conduct experiments on the dev set firstly. We try the ratio of top k correlated dimensions to the original dimensions from $\{0.3, 0.4, 0.5, 0.6, 0.7, 0.8, 0.9, 1.0\}$. According to the performance on the dev set in the Fig. 1, the performance at ratio 0.5 is achieved best, so we set the ratio 0.5, i.e. the top 100 correlated dimensions are selected.

We choose the approach in [12] as our baseline which uses the bilingual word embedding directly. The performance is measured by precision of translation

Table 3. The precision at the Top-1, Top-5 and Top-10 achieved by the Baseline and CCA. The approach [11] is the Baseline, and CCA row means our methods.

Method	P@1	P@5	P@10
Baseline	0.117	0.223	0.282
CCA	**0.190**	**0.349**	**0.417**

Table 4. The performance over the test set with the different size of sampling multiple translation for a word from the bilingual lexicon.

Sampling ratio	P@1	P@5	P@10
1	0.173	0.326	0.396
30 %	0.183	0.336	0.402
60 %	0.189	0.341	0.414
100 %	0.190	0.349	0.417

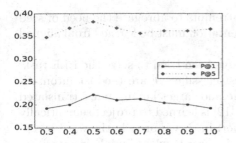

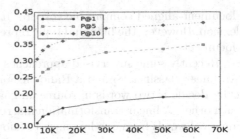

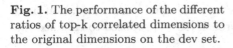

Fig. 1. The performance of the different ratios of top-k correlated dimensions to the original dimensions on the dev set.

Fig. 2. The performance of the different sizes of bilingual lexicon for training over the test set.

retrieval list of the test set at Top-k, where k is set from $\{1, 5, 10\}$. The results obtained by our method are presented in Table 3. From the Table 3, we can see the CCA based method gains the significant improvement over the baseline method. The precision at Top-1 is increased from 0.117 to 0.190. The same significant improvement can be seen from the precision at Top-5 and Top-10. So it confirms that a better transformation matrix is learned after projecting the original word embedding into the new space with CCA. Also we investigate how the size of the training data of CCA affects the final performance for BLE. We randomly select word pairs with the size of $\{6\,\mathrm{K}, 8\,\mathrm{K}, 10\,\mathrm{K}, 15\,\mathrm{K}, 30\,\mathrm{K}, 50\,\mathrm{K}\}$ from the training bilingual lexicon to learn the correlation. The final performance for BLE is reported in Fig. 2. From the Fig. 2, we can see the larger training data for CCA, the better transform matrix can be learned.

Because our bilingual lexicon consists of many words which have multiple translations, it's interesting to investigate how the number of translation for a word affects the quality of the learned transform matrix. So we select the three subsets of the whole training bilingual lexicon where words in the source language have 1 translation, 30 % translations and 60 % translations to train the CCA. For any subset, we sample 4 times and the final average performance is reported in Table 4. Although the CCA is assumed to learn an one-to-one correlation, from the Table 4 we can see that it's still effective for the BLE to use the bilingual lexicon with the multiple translations to train the CCA.

5 Related Works

In the BLE task from comparable corpora, most of the previous methods [3, 5, 7, 13] are based on the distributional hypothesis that a word and its translation tend to appear in similar contexts across languages. Based on this assumption, generally an unsupervised standard approach [2, 10] which uses the co-occurred context words to represent the target word calculates the context similarity and then extract word translation pairs with the highest similarity.

Another unsupervised approach in [15] uses the bilingual topic distribution to represent the target word. The authors train a bilingual topic model on the document-aligned comparable corpora. It attempts to abrogate the need of seed lexicon. However, the bilingual topic representation must be learned from aligned documents.

Recently some supervised approaches have been tried to solve the BLE task. An linear classifier [8] and a Random Forest classifier [9] are used to automatically decide if two words in source language and target language are translated each other. A linear transformation matrix [12] is learned to project semantically identical words from one language to another. In this approach, the word is represented with a continues and dense vector i.e. word embedding. It is surprising that this approach achieved a high accuracy on a bilingual word translation than the standard approach. An normalized approach [16] is proposed to enhance the word embedding.

The most similar work with us is proposed by Faruqui et al. [6] which also learns the linear transformation that project word vectors of all languages to a common low-dimensional space, where the correlation of the multilingual word pairs is maximized the canonical correlation analysis. However, their goal is to learn better monolingual word embedding other than the better bilingual transformation matrix.

6 Conclusions

We presented a simple but effective method based on canonical correlation analysis (CCA) to force the train data for linear transformation to be correlated linearly. After projecting the original word embedding into a new vector space along the correlation dimensions in two languages, a better transform matrix is again learned with the new projected word embeddings. Experiments conducted on an English-Chinese comparable corpora indicate that our method can improve the baseline method significantly.

Acknowledgments. This work is supported by the project of National Natural Science Foundation of China (61173073, 61272384) and International Science and Technology Cooperation Program of China (2014DFA11350).

References

1. Bengio, Y., Courville, A., Vincent, P.: Representation learning: a review and new perspectives. IEEE Trans. Pattern Anal. Mach. Intell. **35**(8), 1798–1828 (2013)
2. Bouamor, D., Semmar, N., Zweigenbaum, P.: Context vector disambiguation for bilingual lexicon extraction from comparable corpora. In: ACL, vol. 2, pp. 759–764 (2013)
3. Chiao, Y.C., Zweigenbaum, P.: Looking for candidate translational equivalents in specialized, comparable corpora. In: Proceedings of the 19th International Conference on Computational Linguistics, vol. 2, pp. 1–5. Association for Computational Linguistics (2002)

4. Collobert, R., Weston, J., Bottou, L., Karlen, M., Kavukcuoglu, K., Kuksa, P.: Natural language processing (almost) from scratch. J. Mach. Learn. Res. **12**, 2493–2537 (2011)
5. Emmanuel, M., Hazem, A.: Looking at unbalanced specialized comparable corpora for bilingual lexicon extraction. In: Proceedings of the 52nd Annual Meeting of the Association for Computational Linguistics (ACL), pp. 1284–1293 (2014)
6. Faruqui, M., Dyer, C.: Improving vector space word representations using multilingual correlation. Association for Computational Linguistics (2014)
7. Fung, P., Yee, L.Y.: An IR approach for translating new words from nonparallel, comparable texts. In: Proceedings of the 17th International Conference on Computational Linguistics, vol. 1, pp. 414–420. Association for Computational Linguistics (1998)
8. Irvine, A., Callison-Burch, C.: Supervised bilingual lexicon induction with multiple monolingual signals. In: HLT-NAACL, pp. 518–523. Citeseer (2013)
9. Kontonatsios, G., Korkontzelos, I., Tsujii, J., Ananiadou, S.: Using a random forest classifier to compile bilingual dictionaries of technical terms from comparable corpora. In: Proceedings of the 14th Conference of the European Chapter of the Association for Computational Linguistics: Short Papers, vol. 2, pp. 111–116 (2014)
10. Laroche, A., Langlais, P.: Revisiting context-based projection methods for term-translation spotting in comparable corpora. In: Proceedings of the 23rd International Conference on Computational Linguistics, pp. 617–625. Association for Computational Linguistics (2010)
11. Mikolov, T., Chen, K., Corrado, G., Dean, J.: Efficient estimation of word representations in vector space (2013). arXiv preprint arXiv:1301.3781
12. Mikolov, T., Le, Q.V., Sutskever, I.: Exploiting similarities among languages for machine translation (2013). arXiv preprint arXiv:1309.4168
13. Rapp, R.: Automatic identification of word translations from unrelated english and german corpora. In: Proceedings of the 37th Annual Meeting of the Association for Computational Linguistics on Computational Linguistics, pp. 519–526. Association for Computational Linguistics (1999)
14. Turian, J., Ratinov, L., Bengio, Y.: Word representations: a simple and general method for semi-supervised learning. In: Proceedings of the 48th Annual Meeting of the Association for Computational Linguistics, pp. 384–394. Association for Computational Linguistics (2010)
15. Vulić, I., Moens, M.F.: Detecting highly confident word translations from comparable corpora without any prior knowledge. In: Proceedings of the 13th Conference of the European Chapter of the Association for Computational Linguistics, pp. 449–459. Association for Computational Linguistics (2012)
16. Xing, C., Wang, D., Liu, C., Lin, Y.: Normalized word embedding and orthogonal transform for bilingual word translation. In: Proceedings of the 2015 Conference of the North American Chapter of the Association for Computational Linguistics: Human Language Technologies, pp. 1006–1011. Association for Computational Linguistics, Denver, May-June 2015. http://www.aclweb.org/anthology/N15-1104

Enhancing the Mongolian Historical Document Recognition System with Multiple Knowledge-Based Strategies

Xiangdong Su[✉], Guanglai Gao, Hongxi Wei, and Feilong Bao

College of Computer Science, Inner Mongolia University, Hohhot 010021, China
{cssxd,csggl,cswhx,csfeilong}@imu.edu.cn

Abstract. This paper describes recent work on integrating multiple strategies to improve the performance of the Mongolian historical document recognition system which utilize the segmentation-based scheme. We analyze the reasons why the recognition errors happened. On such basis, we propose three strategies according to the knowledge of the glyph characteristics of Mongolian and integrate them into glyph-unit recognition. The strategies are recognizing the under-segmented and over-segmented fragments (RUOF), glyph-unit grouping (GG) and incorporating the baseline information (IBI). The first strategy helps in correcting the segmentation error and the remaining two strategies further improve the classifiers accuracies. The experiment on the historical Mongolian Kanjur demonstrates that utilizing these strategies could effectively increase the accuracy of word recognition.

Keywords: Mongolian · Historical document · Knowledge-based strategy · Glyph-unit recognition

1 Introduction

Recently, the historical document digitalization has gained much attention. Many methods have been proposed in the published literatures. In Inner Mongolian Autonomous Region of China, there is a famous historical document called Mongolian Kanjur. This book is produced with the woodblock printing technology in Qing Dynasty. It involves history, literature, religion, sociology and many other subject. To make this precious document easier to be preserved and accessed, an efficient and effective way is to convert it into text by means of optical character recognition (OCR).

However, a few studies have been focused on Mongolian OCR. Most of them deal with the machine-printed Mongolian [1–3], which generally took a segmentation-based approach to recognize the words. These approaches are inel-igible to recognize the classical Mongolian words in the Mongolian Kanjur which is produced in woodblock printing technologies and is considered to be a cursive script. Although Gao et al. in [4] investigate the characteristics of the histori-cal words and design segmentation-based method to recognize them, it cannot

© Springer International Publishing Switzerland 2015
S. Arik et al. (Eds.): ICONIP 2015, Part II, LNCS 9490, pp. 536–544, 2015.
DOI: 10.1007/978-3-319-26535-3_61

dispose of the character heavily overlapped cases and results in many under-segmented fragments. Therefore, we proposed a hybrid system to recognize the words in the documents. A few words which are infeasible to be segmented into glyph-units are recognized by holistic scheme. The rest will be recognized by segmentation-based scheme. In the segmentation-based scheme, the words are segmented into glyph-units through contour analysis and these segments are recognized with convolutional neural network (CNN). Glyph-units are the smallest indivisible units which make up of the glyphs (visual forms of characters) in Mongolian.

The motivation of our work in this paper is to improve the performance of the above-mentioned systems through making use of the knowledge concerning the glyph characteristics of Mongolian. It is derived from the analysis of the recognition errors. In the segmentation-based scheme, the errors of word recognition are mainly attributed to two points. The first one is the segmentation errors resulted from the segmentation algorithm. Since the glyph-units in each word are intrinsically connected together and possess remarkable overlapping and variation, it is inevitable that the segmentation algorithm introduces some unexpected segmentation paths and/or misses some expected segmentation paths. The resultant segments are not glyph-units and can not be recognized in subsequent section. The second one is the recognition errors resulted from the classifier in the glyph-unit recognition since it may assign the wrong labels to the glyph-units in some cases. In short, these two reasons lower the performance of the Mongolian historical document recognition system.

To solve the problems in word recognition and further improve the systems performance, we integrate three knowledge-based strategies into segmentation-based scheme, including recognizing the under-segmented and over-segmented fragments (RUOF), glyph-unit grouping (GG) and incorporating the baseline information (IBI). The experiment demonstrates that utilizing these strategies could effectively improve the word accuracy.

The methodology proposed in this paper is different from those used in post-processing section. It improves the systems performance in the recognition section. It can be used with the latter simultaneously. These strategies also make sense to the OCR systems which process the accumulative languages (i.e. Uygur, Arabic and Czech).

The remainder of this paper is organized as follows. Section 2 presents the Mongolian historical document recognition system. Section 3 describes the methodology in details. Section 4 demonstrates the experiment. Section 5 draws the conclusion finally.

2   Mongolian Historical Document Recognition System

Mongolian Kanjur was made through woodblock printing 300 years ago. This ancient printing technology brings several challenges. To begin with, multiple xylographers created many variants of the same word. Even the instances in the same page also differ from each other. Figure 1 shows some variants of the word which means have done in English. Secondly, ink spreading caused deformation of words and spur noise in the documents. In addition, the phenomena of character overlapping took place within most of words. In some cases, a few characters overlapped with non-adjacent characters. The same is true for glyph-unit overlapping.

Fig. 1. The variants of classical Mongolian word (means "have done").

Figure 2 shows the Mongolian historical document recognition system. In the segmentation-based scheme, the words are segmented into glyph-units and recognized with CNN. Instead of segmenting the words into glyphs as the method in [5], our system segments the words into glyph-units in order to reduce the complexity and improve the segmentation accuracy. After segmentation, we recognize the glyph-units with the CNN. CNN is adopted because it and its variants gain the state-of-the-art performance in many applications, such as house number recognition [6] and image classification [7], and so on. It also proves to be efficient in object detection [8,9] and feature extraction [10]. The success of CNN is credited with that it provides some degree of shift and translation invariance and is free of feature extraction before classification.

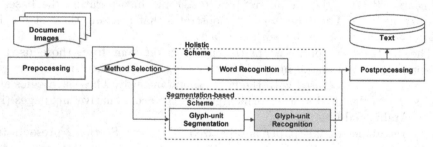

Fig. 2. Mongolian historical document recognition system.

3   Multiple Knowledge-based Strategies Integration

Although the performance of our original system satisfies the basic requirement of document indexing and retrieval, there is still room for improvement. According to our analysis, we find that the errors of word recognition in the segmentation-based scheme are attributed to the following two points. One is the segmentation errors in glyph-unit segmentation. The other is the recognition error in glyph-unit recognition. Therefore, we propose three knowledge-based strategies and integrate them into glyph-unit recognition to enhance the system.

3.1 Recognizing the Under-Segmented and Over-segmented Fragments

In glyph-units segmentation, two kinds of scenarios may occur. One is the over-segmentation when an unexpected segmentation line is introduced. This leads to a situation that a glyph-unit is cut into two segments. The other is the under-segmentation when an expected segmentation line is missed. It results in a segment which consists of two or more of glyph-units. The segments in such cases are not true glyph-units and cannot be correctly recognized because the training set does not include the samples of their categories. Figure 3 lists some samples of under segmented fragments. Owning to word variation and noise, it is unlikely to completely avoid these two scenarios through improving the segmentation algorithm.

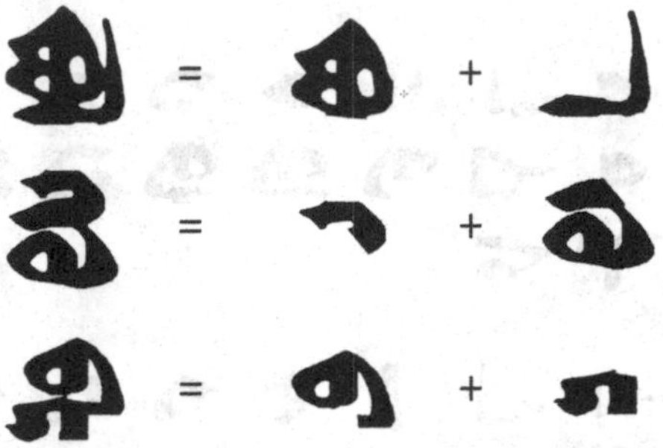

Fig. 3. Some samples of under segmented fragments.

To cope with the problem, we recognize the under-segmented and over-segmented fragments which are frequently occurring, and decode them into correct glyph-unit(s) later. These fragments are treated in the same way as the true glyph-units. This could reduce the recognition errors caused by the segmentation errors and thus improve the recognition performance. In the experiment, the training set includes samples of these fragments. This is equivalent to extending the glyph-unit set.

This is inspired by the works [11,12], which use the recognition feedback of the fragments to rectify the character segmentation. However, as to segmentation-based recognition, it is unnecessary to use the recognition result of the under-segmented and over-segmented fragments to rectify the segmentation result. When they are recognized, we could directly decode them into correct glyph-units.

3.2   Glyph-unit Grouping

In segmentation-based scheme, the resultant glyph-units after segmentation are generally recognized with the same classifier. However we could make an alteration in Mongolian historical document recognition. In classical Mongolian words, each character has as many as three different glyphs depending on whether the character appears in an initial, medial, or final position of a word. It means that among all the glyph-units, some only appear in the initial position of the words, others only appear in the medial position, and the rest only exist in the final position. Therefore, the glyph-units are divided into three groups according to their position in the word, and then recognized in different classifiers. This strategy is called glyph-unit grouping (GG). The glyph-units in the three groups are shown in Fig. 4.

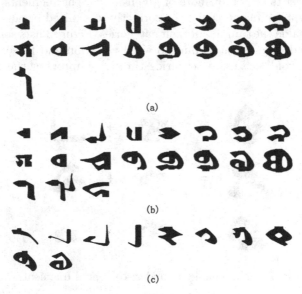

Fig. 4. The glyph-units of classical Mongolian.

The strategy GG could decrease in number of categories recognized by the classifiers and increase recognition accuracy. This practically amounts to adding a decision tree classifier before the CNN classifiers. In the experiment, the first segment of a word is classified with the first CNN, the last segment is classified with the third CNN, and the rest are classified with the second CNN.

3.3   Incorporating the Baseline Information

In Mongolian, glyph-units are connected along the baseline. The width of the baseline approximates the width of the stroke. The position of the baseline is different in different glyph-unit. Thus, the baseline information helps in discriminating the glyph-units.

To utilize the baseline, we incorporate it into the glyph-unit images which will be recognized. In glyph-unit segmentation, the baselines will be detected to facilitate the segmentation. After segmentation, we generate a variant for the binary image of each segment by replacing the value of each black pixel at the baseline with the medium gray level between black and white, as shown in Fig. 5. Figure 5c is the glyph-units and Fig. 5d is the variants of the glyph- units. These variants carry the baseline information. We use these variants as the inputs of the classifiers. Certainly, the classifiers are also trained with the samples of the variants of the glyph-units.

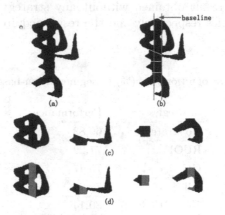

Fig. 5. An example of IBI. **a** is the word ("have done"), **b** is its segmentation lines and baseline, **c** is the glyph-units and **d** is the variants of the glyph-units.

4 Experiment

4.1 Experiment Setting

To check the effectiveness of the proposed strategies, we conduct the test experiment on 5000 words, which are extracted from the Mongolian Kanjur images. Among them, 857 words are recognized by the holistic scheme, and 4143 words are recognized by the segmentation-based scheme. In holistic scheme, the word are directly recognized by CNN. The rest 4143 words are split into 21668 segments in the glyph-unit segmentation. Then the segments are normalized and recognized with CNNs. Each CNN has nine layers except for the input layer. The kernel size and the number of the output feature maps are set experimentally. The experiment is carried out with the help of DeepLearnToolbox provided by Rasmus Berg Palm [13].

4.2 Result and Analysis

It is worth noting that, the proposed three strategies only improves the word accuracy in segmentation-based scheme. Hence, their effect is compared on the

4143 words which are recognized by the segmentation-based scheme. Actually, the strategy RUOF improves the word recognition rate rather than the glyph-unit recognition rate because it recognizes the fragments which are not glyph-units. After glyph-unit recognition, we make use of the result to form the words. It is worth noting that the accuracy of glyph-unit recognition is not the same as the accuracy of word recognition, and the reason is that if a glyph-unit in the word is recognized incorrectly, the whole word will be recognized incorrectly. For the sake of simplicity and consistency, this paper directly compares the accuracies of word recognition rather than the accuracies of glyph-unit recognition. Table 1 presents the result obtained without any strategy, the results gained with the three strategies respectively and the result acquired with three strategies simultaneously.

Table 1. Word recognition rate (%) in segmentation-based recognition.

Strategies	Performance
No strategy	69.83
RUOF	77.29
GG	75.47
IBI	72.60
GG+RUOF+IBI	80.14

From Table 1 we can clearly see that the result obtained with any one of the three strategies is much better than the result gained without any strategy. It implies that each strategy is effective in improving the performance of the recognition system.

When the three strategies are used simultaneously, the accuracy of word recognition in the segmentation-based scheme is improved by more than 10 percent from 69.83 percent to 80.14 percent, which leads to a 34.17 percent reduction in error rate. This result is better than the result acquired with individual strategy.

We also compare the improved system with the method proposed in literature [4] on all the 5000 words. Table 2 lists their performance individually. Meanwhile, the performance of our original system (without any strategy in segmentation-based scheme) is also introduced. In the table, the first row is Gao system proposed in literature [4], the second row is our original system, and the third row is the improved systems when the strategies are used in glyph-unit recognition.

It is clear that the enhanced system significantly outperforms the method [4] in the historical Mongolian document recognition. Another advantage of our system is that it is free of feature extraction.

Table 2. Word recognition rate (%) comparison.

System	Performance
Gao system	71.00
Original system	72.78
Improved system	81.32

5   Conclusion

This paper proposes three strategies in glyph-unit recognition to improve the performance of the Mongolian historical document recognition system. The strategies are designed on the basis of the in-depth analysis of recognition errors and fully utilization of the knowledge of glyph characteristics of Mongolian. The strategy RUOF could correct parts of the recognition errors caused by the segmentation errors. The strategies GG and IBI improve the classifier's accuracies. The experiment demonstrates that utilizing these strategies could effectively increase the accuracy of word recognition. These strategies can also be applied to other accumulative languages, such as Uygur, Arabic and Czech. Provided that there are many segmentation errors, the strategy RUOF can be used. For languages (i.e. Devanagari script) exhibiting zone feature, the strategy GG is applicable.

In the future, we will integrate language-related knowledge in the postprocessing to further improve the system's performance.

Acknowledgements. This work is funded by National Natural Science Foundation of China (Grant No. 61263037, No. 61463038, and No. 61563040) and the Research Project of Higher Education School of Inner Mongolia Autonomous Region of China (Grant No. NJZY14007).

References

1. Wei, H.: Study of key techniques in the printed Mongolian character recognition. Master thesis, Inner Mongolia University (2006)
2. Wei, H., Gao, G.: Machine-printed traditional mongolian characters recognition using bp neural networks. In: Proceedings of International Conference on Computational Intelligence and Software Engineering (CiSE), pp. 1–7 (2009)
3. Peng, L., Liu, C., Ding, X., Jin, J., Wu, Y., Wang, H., Bao, Y.: Multi-font printed mongolian document recognition system. IJDAR **13**(2), 93–106 (2010)
4. Gao, G., Su, X., Wei, H., Gong, Y.: Classical mongolian words recognition in historical document. In: Proceedings of the 11th International Conference on Document Analysis and Recognition (ICDAR), pp. 692–697. IEEE Computer Society (2011)
5. Su, X., Gao, G., Wang, W., Bao, F., Wei, H.: Character segmentation for classical mongolian words in historical documents. In: Li, S., Liu, C., Wang, Y. (eds.) CCPR 2014, Part II. CCIS, vol. 484, pp. 464–473. Springer, Heidelberg (2014)

6. Sermanet, P., Chintala, S., LeCun, Y.: Convolutional neural networks applied to house numbers digit classification. In: Proceedings of 21st International Conference on Pattern Recognition (ICPR), pp. 3288–3291 (2012)
7. Krizhevsky, A., Sutskever, I., Hinton, G.E.: Imagenet classification with deep convolutional neural networks. Adv. Neural Inf. Process. Syst. **25**, 1097–1105 (2012)
8. Szegedy, C., Toshev, A., Erhan, D.: Deep neural networks for object detection. Adv. Neural Inf. Process. Syst. **26**, 2553–2561 (2013)
9. Sermanet, P., Eigen, D., Zhang, X., Mathieu, M., Fergus, R., LeCun, Y.: Overfeat: integrated recognition, localization and detection using convolutional networks. In: Proceedings of International Conference on Learning Representations (ICLR) (2014)
10. Kim, H.-J., Lee, J.S., Yang, H.-S.: Human action recognition using a modified convolutional neural network. In: Liu, D., Fei, S., Hou, Z., Zhang, H., Sun, C. (eds.) ISNN 2007, Part II. LNCS, vol. 4492, pp. 715–723. Springer, Heidelberg (2007)
11. Xiao, X., Leedham, G.: Knowledge-based English cursive script segmentation. Pattern Recogn. Lett. **21**(10), 945–954 (2000)
12. Sulong, G., Rehman, A., Saba, T.: Improved offline connected script recognition based on hybrid strategy. Int. J. Eng. Sci. Technol. **2**(6), 1603–1611 (2010)
13. Palm, R.B.: Prediction as a candidate for learning deep hierarchical models of data. Master thesis, Technical University of Denmark (2012)

Empirical Analysis of Sampling Based Estimators for Evaluating RBMs

Vidyadhar Upadhya[✉] and P.S. Sastry

Indian Institute of Science, Bangalore, India
sastry@ee.iisc.ernet.in

Abstract. The Restricted Boltzmann Machines (RBM) can be used either as classifiers or as generative models. The quality of the generative RBM is measured through the average log-likelihood on test data. Due to the high computational complexity of evaluating the partition function, exact calculation of test log-likelihood is very difficult. In recent years some estimation methods are suggested for approximate computation of test log-likelihood. In this paper we present an empirical comparison of the main estimation methods, namely, the AIS algorithm for estimating the partition function, the CSL method for directly estimating the log-likelihood, and the RAISE algorithm that combines these two ideas.

1 Introduction

The Restricted Boltzmann Machines (RBM) are among the basic building blocks of deep learning models. They can be used as classifiers as well as generative models. The quality of the learnt generative RBM is evaluated based on the average log-likelihood which, for N test samples, is given by,

$$\mathcal{L} = \frac{1}{N} \sum_{i=1}^{N} \log p(\mathbf{v}^{(i)}) \tag{1}$$

where $p(\mathbf{v}^{(i)})$ is the probability (or the likelihood) of the ith test sample, $\mathbf{v}^{(i)}$. The model with higher average test log-likelihood is better. The test log-likelihood can also be used in devising a stopping criterion for the learning and for fixing the hyper-parameters through cross validation.

The likelihood $p(\mathbf{v})$ can be written as $p^*(\mathbf{v})/Z$. While $p^*(\mathbf{v})$ is easy to evaluate, the normalizing constant Z, called the partition function, is computationally expensive. Therefore, various sampling based estimators have been proposed for the estimation of the log-likelihood. In this paper we present an empirical analysis of the two main approaches for estimating the log-likelihood.

The first approach is to approximately estimate the partition function using the samples obtained from the model distribution. Since sampling from the model distribution of an RBM is complicated, a useful sampling technique is the importance sampling where samples obtained from a simple distribution, called the proposal distribution, are used to estimate the partition function. Since choosing a good proposal distribution is difficult, a sequence of intermediate distributions

© Springer International Publishing Switzerland 2015
S. Arik et al. (Eds.): ICONIP 2015, Part II, LNCS 9490, pp. 545–553, 2015.
DOI: 10.1007/978-3-319-26535-3_62

is used. The annealed importance sampling (AIS) is one such estimator [7]. The second approach is to estimate the average test log-likelihood directly by marginalizing over the hidden variables from the model distribution. However, the computational complexity grows exponentially with the number of hidden units present in the model. Hence, an approximate method which uses a sample based estimator called conservative sampling based likelihood estimator (CSL) is proposed in [1]. A more efficient method called reverse annealed importance sampling estimator (RAISE) implements CSL by formulating the problem of marginalization as a partition function estimation problem.

In this paper, we present an empirical comparison of the performance of the sampling based estimators, namely, annealed importance sampling estimator (AIS) [7], conservative sampling based likelihood estimator (CSL) [1], reverse annealed importance sampling estimator (RAISE) [2]. Initially we learn RBM models (with different number of hidden units) using the standard CD algorithm on the MNIST dataset with suitably chosen hyperparameters. We evaluate the average test log-likelihood of each of the learnt models using the estimators mentioned above and compare their performance.

The rest of the paper is organized as follows. Section 2 briefly describes the RBM model. Different sampling methods considered in this study are described in Sect. 3. Section 4 discusses results of our empirical study and Sect. 5 concludes the paper.

2   Restricted Boltzmann Machines

The Restricted Boltzmann Machine (RBM) is a special case of the Boltzmann Machine where the intra-layer connections are restricted [4,5,9]. The visible stochastic units ($\mathbf{v}$) are connected to the hidden stochastic units ($\mathbf{h}$) through bidirectional links. We restrict our study to the binary RBM with $\mathbf{v} \in \{0,1\}^m$ and $\mathbf{h} \in \{0,1\}^n$, being the states of visible and hidden units respectively. For the binary RBM, the energy function and the probability that the model assigns to $\{\mathbf{v}, \mathbf{h}\}$ is given by,

$$E(\mathbf{v}, \mathbf{h}; \theta) = -\sum_{i,j} w_{ij} h_i v_j - \sum_{j=1}^{m} b_j v_j - \sum_{i=1}^{n} c_i h_i, \text{and} \quad p(\mathbf{v}, \mathbf{h}|\theta) = \frac{e^{-E(\mathbf{v},\mathbf{h};\theta)}}{Z}$$

$$(2)$$

where $Z = \sum_{\mathbf{v},\mathbf{h}} e^{-E(\mathbf{v},\mathbf{h};\theta)}$ is the partition function and $\theta = \{\mathbf{w} \in \mathbb{R}^{n \times m}, \mathbf{b} \in \mathbb{R}^m, \mathbf{c} \in \mathbb{R}^n\}$ is the set of model parameters.

The log-likelihood of (learnt) RBM on a test data $\mathbf{v}$ is given as, (see [3] for details),

$$\log p(\mathbf{v}) = \log \sum_{\mathbf{h}} e^{-E(\mathbf{v},\mathbf{h};\theta)} - \log Z = \sum_{j} b_j v_j + \sum_{i} \log(1 + e^{c_i + \sum_j w_{ij} v_j}) - \log Z$$

$$(3)$$

The difficulty in evaluating the above equation is due to the presence of the intractable log partition function. The log partition function can be expanded as [3],

$$\log Z = \log \sum_{\mathbf{v},\mathbf{h}} e^{-E(\mathbf{v},\mathbf{h};\theta)} = \log \sum_{\mathbf{v}} e^{\sum_j b_j v_j} \prod_i \left(1 + e^{c_i + \sum_j w_{ij} v_j} \right) \quad (4)$$

The above equation can also be written in terms of $\mathbf{h}$. Hence, to evaluate the partition function we have to sum 2^L terms, where $L = \min\{m,n\}$, which is computationally expensive for large models. However, if we can estimate Z by some other method then we can efficiently estimate the log-likelihood using Eq. (3).

The test log-likelihood can also be estimated directly by marginalizing the model distribution over the latent variable $\mathbf{h}$ through a sample average as,

$$\log p(\mathbf{v}) = \log \sum_{\mathbf{h}} p(\mathbf{h})p(\mathbf{v}|\mathbf{h}) \approx \log \frac{1}{N} \sum_{i=1}^{N} p(\mathbf{v}|\mathbf{h}^{(i)}) \text{ where } \mathbf{h}^{(i)} \sim p(\mathbf{h}), \forall i. \quad (5)$$

3   Sampling Based Estimators for Evaluating RBMs

3.1   Annealed Importance Sampling for Estimating Partition Function

Suppose two distributions $f_A(\mathbf{x}) = f_A^*(\mathbf{x})/Z_A$ and $f_B(\mathbf{x}) = f_B^*(\mathbf{x})/Z_B$ are given such that $f_A(\mathbf{x}) \neq 0$ when $f_B(\mathbf{x}) \neq 0$ and it is possible to obtain independent samples from $f_A(\mathbf{x})$. It is also assumed that $f_A^*(\mathbf{x})$, $f_B^*(\mathbf{x})$ and Z_A are easy to evaluate. Then the ratio of partition functions can be written as,

$$\frac{Z_B}{Z_A} = \int \frac{f_B^*(\mathbf{x})}{Z_A} d\mathbf{x} = \int \frac{f_B^*(\mathbf{x})}{f_A^*(\mathbf{x})} f_A(\mathbf{x}) \, d\mathbf{x} \approx \frac{1}{M} \sum_{i=1}^{M} \frac{f_B^*(\mathbf{x}^{(i)})}{f_A^*(\mathbf{x}^{(i)})} \approx \frac{1}{M} \sum_{i=1}^{M} w_{\text{imp}}^{(i)} \quad (6)$$

where $w_{\text{imp}}^{(i)} = f_B^*(\mathbf{x}^{(i)})/f_A^*(\mathbf{x}^{(i)})$ is termed as the ith importance weight and $\mathbf{x}^{(i)}$ are independent samples from the distribution f_A. However, the variance of this estimator is very high unless the proposal distribution, f_A, is a good approximation of the target distribution, f_B [6]. In order to overcome this, in annealed importance sampling a sequence of intermediate probability distributions is used to move gradually from the proposal to the target distribution [7]. Suppose there are $K - 1$ intermediate distributions: $f_A = f_0, f_1, \ldots, f_{K-1}, f_K = f_B$, given by $f_i(\mathbf{x}) = f_i^*(\mathbf{x})/Z_i$, $i = 0, 1, \ldots, K$. Then using Eq. (6), we have, for every k, $0 \leq k \leq K - 1$,

$$\frac{Z_{k+1}}{Z_k} \approx \frac{1}{M} \sum_{i=1}^{M} \frac{f_{k+1}(\mathbf{x}_k^{(i)})}{f_k(\mathbf{x}_k^{(i)})}, \text{ where } \mathbf{x}_k^{(i)} \sim f_k \quad (7)$$

Further, we can write,

$$\frac{Z_K}{Z_0} = \frac{Z_1}{Z_0}\frac{Z_2}{Z_1}\cdots\frac{Z_K}{Z_{K-1}} = \prod_{k=0}^{K-1}\frac{Z_{k+1}}{Z_k} \tag{8}$$

The above gives us a good estimate for Z_K (assuming we can easily calculate Z_0.

A standard method to sample from each of the f_k is as follows. Initially a sample $\mathbf{x}_1$ is drawn from $f_0 = f_A$. Then for $1 \leq k \leq K-1$, the sample $\mathbf{x}_{k+1}$, is obtained by sampling from $T_k(\mathbf{x}_{k+1}|\mathbf{x}_k)$ where T_k is a transition function of a (reversible) Markov chain for which f_k is the invariant distribution. This sequence of states, $\mathbf{x}_1, \mathbf{x}_1, \ldots \mathbf{x}_K$ is one sample to estimate Z_K using Eqs. (7) and (8). Then we can get M samples by repeating this M times. (For the detailed analysis of this estimator see [7]).

The choice of intermediate probability distributions is problem specific. The geometric average of $f_A(\mathbf{x})$ and $f_B(\mathbf{x})$ is a popular choice in literature. Here one keeps, $f_k(\mathbf{x}) \propto f_A(\mathbf{x})^{1-\beta_k} f_B(\mathbf{x})^{\beta_k}$, where β_k's are chosen such that $0 = \beta_0 < \ldots < \beta_K = 1$. There are many ways of selecting the so called annealing path $\beta = \{\beta_0, \beta_1, \ldots, \beta_K\}$.

The AIS yields an unbiased estimate of Z and hence, by Jensen inequality, it underestimates the log partition function, in an expected sense [2]. This implies that, on the average, we overestimate the log likelihood.

Estimating the Partition Function of an RBM [8]: Consider the two RBMs with parameters $\theta^A = \{\mathbf{w}^A, \mathbf{b}^A, \mathbf{c}^A\}$ and $\theta^B = \{\mathbf{w}^B, \mathbf{b}^B, \mathbf{c}^B\}$ corresponding to the proposal (f_A) and the target distribution (f_B) respectively.

The two choices for the proposal RBM used in the present study are the *uniform* and the *base rate* RBM. For the uniform case we take $\mathbf{w}^A = 0, \mathbf{b}^A = 0, \mathbf{c}^A = 0$. If we take only the weight matrix to be zero, the proposal distribution is given as $f_A(\mathbf{v}) = e^{\sum_j b_j v_j}/\prod_j(1 + e^{-b_j})$. The ML solution for b_j turns out to be $\log(\bar{\mathbf{v}}) - \log(1 - \bar{\mathbf{v}})$ where $\bar{\mathbf{v}}$ is the mean of the training samples. This is termed as the 'base rate' RBM.

The intermediate distributions are constructed using the geometric average path. Then the kth intermediate distribution is an RBM with the energy function given by,

$$E_k(\mathbf{v}, \mathbf{h}) = -\left((1 - \beta_k)\, E(\mathbf{v}, \mathbf{h}^A; \theta^A) + \beta_k\, E(\mathbf{v}, \mathbf{h}^B, \theta^B)\right) \tag{9}$$

where $\mathbf{h} = \{\mathbf{h}^A, \mathbf{h}^B\}$. Since each of the f_k is defined by the RBM, the samples are obtained through Gibbs sampler. More details of the method are available in [8].

3.2 Conservative Sampling Based Log-likelihood Estimator

Given the conditional probability $p(\mathbf{v}|\mathbf{h})$ and a set of samples $\{\mathbf{h}^{(1)}, \mathbf{h}^{(2)} \ldots \mathbf{h}^{(N)}\}$ from the distribution $p(\mathbf{h})$, the log-likelihood on a test sample using the CSL estimator is given in Eq. (5).

The hidden units samples are obtained using the Gibbs sampler for the RBM. The state of the visible units can be initialized randomly (called the unbiased CSL) or by using the training samples (called the biased CSL). Then, the hidden units and the visible units are sampled alternately from the corresponding conditional distributions for a large number of iterations. The Jensen's inequality shows that the CSL estimator underestimates the test log-likelihood [1], i.e.,

$$\mathbb{E}_{\mathbf{h}}[\log p(\mathbf{v})] \leq \log \mathbb{E}_{\mathbf{h}}[p(\mathbf{v}|\mathbf{h})] = \log p(\mathbf{v}).$$

3.3 Reverse AIS Estimator

The RAISE estimates the log-likelihood similar to the CSL approach while implementing it through the AIS approach [2]. Consider the likelihood on a given test data $\mathbf{v}$ as in Eq. (5), $p(\mathbf{v}) = \sum_{\mathbf{h}} p(\mathbf{h})p(v|\mathbf{h})$. Finding this $p(\mathbf{v})$ is equivalent to finding the partition function of a distribution given by $f(\mathbf{h}) = p(\mathbf{h})p(\mathbf{v}|\mathbf{h})$ for a fixed $\mathbf{v}$. Hence, the AIS estimator (with suitable modifications) can be used. For the analysis and the implementation details of the method refer [2].

The test data $\mathbf{v}$ is assumed to be a sample from the target distribution. This test sample is used as the initial state of the AIS Markov chain which is defined to move gradually from the target to the proposal distribution as opposed to moving from the proposal to the target distribution as in the AIS estimator discussed in Sect. 3.1. The visible and hidden units, $\mathbf{h}$ and $\mathbf{v}$, of intermediate distributions are sampled alternately from this chain using the transition operators similar to the ones defined for the AIS estimator. Once the samples are obtained the estimate of the likelihood is given as [2],

$$p(\hat{\mathbf{v}}) = \frac{f_K(\mathbf{v})}{Z_0} \prod_{k=1}^{K} \frac{f_{k-1}(\mathbf{x}_k)}{f_k(\mathbf{x}_k)}. \tag{10}$$

4 Experimental Results and Discussions

In this section we present the results of estimating the partition function and the average test log-likelihood using the different methods discussed in the previous section. We first learn the RBMs with different number of hidden units (20, 200 and 500, denoted as RBM20, RBM200, RBM500) using the standard CD-20 algorithm on the MNIST dataset. We fix the learning rate, $\eta = 0.1$, batch size $= 100$, weight decay $= 0.001$, initial momentum $= 0.5$, final momentum $= 0.9$ and change the momentum at 5th step. Initial weights are sampled from $U[-1/\sqrt{L}, 1/\sqrt{L}]$ where $L = \min\{m, n\}$.

Then, for each of the learnt RBMs, we estimate the test log-likelihood on the MNIST test dataset using the different estimators. Note that, even though learning the RBM with different hyperparameter settings produce (significantly) different models, we have observed that the behaviour of estimators are similar on these models. Therefore we present results for one such RBM which is learnt using the above hyperparameter settings. For the case of RBM20, the ground truth is calculated by summing over all 2^{20} states so that we can assess accuracy of the estimates.

AIS: The performance of the AIS estimator depends on the distribution of the proposal RBM, the number of intermediate distributions (K) and the number of samples $(M$, also called AIS runs). We fix $M = 500$ and use uniform or baserate as the proposal distribution. The value of K is chosen from $\{100, 1000, 10000\}$ and we use the linear path $\beta = [0 : 1/K : 1]$ for all the experiments. The handcrafted schedule having $K = 14500$ given in [8] with $\beta = [0 : 1/1000 : 0.5, 0.5 : 1/10000 : 0.9, 0.9 : 1/100000 : 1]$ is also implemented for comparison. In order to estimate the variance of the estimator we repeat the experiment 50 times with a random initial state each time. Table 1 gives the estimate of $\log Z$, $\mathscr{L}$ and σ (the standard deviation of the estimate of $\widehat{\log Z}$).

Based on the ground truth available for RBM20, we observe that the AIS estimator, on the average, overestimates the test log-likelihood. The use of base rate proposal distribution gives slightly better estimate with less variance compared to using the uniform proposal distribution for the RBM20. However, for the RBM200 and RBM500 the proposal distribution has no significant effect on the estimated value though it affects the variance of the estimate. On the whole, the linear annealing schedule with $K = 10,000$ seems to perform as well as the hand-crafted annealing schedule with $K = 14,500$.

CSL: The samples required for the CSL estimator are obtained through the Gibbs sampler which alternately samples the hidden and the visible units. We ignore the first B samples to allow burn-in and then collect samples after every T (called 'Thin' parameter) steps, discarding the samples in between, to avoid correlation. We observed that the estimates obtained with a single chain is poor even if we run the Gibbs sampler for a large number of steps. This possibly indicates the poor mixing rate of the Gibbs chain. Therefore we experiment with many parallel Gibbs chains with different initial states.

We first simulate S_M parallel Gibbs chains with S_T steps for both the biased and the unbiased CSL. Under various values of the parameters M and T, the

Table 1. AIS estimate of $\mathscr{L}$, **Ground truth:** $\log \mathbf{Z} = \mathbf{230.61}$, $\mathscr{L} = \mathbf{-141.24}$ for **RBM20**

		Uniform proposal			*Baserate* proposal		
n	K	$\widehat{\log Z}$	$\hat{\mathscr{L}}$	σ	$\widehat{\log Z}$	$\hat{\mathscr{L}}$	σ
20	1000	229.2660	−139.8933	0.2905	230.5353	−141.1626	0.2114
	10000	229.5245	−140.1518	0.2423	230.6192	−141.2465	0.0629
	14500	229.8978	−140.5250	0.6095	230.6082	−141.2355	0.0431
200	1000	174.1720	−111.2346	0.6724	173.5514	−110.6140	1.2344
	10000	174.7664	−111.8290	0.1528	174.6819	−111.7445	0.2408
	14500	174.7424	−111.8050	0.2623	174.7512	−111.8138	0.1300
500	1000	173.4807	−114.8451	0.3071	172.2157	−113.5801	1.1597
	10000	173.4802	−114.8446	0.1207	173.4767	−114.8411	0.2244
	14500	173.4567	−114.8211	0.1902	173.4240	−114.7883	0.1717

samples of **h** (to estimate the log-likelihood) are selected from these simulated chains. The experiment is repeated 100 times to find the variance of the estimate, where for each experiment we select a random burn-in value and select chains randomly. We consider $S_M = 5000$ and $S_T = 25000$ for RBM20 and $S_M = 2500$ and $S_T = 50000$ for the other two RBMs. We consider both small and large burn-in setup for the biased CSL estimator.

We fix the value of T to 100 and vary the number of chains, M, keeping the number of samples, N, constant and these results are presented in Table 2. We observe from the table that better estimates with less variance can be obtained by increasing the number of chains. Note that we are keeping the total number of samples fixed even when we vary the number of chains. Thus, for example, for RBM500, we get lower variance by having a total of 100,000 samples from 1000 chains rather than having 200,000 samples but from only 500 chains. Thus

Table 2. CSL estimate of $\mathscr{L}$ for $T = 100$ and for various values of M, keeping N constant. **Ground truth: log Z = 230.61, $\mathscr{L} = -141.24$ for RBM20**

		$n = 500$		$n = 200$		$n = 20$		
N	M	$\hat{\mathscr{L}}$	σ^2	$\hat{\mathscr{L}}$	σ^2	M	$\hat{\mathscr{L}}$	σ^2
a. Unbiased CSL								
100×10^3	250	-296.18	5.90	-183.67	2.31	500	-153.83	0.83
	500	-284.25	5.87	-175.54	1.65	1000	-150.39	0.58
	1000	-272.62	3.92	-167.70	1.45	2000	-147.94	0.31
200×10^3	500	-270.14	4.56	-172.41	1.65	1000	-148.61	0.35
	1000	-257.29	3.19	-165.04	1.01	2000	-146.31	0.20
	2000	-244.87	2.49	-158.76	0.68	4000	-144.44	0.09
b. Biased CSL with large Burn-in								
100×10^3	250	-219.34	4.64	-176.94	1.89	500	-151.18	0.53
	500	-208.12	3.86	-169.26	1.47	1000	-148.41	0.37
	1000	-197.22	2.17	-162.05	0.92	2000	-146.22	0.25
200×10^3	500	-198.84	2.61	-167.04	1.56	1000	-147.17	0.26
	1000	-188.32	1.65	-160.11	0.93	2000	-145.27	0.17
	2000	-179.69	1.24	-154.05	0.54	4000	-143.52	0.09
c. Biased CSL with small Burn-in								
100×10^3	250	-187.83	2.52	-166.91	1.04	500	-147.82	0.22
	500	-178.12	1.99	-160.23	0.98	1000	-145.54	0.23
	1000	-169.56	1.50	-154.03	0.54	2000	-143.74	0.20
200×10^3	500	-177.09	1.89	-159.31	0.99	1000	-145.31	0.19
	1000	-168.82	1.42	-153.27	0.59	2000	-143.57	0.14
	2000	-161.87	1.45	-148.01	0.39	4000	-142.06	0.13

having more chains also reduces the total computational effort (because we can do with less number of samples).

The final estimates are close to ground truth in case of RBM20, even though the accuracy here is a bit poorer than that of AIS. For RBM200 and RBM500, the estimates differ by a large amount from those obtained with the AIS, even when N is very large. This deviation is smaller for the biased CSL estimates obtained with small burn-in than that of the unbiased CSL estimates and the biased CSL estimates obtained with large burn-in (refer Table. 2). This may be due to high level of correlation among the samples. Further, we observe that, for the RBM20, the biased CSL estimate gives a lower bound.

RAISE: The RAISE estimator requires implementation of the AIS chain for each test sample. This makes the estimator computationally expensive because the MNIST dataset contains $10,000$ test samples. Therefore, for the experiment we randomly select a subset of size 500 (50 from each class) from the test dataset and then estimate the average test log-likelihood on this subset. We experiment with both uniform and base rate proposal distribution by fixing the number of AIS runs and varying the number of intermediate distributions, K. We also estimate the test log-likelihood on the chosen test subset using the AIS and CSL estimators, for comparison with the RAISE estimates. We keep the number of Gibbs steps used in the AIS and CSL estimators equal.

We observe that, only when K is very large and proposal distribution is uniform, the estimator provides conservative estimates. However, the baserate proposal gives overestimates even when $K = 10000$. The lower bound on the test log-likelihood is similar to that of CSL estimate for RBM20. However, unlike the case for CSL, for larger RBMs, the RAISE estimates matches closely with the AIS estimates (Table 3).

Table 3. RAISE estimate of $\mathscr{L}$ with 500 test set, **Ground truth:** $\mathscr{L} = -142.39$ for **RBM20**. The baserate AIS with $K = 10000$, $M = 500$ is used. For the CSL, 5000×10^3 samples are used to make the number of Gibbs steps equal to that of AIS.

n	Proposal \ K	1000	10000	AIS	CSL
20	Uniform	-147.25	-145.99	-142.38	-143.58
	Baserate	-146.97	-144.14		
200	Uniform	-109.29	-112.46	-112.96	-142.64
	Baserate	-110.42	-109.01		
500	Uniform	-114.75	-118.02	-116.46	-154.76
	Baserate	-108.75	-112.04		

5   Conclusion

Calculating the average test log-likelihood of a learnt RBM is important for evaluating different learning strategies. In this paper we present extensive empirical analysis of the sampling based estimators for average test log-likelihood to gain insight into the performance of these methods. We experiment with RBMs with 20, 200 and 500 hidden units. We observed that the AIS estimator delivers good estimates with low variance. We also observe that the proposal distribution does not seem to have much influence on the estimate.

Compared to the AIS estimate the CSL estimate is poorer and its variance is high especially for the RBM200 and RBM500. The estimated value also differ significantly with that of AIS. However, CSL is a much simpler estimator computationally. We also showed that better estimates can be obtained with less computational effort by using multiple independent chains to generate samples.

Unlike AIS, the RAISE gives conservative estimates. More importantly, for large RBMs, the deviation of RAISE estimate from the AIS estimate is not very large compared to that of the deviation of CSL estimate from the AIS estimate. It means, for large K, the RAISE estimate will have tighter lower bound than the CSL estimate. However it is computationally much more expensive. The conservativeness of RAISE may not be enough to justify the high computational cost.

Since large hidden unit RBMs are an important part of deep networks such as stacked-RBMs and DBNs, one needs efficient estimators to evaluate the learnt networks. Our empirical study indicates that there may be much scope for improving CSL like estimators to come up with computationally simple methods to get good estimates of the average test log-likelihood.

References

1. Bengio, Y., Yao, L., Cho, K.: Bounding the test log-likelihood of generative models (2013). arXiv preprint arXiv:1311.6184
2. Burda, Y., Grosse, R.B., Salakhutdinov, R.: Accurate and conservative estimates of MRF log-likelihood using reverse annealing (2014). arXiv preprint arXiv:1412.8566
3. Fischer, A., Igel, C.: An introduction to restricted boltzmann machines. In: Alvarez, L., Mejail, M., Gomez, L., Jacobo, J. (eds.) CIARP 2012. LNCS, vol. 7441, pp. 14–36. Springer, Heidelberg (2012)
4. Freund, Y., Haussler, D.: Unsupervised learning of distributions on binary vectors using two layer networks. In: Moody, J., Hanson, S., Lippmann, R. (eds.) Advances in Neural Information Processing Systems 4, pp. 912–919. Morgan-Kaufmann, USA (1992)
5. Hinton, G.E.: Training products of experts by minimizing contrastive divergence. Neural Comput. 14(8), 1771–1800 (2002)
6. MacKay, D.J.: Information Theory, Inference, and Learning Algorithms, vol. 7. Cambridge University Press, Cambridge (2003)
7. Neal, R.: Annealed importance sampling. Stat. Comput. 11(2), 125–139 (2001)
8. Salakhutdinov, R., Murray, I.: On the quantitative analysis of deep belief networks. In: Proceedings of the 25th International Conference on Machine Learning, pp. 872–879. ACM (2008)
9. Smolensky, P.: Information processing in dynamical systems: foundations of harmony theory (1986)

Autonomous Depth Perception of Humanoid Robot Using Binocular Vision System Through Sensorimotor Interaction with Environment

Yongsik Jin, Mallipeddi Rammohan, Giyoung Lee, and Minho Lee[✉]

School of Electrics Engineering, Kyungpook National University, 1370 Sankyuk-Dong, Puk-Gu, Taegu, 702-701, South Korea
{specialone.ys,mallipeddi.ram,giyoung0606,mholee}@gmail.com

Abstract. In this paper, we explore how a humanoid robot having two cameras can learn to improve depth perception by itself. We propose an approach that can autonomously improve depth estimation of the humanoid robot. This approach can tune parameters that are required for binocular vision system of the humanoid robot and improve depth perception automatically through interaction with environment. To set parameters of binocular vision system of the humanoid robot, the robot utilizes sensory invariant driven action (SIDA). The sensory invariant driven action (SIDA) gives identical sensory stimulus to the robot even though actions are not same. These actions are autonomously generated by the humanoid robot without the external control in order to improve depth perception. The humanoid robot can gather training data so as to tune parameters of binocular vision system from the sensory invariant driven action (SIDA). Object size invariance (OSI) is used to examine whether or not current depth estimation is correct. If the current depth estimation is reliable, the robot tunes the parameters of binocular vision system based on object size invariance (OSI) again. The humanoid robot interacts with environment so as to understand a relation between the size of the object and distance to the object from the robot. Our approach shows that action plays an important role in the perception. Experimental results show that the proposed approach can successfully and automatically improve depth estimation of the humanoid robot.

Keywords: Autonomous development robot · Depth estimation · Robot learning · Interactive robot · Binocular vision system · Humanoid robot · Action-perception cycle learning

1 Introduction

How human being can learn and understand to transform visual stimuli into accurate distance information? Visual information surrounding humans is much noisy, messy and provides very little clear meaningful information to them [1, 2]. This complex perception procedure still remains unknown. The human being is not born having binocular depth perception ability. Infants learn visual ability after birth [3, 4]. Humans should learn to process brain signals from external stimuli in order to understand information

© Springer International Publishing Switzerland 2015
S. Arik et al. (Eds.): ICONIP 2015, Part II, LNCS 9490, pp. 554–561, 2015.
DOI: 10.1007/978-3-319-26535-3_63

from environment (see Fig. 1). Several studies show that actions play an important role when humans and animals learn perception ability [4, 5]. Results of Held and Hein' study [5] show that action is critical in the development of visual processing of kitty. This result of study means that self-driven action largely concerns development of perception ability. In case of robots that are not programmed with depth estimation or don't have tuned suitable parameters, the robots have problem similar to infants. Previous research proposed different ways to develop robot perception ability by robots itself [6, 7]. Timothy A. Mann et al. [6] proposed how the robots can learn to improve binocular depth perception ability. Their approach shows that some actions of the robot are helpful to get information regarding parameters and find proper parameter for binocular depth estimation.

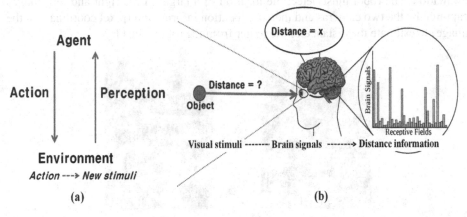

Fig. 1. Overall concept for autonomous depth perception: (a) (1) Action generates new stimuli in environment. (2) Agent gets new stimuli taking actions in environment. (3) Agent learns from stimuli. (b) Humans are not born with binocular depth estimation ability but they should learn to estimate distance from visual stimuli after birth.

In this study, we propose an approach that can improve binocular depth estimation ability by a robot itself through interaction with environment (see Fig. 1 (a)). Environment gives stimuli to robots according to actions as shown Fig. 1 (a). The robot interacts with environment taking some actions. The sensory invariant driven action (SIDA) gives identical sensory stimuli to the robot even though actions are not same. The robot can gather training data to tune parameters of binocular depth estimation from these actions. We use the object size invariance (OSI) to improve depth estimation ability [8]. Distance from object is derived by the relation between perceived size of object and distance [8, 9]. The robot utilizes object size invariance (OSI) to examine whether or not current depth estimation is correct. If the current depth estimation is reliable, the robot tunes the parameters of binocular vision system based on object size invariance (OSI). The robot can improve depth estimation ability walking around the object based on sensory invariant driven action (SIDA) and object size invariance (OSI).

The rest of the paper is organized as follows: in the next section, we explain our proposed approach for autonomous learning and improvement of binocular vision system of the humanoid robot in details. Experimental results are reported in Sect. 3. In Sect. 4, we conclude our results and present our future plans.

2   Proposed Method

2.1   Experimental Setting of Robot

A modified Aldebaran Nao humanoid robot is used in our experiment (Fig. 2 (a)). We mounted two cameras on the head of the robot. We attached a yellow sticker to the hand of the robot in order to easily detect the finger (Fig. 2 (b)). Yellow sticker on the finger of the robot is used as a target to gather training data for the improvement of depth estimation. The robot must detect the location of a target in the right and left images captured by the two cameras and then use location information (pixel coordinate in the image) to estimate the distance of the finger from the robot's head.

(a)   **(b)**

Fig. 2. Experimental setup using Nao robot with stereo cameras: (a) A modified Aldebaran Nao humanoid robot (b) A target on the hand of the robot (Color figure online)

In case of actions of the robot, we use pre-implemented control functions for generating behavior of robot. Owing to the libraries of Nao humanoid robot, we can easily use low level control functions.

2.2   Visual Attention Saliency Map

Correspondence is one of problem of binocular vision system. It is difficult to match objects in right and left images exactly because of some troubles such as noise, illumination. We use biologically inspired attention mechanism proposed by others [10] to detect object and solve correspondence problem. This attention model is useful to detect the same target object from the left and right images using limited top-down signals (e.g. color). We can get angles and size information of object in the right and left images. The robot estimates distance to object from robot's head using size and angles information of detected object.

2.3 Object Size Invariance (OSI)

Perceived object size is used to estimate distance to target object from robot's head. The relation between perceived size of object and distance is inverse proportional [8, 9]. Previous studies suggest some approaches to predict distance from perceived size of object [9]. We use a regression analysis method to apply to real world problem such as robot. We introduce fractional function to fit data distribution between perceived size of object and distance.

$$d = \frac{1}{a_1 s + a_0} \tag{1}$$

Where d is distance and where s is perceived size of object. We change Eqs. (1) and (2) to make linear relationship between perceived object size and distance to object from robot's head.

$$\frac{1}{d} = a_1 s + a_0 \tag{2}$$

We use follow linear regression equation to calculate a_1 and a_0.

$$a_1 = \frac{n \sum s_i \frac{1}{d_i} - \sum s_i \sum \frac{1}{d_i}}{n \sum s_i^2 - (\sum s_i)^2} \tag{3}$$

$$a_0 = \left(\overline{\frac{1}{d}}\right) - a_1 \bar{s} \tag{4}$$

Where n is the number of sample data. Where $\bar{s}$ and $\left(\overline{\frac{1}{d}}\right)$ are mean of perceived size of object and mean of inverse of distance.

2.4 Binocular Depth Estimation Algorithm

In our case, the robot has innate algorithm to estimate distance, but the agent do not know the parameters of binocular depth estimation system. We simplify depth estimation algorithm to reduce parameters. To calculate distance from binocular system, the two angles θ_L and θ_R in radians of the center coordinate (x_L, x_R) of the object in images should be estimated by simple triangular equation with focal length of two cameras [10].

$$\theta_L = \mathrm{atan}\left(\frac{f}{\frac{P_{width}}{2} + x_L}\right) \tag{5}$$

$$\theta_R = \mathrm{atan}\left(\frac{f}{\frac{P_{width}}{2} - \left(P_{width} - x_R\right)}\right) \tag{6}$$

Where f is focal length of the cameras and P_{width} denotes the pixel width of images. The robot must find focal length automatically because focal length remains unknown in our binocular vision system. From the two angles (θ_L, θ_R) of the center coordinate (x_L, x_R) of the object in images, we can calculate distance to target object from baseline. Where B is the baseline or distance between the two cameras as shown Fig. 3. The distance, y is calculated as the following Eq. (7).

$$y = \frac{B(\tan(\theta_L)\tan(\theta_R))}{\tan(\theta_L) + \tan(\theta_R)} \tag{7}$$

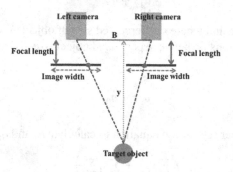

Fig. 3. Binocular depth estimation system

Proper B is given to the robot according to robot's hardware information. We just consider focal length of the camera.

2.5   Sensory Invariant Driven Action (SIDA)

Sensory invariant driven action (SIDA) explains how the brain can learn the meaning of visual stimuli [11]. Self-learning for improvement of binocular depth estimation is proposed by Timothy A. Mann et al. [6]. This approach uses head rotation to gather training data and then the robot adjusts biased parameters. Our approach also uses head rotation to find a proper parameter. Our robot should tune f(focal length) by itself so as to estimate depth accurately because f(focal length) is not given by a supervisor. We can get training data for tuning f(focal length) from sensory invariant driven action (SIDA).

Distance from center of rotation to the target is calculated as follow:

$$x = \frac{y}{tan(\theta_L)} - \frac{B}{2} \tag{8}$$

$$d = \sqrt{x^2 + (y + r)^2} \tag{9}$$

The y is calculated from Eq. (7). In Eq. (8), B is baseline between the two cameras. Equation (8) consider the case where θ_L and θ_R are both less than $\frac{\pi}{2}$ radians. Other cases can be derived similarly.

As shown in Fig. 4 and Eq. (9), x (horizontal distance from the point between the left and right camera and the target point) and y (the vertical distance from baseline of two cameras to the target) are changed when humanoid robot rotates its head but d (the distance from the center of rotation to the target) does not change. From this fact, we can gather training data for tuning focal length in Eqs. (5) and (6) using sensory invariance driven action (SIDA). We use rotating head as sensory invariant driven action (SIDA).

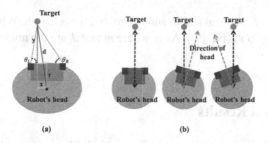

(a)   (b)

Fig. 4. Depth estimation: (a) Aerial view of binocular distance estimation system of humanoid robot. The y quantity (red line) is the vertical distance from baseline of two cameras to the target. The x (yellow line) quantity is the horizontal distance from the point between the left and right cameras and the target point. The d (blue line) is the distance from the center of rotation to the target, with radius r (green line). (b) Distance from center of rotation to target is identical although the direction of face is changed by rotating robot's head (Color figure online).

3   Learning Framework

In this paper, we propose a new depth perception learning framework that gathers training data automatically and incrementally and then retrains depth perception by robots itself. Our robot can know the relation between perceived size of target object and distance from robot's head to target object from Eq. (1) if object size information is known. At first our robot searches focal length from training data gathered by sensory invariant driven action (SIDA). In first stage, target object is a yellow sticker attached on robot's hand. The robot knows the distance from robot's head to robot's hand (yellow sticker attached on the finger of robot). Our humanoid robot can estimate distance using found focal length in first stage but it is not suitable because this depth estimator becomes inaccurate when target object is located apart from robot. To solve this problem, the robot utilizes object size invariance (OSI) information. If results of depth estimation from binocular depth estimator (Eq. (9)) and OSI (Eq. (1)) are similar, the robot can believe this distance information. The robot walks around another target object (The information of size of this object is known by a supervisor. We used red box ($7 \times 7 \times 7$ Cm^3)) after first stage and then checks the results of depth estimation. In case that difference of depth estimation between binocular depth estimator (Eq. (9)) and OSI (Eq. (1)) is less than a threshold, the robot gathers training data using sensory invariant driven action (SIDA) based on reliable this distance (we use OSI distance as reference distance) and then searches focal length again. All of training data is stored in memory and then used in parameter (focal length) search. Stochastic optimization or exhaustive grid-based search such as genetic algorithms or simulated annealing can be used to find focal length [12]. We modify Timothy A. Mann et al.'s approach [6]. The focal length is searched as follow:

$$\hat{f} \approx \arg\min_f \sum_i^m (d_i - D(f, \theta_{Li}, \theta_{Ri}))^2 \qquad (10)$$

Where f and $\hat{f}$ are focal length and optimized focal length. Where D is binocular depth estimation function (Eq. (9)). Also, where m and d_i are the number of samples and real distance at sample i.

4   Experimental Results

Figure 5 shows the experimental results. The red solid line represents the result of our learning framework. After the robot learns to estimate depth from target object (Yellow sticker) on the hand of the robot, the robot gathers training data from another object (We used red box ($7 \times 7 \times 7$ cm³)) and then train again the parameter (focal length). The blue solid line shows the results that are trained by only target object on the hand of the robot one time. As shown in Fig. 5, the experimental results show that the humanoid robot can learn to estimate binocular depth perception as well as improve depth estimation ability using our approach.

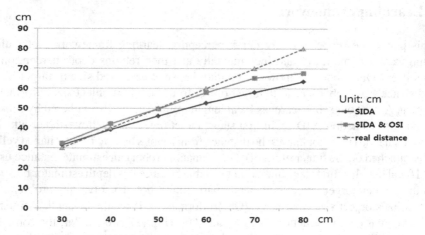

Fig. 5. Experimental result: SIDA (Blue solid line) was trained only one time based on only sensory invariant driven action (SIDA). SIDA & OSI (Red solid line) was trained several times based on sensory invariant driven action (SIDA) and object size invariance (OSI) by another object after training from the object on the robot's hand (Yellow sticker) (Color figure online).

5   Conclusion and Further Work

We propose a learning framework that can learn to estimate depth from object size invariance (OSI) and sensory invariant driven action (SIDA). This proposed approach shows that the robot can gather training data automatically and incrementally. Also, the robot can learn and improve binocular depth perception autonomously.

As a further work, we are considering more accurate and complex depth estimator and learning frameworks so as to solve intrinsic problems of binocular vision system or multiple parameters search problems.

Acknowledgements. This work was partly supported by the Industrial Strategic Technology Development Program (10044009) funded by the Ministry of Trade, Industry and Energy (MOTIE, Korea) (50 %) and Basic Science Research Program through the National Research Foundation of Korea (NRF) funded by the Ministry of Science, ICT and future Planning (2013R1A2A2A01068687) (50 %).

References

1. Mugan, J., Kuipers, B.: Autonomous learning of high-level states and actions in continuous environments. IEEE Trans. Auton. Ment. Dev. **4**(1), 70–86 (2012)
2. Kuipers, B.J., Beeson, P., Modayil, J., Provost, J.: Bootstrap learning of foundational representations. Connection Sci. **18**(2), 145–158 (2006)
3. Sokol, S.: Measurement of infant visual acuity from pattern reversal evoked potentials. Vis. Res. **18**(1), 33–39 (1978)
4. Nawrot, E., Mayo, S.L., Nawrot, M.: The development of depth perception from motion parallax in infancy. Percept. Psychophys. **71**(1), 194–199 (2009)
5. Held, R., Hein, A.: Movement-produced stimulation in the development of visually guided behavior. J. Comp. Physiol. Psychol. **56**(5), 872 (1963)
6. Mann, T.A., Park, Y., Jeong, S., Lee, M., Choe, Y.: Autonomous and interactive improvement of binocular visual depth estimation through sensorimotor interaction. IEEE Trans. Auton. Ment. Dev. **5**(1), 74–84 (2013)
7. Mansard, N., Stasse, O., Chaumette, F., Yokoi, K.: Visually-guided grasping while walking on a humanoid robot. In: 2007 IEEE International Conference on Robotics and Automation. IEEE (2007)
8. McCready, D.: On size, distance, and visual angle perception. Percept. Psychophys. **37**(4), 323–334 (1985)
9. Gilinsky, A.S.: Perceived size and distance in visual space. Psychol. Rev. **58**(6), 460 (1951)
10. Choi, S.-B., Jung, B.-S., Ban, S.-W., Niitsuma, H., Lee, M.: Biologically motivated vergence control system using human-like selective attention model. Neurocomputing. **69**(4), 537–558 (2006)
11. Choe, Y., Yang, H.-F., Eng, D.C.-Y.: Autonomous learning of the semantics of internal sensory states based on motor exploration. Int. J. Humanoid Rob. **4**(02), 211–243 (2007)
12. Whitley, D.: A genetic algorithm tutorial. Stat. Comput. **4**(2), 65–85 (1994)

An Effective Resolution Method of Chinese Multi-category Words with Conditional Random Field in Electronic Commerce

Fan Fei, Yanqin Yang[✉], Wenchao Xu, and Yanfeng Yang

School of Information Science Technology, East China Normal University,
500 Dongchuan Road, Shanghai 200241, China
yqyang@cs.ecnu.edu.cn

Abstract. Most existing recognition methods of multi-category words merely focus on the area of traditional Chinese words rather than the area of electronic commerce. In this paper, we propose an effective method on how to recognize Chinese multi-category words in electronic commerce using Conditional Random Field. Experimental results show that our method remarkably enhances the accuracy of the recognition, reduces the misunderstanding and improves the user experience of electronic commerce retrieval.

Keywords: Electronic commerce · Chinese multi-category words · Conditional random field

1 Introduction

In recent years, electronic commerce has become more and more popular and played more important influence on our working and life than ever before. Ambiguous word, the word or phrase has no less than two kinds of meanings, is widely used in electronic commerce. In Chinese Word, ambiguous word includes polyphone, homophone, polysemant, enantiosemy, multi-category words etc. Especially, Chinese multi-category word is the most representative form of ambiguous word. The more accurate recognition method of multi-category word is used, the more effective processing and searching results of the texts can be reached. The accuracy of part-of-speech tagging will directly affect recognition performance of multi-category words and results of analysis and processing.

Currently, there are two frequently used recognition methods for part-of-speech tagging: rule-based method and statistical-based method. Rule-based method focuses on string matching algorithm [9, 15] or the shortest path algorithm [11, 13]. Rule-based method relies on the set of rules which need to be completeness and reasonable [7]. Because the subjectivity cannot be avoided, rule-based method is difficult to guarantee the consistency of the rules and not applicable for any corpus. When dealing with the multi-category words, it's time-consuming and toilsome. Statistical-based method is the most widely used approach of part-of-speech tagging [6, 14] currently. This method counts the frequency weight for each word or phrase on the basis of composition, characteristics of information and related context information [10, 12]. Then it will select

© Springer International Publishing Switzerland 2015
S. Arik et al. (Eds.): ICONIP 2015, Part II, LNCS 9490, pp. 562–570, 2015.
DOI: 10.1007/978-3-319-26535-3_64

the most optimal state transition probability to determine the word and word class. The Most typical models of the statistical-based measure are Hidden Markov Model [5], Maximum Entropy Markov Model [3] and Conditional Random Field [2]. The Drawback of Hidden Markov Model is that the observed value only depends on the state which makes each observation element exist independently under the condition of a given observation sequence [4]. However in the practical context, the center word is not only related to its close neighbors but also associated with words far from which has some close characteristic information. This algorithm merely provides a locally optimal method. Maximum Entropy Markov Model takes the associated characteristic information between the current word and the relevant words into account [3]. But at the time of probability state transition, due to the different branch number and the imbalance probability distribution, it will lead to the problem of label bias: during the probability state transition, the current state resides in a particular probability. Conditional Random Field is an undirected non-cyclic graphical model [8, 9]. The characteristic takes associated information between the current word and relevant words into consideration, thus it can avoid the problem of strict independence assumption, and also overcome the problem of label bias. The conditional random filed propose an efficient algorithm which obtains a globally optimal solution.

In the area of traditional Chinese words, applying recognition method into the area of electronic commerce is rarely involved. In our previous study [1], we have accomplished the out-of-vocabulary words recognition. Furthermore, we find the processing performance of ambiguous word is of great significance in the area of electronic commerce. In this paper, we propose a method based on Conditional Random Field [2] to recognize Chinese multi-category words in electronic commerce. According to the text characteristic in electronic commerce, we optimize the estimation parameter and modify the feature template, then we extract experimental data from two working electronic commerce website. Experimental results show that our method remarkably enhances the accuracy of the recognition, reduce the misunderstanding and improve the user experience of electronic commerce retrieval.

The rest of the paper is organized as follows. Section 2 proposes Conditional Random Field and recognition method of Chinese multi-category words in electronic commerce. The results and analysis of comparison experiment are given in Sects. 3 and 4 comes to the conclusion.

2   Conditional Random Field and Recognition of Chinese Multi-category Words in Electronic Commerce

2.1   Conditional Random Field

Conditional Random Field is originally proposed in 2001 [2], which comes from the model of maximum entropy. It is used as a statistical framework model for labeling and segmentation of sequence data. We can take the Conditional Random Field as an undirected graph model or Markov Random Field. Assuming that X, Y represents the joint distribution of random variables of observed sequence and the relevant tag sequence respectively, Conditional Random Field (X, Y) is an undirected graph model which takes observed sequence x as the condition. The definition of G = (V, E) is an undirected graph. In theory, the structure of the graph G can be flexible which describes conditional independence of tag sequence. The simplest and most common undirected graph structure is a linear chain structure as shown in Fig. 1.

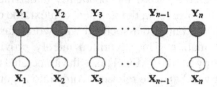

Fig. 1. Linear chain structure

We assume that each output node has first-order Markov independence. There exists no graph structure between the elements of X, because we just take the observed sequence as the condition and don't make independence assumption for any elements of X. The chain structure of Conditional Random Field can also be expressed in Fig. 2.

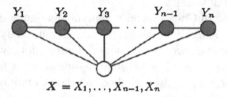

Fig. 2. Another chain structure of an undirected graph

Each state transition of Conditional Random Field corresponds to a non-normalized weight. Any state sequence weight is given by the global normalization factor, thus Conditional Random Field has avoided occurrence of label bias problem.

2.2 Optimization of Parameter Estimation

For Conditional Random Field and other maximum entropy model, the gradient vector of each element is given by the following formula:

$$\frac{\partial L(\Theta)}{\partial \lambda_k} = E_{p(x,y)}[f_k] - E_{p(y|x,\Theta)}[f_k] \tag{1}$$

Compared with the direct method using the gradient iteration method, iterative scaling method doesn't have any benefits. On one hand, when estimating parameters of conditional Maximum Entropy Model, the performance of first-order, second-order numerical optimization technique is superior to iterative scaling algorithm. On the other hand, under the condition of a given observation sequence, we find the tag sequence of conditional distribution formula for p (y|x):

$$p(y|x) = \frac{1}{z(x)}\exp\left(\sum_i \sum_k \lambda_k f_k(y_{i-1}, y_i, x) + \sum_i \sum_k \mu_k g_k(y_i, x)\right) \qquad (2)$$

The form of above formula is similar to the form of function of Maximum Entropy Model:

$$p(y|x) = \frac{1}{z(x)}\exp\left(\sum_k \lambda_k f_k(x,y)\right) \qquad (3)$$

Numerical optimization methods show good performance of parameter estimation in Conditional Random Field.

2.3 The Recognition of Chinese Multi-category Words in Electronic Commerce

In this part, we apply the Conditional Random Field to the recognition of Chinese multi-category word in electronic commerce, and furthermore compare the experiment result generated by the improved feature template to the result obtained from the original feature template. The model is shown in Fig. 3.

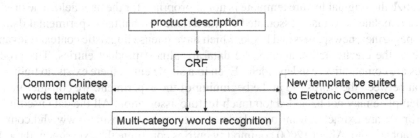

Fig. 3. The model of multi-category words with Conditional Random Field

The recognition steps of Chinese multi-category words in electronic commerce are followed:

(1) Take the data whose brand name is consistent with the title on commodity packaging picture as corpus for experiment and meanwhile revise the wrong entries according to the corresponding commodity.
(2) Cut the commodity entries data into manufacturer block, origin block, brand block, and net content block. Then divide each separate lexical block into characters, and use B to mark the separate lexical block of the first text features for the initial term of lexical block, and use I to mark all the text features of lexical block as the following words. If the lexical block only contains one word, O is described in the text features of lexical block as the independent block.
(3) Attach the label to each character in this given corpus. The specific corresponding part-of- speech tagging is shown in Table 1.

(4) Sort out the commodity entries which contains the same multi-category word from the corpus, a part of which are chosen as the training set and rest are as the test set. The remaining commodity entries in this corpus are processed in the same way. Finally, we will evaluate the recognition performance of multi-category word.

Table 1. Part-of-speech tagging on Chinese multi-category words in electronic commerce

Tab	Block	Tab	Block	Tab	Block	Tab	Block
M	Package	M	Bottle	M	Bucket	M	Slice
M	Bowl	Z	Taste	N	Brand	NP	Category
NP	Goods	NJ	Merchants	NS	Place	T	Time
Q	Quantity	NQ	Quality	NG	Specification	NM	money
A	Adjectives	AD	Shape	TG	Season	W	Symbol

3 Experiment Results and Analysis

When applying Conditional Random Field into the new area of electronic commerce, we realize the original feature template is not appropriate for the new field. The original feature template adds many associated characteristics in that the experimental data are from magazines, newspapers and books which have intense linguistic context relevance. While in the electronic commerce, the data are almost product entries. The product entries are comparatively independent. Even in a single entry, there exists no tight association between the words. So in the beginning of the experiment, we consider that the new template may not be added too much feature association. All the data used in this experiment are extracted from two electronic commerce website: www.yhd.com and www.taobao.com. About 12000 commodity entries constitute the experiment data. The corresponding training set of commodity entries which contains the same multi-category word probably involves 1000 entries, and the corresponding test set includes approximately 500 entries. In this experiment, the feature template set we select for this corpus is shown in Fig. 4.

Unigram U00:%x[-3,0] U03:%x[0,0] U06:%x[3,0] U07:%x[-2,0]/%x[0,0] U10:%x[-2,0]/%x[-1,0]

U01:%x[-2,0] U04:%x[1,0]   U08:%x[-1,0]/%x[0,0] U11:%x[1,0]/%x[2,0]

U02:%x[-1,0] U05:%x[2,0]   U09:%x[0,0]/%x[1,0] U12:%x[0,0]/%x[2,0]

Fig. 4. Modified template

The evaluation criteria of the experiment result are precision, recall rate and F-measure. They can be computed as followed:

$$\text{precision} = \frac{A}{A+C}, \text{ recall rate} = \frac{A}{A+B}, \text{ F-measure} = \frac{2 \times \text{precision} \times \text{recall}}{\text{precision} + \text{recall}} \times 100\%$$

where A, B, C and D is the quantity of related words the system can extract, related words the system cannot extract, unrelated words the system can extract, and unrelated words the system cannot extract respectively.

The total precision reflects the probability that the predicted mark is equal to the real mark. Therefore, the accuracy of the word recognition largely depends on the quantity of words which the state field contains. Generally, the more the quantities of the words are, the higher the accuracy of the experiment result is.

The results of two comparison experiments are shown in Table 2 and Figs. 5, 6, 7 and 8. One of the templates is common Chinese words template and the other is a modified template which is fit for Chines multi-category words in electronic commerce. In electronic commerce, the form of Chinese multi-category word is commonly noun and adjective. We consider two kinds of words accordingly. When recognizing the form of noun, we find the precision from modified template is up to 77.49 % and raise by 4.23 % than common template, recall rate is up to 71.22 % and raise by 2.51 %, f-measure is up to 74.22 % and raise by 3.31 %. When recognizing the form of adjective, we find the precision from modified template is up to 75.36 % and raise by 10.33 %, recall rate is up to 70.09 % and raise by 6.94 %, f-measure is up to 72.63 % and raise by 8.55 %.

Table 2. The recognition results based on common template and modified template

common template	precision	recall	f-measure	modified template	precision	recall	f-measure
noun	73.26%	68.71%	70.91%	noun	77.49%	71.22%	74.22%
adjective	65.03%	63.15%	64.08%	adjective	75.36%	70.09%	72.63%

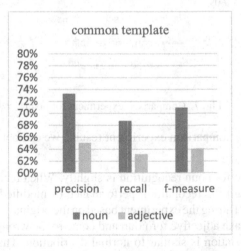

Fig. 5. Common template results

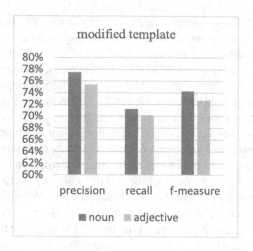

Fig. 6 Modified template results

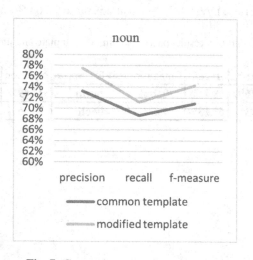

Fig. 7 Comparison experiment for noun

On the basis of the comparison experiment results, we can obtain some conclusion as followed:

(1) The improvement for noun recognition is slightly. When recognizing the form of adjective, we learn the recognition performance of modified template is significantly promoted. During the experiment based on the original template, the machine will usually mistake adjective for noun and come to the wrong meaning.

(2) The feature association is similar to normal distribution. That is to say, the new template may not be added too much or too little feature association. The proper feature association will help the machine decrease the error range.

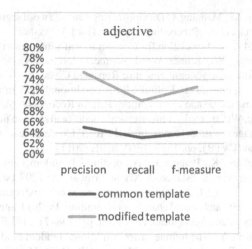

Fig. 8 Comparison experiment for adjective

4 Conclusion

Due to the unique property of electronic commerce, the introduction of the original common template for Chinese multi-category words recognition will appear deviation and even cause error. This paper presents a modified template based on Conditional Random Field for the recognition of Chinese multi-category words in electronic commerce. The results show that the modified template remarkably enhances the accuracy of the recognition, reduce the misunderstanding and improve the user experience of electronic commerce retrieval.

Acknowledgment. This work was supported by the National Natural Science Foundation of China under Grant No. 61300043.

References

1. Yang, Y., Yang, Y., Guan, H., Xu, W.: Out-of-vocabulary words recognition based on conditional random field in electronic commerce. In: Loo, C.K., Yap, K.S., Wong, K.W., Teoh, A., Huang, K. (eds.) ICONIP 2014, Part II. LNCS, vol. 8835, pp. 532–539. Springer, Heidelberg (2014)
2. Lafferty, J., McCallum, A., Pereira, F.C.N.: Conditional random fields: Probabilistic models for segmenting and labeling sequence data (2001)
3. Sutton, C., McCallum, A., Rohanimanesh, K.: Dynamic conditional random fields: Factorized probabilistic models for labeling and segmenting sequence data. J. Mach. Learn. Res. **8**, 693–723 (2007)
4. McCallum, A., Bellare, K., Pereira, F.: A conditional random field for discriminatively-trained finite-state string edit distance. ArXiv preprint arXiv: 1207.1406 (2012)
5. Levow, G.A.: Automatic prosodic labeling with conditional random fields and rich acoustic features. In: IJCNLP, pp. 217–224 (2008)

6. Chang, P.C., Galley, M., Manning, C.D.: Optimizing Chinese word segmentation for machine translation performance. In: Proceedings of the Third Workshop on Statistical Machine Translation, pp. 224–232. Association for Computational Linguistics (2008)

7. Van, C., Kameyama, W.: Khmer Word Segmentation and Out-of-Vocabulary Words Detection Using Collocation Measurement of Repeated Characters Subsequences (2013)

8. Chatzis, S.P., Kosmopoulos, D.I., Doliotis, P.: A conditional random field-based model for joint sequence segmentation and classification. Pattern Recogn. **46**(6), 1569–1578 (2013)

9. Huang, Q., Han, M., Wu, B., et al.: A hierarchical conditional random field model for labeling and segmenting images of street scenes. In: 2011 IEEE Conference on Computer Vision and Pattern Recognition (CVPR), pp. 1953–1960. IEEE (2011)

10. McCallum, A., Bellare, K., Pereira, F.: A conditional random field for discriminatively-trained finite-state string edit distance. ArXiv preprint arXiv: 1207.1406 (2012)

11. Zhang, H., Liu, C., Yang, C., et al.: An improved scene text extraction method using conditional random field and optical character recognition. In: 2011 International Conference on Document Analysis and Recognition (ICDAR), pp. 708–712. IEEE (2011)

12. Chatzis, S.P., Demiris, Y.: The infinite-order conditional random field model for sequential data modeling. IEEE Trans. Pattern Anal. Mach. Intell. **35**(6), 1523–1534 (2013)

13. Yang, H.D., Lee, S.W.: Robust sign language recognition by combining manual and non-manual features based on conditional random field and support vector machine. Pattern Recogn. Lett. **34**(16), 2051–2056 (2013)

14. Wang, Y.R., Liao, Y.F., Wu, Y.K., et al.: Conditional random field-based parser and language model for traditional chinese spelling checker. In: Proceedings of the 7th SIGHAN Workshop on Chinese Language Processing (SIGHAN 2013), pp. 69–73 (2013)

15. Yu, D., Deng, L., Wang, S.: Deep-structured conditional random fields for sequential labeling and classification: U.S. Patent 8,473,430, 25 June 2013

A Novel Hybrid Modelling for Aggregate Production Planning in a Reconfigurable Assembly Unit for Optoelectronics

Francesco G. Sisca[✉], Maurizio Fiasché, and Marco Taisch

Department of Management, Economics and Industrial Engineering, Politecnico di Milano,
Milan, Italy
francescogiovanni.sisca@polimi.it

Abstract. In this work it is presented a novel fuzzy multi-objective linear programming (FMOLP) model based on hybrid fuzzy inference systems for solving the general management framework fort the integration of a self-contained robotic reconfigurable assembly unit in a pre-existing shop-floor. It is developed an aggregate production planning in a fuzzy environment where the product price, unit inventory cost, unit assembly cost, resource-product suitability cost, availability of the assembly line, the assembly time and the market demands are fuzzy in nature. The proposed model attempts to minimize total production costs, maximizing the shop floor resources utilization and the profits, considering nominal capacity and inventory towards zero. Pareto solutions optimization is computed with different techniques and results are presented and discussed with interesting practical implications.

Keywords: Aggregate production planning · APP · Hybrid fuzzy systems · Fuzzy Multi-objective programming · MOLP

1 Introduction

Nowadays world markets are driven by demand for personalization, products life cycles shrinking and customers grown awareness. The manufacturing systems are driven towards the deployment of reconfigurable machinery and robots in order to support mass customized highly personalized products and fast reactions to variability of market demands.

Despite this worldwide trend there are still many cutting edge technologies industries where high handy level is still the main drive in manufacturing activities.

Sectors as the ones producing high precision content devices for industrial and professional application, where often dimensions of the devices to be produced and handled are very small are still hugely employing manual activities in their production processes.

Markets push industries to deliver customized products often characterized by a high rate of complexity: extremely complex shape, fragile and delicate materials, highly degradable, high variants in relation to product geometry, technological features and

© Springer International Publishing Switzerland 2015
S. Arik et al. (Eds.): ICONIP 2015, Part II, LNCS 9490, pp. 571–582, 2015.
DOI: 10.1007/978-3-319-26535-3_65

performance ranges, frequent features evolution over time to match the market dynamics. All of these characteristics come along often with low production volumes.

These type of products are called in literature as HMLV, acronym that stands for High Mix Low Volume [1].

Manufacturing worldwide trends suggest that for the HMLV products novel production systems able to provide comprehensive automated assembly operations, overall better saturation, reduced set-up and ramp-up times and equipment reusability are needed. Eventually the utilization of automated industrial equipment whose flexibility would be capable of matching the various production requirements across their lifecycles would globally impact on the production management system.

In this paper hence it is developed and detailed the management integration framework proposed in [2] for the integration of the self-contained re-configurable assembly island developed in the F7 EU Project white'R. The white'R described in [2, 3] will be in this paper seen as a generic Reconfigurable Robotic Assembly Unit (RRAU). Furthermore the fuzzy multi-objective linear programming model presented in [3] is customized to the general case of the integration of a RRAU in a fuzzy environment.

The study is organized as follow: In Sect. 2 it is analysed the general shop floor suitable to integrate a RRAU. In Sect. 3 general issues incurring in the integration of the RRAU are discussed. Afterwards it is presented the general framework allowing the integration of the RRAU in a pre-existing shop floor in Sect. 4. In Sect. 5 it is proposed the fuzzy multi-objective linear programming model solving the aggregate production planning decision problem presented in the general framework. Then in Sect. 6 the model is solved and a numerical example is presented in Sect. 7. Finally in Sect. 8 it is inferred conclusions and discussed the important topics for the future research.

2   Typical Shop Floor

The typical production Environment of a shop floor suitable for integration of a self-contained reconfigurable assembly unit is a high handy rate flow shop where despite the fact activities are carried on manually by operators the logical production phases are neat and scoped. Generally the machinery used in these production environments is most likely dedicated to one or few more operations and thus it is easy to be integrated with the reconfigurable robotic assembly unit (hereafter called simply RRAU). The RRAU can be simplified as a box where a storage of raw material is designed to grant a certain autonomy during a fixed time horizon and a set of specific assembly processes which are designed and customized for each diverse application (Fig. 1).

The whole production cycle goes through automated activities intersected with manual activities. The shop floors are carried on by law populated work teams, whose operators' interoperability rate is high. The layout of the two plants is pretty contained, and develops on a single floor where both preparation of components and final assembly is carried on.

The products are basically rolled into one, starting from a range of components which lay on few levels of the BOM. It is desirable that the raw material storage of the shop floor and the RRAU one would be on the same level and easy to be connected each-others.

Robotic Reconfigurable Assembly Unit (RRAU)

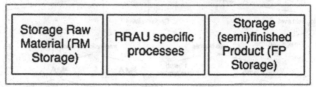

| Storage Raw Material (RM Storage) | RRAU specific processes | Storage (semi)finished Product (FP Storage) |

Fig. 1. Robotic Reconfigurable Assemble Unit (RRAU) macro-functions

Due to the nature of the business of the companies that would be interested in employing a RRAU in order to automatize most of its assembly phases the finished product storage of the shop floor H_{it} is assumed to be nearly null.

3 RRAU Integration in a Pre-existing Shop-Floor

The RRAU is located in a way that the material flow goes along with the technological cycle. Consequently from the preparation of raw material in the RRAU storage till the final assembly operations of the final product a flow shop is constituted. The flow shop may be internal the RRAU, or constituted by the island and one or more complement manual assembly processes (Fig. 2), downstream (a) or upstream (b), or both (c).

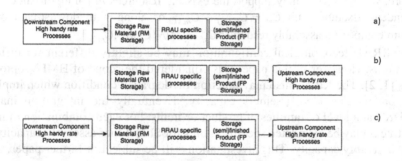

Fig. 2. The different integration scenarios

The integration issue it's important because different integration scenarios imply different assembly capacity of the shop-floor and even of the RRAU. The integration scenarios may be graphically visualized within a tree graph with different combination of the nodes. There are, basically two main decisions: the way the RRAU will be employed to assemble products (Use choice) and the way it will be integrated in the already existing production environment (Integration choice) (Fig. 3).

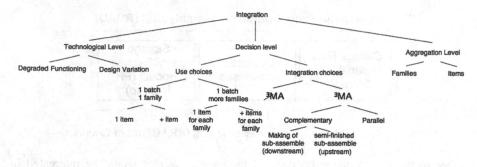

Fig. 3. The integration framework

Regarding the use choice RRAU may be used to deliver in output mono-product batches (batches of only one product family in which it would be assembled just one item or multiple items), or multi-products batches (batches of more than one family of products in which two further options may be followed such as one item for each product family or more than one item for each family of products).

Regarding the Integration choice it may be assumed that:

RRAU substitutes wholly the previous resources concentrating within itself all the production/assembly phases ($\nexists MA$ stands for "don't exist Manual Assembly"). In this case the shop floor assembly capacity is equal to RRAU assembly capacity: $C_{sf} = C_{RRAU}$

Moreover the RRAU may support the existing resources in parallel to an existing independent assembly unit $C_{sf} = C_{RRAU} + C_{mAL}$ or along with downstream or/and upstream component assembly resources $C_{sf} = MIN[C_{RRAU}; C_{mAL}]$.

The RRAU technological configuration may be slightly different for different applications. However in this paper it is considered the scope of HMLV optoelectronics [1, 2]. The RRAU also may function on degraded condition which implies a correspondent degraded assembly capacity. Eventually the integration may be considered at a level of families of product or items. For each combination of nodes of the tree it may be calculated specific values for Shop Floor assembly capacity and RRAU assembly capacity. This investigation will be subject of further paper.

4   The Management Integration Framework for a RRAU

The plane framework presented in [2] showed simply the steps of the process and the flows of inputs and outputs in order to give basis through which the framework a decision maker can implement proper LP models to solve the specific production planning problem [2]. In this section it will be discussed an extended version of the production planning and scheduling framework presented in [2]. The aim of this extended version is besides supporting the management of a company employing RRAU visualizing the production planning and scheduling decisional problem [2] to underpin the main issues in integration of such a system. The whole framework is developed in a perspective of integration of each processes within the production planning system and for Aggregate,

Disaggregate Planning and Material Planning with the ERP system of the company. Simulation can be adopted for the lower level of the framework in order to understand the typical dynamics of operational scheduling and capacity planning (Fig. 4).

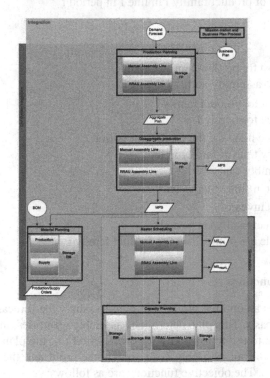

Fig. 4. Production planning and scheduling framework

5 Problem Formulation

In this section, the model formulation for aggregate planning is presented.

The sign "~" and "^" indicates the fuzzy constraint and parameter.

Decision maker has to decide if using assembly line mainly manual or the RRAU using assembly line considering different scenarios of matching resources to different family of products.

5.1 Indices

Here below are listed the indices used for decision variables and parameters.

t period: 1,..., T
i product family: 1,..., N
l line: 1,..., L
j item: 1,..., J

5.2 Decision Variables

Decision variables of the model presented in this work are as follow:
$\hat{q}_{ilt}$ Assembly units of product family i in line l in period t   [units]

5.3 Parameters

$\hat{z}_{il}$ unit assembly cost for product family i in line l	[euro/unit]
$\hat{a}_{il}$ suitability cost to assembly family i in line l	[euro/set-up]
$\overline{h}_l$ unit inventory cost for product family i	[euro/ stocked unit]
$\hat{p}_i$ average unit price for family i	[euro/unit]
$\hat{b}_{ilt}$ availability of the line l for the family i at period t	[%]
$\hat{g}_{il}$ average assembly time of family in line l	[min/unit]
C_{sf} shop floor assembly capacity	[units/day]
C_{fp} finished product inventory capacity	[units/month]
H_{it} finished product inventory	[units/month]
$\hat{D}_{it}$ demand of product family i at period t	[units]
st_{ilt} Boolean variable use of the line l for product family i	[0; 1]

5.4 Objective Functions

This work develops an optimal production plan to satisfy the forecasted demand on a mid-term horizon basis. This plan determines assembly quantities, inventory level and the proper resource to use to assembly for each period t of the planning horizon T in order to minimize overall assembly costs, maximize profit and the utilization of the shop-floor resources. The objective functions are as follows:

MINIMIZATION OF OPERATIONAL SHOP-FLOOR COSTS

$$\min \sum_{i=1}^{N} \sum_{l=mAL}^{RRAUAl} \sum_{t=1}^{T} [\hat{z}_{il} * \hat{q}_{ilt} + \hat{a}_{il} * st_{il} + (\hat{D}_{ilt} - \hat{q}_{ilt}) * \hat{c}_i] \tag{10}$$

This objective is the overall shop-floor management cost which is seen as constituted by a factor regarding the overall assembly costs given as:

$$\sum_{i=1}^{N} \sum_{l=mAL}^{RRAUAl} \sum_{t=1}^{T} [\hat{z}_{il} * \hat{q}_{ilt}],$$

A factor considering the cost given by the use of manual assembly station and of RRAU for families of product not properly designed for them. Thus it is assumed that there's a suitability of different resources for specific families of products: proper allocation of the resource for each family of product avoids this cost:

$$\sum_{i=1}^{N} \sum_{l=mAL}^{RRAUAl} [\hat{a}_{il} * st_{it}]$$

A factor considering the inventory cost thought as the surplus of batch and manual production at the end of the period:

$$\sum_{i=1}^{N} \sum_{l=mAL}^{RRAUAl} \sum_{t=1}^{T} (\hat{D}_{ilt} - \hat{q}_{ilt}) * \hat{c}_i$$

MAXIMIZATION OF THE PROFIT

$$\max \sum_{i=1}^{N} \sum_{l=mAL}^{RRAUAl} \sum_{t=1}^{T} [(\hat{p}_{il} * \hat{q}_{ilt}) - (\hat{z}_{il} * \hat{q}_{ilt}) - (\hat{D}_{ilt} - \hat{q}_{ilt}) * \hat{c}_i] \tag{11}$$

This objective is to contribute to profit seen as revenues $\sum_{i=1}^{N} \sum_{l=mAL}^{RRAUAl} \sum_{t=1}^{T} [\hat{p}_{il} * \hat{q}_{ilt}]$ minus production and inventory costs $\sum_{i=1}^{N} \sum_{l=mAL}^{RRAUAl} \sum_{t=1}^{T} [\hat{z}_{il} * \hat{q}_{ilt} + (\hat{D}_{ilt} - \hat{q}_{ilt}) * \hat{c}_i]$.

MAXIMIZATION OF SHOP-FLOOR RESOURCE UTILIZATION

$$\max \sum_{i=1}^{N} \sum_{l=mAL}^{RRAUAl} \sum_{t=1}^{T} (b_{ilt} * g_{il} * \hat{q}_{il})/(g_{il} * \hat{q}_{ilt}) \tag{12}$$

This objective is to maximize the shop-floor utilization level thought as the actual time employed to assembly a certain quantity of product i in line l at period t divided by the nominal assembly time. The actual time is given by the factor $\sum_{i=1}^{N} \sum_{l=mAL}^{RRAUAl} \sum_{t=1}^{T} [b_{ilt} * g_{il} * \hat{q}_{ilt}]$ where b_{ilt} is a number among 0 and 1 and it is a rate of availability of the line l.

5.5 Constraints

For each period, the following constraints apply:

The demand satisfaction constraint

$$\sum_{i=1}^{N} \hat{q}_{ilt} + H_{ti} \stackrel{\leq}{=} \hat{D}_{it}$$

The inventory capacity constraint

$$\sum_{i=1}^{N} H_{it} \stackrel{\leq}{=} 0 \quad \text{And} \quad \sum_{i=1}^{N} H_{it} \stackrel{\leq}{=} C_{fp}$$

The Shop Floor Assembly capacity constraint

$$\sum_{i=1}^{N} \hat{q}_{ilt} \; \hat{\leq} \; C_{sf}$$

The non-negativity constraint on decision variables

$$q_{itl}, H_{it} \geq 0$$

The constraint on the Boolean nature of set-up variable

$$S_{ilt} \in \{0, 1\}$$

The constraint on the nature availability variable

$$0 < b_{ilt} < 1 \in R$$

6 Model Development

6.1 FMOLP Solving Method

The proposed FMOLP model is constructed using the piecewise linear membership function of Hannan [4] to represent the fuzzy goals of the Decision Maker (DM) in the MOLP model, together with the minimum operator of the fuzzy decision-making of Bellman and Zadeh [5].

Moreover, the original fuzzy MOLP problem can be converted into an equivalent crisp LP problem and it is easily solved by the standard simplex method. There are different fuzzy goal programming models [4, 6–8].

The important differences among these models result from the types of membership functions and aggregation operators they apply. In general, aggregation operators can be roughly classified into three categories: intersection, union, and averaging operators [9].

The minimum operator is preferable when the DM wishes to make the optimal membership function values approximately equal or when the DM feels that the minimum operator is an approximate representation. The application of the aggregation operator to draw maps above the maximum operator and below the minimum operator may be important in some practical situations.

Alternatively, averaging operators consider the relative importance of fuzzy sets and have the compensative property so that the result of combination will be medium. The g-operator [10], which yields an acceptable compromise between empirical fit and computational efficiency, seems to be the convex combination of the minimum and maximum operators [11, 12]. Zimmermann [9] pointed out that the following eight important criteria must be applied selecting an adequate aggregation operator-axiomatic strength, empirical fit, adaptability, numerical efficiency, compensation, range of compensation, aggregating behavior, and required scale level of membership function.

In order to solve the FMOLP problem involving fuzzy parameters it has been adopted the algorithm proposed by Wang and Liang [13]. The Block diagram of the interactive FMOLP is represented in Fig. 5.

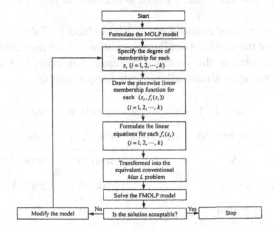

Fig. 5. Block diagram of the interactive FMOLP

Step 1. Formulate the original fuzzy MOLP model for the considered problem.

Step 2. Given the minimum acceptable membership level, α, and then convert the fuzzy inequality constraints with fuzzy available resources (the right-hand side) into crisp ones using the weighted average method.

Step 3. Specify the degree of membership for several values of each objective function.

Step 4. Draw the piece-wise linear membership functions for each objective function.

Step 5. Formulate the piece-wise linear equations for each membership function.

Step 6. Introduce an auxiliary variable, thus enabling the original fuzzy multi-objective problem to be aggregated into an equivalent ordinary LP form using the minimum operator.

Step 7. Solve the ordinary LP problem, and execute the interactive decision process. If the decision maker is dissatisfied with the initial solutions, the model must be adjusted until a set of satisfactory solutions is derived.

6.2   The Multi-objective Linear Programming Associated– Results and Comments

The approach proposed above has been tested on a synthetic data-set built from simulations models of the RRAU environment (made in SIMIO®). Mathematical programming and Evolutionary algorithms have been used, as well as the evolutionary multi-objective optimization (EMO) algorithms for solving the MOLP associated. Thus Normal Boundary Intersection (NBI) [14], Modified Normal Boundary Intersection (NBIm) [15], Normal Constraint (NC), [16, 17] Successive Pareto Optimization (SPO) [18] and for the seconds Non-dominated Sorting Genetic Algorithm-II (NSGA-II) [19] and Strength Pareto Evolutionary Algorithm 2 (SPEA-2) [20]. Also a hybrid approach incorporating Multi-Criteria Decision Making (MCDM) approaches into EMO algorithms as a local search operator and to lead a DM to the most preferred solution(s). Here the local search operator is mainly used to enhance the rate of convergence of EMO algorithms [21].

A comparison on several running (fi) is highlighted in Table 1. Here the Hybrid approach can achieve the same subset Pareto solutions faster than other techniques and when there is not a solution the search stop faster than other techniques. The other methods, performs worst on almost all our tests for our database.

Table 1. Comparison among methods tested

	Hybrid MCDM-EMO	NSGA-II	NBIm
$f1$	3500	40000	300000
$f2$	29500	30000	250000
$f3$	55000	103000	70000
$f4$	20000	90000	86000

7 Conclusions

In order to develop a model for production planning in a production environment of HMLV products employing a Re-configurable Robotic Assembly Unit (RRAU) it is important to identify all different scenarios of integration. This is needed in order to distinguish different cases in terms of production capacity of the shop-floor. The production planning system developed along with the integration of the RRAU has to be integrated in all its components as well as with the management system already existing in the company. The production planning model developed on the management frameworks proposed will present fuzzy variables and fuzzy constraints because the production environment is fuzzy in nature. Following this idea, the need was to solve a Multi-objective Linear Programming with Fuzzy variables and constraints and used an algorithm where an important step was the de-fuzzyfication of the MOLP proposed. A comparison of different methods has been run and interesting results have been reported. In the future authors would propose some EMO algorithm with quantum Principle [22, 23] in the method proposed above.

References

1. Mahoney, R.M.: High-Mix Low-Volume. Hewlett Packard (1997)
2. Fiasché, M., Ripamonti, G., Sisca, F.G., Taisch, M., Valente, A.: Management integration framework in a Shop-Floor employing self-Contained Assembly Unit for optoelectronic products. Accepted at 1st IEEE international forum on RTSI 2015. In press on IEEE Xplore digital library

3. Sisca, F.G., Fiasché, M., Ripamonti, G., Taisch, M.: A novel hybrid fuzzy multi-objective linear programming model for APP in a shop-floor employing white'R for HMLV optoelectronics. Accepted at WIRN 2015, Springer Smart Innovation, Systems and Technologies, in Press

4. Kahraman, C., Kaya, I.: Fuzzy multiple objective linear programming. In: Kahraman, C. (ed.) Fuzzy Multi-Criteria Decision Making. Springer Optimization and Its Applications, vol. 16, pp. 325–337. Springer, US (2008)

5. Zimmerman, H.-J.: Description and optimization of fuzzy systems. Int. J. Gen. Syst. **2**, 209–215 (1976)

6. Hannan, E.L.: Linear Programming with multiple fuzzy goals. Fuzzy Sets Syst. **6**, 235–248 (1981)

7. Zimmerman, H.-J.: Fuzzy programming and linear programming with several objective functions. Fuzzy Sets Syst. **2**, 209–215 (1978)

8. Leberling, H.: On finding compromise solutions in multicriteria problems using the fuzzy min-operator. Fuzzy Sets Syst. **6**, 105–118 (1981)

9. Sakawa, M.: An interactive fuzzy satisficing method for multiobjective linear programming problems. Fuzzy Sets Syst. **28**, 129–144 (1988)

10. Zimmermann, H.J.: Fuzzy Set Theory and its Application. Kluwer, Boston (1996)

11. Zimmermann, H.J., Zysno, P.: Latent connectives in human decision making. Fuzzy Set Syst. **4**, 37–51 (1980)

12. Zimmermann, H.-J.: Fuzzy linear programming. In: Gal, T., Greenberg, H.J. (eds.) Advances in Sensitivity Analysis and Parametric Programming, pp.15.1–15.40. Kluwer, Boston (1997)

13. Wang, R.-C., Liang, T.-F.: Application of fuzzy multi-objective linear programming to aggregate production planning. Comput. Ind. Eng. **46**, 17–41 (2006)

14. Das, I., Dennis, J.E.: Normal-boundary intersection: a new method for generating the pareto surface in nonlinear multicriteria optimization problems. SIAM J. Optim. **8**(3), 631 (1998). doi:10.1137/S1052623496307510

15. S. Motta, R., Afonso, S.M.B., Lyra, P.R.M.: A modified NBI and NC method for the solution of N-multiobjective optimization problems. Structural and Multidisciplinary Optimization, 8 January 2012. doi:10.1007/s00158-011-0729-5

16. Messac, A., Ismail-Yahaya, A., Mattson, C.A.: The normalized normal constraint method for generating the Pareto frontier. Struct. Multi. Optim. **25**(2), 86–98 (2003). doi:10.1007/s00158-002-0276-1

17. Messac, A., Mattson, C.A.: Normal constraint method with guarantee of even representation of complete Pareto frontier. AIAA J. **42**(10), 2101–2111 (2004). doi:10.2514/1.8977

18. Mueller-Gritschneder, D., Graeb, H., Schlichtmann, U.: A successive approach to compute the bounded pareto front of practical multiobjective optimization problems. SIAM J. Optim. **20**(2), 915–934 (2009). doi:10.1137/080729013

19. Deb, K., Pratap, A., Agarwal, S., Meyarivan, T.: A fast and elitist multiobjective genetic algorithm: NSGA-II. IEEE Trans. Evol. Comput. **6**(2), 182 (2002). doi:10.1109/4235.996017

20. Zitzler, E., Laumanns, M., Thiele, L.: SPEA2: Improving the Performance of the Strength Pareto Evolutionary Algorithm, Technical Report 103, Computer Engineering and Communication Networks Lab (TIK), Swiss Federal Institute of Technology (ETH) Zurich (2001)

21. Sindhya, K., Deb, K., Miettinen, K.: A local search based evolutionary multi-objective optimization approach for fast and accurate convergence. In: Rudolph, G., Jansen, T., Lucas, S., Poloni, C., Beume, N. (eds.) PPSN 2008. LNCS, vol. 5199, pp. 815–824. Springer, Heidelberg (2008)

22. Fiasché, M.: A quantum-inspired evolutionary algorithm for optimization numerical problems. In: Huang, T., Zeng, Z., Li, C., Leung, C.S. (eds.) ICONIP 2012, Part III. LNCS, vol. 7665, pp. 686–693. Springer, Heidelberg (2012). doi:10.1007/978-3-642-34487-9_83
23. Fiasché, M., Taisch, M.: On the use of quantum-inspired optimization techniques for training spiking neural networks: a new method proposed. In: Bassis, S., Esposito, A., Morabito, F.C. (eds.) Recent Advances of Neural Networks Models and Applications. SIST, vol. 37, pp. 359–368. Springer, Heidelberg (2015). doi:10.1007/978-3-319-18164-6_35

Design of Distributed Adaptive Neural Traffic Signal Timing Controller by Cuckoo Search Optimization

Sahar Araghi$^{(\boxtimes)}$, Abbas Khosravi, and Douglas Creighton

Center for Intelligent Systems Research (CISR),
Deakin University, Geelong, VIC 3216, Australia
saraghi@deakin.edu.au

Abstract. This paper focuses on designing an adaptive controller for controlling traffic signal timing. Urban traffic is an inevitable part in modern cities and traffic signal controllers are effective tools to control it. In this regard, this paper proposes a distributed neural network (NN) controller for traffic signal timing. This controller applies cuckoo search (CS) optimization methods to find the optimal parameters in design of an adaptive traffic signal timing control system. The evaluation of the performance of the designed controller is done in a multi-intersection traffic network. The developed controller shows a promising improvement in reducing travel delay time compared to traditional fixed-time control systems.

1 Introduction

Urban traffic is one of the sensible problems in most of the modern cities. Different solutions have been proposed to handle urban traffic during the years. Among them using traffic signal lights at intersections is one of the successful approaches in this regard.

In one category, three generations are considered for traffic signal's timing controllers. The first generation fixed-time method requires pre-set signal sequences. The second generation adjusts the signal timing based on the information transformed from traffic detectors. The third generation is characterized by the ability of dynamic decision making and distributed control.

Traditional traffic control methods are not efficient for today's traffic condition. Traffic flows in urban area are complex and random. In this situation, proposing appropriate pre-defined formula for traffic controlling is challenging. Artificial Intelligence (AI) techniques, which have the ability of learning and decision making, are useful to control traffic signals at an intersection.

The first research used fuzzy logic systems (FLS) for isolated traffic signal control was published in 1977 [5]. Following that NN and genetic algorithm (GA) were applied in design of the intelligent traffic signal controllers. Using hybrid systems was the next step in this regard. Hybrid controllers use the combination of the aforementioned methods and enable the controllers to adapt traffic patterns [3].

© Springer International Publishing Switzerland 2015
S. Arik et al. (Eds.): ICONIP 2015, Part II, LNCS 9490, pp. 583–590, 2015.
DOI: 10.1007/978-3-319-26535-3_66

Traffic signal timing controllers who apply artificial intelligence techniques have high performance. Their performance will be increased if they obtain their parameters through training instead of using manual parameters. In this paper, we applies a feed-forward NN in design of the traffic signal controller. The NN gets its optimal weights by using CS, which its high performance have been demonstrated in finding optimal parameters and not being trapped in local optima. Designing a distributed control system extremely reduces the computational cost of the training process for NN controller.

The rest of this paper is organized as: Sect. 2 is talking about NN in traffic signal controller, Sect. 3 is about the testbed, then the proposed traffic signal controller is introduces in Sect. 4, finally we have conclusion in Sect. 5.

2   Neural Network in Traffic Signal Controllers

NN, as a subfield of AI, is a strong method to be used as a controller in different applications. Spall and Chin [6] employed simultaneous perturbation stochastic approximation (SPSA) [7] based gradient estimates with an NN controller for optimizing a system. In their research, SPSA was used to model the weight update process of a NN. The current traffic information are used to generate the signal timings. Actually, the current traffic information was used to solve the current instantaneous traffic issue. The proposed system called S-TRAC and has advantages: (1) Didn't need system-wide traffic flow model; (2) Automatically adaptation to long-term changes in the system such as seasonal variations while providing real-time responsive signal commands; and (3) Be able to work with existing hardware and sensor configurations within the network of interest while additional sensors may help the overall control capability. S-TRAC included a feed-forward NN with 42 inputs and two hidden layers. Inputs are composed of: (1) the queue at each cycle termination for 21 traffic queues of the simulation; (2) eleven nodes for per-cycle vehicle arrivals in the system; (3) simulation start time; and (4) the nine outputs from the previous control solution. To evaluate the performance of S-TRAC, a simulation of a nine-intersection network is considered and they obtained about 11 % improvement against fixed-time method.

In a study done by Srinvasan et al. [8], the authors presented an enhanced version of the SPSA-NN system for a multi-agent system and they tested that in more complicated scenarios. Authors claimed that although the SPSA algorithms is a useful method for updating the weight online, the model proposed by Spall et al. [6] had some limitations influence its performance.

Spall et al. designed a system by a three-layer NN and relevant traffic variables were considered as inputs. In [8], it is claimed that the proposed system by Spall et al. has two shortcomings: First of all, the system used heuristic methods to identify the general traffic patterns (morning and evening peaks) and assignment of time periods for patterns. This system is not robust for non-periodic traffic patterns. As the second issue, a NN was considered for each time period, and the weight were updated only whenever the same traffic pattern and time period was arisen. Therefore, it may not be possible to respond appropriately to changes of the traffic inside the same time period.

Srinivasan *et al.* improved aforementioned method and compared it with the hybrid multiagent architecture presented by Choy *et al.* [2] and Green Link Determining (GLIDE), which was the excising traffic signal control for the city and is the local version of the SCATS. Evaluation is done in a large traffic network in Singapore Central Business District with 25 intersections. After 15 separate simulations with different seeds, which each was set for three hours, the lowest mean delay belonged to SPSA-NN, Hybrid NN and GLIDE respectively.

The study done by Nagare and Bhatia [4] was another work, in which traffic flow is predicted for controlling traffic congestion. It was mentioned that NN introduces some flaws such as flow convergence and the obtained solution is usually local optimal. It is explained that by using a combination of optimization methods better optimization results can be reached.

3   Introducing the Testbed and Conditions

In order to evaluating the experiments, a network of nine intersection is designed in a traffic simulator called PARAMICS. In this network, each intersection is consist of four approaching links and four phases (A, B, C, D), as shown in Fig. 1a. The cycle times are divided between these four phases. Zones are the areas that vehicles are released into the intersection. Figure 1b, is a snapshot of the designed intersection in PARAMICS. Here, four different entrances to the intersection create four zones.

The simulation is modelled in PARAMICS version 6.0. and all controllers are implemented in Matlab R2011b.

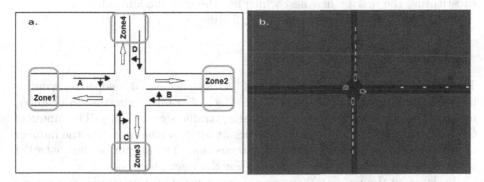

Fig. 1. a. Intersection with four approaching links; b. Snapshot of the isolated intersection in PARAMICS

4   Proposed Neural Traffic Signal Timing Controller

The basic concept of NN were originally obtained from the way biological nervous systems work. A NN is an information processing technique well-known because of its excellent approximation and learning capabilities. A NN is a universal approximator that for any nonlinear mapping can approximate various degree of accuracy [1], using this feature NNs are able to recognize hidden patterns from imprecise and complicated data. In a better word for a problem that is too complicated to be considered by either traditional data mining methods or humans, NN can be a suitable option which is broadly used for control, modeling, prediction and classification problems.

In supervised learning, NNs are usually trained by the minimizing an error-based cost function. The parameters of the NN can be optimally set for situations with unknown expected values, by the using and minimizing the cost function. To reach the optimal set of parameters, the global optimization methods are suitable. Here, we use CS to find the optimal parameters for the developed NN controller.

4.1 Cuckoo Search Optimization Algorithm via Lévy Flight

CS is a heuristic search algorithm. In this method three different rules are considered:

First: Each cuckoo lays an egg, choose a host nest and dispose it there.

Second: The best nest, nest with higher quality of eggs, will carry to the next generation.

Third: The number of host nests are fixed. The probability of egg-discovery by host bird is $p_a \in [0, 1]$. The host bird in cuckoo search can throw out the parasitic egg or abandon that nest for a better new nest. In fact, p_a is the probability of substituting the nest by new nest which are the new random solutions.

Pseudo code for CS presented in Fig. 2 [10].

4.2 Design of CS-NN Controller

The proposed CS-NN controller (CS-NNC) is designed and developed for a multi-intersection network. In this regard, the effect of traffic congestion in neighbor intersections are also considered in adjusting traffic signal timing. The inputs of CS-NNC are the length of vehicles' queues at approaching links and the number of vehicles coming from the neighbor intersection. The output of the controller is the appropriate green time calculated for the next phase.

In designed timing control systems, against previous introduced systems, we do not use a central NN controller for timing the traffic phase in the whole traffic network. The proposed system is a distributed control system, in which each intersection use its own controller for adjusting its traffic congestion, however by accounting the number of vehicles arriving from neighbor intersections we balance the traffic in the whole traffic network. In addition, considering travel time of all the vehicles in the network as the cost function of the optimization, is the other factor to balance the traffic in whole system while each intersection has its own traffic timing control system.

```
1: begin
2: Objective function f(x), x = (x₁, ..., xₘ)ᵖ
3: Generate initial population of n host nests Xᵢ(i = 1, 2, ..., n)
4: while (t < MaxGeneration) or (stopcriterion) do
5:     Get a cuckoo randomly by Lévy flight
6:     evaluate its quality / fitness Fᵢ
7:     Choose a nest among n (say,j) randomly
8:     if (Fᵢ < Fⱼ) then
9:         replace j by the new solution;
10:    end if
11:    A fraction (pₐ) of worse nests are abandoned and new ones are built;
12:    Keep the best solutions (or nests with quality solutions);
13:    Rank the solutions and find the current best
14: end while
15: Postprocess results and visualization
16: end
```

Fig. 2. Cuckoo search via Lévy flights

Separate NNC for each intersection reduces the number of inputs that leads to having faster convergence during training of NNC. In the case of central controlling system the NNC needs the traffic data of the whole network, here NNC needs their own traffic information and their neighbors.

The developed CS-NNC for each intersection uses a feed-forward neural network. It consists of four input neurons, ten neurons in the hidden layer, and one neuron in the output layer. Considering four neurons as the inputs is because of 4-way intersections and having detected traffic information for four approaching links. The output neuron provides the green time for the next green phase in the intersection.

Cuckoo Search (CS) is applied for optimizing the parameters of the controller. Average delay time of vehicles in a complete run of a simulation is considered as the cost function. Therefore, CS aims at reducing the average delay time in whole traffic network for finding optimal parameters. Mathematically, it is defined as:

$$cost\ function = \frac{\sum_{i=1}^{m} \sum_{j=1}^{n} d_j}{\sum_{i=1}^{m} k_i} \qquad (1)$$

where $i = 1, \ldots, m$ is the number of intersections, $j = 1, \ldots, n$ is the number of phases executed during the simulation time, d is the calculated delay time for each phase, and k is the number of cars released in each simulation scenario.

During the training the following process are repeat until finding the optimal parameters or after reaching the limitation in number of iterations.

1. Generating some random parameters as the weights of the CS-NNC.
2. Proposing green time by new weights for next green phases of the network until the end of the simulation.
3. Calculating the average delay time of the vehicles for traveling in the simulated traffic network.

4. proposing the new parameters based on the efficiency of the previous proposed parameters and CS algorithm.

The diagram of training process is presented in Fig. 3. These process are done for each controller at each traffic junction. For a network with nine intersections there will be nine controllers. The training process is done for all of controller simultaneously.

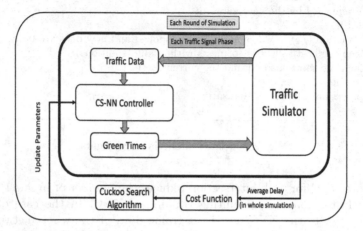

Fig. 3. Training process of NNC by CS to obtain the optimal weights of the traffic controller.

5   Experiments Results and Discussion

Three different scenarios are considered for evaluating the performance of the controllers:

1. fifteen minutes simulation with 700 vehicles;
2. fifteen minutes simulation with 1,000 vehicles; and
3. fifteen minutes simulation with 1,400 vehicles.

All the intelligent controllers are implemented for a network of intersections. The proposed controllers do not need a central controller for cooperation. Each controller receives the information from its neighbors and adjusts the appropriate green time for the next phase based on its current traffic condition and the information received from the neighbors.

Here, to set the weights of the NN controller, instead of doing manually, we apply CS [9,11] to find the optimal parameters as the weights of the NN controller.

The CS-NNC with defined structure in previous section (3 layers and [4, 10, 1] neurons for each layer respectively), have 61 parameters need to be optimized. This number of parameters is for each controller at an intersection and for nine controller we will have 549 parameters. The optimization with CS is done for

Table 1. Average Delay Time per Vehicle in the Network for four developed controllers for seed-1 (times are presented in second)

Controllers	Scenario-1	Scenario-2	Scenario-3
$CS - NNC$	48.7	74.3	101.9
$Fixed - time20$	55.5	69.5	99.8
$Fixed - time50$	69.7	78.6	103.5
$Fixed - time80$	84.7	90.6	110.9

100 generations. The step size is set to 0.5 and the number of population is 20, P_a is considered as 0.25. The outputs of these controllers which are the proposed green times for next phase are set between 0 and 100 s.

Three fixed-time controllers are also developed as benchmarks. Their phase green times are set to 20, 50, and 80 s respectively. Fixed-time controllers are a common benchmark for evaluating the performance of traffic signal controllers. In developed fixed-time controllers all green phases have same predefined values (Table 1).

The results show the average delay per vehicle in a nine-intersection network for each CS-NNC and three fixed time controllers. The better performance, or the less amount of delay, in most of the cases is for CS-NNC. Fixed-time controller with different setting performs differently. This shows fixed-time controllers are not robust for unpredictable traffic conditions. Here, we can see that fixed-time with 20 s works almost well, however the results for fixed-time with 80 s is not suitable compared to CS-NNC. Finding the appropriate value as the fixed-time value is not straightforward and depends on traffic conditions.

6   Conclusion

In this paper, CS is used in design of the traffic signal controller. Instead of setting the parameters of the NN controller manually, the developed system obtain the optimal parameters during a training and using CS optimization method. To evaluate the performance of the distributed CS-NNC the controller tested on a multi-intersection network in three different scenarios. The results show a better performance of the CS-NNC controller against benchmarks in most of the cases.

CS is a quit new optimization method that revealed its high performance in different applications. Here, we evaluate its performance in combination with NN and as the future work, CS will be used in combination with other intelligent techniques. It is aimed to improve its performance especially for the cases with huge number of parameters.

References

1. Bishop, C.M.: Neural Networks for Pattern Recognition. Oxford University Press, New York (1995)
2. Choy, M.C., Srinivasan, D., Cheu, R.L.: Cooperative, hybrid agent architecture for real-time traffic signal control. IEEE Trans. Syst. Man Cybern. Part A Syst. Hum. **33**(5), 597–607 (2003)
3. Malej, A., Brodnik, A.: Fuzzy urban traffic signal control - an overview. Elektrotehniski Vestn./Electrotechnical Rev. **74**(5), 291–296 (2007). www.scopus.com
4. Nagare, A., Bhatia, S.: Traffic flow control using neural network. Int. J. Appl. Inf. Syst. **1**, 50–52 (2012). New York, USA
5. Pappis, C.P., Mamdani, E.H.: A fuzzy logic controller for a traffic junction. IEEE Trans. Syst. Man Cybern. **7**(10), 707–717 (1977)
6. Spall, J.C., Chin, D.C.: Traffic-responsive signal timing for system-wide traffic control. Transp. Res. Part C Emerg. Technol. **5**(3–4), 153–163 (1997)
7. Spall, J.: Multivariate stochastic approximation using a simultaneous perturbation gradient approximation. IEEE Trans. Autom. Control **37**(3), 332–341 (1992)
8. Srinivasan, D., Choy, M.C., Cheu, R.: Neural networks for real-time traffic signal control. IEEE Trans. Intell. Transp. Syst. **7**(3), 261–272 (2006)
9. Yang, X.S.: Chapter 9 - cuckoo search. In: Nature-Inspired Optimization Algorithms, pp. 129–139. Elsevier, Oxford (2014)
10. Yang, X.S., Deb, S.: Cuckoo search via levy flights. In: Nature Biologically Inspired Computing, pp. 210–214, Dec 2009
11. Yang, X., Deb, S.: Cuckoo search: recent advances and applications. Neural Comput. Appl. **24**(1), 169–174 (2014)

A Causal Model Using Self-Organizing Maps

Younjin Chung$^{(\boxtimes)}$ and Masahiro Takatsuka

ViSLAB, The School of IT, The University of Sydney, Sydney, NSW 2006, Australia
jina1407@gmail.com, jina@vislab.net

Abstract. Understanding causality is very important for problem solving in many areas. However, conducting causal analysis for multivariate and nonlinear data, unlabeled in nature, still faces many problems with existing methods. Artificial Neural Networks have been developed for such data analyses and Backpropagation Network has been most used to learn the relationships between causes and effects. However, its supervised and black-boxed learning structure for labeled data often limits modeling and reasoning of uncertain, graded and fuzzy causality through the network. In this paper, an approach for analyzing such causality is proposed by networking Self-Organizing Maps handling unlabeled data. A new weighting scheme on connection weight vector similarity is developed to approximate conditional distributions of data in order to capture the indeterminate nature of causality. The experiments demonstrate that the method well approximates conditional output distributions for given inputs and allows estimating causal effects based on the similarity weight distributions.

Keywords: Unlabeled multivariate and nonlinear data · Self-Organizing Map · Causal analysis · Connection weight vector similarity distribution

1 Introduction

The diverse study areas such as health, social and ecological sciences deal with multivariate and nonlinear data, unlabeled in nature. They share many analytical problems and the analytical questions are not just associational but more likely causal in solving their domain problems. This directs most research to understanding causality as it plays an important role in decision makings [7].

According to the current conception of causality [13], there is a causality between two events if changes of one (cause) trigger changes of the other (effect). Causal analysis is then explained by hypothetical assumptions and changing actions on such events, and it involves modeling and reasoning of causality [7]. The most used methods are reviewed to understand the current state of causal analysis for multivariate and nonlinear data in this study. Bayesian Network has been most used with the structural and probabilistic approach for causal analysis [3,17]. However, its causal modeling and reasoning by the classification of continuous data have been the significant bottleneck to assess probabilistic parameters for large data structure [9]. Artificial Neural Networks have been

© Springer International Publishing Switzerland 2015
S. Arik et al. (Eds.): ICONIP 2015, Part II, LNCS 9490, pp. 591–600, 2015.
DOI: 10.1007/978-3-319-26535-3_67

developed to solve such data analysis problems. Among them, Backpropagation Network has been used to learn the relationships between causes and effects. However, its supervised and deterministic learning structure through hidden layers limits modeling and reasoning of graded, uncertain and fuzzy causality through the network [12].

In this paper, an approach for causal analysis is proposed by using Self-Organizing Maps (SOMs), an unsupervised learning algorithm, handling unlabeled continuous data. The framework is designed based on the basic structure of SOM Tree (SOMT) [2]. The causal structure among multiple domain events can be modeled by networking multiple SOMs. A new weighting scheme on connection weight vector similarity is developed to approximate conditional distributions of data. The scheme assigns similarity weights to the neurons of a SOM for given data to process causal information for possible solutions. Unlabeled ecological data in multiple domains are applied to the experimental evaluation of the method. The experiments demonstrate that this approach allows modeling various causal assumptions using SOMs between multiple data sets. The experimental results also show that the method well approximates conditional output distributions for given inputs and allows estimating causal effects based on the similarity weight distributions.

The following section reviews the most used causal analysis methods and identifies issues of them by defining causal analysis for the current research. Our approach is described in Sect. 3 in order to address the issues. Experimental results are given in Sect. 4, followed by the conclusion in Sect. 5.

2   Background

2.1   Causal Analysis

The conceptual approach of causality has been evolved in a long history and the conception given by Hume [13] has been widely used for the current definition of causality. According to the conception, causality is defined as the analytical relationship between observable events (causes and effects) by changes. The causal changes make causality as a function and different from correlation [19]. The hypothetical assumptions and changing actions are required for modeling and reasoning of causality [7]. Thus, causal analysis is defined as the process by which one can use data to make claims about causality.

The causal inferences should not be based on deterministic interpretation of causality as our knowledge in reality is probabilistic, uncertain, graded and fuzzy [20]. Thus, all possible effects for a given cause must be considered in the inference processes for better decision makings. The mathematical expression of causal analysis with the probabilistic approach is well described by Structural Equation Model (SEM) [17]. In SEM, the general expression for causal effect is described as $P(Y|do(X))$, where P is probability distribution of event Y changed when event X is changed. The value of Y to estimate the effect is based on the probability information of all possible outputs in Y. Based on the conditional probability distributions, some inferential queries such as Most Probable Estimation (MPE) and the estimation of relative contributions of causes on effects can be answered in the process of causal reasoning [4]. The beliefs of the underlying causality can then be better interpreted if the information of all possible solutions is provided when answering such queries.

2.2 Methods for Causal Analysis

The contemporary approach has required structural causal assumptions and probabilistic estimation of causality in the process of causal analysis [18]. It brings Bayesian Network (BN) to be the most used method for causal analysis today [3,17]. A BN locally represents a causal structure and compactly represents all joint probability distributions between variables [3]. The conditional relationships based on probability distributions among variables are measured using Bayes' rule and inference techniques such as Jointree [4,15]. However, the use of BN for causal analysis has had several common limitations [1,6,14,21]. Its local and probabilistic modeling approach in learning data is computationally complex and inflexible to modify hypothetical structures in the causal analysis processes. Discretizing of continuous data has been another issue with the probabilistic approach. A smaller number of bins are more effective for modeling although a large number of bins are better to present complex data distributions [3]. Therefore, the size of bins and the sparse topology of data are the significant bottleneck of BN being infeasible to assess probabilistic parameters for a large number of joint interactions of large data structures [9].

Artificial Neural Network (ANN) models have also recently been proposed for causal analysis when dealing with multivariate and nonlinear data. The properties of ANNs such as massive parallelism, generalization, noise tolerance and adaptability have been interested in accounting for causal analysis [20]. ANNs are widely used for classification problems in nonlinear systems by learning the structure of dependencies between data [8,12,16]. Among ANNs, Backpropagation Network (BPN) is most used in conducting causal analysis with its nonlinear regression approach. BPN is a supervised learning algorithm that learns deterministic relationships between inputs and outputs by classification or maximum probability adjustment. Based on the network relationships, the relative effect could be estimated by comparing the similarity between different classes of resulting factors [8]. However, the BPN approach is not appropriate in situations where the same input may have more than one acceptable outputs. Through its joint activations, a weight cannot also be given to the similar input but different outputs using classes or clusters. Therefore, the information about the actual joint distribution of allowed output values can be lost. Furthermore, BPN approach has been discussed that such class values are not sufficient and significant in interpreting the underlying causality as they are black-boxed [12].

3   Self-Organizing Map Network for Causal Analysis

3.1   SOM Tree (SOMT)

For our study, Self-Organizing Map (SOM) [10] has been studied to conduct causal analysis for unlabeled continuous data. SOM projects high dimensional continuous data space on its discretized topological map and represents relational data value distributions. Based on such properties of SOM, SOM Tree [2] has been introduced for various predictive analyses among multiple data sets.

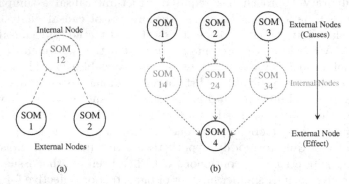

Fig. 1. (a) The basic tree structure of SOMT. The internal node SOM jointly associates two external node SOMs sharing two different domain data information. (b) A network structure using SOMs. The analytical processes flow from three input SOMs (SOM_1, SOM_2 and SOM_3) to an output SOM (SOM_4) through their link SOMs.

In the previous work of Chung and Takatsuka [2], SOM Tree (SOMT) was designed by utilizing multiple SOMs to associate multiple data sets. The basic unit of SOMT consists of one internal node SOM and two external node SOMs in a binary tree fashion (see Fig. 1(a)). Each external SOM represents each of two data sets and the internal SOM is trained with the combined data set of the two data sets. The role of the internal SOM is to link two external SOMs by associating their connection weight vectors. Therefore, the weight vector association method was developed to define an optimized linking range for all links between SOMs. Each of the given multiple data sets is from different data domains but associated with each other through some related factors. SOMT learns the joint factors, which are not explicitly expressed in the data space, by the linking procedure through the internal SOMs. Using a SOMT, one could select a structure of associations, like the one in Fig. 1(b), and generate analytical hypotheses for data predictions.

3.2   Modeling Causality by Networking SOMs

Once a method is feasible for data predictions it plays an important role in assisting causal analysis. Using the linking process of SOMT, one could predict output patterns for given inputs. However, the associational information alone cannot be used for causal reasoning. Based on the predictive functions of SOMT, we propose a method for causal analysis using SOMs. The network structure in Fig. 1(b) can be used to model causalities between multiple data sets. The basic building block of SOMT can allow one to simply configure hypothetical assumptions by changing the causal link between SOMs. With the network structure, some inferential queries are expected to be answered in the process of causal reasoning. Thus, a new weighting scheme is developed for causal modeling and reasoning in this study.

3.3   A Weighting Scheme for Conditional Similarity Distribution

From Fig. 1(b), let SOM_1, SOM_2 or SOM_3 as a cause SOM (S_C) and SOM_4 as the effect SOM (S_E). S_C and S_E have their weight vectors as $W_{C_i} = [wc_{i_1}$ $wc_{i_2} \ ... \ wc_{i_a}] \in R^a$ ($i = 1, 2, ..., s$: where s is the number of S_C neurons) and $W_{E_j} = [we_{j_1} \ we_{j_2} \ ... \ we_{j_b}] \in R^b$ ($j = 1, 2, ..., t$: where t is the number of S_E neurons) respectively. Their internal SOM S_{CE}, which links S_C and S_E, has joint weight vectors as $W_{L_k} = [wl_{k_1} \ wl_{k_2} \ ... \ wl_{k_a} \ wl_{k_{(a+1)}} \ ... \ wl_{k_{(a+b)}}]$ ($k = 1, 2, ..., r$: where r is the number of S_{CE} neurons). The linking range between the SOMs is defined by the weight vector association method of SOMT. The connection weight vectors of the linked joint weight vectors on S_{CE} for an input vector from S_C become the input vectors when linking to S_E.

The connection weight vectors take a new weighting scheme into account for the conditional similarity distributions of data as a function for causal analysis. The similarity ratios of connection weight vectors are measured to assign a weight to their conditional similarity. The weights are assigned to describe how similar the weight vectors are distributed for given data in the event space. When the neurons are linked on S_{CE} from S_C, each connection neuron (W_{L_m}; $m = 1, 2, ..., n$) among n linked ones on S_{CE} is given its similarity weight (SW) for the given input ($W_{C_{bmu}}$) on S_C (bmu: the best matching unit). The SW of W_{L_m} is calculated using Euclidean distance as:

$$SW(W_{L_m}|W_{C_{bmu}}) = \frac{\|W_{C_{bmu}} - W_{L_m}\|^{-1}}{\sum_{m=1}^{n} \|W_{C_{bmu}} - W_{L_m}\|^{-1}}. \tag{1}$$

The sum of the weights for all linked connection neurons is 1. Each of the neurons on S_{CE} now holds its SW for the linked connection weight vector and the SW is given to the joint connection weight vector when they link to S_E. The SW of every neuron of S_E is then calculated for a given connection weight vector W_{L_m} on S_{CE} as:

$$SW(W_{E_j}|W_{L_m}) = SW(W_{L_m}|W_{C_{bmu}}) \frac{\|W_{L_m} - W_{E_j}\|^{-1}}{\sum_{j=1}^{t} \|W_{L_m} - W_{E_j}\|^{-1}}. \tag{2}$$

The similarity weights are distributed over S_E for the given connection weight vector from S_{CE} by Eq. (2). The sum of all weights on S_E is the same as the SW of the given connection weight vector. After the individual calculations, the total distribution of the weights is applied to every neuron on S_E for all given connection weight vectors from S_{CE}. Thus, the sum of the total similarity weights distributed on S_E becomes 1. The finally distributed conditional similarity weight of each neuron on S_E is calculated as:

$$SW(W_{E_j}|W_{C_{bmu}}) = \sum_{m=1}^{n} SW(W_{E_j}|W_{L_m}).$$ (3)

With such similarity weight distributions between events, a MPE can be carried out by taking the event condition with the highest weight ($MPE(S_C|S_E) =$ arg max$_{W_C \in S_C} SW(S_C|S_E)$). The output changes by the input changes can also be measured using relative entropy or KullbackLeibler Divergence (D_{KL}) with the distribution information. The D_{KL} of Q (changed distribution) from P (initial distribution) on S_E is defined as [11]:

$$D_{KL}(P||Q) = \sum_{j=1}^{t} P(j) \ln \frac{P(j)}{Q(j)}.$$ (4)

The small D_{KL} indicates the change of the effect on S_E is small. Thus, it can be used to explain the relative contributions of causes on the effect.

4   Experiment and Discussion

Ecosystem is influenced by interactions of various types of environmental factors. Understanding causalities between environmental and biological data is significant for solving ecological problems. In this study, unlabeled 130 ecological data in 4 data sets with 5 variables each data set [5] are used to conduct a causal analysis. Each data set is formed with 5 components[1] by considering the most influential factors and indicators for ecological data analyses.

4.1   Processes for Causal Modeling and Reasoning

All data sets were normally standardized. The basic causalities were modeled from three environmental data sets (causes) to the biological data set (effect) as depicted in Fig. 1(b). Thus, all seven SOMs were produced by selecting each SOM size with the minimum value of quantization and topological errors. The initial learning rate of 0.05 and 1000 learning iterations with Gaussian neighborhood function were applied to all SOMs. Each external SOM for each environmental data set (S_P for physical, S_C for chemical and S_L for land use data set) was hence networked with the external SOM for biological data set (S_B) through

---

[1] Shredders(%), Filtering-Collectors(%), Collector-Gathers(%), Scrapers(%) and Predators(%) for Biological data set; Elevation(m), Slope(%), Stream Order, Embeddedness(%) and Water Temperature($°C$) for Physical data set; Dissolved Oxygen(DO, mg/l), PH, Nitrates(NO3, mg/l), Organic Carbon(DOC, mg/l) and Sulfate(SO4, mg/l) for Chemical data set; Developed Land(%), Forest(%), Herbaceous Up Land(%), Crop & Pasture Land(%) and Wetlands(%) for Land Use data set.

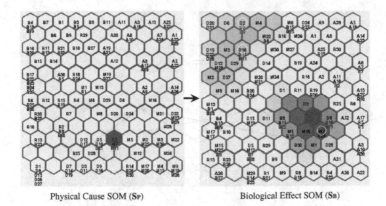

Physical Cause SOM (S_P)   Biological Effect SOM (S_B)

Fig. 2. Conditional similarity distribution of biological outputs for the given physical input of a sample data, '$M7$'. The darker color indicates the more similar output to be for the given input. The neurons on S_B for the sample data '$D2$', '$M7$' and '$M11$' are circled as they have the same best matching unit on S_P (Color figure online).

their internal SOMs. The linking range by the SOMT was approximated with 1.5 standard deviation (SD) for all links.

Using the model of networking the SOMs, we evaluated the method for causal inferences by the following processes.

- The weighting scheme was examined for its approximation of conditional output distribution for a given input. As shown in Fig. 2, a causal link from S_P to S_B was used to demonstrate the similarity distribution of outputs conditional on a given input ('$M7$').
- A MPE was carried out to estimate the most likely environmental conditions for a biological entity based on the similarity distributions. As shown in Fig. 3, the causal links between each of S_P, S_C, S_L and S_B were used to demonstrate the MPE process. The estimated environmental conditions were also tested for the biological effect validation.
- The relative contribution of each environmental data set was estimated by measuring the changes of the similarity distribution of biological outputs on the changes of environmental inputs using D_{KL} calculation. The initial values of environmental inputs were given and the value was changed by 0.5 SD interval to every variable while others were invariant. A BPN[2] was applied to examine the estimated result by the same changes of environmental inputs but calculating the mean square errors on its target biological responses.

---

[2] The BPN was trained with the same 130 sample data and tested with 16 new data for environmental inputs and biological outputs. The learning process was converged after 1000 iterations with the mean square error of 0.091. It showed the high accuracy in predicting the biological data on the basis of the environmental data ($R = 0.95, P < 0.05$ for trained data and $R = 0.56, P < 0.05$ for test data).

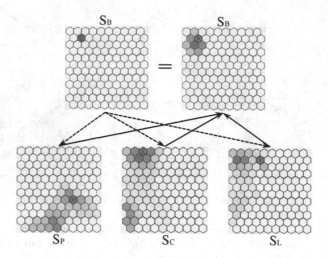

Fig. 3. A MPE process for S_P, S_C and S_L linked to S_B. The neuron, which has the highest weight in the conditional similarity distribution is circled on each SOM.

4.2   Results and Discussion

- From the weighting scheme examination, we could inspect conditional biological output distribution for a given physical input as seen in Fig. 2. The input neuron on S_P for '$M7$' showed the other commonly valued inputs ('$D2$' and '$M11$'), which had differently valued outputs. The similarities of the different outputs for the common input neuron were well distributed over S_B with higher weights than others. Those outputs were highly weighted based on their joint information with the inputs in the distribution.
- According to the weighting scheme, the most likely environmental conditions for a given biological entity were estimated through the MPE process as seen in Fig. 3. The neuron circled on each of S_P, S_C, and S_L showed the highest weight of conditional similarity for the given effect on S_B and was selected as the most likely causal condition. The estimation was validated since the neuron of the initially given effect on S_B also showed the highest similarity weight when the estimated causal conditions were given.
- The relative contributions of environmental data sets were also estimated with the distributed similarity information on the changes of biological outputs. The graphs in Fig. 4 show the results estimated by (a) BPN and (b) our approach respectively. As seen in the graphs, the result of our approach showed an overall agreement with the result of the BPN, used for the estimation in many applications.

The evaluational results demonstrate that our approach is highly acceptable and useful for analyzing uncertain, graded and fuzzy causality of unlabeled multivariate and nonlinear data. As each external SOM represents each continuous data structure, one can examine conditional distributions of data in the data

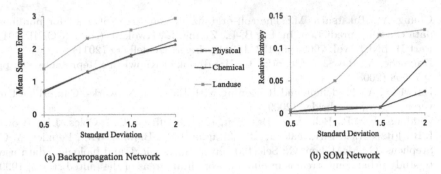

Fig. 4. Sensitivity analyses of the biological data changes by three environmental data changes using (a) Backpropagation Network and (b) SOM Network.

space using the weighting scheme on connection weight vector similarity. The network connection between external SOMs transfers the weighted joint information from one to another through their internal SOM. Thus, the mapping of the causal changes to the effect changes can be investigated through the linked path. This can help one to reason complex causality through various inferential processes detecting all possible alternatives. Therefore, empirical experiments involving domain specialists are required in the future to measure the integrated quality of the method for causal analysis.

5   Conclusion

We have presented a new approach networking SOMs for causal analysis when dealing with unlabeled nonlinear and multivariate data from multiple sources. The issues of the most used existing methods for causal analysis were identified on modeling and reasoning of uncertain, graded and fuzzy causality for such data. In our framework, a new weighting scheme has been developed to obtain conditional distributions of data based on connection weight vector similarity between SOMs. The potential to be outputs for the given input could be inspected over the similarity weight distributions on SOMs. Some causal analysis processes, a MPE and the relative contribution estimation, were demonstrated using the distributional similarity information for multiple domains of ecological data. The experimental results showed that our approach is highly acceptable for causal information processing. This approach with the new weighting scheme on connection weight vector similarity can provide a new opportunity in causal modeling and reasoning for such data.

References

1. Adriaenssens, V., Goethals, P.L.M., Charles, J., Pauw, N.D.: Application of bayesian belief networks for the prediction of macroinvertebrate taxa in rivers. Ann. Limnol. Int. J. Limnol. **40**, 181–191 (2004)

2. Chung, Y., Takatsuka, M.: The self-organizing map tree (SOMT) for nonlinear data causality prediction. In: Lu, B.-L., Zhang, L., Kwok, J. (eds.) ICONIP 2011, Part II. LNCS, vol. 7063, pp. 133–142. Springer, Heidelberg (2011)
3. Darwiche, A.: Bayesian networks. In: Handbook of Knowledge Representation, pp. 467–508 (2008)
4. Darwiche, A.: Modelling and Reasoning with Bayesian Networks. Cambridge University Press, Cambridge (2009)
5. Giddings, E.M.P., Bell, A.H., Beaulieu, K.M., Cuffney, T.F., Coles, J.F., Brown, L.R., Fitzpatrick, F.A., Falcone, J., Sprague, L.A., Bryant, W.L., Peppler, M.C., Stephens, C., McMahon, G.: Selected physical, chemical, and biological data used to study urbanizing streams in nine metropolitan areas of the united states, 1999–2004. Technical report Data Series 423, National Water-Quality Assessment Program, U.S. Geological Survey (2009)
6. Giles, H.: Using bayesian networks to examine consistent trends in fish farm benthic impact studies. Aquaculture **274**, 181–195 (2008)
7. Hidalgo, F.D., Sekhon, J.S.: Causality. In: Badie, B., Berg-Schlosser, D., Morlino, L. (eds.) International Encyclopedia of Political Science, pp. 204–211. SAGE Publications, Thousand Oaks (2011)
8. Hillbrand, C., Karagiannis, D.: Using artificial neural networks to prove hypothetic cause-and-effect relations: a metamodel-based approach to support strategic decisions. In: The Fourth Conference on Enterprise Information Systems. Proceedings of ICEIS 2002, vol. 1, pp. 367–373 (2002)
9. Jurgelenaite, R., Lucas, P.J.F.: Exploiting causal independence in large bayesian networks. Knowl.-Based Syst. **18**, 153–162 (2005)
10. Kohonen, T.: Self-Organizing Maps. Information Sciences, 3rd edn. Springer, Heidelberg (2001)
11. Kullback, S.: Information Theory and Statistics. Courier Corporation, Mineola (2012)
12. Lee, C., Rey, T., Mentele, J., Garver, M.: Structured neural network techniques for modeling loyalty and profitability. In: Data Mining and Predictive Modeling Paper 082–30. Proceedings of SAS SUGI, pp. 1–13 (2005)
13. May, W.E.: Knowledge of causality in hume and aquinas. Thomist **34**, 254–288 (1970)
14. McCann, R.K., Marcot, B.G., Ellis, R.: Bayesian belief networks: applications in ecology and natural resource management. Can. J. For. Res. **36**, 3053–3062 (2006)
15. Murphy, K.P.: The bayes net toolbox for matlab. In: Computing Science and Statistics. Proceedings of the Interface (2001)
16. Palaneeswaran, E., Love, P.E.D., Kumaraswamy, M.M., Ng, T.S.T.: Mapping rework causes and effects using artificial neural networks. Build. Res. Inf. **36**(5), 450–465 (2008)
17. Pearl, J.: Causal inference in statistics: An overview. Technical report R350, Statistics Surveys (2009)
18. Peyrot, M.: Causal analysis: theory and application. J. Pediatr. Psychol. **21**(1), 3–24 (1996)
19. Simon, H., Rescher, N.: Cause and counterfactual. Philos. Sci. **33**, 323–340 (1966)
20. Sun, R.: A neural network model of causality. IEEE Trans. Neural Networks **5**(4), 604–611 (1994)
21. Uusitalo, L.: Advantages and challenges of bayesian networks in environmental modelling. Ecol. Model. **203**, 312–318 (2007)

In-Attention State Monitoring Based on Integrated Analysis of Driver's Headpose and External Environment

Seonggyu Kim, Mallipeddi Rammohan, and Minho Lee[✉]

School of Electronics Engineering, Kyungpook National University,
1370 Sankyuk-Dong, Puk-Gu, Taegu, 702-701, South Korea
{kimsg99550,mallipeddi.ram,mholee}@gmail.com

Abstract. In Advanced Driving Assistance Systems (ADASs), for traffic safety, one of main application is to notify the driver regarding the important traffic information such as presence of a pedestrian or information regarding traffic signals. In a particular driving scenario, the amount of information related to the situation available to the driver can be judged by monitoring the internal information (for example driver's gaze) and external information (for example information regarding forward traffic). Therefore, to provide sufficient information to the driver regarding a driving scenario it is essential to integrate the internal and external information which is lacking in the current ADASs. In this work, we employ 3D pose estimate algorithm (**POSIT**) for estimation of driver's attention area. In order to estimate the distributions corresponding to the forward traffic we employ both bottom-up saliency map model and a top-down process using HOG pedestrian detection. The integration of internal and external information is done using the mutual information.

Keywords: Advanced driving assistance systems (ADASs) · Face landmark detection · Mutual information · Saliency map

1 Introduction

Currently, more than 90 % of the automotive accidents occur due to the driver's lack of situation understanding, according to the "European accident research and safety report 2013". In other words, most of the accidents occur due to the lack of sufficient information regarding the driving situation due to in-attentional blindness. Therefore, recently the use of Advanced Driver Assistance Systems (ADASs) is gaining prominence to achieve traffic safety. In [6], the authors proposed driver's distraction level modeling using head pose and eye gaze while in [6] a driver's in-attention state monitoring system based on the classification of eye-blinking pattern, head pose and OCSVM classifier was proposed. Also, in [4], the driver's lane change intent was predicted using driver behavioral cues such as, head dynamics, eye data etc. In addition, lane departure warning, vehicle detection, advanced cruise control systems, etc. exist in the literature. However, most of the works in the literature do not consider the integration of internal and external information. In other words, in a driving situation an efficient ADAS should monitor the driver's awareness in addition to traffic information distributions.

© Springer International Publishing Switzerland 2015
S. Arik et al. (Eds.): ICONIP 2015, Part II, LNCS 9490, pp. 601–608, 2015.
DOI: 10.1007/978-3-319-26535-3_68

In this work, we will focus on integrating the internal and external information. First, in order to analyze the internal status the driver's frontal face image is captured using a monocular infrared camera. In the current work, the driver's attention area is recognized using regression tree based face landmark detector [8] and POSIT-algorithm [3] which is a 3D-pose estimate method. Secondly, the external road information distributions are analyzed by bottom-up and top-down processing. The bottom-up model employed is a saliency map based on Gestalt principle [1] while the top-down process uses HOG cascade based pedestrian detector [2]. Finally, the internal and external information are integrated to recognize the driver's cognitive status.

The remaining part of this paper is organized as follows: Sect. 2 explains the proposed integration system. Section 3 discusses the experimental result. Conclusion is presented in Sect. 4.

2   Integration System of Internal and External Information

Generally, in a driving situation a driver is exposed to too much of information. Therefore, from the information available, it is difficult to segregate the necessary and sufficient information for a safe driving. Motivated by the above, we propose a model to integrate the internal and external situations of a driver as shown by the block diagram in Fig. 1.

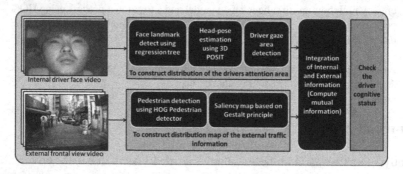

Fig. 1. The flow chart of proposed ADASs

As shown in Fig. 1, the proposed model consists of two parts: (1) obtaining internal information and (2) obtaining external information through bottom-up and top-down processes. To extract internal information, we use monocular infrared camera and obtain driver's attention area based on head-pose estimator. To extract external information, we calculate the road information distributions based on both bottom-up and top-down processes. In other words, the bottom-up saliency map model alone may not be sufficient to acquire the external information. Therefore, we consider the top-down process where we employ pedestrian detector based on HOG cascade method. Then the information available to the driver can be evaluated by the mutual information extracted using internal and external information. Internal information such as driver's frontal face images are obtained by the camera placed inside the vehicle. In the driver face image, we detect face landmarks such as the left face landmark, right face landmark, nose point and chin point using regression tree based face landmark detector. These landmark points are used to estimate driver's head-pose orientation by POSIT algorithm [3]. And, we determine the driver's attention area by head-pose orientation. In addition, the external road traffic images are obtained by the camera placed on the front side. We calculate the traffic information distribution based on the Gestalt saliency map model [7, 9] and HOG cascade pedestrian detector [2] as bottom-up and top-down process. Finally, in order to recognize the driver's cognitive status, we integrate the output of internal image and external image using mutual information, which indicates the amount of information regarding the current traffic scenarios is acquired by the driver.

2.1   Construction of External Traffic Information Distribution Map

Gestalt Saliency Map: The saliency map model employed in the current work is based on Gestalt principles. Conventional saliency map is inspired by the human biological bottom-up visual process. Human subjects react to the contrast of color and intensity by center-surround effects. Also on the road, most of the important parts (such as signs, traffic lights, pedestrians, etc.) comprise of maximized color contrast and regular patterns of shape and arrangement. In order to take the advantage of this we applied the Gestalt principle that includes concepts such as similarity, continuation, closure and proximity [1]. These theories are the basis for emphasizing the recognition area of human view of the bottom-up saliency map. Figure 2 presents the proposed external traffic information distribution map block diagram. Proposed model includes similarity, continuity, symmetry, closure theory based Gestalt principles. In the first step, we separate color image to R, G, B color and intensity image map. Next, we have coded the color opponent, such as red and green color, blue and yellow color. Also, the orientation and edge maps are extracted from intensity map. In the third step, we applied the Gestalt principle as shown in the Fig. 2 using similarity, continuity, symmetry and closure. The similarity feature map can be constructed by modified center-surround difference and normalization (CSDN) algorithm. The symmetry feature map can be extracted by symmetry extraction mechanism [5]. The continuity case is opposite to the symmetry case. Finally, in order to compose the closure feature map, we check parallel area in more than four directions and calculate the value. The final step is center-surround difference normalization. In the current work, we applied Gaussian kernels with different sizes to extract feature maps in order to implement the center-surround difference. In addition, we normalize each feature map and integrated all the feature maps.

HOG Pedestrian Detector: As discussed earlier, the use of Gestalt saliency map to construct bottom-up information is not sufficient to extract the external information. In other words, we need to reflect the top-down process to detect specific information such as pedestrian, which is the most important information for safety driving. The failure to recognize the pedestrians on the frontal side can lead to fatal accidents. Therefore, the recognition of pedestrians on the front side is essential in ADASs. The proposed system contains the HOG cascade pedestrian detector developed in OPENCV version 2.4.9. As

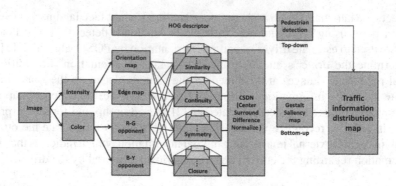

Fig. 2. The structure of traffic information distribution map

shown in Fig. 2, the HOG descriptor is extracted in the intensity map. Finally, we detect pedestrian using pre-learned detector.

Construction of External Information Distribution Map: In order to construct the external traffic information distribution map, we consider both Gestalt saliency map and the result of pedestrian detection. Figure 3(a) is a sample of external road image. Figure 3(b) is the result of proposed Gestalt saliency map. And, Fig. 3(c) is the result of HOG pedestrian detector. Figure 3(d) is the traffic information distribution map based on Fig. 3(b) and (c). As shown in Fig. 3(d), the distribution map is divided into 6 regions. And, we have allocated different weights to different regions in the distribution map according to the position of pedestrian to the 6 regions. The second and fifth regions are given larger weights than others because the frontal pedestrian is more important than side pedestrian.

(a)   (b)   (c)   (d)

Fig. 3. Example image for external environment and preprocessing stages for driving monitoring (a) Outdoor image (b) Gestalt saliency map (c) Result of HOG cascade pedestrian detector (d) Traffic information distribution map

2.2   Construction of Driver's Attention Area Distribution Map

A driver attention area is estimated by POSIT-algorithm [3]. To apply this method, it is essential that the driver's face the landmark points are detected. In this paper, we extract 68 face landmark points using regression tree landmark detector [8]. The acquired images of a driver in driving situation have large illumination deviation and difference of face pose. So, we use an infrared camera and regression tree based face landmark detector. Figure 4 shows the process of extracting driver attention area.

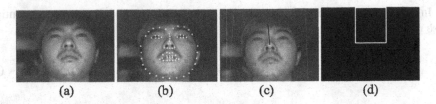

(a)   (b)   (c)   (d)

Fig. 4. Example image for driver's sate monitoring (a) Face image (b) 68 detected face landmark points (c) Head-pose estimation (d) Driver attention area

Driver's Face Land-mark Detection: In our system, we detect the 68 face landmark points using **dlib-18.10** C++ library [8]. Figure 4 (b) is the result of regression tree based face landmark detector. We employ the 4 points among 68 face landmarks for the estimation of driver head pose.

3D Head Pose Estimation: Figure 4(c) is a driver head pose estimation. As shown in this figure, we select each of the four driver's face landmark points; the left face landmark, right face landmark, nose point and chin point.

To estimate driver's head-pose, the proposed system employs 3D-pose estimation, POSIT (POS with Iterations), method [3]. POSIT is a method to reduce the error rate through POS (Pose form Orthography) iteratively.

$$P = K[R|t] \tag{1}$$

In Eq. (1), the projection matrix P can be calculated by camera parameter matrix K, rotation matrix R and motion-vectors t. Head pose is based on a 3D model nose coordinates. And it is normalized by dividing with the distance between landmark points on the either side of the face.

$$\text{HeadPose} = \frac{nose(x, y, z)}{distance_{side\ landmarks}} \tag{2}$$

$$distance_{landmarks} = landmark_l(x, y) - landmark_r(x, y) \tag{3}$$

2.3   Integration of Internal and External Information

In this paper, in order to identify the driver' cognitive status we analyze the mutual information between internal and external information. It represents the amount of traffic information acquired by the driver's during driving. The external image and the driver attention region are divided into 6 equal parts to calculate the mutual information.

The bottom-up saliency map model reacts strongly about traffic signals, warning lights and road sign. If a driver sees these areas, the amount of obtained information increases.

$$A = \{a_1, a_2, a_3, \dots, a_i\} \tag{4}$$

$$B = \{b_1, b_2, b_3, \dots, b_j\} \tag{5}$$

In Eqs. (4) and (5), **A** represents vehicle's outside information, **B** represents attention gaze ratio per unit time of the driver.

$$Information_{trans} = \sum_j P(b_j \backslash a_i) log \frac{P(a_i, b_j)}{P(a_i)P(b_j)} \tag{6}$$

Equation (6) represents the average amount of transmitted information from **A** to **B**. The driver attention gaze ratio is the mean of 15 frame units (Fig. 5).

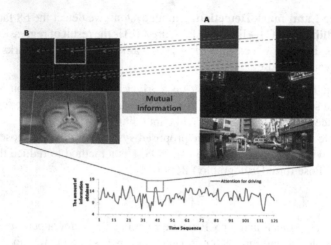

Fig. 5. Information integration process

3   Experiment and Result

In this paper, to integrate the internal and external information, we use desktop PC with Intel Core i7-4770 k processor and 16 GB main memory. And Pointgrey Ltd. FL3-U3-13E4 M-C model camera (resolution 1280 * 1024) is used to obtain a driver's face image. The camera can capture images at a speed of up to 60 fps. External road images are obtained using black box video image (resolution 640 * 360). Both the images are synchronized over each frame. Test videos contain a total of 1800 frames. In the videos, the average of 15 frames is used to compute the mutual information. The external video includes traffic information and pedestrian. As shown in Fig. 6, the result is depicted in two plots. Figure 6(a) shows the road images which are used as samples to test the proposed method. Figure 6(b) depicts the mutual information obtained using internal information and external information using both bottom-up (Gestalt saliency map) and top-down processes (pedestrian detection). Figure 6(c) also depicts the mutual information but the external information is based only on bottom-up process (Gestalt saliency map).

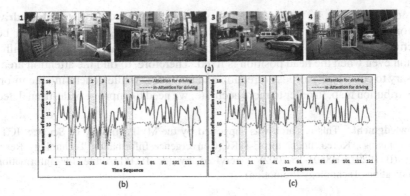

Fig. 6. Experimental results: (a) The road images with pedestrians (b) Mutual information where external information is obtained using both Bottom-up and Top-down processes (c) Mutual information where external information is obtained using only bottom-up process

In Fig. 6(b) and (c), the orange areas correspond to the detection of the pedestrians in Fig. 6(a). From Fig. 6(b), it can be observed that during an in-attentional driving the mutual information (red line) goes down (see the shaded regions) if the driver fails to detect the pedestrian. However, during attentional driving (blue line) the detection of pedestrians results in the increase of mutual information as shown in the shaded regions. The increase in the mutual information if the driver is attentive and the decrease if the driver is in-attentive will allow the ADAS to prompt appropriate signals to the driver. However, such kind of facility is not available if the top-down process such as the pedestrian detection is not employed as shown in Fig. 6(c). Therefore, the proposed system that employs both top-down and bottom-up processes to acquire mutual information can better classify a driver to be attentive and inattentive during a driving scenario. If the driver is found to be inattentive, an alarming signal can be given to alert the driver. Therefore, by employing the proposed system it would be possible to avoid traffic accidents that results due to the driver inattentiveness.

4   Conclusion

In this paper, we proposed an integrated system to analyze driver's cognitive status into attentive driving and inattentive driving. In this work, we proposed to estimate the driver's cognitive status through mutual information which was obtained by combining the internal and external information. In the proposed system, we estimated the driver attention area through internal information such as driver's face image. It was determined by POSIT algorithm [3]. The landmark points were detected by regression tree based face landmark detector [8]. Secondly, we calculated the distributions corresponding to the external road information using bottom-up saliency map (Gestalt Saliency map) and HOG cascade pedestrian detector. Third is the integration of both internal and external information unlike conventional ADASs. From the results, we demonstrated that integration of internal and external information was able to generate meaningful results in estimating the cognitive status of the driver.

In our future work, we will consider the eye gaze point in order to determine a driver's attention area regarding external traffic images. The eye gaze will help in better predicting the driver's cognitive status since the eye gaze can be pointed in different direction even when the head position is fixed. Therefore, to find the attention area it is necessary to consider head-pose direction and eye-gaze. While generating the information distribution map, in order to more accurate, we need to improve pedestrian detector.

Acknowledgment. This research was supported by the MSIP(Ministry of Science, ICT and Future Planning), Korea, under the C-ITRC(Convergence Information Technology Research Center) (IITP-2015-H8601-15-1002) supervised by the IITP(Institute for Information & communications Technology Promotion)

References

1. Chang, D., Nesbitt, K.V., Wilkins, K.: The Gestalt principles of similarity and proximity apply to both the haptic and visual grouping of elements. In: Proceedings of the Eight Australasian Conference on User Interface, vol. 64, pp. 79–86 (2007)
2. Dalal, N., Triggs, B.: Histograms of oriented gradients for human detection. In: IEEE Computer Society Conference on Computer Vision and Pattern Recognition, CVPR 2005, pp. 886–893 (2005)
3. Dementhon, D.F., Davis, L.S.: Model-based object pose in 25 lines of code. Int. J. Comput. Vis. **15**(1–2), 123–141 (1995)
4. Doshi, A., Trivedi, M.M.: On the roles of eye gaze and head dynamics in predicting driver's intent to change lanes. IEEE Trans. Intell. Transp. Syst. **10**(3), 453–462 (2009)
5. Fukushima, K.: Use of non-uniform spatial blur for image comparison: symmetry axis extraction. Neural Netw. **18**(1), 23–32 (2005)
6. Jo, H., Lee, M.: In-attention state monitoring for a driver based on head pose and eye blinking detection using one class support vector machine. In: Neural Information Processing, pp. 110–117 (2014)
7. Jo, H., Ojha, A., Lee, M.: A study on region of interest of a selective attention based on gestalt principles. In: Lee, M., Hirose, A., Hou, Z.-G., Kil, R.M. (eds.) ICONIP 2013, Part III. LNCS, vol. 8228, pp. 41–48. Springer, Heidelberg (2013)
8. King, D.E.: Dlib-ml: a machine learning toolkit. J. Mach. Learn. Res. **10**, 1755–1758 (2009)
9. Shen, J., Ojha, A., Lee, M.: Role of gestalt principles in selecting attention areas for object recognition. In: Neural Information Processing, pp. 90–97 (2013)

Item Category Aware Conditional Restricted Boltzmann Machine Based Recommendation

Xiaomeng Liu[1,2], Yuanxin Ouyang[1,2(✉)], Wenge Rong[2,3], and Zhang Xiong[2,3]

[1] State Key Laboratory of Software Development Environment,
Beihang University, Beijing, China
xiaomeng.liu@buaa.edu.cn

[2] School of Computer Science and Engineering, Beihang University, Beijing, China
oyyx@buaa.edu.cn

[3] Research Institute of Beihang University in Shenzhen, Shenzhen, China
{w.rong,xiongz}@buaa.edu.cn

Abstract. Though Collaborative Filtering is one of most effective recommendation technique, the problem of dealing sparsity brings traditional collaborative filtering recommendation systems great challenge. In this paper, we propose an improved Item Category aware Conditional Restricted Boltzmann Machine Frame model for recommendation by integrating item category information as the conditional layer, aiming to optimise the model parameters, so as to get better recommendation efficiency. Experimental studies on the standard benchmark datasets of MovieLens 100 k and MovieLens 1 M have shown its potential in improving recommendation accuracy.

Keywords: Recommendation system · Collaborative filtering · Conditional restricted boltzmann machine · Item category

1 Introduction

Nowadays, recommender systems have become indispensable to assist people to make decisions and one of the most important recommendation techniques is collaborative filtering (CF) due to its simpleness and robustness [7]. Though achieving great achievements, CF also faced several challenges among which an important one is to deal with large and highly sparse datasets [3].

One possible effective solutions to sparsity problem is matrix dimensionality reduction and Singular Value Decomposition (SVD) is the representative technique [13]. Its main idea is that the low-dimensional vectors lived in a reduced space with increased density can reduce data sparseness and then allow CF systems to handle sparse datasets. Another promising approach in this respect is to use of Restricted Boltzmann Machines (RBM), which can also infer lower-dimensional representations automatically [5]. With the advent of a fast learning algorithm for RBM, it has been proven that RBM models are potential and effective in handling hundreds of millions of user-item ratings [9].

© Springer International Publishing Switzerland 2015
S. Arik et al. (Eds.): ICONIP 2015, Part II, LNCS 9490, pp. 609–616, 2015.
DOI: 10.1007/978-3-319-26535-3_69

Apart from proposing sophisticated models for sparisity challenge, an alternative solution is to employ external available information into CF systems, e.g., demographic statistical information [11], social network information [6], item category information [12] and etc. These methods tried to integrate rich external knowledge information and also proven the effectiveness to some extent.

Inspired by the previous researches, in this paper we proposed Item Category aware Conditional Restricted Boltzmann Machine Frame model (IC-CRBMF) which employed both RBM and item category information for recommendation. Experimental studies on standard benchmark datasets of MoiveLens 100 k and MoiveLens 1 M have shown its poential in improving recommendation accuracy.

The remainder of the paper is organised as follows. Section 2 will present related works and Sect. 3 will elaborate the proposed IC-CRBMF model in detail. The experimental results will be illustrated in Sect. 4 and Sect. 5 will conclude this paper and point out possible future work.

2   Related Work

Sparsity problem is one of the most important challenges in the development of CF algorithms. To overcome this limitation, a lot of techniques have been proposed and one of most effective solutions is dimension reduction techniques. A representative work by Bokde et al. utilised matrix factorisation to overcome sparseness constraints [1]. Paterek et al. further proposed an improved regularized SVD for CF [13].

As an another promising approach to overcome this problem, RBM has also been successfully applied into CF to model highly sparse ratings by Salakhutdinov et al. [9]. Similarly, Georgiev et al. also proposed a unified hybrid non-IID framework based on RBM [2]. Our team has made the autoencoders for CF get better local optimum with pretraining RBMs [8].

Besides, some researchers proposed several methods which imported external information into CF models to improve recommendation efficiency. For example, Vozalis et al. combined demographic information to enhance the performance of traditional CF algorithms [11]. Kaya et al. used social networks to solve data sparsity problem in CF [6]. Specially, Yao et al. integrated item category information into CF recommendation algorithm and provided a better performance than other item-based algorithms [12].

3   IC-CRBMF model

3.1   RBM and Conditional RBM Model

RBM is a probabilistic undirected graphical model, which consists of m visible units $V = (V_1, ..., V_m)$ representing observable data and n hidden units $H = (H_1, ..., H_n)$ capturing dependencies between observable variables. When bsasic RBM model is employed for CF, each user (or item) can be considered as a single training case for an RBM. Each RBM may have different number of visible softmax units but has the same number of hidden units, namely, all of the connected weights and bias are shared among the RBM Model.

Conditional RBM (CRBM) model is developed from basic RBM model, which contains three layers including one visible layer, one hidden layer and one conditional layer. The conditional layer units $R = (R_1, ..., R_p)$ are modelled with an additional source of information.

3.2 IC-CRBMF_UserBased and IC-CRBMF_ItemBased

In this paper we proposed an IC-CRBMF model by integrating item category information for CF, as shown in Fig. 1. The proposed IC-CRBMF model takes use of item category feature vector as the conditional layer. According to the difference of visible layer representation, IC-CRBMF model is further divided into two parts, i.e., IC-CRBMF_UserBased model and IC-CRBMF_ItemBased model.

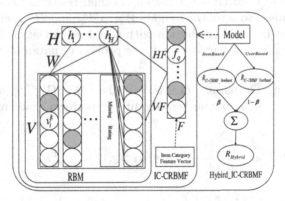

Fig. 1. IC-CRBMF model

IC-CRBMF_UserBased model builds on each user' rating vectors on all items and the conditional layer is these items's category or genre. Different from IC-CRBMF_UserBased model, IC-CRBMF_ItemBased model builds on each item's rating vectors by all users and the conditional layer is just the given item's genre feature vector. As such the most important difference between the two IC-CRBMF models is the conditional layer representation about genre feature vector, while their modelling process is similar and interlinked.

3.3 Modelling and Model Learning

Modelling. IC-CRBMF model includes three parts, a visible layer consists of a rating matrix V, a hidden layer contains a hidden feature vector H and a conditional layer consists of given category feature vector F. We use conditional multinomial distribution (softmax units) for modelling each column of the visible layer's binary rating matrix V and a Bernoulli distribution for modelling hidden feature vector H, while the category feature vector F for conditional layer is constant for each RBM. The visible layer units and the conditional layer units are activated as follows:

$$p(v_i^k = 1|H, F) = \frac{\exp(b_i^k + \sum_{j=1}^{H} W_{ij}^k h_j + \sum_{q=1}^{F} f_q V F_{qi})}{\sum_{k=1}^{K} \exp(b_i^k + \sum_{j=1}^{H} W_{ij}^k h_j + \sum_{q=1}^{F} f_q V F_{qi})} \tag{1}$$

$$p(h_j = 1|V, F) = \sigma(b_j + \sum_{i=1}^{V} \sum_{k=1}^{K} v_i^k W_{ij}^k + \sum_{q=1}^{F} f_q H F_{qj}) \tag{2}$$

where $\sigma(x) = \frac{1}{1+e^{-x}}$ is the logistic function, W_{ij}^k is the connected weight between hidden feature j and rating k by user i on this movie, b_i^k is the bias of rating k by user i, b_j is the bias of feature j, $V F_{qi}$ is the connected weight between rating term by user i and item genre feature q, $H F_{qj}$ is the connected weight between hidden feature j and item genre feature q.

Note that the b_i^k is initialised to the logs of their respective base rates over all users; b_j is initialised to zero; the weights W_{ij}, $V F_{qi}$ and $H F_{qj}$ are all initialised with a zero-mean normal distribution with standard deviation 0.01; and genre feature vector f does not participate in the process of parameters update.

Model Learning. The parameter Θ is defined as: $\Theta = \{W_{ij}^k, b_i^k, b_j, V F_{qi}, H F_{qj}\}$, and update of all of the parameters require to perform gradient ascent in the log-likelihood:

$$\Delta W_{ij}^k = \frac{\partial \log p(V)}{\partial W_{ij}^k} = < v_i^k h_j >_{data} - < v_i^k h_j >_{model} \tag{3}$$

$$\Delta b_i^k = \frac{\partial \log p(V)}{\partial b_i{}^k} = < v_i^k >_{data} - < v_i^k >_{model} \tag{4}$$

$$\Delta b_j = \frac{\partial \log p(V)}{\partial b_j} = < h_j >_{data} - < h_j >_{model} \tag{5}$$

$$\Delta V F_{qi} = \frac{\partial \log p(V)}{\partial V F_{qi}} = < v_i^k f_q >_{data} - < v_i^k f_q >_{model} \tag{6}$$

$$\Delta H F_{qj} = \frac{\partial \log p(V)}{\partial H F_{qj}} = < h_j f_q >_{data} - < h_j f_q >_{model} \tag{7}$$

where $< \cdot >_{data}$ is the expectation defined from the training datasets, and $< \cdot >_{model}$ is an expectation of the distribution defined by the IC-CRBMF model.

However, $< \cdot >_{model}$ is difficult to obtain, as such we use the gradient of the Contrastive Divergence (CD) objective function, as indicated in [4,9].

$$\Delta W_{ij}^k = \frac{\partial \log p(V)}{\partial W_{ij}^k} = < v_i^k h_j >_{data} - < v_i^k h_j >_T \tag{8}$$

where expectation $< \cdot >_T$ represents a distribution of samples by running the Gibbs samper, initialising at the data, for T steps [9]. In accordance with the same method like the Eq. (8) to simplify solving process of Δb_i^k, Δb_j, ΔVF_{qi}, and ΔHF_{qj}. In the practical process of updating parameters, we also introdue momentum method [10] in order to take into accound the iteration $t - 1$ when computing the parameters update at interation t:

$$W_{ij}^k(t) = k \cdot W_{ij}^k(t-1) + \epsilon \cdot \Delta W_{ij}^k \tag{9}$$

where the k is the learning rate of the momentum and the ϵ is the learning rate of parameters updated. The biases b_i^k and b_j, connected weights VF_{qi} and HF_{qj} are learned by the simplified Eqs. (4)–(7) with momentum like the Eq. (9).

3.4 Rating Prediction

To make rating prediction for given user, we just simply clamp the given ratings vector on the visible layer. Afterwards we need to calculate the probabilistic distribution of the hidden layer and let the hidden units be activated through the connected weights between them and the visible layer. Finally, we calculate the probabilistic distribution of the visible units to make predictions. Let unit v_i^k in the visible layer represent the probability of unit i with rating k, we use the expectation as the final predicted value, as follows:

$$R = \sum_{k=1}^{K} k \cdot p(v_i^k = 1 | H, F) \tag{10}$$

According to the difference of visible layer representation, the IC-CRBMF model includes IC-CRBMF_ItemBased and IC-CRBMF_UserBased. In order to simplify and reduce the complexity of optimising hybrid model, the Hybrid IC-CRBMF model just is implemented by weighing linear combination on scores between IC-CRBMF_ItemBased and IC-CRBMF_UserBased model.

$$R_{Hybrid} = \beta \cdot R_{IC-CRBMF\_ItemBased} + (1 - \beta) \cdot R_{IC-CRBMF\_UserBased} \tag{11}$$

where β is the empirically evaluated weight coefficient.

4 Experimental Study

4.1 DataSet and Evaluation Metrics

We evaluated the proposed IC-CRBMF model on MovieLens datasets (MovieLens 100 k and MovieLens 1 M), which are commonly used for evaluating collaborative filtering algorithms. MovieLens 100 k consists of 100,000 ratings for 1,682 movies assigned by 943 users, and MovieLens 1 M contains 1,000,000 ratings for 3,952 movies by 6,040 users. Each rating is an integer between 1 (worst) to 5 (best). In this paper we follow previous reseach methods and the training and testing datasets partition is set to 80 % and 20 %.

To verify the recommendation efficiency of our proposed methods, Mean Absolute Error (MAE) and Root Mean Squared Error (RMSE) are employed:

$$MAE = \frac{\sum_{j=1}^{L} |r_j - p_j|}{L} \qquad RMSE = \sqrt{\frac{\sum_{j=1}^{L} (r_j - p_j)^2}{L}} \qquad (12)$$

where L is the size of testing datasets.

4.2 Experimental Results and Discussion

Firstly, we compared prediction quality about the dependence relationship of IC-CRBMF model on different numbers of hidden units and different numbers of iterations of training epochs. Figure 2 shows the dependency of MAE & RMSE on the number of hidden units when the proposed IC-CRBMF model is trained for 10 epochs. Figure 3 shows the dependency of MAE & RMSE on the number of epochs when IC-CRBMF model is trained with 90 hidden units.

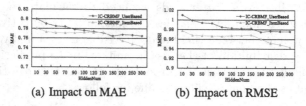

(a) Impact on MAE   (b) Impact on RMSE

Fig. 2. Dependency of MAE & RMSE on number of hidden units

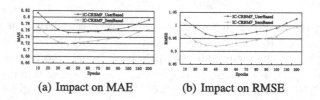

(a) Impact on MAE   (b) Impact on RMSE

Fig. 3. Dependency of MAE & RMSE on number of epochs

Secondly, we implement Hybrid IC-CRBMF model via linear combination on scores between IC-CRBMF_ItemBased and IC-CRBMF_UserBased. The dependency relationship between MAE & RMSE and the values of β is shown in Fig. 4. Experimental results show that Hybrid RBM and Hybrid IC-CRBMF can reach lowest values of MAE & RMSE are equal to 0.45, 0.70 on MovieLens 100 k and 0.55, 0.45 on MovieLens 1 M respectively.

Finally, we calculate and evaluate the prediction quality of IC-CRBMF models (IC-CRBMF_UserBased, IC-CRBMF_ItemBased, Hybrid IC-CRBMF) against other CF algorithms including Basic UserBased and ItemBased, SVD, RBM, AutoEncode [8], Stack_AutoEncode [8] and Hybrid RBM. The result is shown in Table 1.

After investigating the experimental results, some interesting points can be revealed:

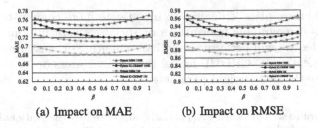

(a) Impact on MAE   (b) Impact on RMSE

Fig. 4. Dependency of MAE & RMSE on the values of β

(1) IC-CRBMF model is efficient and can be preferably adapted to the larger and more highly sparse datasets.
(2) IC-CRBMF_UserBased model has more advantages than IC-CRBMF_Item Based model when working on larger and more sparse datasets.
(3) Due to overfitting of the IC-CRBMF model parameters, the recommendation efficiency of IC-CRBMF model will be restricted with continuously increasing of epochs.
(4) Hybrid IC-CRBMF model is an effective and conveniently implemented combination model for CF recommendation algorithms.

Table 1. Prediction quality on MovieLens dataSets

CF models	MovieLens 100 k		MovieLens 1 M	
	RMSE	MAE	RMSE	MAE
Basic UserBased CF	0.936	0.735	0.914	0.708
Basic ItemBased CF	0.933	0.733	0.902	0.699
SVD	0.907	0.717	0.888	0.704
AutoEncode	0.939	0.737	0.892	0.688
Stack_AutoEncode	0.933	0.728	0.890	0.684
RBM_UserBased	0.970	0.763	0.921	0.726
RBM_ItemBased	0.966	0.770	0.910	0.722
Hybrid RBM Model	0.936	0.748	0.894	0.710
IC-CRBMF_UserBased	**0.950**	**0.749**	**0.889**	**0.698**
IC-CRBMF_ItemBased	**0.924**	**0.725**	**0.894**	**0.704**
Hybrid IC-CRBMF Model	**0.910**	**0.719**	**0.867**	**0.681**

5   Conclusion and Future Work

In this paper, we take full advantage of item category information and proposed an improved IC-CRBMF model. In order to achieve better recommendation efficiency and accuracy, we furhter implement Hybrid IC-CRBMF model by linear combination user and item based on RBM models. Experimental study has shown its potential in improving overal performance.

As to the future plan, it deserve to investigate more content-based features such as user's demographic information or social network information into the models. Besides, we will perform similarity computing or clustering information before implementing IC-CRBMF model that similar users or items in the same category share the same weight parameters.

Acknowledgments. This work was partially supported by the National Natural Science Foundation of China (No. 61472021), the National High Technology Research and Development Program of China (No. 2013AA01A601), and SKLSDE project under Grant No. SKLSDE-2015ZX-17. We are grateful to Shenzhen Key Laboratory of Data Vitalization (Smart City) for supporting this research.

References

1. Bokde, D.K., Girase, S., Mukhopadhyay, D.: Role of matrix factorization model in collaborative filtering algorithm: a survey (2015). CoRR abs/1503.07475
2. Georgiev, K., Nakov, P.: A non-IID framework for collaborative filtering with restricted boltzmann machines. In: Proceedings of 30th International Conference on Machine Learning, pp. 1148–1156 (2013)
3. Grčar, M., Mladenič, D., Fortuna, B., Grobelnik, M.: Data sparsity issues in the collaborative filtering framework. In: Nasraoui, O., Zaïane, O.R., Spiliopoulou, M., Mobasher, B., Masand, B., Yu, P.S. (eds.) WebKDD 2005. LNCS (LNAI), vol. 4198, pp. 58–76. Springer, Heidelberg (2006)
4. Hinton, G.E.: Training products of experts by minimizing contrastive divergence. Neural Comput. **14**(8), 1771–1800 (2002)
5. Hinton, G.E.: A practical guide to training restricted boltzmann machines. In: Montavon, G., Orr, G.B., Müller, K.-R. (eds.) Neural Networks: Tricks of the Trade, 2nd edn. LNCS, vol. 7700, pp. 599–619. Springer, Heidelberg (2012)
6. Kaya, H., Alpaslan, F.N.: Using social networks to solve data sparsity problem in one-class collaborative filtering. In: Proceedings of 7th International Conference on Information Technology, pp. 249–252 (2010)
7. Langseth, H., Nielsen, T.D.: Scalable learning of probabilistic latent models for collaborative filtering. Decis. Support Syst. **74**, 1–11 (2015)
8. Ouyang, Y., Liu, W., Rong, W., Xiong, Z.: Autoencoder-based collaborative filtering. In: Loo, C.K., Yap, K.S., Wong, K.W., Beng Jin, A.T., Huang, K. (eds.) ICONIP 2014, Part III. LNCS, vol. 8836, pp. 284–291. Springer, Heidelberg (2014)
9. Salakhutdinov, R., Mnih, A., Hinton, G.E.: Restricted boltzmann machines for collaborative filtering. In: Proceedings of 24th International Conference on Machine Learning, pp. 791–798 (2007)
10. Takács, G., Pilászy, I., Németh, B., Tikk, D.: Scalable collaborative filtering approaches for large recommender systems. J. Mach. Learn. Res. **10**, 623–656 (2009)
11. Vozalis, M.G., Margaritis, K.G.: Using SVD and demographic data for the enhancement of generalized collaborative filtering. Inf. Sci. **177**(15), 3017–3037 (2007)
12. Yao, Z., Lai, F.: Integrating item category information in collaborative filtering recommender algorithm. In: Proceedings of 4th International Conference on Natural Computation, pp. 33–38 (2008)
13. Zhang, S., Wang, W., Ford, J., Makedon, F., Pearlman, J.D.: Using singular value decomposition approximation for collaborative filtering. In: Proceedings of 7th IEEE International Conference on E-Commerce Technology, pp. 257–264 (2005)

Figure-Ground Segregation by a Population of V4 Cells

A Computational Analysis on Distributed Representation

Masaharu Hasuike[1], Shuto Ueno[1], Dai Minowa[1], Yukako Yamane[2],
Hiroshi Tamura[2], and Ko Sakai[1(✉)]

[1] Department of Computer Science, University of Tsukuba, 1-1-1 Tennodai
Tsukuba, Ibaraki, 305-8573, Japan
{hasuike,ueno,sakai}@cs.tsukuba.ac.jp
http://www.cvs.cs.tsukuba.ac.jp/
[2] Graduate School of Frontier Biosciences, Osaka University, 1-4 Yamadaoka
Suita-Shi, Osaka, 560-0871, Japan
{yukako,tamura}@bpe.es.osaka-u.ac.jp

Abstract. Figure-ground (FG) segregation is a crucial function of the intermediate-level vision. Physiological studies have suggested that monkey V4 cells
are sensitive to FG organization, although their selectivity to FG and what they
represent have not been clarified. We investigated computationally whether a
subpopulation of cells in V4 represents FG. Specifically, we applied support
vector machine to the responses of 1–100 cells to natural-image patches that were
recorded from V4 of two macaque monkeys (*Macaca fuscata*) by multi-electrode
arrays. Our results showed that the responses of a group of cells were capable of
segregating FG while the responses of single-cells were not, suggesting a distributed representation of FG in V4.

Keywords: Visual cortex · V4 · Figure ground segregation · Natural image ·
Electrophysiology · Population coding · Support vector machine

1 Introduction

Figure-ground segregation, a process to separate an object from background in a visual
scene, is a crucial step toward object recognition in the visual cortex. Recent physiological studies [e.g., 1, 2] have reported that a number of cells in early-to-intermediate-
level visual areas are selective to the local direction of figural region along a contour
segment. These neurons increase their activity when a figural region exists on one side
(say, left) with respect to their classical receptive field (CRF). In such a case, the contour
on the CRF is owned by that side (left) so that these cells are defined as border-ownership
(BO) selective and thought to signal the local direction of figure at the location of the
CRF. Recent computational studies reported that model BO-selective cells based on
surround modulation are capable of determining BO in natural contours with 66 %
success rate [3, 4]. Our computational study has shown that an integration of 10 BO-
selective cells is capable of yielding 85 % success with the natural contours [5]. Since
BO represents the direction of figure *on contours*, a mechanism that represents FG

S. Arik et al. (Eds.): ICONIP 2015, Part II, LNCS 9490, pp. 617–622, 2015.
DOI: 10.1007/978-3-319-26535-3_70

organization of *surfaces* is expected to follow the BO determination. Specifically, we expect that a subpopulation of V4 cells modulate their responses depending on whether their CRF is located on a figural region or a ground region. Physiological studies on monkeys have reported that V4 cells appear to be sensitive to FG organization [6, 7] although their selectivity to FG has not been clarified. We investigated computationally whether the intermediate-level visual areas represent information capable of determining FG.

Our analysis of single-cell responses from electrophysiolocial recordings showed that individual cells in monkey V4 are not capable of segregating FG. We investigated what happens if multiple cells work together in the challenging problem of FG determination in natural scenes. We used support vector machine (SVM) as an excellent machine for the integration of the responses of multiple cells. If the machine was capable of segregating correct FG from the cellular responses, it indicates that the responses of the cells included the sufficient information for judging FG. Our simulations showed that the responses of fifty V4 cells were capable of achieving 66 % correct in the FG discrimination of natural contours, indicating the distributed representation of FG in the intermediate-level visual area.

2 Methods

Electrophysiological recordings from the visual area V4 of two macaque monkeys (*Macaca fuscata*) were provided from the Laboratory for Cognitive Neuroscience of Osaka University. During experiment sessions, the monkey were anesthetized and seated in front of a computer monitor. The neural signals from multiple-electrode arrays with 32 channels were sorted to obtain the responses of individual cells. The experiments were approved by the ethics committee of the institute and performed in accordance with The Code of Ethics of the World Medical Association (Declaration of Helsinki).

Stimuli presented to the monkeys were comprised of three sets of small patches. The first set consisted of single squares and overlapping two squares, which were similar to those used in previous studies on BO [1, 2]. With a variation in contrast, translation and rotation, a total of 32 square stimuli were presented, as examples are shown in Fig. 1. The second set consisted of natural contours with one side filled with black and the other with white. We used Human Marked Contour (HMC) available in Berkeley Segmentation Dataset [9, 10] that were drawn by 10 human participants. We chose pseudo-randomly 104 subregions (69 × 69 pixel) from the HMC, as similar to our previous experiment [5]. Figure 1 shows a few examples of filled patches. Note that purely-random choice of image patches end up with a set of similar contours such as straight lines because the distribution of contour characteristics are highly non-uniform in natural scenes. To assure the diversity of stimuli (not to chose similar contours), we controlled the distribution of the degree of convexity, closure and parallel of stimuli (uniformly chosen from each range of these characteristics) [5]. With a variation in contrast, translation and mirror image, a total of 832 filled patches were presented. The patches in the third set were the original natural-image patches of the second set (a total of 416). To obscure the boundary of stimulus and background, we attenuated contrast towards the periphery with a Gaussian. The patches were scaled to match the typical dimension of the receptive fields of V4 cells under examination. We performed a psychophysical experiment to obtain the veridical label of BO direction for image patches (stimuli) taken from natural scenes [5]. The stimuli were presented for 200 ms with a blank interval of 200 ms in a random order with the repetition of 10.

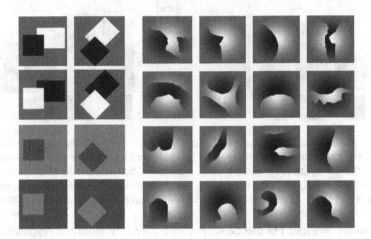

Fig. 1. Examples of square stimuli (left) and filled patches whose borders represent natural contours (right).

3   Single Cell Responses to Natural Images

We analyzed whether the responses of single cells in monkey V4 depend on FG organization of stimuli with respect to their CRF center. The electrophysiological recording included the responses to 2430 neurons of which 1487 cells showed their CRF center located within stimulus patches. To examine whether FG and contrast are significant factors for the responses, we performed a two-way ANOVA with the two factors at the CRF center. Nineteen percent (275/1487) of cells showed the significance to FG ($p < 0.05$), and 37 % to contrast with 19 % to the interaction. Figure 2 shows the responses of a single cell that showed significance to both FG ($p = 0.030$) and contrast ($p = 0.002$). We observed that the number of cells that showed the significance to square stimuli (51 %) was larger than that to filled-contour patches (33 %) and to natural patches (44 %).

The results indicate that a part of cells in monkey V4 show the responses that depend on FG at their CRF center. This result means that the responses include information related to FG, however, it does not directly mean that the cells are capable of discriminating FG correctly for the patches. Traditional physiological indices often use a difference such as $R_{\text{BO-right}} - R_{\text{BO-left}}$ in which the response to an original patch (say, a square located on the left side of the CRF) is subtracted from that to the mirror patch (a square on the right). However, in the discrimination of natural-image patches, neurons need to judge its FG without knowing the counterpart. To examine whether the FG-significant cells are capable of segregating FG in natural-image patches and their filled patches, construction of a model is required.

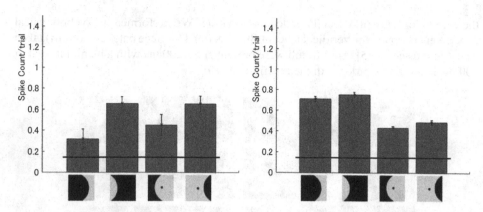

Fig. 2. The responses of a single cell to the square stimuli (left) and the filled natural-contour patches (right). The mean responses per trial (200 ms) are indicated for four categories; black-figure (B-F, the left most), black-ground (B-G, the second from the left), white-figure (W-F, the third), and white-ground (W-G, the right most). The red dots in icons illustrate the location of the CRF center. This cell shows stronger responses to ground compared to figure (Color figure online).

To examine whether FG can be determined correctly from the responses of FG-significant cells, we used a threshold model based on rate coding. If a response exceeded a threshold, the location of CRF center was judged as figure or ground. We computed the correct rate of the model for the natural-contour stimuli presented in electrophysio-logical recording (set 2). The mean correct rate was 49 % with 6 % SD, indicating that the single cells were not capable of segregating FG with natural contours.

4 Discrimination of FG from a Group of Cellular Responses

The determination of FG in natural image patches was difficult for single cells, leading to the idea that multiple cells work simultaneously as a population for the veridical determination of FG. To examine this hypothesis, we applied SVM to the responses of cells so as to maximize the correct rate of FG discrimination. In other words, we used SVM as an ideal integrator of the responses of multiple cells. This procedure is equiv-alent to the computation of theoretical upper limit given the responses of the cells.

Specifically, we used SVM to discriminate FG at the center of CRF from the cellular responses to filled-contour stimuli (set 2). We used five-fold cross-validation so that 1/5 of the responses to patches were reserved for test in each trial. The number of cell (N) was varied from 1 to around 50. For each N, we repeated for 50 times (1) randomly choosing N cells from the total of 275, (2) randomly choosing 4/5 of the responses to image patches, (3) learning with the 4/5 responses of the N cells to yield a classifier, and (3) testing the classifier with the remaining 1/5 of the responses. The veridical FG for each patch was given from the psychophysical experiment [5]. We implemented this procedure by using LIBSVM software package [8] with Gaussian kernels. The learning was successful with more than 99 % correct.

The mean corect rates for the test sets are shown in Fig. 3 as a function of the number of cells. The correct rate increases as the number of cells increases. With 50 cells, the mean correct rate was 66 % and the SD was 5.2 %. As a control, we generated the SVM models that consisted of randomly-selected cells with or without FG significance, and computed their discrimination rate for the test as shown in Fig. 3. The mean correct rates were in the range of chance for the entire range of the number of cells examined (up to 100). These results indicate that a subpopulation of V4 cells carry the information relevant to FG, and that a group of a few tens of the FG-significant cells are capable of discriminating FG at their CRF location.

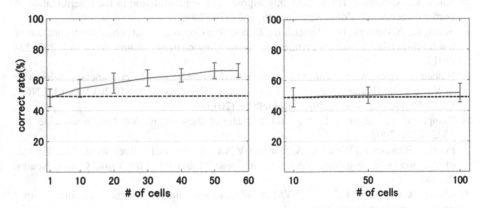

Fig. 3. The correct rates of the SVM model as a function of the number of cells. The error bars indicate SD. With FG-significant cells (left), the correct rate increases as the number of cells increases. The mean correct rate of 50 cells was 66 % while that of single cells was chance. The results of randomly-chosen cells (right) were in the range of chance independent of the number of cells.

5 Conclusion and Discussion

We proposed that a few tens of V4 cells were capable of discriminating FG from various natural-image patches. We applied SVM to the responses of individual cells so as to maximize the discrimination rate. The results of the computation showed that the responses of 50 cells achieved 66 % correct for natural contour patches (set 2, filled contours), suggesting that a group of V4 cells are capable of determining FG in natural images.

Note that the spatial extent of the receptive field of V4 cells is typically several to ten degrees in central vision. Therefore, FG-selective cells determine a *local* FG along a contour. A psychophysical study [9] reported that human subjects are able to achieve 70 % correct from local patches with respect to the global inspection of the scene. The present study focuses on the local determination of FG by a group of V4 neurons. The global figure-ground determination of an entire scene requires additional mechanisms probably residing throughout the visual pathway [e.g., 11].

Acknowledgment. This work was supported by grant-in-aids from JSPS and MEXT of Japan (KAKENHI 26280047, 25135704 (Shitsukan)), and from RIEC of Tohoku University (H25-A09).

References

1. Zhou, H., Friedman, H.S., von der Heydt, R.: Coding of border ownership in monkey visual cortex. J. Neurosci. **20**, 6594–6611 (2000)
2. Zhang, N.R., von der Heydt, R.: Analysis of the context integration mechanisms underlying figure-ground organization in the visual cortex. J. Neurosci. **30**, 6482–6496 (2010)
3. Sakai, K., Nishimura, H.: Surrounding suppression and facilitation in the determination of border ownership. J. Cogn. Neurosci. **18**, 562–579 (2006)
4. Sakai, K., Nishimura, H., Shimizu, R., Kondo, K.: Consistent and robust determination of border-ownership based on asymmetric surrounding contrast. Neural Netw. **33**, 257–274 (2012)
5. Nakata, Y., Sakai, K.: Structures of surround modulation for the border-ownership selectivity of V2 cells. In: Huang, T., Zeng, Z., Li, C., Leung, C.S. (eds.) ICONIP 2012, Part I. LNCS, vol. 7663, pp. 383–391. Springer, Heidelberg (2012)
6. Pasupathy, A., Connor, C.E.: Population coding of shape in area V4. Nat. Neurosci. **5**(12), 1332–1338 (2002)
7. Poort, J., Raudies, F., Wanning, A., Lamme, V.A.F., Neumann, H., Roelfsema, P.R.: The role of attention in figure-ground segregation in Areas V1 and V4 of the Visual Cortex. Neuron **75**, 143–156 (2012)
8. Chang, C.C., Lin, C.J.: LIBSVM-A library for support vector machines. http://www.csie.ntu.edu.tw/~cjlin/libsvm/
9. Fowlkes, C.C., Martin, D.R., Malik, J.: Local figure-ground cues are valid for natural images. J. Vis. **7**(8), 2, 1–9 (2007)
10. Berkeley Segmentation Dataset; Figure-Ground Assignments in Natural Images. http://www.eecs.berkeley.edu/Research/Projects/CS/vision/grouping/fg/
11. Jones, H.E., Andolina, I.M., Shipp, S.D., Adams, D.L., Cudeiro, J., Salt, T.E., Sillito, A.M.: Figure-ground modulation in awake primate thalamus. PNAS **112**(22), 7085–7090 (2015)

Weighted Probabilistic Opinion Pooling
Based on Cross-Entropy

Vladimíra Sečkárová[✉]

Institute of Information Theory and Automation of the CAS,
Pod Vodárenskou věží 4, 182 08 Prague 8, Czech Republic
seckarov@utia.cas.cz

Abstract. In this work we focus on opinion pooling in the finite group of sources introduced in [1]. This approach, heavily exploiting Kullback-Leibler divergence (also known as cross-entropy), allows us to combine sources' opinions given in probabilistic form, i.e. represented by the probability mass function (pmf). However, this approach assumes that sources are equally reliable with no preferences on, e.g., importance of a particular source. The discussion about the influence of the combination by preferences among sources (represented by weights) and numerical demonstration of the derived theory on an illustrative example form the core of this contribution.

Keywords: Minimum cross-entropy principle · Kullback-Leibler divergence · Linear opinion pooling · Combining probability distributions

1 Introduction

In this work we focus on decision making in the finite group (set) of sources, providing their opinions about the underlying (studied) problem to each other. The problem relates to a hidden (stochastic) phenomenon, that is not directly observable, but about which an opinion can be formulated. Examples of such phenomena are anticipated elections results, companies contracts and many others. The assumption on obtaining opinions yields a specific decision making process commonly known as opinion pooling. Due to the complexity of the space of possible decisions we consider the probability distributions over this set rather than single values. The final decision (result of pooling) is then a combination of probability distributions provided by sources.

Combining probabilistic information within a group of sources has been of interest for a long time, see, e.g., [2]. Different approaches consider different levels of cooperation among sources and different exploitations of their proposed combination. One can assume the sources are represented as a group of individuals "who must act together as a team and reach consensus" [3]. Or, one can consider that sources are "perfectly coherent, rational as decision makers and cooperate in agreeing to adopt a group utility function" [4].

Many probability combining approaches exploit information theory, namely, information divergences such as the Kullback-Leibler (KL) divergence [5] (the

© Springer International Publishing Switzerland 2015
S. Arik et al. (Eds.): ICONIP 2015, Part II, LNCS 9490, pp. 623–629, 2015.
DOI: 10.1007/978-3-319-26535-3_71

term cross-entropy is also used). Based on the order of its arguments we arrive at two basic classes - linear pools and log-linear pools, see, e.g., [6].

Linear pools consider, e.g., in [7], where "expert opinion is represented as a probability and associated with a confidence level that expresses the conviction of the corresponding expert on its own judgement".

The examples of KL-divergence based log-linear pools include, e.g., estimation [8], or determination of weights for prior distributions pooling [9].

Here, we follow the idea of combination derived under certain assumptions on cooperation between sources [10]. In particular, we focus on approach recently introduced in [1], which heavily exploits cross-entropy and yields a linear combination of discrete probability distribution. The formula for this combination was derived under the *wise-selfish* cooperation scenario assumption. *Wise* source is willing to cooperate and share its information with other sources in the group, but *selfishly* requires the result of combining to be "close" in the sense of bounded KL-divergence. There were no assumptions on preferences among sources reflecting their reliability or importance.

In this contribution we study how the values of this opinion pool changes when preferences about sources reliability or importance are known prior to combining. First, we briefly summarize the combining introduced in [1]. Then, we discuss the results when weights are included. To demonstrate the proposed idea we give an illustrative example.

2   Opinion Pooling of Discrete Probability Distributions Based on Cross-Entropy

As considered in [1], let us have a finite number of sources, $j = 1, \ldots, s < \infty$. Assume also that each source provides its opinion in the probabilistic form: as a probability mass function (pmf) assigning probability to each of $n < \infty$ outcomes of stochastic phenomenon:

$$\boldsymbol{p}_j = (p_{j1}, \ldots, p_{jn}): \quad p_{ji} > 0, \quad \sum_{i=1}^{n} p_{ji} = 1, \quad j = 1, \ldots, s. \tag{1}$$

Let $\boldsymbol{q}$ represent the combination of $\boldsymbol{p}_1, \ldots, \boldsymbol{p}_s$. To obtain $\hat{\boldsymbol{q}}$, the estimator of $\boldsymbol{q}$, we search for the minimizer of the expected loss [11]. In particular, we search for the minimizer of the conditional expected value of the KL-divergence (KLD) with respect to the conditional probability density function (pdf) $\pi(\boldsymbol{q}|\boldsymbol{p}_1, \ldots, \boldsymbol{p}_s)$ conditioned on $\boldsymbol{p}_1, \ldots, \boldsymbol{p}_s$

$$\mathrm{E}_{\pi(\boldsymbol{q}|\boldsymbol{p}_1, \ldots, \boldsymbol{p}_s)} \mathrm{KLD}(\boldsymbol{q}||\hat{\boldsymbol{q}}) = \mathrm{E}_{\pi(\boldsymbol{q}|\boldsymbol{p}_1, \ldots, \boldsymbol{p}_s)} \sum_{i=1}^{n} q_i \ln \frac{q_i}{\hat{q}_i}. \tag{2}$$

The minimizer is the conditional expected value of $\boldsymbol{q}$ with respect to $\pi(\boldsymbol{q}|\boldsymbol{p}_1, \ldots, \boldsymbol{p}_s)$

$$\hat{\boldsymbol{q}} = \mathrm{E}_{\pi(\boldsymbol{q}|\boldsymbol{p}_1, \ldots, \boldsymbol{p}_s)}[\boldsymbol{q}|\boldsymbol{p}_1, \ldots, \boldsymbol{p}_s]. \tag{3}$$

The estimator (3) heavily depends on the form of the unknown conditional pdf $\pi(q|p_1,\ldots,p_s)$. To obtain this pdf we introduce the notion of the source's *selfishness* represented by the constraints on the expected KL-divergences from p_j to q with respect to the conditional pdf $\pi(q|p_1,\ldots,p_s)$. Moreover, we assume the sources consider the following equalities among the expected values of the KL-divergence:

$$\mathrm{E}_{\pi(q|p_1,\ldots,p_s)}[\mathrm{KLD}(p_s\|q)|p_1,\ldots,p_s] = \mathrm{E}_{\pi(q|p_1,\ldots,p_s)}[\mathrm{KLD}(p_j\|q)|p_1,\ldots,p_s], \tag{4}$$

$j = 1,\ldots,s-1$.

We exploit the minimum cross-entropy principle [12] and choose the conditional pdf $\pi(q|p_1,\ldots,p_s)$ that solves:

$$\min_{\pi(q|p_1,\ldots,p_s)} \mathrm{KLD}(\pi(q|p_1,\ldots,p_s)\|\pi_0(q)), \tag{5}$$

where $\pi_0(q)$ is the prior guess on the conditional pdf $\pi(q|p_1,\ldots,p_s)$.

The prior pdf is chosen as the pdf of the Dirichlet distribution for its computationally advantageous properties. The KL-divergence in (5) and the constraints in (4) can be then rewritten as follows

$$\ln\frac{\Gamma(\sum_{i=1}^n \nu_{0i})}{\Gamma(\sum_{i=1}^n \nu_{0i})} + \sum_{i=1}^n \ln\frac{\Gamma(\nu_{0i})}{\Gamma(\nu_i)} + \sum_{k=1}^n (\nu_k - \nu_{0k})\left(\psi(\nu_k) - \psi(\sum_{l=1}^n \nu_l)\right) \tag{6}$$

and

$$-H(p_j) + H(p_s) = \sum_{i=1}^n (p_{ji} - p_{si})\left(\psi(\nu_i) - \psi\left(\sum_{i=1}^n \nu_i\right)\right), \quad j = 1,\ldots,s-1, \tag{7}$$

where $\Gamma(.)$ is the Gamma function, $\psi(.)$ is the digamma function [13] and $\nu_{01},\ldots,\nu_{0n}$ is the prior guess on parameters of the Dirichlet distribution. It can be shown that the conditional pdf $\pi(q|p_1,\ldots,p_s)$ minimizing (5) is also the pdf of the Dirichlet distribution. Thus, it is satisfactory to perform the minimization in the set of possible values of parameters $\nu_1,\ldots,\nu_n$ instead of searching in the set of all possible probability distributions. The values of parameters, for which the value (5) is minimal, are denoted by $\hat{\nu}_1,\ldots,\hat{\nu}_n$.

The values of the parameters $\hat{\nu}_1,\ldots,\hat{\nu}_n$ are theoretically expressed by the following formula:

$$\hat{\nu}_i = \nu_{0i} + \sum_{j=1}^s \lambda_j(p_{ji} - p_{si}), \quad i = 1,\ldots,n, \tag{8}$$

where λ_j are the Lagrange multipliers resulting from minimization of (5) with respect to $(s-1)$ equations in (4).

The final weighted combination of $p_1,\ldots,p_s$, represented by $\hat{q}$ in (3), has then the following form:

$$\hat{q}_i = \frac{\hat{\nu}_i}{\sum_{k=1}^n \hat{\nu}_k} = \frac{\nu_{0i}}{\sum_{k=1}^n \nu_{0k}} + \frac{\sum_{j=1}^s \lambda_j(p_{ji} - p_{si})}{\sum_{k=1}^n \nu_{0k}}, \quad i = 1,\ldots,n, \tag{9}$$

since, based on (8), it holds that $\sum_{k=1}^{n} \hat{\nu}_k = \sum_{k=1}^{n} \nu_{0k}$.

To perform the numerical part of the optimization we exploited Matlab.

3   Cross-Entropy Based Opinion Pooling Influenced by Preferences Among Sources

The opinion pooling of sources' $p_1, \ldots, p_s$ described above assumed no preferences among sources. If the preferences, represented by the weights $w_1, \ldots, w_s$ about sources $1, \ldots, s$, are available, they can influence the prior information about $(\nu_{01}, \ldots, \nu_{0n})$ and selfishness restrictions in (4). Both cases are discussed below.

For the guess on values of $(\nu_{01}, \ldots, \nu_{0n})$ it is optional to exploit the information known prior to combining. Since we focus on combining opinions about hidden (stochastic) phenomena with unavailable prior knowledge, we suggest to use the weighted arithmetic mean of $p_1, \ldots, p_s$ as the prior guess:

$$\nu_{0i} = \sum_{j=1}^{s} w_j p_{ji}. \tag{10}$$

Since

$$\sum_{i=1}^{n} \nu_{0i} = \sum_{i=1}^{n} \sum_{j=1}^{s} w_j p_{ji} = \sum_{j=1}^{s} w_j, \tag{11}$$

we obtain that the part of combination (9) representing the part induced by the prior $\pi_0(q)$ is

$$p_{0i} = \frac{\nu_{0i}}{\sum_{k=1}^{n} \nu_{0k}} = \sum_{j=1}^{s} \frac{w_j}{\sum_{l=1}^{s} w_l} p_{ji}. \tag{12}$$

It is somewhat surprising that the Eq. (9) combines simultaneously both, the parameters of the Dirichlet distribution and pmfs $p_1, \ldots, p_s$. Provided pmfs can be viewed as individual guess for $(\nu_{01}, \ldots, \nu_{0n})$ when $\sum_{i=1}^{n} \nu_{0i} = 1$ (yielding $\sum_{j=1}^{s} w_j = 1$).

Besides the prior values $\nu_{01}, \ldots, \nu_{0n}$, it is also desirable that the constraints (4) will be affected by these weights. In particular, we approach the combination to more important sources by requiring

$$w_j \mathrm{E}\left[\mathrm{KLD}(p_j \| q) | p_1, \ldots, p_s\right] = w_s \mathrm{E}\left[\mathrm{KLD}(p_s \| q) | p_1, \ldots, p_s\right], \tag{13}$$

$j = 1, \ldots, s-1$, yielding the following weighted counterpart of constraints (7):

$$-w_j H(p_j) + w_s H(p_s) = \sum_{i=1}^{n} (p_{ji} w_j - p_{si} w_s) \left[\psi(\nu_i) - \psi\left(\sum_{k=1}^{n} \nu_k\right)\right]. \tag{14}$$

Then, the final combination $\hat{q}$ in (9) is

$$\hat{q}_i = p_{0i} + \sum_{j=1}^{s-1} \frac{\lambda_j}{\sum_{l=1}^{s} w_l} (w_j p_{ji} - w_s p_{si}), \quad i = 1, \ldots, n. \tag{15}$$

with p_{0i} given in (12).

In the presented discussion, we have assumed that the elements of pmf provided by source j, $j = 1, \ldots, s$, have the same weights $w_{ji} = const$, $i = 1, \ldots, n$. If needed, also an element-dependent version w_{ji}, $i = 1, \ldots, n$ can be developed.

Let us now consider the following two sets of weights

$$\sum_{j=1}^{s} w_j = d \quad \text{and} \quad \sum_{j=1}^{s} w_j^* = \sum_{j=1}^{s} k w_j = d^*. \tag{16}$$

According to (10), the prior guesses on the parameters of the Dirichlet distribution look as follows:

$$\text{for } w_1, \ldots, w_s : \quad \nu_{0i} = \sum_{j=1}^{s} w_j p_{ji},$$
$$\text{for } w_1^*, \ldots, w_s^* : \quad \nu_{0i}^* = \sum_{j=1}^{s} w_j^* p_{ji}.$$

From the combination $\hat{q}$ derived for weights w^*

$$\hat{q}_i = \sum_{j=1}^{s} \frac{k w_j p_{ji}}{\sum_{l=1}^{s} k w_l} + \sum_{j=1}^{s-1} \frac{\lambda_j}{\sum_{l=1}^{s} k w_l} (k w_j p_{ji} - k w_s p_{si}), \tag{17}$$

where $k = \frac{d^*}{d}$, it might seem that both optimal estimators, based on w and w^*, coincide. Recall the KL-divergence in (6) leading to $\hat{q}$ by minimization with respect to $\nu_1, \ldots, \nu_n$. Due to non-linearity of this function in prior guess $\nu_{01}, \ldots, \nu_{0n}$, we can not expect the combinations based on $w_1, \ldots, w_s$ and $w_1^* = k w_1, \ldots, w_s^* = k w_s$ to be equal.

In the following illustrative example we demonstrate the change in values of $\hat{q}$ for different set of weights $w_1, \ldots, w_s$.

4   Illustrative Example

Let us have 2 sources ($s = 2$) which provided the following pmfs ($n = 3$)

$$\boldsymbol{p}_1 = [0.75, 0.05, 0.2],$$
$$\boldsymbol{p}_2 = [0.3, 0.1, 0.6].$$

Consider the following weights:

- first set of weights: arithmetic mean (equal weights summing to one),
- second set of weights: first source fixed $w_1 = 0.1$, weight of the second source is rising,
- third set of weights: weights of both sources are rising ($w_2 = 2 \times w_1$).

Values of the combination $\hat{q}$ of $\boldsymbol{p}_1$ and $\boldsymbol{p}_2$ in the last instant, see Fig. 1, are given in Table 1.

In Fig. 1 we see that the resulting combination stabilizes quickly and with rising values of the weights (second set and third set of weights) the combination of $\boldsymbol{p}_1$ and $\boldsymbol{p}_2$ based on (15) tends towards the arithmetic mean of $\boldsymbol{p}_1$ and $\boldsymbol{p}_2$. This might be a consequence of the considered *selfish* scenario – the equality in the constraints in (14).

Table 1. Values of the combination $\hat{q} = (\hat{q}_1, \hat{q}_2, \hat{q}_3)$ for different sets of weights (value given for the last considered value of weights).

Arithmetic mean:	(0.525, 0.075, 0.4)
First set of weights:	(0.403, 0.089, 0.508)
Second set of weights $(w_1 = 0.1,\ w_2 = 27.2)$:	(0.519, 0.076, 0.405)
Third set of weights $(w_1 = 13.6,\ w_2 = 27.2)$:	(0.522, 0.075, 0.403)

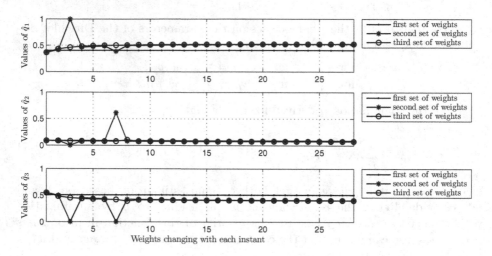

Fig. 1. Resulting combination of p_1 and p_2 for different sets of weights. *First set of weights*: equal weights summing to one. *Second set of weights*: fixed weight of the first source ($w_1 = 0.1$), weight of the second source rising. *Third set of weights*: $w_2 = 2 \times w_1$, both weights rising.

5 Conclusion and Future Work

We focused on combining opinions provided in the probabilistic form based on the cross-entropy introduced in [1]. We discussed theoretically and numerically the change in the value of the combination when preferences among sources, represented by the weights, were included. The results of the illustrative example showed that, with rising values of the weights, the considered combining approach is stable and tends towards the arithmetic mean. Thus, the properties of suggested weighted version of the combination have to be properly investigated.

The further research includes development of another way how to involve preferences among sources, e.g., by allowing inequality in constraints (14).

Acknowledgments. This research has been supported by GAČR 13-13502S.

References

1. Sečkárová, V.: Minimum cross-entropy based weights in dynamic diffusion estimation in exponential family. Pliska Studia Math. Bulg. **24**, 181–188 (2015)
2. von Neumann, J., Morgenstern, O.: Theory of Games and Economic Behavior (60th Anniversary Commemorative Edition). Princeton Classic Editions. Princeton University Press, Princeton (2007)
3. DeGroot, M.H.: Reaching a consensus. J. Am. Stat. Assoc. **69**(345), 118–121 (1974)
4. West, M.: Bayesian aggregation. J. R. Stat. Soc. (Ser. A) **147**, 600–607 (1984)
5. Kullback, S., Leibler, R.A.: On information and sufficiency. Ann. Math. Stat. **22**, 79–86 (1951)
6. Abbas, A.E.: A Kullback-Leibler view of linear and log-linear pools. Decis. Anal. **6**(1), 25–37 (2009)
7. Garcia, M., Puig, D.: Robust aggregation of expert opinions based on conflict analysis and resolution. In: Conejo, R., Urretavizcaya, M., Pérez-de-la-Cruz, J.-L. (eds.) CAEPIA/TTIA 2003. LNCS (LNAI), vol. 3040, pp. 488–497. Springer, Heidelberg (2004)
8. Dedecius, K., Sečkárová, V.: Dynamic diffusion estimation in exponential family models. IEEE Sig. Process. Lett. **20**(11), 1114–1117 (2013)
9. Rufo, M.J., Martín, J., Pérez, C.J.: Log-linear pool to combine prior distributions: a suggestion for a calibration-based approach. Bayesian Anal. **7**(2), 411–438 (2012)
10. Kárný, M., Guy, T.V., Bodini, A., Ruggeri, F.: Cooperation via sharing of probabilistic information. Int. J. Comput. Intell. Stud. **1**(5), 139–162 (2009)
11. Bernardo, J.M.: Expected information as expected utility. Ann. Stat. **7**, 686–690 (1979)
12. Shore, J.E., Johnson, R.W.: Axiomatic derivation of the principle of maximum entropy and the principle of minimum cross-entropy. IEEE Trans. Inf. Theory **26**, 26–37 (1980)
13. Abramowitz, M., Stegun, I.A.: Handbook of Mathematical Functions: With Formulas, Graphs, and Mathematical Tables. Applied Mathematics Series. Dover Publications, New York (1964)

Tailgating Enforcement based on Back-Tracking in Intersection

Su-Hyoung Choi, Jong-Pil Ahn, Jee-Hyung Rheu, and Young-Mo Kim[✉]

School of Electronics Engineering, Kyungpook National University,
Sankyuk-dong, Buk-gu, Daegu, Korea
suhyoung89@knu.ac.kr, ymkim@ee.knu.ac.kr

Abstract. In this paper, we propose the new method to enforce the tailgating violations that occurs at the intersection. The tailgating enforcement system is composed of the wide-view camera for tracking a vehicle and the narrow-view camera for recognizing a license plate. The images from the wide-view camera are sequentially stored in the stack in order to monitor the ROI (Region of Interest) at the intersection area. The narrow-view camera recognized the license plate. And the image coordinate of the license plate from the narrow-view camera was converted into the image coordinate from the synchronized wide-view camera. The feature points of the vehicle were extracted the image from wide-view camera using the converted image coordinate of the license plate from the narrow-camera. The images from the wide-view camera are searched in reverse order, and the vehicle trajectories are tracked. As a result, the success rate of the back-tracking is 98 % at day and 93 % at night. And the success rate of the sequential tracking is 90 % at day and 76 % at night. The back-tracking is the better results than the sequential tracking. Therefore the back-tracking is the appropriate to enforce the tailgating violation at the intersection.

Keywords: ITS · Vehicle · Intersection · Back-tracking · Optical flow · Template matching

1 Introduction

The vehicle tracking in the intelligent traffic monitoring system is the important technology to control the traffic flow and to detect the violation vehicles. The proposed vehicle tracking methods are KLT based tracking [1], extended Kalman filter based tracking [2], and particle filter based tracking [3] at the present. ROI (Region of interest) should be set in the space to be monitored; the vehicle entering the ROI can be recognized. And then the vehicle recognized is traced by the normal tracking algorithm. These normal tracking methods are difficult to track the vehicle when the vehicle is far from the camera.

This paper explains that the tailgating enforcement system is composed of the wide-view camera for tracking a vehicle and the narrow-view camera for recognizing a license plate. The wide view camera monitors all region of the cross in order to track vehicles and the images from the wide-view camera are sequentially stored in the stack.

© Springer International Publishing Switzerland 2015
S. Arik et al. (Eds.): ICONIP 2015, Part II, LNCS 9490, pp. 630–637, 2015.
DOI: 10.1007/978-3-319-26535-3_72

The narrow-view camera recognize the position of the license plate as coordinate system and this coordinate value can be converted into one on the image from the wide-view camera. The coordinate value can be used in order to extract feature point. The images from the wide-view camera are searched in reverse order, and the vehicle trajectories are tracked using Template Matching [4] and Optical Flow [5].

The back-tracking method easily obtains the unique feature points on the vehicle because the license plate image comes from the vehicle in close range. The bigger license plate image from close range explains us detail and clear information of coordinate value. The variety feature points can also be extracted using this method. Template Matching and Optical Flow has a real-time processing and a high success rate of vehicle tracking at a long distance.

This paper is organized as follows. Chapter 2 explains the back-tracking in order to enforce the tailgating. Section 1 is the analysis of the tailgating violation. Section 2 is the back-tracking method. Section 3 is the method of mapping the image coordinates of a wide-view camera and a narrow-view camera. Section 4 is the stack structure for the back-tracking. Chapter 3 is the experiment about normal tracking and back-tracking. Finally we conclude the tailgating enforcement using the back-tracking in Chapt. 4.

2   The Back-tracking to Enforce the Tailgating

The camera as road surveillance system diagonally installed more than 6 m height in order to monitor the entire road. The image from the surveillance camera is gradually decreased by the perspective ratio along the distance between the camera and the vehicle. So the far vehicle is relatively small than the closer vehicle. It is difficult to extract the feature points on image came from far vehicle in order to track it. And the bottom of the image variation is greater than the top. Because the camera obtains the images diagonally. The image variation according to the vehicle location is very big by this reason. As these properties of surveillance camera, it is difficult to track the vehicle exactly. Chapter 2, we propose the method of the back-tracking using Template Matching and Optical Flow to solve these problems.

2.1   The Analysis of the Tailgating Violation

The tailgating is that the vehicle shouldn't pass through the intersection during green light signal. A driver ignored the traffic congestion and entered the intersection, consequently vehicle cannot pass through the intersection during green light signal. This action makes interrupt the traffic flow. Figure 1 is the image of the tailgating violation in the intersection. The interval between each vehicle in the intersection is very narrow. And the lane-change of the vehicle happens frequently. So, the occlusions are generated by each vehicle and the license plate disappeared on the image. However, the bigger plate images of vehicles are obtained on the bottom of the image without occlusion. We implement the method of the back-tracking using the above mechanism.

A. Red signal 5 seconds elapse

B. Red signal 10 seconds elapse

C. Red signal 15 seconds elapse

D. Red signal 20 seconds elapse

Fig. 1. The image of the tailgating violation in the intersection.

2.2 The Back-tracking Method

The tailgating enforcement system is composed of the wide-view camera for tracking a vehicle and the narrow-view camera for recognizing a license plate. The images from the wide-view camera are sequentially stored in the stack in order to monitor the ROI (Region of Interest) at the intersection area. The narrow-view camera recognizes the license plate. And the image coordinate of the license plate from the narrow-view camera is converted into the image coordinate from the synchronized wide-view camera. The feature points of the vehicle is extracted the image from wide-view camera using the converted image coordinate of the license plate from the narrow-camera. The images from the wide-view camera are searched in reverse order, and the vehicle trajectories are tracked using Template Matching [4] and Optical Flow [5]. The close vehicle tracked using the Template Matching because the close vehicle can be extracted the number of feature points. And the distance vehicle tracked using the Optical Flow. When vehicle tracking is complete, determine the violation. And the violation vehicles are to be classified separately. Figure 2 is the flowchart of the back-tracking method. The License Plate Recognizer was used for vehicle detection.

2.3 The Mapping Method between Each Coordinate Values of Narrow-view Image and Wide-View Image

The FOV (field of view) of the narrow-view camera is located in the bottom on the wide-view image. The images from the wide-view camera are sequentially stored in the stack

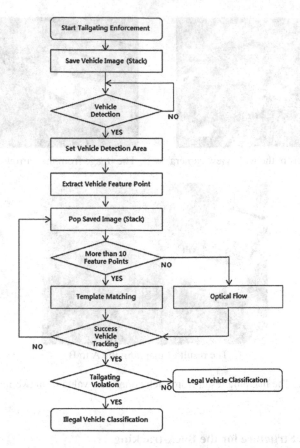

Fig. 2. The flowchart of the back-tracking method

in order to monitor the ROI at the intersection area. The narrow-view camera recognizes the coordinate value of the license plate. The coordinate value of the synchronized wide-view camera came from narrow-view image. The converted coordinate value on narrow-view image is stored on 320×240 coordinate table. The converted coordinate value of license plate from narrow-view image is use for extracting feature point of vehicle on wide-view image. And the images from the wide-view camera are searched in reverse order, and the vehicle trajectories are tracked. Figure 3 is the mapping result from each coordinate value from two images. The setting of the correct vehicle area on the wide-view image is required to improve the success rate of the vehicle tracking. In the case of the tilted vehicle from camera standing, it is difficult to set the correct vehicle area. In other words, although the position of the plates is the same, but the position of the vehicle is different. However, the vehicle area can be set by using tilted angle and the trajectory of the license plate.

A. The image from the wide-view camera   B. The image from the narrow-view camera

C. The result of mapping the A to B

Fig. 3. The mapping result from each coordinate value from two images

2.4   The Stack Structure for the Back-tracking

The feature point of the vehicle is extracted from the vehicle detection area on the wide-view image, and then the feature point can be tracked by reading stored images from the stack. Figure 4 explains the stack structure method for vehicle tracking.

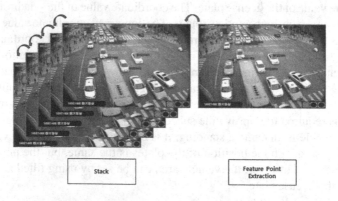

Fig. 4. The stack structure method for vehicle tracking

As processing the back-tracking, the size of the vehicle is reduced by the perspective. And also the feature points converged at the vehicle viewed as point because the image size of the vehicle reduced. The feature point should be excluded in tracking on the next images when the dispersion is exceeded the threshold value. Figure 5 explains the vehicle tracking at the intersection.

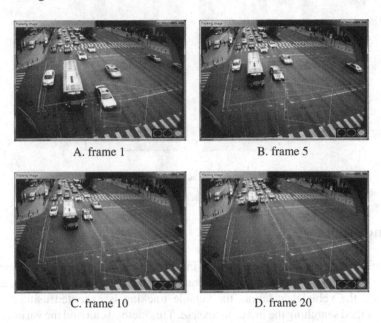

A. frame 1   B. frame 5

C. frame 10   D. frame 20

Fig. 5. The vehicle tracking in the intersection.

3   Experiments: Back-tracking and Sequential Tracking

The experiments consist of the back-tracking and sequential tracking. Figure 6 is a result of the back-tracking and sequential tracking. We obtained successful experiment results at finding various feature point and tracking vehicles. The back-tracking has better results than the sequential tracking in the experiment. As the reason why the back-tracking is better, the more feature points can be found compared to the sequential tracking. As a result, the success rate of the back-tracking is 98 % at day and 93 % at night. And the success rate of the sequential tracking is 90 % at day and 76 % at night. The back-tracking is the better results than the sequential tracking. Therefore the back-tracking is the appropriate to enforce the tailgating violation at the intersection. In this paper, the vehicle tracking experiment is only executed at day and night conditions. And camera shaking, abnormal weather (snow, rain, fog, etc.), reflective light at the road (sun, rain, buildings, etc.), shadow (vehicles, buildings, trees, etc.) were excluded in the experiments.

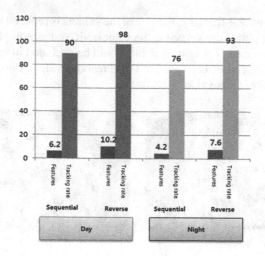

Fig. 6. A result of the back-tracking and sequential tracking.

4 Conclusion

In this paper, we propose the new method in order to enforce the tailgating violations that occur at the intersection. The images from the camera are sequentially stored in the stack. After the vehicle detection, the vehicle tracking is executed using the back-tracking method searching the image in reverse. This method can find the various feature point of the vehicle easily because the closer or bigger image is used at the start point for tracking. Normally bigger or closer images give us more clear and correct information. And because the feature points are converged from big image to image of vehicle, we can get higher success rate of tracking compare to sequential tracking.

We have to overcome the hard environment, narrow enforcement range and complicated tracking condition, at the intersection for taking good result at tailgating enforcement. The proposed method of the back-tracking overcomes the hard environment like the intersection for processing the vehicle detection and vehicle tracking well.

Acknowledgements. "This research was supported by the MSIP (Ministry of Science, ICT and Future Planning), Korea, under the C-ITRC (Convergence Information Technology Research Center) (IITP-2015-H8601-15-1002) supervised by the IITP (Institute for Information & communications Technology Promotion)"

References

1. Lim, Y.-C., et al.: A fusion method of data association and virtual detection for minimizing track loss and false track. In: 2010 IEEE Intelligent Vehicles Symposium (IV). IEEE (2010)
2. Barth, A., Franke, U.: Estimating the driving state of oncoming vehicles from a moving platform using stereo vision. IEEE Trans. Intell. Transp. Syst. **10**(4), 560–571 (2009)

3. Taekyu, Y., Sukbum, K.: Tracking for moving object using invariant moment and particle filter. In: 27th Chinese Control Conference, CCC 2008. IEEE (2008)
4. Bouguet, J.-Y.: Pyramidal implementation of the affine lucas kanade feature tracker description of the algorithm. Intel Corporation **5**, 1–10 (2001)
5. Nguyen, H.T., Worring, M., Van Den Boomgaard, R: Occlusion robust adaptive template tracking. In: Proceedings of the Eighth IEEE International Conference on Computer Vision, ICCV 2001, vol. 1. IEEE (2001)

Virtual Reality Based GIS Analysis Platform

Weixi Wang[1,2], Zhihan Lv[3](✉), Xiaoming Li[1,2,3], Weiping Xu[4],
Baoyun Zhang[5], and Xiaolei Zhang[6]

[1] Shenzhen Research Center of Digital City Engineering, Shenzhen, China
[2] Key Laboratory of Urban Land Resources Monitoring and Simulation,
Ministry of Land and Resources, Shenzhen, China
[3] SIAT, Chinese Academy of Science, Shenzhen, China
lvzhihan@gmail.com
[4] Wuhan University, Wuhan, China
[5] Jining Institute of Advanced Technology(JIAT), Jining, China
[6] Ocean University of China, Qingdao, People's Republic of China

Abstract. The proposed platform supports the integrated VRGIS functions including 3D spatial analysis functions, 3D visualization for spatial process and serves for 3D globe and digital city. The 3D analysis and visualization of the concerned city massive information are conducted in the platform. The amount of information that can be visualized with this platform is overwhelming, and the GIS-based navigational scheme allows to have great flexibility to access the different available data sources.

Keywords: WebVRGIS · WebVR · Big data · 3D globe

1 Introduction

Virtual Reality Geographical Information System (VRGIS), a combination of geographic information system and virtual reality technology [4] has become a hot topic. By integrating the friendly interactive interface of Virtual Reality System and spatial analysis specialty of Geographical Information System, WebVRGIS [7,11] based on WebVR [13] is preferred in practical applications, especially by the geography and urban planning. Accordingly, '3-D modes' has been proved as a faster decision making tool with fewer errors [15]. A parallel trend, the utilize of bigdata is becoming a hot research topic rapidly recently [1]. GIS data has several characteristics, such as large scale, diverse predictable and real-time, which falls in the range of definition of Big Data [2]. As a practical tool, most commonly used functions of VRGIS are improved according to practical needs [9]. For our platform, the customer are the employees of the governmental public service or social service agencies. The junior version of our platform is also planed to open the right to use to public. All the presented functions are extracted from the practical customer needs [8,10,12,16].

This research provides a new effective model of three-dimensional spatial information framework and application for urban construction and development directly, which must significantly improve the technical level and efficiency of

© Springer International Publishing Switzerland 2015
S. Arik et al. (Eds.): ICONIP 2015, Part II, LNCS 9490, pp. 638–645, 2015.
DOI: 10.1007/978-3-319-26535-3_73

urban management and emergency response and bring revolutionary changes to the engineering design and construction management field from two-dimensional drawing to three-dimensional collaborative design and construction.

2   System Information Process

Under the single-computer environment, the three-dimensional space analytical components have an access to interface access space to analyze the data which will be treated through the uniform data provided by three-dimensional spatial data engine; according to analysis requirement, the access is made to the interface of general spatial analytical components or comprehensive spatial analytical components, and the analysis result can be returned to database or applied into three-dimensional visualization and professional application through the uniform data access interface. Some key technologies are used to tackle the key issues, etc. massive geographical data storage [5, 6, 14, 17–29]. The information processing process is shown in figure below (Fig. 1).

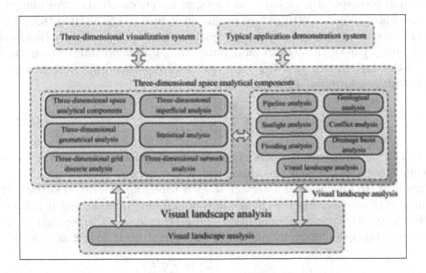

Fig. 1. Information process of three-dimensional space analytical components

2.1   Three-Dimensional Superficial Analytical Module

The surface is an area composed of many points and it contains large quantity of useful information. Through superficial analysis on three-dimensional network, it is able to know the superficial features as a whole, and carry out some specific calculations on existing surfaces, and generate into new data and recognition mode to extract more information. Topographic factor analysis: It mainly includes the calculation of gradient, slope aspect, and curvature. The different topographic factors reflect topographic features from different sides. The gradient and slope aspect are two mutually related parameters: the gradient reflects the inclining degree of slope, and the slope aspect reflects the direction that the slope faces. As the basic elements of topographic feature analysis and visualization, the gradient and slope aspect play a very important role in the research of drainage basin unit, landscape unit, and morphologic measurement. The gradient and slope aspect are used together with other parameters, which can help to solve the application problems such as estimation of forest content, conservation of water and soil, site analysis, and land utilization. Calculation formula of gradient: $\beta = \arctan \sqrt{p^2 + q^2}$ Calculation formula of slope aspect:

$$\alpha = 180° - \arctan \frac{q}{p} + 90° \frac{p}{|p|},$$

where $p = \frac{\partial H}{\partial x}$, which is the change rate of elevation at x direction; $q = \frac{\partial H}{\partial y}$, which is the change rate of elevation at y direction.

Topographic curvature: It is the reflection of change in shape and concave-convex of topographic curved surface at each section direction, and it is the function of plane point location. The topographic curvature includes plane curvature (contour line curvature) and profile curvature (vertical curvature).

(1) The plane curvature refers to the curvature of contour surface (horizontal plane) which passes through ground point and superficial intersection, and it reflects the convergence and divergence mode of superficial physical movement. The calculation formula is shown as below:

$$C_c = -\frac{q^2 r - 2pqs + p^2 t}{(p^2 + q^2)^{\frac{3}{2}}}$$

(2) The profile curvature refers to the curvature of normal vector which passes through ground point and the curve which the normal section parallel with the gradient at this point is intersected with topographic curved surface. The profile curvature describes the change in topographic gradient, and it influences the acceleration, slow-down, deposition and flowing state of superficial physical movement. The calculation formula of profile curvature is shown as below:

$$C_p = -\frac{p^2 r + 2pqs + q^2 t}{(p^2 + q^2)(1 + p^2 + q^2)^{\frac{3}{2}}}.$$

2.2 Three-Dimensional Statistical Analysis Module

The spatial correlation analysis mainly focuses on determining the correlation of two or more variables, and the main purpose is to calculate out the degree of correlation and property of relevant variables. The trend-surface analysis is a method in which the spatial distribution and time process of entity features are simulated through mathematical model to predict partial interpolations among actually measured data points of geographic elements under spatial and temporal distribution. Spatial fitting analysis: The so-called fitting refers to the situation that several discrete function values f1,f2,fn of one function are known, and then several undetermined coefficients f(1, 2, 3) in this function are adjusted to realize minimum difference (significance of least squares) between this function and the known point set. If the undetermined function is linear, this process is called linear fitting or linear regression (in statistics); otherwise this process is called nonlinear fitting or nonlinear regression. The expression can be piecewise function; under this condition, the process is called spine fitting. The spatial interpolation mainly includes Kring and inverse distance weighted method. The Kring interpolation method is one of important contents of spatial statistical analysis method; it is established on the basis of theoretical analysis of semi-variable function, and it is a kind of method of carrying out unbiased optimal estimation for regionalized variable value with finite region. The inverse distance weighted (IDW) is based on similarity principle, that is, the closer two objects are, the more similar their property is; on the contrary, the further two objects are, the less similar their property is.

2.3   Three-Dimensional Network Analysis Module

The network analysis is one of core problems of GIS spatial analysis function, and the main task and purpose of GIS network analysis function is to carry out geographic analysis and modeling on geographic network (such as traffic network) and urban infrastructure network (such as various kinds of reticle, power lines, telephone lines, and water supply and drainage lines). The network analysis can be used to research and plan how to arrange a network engineering and realize the best operation effect, such as the optimum allocation of certain resource, the shortest operation time or the least consumption from one place to another place.

Network measurement: It is mainly used to measure the incidence relation between the peak and side or the degree of connectivity between peaks in network diagram. The common measurement indexes include: β index, number of loops k, α index, and γ index. As for any three-dimensional network diagram, there exist three kinds of common basic indexes: (1) Number of lines (sides or arcs); (2) Number of nodes (peaks); (3) Number of sub-graphs in three-dimensional network.

β index is also called the rate of line points, and it is the number of average lines for each node in three-dimensional network; the calculation formula is $\beta = \frac{m}{n}$. The loop is a kind of closed path, and its starting point is also the ending point. The number of loops k is the value obtained via subtracting the number of lines $(n - p)$ under minimum-degree connection from the number of actual lines, that is $k = m - n + p$. α index refers to the ratio between the number of actual loops and the possible maximum number of loops in the network. The possible maximum number of loops in the network is obtained via subtracting the number of lines under minimum-degree connection from the possible maximum number of lines. Then, the α index is

$$\alpha = \frac{m - n + p}{\frac{n(n-1)}{2} - (n - 1)} \times 100\,\%.$$

γ index refers to the ratio between the actual number of lines and the possible maximum numbers of lines in the network, and the calculation formula is $\gamma = \frac{m}{n(n-1)/2} \times 100\%$ (Fig. 2).

Fig. 2. Effect diagram of application of Kring in mine

Optimal path analysis: It refers to realizing the best path selection in three-dimensional network model according to the given parameters. Through establishing three-dimensional path network mode, the users assign starting point and ending point to seek for the nearest path on the network [3].

Indoor and outdoor integrated three-dimensional routing and navigation: A whole process not only includes the route planning of road traffic, but also includes the route planning of indoor architecture. Through inputting the starting point and ending point, the users can analyze the indoor and outdoor integrated three-dimensional route; as for analysis result, it is able to adopt highlighting way to show the found route in the three-dimensional network model (Fig. 3).

Fig. 3. Indoor and outdoor integrated three-dimensional routing and navigation

Connectivity analysis: It refers to analyzing the ability of keeping connection between nodes in the network. Through connectivity analysis on the three-dimensional network model established, the users can obtain the network connectivity and which nodes are adjacent to the node. In this way, it is able to provide network structure data for actual geographic network such as power distribution network, network reconstruction such as pipeline network, state estimation, and safety analysis.

Network address matching: In essence, it refers to inquiry on geographic position and it involves the address coding. The users can enter address list, street network which contains the range of address, and the property value of address to be inquired; through address matching technology, it is able to carry out contrast and matching for the address information entered by users and the address in standard address library, carry out relevance for matched address data, and show it on the map. The network address matching shall be combined with other network analysis functions to meet the complicated analysis required in actual work.

3 Analysis on Spatial Trend Surface

The spatial distribution and time process of the solid feature were simulated through the mathematical model of trend surface, and the local interpolation or trend between the measured data point of spatial and temporal distribution of geographical elements was used for analysis.The purpose was to count and analyze the variable trend surface and nature of data set. During use, a group of 3D discrete points were selected as the analysis target, and trend surface was selected in the trend surface analysis dialogue box to analyze the function type and conduct calculation, and in this way, it is possible to give the trend surface analysis coefficient of discrete point set and the trend surface fitness. The figure below is a binary cubic polynomial trend surface generated based on the 3D discrete point coordinate. Aimed at the spatial discrete point set, 3D Nurbs curved surface was used for spatial fitting to analyze the spatial distribution of these discrete points. Nurbs curved surface can be used for the fitting analysis of spatial discrete point set, so as to analyze the spatial distribution of the discrete points. As shown in Fig. 4, a group of spatial discrete points to be analyzed were selected for spatial fitting analysis, and in this way, it is possible to generate the Nurbs fitting curved surface of these spatial discrete points.

Fig. 4. Effect diagram of trend surface analysis

4   Conclusion

3D city visualization and analysis platform is a useful tool for the social service agencies and citizens for browsing and analyzing city big data directly, and is agreed upon as being both immediately useful and generally extensible for future applications. The user-need-oriented 3D GIS based smart government portal makes rapid response by real-time and thorough perception of users needs, so as to make timely improvement to the short service board, actively provide convenient, accurate and high-efficiency service to the public and enterprises in online public service.

Acknowledgments. The authors are thankful to the National Natural Science Fund for the Youth of China (41301439).

References

1. Big data specials. Nature, 455(7209), Sept 2008
2. Briggs, F.: Large data - great opportunities. Presented at IDF 2012, Beijing (2012)
3. Dang, S., Ju, J., Matthews, D., Feng, X., Zuo, C.: Efficient solar power heating system based on lenticular condensation. In: 2014 International Conference on Information Science, Electronics and Electrical Engineering (ISEEE), vol. 2, pp. 736–739. IEEE (2014)
4. Huang, B., Jiang, B., Li, H.: An integration of gis, virtual reality and the internet for visualization, analysis and exploration of spatial data (2001)
5. Li, T., Zhou, X., Brandstatter, K., Raicu, I.: Distributed key-value store on hpc and cloud systems. In: 2nd Greater Chicago Area System Research Workshop (GCASR), Citeseer (2013)
6. Li, T., Zhou, X., Brandstatter, K., Zhao, D., Wang, K., Rajendran, A., Zhang, Z., Raicu, I.: Zht: a light-weight reliable persistent dynamic scalable zero-hop distributed hash table. In: 2013 IEEE 27th International Symposium on Parallel and Distributed Processing (IPDPS), pp. 775–787. IEEE (2013)
7. Li, X., Lv, Z., Hu, J., Zhang, B., Shi, L., Feng, S.: Xearth: A 3d gis platform for managing massive city information. In: 2015 IEEE International Conference on Computational Intelligence and Virtual Environments for Measurement Systems and Applications (CIVEMSA), pp. 1–6. IEEE (2015)
8. Li, X., Lv, Z., Hu, J., Zhang, B., Yin, L., Zhong, C., Wang, W., Feng, S.: Traffic management and forecasting system based on 3d gis. In: 15th IEEE/ACM International Symposium on Cluster, Cloud and Grid Computing (CCGrid). IEEE (2015)
9. Lin, H., Chen, M., Lu, G., Zhu, Q., Gong, J., You, X., Wen, Y., Xu, B., Hu, M.: Virtual geographic environments (vges): a new generation of geographic analysis tool. Earth Sci. Rev. **126**, 74–84 (2013)
10. Lv, Z., Li, X., Zhang, B., Wang, W., Feng, S., Hu, J.: Big city 3d visual analysis. In: European Association for Computer Graphics 36th Annual Conference of the European Association for Computer Graphics (Eurographics2015) (2015)
11. Lv, Z., Réhman, S.U., Chen, G.: WebVRGIS: a P2P network engine for VR data and GIS analysis. In: Lee, M., Hirose, A., Hou, Z.-G., Kil, R.M. (eds.) ICONIP 2013. LNCS, vol. 8226, pp. 503–510. Springer, Heidelberg (2013)

12. Lv, Z., Su, T.: 3d seabed modeling and visualization on ubiquitous context. In: SIGGRAPH Asia 2014 Posters, p. 33. ACM (2014)
13. Lv, Z., Yin, T., Han, Y., Chen, Y., Chen, G.: Webvr-web virtual reality engine based on p2p network. J. Netw. **6**(7), 990–998 (2011)
14. Pintus, R., Pal, K., Yang, Y., Weyrich, T., Gobbetti, E., Rushmeier, H.: A survey of geometric analysis in cultural heritage. Computer Graphics Forum (2015)
15. Porathe, T., Prison, J.: Design of human-map system interaction. In: CHI 2008 Extended Abstracts on Human Factors in Computing Systems, CHI EA 2008, pp. 2859–2864. ACM, New York (2008)
16. Su, T., Lv, Z., Gao, S., Li, X., Lv, H.: 3d seabed: 3d modeling and visualization platform for the seabed. In: 2014 IEEE International Conference on Multimedia and Expo Workshops (ICMEW), pp. 1–6. IEEE (2014)
17. Wang, J.J.-Y., Gao, X.: Partially labeled data tuple can optimize multivariate performance measures. In: 24th ACM International Conference on Information and Knowledge Management (CIKM 2015) (2015)
18. Wang, J.J.-Y., Wang, Y., Jing, B.-Y., Gao, X.: Regularized maximum correntropy machine. Neurocomputing **160**, 85–92 (2015)
19. Wang, K., Kulkarni, A., Zhou, X., Lang, M., Raicu, I.: Using simulation to explore distributed key-value stores for exascale system services. In: 2nd Greater Chicago Area System Research Workshop (GCASR) (2013)
20. Wang, K., Liu, N., Sadooghi, I., Yang, X., Zhou, X., Lang, M., Sun, X.-H., Raicu, I.: Overcoming hadoop scaling limitations through distributed task execution
21. Wang, K., Zhou, X., Chen, H., Lang, M., Raicu, I.: Next generation job management systems for extreme-scale ensemble computing. In: Proceedings of the 23rd International Symposium on High-performance Parallel and Distributed Computing, pp. 111–114. ACM (2014)
22. Wang, K., Zhou, X., Li, T., Zhao, D., Lang, M., Raicu, I.: Optimizing load balancing and data-locality with data-aware scheduling. In: 2014 IEEE International Conference on Big Data (Big Data), pp. 119–128. IEEE (2014)
23. Wang, K., Zhou, X., Qiao, K., Lang, M., McClelland, B., Raicu, I.: Towards scalable distributed workload manager with monitoring-based weakly consistent resource stealing. In: Proceedings of the 24rd International Symposium on High-performance Parallel and Distributed Computing, pp. 219–222. ACM (2015)
24. Wang, Y., Bicer, T., Jiang, W., Agrawal, G.: Smart: a mapreduce-like framework for in-situ scientific analytics (2015)
25. Wang, Y., Su, Y., Agrawal, G.: A novel approach for approximate aggregations over arrays. In: Proceedings of the 27th International Conference on Scientific and Statistical Database Management, p. 4. ACM (2015)
26. Yang, Y., Ivrissimtzis, I.: Mesh discriminative features for 3d steganalysis. ACM Trans. Multimedia Comput. Commun. Appl. **10**(3), 1–27 (2014)
27. Zhang, S., Zhang, X., Ou, X.: After we knew it: empirical study and modeling of cost-effectiveness of exploiting prevalent known vulnerabilities across iaas cloud. In: Proceedings of the 9th ACM Symposium on Information, Computer and Communications Security, pp. 317–328. ACM (2014)
28. Zhao, D., Zhang, Z., Zhou, X., Li, T., Wang, K., Kimpe, D., Carns, P., Ross, R., Raicu, I.: Fusionfs: toward supporting data-intensive scientific applications on extreme-scale high-performance computing systems. In: 2014 IEEE International Conference on Big Data (Big Data), pp. 61–70. IEEE (2014)
29. Zhou, X., Chen, H., Wang, K., Lang, M., Raicu, I.: Exploring distributed resource allocation techniques in the slurm job management system. Illinois Institute of Technology, Department of Computer Science, Technical Report (2013)

Hierarchical Nearest Neighbor Graphs
for Building Perceptual Hierarchies

Gi Hyun Lim[✉], Miguel Oliveira, S. Hamidreza Kasaei, and Luís Seabra Lopes

University of Aveiro, 3810-193 Aveiro, Portugal
hhmetal@gmail.com, {mriem,seyed.hamidreza,lsl}@ua.pt

Abstract. Humans tend to organize their knowledge into hierarchies, because searches are efficient when proceeding downward in the tree-like structures. Similarly, many autonomous robots also contain some form of hierarchical knowledge. They may learn knowledge from their experiences through interaction with human users. However, it is difficult to find a common ground between robots and humans in a low level experience. Thus, their interaction must take place at the semantic level rather than at the perceptual level, and robots need to organize perceptual experiences into hierarchies for themselves. This paper presents an unsupervised method to build view-based perceptual hierarchies using *hierarchical Nearest Neighbor Graphs* (hNNGs), which combine most of the interesting features of both *Nearest Neighbor Graphs* (NNGs) and self-balancing trees. An incremental construction algorithm is developed to build and maintain the perceptual hierarchies. The paper describes the details of the data representations and the algorithms of hNNGs.

Keywords: Hierarchical nearest neighbor graphs · Incremental learning · Agglomerative clustering · Open-ended environments

1 Introduction

Humans tend to organize their knowledge into hierarchies, because searches are efficient when proceeding downward in the tree-like structures. It's said that the same categorical structure may be inherent in humans [11]. Cognitive psychologists say that representations are one of key devices to understand human behaviors [1]. Taxonomies which show classification of things or concepts are highly-abstracted and well-organized knowledge structures. Similarly, many intelligent service robots also contain some form of hierarchical semantic knowledge used to support interaction with humans, execution of service tasks and acquiring experiences for learning [15,18].

Semantic representations such as taxonomies and hierarchical semantic knowledge are easy to share across not only human-robot but also robots. Otherwise, in the case that two agents have different types of sensors or use different kinds of processing methods, their perceptual experiences are almost useless to each other. In the same manner, the perceptual information of a robot is meaningless

© Springer International Publishing Switzerland 2015
S. Arik et al. (Eds.): ICONIP 2015, Part II, LNCS 9490, pp. 646–655, 2015.
DOI: 10.1007/978-3-319-26535-3_74

to a human. Thus, human-robot interaction is commonly performed via symbolic ways, and robots need to organize perceptual experiences into hierarchies for themselves. *Grounding* symbols is one of the primary challenges in service robotics [22].

This paper addresses the problem of building a perceptual hierarchy of object views from object categories. It is here assumed that an object category may have several instances and each instance has several views according to its view point changes. An object might be recognized and be assigned a symbol by matching between a newly observed view and stored views [17] in the feature level, where views are composed of a set of perceptual features. This recognition is a start point of grounding of the perceptual observations to the semantic representations. When a hierarchical structure is built by using the similarities between views of a specific feature, it might be independent with semantic hierarchy and perceptual hierarchies of other features.

Intelligent robots need to continuously improve their knowledge to adapt to dynamic environments which they do not know in advance. This problem is referred to as *open-ended learning* [20]. To handle this in the perceptual level, an incremental clustering method is implemented to learn a perceptual hierarchy with user mediation. A user triggers the learning procedure and the perceptual system corrects key views of the object to collect noiseless views and reduce the number of views [14].

Recently, many researchers in the area of intelligent robotics and machine learning have increasingly focused on building a variety of perceptual hierarchies. In [21], an approach to discover visual object class hierarchies is described. The authors have adapted a generative hierarchical latent Dirichlet allocation (hLDA) method to adapt to the visual domain with unsupervised manner. Their results show perceptual object hierarchies and object segmentation, while their learning considers static environments. Collet et al. [3] proposed a graph-based framework to solve the problem of Lifelong Robotic Object Discovery (LROD). They used Constrained Similarity Graphs (CSGs) to encode object and domain knowledge as constraints into nodes, edges, node weights and edge weights. Several types of constraints including motion and object shape into nodes, visual similarity and shape similarity into edges are provided to address their problem domain.

2   View-Based Perceptual Hierarchy

In this paper, a perceptual hierarchy is considered as a rooted tree from an object view perspective. An object category is represented with a set of object views, which are image regions for particular viewpoints and consist of a set of perceptual features such as spin-images [9]. In the bottom level, each view is associated to a *Perceptual Element* (PE). Several elements with similar views are organized into a *Perceptual Cluster* (PC) using the *Nearest Neighbor Graph* (NNG) clustering method [4,7]. For each PC, one PE is selected to be the representative element of the cluster, and relocates to the upper level. As moving up the hierarchy, groups of elements are merged until only one cluster remains in the level.

To represent PCs and PEs, two kinds of data structures are defined in terms of cluster (*Clst*) and element (*Elem*), respectively. In the implementation, these are represented as *Robot Operating System* (ROS) messages that consist of sets of key-value pairs to store into the perceptual memory system [16] of *Robustness by Autonomous Competence Enhancement* (RACE) [8]. *The details of hierarchical Nearest Neighbor Graphs* (hNNGs) and their incremental construction algorithm are described in the Sects. 3 and 4, respectively.

The followings are formal definitions for hNNG and NNG.

Definition 1. *A hNNG is a set of NNGs,*

$$hNNG = \{NNGs(lvl)|lvl \in levels\}, \tag{1}$$

$$NNGs(lvl) = \{Clst|level(Clst) = lvl\}, \tag{2}$$

where *levels* is a list of the existing levels from the bottom to the root of a hNNG, and *level*() is a function returning the level of a given *Clst*.

Definition 2. *A cluster (Clst) in a level is a tuple,*

$$Clst = (lvl, rep, p, Elem). \tag{3}$$

It contains information about a level of hierarchy (*lvl*), a representative element (*rep*), a key pointing at its parent cluster (*p*) and a list of member elements (*Elem*). Its member elements are grouped by NNG clustering, and one of them is selected as the representative of the cluster.

Definition 3. *An element (Elem) in a level is a tuple,*

$$Elem = (lvl, vk, outlink, dist, Inlinks). \tag{4}$$

Its fields contain a level of hierarchy (*lvl*), a view key (*vk*), which points at a set of features for the object view, its outlink key (*outlink*) pointing to the nearest neighbor, the distance (*dist*) between its the nearest neighbor, and a list of inlink keys (*Inlinks*) which are incoming links to this element as their nearest neighbor. The representation of *Inlinks* makes it easy to navigate between links without queries or matching.

3   Hierarchical Nearest Neighbor Graphs

In order to make a cluster, we model the relationship between nodes using a *Nearest Neighbor Graph* (NNG), which is a directed graph whose nodes have a directed edge to point out their nearest neighbor. Each node in a graph represents a *Elem* and is linked to its most similar *Elem*. All the linked nodes are grouped into a *Clst*, which represents a *Perceptual Cluster* (PC) in a perceptual hierarchy. One of the most important properties of NNG is scale and rotation invariant [7]. Additionally, NNG is applicable to asymmetric relations. Note that not only feature distance metrics but also nearest neighbor relations are asymmetric.

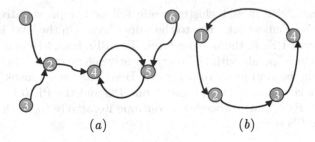

Fig. 1. Examples of NNG.

Figure 1 shows examples of two NNGs. One interesting characteristic is a NNG always has one loop. Assume that a cluster grouped into a NNG has n elements. The cluster also has n edges, since each element has one link to point at its nearest neighbor. However, the cluster needs $n - 1$ edges to connect all elements without loop. Thus, one additional edge makes one loop. Even though the right figure is a special case of NNG, each example has one loop.

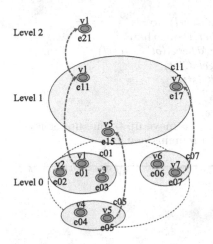

Fig. 2. An example of three level hNNG.

In this paper, NNG representation is extended in a hierarchical way, coined *hierarchical Nearest Neighbor Graph* (hNNG). hNNGs combine most of the interesting features of both NNGs and self-balancing trees. From NNGs, they are applicable in a more general mathematical space. While all the elements in r-trees [6,19] and k-d trees [2,5] are defined in a fixed dimensional space, hNNGs are applicable in a topological space, in which a set of elements are defined on their neighbourhood relationships. Therefore a hNNG can represent the distance between views, which are commonly represented as a set of features. The number of features, which depends on its feature selection algorithm, might be various so that the distance between views is metric between different size of sets. From self-balancing trees whose all leaf nodes are at the same height from root through the same number of intermediate nodes, and keep the height automatically, hNNGs guarantee same worst-case performance as average performance. While hNNGs do not require balancing, r-trees and k-d trees require more complex balancing techniques, because they are sorted in multiple dimensions. Anyway, these hierarchical or tree-like structures allow fast enumeration to find the nearest neighbor.

Figure 2 shows an example of a three level hNNG. On the bottom level, there are seven PEs which are associated with all the stored object views and three PCs which are grouped using NNG clustering method as described in the following section. A PE in each cluster is selected as its representative, and only the representative element relocates to the upper level. On the level 1, three PEs remain from seven PEs in the bottom level. The PEs ($e11$, $e15$ and $e17$) in the level 1 exactly corresponds with the representative elements ($e01$, $e05$ and $e07$) of all PCs in the bottom level, respectively. Here, the same clustering method is applied to make clusters. Thus, there is one PC and the PE ($e11$) is selected as its representative. These procedures continue iteratively to reach a top level where only one PE remains.

Algorithm 1. Agglomerative Incremental hNNG Construction.

Input:
 list of new elements **NewElem**
1: **procedure** AGGLOMERATIVE INCREMENTAL hNNG CONSTRUCTION
2:    $level \longleftarrow 0$ ▷ start clustering from the bottom level
3:    **repeat**
4:       **for** $elem \in$ **NewElem do**
5:          $nn \longleftarrow getNearestNeighborAtLevel(elem, level)$
6:          $nnClst \longleftarrow getClusterAtLevel(nn, level)$
7:          $addElement(elem, nnClst)$
8:          **for** $other \in getElementsAtLevel(level)$ **do**
9:             **if** $other.dist < distance(other, elem)$ **then**
10:                $oldClst \longleftarrow getClusterAtLevel(other, level)$
11:                $removeElement(other, oldClst)$
12:                $selectRepresentive(oldClst)$
13:                $addElement(other, nnClst)$
14:          $selectRepresentive(nnClst)$
15:       $++level$ ▷ move up to one upper level
16:    **until** number of ElementsAtLevel(level) remains one
17:    **return** hNNG

---

4   Incremental Construction

An intelligent robot needs to learn continuously to adapt itself to a dynamic environment [13]. Such robot is evidently an open-ended learning system [10], whose learning targets are not known in advance and experiences that are acquired for the learning are also not known in advance. In order to learn a perceptual hierarchy under an open-ended environment, an incremental algorithm was implemented that works in an unsupervised manner, as shown in Algorithm 1. When a new object instance is provided, its views are collected to add into the perceptual hierarchy. In this paper, these views are input data of *Agglomerative Incremental hNNG Clustering* (AIhNNGC), and each of them becomes an element of a hNNG. The algorithm continues until only one element remains at the highest level. In the bottom level, elements are inserted into a NNG by searching their nearest neighbors, with respect to the distance measure between views.

The algorithm also updates the other elements whose distances from the added element are nearer than previous ones. In order to build an efficient tree, it is implemented with two lists: an adding list (*addList*) and a deleting list (*delList*). The first one is for adding elements and the other for deleting elements in the corresponding level. Whenever a cluster is changed, the *selectRepresentive* procedure follows. A representative element of the cluster is selected by the following priorities: the number of links of element, the average distance of *inlinks*, the distance of *outlink* and random selection.

5   Experimental Results

Figure 3 shows an evolution of a perceptual hierarchy generated from 51 object categories. The object categories were randomly selected from the Washington RGB-D object dataset [12] which contains 51 object categories with 300 instances. In the case of center one, which were trained from 100 instances, the actual number of selected instances per category varies between 0 and 5 and the actual number of categories included in the example is 47. For each selected instance, the first 3 views are selected. Thus, a total of 300 views are used for the perceptual hierarchy. Spin-image features are extracted from each view using the configuration proposed in [10]. The obtained perceptual hierarchy is structured in four levels, containing 4 clusters in the third level (below the root), 15 clusters in the second level, and 62 clusters in the first level (leaf level). The last perceptual hierarchy from 900 views of 300 instances in the Fig. 3 is structured in five levels, containing 2 clusters in the fifth level, 7 clusters in the third level, 31 cluster in the second level, and 143 clusters in the bottom level.

Cluster CLST_0_42_1 in the first level (level 1) includes views from 4 object categories: *bell pepper, onion, pear* and *potato*[1], as shown in Fig. 4. These views have very similar shape, as shown in Fig. 5. This cluster is united with the cluster CLST_0_67_1, which includes views of *orange, onion* and *peach*, resulting in the cluster CLST_1_38_2 at the immediately upper level (level 2). In another example, views of *food box, kleenex* and *cereal box* are grouped in the same cluster (CLST_0_18_1). These results show that, as expected, the perceptual hierarchy groups together object views with similar shape. To evaluate clustering results, the cluster purities, which are defined as the percentage of the most frequent class in a cluster, are measured according to an evolution of leaf clusters, as shown in Fig. 3. The results show the purity value converges around 0.68.

Figure 6 shows the processing time to add a new instance with 3 views up to 300 instances from 51 categories. In Algorithm 1, inserting a new element into a hNNG takes time $\mathcal{O}(2n \log_{avg(cs)} n)$, where $avg(cg)$ is an average cluster size, that is $\mathcal{O}(n \log_{avg(cs)} n)$, because the value of $avg(cg)$ converges a certain value. It is approximately $\mathcal{O}(n)$, as shown in Fig. 6 (right).

Figure 7 shows the evolution of the number of clusters and cluster size at each level. The numbers of clusters seem to increase linearly according to the number of views, since the average cluster size at each level, that is a number of

---

[1] The labels are provided just for evaluation.

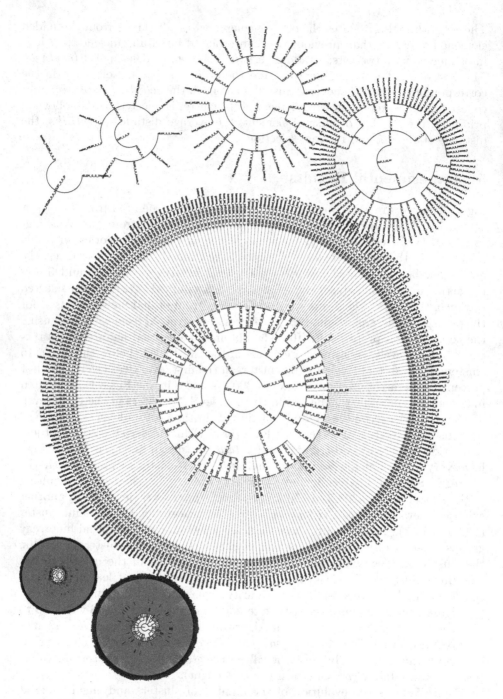

Fig. 3. Evolution of a perceptual hierarchy trained from the numbers of instances (clockwise from top left): 1, 3, 10, 30, 100, 200 and 300 with the number of views 3, 9, 30, 90, 300, 600 and 900, respectively.

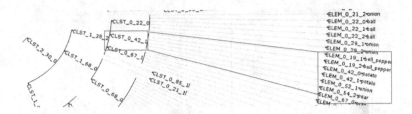

Fig. 4. A detail of the perceptual cluster (CLST_0_42_1).

(a) bell pepper   (b) onion   (c) pear   (d) potato

Fig. 5. Object instances and point-cloud data of the perceptual cluster (CLST_0_42_1).

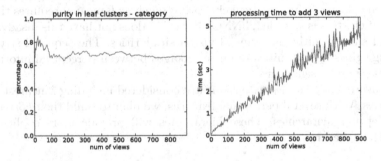

Fig. 6. Evolution of purity (left) and processing time (right).

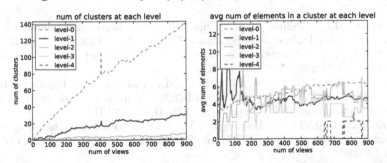

Fig. 7. Evolution of number of clusters (left) and cluster size (right).

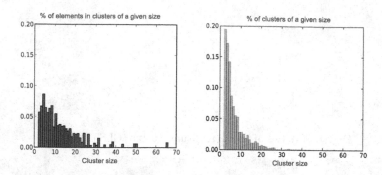

Fig. 8. Distribution of cluster size.

elements in a cluster, converges around 6. It means that a 5-levels hNNG may have $6^5 = 7776$ elements.

Finally, Fig. 8 shows distributions of cluster size in terms of elements and clusters. Figure 8 (right) suggests there are many clusters of small size, such as 2, 3 or 4. However, the proportion of elements in clusters looks evenly distributed.

6   Conclusions

We proposed an agglomerative and incremental clustering method that works in an unsupervised manner. hNNG clustering is also applicable to non fixed dimensional space. While classical clustering (such as k-means) requires that the number of clusters is specified, hNNG clustering does not have this disadvantage and results are intuitive in terms of visual similarities. The acquired structural knowledge gives opportunities to make linkages between perceptual information and semantic representations.

For future work, several extensions are considered including k nearest neighbor queries. With several perceptual features, we plan to build their corresponding perceptual hierarchies. These hierarchies will provide more sophisticated knowledge structure.

References

1. Austerweil, J.L., Griffiths, T.L.: A nonparametric bayesian framework for constructing flexible feature representations. Psychol. Rev. **120**(4), 817 (2013)
2. Bentley, J.L.: Multidimensional binary search trees used for associative searching. Commun. ACM **18**(9), 509–517 (1975)
3. Collet, A., Xiong, B., Gurau, C., Hebert, M., Srinivasa, S.S.: Exploiting domain knowledge for object discovery. In: 2013 IEEE International Conference on Robotics and Automation (ICRA), pp. 2118–2125. IEEE (2013)
4. Franti, P., Virmajoki, O., Hautamaki, V.: Fast agglomerative clustering using a k-nearest neighbor graph. IEEE Trans. Pattern Anal. Mach. Intell. **28**(11), 1875–1881 (2006)
5. Friedman, J.H., Bentley, J.L., Finkel, R.A.: An algorithm for finding best matches in logarithmic expected time. ACM Trans. Math. Softw. (TOMS) **3**(3), 209–226 (1977)

6. Guttman, A.: R-trees: A dynamic index structure for spatial searching, vol. 14. ACM (1984)
7. Hautamäki, V., Kinnunen, T., Fränti, P.: Text-independent speaker recognition using graph matching. Pattern Recogn. Lett. **29**(9), 1427–1432 (2008)
8. Hertzberg, J., Zhang, J., Zhang, L., Rockel, S., Neumann, B., Lehmann, J., Dubba, K.S.R., Cohn, A.G., Saffiotti, A., Pecora, F., Mansouri, M.: The RACE project. KI - Künstliche Intelligenz **28**(4), 297–304 (2014). http://dx.doi.org/10.1007/s13218-014-0327-y
9. Johnson, A., Hebert, M.: Using spin images for efficient object recognition in cluttered 3D scenes. IEEE Trans. Pattern Anal. Mach. Intell. **21**(5), 433–449 (1999)
10. Kasaei, S.H., Oliveira, M., Lim, G.H., Lopes, L.S., Tomé, A.M.: Interactive open-ended learning for 3d object recognition: An approach and experiments. J. Intell. Robotic Syst., 1–17 (2015)
11. Kriegeskorte, N., Mur, M., Ruff, D.A., Kiani, R., Bodurka, J., Esteky, H., Tanaka, K., Bandettini, P.A.: Matching categorical object representations in inferior temporal cortex of man and monkey. Neuron **60**(6), 1126–1141 (2008)
12. Lai, K., Bo, L., Ren, X., Fox, D.: A large-scale hierarchical multi-view rgb-d object dataset. In: 2011 IEEE International Conference on Robotics and Automation (ICRA), pp. 1817–1824. IEEE (2011)
13. Lim, G.H., Kim, K.W., Suh, H., Suh, I.H., Beetz, M.: Knowledge-based incremental bayesian learning for object recognition. In: Autonomous Learning Workshop, ICRA (2013)
14. Lim, G.H., Oliveira, M., Mokhtari Hassanabad, V., Kasaei, S.H., Seabra Lopes, L., Maria Tomé, A.: Interactive teaching and experience extraction for learning about objects and robot activities. In: RO-MAN 2014. IEEE (2014)
15. Lim, G.H., Suh, I.H., Suh, H.: Ontology-based unified robot knowledge for service robots in indoor environments. IEEE Trans. Syst., Man Cybern., Part A: Syst. Humans **41**(3), 492–509 (2011)
16. Oliveira, M., Lim, G.H., Seabra Lopes, L., Kasaei, S.H., Maria Tomé, A., Chauhan, A.: A perceptual memory system for grounding semantic representations in intelligent service robots. In: IROS 2014. IEEE (2014)
17. Riesenhuber, M., Poggio, T.: Models of object recognition. Nature Neurosci. **3**, 1199–1204 (2000)
18. Rockel, S., Neumann, B., Zhang, J., Dubba, K.S.R., Cohn, A.G., Konečný, Š., Mansouri, M., Pecora, F., Saffiotti, A., Günther, M., Stock, S., Hertzberg, J., Tomé, A.M., Pinho, A.J., Lopes, L.S., von Riegen, S., Hotz, L.: An ontology-based multilevel robot architecture for learning from experiences. In: Designing Intelligent Robots: Reintegrating AI II, AAAI Spring Symposium, Stanford, USA, March 2013
19. Roussopoulos, N., Kelley, S., Vincent, F.: Nearest neighbor queries. In: ACM Sigmod Record. vol. 24, pp. 71–79. ACM (1995)
20. Seabra Lopes, L., Chauhan, A.: Open-ended category learning for language acquisition. Connection Sci. **20**(4), 277–297 (2008)
21. Sivic, J., Russell, B.C., Zisserman, A., Freeman, W.T., Efros, A.A.: Unsupervised discovery of visual object class hierarchies. In: IEEE Conference on Computer Vision and Pattern Recognition, CVPR 2008, pp. 1–8. IEEE (2008)
22. Steels, L.: Grounding language through evolutionary language games. In: Steels, L., Hild, M. (eds.) Language Grounding in Robots, pp. 1–22. Springer, US (2012)

Author Index

Printed in the United States
By Bookmasters